*Tropical Ecology and
Physical Edaphology*

Tropical Ecology and Physical Edaphology

R. LAL
Soil Physicist
International Institute of Tropical Agriculture
Ibadan, Nigeria

A Wiley–Interscience Publication

JOHN WILEY & SONS
Chichester ● New York ● Brisbane ● Toronto ● Singapore

S
481
.L34
1987

Copyright © 1987 by John Wiley & Sons Ltd.

All rights reserved.

No part of this book may be reproduced by any means, or transmitted, or translated into a machine language without the written permission of the publisher

Library of Congress Cataloging-in-Publication Data:

Lal, R.
   Tropical ecology and physical edaphology.

   'A Wiley–Interscience publication.'
   Includes indexes.
     1. Agricultural ecology—Tropics. 2. Ecology—Tropics. 3. Soil ecology—Tropics. 4. Man—Influence on nature—Tropics. I. Title.
S481.L34   1986   574.5'2623   85–16906
ISBN 0 471 90815 0

British Library Cataloguing in Publication Data:

Lal, R.
   Tropical ecology and physical edaphology.
   1. Ecology—Tropics
   I. Title
   574.5'0913   QH541.5.T7

   ISBN 0 471 90815 0

Printed and bound in Great Britain   # 12313352

This Book is Dedicated to the Farmers
of the Tropics

Contents

Foreword ix

Preface xi

Part I   Tropical Ecology 1
1   Introduction 3
2   Tropical climate 24
3   Rainfall, vegetation, soil 46
4   Soil physical properties 113
5   Variability in soil physical properties 178

Part II   Ecological factors and soil physical properties 229
6   Vegetation and soil 231
7   Soil fauna and flora 260
8   Earthworms 285
9   Termites 337
10 Ants 423

Part III   Man as an ecological factor 443
11 Man and soil 445
12 Fire 452
13 Conversion of tropical rain forests 503
14 Tillage 565
15 Farming systems 618

Part IV   Towards improvement in tropical agriculture 685
16 An ecological approach to tropical agriculture 687

Author index 711

Subject index 725

Foreword

Throughout the history of agriculture, farmers have used practical experience to settle on the naturally productive soils and in environments that favour crop growth. Now the major unsettled agricultural land resources of the world lie in the humid and sub-humid tropics. These resources are in regions where human populations are increasing at a rate approximating three per cent annually. It is inevitable that large acreages will be placed under cultivation within the next two decades. Unless appropriate technologies are used, many cultivation schemes will result in unacceptably low crop production and considerable land degradation. In this book the author has sought to identify and quantify the unique ecological relationships of tropical environments and to apply science to overcome contraints to food production through management systems that are stable, sustainable and safe.

The contribution of science to increased agricultural food production in temperate climates is well documented, with yield increases for several grain crops being two to three fold. In contrast the scientific knowledge base of the complex tropical ecology is considerably less, and food production has generally not increased. Thus a great challenge for scientists is critically to evaluate tropical environments and determine how they may be managed for food crop production.

Newcomers to the tropics often view the lush tropical growth as a veritable Garden of Eden and ask 'How can food production be a problem here?' Little do they realize the basic chemical and physical limitations of both soil and above-ground environments. High rainfall and temperature usually bring about intense soil weathering and render soils susceptible to a range of degradative processes. In addition to being inherently low in major plant nutrients, most soils have severe physical restraints that limit the uptake and effective utilization of available nutrient reserves. Other soil physical restraints, especially in tropical soils under intensive cultivation, are characterized by supra-optimal temperatures, low water reserves, restricted rooting depth, and susceptibility to compaction and accelerated soil erosion. Conversely, these soil physical properties are generally at optimum levels when under forest cover by intense activity of soil macro-fauna, e.g. earthworms, termites, ants and other soil animals; the recycling of nutrients by tree roots; and a protective cover to reduce soil erosion. These beneficial biotic and biophysical factors are disrupted by deforestation, burning, tillage methods, use of acidifying fertilizers, indiscriminate use of insecticides and heavy farm machinery.

Thus many persons sagely observe that tropical ecosystems and, particularly, rainforests must not be disturbed, else other areas of the earth's environment will be changed. Their solution is to leave them unaltered or compromise by growing tree crops. Like many environmental and food issues, the trade-offs must be carefully considered. An increasingly large population is to be fed and clothed in the tropics. Historically, certain crops provided the means to purchase food. But a casual study of the fluctuating world market prices shows that income based on export crops is hazardous. Secondly, past and present research by the chemical industry led to development of substitutes for plant products at a cheaper price. Occasionally a fortunate country will have other natural resources to sell and purchase food, but often such resources are finite. So the basic questions remain. Should food to sustain increasing populations be imported? If so, how is it to be paid for? Or should increasing efforts be made to apply science to increase in-country food production?

This book deals with the challenge of the latter question, and with the assumption that more food can be produced in the tropics of the world if the ecology is understood and wisely managed. Scientists interested in tropical ecology as a natural body, or resource, and agricultural scientists concerned with using the tropics for increased food production will find valuable information in this volume. It is further hoped that policy makers of tropical countries that are striving for food sufficiency will heed both the potentials and warnings that are presented so concisely.

J.W. Pendleton
Deputy Director General (Research)
International Institute of Tropical Agriculture (IITA)
Ibadan, Nigeria

Preface

Given proper management and necessary inputs, the soils and environment of the tropics can produce high and sustained yields of many crops. However, understanding soil–climate–vegetation interaction and the effects of human intervention on life-supporting processes in various ecological environments is important for proper development, utilization, and restoration of fertility status of soil resources. This book is a state-of-the-art review of the properties, potential and constraints of some major agro-ecological regions of the tropics. Particular emphasis is placed upon soil physical properties under natural conditions, effects of vegetation and soil fauna, and alterations by man's intervention. Although soils and environments of the tropics are called fragile, there is still a great deal to learn about factors and processes responsible for their fragility and how to manage them.

This book is arbitrarily divided into four parts: (a) Tropical Ecology; (b) Ecological Factors and Soil Physical Properties; (c) Man as an Ecological Factor; and (d) Towards Improvements in Tropical Agriculture. The title of the volume has been carefully chosen to reflect the contents and highlight the importance of ecological environments in relation to soil physical properties and crop production in the tropics. Although scientific bases of technologies described are transferable to other regions, it will be oversimplification to expect the universalities of technological packages discussed.

Although it was the intention to draw upon the most current literature available, the book is by no means an exhaustive treatise of all that is known and published. Our knowledge of tropical ecology, soil resources and their management is increasing very rapidly. Because of the perpetual food deficit in parts of tropical Africa, scientists are generating a wealth of new and valuable research information. It is likely that some very important information has been omitted. For example it was only after the proof-reading stage that I came across an important book on earthworms by K.E. Lee.

I have drawn heavily on material and data from many sources. The author took the liberty of citing data of many colleagues and friends around the world. Furthermore, it is impossible to name all those who have very generously helped and contributed in one way or the other towards completing this volume. Nonetheless, the names of a few need a special mention. The book was typed, often more than once, by Mrs Agnes Nwamadi. I am particularly grateful to Mrs Janet Keyser for editing the book. The entire artwork and line drawings were prepared by Mr Olayinka Ogunjimi of the Soil Physics Laboratory, all photographs were prepared by Messrs Benson Fadare and Lawrence Umoh of the communication unit, and subject and author indices were compiled by Mrs E.F. Nwajei. I am grateful to Dr F.E. Caveness for reviewing chapters dealing specifically with soil fauna. My special thanks go to Professor George S. Taylor of the Ohio State University and Dr J.W. Pendleton of IITA, who very kindly reviewed the manuscript.

R. Lal
Ibadan, Nigeria

PART 1

Tropical ecology

Chapter 1
Introduction

I	LAND RESOURCES	3
II	DEMOGRAPHIC PRESSURE	7
III	FOOD SUPPLY AND DEMAND IN THE TROPICS	12
IV	PRODUCTIVE POTENTIAL	17
V	THE PROBLEM	21

The tropics cover approximately 40% of the earth's surface, receive over half of the world's total rainfall, provide for about half of the total and 62% of the world's agricultural population and contain at least 53% of the potentially cultivable land area of the world. With a few exceptions, the tropics are also distinguished by low yield, high rate of soil degradation followed by drastic decline in soil productivity, widespread poverty, malnutrition and low standards of living. In spite of national and international efforts, food production has grossly lagged behind demand, making hunger the most stubborn problem in tropical Africa. And yet, the tropics have the potential to produce more than is required to support its human and cattle populations. The answer to this dilemma lies in understanding soils and their interaction with environments, and socio-economic factors. If the potential is not being realised, it is probably due to the lack of appropriate technology and infrastructure and mismanagement of resources, or both. An evaluation of the land resources, food supply, and demand is necessary to identify constraints and define research and action priorities.

I. LAND RESOURCES

The potential arable land area of the earth is about 24% of its ice-free surface, and about three times the area currently being harvested annually (President's Science Advisory Committee, 1967). An evaluation of the precise potentially cultivable land area of the tropics is a difficult task because of the lack of the baseline data on existing land use and land capability for these regions. There exists, therefore, a wide variation in the estimates of potentially arable land area in the tropics. For example, the data in Tables 1.1, 1.2 and 1.3 compare the potentially cultivable land area assessed by Buringh *et al*. (1976), Norse (1976) and FAO (1981 a, b). These estimates are difficult to compare

Table 1.1 Totals of the production potentials of the tropics and the world (Buringh et al., 1975). *Reproduced by permission of Agricultural University, Wageningen*

	A	PAL	IPAL	MPDM	PIAL	IPALI	MPDMI	MPGE
S. America	1780	616.5	333.6	25224	17.9	340.7	25710	11106
Australia	860	225.6	74.2	5297	5.3	76.1	5462	2358
Africa	3030	761.2	306.5	24162	19.7	317.5	25115	10845
Asia	4390	1083.4	433.5	24966	314.1	581.6	33058	14281
N. America	2420	628.6	320.0	15443	37.1	337.5	16374	7072
Europe	1050	398.7	233.1	8289	75.9	247.1	9653	4168
Antarctica	1310	0	0	0	0	0	0	0
Total	14840	3714.1	1700.9	103381	470.0	1900.5	115372	49830

A,   area of broad soil regions (10^6 ha)
PAL,   potential agricultural land (10^6 ha)
IPAL,   imaginary area of PAL with potential production without irrigation (10^6 ha)
MPDM,   maximum production of dry matter without irrigation (10^6 tonnes/ha)
PIAL,   potentially irrigable agricultural land (10^6 ha)
IPALI,   imaginary area of PAL with potential production including irrigation (10^6 ha)
MPDAMI,   maximum production of dry matter including irrigation (10^6 tonnes/year)
MPGE,   minimum production of grain equivalents, including irrigation (10^6 tonnes/year)

The potential agricultural land (PAL) of the world is 3419 million ha, which is about 25% of the total land area. A maximum of 470 million ha of this area can be irrigated. At present 1406 million ha of land are cultivated, of which 201 million ha are irrigated.

because of the different systems used for land classification. The tropical land area is grossly overestimated if one adds together South America, Africa and Asia from Buringh's assessment in Table 1.1. Obviously not all land areas in these continents are strictly 'tropical'. The estimates in the FAO's (1981) classification of 'developing countries' provide the gross land area without making a distinction between arable and potentially arable land area. On the basis of Norse's classification of wet and dry tropics, the present arable land area of the tropics is estimated to be 510 million ha out of the world total of 1447 million ha. Out of the potentially cultivable land area of about 3 billion ha, at least 53% lies in the wet and dry tropics.

It is estimated that about 64% of the 3 billion hectares of potentially arable land lies in Asia, Africa and South America and only 37% of the land in the tropics has ever been cultivated (FAO, 1969; President's Comm. 1967). The estimates of potentially cultivable area in the developing tropical countries alone range from 1.145 billion ha (FAO, 1969) to 1.648 billion ha (President's Comm., 1967). The FAO's (1974) estimates of arable land area in some tropical countries are shown in Table 1.3. If one includes vast unexploited regions of the Australian tropics and considers the possibilities of multiple cropping that exist in the tropics (Okigbo and Greenland, 1976; Okigbo, 1978), the potential gross arable land in the tropics is several times this estimate.

The data on land resources also indicate that there are large, scarcely inhabited areas in the tropics that can be developed for food crop production (Tables 1.5 and 1.6.). For example, Sub-Saharan Africa has an arable land area

Table 1.2 Land and water resources of the world (Norse, 1979) (areas in 10^6 ha)

Agro-climatic zone	Present arable area		Additional potential arable area	Additional irrigable area in columns 1 and 3	Total potentially cultivable area
	Non-irrigated	Irrigated			
(a) *Developed countries*					
Wet tropics	1	0	2	0	3
Dry tropics	10	1	7	5	18
Warm humid	43	3	20	4	69
Warm dry	69	17	20	23	103
Cool humid	367	13	240	8	630
Cool dry	106	20	46	58	170
Cool temperature and polar	20	0	30	0	50
Total	616	52	315	98	1043
(b) *Developing countries*					
Wet tropics	76	9	414	80	499
Dry tropics	373	40	733	105	1146
Warm humid	28	26	58	6	112
Warm dry	94	44	90	9	228
Cool humid	26	32	12	20	70
Cool dry	20	11	10	4	41
Total	617	162	1317	224	2096
Grand total	1233	214	1632	322	3139

Table 1.3 Estimate of the potentially cultivable land area in 10^6 ha (FAO, 1981a,b). *Reproduced by permission of the Food and Agricultue Organization of the United Nations*

Land use	World	Developing countries	Developed economies
Arable and perm. crops	1413	628	394
Arable land	1326	564	378
Perm. crops	87	63	16
Pasture	3126	1481	881
Forest and woodlands	4209	2191	880
Other land	4325	2140	1003

of about 700 million hectares and the potentially cultivable area in the Amazon is an additional 450 million hectares. Many feasibility surveys indicate that if properly developed and managed, these regions can support a high population (Sanchez and Buol, 1975; Revelle, 1976).

The seventies have in fact witnessed a drastic change in the existing land use. In the developing countries of the tropics and subtropics, surveys of the land use for the period of 1970 to 1980 (FAO, 1981 a, b) have indicated an increase in land area of 6, 10 and 1.1% under arable land use, permanent crops and pastures, respectively, and a decrease of 4.2% in the area under forest and woodlands (Fig 1.1). These percentages correspond with the actual area + 34.9, +6.3, +16.2 and −91.7 million ha. The forest and woodlands in the developing countries of the tropics and subtropics have been shrinking at the rate of 9.2 million ha per annum.

Table 1.4 Land use per person in some tropical countries (Adapted from Crosson and Frederick, 1977). *Reproduced by permission of Resources for the Future, Inc*

Region	Land use 10^6 ha)		Land use per person economically active in agriculture (ha/person)	
	1961–63	1985	1961–63	1985
Africa, South of Sahara				
Chad	6.8	7.1	6.5	4.5
Ethiopia	14.0	19.8	1.6	1.7
Ghana	8.0	8.0	4.6	3.3
Kenya	8.1	9.8	2.7	2.0
Niger	8.8	9.6	9.3	5.9
Nigeria	31.8	31.8	2.3	1.7
Tanzania	14.0	17.4	3.2	2.7
Zaire	7.2	14.0	1.2	2.0
Regional total	151.8	189.3	2.5	2.3
Far East				
India	161.5	164.4	1.1	1.0
Pakistan and Bangladesh	25.5	28.7	1.0	0.7
Philippines	7.9	10.1	0.9	0.8
Thailand	8.6	11.1	0.7	0.7
Regional total	209.8	222.4	1.1	0.9
Near East and N.W. Africa				
Afganistan	9.0	11.1	2.1	1.8
Egypt	2.6	3.2	0.6	0.5
Iran	16.8	19.6	4.8	4.9
Iraq	6.7	6.3	6.7	4.7
Sudan	7.1	10.1	2.0	1.8
Syria	6.6	6.9	9.5	7.0
Regional total	50.7	59.2	2.6	2.2
South America				
Argentina	25.7	41.3	16.8	42.8
Brazil	48.8	48.6	4.1	4.5
Chile	4.5	4.7	5.9	6.2
Colombia	5.1	7.1	2.0	2.0
Peru	2.6	3.7	1.5	1.8
Venezuela	5.2	5.6	6.1	6.6
Regional total	101.0	133.7	4.7	5.5
Grand total	513.3	604.6	1.7	1.6

Table 1.5 Forest areas (10^6 ha) in the tropics that can be developed for agriculture (Barney, 1979)

Region	Closed forest	Closed forest bordering land area
Latin America	680	33
Asia	410	15
Africa	190	6
Oceania	89	10
World total	2657	20

Table 1.6 Estimated sources of new agricultural land (modified Chou et al., 1977)

Region	Area (10^6 ha)	Constraints
1. Latin America (principally Brazil)	450	Al toxicity, P fixation and availability, erosion and compaction
2. Latin America (Peruvian and Chilean Sea Coast)	50	Unavailability of good quality irrigation water
3. Sub-Saharan Africa	500–700	Erosion, compaction, cropping system
4. Australia's semi-arid tropics		Erosion, water conservation

The changes in land use during the decade ending in 1980 for Asia, South America and tropical Africa are presented in Figs. 1.2, 1.3 and 1.4, respectively. Forests and woodlands decreased at the rate of 1.3, 4.7 and 2.7 million per annum, respectively, in Asia, South America and tropical Africa. The corresponding increase was observed to be 4.6, 11.1, 7.0% in arable land area, and 5.7, 8.6 and 14% in permanent crops, respectively. While in tropical Africa and the Far East there was a net decrease of 1.13 million ha (3.2%), forest and woodlands in tropical America were converted to pastures. Most of the increase in arable land area in the tropics over the decade ending in 1980 was brought about by horizontal expansion, whereby the forests and woodlands were brought under arable land use.

II. DEMOGRAPHIC PRESSURE

The world population is estimated to be 4.9, 5.4 and 6.5 billion inhabitants with corresponding annual growth rates of 1.9, 2.0 and 2.0 per cent for the years 1985, 1990 and 2000, respectively (FAO, 1974). The rate of population growth and density varies widely among regions (Table 1.7). For example, the population growth rate among the developed economies of the temperate latitudes averages 0.97% per year, and that of tropical countries averages 2.67%. The estimates of actual population for the decade of 1970 to 1980

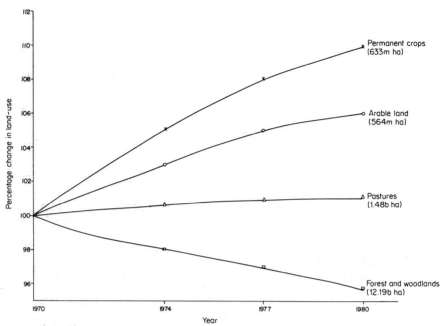

Figure 1.1 Percentage change in land-use in the developing countries of the tropics. Figures in parenthesis refer to the land area in 1970. *Redrawn from FAO (1982) by permission of the Food and Agriculture Organization of the United Nations*

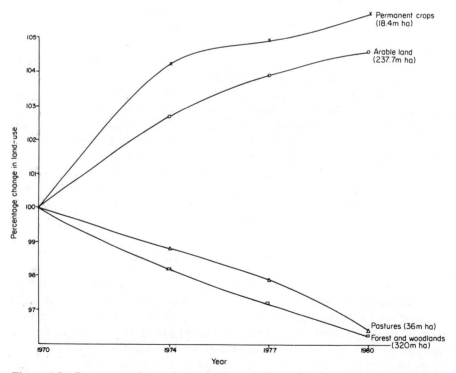

Figure 1.2 Percentage change in the land-use in the tropical regions of the Far East with reference to 1970 (figures in parenthesis). *Redrawn from FAO (1982) by permission of the Food and Agriculture Organization of the United Nations*

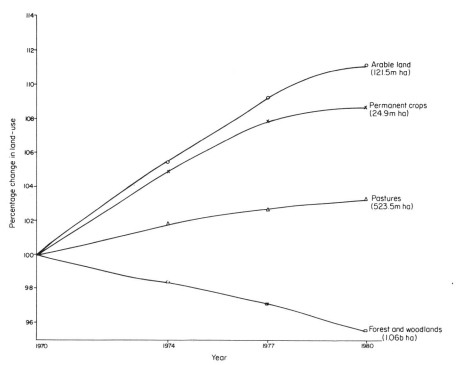

Figure 1.3 Percentage change in land-use in tropical America. Figures in parenthesis refer to the land-use in 1970. *Redrawn from FAO (1982) by permission of the Food and Agriculture Organization of the United Nations*

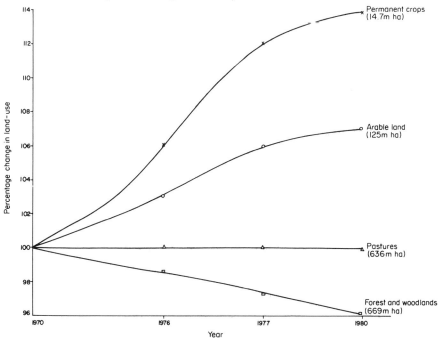

Figure 1.4 The trends in change in land-use in tropical Africa in the decade ending in 1980. The land-use figures in 1970 are shown in parenthesis. *Redrawn from FAO (1982) by permission of the Food and Agriculture Organization of the United Nations*

Table 1.7 World population (millions) and its growth rate (% per year) (Keyfitz, 1966; FAO, 1983). *Reproduced by permission of the Food and Agriculture Organization of the United Nations*

	Actual	Projected		Growth rates		
	1970	1985	1990	1960–70	1970–83	1970–90
Developed countries	1072	1227	1277	1.1	0.9	0.9
Developing countries	2549	3631	4069	2.3	2.4	2.4
Africa	279	427	498	2.5	2.9	2.9
Latin America	284	428	489	2.8	2.8	2.7
Near East	171	262	303	2.7	2.9	2.9
Far East	1021	1506	1707	2.5	2.6	2.6

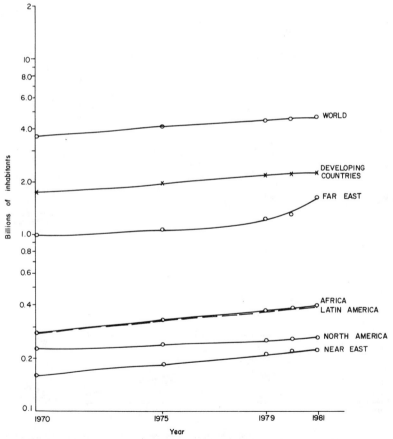

Figure 1.5 Population trends in developing countries of the tropics (FAO, 1981)

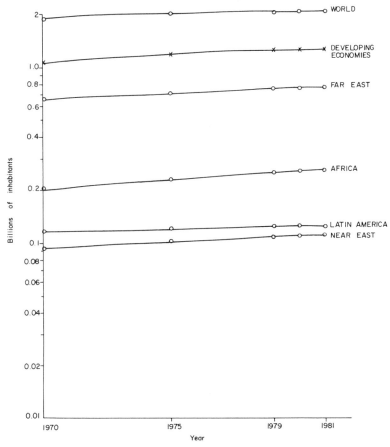

Figure 1.6 Agricultural population of the world (FAO, 1981). *Reproduced by permission of the Food and Agriculture Organization of the United Nations*

shown in Fig. 1.5 indicate an increase in population by 22, 30, 37, 31, 27 and 35% in the world, tropical countries, tropical Africa, Latin America, Far East and Near East, respectively. The corresponding increase in the agricultural population (Fig. 1.6) is 8, 16, 24, 7, 15, and 17 percent for the world, tropical countries, tropical Africa, Latin America, Far East and Near East, respectively. This implies that the relative percentage of agricultural population declined over the past decade. The percentage of population engaged in agriculture ranged from 51.5, 63.7, 73.8, 41.2, 67.9 and 59.5 in 1970 and 45.4, 56.5, 66.5, 33.6, 61.2 and 51.2 in 1981 for world, tropical countries, tropical Africa, Latin America, Far East and Near East regions, respectively. In the tropics, not only the relative proportion of population engaged in agriculture has declined but the mean age of the farmer is higher now than in the previous decade (Ay, Personal communication) because of the rapid rate of urbanization.

The increase in arable land area has not kept pace with the increase in

Table 1.8 Arable area per capita, actual and projected. (Global 2000 Report, Barney, 1979). *Reproduced by permission of the Food and Agriculture Organization of the United Nations*

Region	1951–55	1971–75	1985	2000
Industrialized countries	0.61	0.55	0.50	0.46
Western Europe	0.33	0.26	0.24	0.22
China	0.19	0.16	0.13	0.11
Developing tropical countries	0.45	0.35	0.27	0.19
World	0.48	0.39	0.32	0.25

population. The per capita land area in the developing tropical countries is estimated to decline from 0.45 hectare in 1951-55 to 0.19 hectare in the year 2000 (Table 1.8). Similarly, due to a high increase in the agricultural population, the land use per person economically active in agriculture is also expected to decline (Table 1.4).

III. FOOD SUPPLY AND DEMAND IN THE TROPICS

In spite of the high potential arable land area, per capita food production has not kept pace with the demand in most tropical regions. Crosson and Frederick (1977) analysed the world food situation and reported that in comparison to the 1970 level the world food demand will rise 44% by 1985 and 112% by 2000. This

Table 1.9 Demand for cereals in the world and in tropical countries (Crosson and Frederick, 1977). *Reproduced by permission of Resources for the Future, Inc*

Region	Actual consumption (10^6 tonnes)	Projected demand (10^6 tonnes)		Annual rates of growth (%)	
	1970	1985	1990	1970–1985	1985–1990
Developed countries	617	796	847	1.7	1.2
Developing market economies	386	629	738	3.3	3.2
Asian centrally planned economies	204	300	325	2.6	1.7
All developed countries	590	929	1063	3.1	2.8
World	1207	1725	1910	2.4	2.1

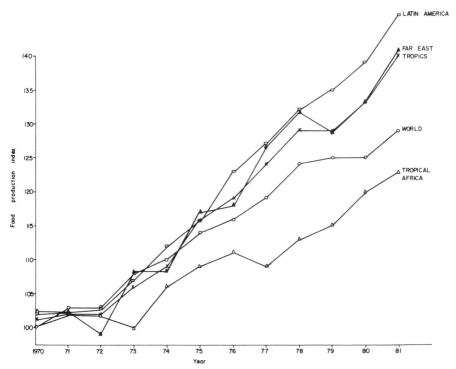

Figure 1.7 Food production index of tropical regions (FAO, 1981). *Reproduced by permission of the Food and Agriculture Organization of the United Nations*

rise in food demand is equivalent to a 2.4 per cent annual increase over the 1970–1985 period and 2.6 per cent over 1985–2000 period. The tropical countries account for 71% of the increase from 1985 to 2000. The data in Table 1.9 show that tropical countries account for 67% of the 518 million tonne increase in annual world cereal demand from 1970 to 1985 and 75% of the 626 million tonne increase from 1985 to 2000.

The FAO's per capita food production index for the period 1970 to 1981 is shown in Fig. 1.7. The index for 1980 showed an increase of 40% in all tropical countries compared with an average increase of only 29% in the world. The increase in agricultural production by the year 1981 in comparison with 1970 was 23, 46, 41 and 42% in tropical Africa, Latin America, Near East and Far East regions respectively. Although these relative increases in agricultural production are rather impressive, these statistics can be misleading because of the very low absolute production levels for most of these countries in the mid-sixties and early seventies.

Whatever the increase in food and agricultural production has been, its effect has been nullified by even greater increase in the population. The data shown in Fig. 1.8 indicate that sub-Saharan Africa is the only region in the world where per capita food production declined over the past two decades,

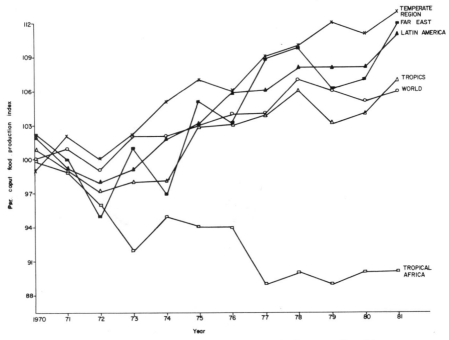

Figure 1.8 Per caput food production index (FAO, 1981). *Reproduced by permission of the Food and Agriculture Organization of the United Nations*

and more so during the seventies. In comparison with 1970, the per capita food production in tropical Africa for 1981 declined by as much as 9.6%. A survey conducted by the USDA (1980) observed that in 1978, per capita food production in Angola, Benin, Ethiopia, Ghana, Nigeria, Senegal, Sierra Leone, Uganda and Upper Volta was less than 90% of the 1961–65 average. In contrast, there was a definite increase in per capita food production index of 12.6, 11.0, 10.4, 6.6 and 6.0% in the temperate zone countries, Far East, Latin America, tropical countries, and the world on the whole, respectively (Fig. 1.8). For the decade ending in 1980, Frére (1983) also noted that the rate of food production in Africa was the lowest among the developing countries and was also less than the annual rate of population growth, i.e. 1.8 versus 2.9% (Table 1.10).

As a consequence of this severe decline in per capita food production in Africa, the per capita calorie intake has fallen below the minimal nutritional standards. A World Bank sponsored survey indicated that in 1975 about 193 million people (more than 60% of Africa's total population) had a seriously inadequate calorie intake. These conclusions are supported by the data in Fig. 1.9, which show that with the exception of oil-rich Gabon and Nigeria, the food supply declined for most countries in sub-Saharan Africa.

The per capita food intake in the mid-seventies was below the average requirement for Africa and Far East. Consequently the number of seriously

Table 1.10 Annual percentage of increase in food production compared to population growth from 1971–80 in some tropical countries (Frére, 1983). *Reproduced by permission of Westview Press*

Region	Food production	Population growth
Africa	1.8	2.9
Far East	3.2	2.5
Latin America	3.9	2.7
Near East	3.3	2.8

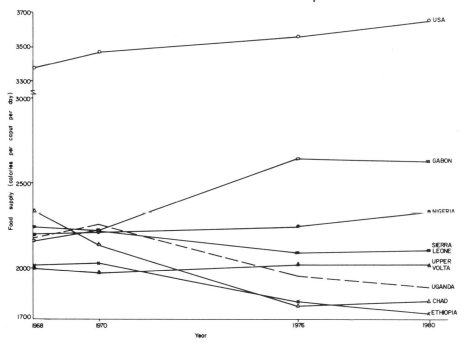

Figure 1.9 Per caput food supply in the world (FAO, 1982). *Reproduced by permission of the Food and Agriculture Organization of the United Nations*

undernourished people in 1975 was estimated to be 436 million. The projected undernourished population in the tropics is estimated at 408 and 390 million persons in 1990 and 2000, respectively (Table 1.11). The food requirements cannot be met in spite of the expected rate of 3.0 to 3.4 per cent for food and agricultural products.

In addition to food scarcity, fuel-wood scarcity is a serious problem in most tropical countries. FAO (1981) projects that by the year 2000, the minimum fuel needs of some 2.7 billion people would not be fully met (Table 1.12). In many tropical countries a typical family spends one-quarter of its earnings on fuel-wood and charcoal (Postel, 1984).

Table 1.11 Mean calorie supplies and the undernourished population in the tropics (FAO, 1981b)

Region	Calories per capita			Number seriously undernourished (millions)			Growth rates of food and agricultural products (%)	
	1975	2000	National average required	1975	1990	2000	1965–80	1980–2000
Africa	2180	2436	2352	72	84	80	1.8	3.4
Far East	2025	2309	2224	304	274	268	2.9	3.0
Near East	2560	2821	2480	19	17	16	3.0	3.0
Latin America	2525	2905	2395	41	33	26	3.0	3.3
Developed countries	3315	3417	—	—	—	—	—	—

Table 1.12 Tropical population (millions) with deficit fuel-wood supplies (Postel, 1984)

	1980		2000
Region	Acute scarcity	Deficit	Acute scarcity or deficit
Tropical Africa	55	146	535
Tropical Asia	31	832	1671
Latin America	26	201	512
Total	112	1179	2718

IV. PRODUCTIVE POTENTIAL

The agricultural potential of the tropics does not seem to be a major constraint, and the vast potential is widely recognised (FAO, 1974; Sanchez et al., 1982). The quality of land resources is also adequate to meet the biological demands of the region (Buringh et al., 1975; Norse, 1979). There is no doubt that considering the moisture regime, the radiation levels and the length of the growing seasons, the agricultural potential of the tropics may be even greater than the temperate latitudes (Table 1.13). For example, Moira (1980), De Vries and De Wit (1983) and Higgins (1984) have estimated that the population carrying capacity of tropical Africa may be as much as 10×10^9 persons (Fig. 1.10). What is lacking is the appropriate technology to develop this potential, and the will-power to attain food self-sufficiency.

In spite of this high potential, the national average yields of most crops in the tropics and sub-tropics are deplorably low. For example, a survey conducted by Cummings (1976) indicated that average yield of rice in Africa was 1387 kg ha^{-1} compared with 5703 kg ha^{-1} in Japan and 5117 kg ha^{-1} in the United States. The national average yield of rice for Zaire was 731 kg ha^{-1}. Similarly, the yield

Table 1.13 Comparison of radiation levels and potential productivity at Samuru, northern Nigeria, and Rothamsted, UK (Kowal, 1972). *Reproduced by permission of ABU Press Ltd*

Samuru			Rothamsted		
Season	Ratiation (cal cm^{-2})	Dry matter (kg ha^{-1})	Season	Radiation (cal cm^{-2})	Dry matter (kg ha^{-1})
6 month rainy season May–October	88 000	22 000	7 month growing season March–Sept.	64 000	16 000
6 month dry season Nov.–April	90 000	22 500	5 month winter Oct.–March	12 000	3000*
Annual total		44 500			16 000

* production limited by low temperatures

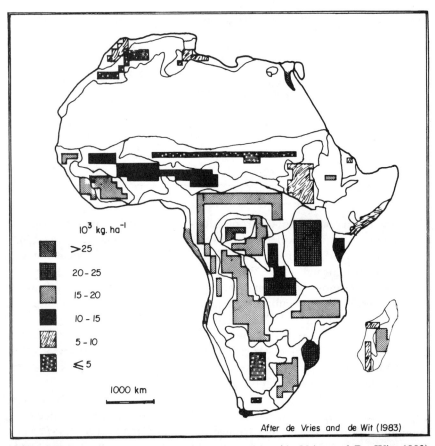

Figure 1.10 Productive potential of tropical Africa (de Vries and De Wit, 1983). *Reproduced by permission of the Food and Agriculture Organization of the United Nations*

of dry beans was 487 kg ha^{-1} in Africa and only 58 kg ha–1 in Niger. The yield of cassava tubers, a staple for most sub-Saharan countries, has been observed to be as low as 4063 kg ha^{-1} for Tanzania (Cummings, 1976). The yield on experimental plots can be as high as 40 t ha^{-1}.

Yield trends during the seventies for most staple crops in the tropics are shown in Table 1.14. In spite of the improved varieties and recommended agronomic practices, the yields per hectare have rarely shown an upward trend. Although the low levels of yields are partly due to higher incidence of disease and pests in the tropics than in the temperate regions (Table 1.15), the overriding influence of soil and environment management in defining the maximum yield potential cannot be over-emphasized.

The low total agricultural production and the low levels of crop yield are only one aspect of this multi-faceted and complex food situation. The variability in food production among seasons and years make it difficult to predict the food supply and plan for the worst situation. The food insecurity is related to both low yields and high variability in production. Valdés and Konandreas (1981) studied food consumption variability among many tropical countries (Table 1.16). The data show that the coefficient of variation ranges from a low of 3 to 4

Table 1.14 Yield (kg ha^{-1}) trends in the seventies (FAO, 1981. *Reproduced by permission of the Food and Agriculture Organization of the United Nations*

Region	1969–71	1979	1980	1981
(a) *Rice*				
World	2331	2676	2770	2885
Tropical Africa	1355	1368	1390	1369
Latin America	1715	1918	1998	1896
Far East	1842	1991	2234	2334
Near East	3712	4576	4240	4219
Developed countries	5388	5733	5091	5588
(b) *Maize*				
World	2472	3326	3060	3370
Tropical Africa	1003	907	903	1044
Latin America	1426	6150	1795	2028
Far East	1106	1203	1275	1334
Near East	2328	2633	2723	2486
Developed countries	4349	5930	5057	5885
(c) *Millet*				
World	709	644	655	686
Tropical Africa	584	590	574	598
Latin America	1049	1303	1032	1270
Far East	520	474	531	581
Near East	1304	833	778	839
Developed countries	1054	1100	962	795
(d) *Sorghum*				
World	1157	1409	1241	1507
Tropical Africa	662	681	672	662
Latin America	2033	2589	2390	3127
Far East	493	702	681	727
Near East	810	788	741	864
Developed countries	3128	3703	2743	3648
Cassava				
Developed countries	8853	8581	8674	9054
Tropical Africa	6639	6292	6369	6456
Nigeria	10592	9130	9167	9167
Zaire	6851	6667	6711	7017
Latin America	13510	11312	11338	11572
Near East	9213	11574	11762	13083
Far East	3402	2541	2689	2687
Dry beans				
World	520	532	519	567
Tropical Africa	565	613	615	618
Latin America	605	576	499	571
Far East	313	283	334	345
Near East	1254	3100	1286	1337
Developed countries	945	1175	1165	1200

Table 1.15 Relative incidence of crop diseases in temperate and tropical regions (Swiminathan, 1980). *Reproduced by permission of the Commonwealth Agricultural Bureau*

Crop	Number of disease outbreaks reported	
	Temperate zone	Tropics
Rice	54	500–600
Maize	85	125
Citrus	50	248
Tomato	32	278
Beans	52	250–280

Table 1.16 Variabilities in staple food consumption and production, 1961–76 (Valdes and Konandreas, 1981). *Reproduced by permission of Westview Press*

Region	Coefficient of variation		Probability of falling below 95% of trend	
	Production	Consumption	Production	Consumption
Asia				
Bangladesh	6.4	7.6	22	26
India	6.4	5.3	22	17
Indonesia	5.4	6.1	18	21
Korea	7.1	6.5	24	22
Philippines	5.7	3.3	19	6
Sri Lanka	9.3	8.3	29	27
Sub-Sahara Africa				
Ghana	5.8	6.1	20	21
Nigeria	5.7	5.6	19	19
Senegal	18.6	15.7	39	37
Tanzania	12.7	14.6	35	37
Upper Volta	9.8	9.5	30	30
Zaire	4.9	4.1	15	11
Latin America				
Brazil	5.2	5.8	17	20
Chile	11.1	14.4	33	36
Colombia	4.4	4.7	13	14
Guatemala	6.5	6.9	22	24
Mexico	7.7	5.3	26	17
Peru	9.8	3.9	30	10

per cent (the Philippines and Peru) to a high of 15 to 16 per cent (Tanzania and Senegal). These authors observed a generally high degree of consumption variability in 44 out of 67 countries, and in 51 out of 67 countries the probability of actual consumption falling below the expected level. In some tropical countries (Bangladesh, India, Indonesia and the Philippines) the coefficient of variation of food production is not high (<6%) and production is relatively stable.

V. THE PROBLEM

The statistics indicate an overwhelming food shortage in the tropics, particularly in Sub-Saharan Africa. Given the land and water resources, this need not be so. One of the research and action priorities for agriculturists should be to develop technology to exploit the vast potential that exists in this region. Basic information is urgently needed to understand the constraints and potential of the resources of the tropics, i.e., soil, water, energy, nutrients and the manpower available. Strengthening of these baseline data regarding the soil resource potential is necessary to dispel some myths about tropical soils being of low fertility, excessively leached, too wet or too dry, lacking response to inputs, and being uninhabitable environments. What it all means is that we do not have the research information nor the technology to overcome and alleviate these soil and environment constraints. It is time that researchers and planners consider climate and soils in the tropics as basic resources to be developed, used and restored rather than as hazards and constraints to be feared and kept away from.

It is this lack of basic understanding of the soil–ecology interaction that have led to the failures of many grandiose agricultural development projects in the tropics. The failures of these otherwise workable and successful models transferred from elsewhere should not come as a surprise in view of our knowledge gaps and the non-availability of appropriate technology. These failures have merely accelerated the widespread pessimistic myths and the negative attitude towards the ability of the tropical countries to feed themselves. It is thus a matter of urgency that basic information be made available so as to develop strategies that improve food production in the tropics and reduce dependence on the vagaries of the climate. The strategies to improve food production should involve understanding of:

(i) the soil and its interaction with ecological environments;
(ii) agricultural systems and their interaction with soil fauna;
(iii) crop requirements and soil's potential.

The traditional agricultural systems are stable and ecologically sound under certain specific conditions. When these conditions are altered, the system is destabilized. The success of high-input farming in temperate regions of Western Europe and North America has led to a widespread implementation of land development schemes in the tropics, without giving serious considerations to the soil-biological processes that stabilize the traditional bush-fallow and shifting cultivation systems. After wasting countless millions of dollars, degrading millions of hectares of fertile lands into barren and unproductive wastes, enhancing malnutrition and suffering of tropical inhabitants, and frustrating scientists and planners alike, we now recognize that it is impossible to sustain and stabilize production by high-input technology *per se*. There is no way to effectively bypass the important soil biological processes that regularly turn over the soil, keeping it alive, porous, aggregated and receptive to high intensity rains, mineralise leaf litter and provide an effective nutrient-recycling mechanism.

This book is written with the objective to collate, review and analyse the existing research information concerning the role of ecological factors in preserving a favourable level of soil physical properties and sustaining soil productivity. Special attention is given to soil fauna, e.g., earthworms, termites and other organisms that influence soil properties and productivity. The role of man is viewed as an important ecological factor. The potential and constraints of ecological environments and their interaction with the soil must be assessed if we are to survive the crises created by burgeoning population and shortage of suitable land to meet its biological needs. The indiscriminate use of agrochemicals (pesticides, herbicides and chemical fertilizers), heavy machinery and other technological innovations of western agriculture have not proven compatible with the 'shifting cultivator' of the tropics and the environments in which he manages to carve out his living. The scientific community should therefore view ecological factors as the vehicle of gradual transformation of traditional agricultural system into commercial but ecologically stable farming.

REFERENCES

Allaby, M., 1977, *World food resources: actual and potential*, Applied Science Publishers Ltd., London, 418 pp.

Barney, G.O., 1979, *The global 2000 report to the United States*, Pergamon Press, New York, Vol. 1–10.

Buringh, P., Van Heemst, H.D.J., and Staring, G.J., 1976, *Computation of the absolute maximum food production of the world*, Agric. Univ. Wageningen, The Netherlands.

Chou, M., Harmon, D.P. Jr., Kahn, H., and Wittwer, S.H., 1977, *World food prospects and agricultural potential*, Praeger Publishers, New York, 316 pp.

Chou, M., and Harmon, D.P., Jr., 1979, *Critical food issues of the Eighties*. Pergamon Press, New York, p. 404.

Clark, C., 1970, *Starvation or Plenty*, Taplinger, New York, 154 pp.

Crosson, P.R., and Frederick, K.D., 1977, *The world situation*, Resources For the Future Inc., Washington D.C., 230 pp.

Cummings, R.W., Jr., 1976, Food crops in the low-income countries: the state of present and expected agricultural research and technology, *Working Papers*, Rockefeller Foundation, New York, 103 pp.

De Vries, P.F.W.T., and De Wit, C.T., 1983, Identifying technological potentials for food production and accelerating progess in productivity. IFPRI conference '*Accelerating Agricultural Growth in Sub-Saharan Africa*', Victoria Falls, Zimbabwe, 29 August–1 September, 1983.

FAO, 1969, *Provisional Indicative World Plan for Agricultural Development*, Vol. 2, FAO, Rome.

FAO, 1974, *Assessment of the world food situation: present and future*, Item 8 of the Provisional Agenda of the United Nations World Food Conference, FAO, Rome.

FAO, 1981a, *Production year book*, FAO, Rome.

FAO 1981b, *Agriculture: Toward 2000*, FAO, Rome,

Frére, M., 1983, *Agroclimate information for development*: Revising the green revolution, D.F. Cusack (ed.), Westview Press, Boulder, Colorado, pp. 18–24.

Hanson, A.D., and Nelson, C.E., 1981, Water: adaptation of crops to drought-prone environments, in Carlson, P.S. (ed.) *The Biology of Crop Productivity*, Academic Press, New York, pp. 78–152.

Higgins, G., 1984, African resources of terrain, land and water for the populations of the future—management for development. *Advancing Agricultural Production In Africa*, CAB conference, Arusha, Tanzania, 12–18 February, 1984.

Keyfitz, N., 1966, How many people have lived on earth?, *Demography*, **3**(2), 581.

Kirit Parikh and Raba'r (eds.), *Food for All in a Sustainable World*: The IIASA Food and Agriculture Program, IIASA, Laxemburg, Austria, 250 pp.

Kowal, J.M. 1972, Radiation and potential crop production at Samaru. Savanna 1, 89–101.

Leakey, C.L.A., and Wills, J.B., 1977, *Food crops of the Lowland tropics*. Oxford Univ. Press, London, pp. 162–167.

Lowenberg, M.E., Todhunter, E.N., Wilson, E.D., Savage, J.R., and Lubawski, J.L., 1968, *Food and Man*, J. Wiley & Sons, New York, 459 pp.

Moira, 1980, *Model of International Relations in Agriculture*, North Holland Publishing Company, Amsterdam, p. 375.

Norse, D., 1976, Development strategies and the world food problem, *J. Agric. Econ.*, **27**, 137–158.

Norse, D., 1979, Natural resources, development strategies and the world food problem, in M.R. Biswas and A.K. Biswas (eds.), *Food, Climate and Man*, John Wiley & Sons, New York.

Okigbo, B.N., and Greenland, D.J., 1976, Intercropping systems in Tropical Africa, in R.I. Papendick, P.A. Sanchez, and G.B. Triplett (eds.), *Multiple Cropping*, ASA Publication 27, Madison, Wisconsin, USA.

Okigbo, B.N., 1978, Cropping systems and related research in Africa. Special Issue *Association for Advancement of Agricultural Sciences in Africa (AAASA)*, Occasional Publication Series—OT—1, Addis Ababa, Ethiopia, pp 81.

Pimental, D., and Pimental, Marcia, 1979, *Resource and Environmental Sciences Series*, Edward Arnold, London, 165 pp.

Postel, S., 1984, Protecting forests, in L.R. Brown et al. (eds.), *State of the World 1984*, Worldwatch Institute, W.W. Norton & Co., New York, pp. 75–92.

President's Science Advisory Committeee, 1967, *The World Food Problem*, Washington, D.C., Vol. 2.

Revelle, R., 1976, The resources available for agriculture, *Scientific American*, **235**, 164–178.

Sanchez, P.A., and Buol, S.W., 1975, Soils of the tropics and the world food crisis, *Science*, **128**, (4188), 598–599.

Swaminathan, M.S., 1980, Past, present and future trends in tropical agriculture, in *Perspectives in World Agriculture*, CAB 1980, pp. 1–47.

USDA, 1980, *Food problems and prospects in Sub-Saharan Africa*: The decade of the 1980's, Foreign Agricultural Economic Report No. 166, 293 pp.

Valdés, A., and Konandreas, P., 1981, Assessing food insecurity based on national aggregates in developing countries, in Valdeés A. (ed.), *Food Security For Developing Countries*, Westview Press, Colorado, pp. 25–51.

Valdés, A., 1981, *Food security for developing countries*, Westview Press, Boulder, Colorado, 251 pp.

WMO No. 537, *World Climate Conference, Geneva, 12—23 Feb. 1979.*

Chapter 2
Tropical Climate

I	INTRODUCTION	24
II	TROPICS	25
III	CLIMATIC CLASSIFICATION	25
	A. Temperature and rainfall	25
	B. Water balance and moisture availability indices	29
	C. Vegetation and ecological indices	37
	D. Duration of the rainy season	39
	E. Agronomic criteria	39
	F. Economic and cultural criteria	40
IV	CHOICE OF AN APPROPRIATE CLASSIFICATION SYSTEM	41
V	REFERENCES	42

I. INTRODUCTION

The impact of climate and its vagaries is more drastic for low-input subsistence farming in the tropics than commercial agriculture in the temperate regions. The adverse effects of unfavourable climate on crops are difficult to avoid in the tropics for two reasons—either the technology has not yet been developed or it is too expensive for the subsistence farming community.

An important prerequisite for the development of agricultural technology that will enable high, economic and sustained production from soils in the tropics is the in-depth understanding of the climatic factors, their magnitude and spatial and temporal variability, and their interaction with one another and with soils and crops. The high productive potential of the tropics has thus far not been realized partly because of the lack of basic climatic information that will dispel misunderstandings and myths about the climatic constraints to crop production, and that will facilitate their alleviation through technological advances.

Even though many planners recognize the importance of the climate as a major factor responsible for successful development of tropical agriculture, they find it difficult to use it for planning because of the lack of a suitable criterion to assess climatic potentials and constraints. In spite of recent advances in atmospheric physics, agroclimatologists and soil physicists have made little progress in developing methods of evaluation and classification of those climatic parameters that affect agriculture in general, and soil-crop interaction in particular. It is rather ironic that there is no one universally acceptable system that is available to evaluate and classify climatic environments and that is indicative of the potentials and contraints of different agro-ecologies of the tropics.

II. TROPICS

The word 'tropic' is derived from Greek and literally means 'turning'. Conventionally, tropical regions are those lying within the Tropics of Cancer and Capricorn, about 23.5° north and south of the equator. The zenithal position of the sun in the sky 'turns' or reverses during its annual rotation, hence the origin of the word. Within these regions, the temperature at low elevations is generally uniformly high with relatively little seasonal or annual variation. Crop growth in the tropics is ususally not limited by sub-optimal temperatures, and rainfall varies from practically zero to 8000 mm per year.

III. CLIMATIC CLASSIFICATION

Attempts have been made to classify tropical climates on the basis of one or more easily measured parameters. The objective of this review is not to describe in detail the different systems proposed, but to highlight the potential and limitations of various classification systems. The result is the lack of generally accepted ecological maps, especially those that relate the vegetation type to climatic factors and indicate the management constraints towards intensive agriculture.

About two dozen classification systems are used, based on one or a combination of the following factors:
  (i) rainfall and temperature regime;
 (ii) water balance;
(iii) moisture availability;
(iv) vegetation;
 (v) length of the growing season; and
(vi) economic and cultural factors.

A. Temperature and rainfall

The temperature regime has long been used as a criterion to delineate tropical regions. As far back as the 19th century, Buck (1829) and Supan (1879) defined tropical environments on the basis of annual precipitation and other climatic factors. Supan defined the upper limit of the tropical regions to coincide with the 20° isotherm of the average annual temperature. Subsequently Philippson (1933) proposed the 24° isotherm as the limit of the inner and the 20° isotherm as a limit for the outer tropics. In addition to temperature, Hettner (1930, 1934) and Flohn (1957) also proposed the use of predominant winds and air masses as factors in delineating ecological limits of the tropics. Troll (1943) put forth an argument based on the vertical differentiation of climate in the tropics and proved that at all altitudes in the tropical zone, the daytime climate is normally constant. Another method that involves the use of temperature in relation to rainfall is that of Gaussen (1955). He defined a dry month in which the total rainfall (P) expressed in mm is equal to or less than twice the minimum temperature (T) expressed in °C.

$$P \leq 2T.$$

Accordingly, the equatorial climates are those that have a mean annual temperature exceeding 20°C. Emberger (1955) suggested a pluviothermic coefficient defined as

$$Q = \frac{100P}{(M-m)(M+m)}$$

where M is the mean maximum temperature (°C) in the hottest month, m is the mean minimum temperature (°C) in the coldest month, and P is the annual rainfall in mm. Emberger related different vegetation zones with this pluviothermic cofficient.

One of the better known and well-founded systems is that proposed by Köppen because it relates climatic factors with the predominant vegetation. Köppen (1931) and Köppen and Geiger (1936) used the ecological concept and related rainfall (amount and distribution) and temperature to the main vegetation types (Fig. 2.1). Köppen (1936) adopted de Candollé's (1874) system of classification of the natural vegetation types, and showed that there exists a good correspondence between vegetation and broad climatic regions. He hypothesized that the delineation of vegetation boundaries can be accomplished by means of a quantitative average of climatic parameters. Köppen distinguished two tropical rainy climates:

(a) tropical rainforest: The humid and hot climate where the minimum rainfall even in the driest month is at least 60 mm; and
(b) savanna: the regions with a distinct dry season and at least one month with a rainfall of less than 60 mm.

Köppen defined those regions as 'dry' in which the average precipitation is deficient in relation to the evaporation demand. The dry climates were further subdivided into hot dry climates and cool dry climates according to whether the mean annual temperature is above or below 18°C (Table 2.1, Fig. 2.2).

More recently, Hargreaves (1971) used temperature regime as a criterion to distinguish between subtropical and tropical climates strictly on the basis of the mean annual temperature:

(i) subtropical climate: regions where the mean annual temperature exceeds 10°C for 10 or more months; and
(ii) tropical climate: regions with mean annual temperature of 17°C or more throughout the year.

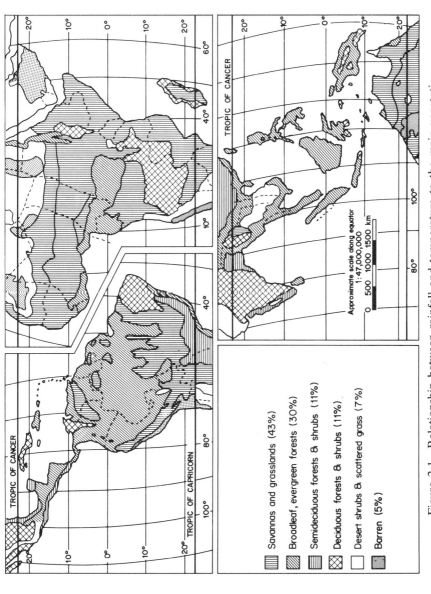

Figure 2.1 Relationship between rainfall and temperature to the main vegetation (Köppen and Geiger, 1936)

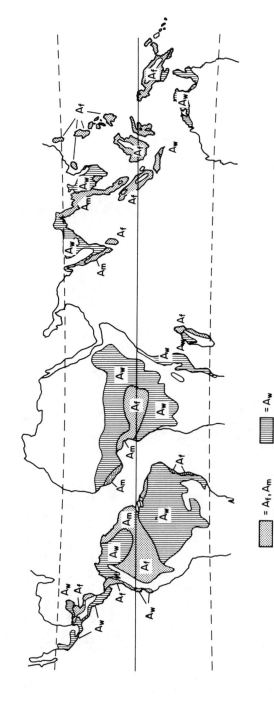

Figure 2.2 Climatic regions on the basis of mean annual temperature (Köppen, 1936)

Table 2.1 Köppen's classification of tropical climates

A – Winterless tropical rainbelt (coldest month averages above 18°C)
    Af — Tropical rainforest climate
    Aw — Periodically dry savannah climate
B – Incomplete dry belt
    BS — Steppe climate
    BW — Desert climate

B. Water balance and moisture availability indices

Water is undoubtedly the most important factor affecting rain-fed agriculture, and has thus rightly been used by many as a criterion to classify tropical climates. For example, each of the broad climatic zones (i.e. tropical rainforest and savanna) are further subdivided depending on the annual rainfall amount. For crop production, however, the rainfall distribution and its effectiveness are more important than the total amount. That has led many climatologists to evaluate rainfall effectiveness by various aridity indices based on rainfall, temperature, and evaporative demand (Table 2.2.).

Thornthwaite (1948) based his classification system on rainfall and evapotranspiration, the factors that affect plant growth. This method relates rainfall to potential evapotranspiration through a 'moisture index' assuming a certain soil storage capacity. He defined climate as 'moist' when rainfall exceeds

Table 2.2 Different indices used to classify moisture regimes

	Climatic parameter	References
(i)	Rainfall and temperature	de Martonne (1926), Emberger (1930, 1955), Köppen (1936), Gaussen (1955), Shanbhag (1956)
(ii)	Rainfall and evaporation	Transeau (1905), Prescott (1934), Trumble (1937), Hosking (1937), Thornthwaite (1948), Capot-Rey (1951), Shanbhag (1956)
(iii)	Rainfall and relative humidity	Mangenot (1951)
(iv)	Rainfall and potential evapotranspiration	Cochémé and Franquin (1967), Hargreaves (1971), UNESCO (1979), Chowdhury and Sarwade (1980)
(v)	Rainfall and actual avapotranspiration	Huke (1982)
(vi)	Length of growing or rainy season	Stefanoff (1930), Thomas (1932), Aubreville (1949), Bagnouls and Gaussen (1953), Troll (1965)
(vii)	Combination of above factors	Papadakis (1961, 1970a,b, 1975)

Table 2.3 Thornthwaite's classification of tropical climates

Humidity provinces		Characteristic vegetation	Precipitation effectiveness index (PE Index)
A	Wet	Rainforest	>128
B	Humid	Forest	64–127
C	Subhumid	Grassland	32–63
D	Semi-arid	Steppe	16–31
E	Arid	Desert	<16

Tropical climates have a temperature effectiveness index (TE Index) >125

$$\text{PE Index} = \sum_{i=1}^{12} 115 \frac{[(P)/(T-10)]}{} n^{10/9}$$ P = mean monthly rainfall (inches), T = mean monthly temperature (F°) and $n = 1$.

$$\text{Or} = \sum_{i=1}^{12} 10\,(P/E)\,n$$ E = evaporation from the free water surface and $n = 1$.

$$\text{TE Index} = \sum_{i=1}^{12} \frac{[T-32]}{4} n$$ T = mean monthly temperature (F°) and $n = 1$.

evapotranspiration and 'dry' when a moisture deficit occurs due to more evaporative demand or less rainfall. He also distinguished between 'actual' and 'potential' evapotranspiration. When rainfall is adequate and meets the evaporative demand, the moisture index is considered equal to zero. The 'moist' or 'wet' climates have positive moisture indices and the 'dry' or 'arid' climates have negative values of these indices (Table 2.3).

Carter (1954) used Thornthwaite's method and classified climates of Africa and India, and Meigs (1953) developed a generalized climatic map of the arid regions. It is important to realize, however, that Thornthwaite's classification system has definite limitations when applied to the tropics. The computation of potential evapotranspiration by this method usually underestimates water losses in the tropics and provides an inadequate estimate of the crop water requirements. One of the reasons for this difference is the differential available water storage capacity of the tropical in comparison with the temperate region soils.

The rainfall effectiveness obviously depends on the temperature regime. A range of aridity indices has been proposed on the basis of temperature and moisture regimes. Lauer (1952) and de Martonne (1926, 1962) developed empirical relations relating to temperature. De Martonne's aridity index, computed for a week, month or year, is based on the following relationship:

$$I = \frac{np}{t + 10}$$

where I is the aridity index, n is number of rainy days, p is mean precipitation per day, and t is the mean temperature of the desired period in °C. When this index is used on an annual basis, the following are the limits to classify different regions by climax vegetation:

<5	desert
10–20	dry steppe
20–30	prairie
>30	forest

Garnier (1961a, b) improved upon this concept of classifying climates on the basis of 'moist' or 'dry' regions by computing potential evapotranspiration using his own method and that of Thornthwaite. He defined a wet tropical month with the following criteria:

(i) mean monthly temperature of over 20°C;
(ii) mean relative humidity of over 65 per cent; and
(iii) mean vapour pressure of over 20 mb.

Garnier (1961) and Budyko (1958) defined a 'humid tropical' region as where these conditions prevail for at least 8 months in a year; and where mean annual rainfall is at least 1000 mm.

In contrast to the humid tropics with generally high available soil moisture, semi-arid tropics are characterized by long dry spells of unpredictable length and timing. Reliability of rainfall delivery is an important consideration in semi-arid regions, where inter-annual variability in rainfall is much more than in humid climates. Different indices have been proposed to assess the adequacy of rainfall by comparing the rainfall received with the potential evapotranspiration, i.e., evaporation from crops when soil water is not limiting. The evaporation of crops fully supplied with water averages about 0.8 of pan evaporation (Jätzold, 1977) and varies widely with soil, climate and crop (Miller, 1977).

Some climatologists (Budyko, 1956; 1974; Lettau, 1969) have partitioned rainfall into evaporation (E) and run-off (N) and have defined the radiation index of dryness as follows:

$$D^* = R/LP$$

where L is latent heat of vaporization (600 cal g^{-1}), P is rainfall (cm), and R is radiation (lyr $^{-1}$).

The asterisk and bars refer to the annual values. Fig. 2.3 shows the bioclimatic zonality on the basis of this aridity index (Budyko, 1974). Lettau (1969) used Budyko's (1956) equations and computed other derived variables as shown below. It is important to remember, however, that Budyko used radiation calculated over the moist surface and Lettau used the one computed over a representative surface (R). The empirical relations of Lettau are:

$P\ \ = R/LD^*$
$A^* = E/P = \tanh D^*$
$E\ \ = R \tanh D^*/LD^*$
$C^* = N/P = 1 - \tanh D^*$
$N\ \ = R(1 - \tanh D^*/LD^*$

where N is run-off, A^* is evaporation ratio and C^* is run-off ratio.

Bailey (1958) developed a 'moisture index' on the basis of mean monthly rainfall and mean monthly temperature. The monthly moisture index (s_i) is defined as

$$s_i = 0.18p/1.045^t$$

where p is the mean monthly precipitation and t is the mean monthly temperature (°C).

The annual moisture index (S) is the summation of mean monthly values

$$S = \sum_{i=1}^{12} s_i.$$

Bailey (1979) classified moisture provinces and moisture realms on the basis of the annual moisture index S (Table 2.4). The moisture provinces for the region 30° north and south of the equator according to this index are shown in Fig. 2.4. According to this, the moisture index for 'semi-arid' climates ranges between 2.5 and 4.7 and the threshold value between dry and moist realms is 6.37. Moisture indices greater than 6.37 indicate surplus and lesser indices indicate moisture deficit. Obviously, the moisture index increases with increase in rainfall but decreases exponentially with increase in temperature. Distribution of tropical climates according to Bailey's annual moisture index is shown in Fig. 2.5. Bailey also computed the relationship between mean annual temperature and mean annual rainfall and compared it with the moisture provinces computed by Thornthwaite's method (Fig. 2.6).

Yet another index of estimated moisture availability has been proposed by Hargreaves (1977). In this system climate is first divided into subtropical and tropical regions on the basis of the mean annual temperature. Within each of these broad categories, climates are further classified to range from very arid to

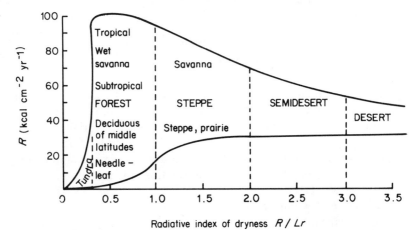

Figure 2.3 The bioclimatic zonality on the basis of Budyko aridity index (Budyko, 1974). *Reproduced by permission of the author*

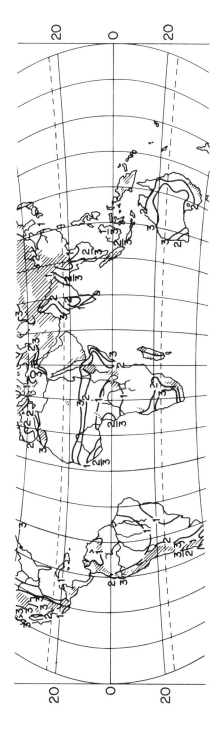

Figure 2.4 The radiational index of dryness for tropical regions (Budyko, 1974). *Reproduced by permission of Mezhdunarodnaya Kniga*

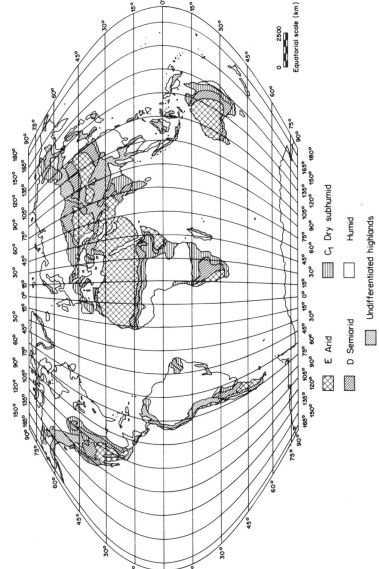

Figure 2.5 Moisture provinces according to Bailey's annual moisture index S (Bailey, 1979). *Reproduced by permission of Springer-Verlag*

Table 2.4 Moisture provinces and realms according to Bailey moisture index (Bailey, 1979)

Annual moisture index	Moisture province	Moisture realm
	Arid	
2.5		
	Semi-arid	Dry
4.7		
	Dry sub-humid	
6.37	-----------------------------------	-----
	Moist sub-humid	
8.7		
	Humid	Wet
16.2		
	Perhumid	

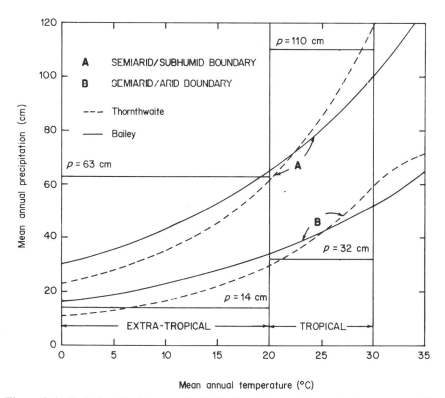

Figure 2.6 Relationship between the Bailey's moisture index and that computed by the Thornthwaite's method (Bailey, 1979). *Reproduced by permission of Springer-Verlag*

Table 2.5 Climatic classification according to Hargreaves' Moisture Availability Index (MAI) (Hargreaves, 1977)

Criteria	Climate classification	Productivity classification
1. All months with MAI in the range of 0.00 to 0.33	Very arid	Not suited for rainfed agriculture
2. 1 or 2 months with MAI of 0.34 or above	Arid	Limited suitability rainfed agriculture
3. 3 or 4 months with MAI of 0.34 or above	Semi-arid	Production possible for crops requiring only 3 to 4 months of growing season
4. 5 or more consecutive months with MAI of 0.34 or above	Wet–dry	Production possible for crops requiring a good supply of water during 5 or more months
5. 1 or more months with MAI above 1.33	Somewhat wet	Natural or artificial drainage required for good production
6. 3 to 5 months with MAI above 1.33	Moderately wet	Good drainage required for normal agricultural production
7. 6 or more months with MAI above 1.33	Very wet	Very good drainage required for normal agricultural production

Table 2.6 Moisture availability index (MAI) during the sorghum-growing season at representative locations in semi-arid Africa (ICRISAT, 1982)

Location	Moisture availability index (MAI)					
North of equator	June	July	Aug.	Sept.	Oct.	Nov.
Ouagadougou (Upper Volta)	0.47	1.00	1.47	0.75	0.06	0.00
Kano (Nigeria)	0.48	1.07	1.76	0.59	0.00	0.00
Geneina (Sudan)	0.07	0.67	1.36	0.29	0.00	0.00
South of equator	Dec.	Jan.	Feb.	Mar.	Apr.	May
Inhambane (Mozambique)	0.37	0.32	0.36	0.36	0.36	0.31
Livingstone (Zambia)	0.62	0.80	0.55	0.29	0.03	0.00

very wet depending on the moisture availability index (MAI). MAI is a water balance evaluation and is calculated according to the following equation:

$$MAI = PD/ETP$$

where MAI is the 75% probability of rainfall occurrence, PD is the amount of rainfall in a given period that is equalled or exceeded during three-quarters of the year, and ETP is the potential evapotranspiration. Using these relations, Hargreaves calculated world-wide values for MAI, PD and ETP for 640 locations, and developed a climatic classification system on the basis of rainfall dependability and crop productivity (Table 2.5). This index has an advantage over other indices because it considers requirements for water management (irrigation or drainage) for obtaining high crop yield. However, the index is still very broad and general and has not been widely tested. The MAI index for some sorghum-growing locations in West Africa is shown in Table 2.6.

C. Vegetation and ecological indices

Both moisture and temperature regimes influence the length of the growing season and the nature of species that can be grown in different environments. The climax vegetation and the predominant species in an ecology are influenced by the climatic variables and soil conditions, as are the crops that can be grown in those environments. The index value of the natural vegetation can, therefore, be used to classify the climatic environment. For example, dry or arid region plants are especially adapted to environments with pronounced water deficits. On the other hand, semi-arid regions receive enough rainfall for at least a few months of the year to bring soil moisture up to the levels to create grasslands or shrubland. The humid and perhumid regions receive enough rainfall to support perennial crops. Köppen (1931) used this concept and distinguished between the 'tropical rainforest' and 'savanna' regions on the basis of the climax vegetation that can be supported in an ecology. Klages (1942) classified tropical natural vegetation as desert, grassland and woodland. The woodland or the forest regions were further subdivided into subtropical rainforest, savanna forest, monsoon forest, and tropical rainforest. Each of the major vegetation types is characterized by distinct requirements. Holdridge (1959) developed a classification system on the basis of the physiognomy of the natural vegetation, and stressed the dependence of an ecosystem on the climate. He divided the climates into natural life zones on the basis of their dependence on temperature, rainfall and evapotranspiration potential. This model has been further developed by Tosi and Voertman (1964), Baker (1966) and Jätzold (1977).

The ecological concept is rather broad, general, and at best a qualitative integrated index of the ecological environment in relation to plant adaptation. The main factor that influences the climatic vegetation is the water available to meet the transpirational requirments. It is difficult, however, to establish a one-to-one quantitative relationship between the available moisture and the vegetation because the condition of growth, plant density, and percentage ground cover all interact with the available moisture regime on the one hand and the soil condition on the other. For example, Ochse *et al* (1961) reported data of Coster (1937) and of Mohr et al (1974) regarding the transpirational needs of different species in Java. Data in Table 2.7 showing wide variations in transpiration among species and due to conditions of their growth support the argument of the qualitative nature of using the climax vegetation as a criterion towards rational climatic classsification.

Mehr-Homji (1960) advocated the use of phytogeograhic criteria to classify tropical climates, and further classified these in the following categories:

Table 2.7 Relationship between species and their growth conditions to the transpirational requirements (mm) (After Ochse et al., 1961). *Reproduced by permission of Macmillan Publishing Co.*

Species	Condition of plant growth		
	Luxuriant	Average	Poor
Imperata cylindrica	1750	1000	300
Eupatorium pallescens	2900	2000–1600	1000
Leucaena glauca	4670	4000–3000	—
Acadia villosa	2400	1600	—
Crotalaria anagyroides	2300	1500	—
Tephrosia maxima	3100	2000	1000
Albizia falcate	2300	—	—
Thea sinensis	900	500	—
Hevea brasiliensis	1200	—	—
Tectona grandis	1200–1100	1000–800	500–400
Mountain forest			
Trees	740 870	—	—
Undergrowth	130	—	—
Bamboo forest	3000	1500	—
Jungle <1000 m.a.s.l.		>1200	—
1000 m.a.s.l.		1200–500	—
25000 m.a.s.l.		600–500	

(i) floristic criteria: some species are exclusively confined to humid or arid climates;
(ii) Morpho-ecological criteria: based on the adaptability of plants to weather/climate;
(iii) agronomic criteria: the rainfall amount is adequate to grow crops of economic importance;
(iv) vegetational criteria: vegetational aspects are the expression of the prevailing ecological conditions.

Some seasonal crops, e.g., rice, can be grown under special ecological conditions with continuous inundation. The soils with this special moisture regime also have a definite place in a landscape or a toposequence, i.e., valley bottoms. Many researchers have attempted to define the moisture regime for rice on the basis of its water requirements and growth habits. Huke (1982) classified climatic environments of southeast Asia in which rice is grown. A monthly rainfall of 200 mm was considered to be the minimum requirement, because 50% of it is generally accounted for by evapotranspiration, and the remainder is percolation losses; dry months were considered to be those in which the potential evapotranspiration exceeded the actual evapotranspiration. It is important to use actual evapotranspiration rather than the rainfall because a water deficit in an agricultural sense does not occur until the soil moisture reserves in the root zone are exhausted. Crop suitability maps on the basis of specific requirements can be made for different crops. The requirements vary not only among crop species but also among cultivars depending on their growth duration, leaf characteristics, and the root system configuration. Jätzold (1977) has developed a similar classification system for coffee in Kenya, and the FAO ecological zonation project has developed production indices for a range of crops in Africa.

D. Duration of the rainy season

Many climatologists have used the duration of the wet or dry periods as a criterion for climatic classification. Troll (1965) proposed a system based on thermal and hygric regimes, and emphasized the importance of the number of the dry or the humid months. He observed that the following values are associated with the main vegetation zones of tropical Africa and South America.

Humid months	*General vegetation*
12 to 9.5	Tropical rainforest and transitional forest
9.5 to 7	Humid savanna
7 to 4.5	Dry savanna (wet/dry semi-arid tropics)
4.5 to 2	Thorn savanna (dry semi-arid tropics)
2 to 1	Semi-desert (arid)
1 to 0	Desert (arid)

A humid month is considered to be one in which rainfall exceeds the potential evapotranspiration. In addition to its simplicity, the system does not consider the average amount of rainfall in the humid month.

E. Agronomic criteria

Moisture availability and the length of the growing season are important criteria for crop production. Two commonly used indices relating climatic factors to crop growth and their geographic distribution are those of Papadakis and Cochémé and Franquin. Papadakis (1961, 1970a; b; 1975) based his classification system on the following criteria:

(i) average daily maximum and minimum tempertures;
(ii) length of the frost-free season;
(iii) evaluation of potential evapotranspiration as a function of the saturation deficit; and
(iv) evaluation of the thermic and hydric regimes.

For example, the tropical and subtropical climates are subdivided into 9 and 7 thermic types, respectively.

Papadakis defined the moisture index (H) as follows:

$$H = \frac{P + W}{E}$$

where P is the monthly rainfall, E is the monthly potential evapotranspiration and W is the water stored in the soil from previous rains. Different climatic types on the basis of this moisture index are as follows:

Moisture index	Climatic type
<0.25	Arid (a)
0.25 to 0.50	Dry (s)
0.50 to 1.00	Intermediate humid (y)
>1.0 (with $P<E$)	Posthumid (p)
1.00 to 2.00 (with $P+W-E<100$ mm)	Humid (g)
>2.00 (with $P+W-E$ 100>mm)	Wet (w)

Papadakis related these indices to crop water requirements and on this basis explained the geographic distribution of individual crops.

In the semi-arid regions, duration of periods during which rainfall (P) exceeds selected levels of evapotranspiration (ET) is a useful index to define the agricultural potential of the region. Brown and Cochémé (1969) and Cochémé and Franquin (1967) used the following limits of water availability for tropical Africa:

$P > $ ET	Humid
ET $>+ P >$ 1/2ET	Moist period
$P = $ 1/2ET to 1/4ET	Moderately dry period
$P = $ 1/4ET to 1/10ET	Dry period
$P <$ 1/10ET	Very dry period

This index considers both the rainfall and the soil water storage in the root zone to develop crop water availability calenders.

The climatic water budgeting approach along with the soil-water storage value has also been used to produce maps of crop-growing seasons (Krishnan and Thanvi, 1972; Lawson, 1977). These maps are useful guidelines to delineate regions of different growing seasons. The combination of moisture index with the major soil types can be used to demarcate broad soil climatic zones of a region similar to those prepared for India by Krishnan and Mukhtar Singh (1968).

F. Economic and cultural criteria

Humans are an important ecological factor because they interact with their environment. Too often, grandiose land development schemes have failed because this important interacting factor, i.e. the socio-economic consideration, has been overlooked. Technological developments have little impact unless supported by political, social, and infrastructure requirements that are conscious of human needs in an environment. Otremba (1950) and others defined tropical regions on the basis of economic geography. The boundaries or limits of economical plants are not fixed and change with the technological advances.

IV CHOICE OF AN APPROPRIATE CLASSIFICATION SYSTEM

Among the different systems discussed, none has a universal applicability from an edaphological point of view. Identification, evaluation and development of soil and water resources of the tropics for sustained productivity requires a system of quantitative evaluation of the climatic environments of the tropics. The successful system must be related to productivity and should identify production constraints. The fact that agricultural production in equatorial tropical environments often falls short of expectations is attributed to the lack of an appropriate system that defines the productive potential of different ecologies in relation to the climatic factors.

As will be discussed in the subsequent chapters, some important edaphological factors must be considered in developing a productivity index for tropical environments. High soil temperature is one such factor that has not been given the attention it deserves. Significant yield reductions are often related to high soil temperatures that adversely affect crop establishment during the seedling stage. A high probability of rainless periods of even 5 to 10 days duration is more critical in determining crop yield in these environments than the use of potential evapotranspiration or the generalised water balance. Although the information is not widely available, the use of 'actual evapotranspiration' is definitely a better indicator of crop moisture availability then potential evapotranspiration.

The choice of an approach to climatic classification is governed by the objectives for which the information is required. A classification system without sharply focused objectives has limited application. It is rather difficult to develop a classification system that has application for a diverse range of environments in the tropics and for varied interests and objectives. A flexible and a problem-solving system can be a useful tool in planning for agricultural development. There is an urgent need to develop a system for planning for new arable land development, such as for the new land development schemes in the Amazon, Sumatra, Central and West Africa, and the Northern Territory of Australia. From the agricultural point of view, the objectives of a suitable classification system are to:

(i) define the productive potential of an ecology and identify the suitability and output of crops that can be grown there;
(ii) delineate the zones of moisture surplus and deficit so that appropriate techniques of water management and resource conservation can be developed;
(iii) identify regions with similar ecological environments (isoclimes) so that technology developed in one environment can be transferred to other locations of similar climatic regimes;
(iv) facilitate the choice of appropriate land use and quantify the levels of risks associated with it; and
(v) promote development of crop calendar and farm planning.

This broad range of objectives can be achieved if the basic climatic data are available for those parameters that have direct bearing on crop growth, development, and economic yields. The range of these parameters can be rather wide and vary among crops and different ecological regions. For rainfed upland agriculture, it is important to have quantitative information about the following factors:

(a) Frequency analysis of long-term rainfall records regarding:

   (i) probability of 5 to 10 day rainless periods during the growing season;
   (ii) maximum amount of rainfall received in a day and its return period;
   (iii) maximum rainfall intensity sustained over a period of 10, 20 or 30 minutes,

(b) Soil temperature in the root zone and its relation to agronomic practices.
(c) Levels of solar radiation during the growing season.
(d) Evapotranspiration, or the basic data required to compute it by different methods.

Adequate assessment of these agroclimatic resources is a prerequitiste for successful implementation of large-scale land development projects. Failures and disappointments of many large-scale projects have been due to oversight in proper assessment of the basic climatic parameters (Chang, 1968).

From the agricultural point of view the most useful index is the one that combines climatic parameters with those of soil resources and defines the integrated yield potential of an ecology. These crop productivity indices in relation to soil and environments are discussed in another chapter.

REFERENCES

Aubreville, A., 1949, Climates, forest et desertification de l' Afrique tropicale, *Paris Société d' Editions Geographiques, Maritimes et Coloniales.*
Bailey, H.P., 1958, A simple moisture index based upon a primary law of evaporation, *Geograf. Ann.*, **3-4**, 196–215.
Bailey, H.P., 1979, Semi-arid climates: Their definition and distribution, in A.E. Hall, G.H. Cannell and H.W. Lawton (eds.), *Agriculture in Semi-Arid Environments*, Springer-Verlag, Berlin, Heidelberg, New York, pp. 73–97.
Bagnouls, F., and Gaussen, H., 1953, Saison seche et regime xerothermique. *Documents pour les Caters des Productions Vegetales, Toulouse*, **3**, 1–49.
Baker, S., 1966, The utility of tropical regional studies, *Prof. Georg.*, **18**, 20–22.
Brown, C.H., and Cochémé, J., 1969, A study of the agro-climatology of the highlands of Eastern Africa, *Technical Note No. 125,* WMO, Geneva.
Buck, L.V., 1829, Quoted by Manshard, W., 1974, *Tropical Agriculture*, Longman, London, 226 pp.
Budyko, M.I., 1956, *The Heat Balance of the Earth's Surface*, Translated by N.A. Stepanova, US Weather Bureau, Office of Technical Services, Washington DC, 1958, MGA 8.5-20 11B25 13E-286.

Budyko, M.I. 1958, *The Heat Balance of the Earth's Surface*, U.S. Dept. of Commerce, Office of the Technical Services, 259 pp.
Budyko, M.I., 1974, *Climate and life*, Academic Press, New York, 508 pp
Candollé, A.P. de, 1874, Geographic botanique raisonnee, *Arch. Sci. Biblio.*, Univ. Génève.
Capot-Rey, R., 1951, Une carte de l'indice d'aridité au Sahara français, *Bull. Assoc. Geog. Francais*, pp. 73–76.
Carter, D.B., 1954, Climates of Africa and India according to Thornthwaite's 1948 classification, Final Report, *Climatology* 7(4), Drexel Inst. Techn., Lab. Climatology, Centerton, N.J.
Chang, J.H., 1968, Progress in agricultural climatology, *Prof. Geog.*, **20,** 317–320.
Chowdhry, A., and Sarwade, G.S., 1980, A simple approach for climatic classification of Index, *Tropical Ecology*, (in Press).
Cochémé, J., and Franquin, P., 1967, An agroclimatological survey of semi-arid area in Africa south of the Sahara, *Technical Note No. 86*, WMO, Geneva.
Coster, Ch., 1937, De verdamping van verschillende vegetatievormen of Java, *Tectona*, **30**, 1–102.
Emberger, L., 1930, La vegetation de la region mediterranéenne. Essai d'une classification des groupements vegetaux. *Revue Gen. Bot.*, 42 641–662, 705–721.
Emberger, L., 1955, Une classification biogéographique des climates. Recueil des travaux des laboratoires des botanique, geologie et Zoologie de la Facultié des Sciences de l' Université de Montpollier, *Serie botanique*, No. 7, pp. 3–45.
Flohn, H., 1957, Zur Frage der Einteilung der Klimazonen, *Erdkunde*, **11**, 161–175.
Garnier, B.J., 1958, Some comments on defining the humid tropics, *Research Notes 2*, Ibadan, 9–25.
Garnier, B.J., 1961a, The idea of humid tropicality, *Proc. Tenth Pacific Sci. Congr.*, Honolulu.
Garnier, B.J., 1961b, Mapping of the humid tropics: Climate criteria, in Delimitation of the Humid Tropics, *Geogr. Rev.*, **51**, 339–346.
Gaussen, H., 1955, Les climats analogues a l' echelle du monde. *Comptes Rendus Hebdomadaires des Seances Acad. Agr. France*, Vol. 41.
Hare, F.K., 1977, Climate and desertification, in *Desertification: its causes and consequences* (UN Conference on Desertification, Nairobi), Pergamon Press, Oxford.
Hargreaves, G.H., 1971, Precipitation dependability and potential for agricultural production in northeast Brazil, Publication No. 74–D–159, EMBRAPA and Utah State Univ. 123 pp.
Hargreaves, G.H., 1975, Water requirements manual for irrigated crops and rainfed agriculture, *Publication* No. 75–D–158. EMBRAPA and Utah State Univ., 40 pp.
Hargreaves, G.H., 1977, *World water for agriculture*, AID, Utah State Univ. Contract No. AID/ta-c-1103, 1977.
Hargreaves, G.H., 1981, Responding to tropical climates —an approach, in L.E. Slater (ed.) *Food and climate review*, Aspen Institute of Humanistic Studies, Colorado, pp. 29–38.
Hettner, A., 1930, Die Klimate der Erde, *Geogr. Schriften*, **5**.
Hettner, A., 1934, Die Klimate der Erde, *Vergleichende Landerkunde 3*, 87–202.
Holdridge, L.R., 1959, Ecological indications of the need for a new approach to tropical land use, in A. Sampler (ed.), *Symposia Interamericana*, Turrialba, Costa Rico 1–12.
Hosking, J.S., 1937, The ratio of precipitation to saturation deficiency of atmosphere in India, *Curr. Sci.*, 422–423.
Huke, R.E., 1982, *Agroclimatic and dry-season maps of South, Southeast and East Asia*, IRRI, Los Banos, Philippines, pp. 15.
ICRISAT, 1982, *Annual Report, International Crops Research Centre for the Semi-Arid Tropics*, Hyderabad, p.234.

Jätzold, R., 1970, Ein Beitrag zur Klassification des Agrarklimas der Tropen, *Tubinger Geogr. Studien*, 57–69.
Jätzold, R., 1977, Humid month isolines as an aid in agricultural planning, *Applied Sci. Develop.*, **9** 140–158.
Klages, K.H.W., 1942, *Ecological Crop Geography*, Macmillan, New York, pp. 615.
Köppen, W. 1931, Die Klimate der Erde, *Geogr. Schriften*, **5**.
Köppen, W., and R. Geiger, 1936, *Handbuch der Klimatologie*, Gebrüdes Bornträge, Berlin.
Köppen, W., 1936, Das geographische System der Klimate, in W. Köppen and R. Geiger (eds.), *Handbuch der Klimatologie* Vol. I, Part C, Gebrüder Bornträge, Berlin.
Krishnan, A., and Mukhtar Singh, 1968, Soil climatic zones in relation to cropping patterns, *Symp. On Cropping Pattern*, ICAR, New Delhi, pp 172–185.
Krishnan, A., and Thanvi, K.P., 1972, Studies on duration of agricultural growing season in India under rainfed farming using climatological water budgeting approach, Annual Report, CARI, ICAR, India.
Lauer, W., 1952, Humide und aride Jahreszei ten in Afrika und Sudamerika und inhre Beziehung Zu den Vegetationsgurtel, *Bonner Geogr. Abhandll.*, **9**, 15–98.
Lawson, T.L., 1977, Possible impact of agroclimatological studies on food production in the humid topics with particular reference to West Africa. *WMO/FAO Conf. on Application of Meteorology to Agriculture*, IITA, Ibadan, Nigeria.
Lettau, H., 1969, Evaportranspiration climatonomy, *Month. Weather Rev.*, **97**, 691–699.
Mangenot, G., 1951, Une formula simple permettant de caracteriser les climats de l' Afrique intertropicale dans leurs rapports avec la vegetation, *Rev. Gen. Bot.*, **58**, 353–372.
Manshard, W., 1979, *Tropical agriculture*, Longman, London and New York, 226 pp.
Martonne, E. de, 1926, Une nouvelle fonction climatologique: l' indice d' aridite, *La Météorologie*, **68**, 449–458.
Martonne, E. de, 1962, Une nouvelle fonction climatologique: l' indice d' aridite, *La Météorologie Nouvelle*, **2**, 449–455.
Mehr–Homji, V.M., 1960, Classification of the semi-arid tropics: Climatic and Phytogeographic approaches, *Proc. Symp. Climatic Classification*, ICRISAT, Hyderabad. 7–16.
Meigs, P., 1953, World distribution of arid and semi-arid homoclimates, *Arid zone programme* **1**, pp. 203–210, UNESCO, Paris.
Miller, D., 1977, Water at the surface of the earth, *Int. Geophys. Ser.* **21**, Academic Press, New York.
Mohr, E.C.J., Van Baren, F.A., and Van Schuylenborgh, J., 1974. *Tropical Soils*. Mouton-Ichtiar Baru-Van Hoeve, The Hague, Netherlands, 481 pp.
Ochse, J.J., Soule, M.J., Dijkmau, M.J., and Wehlburg, C., 1961, *Tropical and Subtropical Agriculture*, Collier-Macmillan, New York, pp. 760.
Otremba, E., 1950, Die wirtschaftsgeographische Ornung der Länder, *Die Erde*, **1**, 216–232.
Papadakis, J., 1961, *Climatic Tables of the World*, Libro De Edicion, Buenos Aires, Argentina.
Papadakis, J., 1966, *Crop Ecology Survey in West Africa*, FAO, Rome: UN Pub. MR/16439/1 vol 2, MR/26939/I vol I.
Papadakis, J., 1970a, *Climates of the world, their Classification, Similitude, Differences and Geographic Distribution*, Libro De Edicion, Argentina, Buenos Aires, 47 pp.
Papadakis, J., 1970b, *Agricultural potentialities of world climates*, Libro De Edicion, Argentina, Buenos Aires, 70 pp.
Papadakis, J., 1975, *Climates of the world and their agricultural potential*, Libro De Edicion, Argentina, Buenos Aires, 200 pp.

Philippson, A., 1933, *Grundzüge der allgemeinen Geographic*, Leipzig, Vol. 1.
Prescott, J.A., 1934, Single value climatic factors, *Trans. Royal Soc. S. Aust.*, **58**, 48–61.
Proshaska, F., 1967, Climatic classifications and their terminology, *Int. J. Biomet.*, **11**, 1–3.
Riehl, H., 1979, *Climate and Weather in the Tropics*, Academic Press, London, 611 pp.
Shanbhag, G.Y., 1956, The climates of India and its vicinity according to a new method of classification, *Indian Geog. J.*, **31**, 20–25.
Slater, L.E., (ed.), 1981, *Food and Climate Review*, Aspen Inst. Humanistic Studies, Colorado, pp. 68.
Stefanoff, B., 1930, A parallel classification of climates and vegetation types, *Sbornik na Bulgarskata Akademiya na Naukite*, 26.
Supan, A., 1879, Die Temperaturzonen der Erde, *Pet. Mitt.*, 13.
Thomas, A.S., 1932, The dry season in the Gold Coast and its relation to the cultivation of Cacao, *J. Ecol.*, **20**, 263–269.
Thornthwaite, C.W., 1948, An approach toward a rational classification of climate, *Geog. Rev.*, **38**, 55–94.
Tosi, J.A., and Voertman, R.F., 1964, Some environmental factors in the economic development of the tropics, Econ. Geogr., **40**, 189–205.
Transeau, E.N., 1905, Forest centres of Eastern America, *Am. Natur.*, **39**, 875–889.
Troll, C., 1943, Thermische Klimatypen der Erde, *Pet. Mitt.*, **89**, 81–89.
Troll, C., 1963a, Landscape, ecology and land development with special reference to the tropics, *J. Trop. Geog.*, **17** 1–11.
Troll, C., 1963b, Quant-Bewässerung in der Alten und Neuen Welt, *Mitt. Osterreich Georgr. Ges.*, **105**, 313–330.
Troll, C., 1965, Seasonal climates of the earth, in Rodenwalt, E., and Jusatz, H., (eds.), *World Maps of Climatology*. Springer Verlag, Berlin.
Trumble, H.C., 1937, The climatic control of agriculture in South Australia, *Trans. Royal Soc. S. Aust.*, **61**, 41–62.
Turc, L., and Lecerf, H., 1972, Indice climatique de potentialité agricole, *Sci. Sol.*, **21**, 81–102.
UNESCO, 1979, Carte de la repartition mondiale des régions arides, Note Technique du MAB 7, UNESCO, Paris.
Virmani, S.M., Huda, A.K.S., Reddy, S.J., Sivakumar, M.V.K, and Bose, M.N.S., 1980, Approaches used in classifying climates with special references to dry climates, *Agroclimatology Progress Report 2*, ICRISAT, Hyderabad, pp. 25.
Walter, H., 1973, *Vegetation of the earth*, Springer-Verlag, Berlin.
Williams, G.D.V., and Masterton, J.M., 1980, Climatic classification, agro-climatic resource assessment, and possibilities for application in the semi-arid tropics, *Symp. Proc. Climatic Classification*, ICRISAT, Hyderabad, pp. 45–57.

Chapter 3
Rainfall, Vegetation, Soil

I	TROPICAL ECOLOGY	46
II	RAINFALL	48
	A. Amount	48
	B. Rainfall Distribution	53
	C. Variability	53
	D. Methods of expressing rainfall variability	54
	E. The choice of method of evaluation of rainfall variability	59
	F. Start of the Growing Season	62
	G. Rainfall Effectiveness	67
	H. Intensity and Energy Load	76
III	SOILS OF THE TROPICS	88
	1. Oxisols	88
	2. Ultisols	90
	3. Alfisols	90
	4. Vertisols	91
	5. Aridisols	91
	6. Andosols	91
	A. Soils of Tropical Africa	91
	B. Soils of Tropical Asia	93
	C. Soils of Tropical Australia	96
	D. Soils of Tropical America	96
	E. Soil Physical Constraints to Crop Production	97
IV	VEGETATION OF THE TROPICS	99
	A. Tropical Forests	99
	B. Tropical Savannas	100
V	POTENTIAL PRODUCTION OF A DIFFERENT TROPICAL ECOLOGIES	102
	A. Comparative Productivity of Different Ecologies	102
	B. Potential cultivable area in different ecologies	106

I. TROPICAL ECOLOGY

The word 'ecosystem' refers to a functional entity of an ecology created by the interaction of all living organisms, or to a community with its physical chemical,

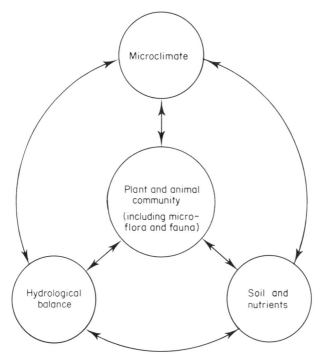

Figure 3.1 An ecosystem

biological and social environments. The ecosystem is, therefore, a region, space or habitat with its unique biophysical and social environments. It is a dynamic entity and is, at a given point in time, at equilibrium with its immediate surroundings (Fig. 3.1). The transition from one ecology to another is gradual in space, but a rapid transition in time is brought about by the alterations of this delicate balance by factors within or outside the ecosystem. All terrestial ecosystems when considered together as one unit are referred to as the biosphere.

The word 'agro-ecology' refers to a sub-region within an ecology that is related to a specific activity of agricultural importance. The concept of ecology can therefore be used at different scales within an ecosystem. For example, one can have an 'upland ecology' or a 'wetland ecology' depending upon the relief and the moisture regimes. Different flora and fauna species are adapted to these ecologies, even though they may occur within short distances of one another. Large regions of a similar vegetation with easily definable characteristics are termed as biomes. The important biomes in the tropics are rainforest, savanna, grassland, etc.

This chapter describes some major climatic variables (the rainfall regime, temperature and solar radiation) that play important roles in delineating different ecologies of the tropics. Also discussed are the major soils and vegetation biomes of economic importance. Subsequently, a brief description is presented of major ecologies of the tropics in relation to one another and in terms of their agricultural importance.

II. RAINFALL

Rainfall is the climatic factor with the most drastic effects on plant and animal growth in the tropics. The success of rain-fed agriculture in particular is dependent on the reliability of rains. No wonder the 'Rain God' has been a powerful diety in many ancient civilizations. The 'God Inder' of Hinduism is but one example of the dependence and helplessness of 'Man' confronted by the vagaries of this powerful ecological factor. It is also for this reason that the attempts to monitor the rainfall regime in tropical Asia and in the Indian subcontinent, for example, date back to 400 BC (Kautilya, 1915; Kurtyka, 1953; Biswas, 1967).

The effectiveness of the rainfall in tropical environments is influenced by the prevalence of high temperature, which causes considerable losses through evaporation and evapotranspiration throughout the year. In addition to temperatures, the important characteristics of rainfall that affect plant growth are: amount, intensity, effectiveness, variability, reliability, length of dry spells and the duration of the rainy season. In this connection both the onset of the rainy season and its termination need to be carefully defined for planning of agricultural operations. Readers are referred to comprehensive reviews of tropical rains elsewhere (Riehl, 1979; Nieuwolt, 1977), and brief summary is presented below.

A. Amount

The mean annual rainfall in tropical regions varies widely, from as low as 250 mm or less to more than 6,000 mm (Fig. 3.2). In some regions of West Africa and northeast India, annual rainfall has been observed to exceed 8000 mm and sometimes even 10 000 or 12 000 mm (Mohr and Van Baren, 1954). A world-wide decrease in tropical rainfall has also been observed since the early twentieth century (Kraus, 1955). Mohr et al. (1972) analysed the relationship between annual rainfall and the number of rainy days for various tropical regions. Regions with monsoonal climates have seasonal rainfall patterns, i.e., rainy seasons alternating with dry seasons. The rainfall events are intense and a small number of rainy days account for most annual rainfall. There are other regions where rainfall is uniformly distributed and the soil remains wet through the year. Rains in these regions are gentle.

The data in Table 3.1 show that the equatorial belt (5°N and 5°S of the equator) receives the maximum rainfall, although there are many exceptions to this general rule. For example, Kowal and Kassam (1978) observed a regular north–south gradient in the rainfall pattern for the African savannah, and reported that rainfall decreased with increasing distance away from the equator. These authors developed a multiple regression equation relating

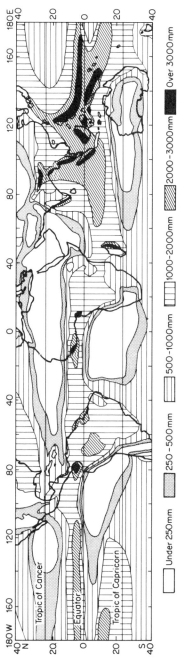

Figure 3.2 Mean annual rainfall in the tropics (Nieuwolt, 1975)

Table 3.1 Rainfall distribution at different latitudes (Schimper, 1903; Ochse *et al.*, 1961). *Reproduced by permission of Macmillan Publishing Co*

Climatic Parameters	Latitude									
	45°N	35°N	25°N	15°N	5°N	5°S	15°S	25°S	35°S	45°S
Rainfall (mm)	570	550	680	950	1970	1890	1230	650	700	1060
Cloudiness (% sunshine)	54	46	40	43	55	59	52	45	49	61
Relative humidity (%)	74	70	71	76	79	81	78	77	79	81

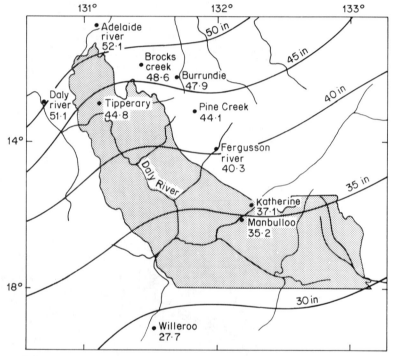

Figure 3.3 Rainfall distribution in northern Australia in relation to latitude Fitzpatrick, 1965)

mean annual rainfall (P) with the latitudinal (LA) and longitudinal (LO) coordinates:

$$P \text{ (mm)} = 2470 - 130.9 \text{ (LA)} - 0.6 \text{ (LO)} \qquad r = 0.91$$

The average decrease in rainfall is 131 mm per degree latitude northward or 1.18 mm per km inland from the ocean. A similar gradient is observed for Northern Territory in Australia by Fitzpatrick (1965). His data in Fig 3.3 show that the mean annual rainfall decreases with the distance away from the equator. There is also evidence that rainfall on the equator is less than in the

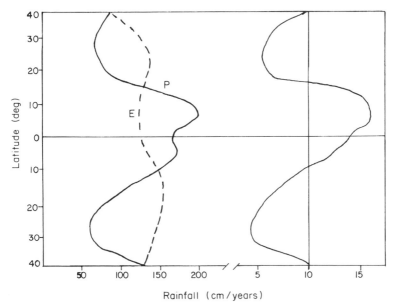

Figure 3.4 Rainfall distribution and location of the meteorological equator (Riehl, 1979). *Reproduced by permission of Academic Press*

region just north or south (Riehl, 1979). Maximum rainfall is received near latitude 5°N, the meteorological equator (Fig. 3.4).

The rainfall patterns in tropical climates are influenced by a zone where the air-flow patterns regularly converge, i.e., where the northeast and southeast trade winds meet and often create a violent confrontation. This zone is referred to as the Inter-Tropical Convergence Zone (ITCZ). The zone has also been termed the equatorial trough. The position of ITCZ during January and July is shown in Fig. 3.5. In the northern hemisphere summer, the ITCZ oscillates to the north, and it migrates to the south during summer in the southern hemisphere. Total annual rainfall varies markedly at different locations in the tropical regions and at the same location with the movement of the equatorial trough. As the trough reaches its extreme latitudes in February and August, these regions experience a single rainfall peak with monomodal distribution. Double rainfall maxima with biomodal distribution appear in those regions where the equatorial trough crosses twice. This is an over-simplification and is not observed everywhere because of differences in the structure and behaviour of the trough at different locations and the effects of monsoons and orographic circulation.

Although the total annual rainfall received in the tropics can be considered adequate, limitations for food crop production throughout the year with rainfed agriculture are imposed by irregularities in the timing, seasonal distribution, and annual variation. It is, therefore, the variability and distribution of rainfall that govern the productive potential of a region.

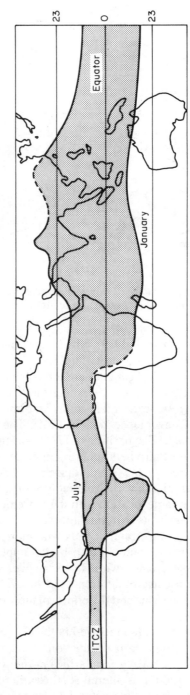

Figure 3.5 Mean position of ITCZ in January and July (Flohn, 1960; Yoshino, 1968; Bouchev, 1975)

Table 3.2 Sub-climate regimes for tropical and equatorial rainfall patterns on the basis of rainfall distribution and the length of rainy season (modified from Le Houérou and Popov, 1981)

Sub-climate	Rainfall amount (mm)	Length of rainy season (days)	Length of growing season (days)
Desert	<100	0	
Very arid	100–200	0–30	0
Arid	200–400	60–90	1–74
Semi-arid	400–600	90–150	75–119
Dry subhumid	600–800	120–150	120–179
Subhumid	800–1200	150–210	180–209
Humid	1200–1500	240–300	210–299
Perhumid	>1500	>300	>300

B. Rainfall distribution:

The water availability to plants depends on rain amount and distribution during the rainy season, and the temperature regime. As described in Chapter 2, the rainy season according to Bagnouls and Gaussens' criterion is defined when rainfall (in mm) exceeds twice the mean monthly temperature in degrees Celsius ($P>2t$). The distribution of rainfall during the rainy season has tremendous edaphological significance because the same amount of rainfall received over different durations results in differential amounts of water available for plants and animals. On the basis of rainfall distribution Le Houerou and Popov (1981) defined four rainfall regimes:

(i) no rainy season/permanent drought,
(ii) a single rainy season/one dry season,
(iii) two rainy seasons/two dry seasons,
(iv) no dry season/permanent rains.

De Martonne (1926) described regions with a monomodal pattern of rainfall distribution as 'tropical' and those with a biomodal pattern as 'equatorial', because the former occur in the vicinity of both tropics and the latter around the equator. This is, however, a misnomer because both rainfall patterns occur near and farther away from equator (i.e. arid and humid varients). Predominant sub-climates on the basis of rainfall distribution for both equatorial and tropical rainfall patterns are shown in Table 3.2.

C. Variability

Tropical rains are believed to have high spatial and temporal variability, and therefore, vary markedly from place to place and year to year (Ojo, 1977). Rainfall varies greatly both in time and space and the importance of variability of rainfall in time is of particular interest in agriculture, because evaporation potential is high throughout the year. Rainfall variability is usually expressed using one of the following methods:

D. Methods of expressing rainfall variability

(i) Relative variability (V_r) is defined as the ratio of the sum of all deviation from the mean, averaged, without respect to sign, to the mean. In other words, it is the ratio between the mean deviation of annual or monthly rainfall over a period of time (P_i) from its arithmetic mean (Schumann and Mostert, 1969):

$$V_r = \frac{\Sigma |P_i - p|}{\Sigma P_i}$$

where P_i is the annual (monthly, daily or hourly) rainfall for a given period and p is the average annual (monthly, daily or hourly) rainfall.

(ii) Relative intersequential variability (V_s) is the ratio between the average absolute differences of successive terms $\frac{(1/}{(\pi - 1)} \Sigma |P_{i+1}|)$ and the arithmetic mean

p (Katsnelson and Kotz, 1957).

$$V_s = \frac{n}{n-1} \frac{\Sigma |P_{i+1}|}{\Sigma P_i}$$

(iii) Coefficient of variation (C_v) is defined as the ratio between standard deviation and the arithmetic mean. Thus.

$$C_v = \frac{\sigma}{p} = n \frac{\Sigma (P_i - p)^2}{P_i}$$

where σ is the standard deviation.

Generally the coefficient of variability (C_v) is used to express the variability from one season to another.

The variability in tropical rains is widely recognized and has been documented by many researchers (Worthington, 1958; Webster and Wilson, 1966). For example, Mohr and Van Baren (1954) observed that annual variations in rainfall in Indonesia can be 60 to 150 per cent for the same location. Ochse et al. (1961) reported annual variations of as much as 20 to 250% for Antilles. Chapman and Kininmonth (1972) reported high variability in annual rainfall received at Darwin (Fig. 3.6).

It is generally believed that variability in rainfall for the tropical belt increases with decreasing annual rainfall (Walker, 1962) (Fig. 3.7). Cochémé and Franquin (1967), from their agroclimatological survey of semi-arid regions of Africa south of the Sahara, reported that the coefficient of variability was about 22.5% at 500 mm and 15% at 1000 mm annual rainfall (Fig. 3.8). The rainfall variability is generally higher in semi-arid than in humid tropics. The

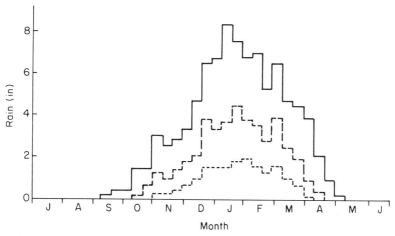

Figure 3.6 Rain variability at Darwin, Australia for the period 1869–1967 (Chapman and Kininmonth, 1972). *Reproduced by permission of Elsevier Science Publishers*

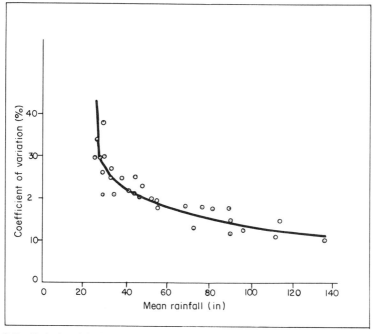

Figure 3.7 Relationship between the coefficient of variation in rainfall and the mean annual rainfall amount (Walker, 1962)

data in Figs. 3.9 (a) and 3.9 (b) show high annual variations in rainfall at Mpwapw, Tanzania, and Maroua, Cameroon, respectively. Also shown is the variation in mean annual rainfall at Maha Illupalma, Sri Lanka, from 1905 to 1957. Le Houerou and Popov (1981) reported that in tropical Africa variability

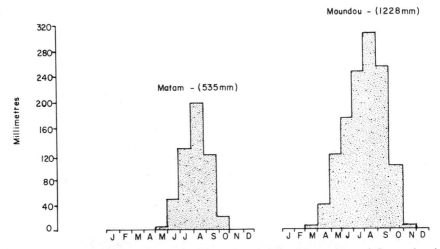

Figure 3.8 Seasonal variation in monthly mean rainfall at Matam in north Senegal and at Moundou in south Chad showing the similarity of distribution pattern (Cochémé and Franquin, 1967). *Reproduced by permission of the World Meteorological Organization*

increases from 10–15% in the rainforest region to more than 50% in the desert. On the basis of the data from the Sudan Meteorological Service, Le Houerou and Popov (1981) reported the following empirical relation between variability (V) and the rainfall amount (P):

$$V = (59.6 + 0.11\ P) \times 100/P$$

This regression is valid for annual rainfall of 100 to 1500 mm with corresponding variability of 70.6 and 14.9%, respectively. This argument is further strengthened if one reviews the variability on a global basis. Regions with annual variability of less than 15% include the Congo and Amazon Basins, and parts of Indonesia receiving high annual rainfall.

Oyebande and Oguntoyinbo (1970) and Balogun (1972) computed the variability of rainfall over Nigeria by using the C_v as a measure of variability and reported a non-linear coefficient of variation between rainfall and C_v. A comparison of the variability in rainfall for two contrasting locations in West Africa (Zaria in northern Nigeria, and Cavalla in southeastern Liberia) was made by Gregory (1969) to indicate the anomaly caused in expressing variability on the basis of mean and standard deviation (Table 3.3). The mean annual rainfalls for Zaria and Cavalla are 1145 and 2637 mm with corresponding length of rainy seasons being 5 and 11 months, respectively. Despite these differences, the annual coefficient of variation is about the same, i.e., 16%. Computing the monthly variability for Zaria for the dry period is, however, difficult because the data have a skewed frequency distribution from November to March, and the percentage variability values are meaningless. Furthermore, the wetter months at drier Zaria are less variable than those at humid Cavalla.

Variations in the annual rainfall distribution of three locations in the humid

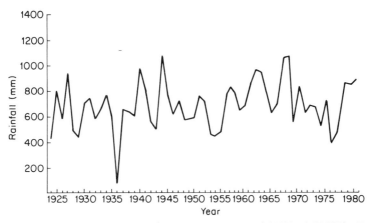

Figure 3.9a Variation in annual rainfall in Mpwapwa (6°20'S, 36°30'E), Tanzania from 1925 to 1980

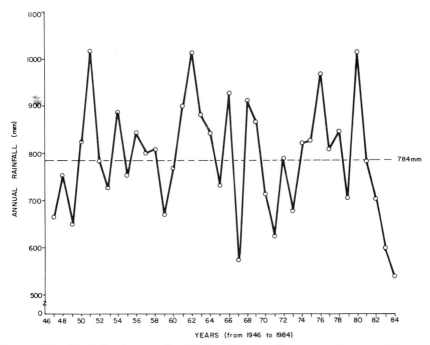

Figure 3.9b Variation in annual rainfall at Maroua, Cameroon (Courtesy Director, IRA, Cameroon)

tropics (Darwin, Honolulu, and Bombay) computed by Riehl (1979) showed that the highest and the lowest rainfall received at Bombay during 110 years of recordings were double and half the annual means, respectively. For Honolulu, with a mean rainfall of 71 cm, 85 years of records indicated a maximum rainfall of 142 cm and a minimum of 25 mm. Often the seasonal variations in monthly

Table 3.3 Monthly and annual average rainfall, standard deviation and coefficient of variation values for Zaria (N. Nigeria), 1904–59 and Cavalla (Liberia) 1928–64. (After Gregory, 1969)

	Monthly												Annual
	J	F	M	A	M	J	J	A	S	O	N	D	
Average (in)													
Zaria	0.02	0.054	0.215	1.72	4.77	6.19	8.97	12.55	8.88	1.60	0.06	0.027	45.09
Cavalla	3.29	4.43	6.81	6.68	14.10	16.72	5.49	4.21	12.67	13.98	9.76	5.69	103.85
Standard deviation (in)													
Zaria	0.010	0.144	0.421	1.26	1.74	1.97	2.71	3.73	2.75	1.24	0.22	0.201	7.32
Cavalla	2.05	2.00	2.93	2.82	6.11	7.60	6.00	3.38	6.17	6.63	3.64	2.50	16.73
Coefficient of variation (%)													
Zaria	500	266	196	73.2	36.5	31.8	30.2	29.8	31.0	77.8	367	743	16.2
Cavalla	62.3	45.2	43.0	42.2	43.3	45.5	109.3	80.3	48.7	47.4	37.3	43.9	16.1

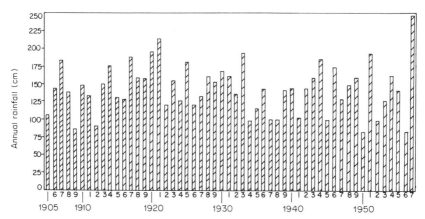

Figure 3.10 Variation in mean annual rainfall at Maha Illupalma in Sri Lanka from 1905 to 1957 (Panabokke, 1955)

rainfall are similar although the total annual rainfall amounts for two locations are very different. The rainfall expectations for sub-saharan Africa computed by Cochémé and Franquin (1967), shown in Fig. 3.7, can be used to evaluate probabilities of receiving annual rainfall within the region where the records of mean annual rainfall are not complete.

Kowal and Kassam (1978) used data from 52 stations in West Africa and developed a multiple regression analysis to describe the distribution of lower and upper limits of rainfall in relation to latitude and longtitude:

P_4:1 Upper limit = 2966 − 154(LA) − 11(LO) mm $r = -0.90$
P_4:1 Lower limit = 2050 − 113(LA) − 7(LO) mm $r = 0.89$

These equations further support the argument of dependence of rainfall amount and its variability on nearness from the source of moisture. Similar probability maps for East Africa have been reported by Manning (1950), Glover and Robinson (1953), Glover et al. (1954), Dow (1955), McCulloch (1961) and Kenworthy (1964), and for West Africa by Sivakumar et al. (1980) and Virmani et al. (1980).

Short-term (monthly, weekly, and even pentad) variations in rainfall have more edaphological significance than the annual or long-term variations. This is particularly true for regions with a growing season of 100 days or less. Long-term records, however, are needed to obtain some reliable estimates of rainfall variability over short time intervals (Figs. 3.8, 3.9 and 3.10).

E. The choice of method of evaluation of rainfall variability reliability:

Maps dealing with rainfall variability are based on the mean deviation as a percentage of the average. The percentage departure from normal mean can be very misleading because many details are often omitted, and the interpretation is rather overgeneralized (Gregory, 1969). It is also perhaps erroneous to evaluate the relationship between annual rainfall and its variability on the basis of this percentage variability. The expression of variability in terms of mean and deviation is more relevant to the problems of crop water availability and agricultural potential of the region than deviation expressed as percentage of the mean. For regions with rainfed agriculture, the variation expressed in absolute terms of the amount of rainfall is more indicative of the crop water needs and possible crop failure than the expression on the basis of percentages.

Gregory (1964, 1969), for Mozambique and Sierra Leone, and Bowden (1980) for Sierra Leone, observed that because the mean values were widely dispersed, direct comparison with standard deviation is not possible. Furthermore, the rainfall data often do not fit the normal frequency distribution particularly in regions of low rainfall. In that case it is not legitimate to use the standard deviation or the coefficient of variation as an expression of variability. However, Kowal and Kassam (1978) reported a normal frequency distribution of rainfall at Kano, northern Nigeria, for the period 1905–73.

For dry regions or in dry seasons, when the rainfall data observes a positive skew, the degree of variability can be expressed by the use of the quartile deviation as a percentage of the medium (Gregory, 1969). If rainfall variability is expressed on this basis, even regions of low rainfall do not show exceptionally high variability. Gregory (1969) reported that rainfall variability for Sierra Leone for the annual rainfall of 2000 to 6250 mm is no more than that observed in Britain. The coefficient of variability for northern Nigeria was also found to be between 13 and 19% (Fig. 3.11).

Shukla (1984) studied the interannual variability of monsoons in India, and observed that high variability could be easily comprehended if analysed in terms of the standardized rainfall anomaly Fig. 3.12. The standardized rainfall anomaly was computed as the difference between actual rainfall and the long-term climatological mean divided by the standard deviation. In some

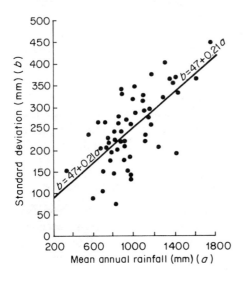

Figure 3.11 Regression of annual standard deviation on mean annual rainfall for northern Nigeria (Gregory, 1969)

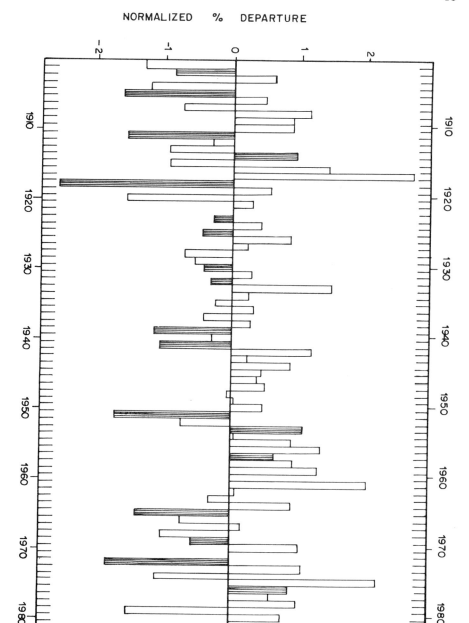

Figure 3.12 Standardized rainfall anomaly (Shukla, 1984), depicting inter annual variation in monsoonal rainfall over India for 1900 to 1980. (The black bars denote the El Niño years.)

years the normalized departure exceeded 200%. Furthermore, he observed that the years with below-average rainfall were 'El Niño' years: 'El Niño' refers to a southward-flowing ocean current that brings warm waters to the normally

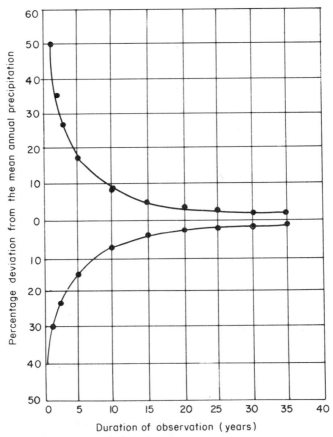

Figure 3.13 Relationship between coefficient of variability and the duration of observations (Ayoade, 1974). *Reproduced by permission of the Singapore Journal of Tropical Geography*

cool coast of Ecuador and Peru. In exceptional years, a catastrophic version of El Niño is associated with a southerly shift in the tropical rainfall, causing widespread disruption to agriculture.

Another major reason for high variability observed in the tropics is the duration and reliability of the records available. The coefficient of variability decreases with the increase in the length of the period for which records are available (Fig. 3.13.)

The probability computations should also be based on data that follow a normal distribution. Transformation of data by square root or log transformation to normality is essential if the data are skewed.

F. Start of the growing season

In the tropics, the springs (at the beginning of the rainy season) are hot and dry rather than cold and wet springs as in the temperature region. Low soil moisture and high soil temperatures are the critical factors that determine the success or failure of early planted crops. If the stresses of low moisture and high temperatures can be alleviated, early planted crops not only yield more than those planted late (Lal, 1973; Kassam and Andrews, 1975; Jones and Stockinger, 1972), but also improve the chance of double cropping by relay or sequential plantings. In regions with a pronounced dry season, the start of the growing season should, therefore, be carefully defined in relation to assured availability of soil water. Otherwise, if assured rains have not set in, crop failures can occur due to poor seed germination and seedling establishment. An unusually dry spell following seeding can necessitate expensive replanting. It is therefore important to define precisely the wet months.

Kenworthy and Glover (1958) distinguished the wet months on the basis of percentage variability in rainfall as a diagnostic criterion. The mean rainfall regime is a poor indication of whether a month is wet or dry. A comparison of the mean monthly rainfall with its standard deviation is a better choice. If the standard deviation is less than 50 per cent of the mean, the month can be considered as a part of the wet season. Dry months generally have a variability exceeding 70 per cent.

Virmani (1975) defined start of the growing season as an event with 20 mm of rain falling on one or two successive days, and advised the pre-planting cultivation to take place between dates d_0 and d_1. After d_1 the event is used for planting. Benoit (1977) defined the start of the growing season as 'the date when accumulated rainfall exceeds and remains greater than one-half of potential evapotranspiration for the remainder of the growing season, provided that no dry spell longer than 5 days occurs immediately after this date'. Periods of rainfall in the early rainy season are often followed by long dry spells leading to crop failure. This 'false start' was observed at IITA, Ibadan, Nigeria in 1983 when two or three rains totalling 6 cm received in the first week of April were followed by 4 weeks of dry weather. Early planted crops suffered severely from drought stress and high soil temperatures.

In order to define the optimum time of planting, it is important to predict the onset of rains. Virmani et al. (1980) have analysed the long-term records of rainfall distribution from India and elsewhere in the semi-arid tropics to define the onset of rains for different ecological zones. Benoit (1977) computed the dates when rainfall was sufficient for planting to begin without risk of a dry spell for the period 1951–75 for three sites in northern Nigeria (Fig. 3.14). It has been observed that in West Africa, the mean start of the growing season delays with increase in distance from the equator (Kowal and Knabe, 1972). There is, however, a considerable variability in the mean start of the growing season from year to year. The range of probable dates also varies with the latitude, as is shown for three locations in northern Nigeria by the data in Table 3.4.

The mean start of the growing season for West Africa was found to be related to the latitude (Kowal and Kassam, 1978) as described in the following equation:

Table 3.4 Probable date for which growing season can be expected to begin in given proportions of years (Benoit, 1977). *Reproduced by permission of Elsevier Science Publishers*

	1 year in 100	1 year in 20	1 year in 10	1 year in 5	3 years in 10	2 years in 5	1 year in 2	3 years in 5	7 years in 10	4 years in 5	9 years in 10	19 years in 20	99 years in 100
P	0.01	0.05	0.10	0.20	0.30	0.40	0.50	0.60	0.70	0.80	0.90	0.95	0.99
Mokwa	5 Apr	14 Apr	19 Apr	25 Apr	29 Apr	3 May	6 May	9 May	13 May	17 May	23 May	28 May	6 June
Samaru	12 Apr	24 Apr	30 Apr	7 May	12 May	17 May	21 May	25 May	30 May	4 June	11 June	17 June	29 June
Kano	3 May	14 May	20 May	27 May	1 June	6 June	10 June	14 June	19 June	24 June	1 July	7 July	18 July

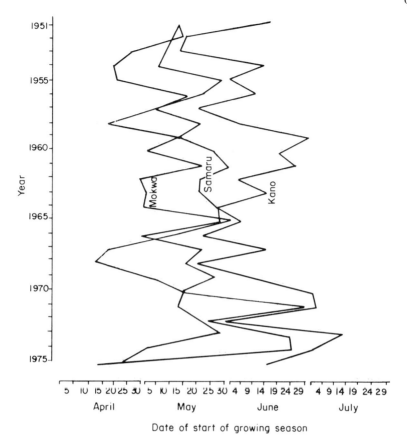

Figure 3.14 Suggested dates of planting in relation to assured rainfall for three sites in northern Nigeria (Benoit, 1977). *Reproduction by permission of Elsevier Science Published*

Start rains (decade) = −1.16 + 1.34(LA) + 0.70(LO) $r = -0.94$

The relationship explained about 88 per cent of the variability by the geographical position described by the latitude (LA) and the longitude (LO). In general, the onset of rains in West African tropics advances at a mean rate of 1.34 decade (13.4 days) per degree latitude. The end of rains, defined as the last 10-day period of the rainy season with at least 1.25 cm of rainfall and E_t in the previous 10-day period not less than rainfall, is also related to the geographical position (Kowal and Kassam, 1978).

End of rains (decades) = 35.18 − 0.57(LA) − 0.07(LO) $r = 0.84$

This regression equation indicates that the retreat of the rains is much faster and occurs at about 5.7 days per degree latitude. Because the advance and the retreat of the rains are related to latitude, it is apparent that the length of the rainy season would follow the same trend. Considering the soil moisture reserve, Kowal and Kassam (1978) computed the end of growing season and the length of the growing season in relation to the geographical location as explained in the following equations:

Length of rainy period (decade) = 36.34−1.91(LA)−0.14(LO)
$$r = -0.94$$
End of growing season (decade) = 40.58−0.82(LA)−0.06(LO)
$$r = -0.89$$
Length of growing season (decade) = 41.74−2.16(LA)−0.01(LO)
$$r = -0.94$$

Similar trends exist in the onset of monsoons in the Indian sub-continent. Fitzpatrick *et al.* (1967) estimated the length of the growing season for Australia's semi-arid and arid regions on the basis of rainfall and soil moisture availability. Soil moisture availability was calculated using climatic models to depict evapotranspiration rates. Two basic models were used. One describes the relationship between an empirical measure of growth stage and the ratio of expected evapotranspiration in the absence of water stress to the potential evaporation. The second model is based on the ratio of actual to expected evapotranspiration to the estimated soil-water storage under conditions of restricted soil water supply. Using these models, the authors observed impor-

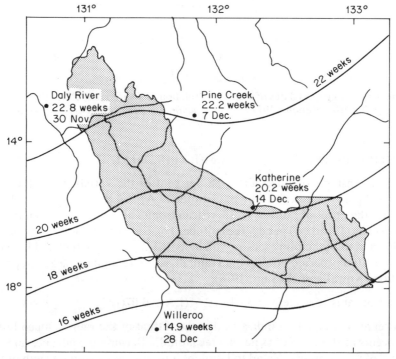

Figure 3.15 Mean length of estimated agricultural growing season in Northern Territory, Australia (Fitzpatrick, 1965)

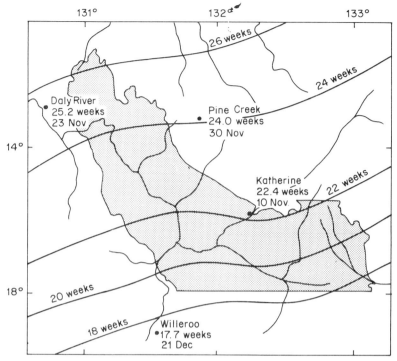

Figure 3.16 Estimated mean total duration of pasture growth in Northern Territory, Australia (Fitzpatrick, 1965)

tant seasonal trends in growth period expectancy that differ markedly from the trends in rainfall expectancy (Figs. 3.15 and 3.16).

G. Rainfall effectiveness

In the tropics much more rainfall is needed to keep the soils at the same level of wetness than in the middle and northern latitudes. This low rainfall effectiveness is due to a multitude of interacting factors. Firstly, tropical rains are in general more intense, and therefore, a relatively high proportion is lost as run-off. Prevalence of high temperatures throughout the year causes considerable losses due to evaporation and evapotranspiration. The growth rates due to favourable day/night temperatures are generally high, leading to high consumptive rates. Soils have low water-holding capacity and the effective root volume is generally less. Seasonal and annual crops, therefore, suffer from drought even within a few days after a heavy rain. If the probability of a 5- to 7-day rainless period is high, as it is for most of the seasonally dry tropics, crop yield potential is adversely affected. For example, Benoit (1977) computed the frequency of dry spells in Nigerian savanna and observed that the probability for dry spells during the early rainy season is very high. This, combined with the

Table 3.5 Frequency of occurrence of various durations of dry top soil for Dodoma (Central Tanzania, 1936–65), and Norfolk and Tifton (USA, 1940–69) (Yao, 1973). *Reproduced by permission of Elsevier Science Publishers*

Duration (days)	Number of occurrences			Frequency (%)		
	Dodoma	Norfolk	Tifton	Dodoma	Norfolk	Tifton
6–10	18	16	27	60	53	90
11–15	19	4	4	63	13	13
16–20	7	0	0	23	0	0
21–25	6	0	0	20	0	0
>26	5	0	0	17	0	0

dry top soil, no moisture reserves in the sub-soil and high soil temperatures, has often led to crop failure.

The famous groundnut scheme of East Africa is often cited as an example of failure due to the occurrence of these dry spells. Yao (1973) computed the frequency of occurrence of various durations of dry topsoil in central Tanzania, the site of this scheme (Table 3.5). During the rainy season from mid-January to mid-April, the probability that the topsoil stays dry for at least 16 consecutive days for Dodoma is 57% and drops only to 37% for 21 or more consecutive days. Comparison of similar analysis for Tifton (Georgia) and Norfolk (Virginia), the major groundnut producing regions in the United States, indicated no occurrence of dry topsoil for as long as 16 consecutive days in 30 years of records analysed. Stern and Coe (1982) computed the probability of dry spells of length, (d) days occurring in the next (m) days by using recurrence relations for Hyderabad, India (Fig. 3.17). The data show that in June, one year in two will experience a dry spell of 10 or more days.

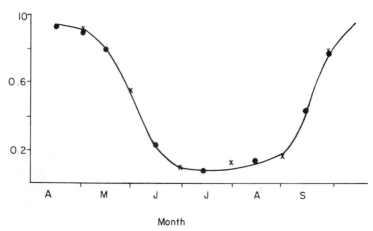

Figure 3.17 The probability of occurrence of dry spells of 10 or more days in next 30 days

Table 3.6 Monthly rainfall (mm) at two rain gauges, A and B, 3.2 km apart, April 1967 – March 1968 (Jackson, 1969b)

Gauge	April	May	June	July	Aug.	Sept.	Oct.	Nov.	Dec.	Jan.	Feb.	March	Total
A	315.2	193.4	31.8	131.2	47.5	89.1	49.7	206.0	134.5	2.6	90.5	145.8	1437.3
B	437.6	236.0	44.3	97.3	59.7	83.1	48.3	180.7	172.4	2.1	82.8	222.2	1666.5

Table 3.7 Annual rainfall (mm) 1946–54 for two stations, Fatemi Sisal Estate (A) and Mgude Sisal Estate (B), 8 km apart, altitude difference 15m (Jackson, 1969b)

Station	1946	1947	1948	1949	1950	1951	1952	1953	1954	Average (1942–1966)
A	754.6	973.1	966.5	790.1	1069.6	1188.7	755.1	1108.7	930.7	943.9
B	859.8	1248.4	1420.9	619.8	1223.3	1127.0	565.4	859.8	869.8	924.8
A–B	−105.2	−275.3	−454.4	+170.3	−153.7	+61.7	+189.7	+248.9	+61.0	+19.1

Tropical storms are often localized and spotty, and the variability among locations within short distances from one another can be high; antecedent magnitude also varies on a day to day basis. That is why a farmer often believes that it has rained everywhere else except on his own farm. This belief originates from a not unusual observation of very high rainfall at one corner of a large field while the adjacent land is absolutely dry.

Spatial variability patterns of rainstorms in East Africa have been reported by Johnson (1962), Sharon (1974), Jackson (1969a, 1969b, 1972, 1974, 1978) and Morth (1968). Among the factors responsible for this high spatial variability are topography and relief, slope aspect, directional storms in relation to the slope and others. Reported rainfall variability can also be due to the use of rain gauges of different sizes and dimensions. Data in Table 3.6 show that even in the absence of topographic differences, the spatial variability in rainfall can be substantial among locations within close proximity of one another. These large differences observed for individual months are due to a few heavy storms. Data in Table 3.7 are also obtained by Jackson (1978) from two stations on relatively flat terrain. These stations are only 8 km apart. Although the mean average rainfall is the same for both stations, there are large year to year variations. Similar records from three stations in Malaysia indicate significant differences in the pattern of variability (Table 3.8). Rainfall variability among two recording rain gauges 3 km apart at IITA are shown in Tables 3.9 and 3.10 for 1982 and 1983, respectively.

Table 3.8 Comparison of monthly rainfall for three Jahore (Malaya) stations (Jackson, 1978). *Reproduced by permission of Elsevier Science Publishers*

Year	Station	January 7	34	40	April 7	34	40	July 7	34	40	October 7	34	40
Average		345	345	337	250	253	256	173	169	182	253	249	240
1935											207	312	119
1939		560	313	238									
1940					156	376	154						
1941					300	445	243						
1948											284	388	173
1950								115	269	204	279	270	162
1952		427	209	247							216	298	229
1957					115	187	315	232	125	204			

Station No.	Name	Altitude (m)	Date opened
7	Kota Tinggi General Hospital	12	Jan. 1932
34	Nan Heng Estate, Kota Tinggi	38	Aug. 1935
30	Malaya Rubber Estate Ltd, Kota Tinggi	12	Jan. 1934

Distance between stations: 7–34 = 0.07 km
7–40 = 4.8 km
34–40 = 4.8 km

Table 3.9 Monthly rainfalls (mm) for two locations at IITA about 3 km apart (1982)

Location	Jan.	Feb.	March	April	May	June	July	Aug.	Sept.	Oct.	Nov.	Dec.	Annual total
A	0	36.3	108.0	50.8	155.5	105.4	138.4	92.7	39.4	69.9	15.2	0	811.6
B	0	12.7	73.7	69.9	132.1	157.5	120.1	101.6	78.7	37.9	0.0	0	784.2
Difference (B–A)	0	–23.6	–34.3	+19.1	–23.4	+52.1	–18.3	+8.9	+39.3	–32.0	–15.2	0	–27.4

A = run-off plots in Block A; B = block D watershed.

Table 3.10 Monthly rainfalls (mm) for two locations at IITA about 3 km apart (1983)

Location	Jan.	Feb.	March	April	May	June	July	Aug.	Sept.	Oct.	Nov.	Dec.	Annual total
A	0	0	0	88.9	242.6	142.2	97.8	25.4	125.7	31.8	29.2	78.7	862.3
B	0	0	3.8	86.4	238.3	166.4	85.1	29.1	148.6	29.2	27.9	63.5	878.3
Difference (B−A)	0	0	+3.8	−2.5	−4.3	+23.8	−12.7	+3.7	+22.9	−2.6	−1.3	−15.2	+16.0

A = run-off plots in Block A; B = block D watershed.

Sreenivasan (1971) computed the spatial variability in monthly rainfall for the 32 stations in Bihar, India, for the 30-year period from 1921 to 1950. He observed that the variations in space were of a much higher order than the variations in time, i.e., spatial variability was more than twice the variability in time for the 30-year period under study. Nwa (1977) reported variability of 10 to 15 per cent among four gauges distributed over a watershed of about 40 hectares. In general, days of high rainfall recorded low values of coefficient of variation (C_v) and days of low rainfall recorded high C_v values. Data in Fig. 3.18 show that average C_v increased as the daily watershed rainfall amount decreased. The coefficient of variation was related to the class mean rainfall, class number and the gauge density as shown in the following equations:

$C_v = 3.710 \quad P^{-0.396} \quad N^{-0.689} \quad R^2 = 0.82$
$C_v = 1.096 \quad P^{-0.396} \quad G^{0.689} \quad R^2 = 0.82$
$C_v = 0.90 \quad e^{-0.206^{1/2}} \quad G^{0.689} \quad R^2 = 0.77$

where P = class mean rainfall (mm), G = gauge density (ha/gauge), N = the number of rain gauges and e is the Napierian constant. The average error also increased as the mean watershed rainfall increased for all rainfall classes (Fig. 3.19). The regression equations relating the magnitude of error term (E) with other variables are as follows:

$E = -1.826 \quad P^{0.557} \quad N^{-0.758} \quad R^2 = 0.90$
$E = -4.70 \quad P^{0.557} \quad G^{0.758} \quad R^2 = 0.90$
$E = -4.50 \quad e^{0.308 P^{1/2}} \quad G^{0.755} \quad R^2 = 0.92$

E is the average error in mm. This type of analysis is useful in providing

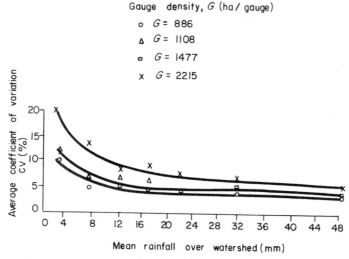

Figure 3.18 The relationship between coefficient of variation and the daily rainfall amount at IITA, Ibadan, Nigeria (Nwa, 1977). *Reproduced by permission of Elsevier Science Publishers*

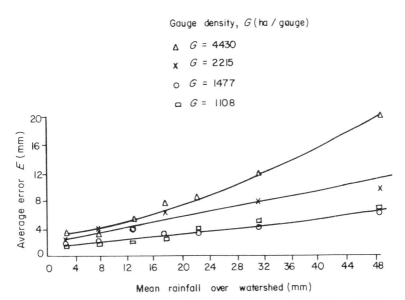

Figure 3.19 The relationship between the average error and the daily rainfall amount at IITA, Ibadan, Nigeria (Nwa, 1977). *Reproduced by permission of Elsevier Science Publishers*

information regarding the required rain gauge density that can give a certain level of accuracy in monitoring rainfall over a region. The optimum rain gauge density, however, varies among geographical regions. Nwa (1977) concluded that for western Nigeria a rain gauge density of 14.77 ha/gauge was sufficient. Similar magnitudes of variability in rainfall received over the watershed were observed at Hyderabad, Central India (Table 3.11). A procedure to compute mean rainfall from spatial interpolation of daily totals of rainfall has been described by Kruzinga and Yperlaan (1978). The readers are referred to more authoritative reviews elsewhere for computation of the mean rainfall over a watershed (Thorpe *et al.*, 1979).

Table 3.11 Range of variation in rainfall (mm) recorded at ICRISAT, India in 1977 (ICRISAT, 1977–78)

Date	Main observatory	Range over the watershed
18 June	19.8	5.2 – 36.7
17 July	54.4	14.4 – 61.0
10 August	74.4	65.0 – 85.0
20 August	10.8	7.2 – 30.6
1 September	14.8	5.6 – 44.0

H. Intensity and energy load

Tropical rains are generally more intense than the temperate zone rains (Lal, 1976; Hudson, 1976; Lal, 1981). High intensities of 150 to 200 mm h^{-1} have been observed for relatively short durations of 5 to 10 minutes. In some exceptional regimes, e.g., Assam in India and a region near Mount Cameroon in West Africa, exceptionally high intensities are commonly observed. An example of the frequency distribution of rainfall intensity for 1972 to 1974 for Ibadan is shown in Fig. 3.20. The 7.5-minute maximum intensities of 61% of the rainstorm in 1972 and 57% of those during 1973 were between 25 and 75 mm h^{-1}; 76% of the storms during 1972 and 66% of those during 1973 had maximum intensities greater than 25 mm h^{-1}. Intensities exceeding 75 mm h^{-1} were recorded for 15% of the rainstorms during 1972 and 10% of those during 1973. Intensities greater than 100 mm h^{-1} were recorded for 7% of the storms

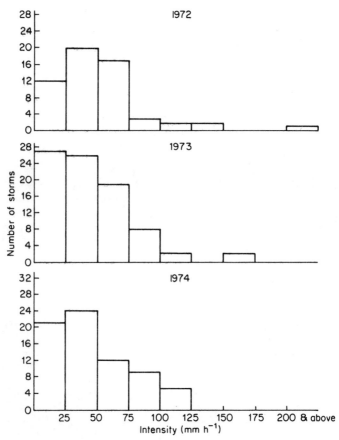

Figure 3.20   The frequency distribution of rainfall intensity for 1972 to 1974 at IITA, Ibadan, Nigeria (Lal, 1976)

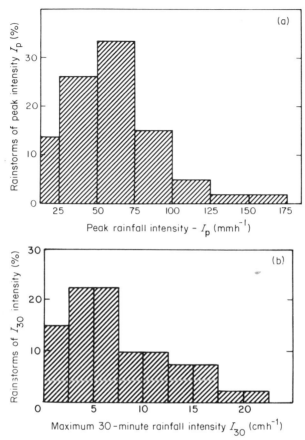

Figure 3.21 The frequency distribution of rainfall intensities in 1975 at IITA. (a) Peak intensity or (b) 30–minute intensity (Aina et al., 1977). *Reproduced by permission of the Soil Conservation Society of America*

during 1972 and 3% of those during 1973. The 7.5-minute intensities of 48.3% of the rainstorms during 1974 were greater than 50 mm h^{-1}; 21% of the storms had maximum intensities greater than 75 mm h^{-1}. About 14% of the rainstorms had intensities greater than 100 mm h^{-1} half of those with intensities of 130 mm h^{-1}.

Aina *et al.* (1977) reported the monthly rainfall distribution and maximum 30-minute intensity of rainstorms at Ibadan, Nigeria (Fig. 3.21). Most of the rainstorms observed were of short duration, reaching their peak intensity within the first few minutes of the onset of rain. About 55% of the storms attained peak intensity in less than 5 minutes. The maximum intensity was between 75 and 100 mm h^{-1} for 16% and more than 100 mm h^{-1} for 5 to 7% of the storms. There was at least one storm with a peak intensity exceeding 200 mm h^{-1}. The data of frequency distribution for 7.5-minute maximum rainfall for 1977 and 1978 reported by Lal *et al.* (1980) are shown in Fig. 3.22; 61% of

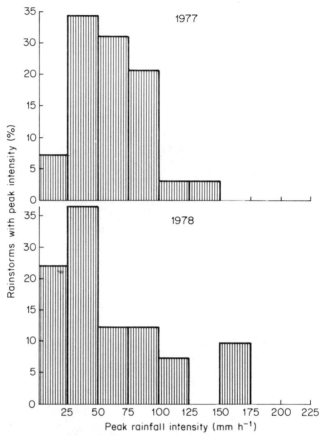

Figure 3.22 The frequency distribution of 7.5-minute maximum rainfall for 1977 and 1978 at IITA, Ibadan, Nigeria (Lal *et al.*, 1980)

Table 3.12 Rainfall intensities recorded at Hyderabad, India, in 1977 (ICRISAT, 1977–78)

Date	Duration (min)	Intensity (mm h^{-1})
14 June	15	36
18 June	15	26
25 June	15	20
3 July	15	58
20 July	15	32
24 July	10	30
28 July	10	24
29 July	7	51
10 August	15	42
21 August	30	38
27 August	15	24
29 August	15	44

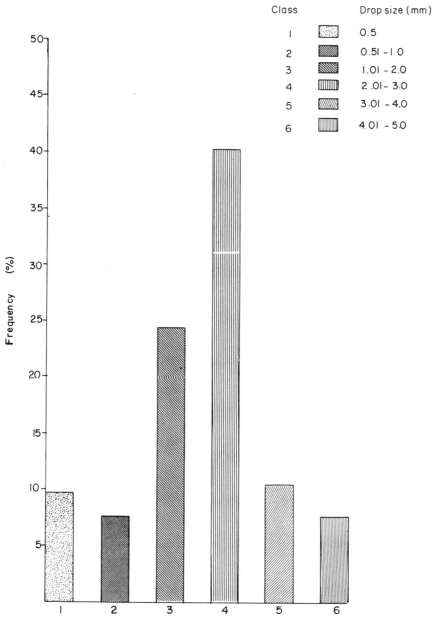

Figure 3.23   The median drop size distribution of rains at Ibadan, Nigeria for the year 1979

the storms had intensity between 25 and 75 mm h^{-1}. In all, 35% of the rains fell at intensities exceeding 50 mm h^{-1}.

An intensity of 58 mm h^{-1} sustained for 15 minutes has been recorded at

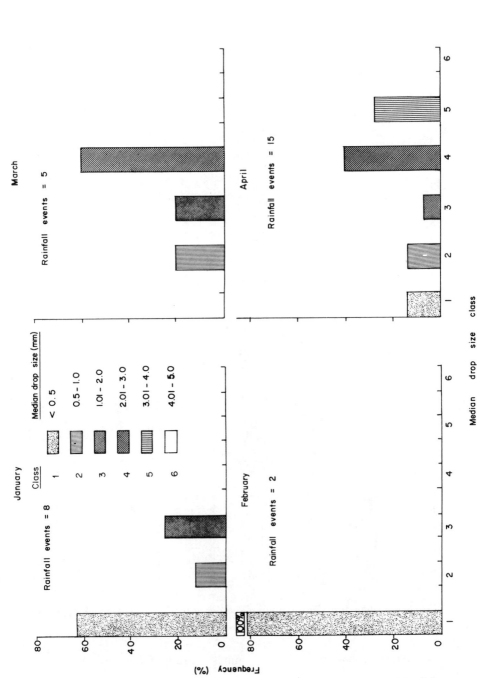

Figure 3.24 The frequency analysis of drop size distribution of rains at IITA, Ibadan for months of January, February, March and April

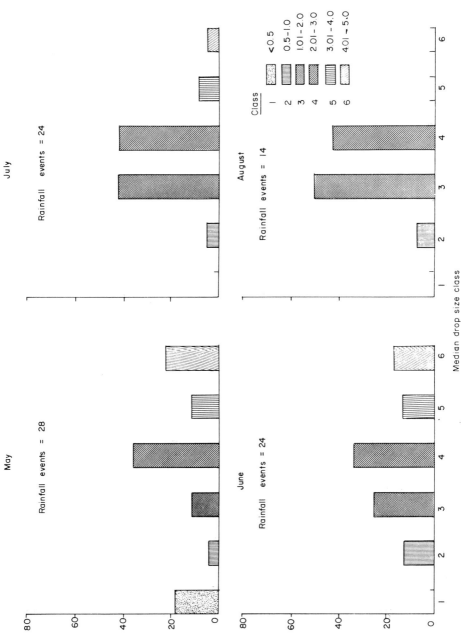

Figure 3.25 The frequency analysis of drop size distribution of rains at IITA, Ibadan for months of May, June, July and August

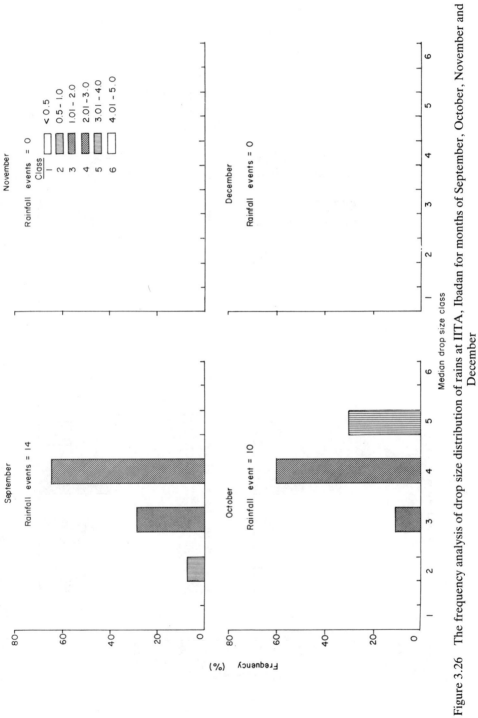

Figure 3.26 The frequency analysis of drop size distribution of rains at IITA, Ibadan for months of September, October, November and December

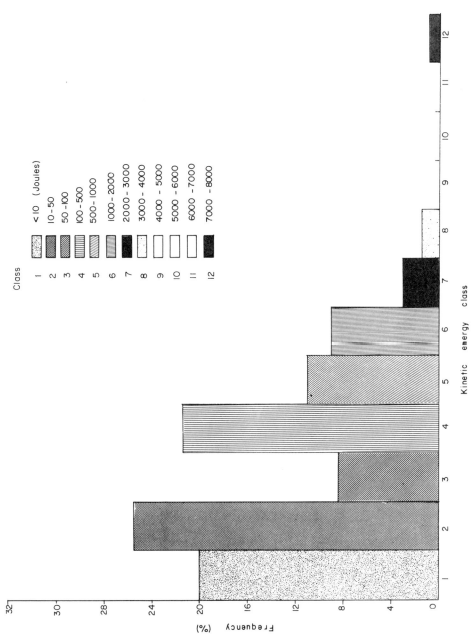

Figure 3.27 The frequency analysis of energy load of rain-storms at Ibadan, Nigeria

Hyderabad, India (Table 3.12). An intensity of 38 mm h^{-1} was reportedly sustained for 30 minutes. Similar observations of high rainfall intensity have been made by Sien and Koon (1971) for southeast Asia, by Roose (1971), Ojo (1977), Walker (1957), Elwell and Stocking (1973), Hudson (1965), Pereira (1973), and C.T.F.T. (1974) for Africa and by Ahmad and Brechner (1974) for the Carribean and South America. Kowal and Kassam (1976) computed the intensity of a rainstorm observed at Samaru, northern Nigeria, during successive 1-minute intervals for a storm of 18 minutes duration. The mean intensity over the 18-minute period was 67 mm h^{-1} but the intensity over 1-minute durations ranged from 36 to 111 mm h^{-1}.

Rainfall intensity and the energy load are influenced and related to the drop size distribution. Information on drop size distribution of rain is available for a few locations only. Measurements made at IITA, Ibadan, indicate that most rains have a median drop size (D_{50}) ranging between 2 mm and 4 mm (Fig. 3.23). The monthly frequent analysis of drop size distribution of rains received at Ibadan during 1979 is shown in Figs 3.24–3.26. The frequency of rains of median drop size exceeding 3 mm is high for the months of April, May, June, July and October.

Big rain drop size results in high cumulative kinetic energy. The frequency of rainstorms with kinetic energy exceeding 500 Joules per storm is usually high. The results of the frequency analysis of energy load of rainstorms at Ibadan are shown in Fig. 3.27. About 6% of the rains had energy loads exceeding 200 Joules. The cumulative monthly kinetic energy for Ibadan shown in Fig. 3.28 indicates that for April to October 1979 the cumulative energy load of the

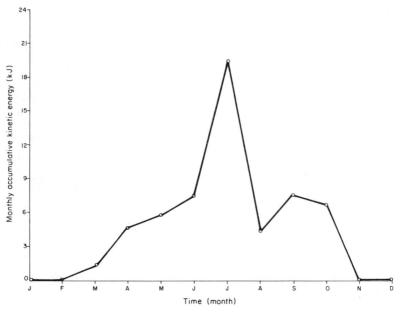

Figure 3.28 The monthly accumulative kinetic energy of rains at Ibadan, Nigeria

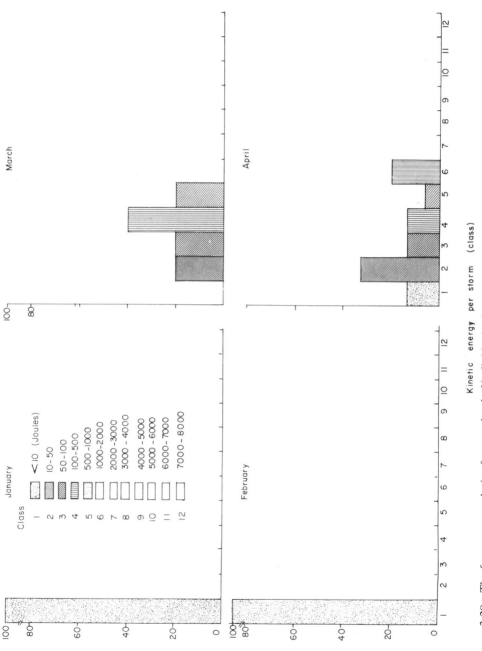

Figure 3.29 The frequency analysis of energy load of individual rainstorms at Ibadan for January, February, March and April

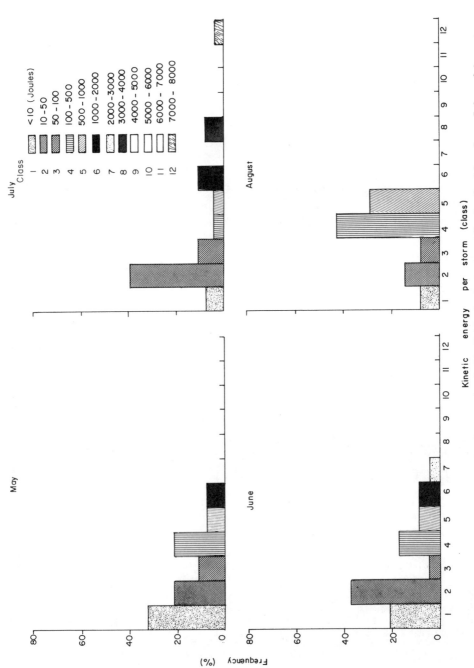

Figure 3.30 The frequency analysis of energy load of individual rainstorms at Ibadan for May, June, July and August

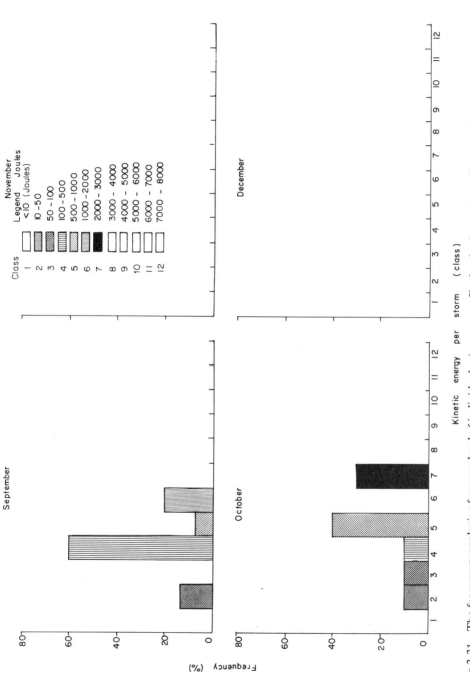

Figure 3.31 The frequency analysis of energy load of individual rainstorms at Ibadan for September, October, November and December

rainfall was 500 Joules m^{-2}. The analysis of data depicting kinetic energy per storm at Ibadan, Nigeria for 1979 is shown in Figs. 3.29–3.31. At least one rain was received at Ibadan in 1979 with accumulative energy of 7000–8000 Joules m^{-2}. Tropical rains are, therefore, more intense than temperate ones; data of this type are still needed for many locations and ecologies in the tropics.

III. SOILS OF THE TROPICS

Tropical soils are supposedly red, old, deeply weathered, leached, acidic, infertile, lateritic and unable to support intensive cultivation. Some soils in the tropics fit this definition, others do not. Highly productive soils of almost neutral pH also occur in the tropics. Neither are all tropical soils lateritic. In fact soils of the tropics are highly diverse, depending on geology, geomorphology, vegetation and the rainfall pattern. To a casual observer most tropical soils are red-coloured, but some agriculturally important soils are brown, yellow, grey, black and even white sands. Soils of the tropics have been described by many (Young, 1976; Mohr et al., 1972; FAO–UNESCO, 1974; Jones and Wild, 1975; Sanchez, 1976). Therefore, only a brief account of some important soils is given here. A majority of tropical soils come under the categories of Oxisols, Aridisols, Alfisols, Ultisols and Vertisols (Table 3.13).

1. Oxisols

Oxisols are old, deep, permeable and well drained soils characterized by an oxic horizon or plinthite forming a continuous phase. The oxic horizon, according to USDA (1975): (i) is at least 30 cm thick; (ii) has a fine earth

Table 3.13 Distribution of major soils of the tropics (Sanchez, 1976)

Soil group	Area (10^6 ha)				
	Humid tropics	Semi-arid	Dry and arid lands	Total	Percent of tropics
Oxidols, Ultisols, Alfisols	920	1540	51	2511	51
Psamments and lithic groups	80	272	482	834	17
Aridisols	0	103	582	685	14
Alluvial (Aquepts, Fluvents)	146	192	28	366	8
Vertisols, Mollisols	24	174	93	291	6
Andepts, Tropepts	5	122	70	207	4
Total area	1175	2403	1316	4896	100
Per cent of tropics	24	49	27	100	

fraction that retains 10 meq or less of NH4-ions/100g clay, (iii) has CEC of less than 16 meq/100g clay; (iv) contains only traces of weatherable primary minerals; (v) contains small amounts of water-dispersible clay in some sub-horizons; (vi) is predominantly sandy loam and contains more than 15% clay; (vii) has gradual or diffuse boundaries between sub-horizons; and (viii) shows rock structure in less than 5% of its volume.

The clay minerals are predominantly kaolinitic with varying amounts of iron and aluminium oxides. The CEC is only partly saturated with bases and the exchangeable Al^{3+} is relatively high. These soils are developed on gentle slopes and are deeply weathered. The soils derived from quartz-rich parent rocks are also characterized by the presence of stonelines. There are Oxisols with and without plinthite or laterite (hardened plinthite). The old term 'laterite' was used to describe Oxisols with hardened plinthite. For details on origin, classification and properties of Oxisols, readers are referred to reviews by Mohr *et al.* (1972) and Bennema (1977). Most Oxisols have low inherent fertility.

Plinthite occurs in only 7% of the tropical world (Sanchez and Buol, 1975) and in less than 2% of the Amazon (Wambecke, 1978). The hardened plinthite is, however, widely observed in the sub-humid and semi-arid regions of West Africa (Fig. 3.32). Fluctuating ground water table is an important factor in the formation of plinthite. In addition high temperatures and an uneven distribution of rainfall over the year, i.e., monsoonal climate with two dry seasons, are required. Vegetation plays an important role in evolving soil properties and

Figure 3.32 The hardened plinthite occurs at shallow depth in some savannah soils of western Nigeria (Photo courtesy of Mr. U. Sabel-Khoschella)

vegetation removal leads to: (a) an increase in effective rainfall; (b) increased erosion; (c) decreased organic matter returned to the soil; (d) a reduced rate of new soil development; (e) decreased biological soil activity; and (f) increased temperatures. Vegetation removal leads to larger fluctuations in subsoil moistures and, therefore, desiccation occurs to greater depths. High temperatures and desiccation accelerate aging and crystallization of amorphous iron and aluminum hydrates, thus cementing other materials imbedded in these hydrates.

The tropical rainforest is the original vegetation of the plinthite soils. If the present soil has a savanna cover, this is the result of forest degradation. In some cases the savanna vegetation contributes to the formation of extremely hard iron crusts (Mohr et al., 1972).

2. Ultisols

Ultisols have a thin A_1 horizon over a leached, light-coloured A_2 horizon underlain by red, yellowish-red or yellow finer-textured B horizons that grade into lighter-coloured, often reticulated, mottled C horizons (Mohr et al., 1972). In the tropics these soils occur at mean annual temperatures of approximately 25°–28°C without marked seasonal fluctuations, with annual rainfall exceeding 2000 mm. The Ultisols are most extensive in warm humid climates that have a seasonal deficit of rainfall. These are younger soils than Oxisols and the kaolinities are relatively more active resulting in somewhat higher CEC. Ultisols may also have layers of plinthite, which often occur higher in the profile due to impeded drainage conditions. Ultisols are characterized by textural B_2 horizons with CEC of 12 meq/100g soil indicating presence of minerals other than kaolinites. The iron has generally been moved from A to B horizon and migration is independent of clay. Desilication is also a characteristic of these soils. Ultisols are found along the coastal plains of Surinam, in the southeastern regions of Nigeria, Central Africa, etc.

3. Alfisols

Alfisols occur in the subhumid and semi-arid tropics, and are characterized by argillic subsurface horizons with high base saturation (>40%), in which water is held at <15 bar tension during at least 3 months each year when the soil is warm enough for plants to grow. These soils have angular blocky and sometimes prismatic structure. The clay minerals are predominantly kaolinitic with some admixture of illite and smectites. Soils developed on basic rocks have much higher base saturation. The 'Terra Roxa estruturada' of Brazil is an Alfisol with high base saturation. Alfisols contain limited amounts of weatherable minerals. These soils occur in West Africa, northeast Brazil, Central India, northeast Thailand, and northern Australia.

4. Vertisols

Vertisols are dark clay soils and are concentrated in the semi-arid tropics. The distribution of Vertisols in millions of hectares is 5.6 in North America, 16.5 in South America, 98.0 in Africa, 63.2 in Asia and 70.5 in Australia (Mohr *et al.*, 1972). These soils have high shrink-swell properties, develop cracks to considerable depths during the dry season, and have a high bulk density between the cracks. The cracks are often filled with surface soil, and with alternate swelling and shrinking, the whole soil is thoroughly churned. The 'self swallowing' or the churning also imparts 'slickensides' to the soils. Most Vertisols contain predominantly montmorillonitic clay but some minerals are also transformed into illites, chlorites and even to traces of kaolinite.

The shrink-swell properties of Vertisols are responsible for the formation of 'gilgai' land surface. The surface layer often has a fine crumb structure but the subsoil develops large prismatic blocks between cracks which are subdivided into angular and sub-angular blocky elements. These elements comprise smaller aggregates. Vertisols, therefore, possess a compound structure. Beneath the layer that is affected by profile churning, the soil has a massive structure.

5. Aridisols

Aridisols are soils of the dry tropics where water is held at tensions >15 bar for more than 9 months a year. The surface horizons are normally light in colour and have a soft consistency when dry.

6. Andosols

Andosols are soils derived from volcanic ash, and cover a relatively small area in the tropics. These soils occur in southeast Asia, Central America and East Africa. Andosols are characterized by very porous, friable, non-plastic and non-sticky A horizons with well-developed crumb or granular structure. These soils have high water-holding capacity.

A. Soils of tropical Africa

A generalized soil map of tropical Africa (Fig.3.33) shows that Alfisols are the predominant soils of the West African savanna, a region with annual rainfall between 800 and 1500 mm received in one or two distinct rainy seasons. The West African Alfisols are derived from crystalline acid rocks and basement complex and are characterized by a gravelly horizon (Fig. 3.34). These soils have a coarse-textured surface horizon, are low in silt content, have low CEC and relatively high base saturation. About 60% of the West African savanna is covered by Alfisols. Alfisols are also found in East Africa, i.e., the east of Lake

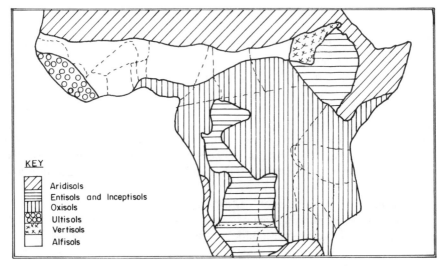

Figure 3.33 A generalized soil map of tropical Africa (Modified from Foth and Schafer, 1980)

Victoria, and in southern Africa. These areas have a much drier climate and the rainfall is generally between 500 and 800 mm per annum.

Ultisols occur in high rainfall regions of West Africa, i.e., the coastal areas of Ivory Coast, Liberia, Sierra Leone, eastern Nigeria and Cameroon. Ultisols are also found in East Africa and occur side by side with Oxisols in central Africa. Some soils of central Africa now classified as Oxisols may in fact be Ultisols. Oxisols are predominant soils of the equatorial zone with high annual rainfall. These soils have low cation exchange capacity, have less than 40% base saturation, and are relatively infertile. These are poorest of the African soils, suitable for some perennial crops. The yields of seasonal crops are low with traditional farming methods. Oxisols occur with and without plinthite, and the rooting depth of Oxisols with hardened plinthite is severely restricted.

Vertisols are widespread in Africa and occur in semi-arid regions with annual rainfall of less than 1000 mm. These soils occupy large areas in southern Sudan, Lake Chad, East Africa and the Accra plains in Ghana. Dudal (1965) estimated that Vertisols cover 98 million hectares in Africa, i.e., 40% of the world's Vertisols. Vertisols occur in Sahel along with Aridisols, i.e., soils of the arid regions with low annual rainfall. Aridisols are often too dry and can only be used for commercial agriculture with supplementary irrigation. In addition to Sahel, Aridisols are also found in the Ethiopian highlands and parts of southern Africa.

Inceptisols and Entisols are the most fertile soils of tropical Africa, developed along the flood plains of major rivers, and valley bottoms. These soils are suited for rice cultivation during the rainy season and for upland crops during the dry season. The African wetlands are under-utilized because of social taboos and health hazards related to their development. Water manage-

Figure 3.34 The occurrence of gravelly horizon in West African Alfisols is a common phenomenon on soils developed on basement complex rocks

ment is one of the major physical constraints toward intensive utilization of these soils. Distribution of major soils in different sub-climatic regions of tropical Africa are shown in Table 3.14. Desert, arid and semi-arid subclimates occupy more than 55% and humid and perhumid only 25% of tropical Africa. Alfisols are the most predominant soils in the dry subhumid and subhumid regions, and Ultisols/Oxisols occupy most areas in the humid and perhumid climates. Lithosols, Regosols and Yermosols, as one would expect, are the most predominant soils in deserts, arid, semi-arid and montane regions.

B. Soils of tropical Asia

The generalized soil map of tropical Asia (Fig. 3.35) shows the predominance of Aridisols in west Asia, Alfisols and Vertisols in south Asia, and predominantly Ultisols in southeast Asia. Inceptisols are major soils along the river valleys and deltas (Ganges and Brahmaputra).

Table 3.14 Distribution of major soils in tropical Africa (modified from Houérou and Popov, 1981)

Major soils		Area (1000 ha) in different subclimate								
FAO	Taxonomy	Desert	Arid	Semi-arid	Dry sub-humid	Sub-humid	Humid	Per-humid	Montane $t<50°C$	Total	Percent
Acrisols	Ultilols	—	5	5358	10 646	16 686	37 749	20 598	11	92 575	3.05
Cambisols	Inceptisols	6272	14 794	12 208	21 118	14 655	24 395	12 266	1425	111 564	3.68
Ferralsols	Oxisols	—	760	2311	15 267	32 495	136 218	121 947	22	335 970	11.08
Gleysols	Entisols	1259	5892	11 367	18 252	20 465	26 906	42 985	—	133 720	4.41
Lithosols	Inceptisols Lithic subgroup	191 321	72 740	26 909	37 988	28 137	30 175	6202	3185	397 667	13.12
Fluvisols	Entisols	30 029	18 092	10 238	9747	9719	12 896	6021	128	101 910	3.36
Luvisols	Inceptisols Alfisols	3058	23 753	25 311	87 845	62 197	51 131	2427	1265	256 988	8.48
Nitosols	Alfisols Ultisols	—	2313	3059	9596	14 678	38 625	—	757	99 316	3.28
Arenosols	Psamment	15 783	108 723	67 394	34 428	19 947	51 065	30 255	2	331 033	10.92
Regosols	Entisol	102 503	83 270	22 371	24 577	11 999	19 202	5055	478	264 707	8.73
Vertisols	Vertisols	1552	21 748	18 439	39 303	13 920	8812	706	487	104 967	3.46
Xerosols	Mollic Aridisols	22 807	52 701	18 202	5461	825	146	—	766	100 908	3.33
Yermosols	Typic Aridisols	312 624	55 305	5283	264	32	—	—	266	373 774	12.33
Solonchcks	Salorthid	20 043	18 783	6541	2833	903	1582	578	62	51 325	1.69
Others	—	194 487	53 214	13 864	23 885	9096	8209	41 505	820	274 699	9.06
Total		901 738	530 093	248 855	341 220	255 754	447 091	220 545	9674	3 031 118	99.98
Percent		29.75	17.49	8.21	11.26	8.44	14.75	9.59	0.32	99.81	

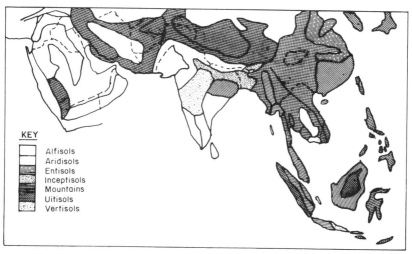

Figure 3.35 The generalised soil map of tropical Asia (Modified from Foth and Schafer, 1980)

Aridisols of west Asia occur in regions of low rainfall and are thinly populated. Some Aridisols are salt-affected soils, and without supplemental irrigation most have low productivity. These soils have, however, supported some of the most ancient civilizations and the earliest recorded agriculture. A major climate change has supposedly occurred since then. With supplementary canal irrigation, some Aridisols of Western India are now highly productive.

In South Asia most uplands are Alfisols, Ultisols or Vertisols. Alfisols occur in central India and in the southern peninsula with mean annual rainfall of 500 to 1000 mm, and in the dry zone of Sri Lanka. Alfisols are also found in eastern Cambodia, northeast Thailand and in southern China. This region receives higher rainfall than the Alfisols of south Asia, i.e., 1000 to 2000 mm. Ultisols are the dominant soils of the humid regions of southeast Asia, i.e., the islands of Indonesia and Malaya Peninsula. The rainfall of the region is 1500 to 2000 mm and the climate is predominantly monsoonal. Ultisols also occur in east India, Burma, and in southern China.

Vertisols are predominant soils of central India (Dudal, 1965) and are potentially productive. These soils, developed on basaltic rocks, occupy an area of about 60 million hectares, and often occur side by side with Alfisols. Vertisols are also found in eastern Java, Indonesia, where they occur along with some Andosols. These soils are more extensively used in Asia than those in Africa.

In west Asia, Inceptisols occur side by side with Aridisols and are less productive. On the other hand, less than 5% of the land surface in tropical Asia is covered by 30% of the world's arable Alluvial soils (Foth and Schafer, 1980). If water is available, these soils are suitable for the most intensive agriculture and the multiple-cropping systems.

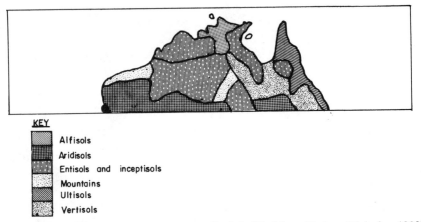

Figure 3.36   Major soils of tropical Australia (Modified from Foth and Schafer, 1980)

C. Soils of tropical Australia

Soils of Tropical Australia are Vertisols, Ultisol, Alfisols, Inceptisols and Aridisols (Isbell, 1978) (Fig. 3.36). Alfisols occur in the semi-arid regions of the Northern Territory, in the monsoonal climate. These soils are similar to the Alfisols of the West African savanna and often contain gravels and iron concretions in the subsoil horizons. The surface soil horizons are often sandy and coarse-textured. Drought stress, high soil temperature, and crusting are some of the management constraints towards intensive utilization. Vertisols occupy extensive regions of semi-arid Australia with mean annual rainfall of 500 to 1000 mm. These soils contain 50 to 80 per cent predominantly montmorillonitic clay. Ultisols are predominant in the humid northeast region of Australia, and are extensively used as rangelands. Entisols and Inceptisols cover extensive regions of northwestern and central Australia.

D. Soils of tropical America

Soils of tropical America are described by Sanchez and Buol (1975), Sanchez (1976), and Tanaka *et al.* (1984). The most predominant soils of tropical America are Oxisols (Table 3.15, Fig. 3.37). The Amazon valley, with a tropical humid climate, is covered predominantly by Oxisols. Similar to Alfisols of the West African savanna, Oxisols of the Amazon are low in silt content. Ultisols are the second most common soils of tropical America and often occur side by side with Oxisols. Some Ultisols also occur in the savanna, i.e., Llanos and the Cerrados (Goosen, 1972)

Inceptisols are found along the flood plains of the Amazon and its tributaries. These are the most fertile soils and can support intensive cropping systems with a minimum of commercial inputs. Inceptisols are also found in the Llanos

Table 3.15 Approximate extent of major soil sub-orders in the tropics (10^6 ha) (Sanchez, 1976)

Order	Sub-order	Africa	America	Asia	Total area	Per cent
Oxisols	Orthox	370	380	0	750	15.0
	Ustox	180	170	0	350	7.5
		550	550	0	1100	22.5
Aridisols	All	840	50	10	900	18.4
Alfisols	Ustalfs	525	135	100	760	15.4
	Udalfs	25	15	0	40	0.8
		550	150	100	800	16.2
Ultisols	Aquults	0	40	0	40	1.0
	Ustults	15	35	50	100	2.2
	Udults	85	125	200	410	8.2
		100	200	250	550	11.2
Inceptisols	Aquepts	70	145	70	285	6.0
	Tropepts	0	75	40	115	2.3
		70	225	110	400	8.3
Entisols	Psamments	300	90	0	390	8.0
	Aquents	0	10	0	10	0.2
		300	100	0	400	8.2
Vertisols	Usterts	40	0	60	100	2.0
Mollisols	All	0	50	0	50	1.0
'Mountain areas'		0	350	250	600	12.2
Total		2450	1670	780	4900	100.0

valley, and in contrast to the Inceptisols of the Amazon, these soils contain more bases (Foth and Schafer, 1980). Alfisols are predominant soils of the semi-arid regions of northeast Brazil. The Terra Roxa soils of the Amazon are Alfisols. Small areas of Alfisols are also located west of the Andes and in Colombia.

E. Soil physical constraints to crop production

The misconception that Oxisols and Ultisols of the wet tropics have excellent soil physical properties is based on the overwhelming effect of nutrient imbalance (Al toxicity, P fixation etc) on crop production. That the economic level of crop production cannot be achieved without prior alleviation of nutrient imbalance does not imply that these soils have no physical constraints. In fact, compaction, erosion, and drought stress are severe problems to mechanized and intensive crop production systems on Ultisols and Oxisols. Similarly, sustained crop yields on Alfisols, though constrained more by severe

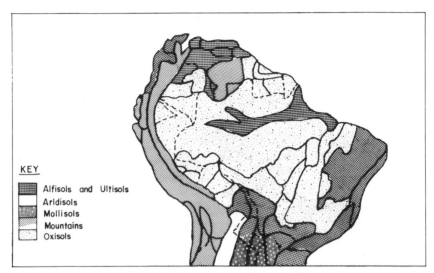

Figure 3.37 Predominant soils of tropical America (Modified from Foth and Schafer, 1980)

soil physical than nutritional constraints, are also limited without frequent and balanced application of plant nutrients. Over-generalizations have often created gross misinterpretations. The problems of soil management for major soil groups are listed in Table 3.16.

Although most tropical soils are deeply weathered, they do not usually have a deep effective rooting depth. Most soils have shallow rooting depth, particularly for seasonal crops, due either to physical, mechanical, or nutritional disorders. Because of the shallow root system, drought stress is often a serious constraint on grain crop production even in the soils of the humid and per humid tropics. The mechanical impedance of the subsoil horizon may be due to:

(i) compact argillic horizons;
(ii) gravel horizons of variable thickness;
(iii) layers of hardened plinthite; and
(iv) layers of fine sand as observed in the loess soils of the semi-arid regions of West Africa.

The mechanical impedance of these soils is often aggravated by mechanized cultivation, lack of soil organic matter, high intensity rains and ultra-desiccation during the dry season. Drought stress and high soil temperatures also inhibit deep root system development.

Poor root development in the subsoil of Ultisols and Oxisols is due both to mechanical impedance and nutritional imbalance. In addition to the physical factors listed above, root growth is inhibited by:

(i) aluminium toxicity;

Table 3.16 Physical constraints towards intensive utilization of upland soils in the tropics

Problems	Oxisols	Ultisols	Alfisol	Inceptisol	Vertisol
Physical constraints					
(i) Erosion	2	2	3	3	3
(ii) Compaction	3	3	3	2	2
(iii) Crusting	1	1	3	3	2
(iv) Drought	2	2	3	1	2
Chemical/nutritional	3	3	2	1	2
Supra-optimal soil temperature	2	2	3	3	2

0 = none; 1= slight; 2 = moderate; 3 = severe.

  (ii) phosphorus deficiency; and
  (iii) absence of calcium.

Nutrient imbalance can inhibit root development even if the physical properties are favourable.

In addition to uplands, there are vast areas of yet-to-be-explored wetland, hydromorphic soils, flood plains and valley bottom lands. Some of the soil physical constraints (e.g. erosion, drought) are not a limitation towards intensive management of these soils. In addition to developing appropriate technology for land development, water control, tillage methods, cropping systems and nutrient management, it will be necessary to deal with the social taboos and health hazards related to tropical wetlands.

IV. VEGETATION OF THE TROPICS

The vegetation of an ecology represents the integrated effects of climate, soil, rainfall distribution, biotic environments and man's intervention. Vegetation surveys of tropical ecosystems have been reported by Richards (1964), Whitmore (1975), UNESCO (1978), Odum and Pigeon (1970), Gomez-Pompa *et al.* (1972), Menaut (1983), Mistra (1983), Gillison (1983), and Sarmiento (1983), and Golley (1983). Major ecological zones in tropical climates are Saharan (rainfall < 100mm), Saharo-Sahelian (100–200mm), Sahelian (200–400mm), Sudano-Sahelian (400–600mm), northern Sudanian and southern Miombo (800–1200mm), northern Guinean (1200–1500mm) and southern Guinean rainforest (>1500mm). The generalized vegetation map of the tropical biomes presented in Fig. 3.38 indicates the two agriculturally important vegetation types, i.e., tropical forest and tropical savanna briefly described below.

A. Tropical forests

Tropical forest vegetation essentially constitutes the 'equatorial green belt'

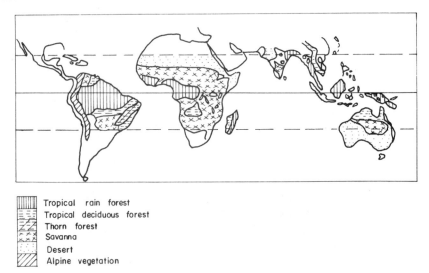

Figure 3.38 Tropical biomass (Vickery, 1984)

with some local modifications. In the forest ecology the rainfall often exceeds evapotranspiration for at least 8 months in a year, resulting in the maximum growth of plant biomass. Most of the nutrient reserves of the ecosystem are immobilized in the plant biomass, and this ecology has the most rapid rate of nutrient recycling (UNESCO, 1978). Tropical forests are also characterized by maximum species diversity. In addition to the water surplus for at least 8 months in a year, the radiation and temperature regimes also play an important role in determining the climax vegetation.

Agricultural crops, improved pastures and shifting cultivators are rapidly transforming tropical forests causing drastic alterations in soils, vegetation, hydrology, and microclimate. In tropical Africa only 5% of the land surface is covered by tropical rainforest (Foth and Schafer, 1980). Many of the forest reserves of tropical Asia have also been denuded. Tropical rainforests exist in Sumatra and other islands of Indonesia, and along the western coast of India. Forests in Australia are confined to the northeastern coast. Tropical South America is still dominated by forests, especially the Amazon Basin. The rainforest in Brazil alone covers 1.7 times the land covered by all the rainforests in Africa. It is estimated that 25% of the world's trees grow in the Amazon Basin and at least 40% of them are inaccessible (Foth and Schafer, 1980). Tropical deciduous forests and thorn forests occur in areas of either marginal rainfall or of marginal soils. In these regions, once the vegetation is removed, it is difficult for the forest to reestablish.

B. Tropical savannas

Savanna vegetation is defined differently for different purposes. The most commonly used definitions are by Beard (1953) and by the Conseil Scientifique pour l'Afrique (1956). Beard has defined the tropical savanna as a 'plant formation comprising a virtually continuous ecologically dominant stratum of more or less xeromorphic herbs, of which the grasses and sedges are the principal components, with scattered shrubs, trees, or palms, sometimes present'. Bourlière and Hadley (1983) considered savannas to have the following characteristics: (i) the grass stratum is continuous and important, occasionally interrupted by trees and shrubs; (ii) bush fires occur from time to time; and (iii) the main growth patterns are closely associated with alternating wet or dry seasons. Some savanna regions have trees and others, drier, are the treeless savanna.

Savanna vegetation occurs extensively in Africa, Asia, South America and

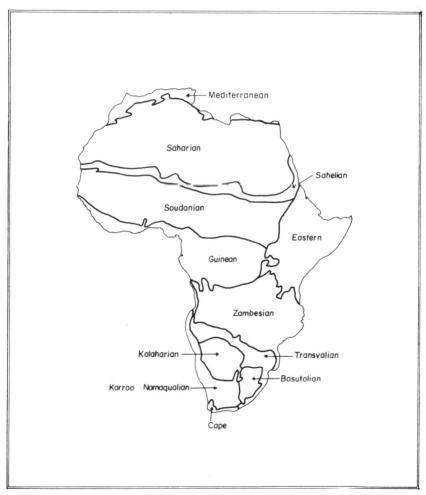

Figure 3.39 A generalized ecological map of Africa (A. de Vos, 1975). *Reproduced by Dr W. Junk Publishers*

Australia. The present-day savanna vegetation over much of humid tropical Africa is due to deforestation. In addition to areas of marginal rainfall, forest vegetation is also not re-established on marginal soils (too shallow, too gravelly, too infertile or too steep). The vegetation of these regions is termed 'derived savanna' or the 'man-made' savanna. Frequent use of fire has destroyed the perennial trees and shrubs, resulting in fire-resistant grass and scrub vegetation. As in Africa, most of the forest vegetation in Asia has been cut and the intensive land-use has changed the vegetation to derived savanna and to arable lands. Tropical America has vast tracts of grasslands and wooded savannas. About 80% of the Llanos in Colombia are covered by natural grasses (Goosen, 1972). Vast tracts of Brazilian Cerrados and grasslands in the Pampas of Uruguay are savanna. These soils are low in chemical fertility and do not support high tree density. Most of Australia is covered by grassland and shrub vegetation, and as agriculture has been relatively recently introduced, the soils and vegetation of the Australian savanna are less disturbed.

These areas are characterized by the seasonality of rainfall pattern, the less effective rainfall during the growing season, and the severity of the dry season. The rainfall of this ecology ranges from 500 to 1500 mm, but the dry season may exceed 5 months. The rainfall is variable in this ecology with a coefficient of variation often exceeding 75% (Nix, 1983). For more details on the savanna ecosystem readers are referred to a review by Bourlière (1983).

On the basis of the soil, climate and predominant vegetation, major biomes of the tropics are rainforest, deciduous forest, thorn forest, savanna, desert and tropical highlands. A detail vegetation and ecological map of Africa (Fig. 3.39) shows that the Guinean, the Sudanian, Sahelian, the Eastern and the Zimbabwean zones are the important agricultural zones of Africa.

V. POTENTIAL PRODUCTIVITY OF DIFFERENT TROPICAL ECOLOGIES

A. Comparative productivity of different ecologies

The net primary* productivity of different ecosystems varies widely depending on rainfall and available soil moisture, radiation, and soil properties. In the savanna and grasslands, water availability is the limiting factor, while in the tropical forest zone, radiation can be a limiting factor. Net primary productivity is the basic amount of energy available within an ecosystem. The potential productivity of tropical ecosystems far exceeds that of the equivalent regions in temperate latitudes. For example, Kowal (1972) compared the radiation level and potential productivity of Samaru, northern Nigeria, with that of Rothamsted, UK (Table 3.17). The potential productivity of Samaru was estimated to be 2.78 times more than that of Rothamsted. Even during the growing season,

* The net primary productivity (NPP) = $GPP - E_r$ where GPP or the gross primary productivity is the rate at which the solar energy is converted into chemical energy, and E_r is the energy or dry matter lost through respiration.

Table 3.17 Comparison of radiation levels and potential productivity at Samaru, northern Nigeria, and Rothamsted, U.K. (Kowal, 1972)

	Samaru			Rothhamsted	
Season	Radiation (cal cm^{-2})	Dry matter (kg ha^{-1})	Season	Ratiation (cal cm^{-2})	Dry matter (kg ha^{-1})
6 months rainy season May–October	88 000	22 000	7 months growing season March–Sept.	64 000	16 000
6 months dry season Nov.–April	90 000	22 500	5 months winter Oct.–March	12 000	3 000*
Annual total		44 500			16 000

* Production limited by low temperatures.

Table 3.18 Comparison of net primary productivity in different territorial ecosystems in tropical ecosystems in tropical and temperate latitudes (Murphy, 1975)

Ecosystem	Net primary productivity (g m^{-2} year^{-1})		
	Tropics	Temperate	Ratio Tropics:Temperate
Forest	2160	1300	1.7
Boreal forest	800	296	2.7
Grasslands	1080	500	2.2
Savannas	890	494	1.8

37.5% more productivity for Samaru than Rothamsted was attributed to better radiation. Although results of such comparisons are rather artificial and are difficult to generalize, it does appear that most tropical ecosystems are more productive on an annual basis than equivalent ecosystems in the temperate zone (Table 3.18).

Net primary productivity also increases with an increase in rainfall amount (Fig. 3.40). The high primary productivity of tropical ecosystems does not, however, mean that productivity of economic importance is also higher in the tropics. For example, in tropical forests leaves form a sizeable portion of the total annual production and the wood production is not necessarily high. Jordan (1981) argued that although the total productivity of natural forests can be high, productivity useful to man, on an area basis, is not greater than in temperate zones.

Within the tropical ecosystems, the net primary productivity of the rainforest on an annual basis is about 2.4 times that of the savanna, and is merely a reflection of the amount of rainfall received. What this essentially means is that the growing season is longer in the moist forest zone than in the drier savanna.

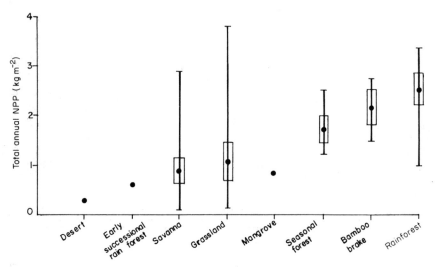

Figure 3.40 Average annual net primary productivity in various tropical ecosystems: ●, mean (or individual values; —, range; and □, standard error of the mean) (After Murphy, 1975)

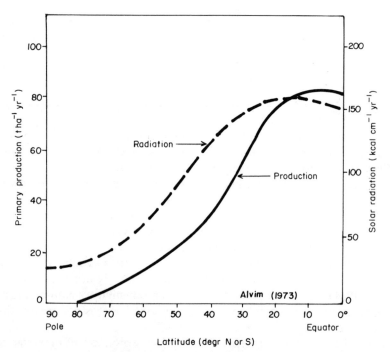

Figure 3.41 Changes in solar radiation and potential primary production for different latitudes (Alvim, 1973)

Table 3.19 Average rate of dry matter production at maximum growth in the savanna and forest zones (Adapted from Kassam and Kowal, 1973)

Crop	Rate of dry matter production (g m^{-2} day^{-1})	
	Savanna zone	Forest zone
Maize	40	21
Kenaf	38	19
Roselle	29	15
Cotton	27	19
Mean	34	18

Table 3.20 Average yields of some crops in savanna and forest zones (Kowal and Kassam, 1973)

Crop	Economic yeilds (kg/ha)		Ratio Savanna: Forest
	Forest	Savanna	
Maize	4140	6000	1.45
Cotton	1374	2556	1.86
Rice	1533	2500	1.63
Soybean	1629	2167	1.33
Groundnut	1136	2000	1.76
Cowpea	1245	1556	1.25
Tomato (t/acre)	6.5	17.1	2.62

The net primary productivity per day, however, may be higher in the savanna than in the forest. The latter has often less radiation due to overcast skies, thereby limiting the photosynthetic rate. The annual solar radiation in the savanna zone is 30% greater than in the forest zone, while the mean monthly radiation during the growing season is 42% greater (Fig. 3.41). Furthermore, the mean daytime and night-time temperatures are 6 and 16% higher in the forest than in the savanna. Consequently, the mean net photosynthesis in the savanna zone is 20 to 40% greater than in the forest zone (Table 3.19). In the forest zone, the gross photosynthesis is lost through the high rate of respiration. Kassam and Kowal (1973) reported that average daily rate of dry matter production of crops in the savanna is higher than in the forest zone by 45% for maize, 86% for cotton, 63% for rice, 33% for soybean, 76% for groundnuts, 25% for cowpea, and by 162% for tomato. Kowal and Kassam observed that economic yields of crops grown in experimental conditions were higher in savanna than in forest by 25 per cent for cowpea and as much as 162 per cent for tomato (Table 3.20).

Table 3.21 Arable land area in the tropics (10^6 ha) (Modifid from Norse, 1979)

Region	Presently cultivated	Additional potential arable area	Total
Humid tropics	86	416	502
Dry tropics	424	740	1164
Total	510	1156	1666

B. Potential cultivable area in different ecologies

The total cultivable land area in the tropics is about 1700 million hectares, only 30% of which is currently being cultivated (Table 3.21) (Norse, 1979). The potentially arable land area in the humid tropics is only 500 million hectares or only 30% of the total cultivable land area. The presently cultivated land area in the humid tropics is only 17% of the total cultivated land area in the tropics. This implies that dry tropics or savanna land provides most of the food grain production and has more potentially cultivable area that can be brought under cultivation.

The low net annual productivity of the savanna is attributed mostly to water deficit. Drought stress is obviously the most important constraint to food crop production in the savanna land. Intensive utilization of the existing agricultural lands and the possibility of bringing new lands under cultivation depends mostly on the availability of supplementary irrigation. The low productivity of the savanna of tropical America (Cerrados, Llanos and Pampas) is also due to severe nutritional problems. In addition to N, these soils are severely deficient in P and Ca. Most savanna soils of tropical America also have Al toxicity limitations.

REFERENCES

Aina, P.O., Lal, R., and Taylor, G.S., 1977, Soil and crop management in relation to soil erosion in the rainforest of western Nigeria, in *Soil Erosion: Prediction and Control*, SCS Special Publication, 21, pp. 75–82.

Ahmad, N., and Brechner, I., 1974, Soil erosion on 3 Tobago soils, *Trop. Agric.*, **51**, 313.

Ahn, P., 1970, *West African Soils*, Oxford University Press, Oxford, pp. 232.

Alvim de, P.T., Los tropicos bajos de América Latina: recursos ambiente para el desarrollo agricole, *Simposio El Potencial del Tropico Bajo*, Octubre, 1973, CIAT, Cali, Colombia, *Serie Simposios* 1, pp. 43–62.

Ayoade, J.P., 1974, A statistical analysis of rainfall over Nigeria, *J. Trop. Geog.*, **39**, 11–23.

Baghirathan, V.R., Shaw, E.M., 1978, Rainfall depth duration–frequency studies for Sri Lanka, *J. Hydrol.*, **37**, 223–240.

Balogun, C., 1972, The variability of rainfall in Nigeria, *Niger. J. Sci.*, **6**, 87–92.

Beard, J.S., 1953, The savanna vegetation of northern tropical America, *Ecol. Monogr.*, **23**, 149–215.

Bennema, J., 1977, Soils, in Alvim, P.T. de, and Kozlowski, T.T. (eds), *Ecolophysiology of Tropical Crops*, Academic Press, New York, pp. 302.
Benoit, P., 1977, The start of the growing season in northern Nigeria, *Agric. Met.*, **18**, 91–99
Biel, E., 1929, Die Veranderlichketi der Jahressumme des Niederschlags auf der Erde, *Geogr. Jber. Öst.*, **XIV**.
Biswas, A.K., 1967, Development of rain gauges, *Am. Soc. Civ. Eng.*, **93** (TR3), 99–124.
Blasco, F., 1983, The transition from open forest to savanna in Continental Southeast Asia, in F. Bourlière (ed.), *Tropical Savannas, Ecosystems of the World* **13**, Elsevier, Amsterdam, pp. 167–181.
Bonell, M., and Gilmore, D.A., 1980, Variation in short-term rainfall intensity in relation to synoptic climatological aspect of the humid tropical northeast Queensland Coast, *Singapore J. Trop. Geog.*, **1**, 16–30.
Bourlière, F., 1983, *Tropical Savannas, Ecosystems of the World* **13**, Elsevier, Amsterdam, 730 pp.
Bourlière, F., and Hadley, M., 1983, Present-day savannas: an overview, in Bourliere, F. (ed.), *Tropical Savannas, Ecosystems of the World*, **13**, Elsevier, Amsterdam, pp. 1–15.
Bowden, D.J., 1980, Rainfall in Sierra Leone, *Singapore J. Trop. Geog.*, **1**, 31–39.
Buringh, P., 1968, *Introduction to the Study of Soils in Tropical and Subtropical Regions*, Agricultural University, Wageningen, The Netherlands, 118 pp.
Chapman, A.L., and Kininmonth, W.R., 1972, A water balance model for rain-grown, lowland rice in northern Australia, *Agric. Met.*, **10**, 65–82.
Clarke, R.T., and Edwards, K.A., 1972, The application of the analysis of variance to mean areal rainfall estimation, *J. Hydrol.*, **15**, 97–112.
Cochémé, J., and Franquin, P., 1967, An agro-climatological survey of a semi-arid area in Africa south of the Sahara, *WMO Tech. Note* No. 26, 136 pp.
Counscil Scientifique pour l'Afrique (CSA), 1956, *Réunion de Spécialistes du CSA en matiere de Phytogéographie, Yangambi*, CCTA, London, Publ. No. 53, 35 pp.
C.T.F.T. Haute-volta, 1974, *Report de Synthese 1973*, C.T.F.T. Ministere de l'Agriculture de Haute Volta, Ouagadougou, 15 pp.
De Arruda, H.V., and Pinto, H.S., 1980, A simplified Gamma probability model for analysis of the frequency distribution of rainfall in the region of Campinas, SP, Brazil, *Agric. Met.*, **22**, 101–108.
De Martonne, E., 1926, *Geographie Physique, III Biogéoraphie* (3e édit), A. Colin, Paris.
Dennett, M.D., Elston, J., and Prasad, P.C., 1978, Seasonal rainfall forecasting in Fiji and the southern oscillation, *Agric. Met.*, **19**, 11–12
D'Hoore, J., 1954, *L' Accumulation des Sesquioxides Libres dans les Tropicaux*, Publ. INEAC, Ser. Sci. 62, 132 pp.
D'Hoore, J.L., 1964, *Soil Map of Africa: Scale 1 to 5,000,000, Explanatory Monograph*, C.C.T.A., Lagos, 205 pp.
Dow, H., Sir, 1955, *East Africa Royal Commission 1953—5 Report*, HMSO, London (Cmd. 9475).
Dudal, R. 1965, *Dark Clay Soils of Tropical and Subtropical Regions*, FAO Development Papers, No. 83, 161 pp.
Elwell, H.A., and Stocking, M.A., 1973, Rainfall parameters for soil loss estimation in a subtropical climate, *J. Agric. Eng. Res.*, **18**, 169–177.
FAO–UNESCO, 1974, *Soil Map of the World 1:5,000,000*, Vol. I. Legeon, FAO/UNESCO, Paris, 59 pp.
Fitzpatrick, E.A., 1965, Climate of the Tipperary Area, in *General Report on Lands of the Tipperary Area*, 1961, CSIRO, Melbourne, Australia, 112 pp.
Fitzpatrick, E.A., Slatyer, R.O., and Krishnam, A.I., 1967, Incidence and duration of

periods of plant growth in central Australia as estimated from climatic data, *Agr. Met.*, **4**, 389–404.
Foth, H.D., and Schafer, J.W., 1980., *Soil Geography and Land-use*, John Wiley & Sons, New York.
Fripiat, J.J., and Gostuche, M.C., 1952, *Etude physico-chemique des surfaces des argiles. Les combinaisons de la kaolinite avec les oxides de fer trivalents*, Publ. INEAC, Sér. Sci. 54.
Gillison, A.N., 1983, Tropical savannas of Australia and the southwest Pacific, in F. Bourliere (ed.), *Tropical Savannas, Ecosystems of the World* **13**, Elsevier, Amsterdam, pp. 183–243.
Glover, J., and Robinson, P., 1953, A simple method of assessing the reliability of rainfall, *J. Agric. Sci.*, **43**, 275–280.
Glover, J., Robinson, P., and Henderson, J.P., 1954, Provisional maps of the reliability of annual rainfall in East Africa, *J. Roy. Met. Soc.*, **80**, 602–609.
Golley, F.B., 1983 (ed.), *Tropical Rain Forest Ecosystems, Ecosystems of the World* **14A**, Elsevier, Amsterdam, 381 pp.
Gómez-Pompa, A., Vazquez-Yanes, C., and Guevara, S., 1972, The tropical rainforest: a non-renewable resource, *Science*, **177**, 762–762.
Gommes, R.A., 1983, Pocket computers in agrometeorology. FAO Plant production and Protection Paper 45, 140 pp.
Goodland, R., and Pollard, R., 1973, The Brazilian cerrado vegetation; a fertility gradient, *J. Ecol.*, **61**, 219–224.
Goosen, D., 1972, *Physiography and Soils of the Llanos Orientales, Colombia*, ITC, Enschede, The Netherlands 198 pp.
Gregory, S., 1964, Annual seasonal and monthly rainfall over Mozambique, in Steel, R.W., and Prothero, R.M. (eds.),*Geographers and the Tropics*, Liverpool Essays, London, pp. 81–109.
Gregory, S., 1969, Rainfall reliability in Thomas, M.F., and Whittington, G.W. (eds.), *Environment and Land-use in Africa*, Methuen, London, pp. 57–82.
Halais, P., 1964, Influence of meteorological fluctuations on sugar production 1955–64, *Mauritius Sugar Industry Research Institute, Ann. Rep.* **1964**, 86–94.
Hills, R.C., 1981, Agrometeorological services in developing countries: aspects of cost-efficiency, *Agric. Met.*, **23**, 292.
Hudson, N.W., 1965, The influence of rainfall on the mechanics of soil erosion, *M.Sc. Thesis*, University of Cape Town.
Hudson, N.W., 1976, *Soil Erosion*, B.T. Batsford Ltd, London.
Isbell, R.F., 1978, Soils of the tropics and sub-tropics, in Andrew, C.S., and Kamprath, E.J. (eds.), *Mineral Nutrition of Legumes in Tropical and Sub-Tropical Soils*, CSIRO, Melbourne, Australia, pp. 1–19.
Jackson, I.J., 1969a, Tropical rainfall variations over a small area, *J. Hydrol.*, **8**, 99–110
Jackson, I.J., 1969b, The persistence of rainfall gradients over small areas of uniform relief, *East Afr. Geogr. Rev.*, **7**, 37–43.
Jackson, I.J., 1972, The spatial correlation of fluctuations in rainfall over Tanzania; a preliminary analysis, *Arch. Met., Geophys. Bioklim. Serv. B.*, **20**, 167–178.
Jackson, I.J., 1974, Inter-station rainfall correlation under tropical conditions, *Catena*, **1**, 235–256.
Jackson, I.J., 1975, Relationships between rainfall parameters and interception by tropical forest, *J. Hydrol.*, **24**, 215–238.
Jackson, I.J., 1977, *Climate, water and agriculture in the Tropics*, Longmans, London, 248 pp.
Jackson, I.J., 1978, Local differences in the patterns of variability of tropical rainfall: some characteristics and implications, *J. Hydrol.*, **38**, 273–288.
Jackson, I.J., 1982, Traditional forecasting of tropical rainy season, *Agric. Met.*, **26**, 167–178.

Johnson, D.H., 1962, Rain in East Africa, *Q.J.Roy. Met. Soc.*, **88**, 1–21.
Jones, M.J., and Stockinger, K.R., 1972, The effect of planting date on the growth and yield of maze at Samaru, Nigeria, *Afr. Soils*, **17**, 27–34.
Jones, M.J., and Wild, A., 1975, *Soils of the West African Savanna*, C.A.B. Tech. Comm. No. 55, 246 pp.
Jordan, C.F. (ed.), 1981, *Tropical Ecology*, Hutchinson Ross, Stroudsburg, Pa., pp. 356.
Kassam, A.H., and Kowal, J., 1973, Productivity of crops in the Savanna and Rainforest zones in Nigeria, *Savanna*, **2**, 39–49.
Kassam, A.H., and Andrews, D.J., 1975, Effects of sowing date on growth, development and yield of photosensitive sorghum at Samaru, northern Nigeria, *Expl. Agric.*, **11**, 227–240.
Katsnelson, J., and Kotz, S., 1957, On the upper limits of some measures of variability, *Arch. Met. Geophys. Bioklim. Band*, 103–107.
Kautilya, Arthasastra, 1915, Government Oriental Library Series, Bibliotheca Sanskrita, No. 37, Bangalore, 321–296 BC.
Kenworthy, J.M., 1964, Rainfalls and water resources of East Africa, in Steel, R.W., and Prothero, R.M., (eds.), *Geographers and the Tropics*, Liverpool Essays, London, pp. 111–127.
Kenworthy, J.M., and Glover, J., 1958, The reliability of the main rains in Kenya. *E. Afr. Agric. J.* 23, 267–272.
Kowal, J.M., 1972, Radiation and potential crop production at Samaru, *Savanna*, **1**, 89–101.
Kowal, J.M., and Knabe, D.T., 1972, *An Agroclimatological Atlas of the Northern States of Nigeria*, ABU Press, Zaria, Nigeria.
Kowal, J.M., and Andrew, D.J., 1973, The pattern of water availability and water requirement for grain sorhgum production at Samaru, Nigeria, *Trop. Agric.*, **50**, 89–100.
Kowal, J.M., and Kassam, A.H., 1976, Energy load and instantaneous intensity of rainstorms at Samaru, northern Nigeria, *Trop. Agric. (Trinidad)*, **53**, 185–197.
Kowal, J.M., and Kassam, A.H., 1978, *Agricultural Ecology of Savanna*, Clarendon Press, Oxford, 403 pp.
Kowal, J.M., Kijewski, W., and Kassam, A.H., 1973, A simple device for analysing the energy load and intensity of rainstorms, *Agric. Met.*, **12**, 271–280.
Kraus, E.B., 1955, Secular changes of tropical rainfall regimes, *Q. J. Roy. Met. Soc.*, **81** 198–210.
Kruzinga, S., and Yperlaan, G.J., 1978, Spatial interpolation of daily totals of rainfall, *J. Hydrol.*, **36**, 65–74.
Kurtyka, J.C.., 1953, *Precipitation Measurement Study. III*, State Water Surv. Div. Rep. Invest. No. 20, 178 pp.
Lal, R., 1973, Effects of seedbed preparation and time of planting on maize in western Nigeria, *Expl. Agric.*, **9**, 303–313.
Lal, R., 1976, Soil erosion problems on an Alfisol in western Nigeria and their control, *IITA Monograph*, 208 pp.
Lal, R., 1981, Analyses of different processes governing soil erosion by water in the tropics, In *Erosion and Sediment Transport Measurement*, IAHS Publ. No. 113, 351–364.
Lal, R., Lawson, T.L., and Anastase, A.H., 1980, Erosivity of tropical rainfalls, in DeBadt, M. and Gabriels, M., (eds.), *Assessment of Erosion*, J. Wiley & Sons, Chichester, U.K., pp. 143–151.
Le Houérou, H.N., and Popov, G.F., 1981, An ecoclimatic classification of intertropical Africa, *FAO Plant Production and Protection Paper 31*, FAO, Rome, 40 pp.
Ling, A.H., and Robertson, 1982, Reflection coefficients of some tropical vegetation, *Agric. Met.*, **27**, 141–144.

Maignien, R., 1961, Les passage des sols ferrugineux tropicaux aux sols ferrallitique dans les regions sud-ouest du Sénegal, *Soils Afrique*, **6**, 113–172.

Manning, H.L., 1949, Planting date and cotton production in the Buganda Province of Uganda, *Emp. J. Exp. Agric.*, **17**, 245–258.

Manning, H.L., 1950, Confidence limits of expected monthly rainfall, *J. Agric. Sci.*, **40**, 169–176.

Mavi, H.S., and Newman, 1976, Crop production potentials in India – A water availability based analysis, *Agric. Met.*, **17**, 387–395.

McCown, R.L., 1973, An evaluation of the influence of available soil water storage capacity on growing season length and yield of tropical pastures using simple water balance models, *Agric. Met.*, **11**, 53–63.

McCulloch, J.S.G., 1961, The statistical assessment of rainfall, *Inter-African Conference on Hydrology*, Nairobi, pp. 108–114.

Menaut, Jean-Claude, 1983, The vegetation of African Savannas, in F. Bourliere (ed.), *Tropical Savannas Ecosystems of the World* **13**, Elsevier, Amsterdam, pp. 109–149.

Mistra, R., 1983, Indian Savannas, in F. Bourliere (ed.), *Tropical Savannas, Ecosystems of the World* **13**, Elsevier, Amsterdam, pp. 151–166.

Mohr, E.C.J., and Van Baren, F.A., 1954, *Tropical soils*, Uitgeverij W. Van Hoeve, The Hague. Interscience, New York, 498 p.

Mohr, E.C.J., Van Baren, F.A., and Van Schuylenborgh, J., 1972, *Tropical Soils*, Mouton-Ichtiar Barn Van Hoeve, The Hague, 481 pp.

Montgomery, R.F., and Askew, G.P., 1983, Soils of tropical savannas, in F. Bourliere (ed.), *Tropical Savannas Ecosystems of the World* **13**, Elsevier, Amsterdam, pp. 63–78.

Morth, H.T., 1968, A study of the areal and temporal variations of rainfall in East Africa, *Proc. 4th Spec. Meet. on Applied Meteorology*, Nairobi, Kenya.

Murphy, P.G., 1975, Net productivity in tropical terrestrial ecosystems, in Lieth, H., and Whittaker, R.H., (eds.), *Primary Productivity of the Biosphere*, Springer-Verlag, New york, pp. 217–231.

Nieuwolt, S., 1977, *Tropical Climatology*, J. Wiley & Sons, Chichester, U.K.

Nix, H.A., 1983, Climate of tropical savannas, in Bourliere, R. (ed.), *Tropical Savannas, Ecosystems of the World* **13**, Elsevier, Amsterdam, pp. 37–62.

Norse, D., 1979, Natural resources development strategies, and the world food problem, in Biswas, M.R., and Biswas, A.K. (eds.), *Food, Climate and Man*, Wiley-Interscience J. Wiley & Sons, New York, pp. 12–51.

Nwa, E.U., 1977, Variability and error in rainfall over a small tropical watershed, *J. Hydrol.*, **34**, 161–170.

Ochse, J.J., Soule, M.J. Jr., Dijkman, M.J., and Wehlburg, C., 1961, *Tropical and Subtropical Agriculture*, Macmillan, New York, 760 pp.

Odum, H.T., and Pigeon, R.F. (eds.), 1970, *A Tropical Rain Forest*, US Atomic Energy Commission, Washington, D.C., 1678 pp.

Ojo, O., 1977, *The Climates of West Africa*, Heinemann, London, 218 pp.

Oyebande, O., and Oguntoyinbo, J.S., 1970, Statistical analysis of rainfall patterns in the Southwestern States of Nigeria, *Niger. Geog. J.*, **13**, 141–162.

Papadakis, J., 1970, *Climates of the World: their Classification, Similitudes, Differences and Geographic Distribution*, Libro De Edicion Argentina, Cordoba 4564, Buenos Aires, 47 pp.

Pereira, H.C., 1973, *Land-use and Water Resources*, Cambridge University Press, Cambridge, 246 pp.

Richards, P.W., 1964, *The Tropical Rain Forest: An Ecological Study*. Cambridge University Press, Cambridge, 450 pp.

Riehl,H., 1954, 1979, *Tropical Meteorology*. McGraw Hill Book Company, New York, 392 pp.

Roose, E.J., 1977, Use of the USLE to predict erosion in West Africa. in: *Soil Erosion: Prediction and Control*. SCS Special Publication 21, 60–74.

Sanchez, P.A., 1976, *Properties and Management of Soils in the Tropics*. J. Wiley & Sons, New York.

Sanchez, P.A., and Buol, S.W., 1975, Soils of the tropics and the World Food Crisis, *Science*, **188**, 598–603.

Sarmiento, G., 1983, The Savannas of Tropical America. in Bourliere, F. (ed.), *Tropical Savannas, Ecosystems of the World* **13**, Elsevier, Amsterdam, pp. 245–288.

Scarf, F., 1977, Estimating potential evapotranspiration, using the Penman procedure, *Hydrol. Proc.* **17**, (Drainage & Irrig. Dept., Malaysia).

Schimper, A.F.W., 1903, *Plant Geography upon a Physiological Basis*, Clarendon Press, Oxford.

Schumann, Je., W., and Mostert, J.S., 1969, On the variability and reliability of precipitation, *Bull. Amer. Met. Soc.*, **30L**, 100.

Sharon, D., 1972, The spottiness of rainfall in a desert area, *J. Hydrol.*, **17**, 161–176.

Sharon, D., 1974, The spatial pattern of corrective rainfall in Sukumaland, Tanzania — a statistical analysis, *Arch. Meteor., Geophys. Bioklim. Ser. 3*, **22**, 201–218.

Sharon, D., 1980, The distribution of hydrological effective rainfall incident on sloping ground, *J. Hydrol.*, **46**, 165–188.

Shukla, J., 1984, Interannual variability of monsoons, in J.S. Fein and P. Stephens (eds.), *Monsoons*, John Wiley & Sons, New York.

Sien, C.L., and Koon, C.K., 1971, The record floods of 10th December, 1969 in Singapore, *J. Trop. Geog.*, **33**, 9–19.

Sivakumar, M.V.K., Virmani, S.M., and Reddy, S.J., 1980, Rainfall climatology of West Africa: Niger ICRISAT, Hyderabad, India, *Information Bulletin* No. 5, 66 pp.

Smyth, A.J., and Montgomery, R.F., 1962, *Soils and Land Use in Central Western Nigeria*, Govt. Printer, Ibadan, Nigeria, 265 pp.

Stern, R.D., and Coe, R., 1982, The use of rainfall models in agricultural planning, *Agric. Met.*, **26**, 35–50.

Sreenivasan, P.S., 1971, A study of districtwise distribution of rainfall in Bihar, *J. Hydrol.*, **13**, 41–53.

Sreenivasan, P.S., and Banerjee, J.R., 1973, The influence of rainfall on the yield of rainfed rice at Karjat (Colaba Dist), *Agric. Met.*, **11**, 285–292.

Stigter, C.J., and Hyera, T.M., 1979, Agro-meteorological research needs in Tanzania, *Agric. Met.*, **20**, 375–280.

Subbaramanya, I., and Rupa Kumar, K., 1980, Crop-weather relationships of sugarcane and yield prediction in northeast Andhra Pradesh, India, *Agric. Met.*, **20**, 265–279.

Subrahmangan, V.P., and Ratnam, B.P., 1969, Albedo Studies in a sugarcane field, *Indian J. Agric. Sci.*, **39**, 774–779.

Tan, B.C., and Rajaratnam, A., 1975, Preliminary results of reflection coefficients of some typical terrains of short wave radiation, *Malays. Agric. Res.*, **4**, 221–225.

Tanaka, A., Sakuma, T., Okagawa, N., Imai, H., and Ogata, S., 1984, *Agro-ecological condition of the Oxisol–Ultisol Area of the Amazon River System*, Faculty of Agric., Hokkaido University, Sapporo, Japan, 101 pp.

Thorpe, W.R., Rose, C.W., and Simpson, R.W., 1979, Areal interpolation of rainfall with a double Fourier series, *J. Hydrol.*, **42**, 171–178.

Trewartha, G.T., 1954, *An Introduction to Climate*, McGraw Hill Book Co., New York, 272 pp.

Tricart, J., 1972, *The Landforms of the Humid Tropics, Forests and Savannas*. Longmans, London, 306 pp.

UNESCO, 1978, *Tropical Forest Ecosystems:* UNESCO/FAO/UNEP Nat. Resour. Res., **14**, 683 pp.

USDA, 1975, *Soil Taxonomy*, Soil Conservation Service, USDA Agric. Handbook, No. 436, Washington, D.C., 754 pp.
Vickery, M.L., 1984, *Ecology of Tropical Plants*, J. Wiley & Sons, New York, 170 pp.
Virmani, S.M., 1975, The agricultural climate of the Hyderabad region in relation to crop planning (a sample analysis), ICRISAT, Hyderabad, India.
Virmani, S.M., Sivakumar, M.V.K., and Reddy, S.J., 1980, Climatic classification of semi-arid tropics in relation to farming systems research, in *Climatic Classification*, A Consultants Meeting 14–16 April 1980, ICRISAT, Hyderabad, India, pp. 27–44.
Virmani, S.M., Reddy, S.J., et Buse, M.S.S., 1980, Manuel de climatologie pluviale de l'Afrique Occidentale. Données pour des stations sélectionnées. ICRISAT Inf. Bull. 7, Hyderabad, India, 52 pp.
Walker, H.O., 1957, The Weather and Climate of Ghana, Dept. Note, No. 5, *Ghana Met. Dept.*, Accra.
Walker, H.O., 1962, quoted by Gregory, S., 1969. Rainfall Reliability, in Thomas, M.F., and Whittington, G.W. (eds) Environment and Landuse in Africa, Methuen, London, pp. 57–82.
Wambeke, A. Van, 1978, Properties and Potentials of soils in the Amazon Basin, *Interciencia*, **3**, 233–242.
Webster, C.D., and Wilson, P.N., 1966, *Agriculture in the Tropics*, Longmans, London.
Wernstedt, F.L., 1972, *World climatic data*. Climatic Data Press, Lement, PA, 522 pp.
Whitmore, T.C., 1975, *Tropical Rain Forests of the Far East*, Clarendon Press, Oxford, 281 pp.
Worthington, E.B., 1958, quoted in Gregory, S., 1969, Rainfall Reliability, in Thomas, M.F., and Whittington, G.W. (eds) 'Environment and Landuse in Africa' Methuen, London, 57–82.
Yao, A.Y.M., 1973, Evaluating climatic limitations for a specific agricultural enterprise, *Agric. Meteorol.*, **12**, 65–73.
Young, A., 1976, *Tropical Soils and Soil Survey*. Cambridge Geographical Studies, 9. Cambridge University Press, London, 468 pp.

Chapter 4
Soil Physical Properties

I	ROLE OF SOIL PHYSICAL PROPERTIES	114
II	MAJOR SOILS OF THE TROPICS	114
III	SOIL PHYSICAL PROPERTIES AND CLAY CHARACTERISTICS	116
IV	PHYSICAL PROPERTIES OF LAC SOILS	117
	A. Texture and mechanical properties of soils	117
	1. Particle size distribution	117
	(i) Tropical America	117
	(ii) Africa	120
	(iii) Asia and Australia	123
	2. Bulk density	125
V	SOIL STRUCTURES	129
	A. Factors Affecting Soil Structure	131
	B. Micro-Aggregation in Tropical Soils	132
	C. Structural Stability and Soil Erodibility	132
VI	SWELL–SHRINK PROPERTIES	135
VII	pF CURVES AND THE AVAILABLE WATER-HOLDING CAPACITY	135
VIII	WATER TRANSMISSION PROPERTIES	150
	A. Water Infiltration	150
	B. Hydraulic Conductivity	152
IX	AERATION	155
X	THERMAL PROPERTIES	157
XI	SOIL PHYSICAL PROPERTIES OF TROPICAL VERSUS TEMPERATE ZONE ALFISOLS	159
XII	PHYSICAL PROPERTIES OF HAC SOILS	160
	A. Vertisols	160
	B. Mollisols	165
XIII	PHYSICAL PROPERTIES OF TROPICAL ANDOSOLS	167
XIV	PHYSICAL PROPERTIES AND SOIL CLASSIFICATION	168
XV	CONCLUSIONS	170

I. ROLE OF SOIL PHYSICAL PROPERTIES

The importance of soil physical properties in sustaining high levels of production and in preserving stability of ecological environments has not been given the recognition it deserves. While the hazards of soil degradation, desertification, accelerated soil erosion and rapid decline in productivity are widely recognized, the underlying causes of these problems are hardly understood. Accelerated soil erosion, a major factor responsible for degradation of soil quality, is a direct result of low structural stability and its interaction with rainfall characteristics and soil mismanagement. Severe constraints of soil compaction and frequent drought often associated with intensive and mechanized arable land-use can be attributed to the collapse of soil structure and decline in water transmission and retention pores. The supra-optimal soil temperatures commonly observed in the surface soil horizons are created by a combination of coarse texture, low organic matter content and low available water holding capacity on the one hand and high insolation on the other. The net result is the severe reduction in crop production caused by accelerated soil erosion, erosion-induced changes in soil properties, frequent occurrence of drought and high soil temperature stresses.

It is a misconception that economic productivity of most tropical soils can be sustained by the addition of the balanced fertilizers and essential plant nutrients in which these soils are grossly deficient. Undoubtedly some soils respond favourably to chemical inputs. In others, however, chemical inputs are easily wasted if soil physical properties are not managed adequately for good crop growth. Fertilizers can easily be leached out of the root zone, washed away in surface run-off, or volatilized by continuously high temperatures. The adverse effects of drought stress are exaggerated on a fertilized soil with inadequate amounts of available soil-water in the root zone. Capital investments in farm equipment and heavy machines are mixed blessings for soils prone to compaction and erosion.

An understanding of the physical properties of soils of the tropics is, therefore, essential for high economic returns, a favourable ecological balance and environmental protection. Some severe and costly errors made in the past in land management and development schemes could have been avoided if soil physical properties and processes were considered in planning and implementation.

II. MAJOR SOILS OF THE TROPICS

The geographical distribution of major soil orders shown in Table 4.1 indicates three major groups of soils: (i) soils containing predominantly low activity clays; (ii) soils of mixed mineralogy and those containing predominantly high activity clays; and (iii) soils of volcanic origin.

Soils of the humid and subhumid tropics are those containing low activity clays (LAC). Low activity clays are those with an effective cation exchange

Table 4.1a Land of different soil orders (modified from Buringh, 1979)

Soil order		World land area		Land area in the tropics	
		(10^6 ha)	%	(10^6 ha)	%
(a)	Low activity clays				
	Alfisols	1730	13.1	800	16.2
	Ultisols	730	5.6	550	11.2
	Oxisols	1120	8.5	1100	22.5
(b)	High activity clays				
	Aridisols	2480	18.8	900	18.4
	Entisols	1090	8.2	400	8.2
	Inceptisols	1170	8.9	400	8.3
	Vertisols	230	1.8	100	2.0
	Mollisols	1130	8.6	50	1.0
	Histosols	120	0.9	—	—
	Spodosols	560	4.3	—	—
(c)	Mountain soils	2810	21.3	600	12.2
	Total	13 170	100.0	4900	100.0

Table 4.1b Soils of the semi-arid tropics

Soil order	Area in each continent (10^6 km^2)			Total
	Africa	Latin America	Asia	
Alfisols	4.66	1.07	1.21	6.94
Aridisols	4.40	0.33	0.47	5.20
Entisols	2.55	0.17		2.72
Inceptisols	0.38	—	0.28	0.66
Mollisols	—	0.78		0.78
Oxisols	1.88	—		1.88
Ultisols	0.24	0.08	0.20	0.52
Vertisols	0.51	—	0.80	1.31
Others	—	0.70	0.23	0.93
Total	14.62	3.13	3.19	20.94

capacity (e.c.e.c.) of 16 meq per 100 g clay or less or determined in ammonium acetate of 24 meq per 100 g or less. These soils are characterized by a clay fraction mostly composed of kaolinite and halloysite with hydrous oxides of iron and aluminium. These soils are mostly Alfisols, Ultisols and Oxisols. Together these three soil orders comprise about 50 per cent of the land surface of the tropics, but only 27 per cent of the world land area (Table 4.1).

Major orders of the soil containing high activity clays (HAC) and those with mixed mineralogy are Aridisols, Vertisols, Mollisols, Inceptisols, and Entisols. Aridisols and Vertisols are predominant soils of the semi-arid and arid tropics and together comprise about 20 per cent of the tropical land area. Alfisols and Vertisols combined comprise 63% of the Asian semi-arid tropics (El-Swaify et al, 1984). The sub-orders Inceptisols and Entisols are relatively fertile soils and cover about 16.5 per cent of the land surface. Exceptionally fertile Mollisols are relatively scarce.

The order Andosols are the soils of volcanic origin and cover vast regions of Central America and parts of southeast Asia, i.e., the Indonesian island of Java. These are very fertile soils and capable of supporting high population density.

Soils containing predominantly low activity clays, in general, have poorer physical properties than those with high activity clays and of volcanic origin. More specifically Alfisols and Ultisols are easily dispersed, prone to crusting and compaction, and susceptible to erosion. In contrast Mollisols, Inceptisols and Entisols are better structured, with generally favourable physical properties. Exception to these general rules are LAC Oxisols with clays resistant to dispersion and HAC Vertisols with easily dispersed clays of unfavourable physical attributes. Also, some Alfisols and Ultisols are characterized by very stable aggregation, e.g., Rhodic Paleustults, also classified as Nitosols according to the FAO system of classification. In Central and East Africa Nitosols cover as much as 200 million hectares of the land surface. In West Africa, however, especially in regions where rainfall exceeds 1200 mm per annum, landscapes are generally dominated by Oxic Alfisols and Ultisols with less favourable soil physical characteristics. The majority of LAC soils in south and southeast Asia and in tropical Australia are similar to the soils of West Africa. Soils of Latin America and particularly those of the Amazon basin, however, are Oxisols with relatively favourable physical properties.

III. SOIL PHYSICAL PROPERTIES AND CLAY CHARACTERISTICS

HAC soils are less easily dispersed than LAC soils because they are flocculated by solutions of higher electrolyte concentration. The ease of dispersion of LAC soils is attributed to the influence of oxides and hydrous oxides of iron and aluminium (Koenigs, 1961). The oxide and hydrous oxide surfaces, normally with net positive charge within the pH ranges of most Alfisols and Ultisols, are often inactivated through adsorption of some anions, e.g., silicate, phosphate and organic anions (Deshphande et al., 1968; Bowden et al., 1980).

In contrast, soils with clays highly resistant to dispersion e.g. Nitosols and Oxisols, are strongly aggregated into fine sand and silt sized particles so that their field texture is often loamy rather than clayey. Even after treatment with hydrogen peroxide and prolonged shaking, the aggregates are usually not destroyed. The aggregates are broken into primary particles only after addition of an anionic dispersing agent (Table 4.2).

Self-dispersing Vertisols have high specific surface area and their exchange complex is mostly dominated by monovalent cations e.g. Na^+ These soils slake readily on quick wetting and are difficult to manage.

Table 4.2 Clay aggregation in soils with low activity clays (Ahn, 1979)

Soil	Horizon (Depth, cm)	Clay content (% of soil)			Proportion of clay in stable aggregates
		Before dispersion	After dispersion	Difference	
More stable					
Rhodic Paleudult	A 11 (5)	20.5	58.4	37.9	0.65
or Nitosol	A 12 (15)	18.9	51.5	32.6	0.63
(Kikuyu friable clay, Kenya)	Upper B 2 (30)	4.2	69.8	65.6	0.94
	Lower B 2 (90)	1.8	82.8	81.0	0.98
Alfic Eutrorthox	A 11 (5)	7.7	45.2	37.5	0.83
(Akamadon	A 12 (13)	7.4	45.2	37.8	0.84
series, Ghana)	B 1 (38)	7.2	48.0	40.8	0.85
	B 21 (80)	11.3	64.7	53.4	0.83
Less stable					
Plinthic Paleudult	A 11 (5)	18.6	19.0	0.4	0.02
(Bekwei series,	A 12 (13)	16.4	20.4	4.0	0.20
Ghana)	B 21 (30)	19.0	43.2	24.2	0.56
	B 23 (101)	31.9	68.1	36.2	0.53
Plinthic Paleudult	A 11 (5)	8.8	12.2	3.4	0.28
(Asuansi series,	A 12 (13)	9.6	12.4	2.8	0.23
Ghana)	B 1 (28)	10.1	22.2	12.1	0.55
	B 23 (82)	21.5	49.7	28.2	0.57

IV. PHYSICAL PROPERTIES OF LAC SOILS

The general physical properties of LAC soils have been described by Lal (1979, 1980) and Lal and Greenland (1984). These are important soils of the humid and subhumid tropics, and judicious management of physical properties is the key towards their intensive utilization and for high and sustained production.

A. Texture and mechanical properties of soils

Sedentary upland soils in the humid and subhumid tropics have coarse-textured surface horizons containing 60 to 70 per cent sand. The fine fraction has long been eluviated to the clayey subsoil horizon or has been preferentially removed downslope by surface run-off or lateral flow. The existing clay and silt content in the surface horizon is primarily due to soil turnover by macrofauna, e.g., termites and earthworms. The clay content of the surface horizon generally decreases with the increase in rainfall amount.

1. Particle size distribution

(i) Tropical America: Textural and mechanical properties of soils of tropic-

Table 4.3 Mechanical properties of a Cerrado Ultisol in Brazil (Stone and da Silveira, 1978). *Reproduced by permission of Pesquisa Agropecuaria Brasileira*

Horizon	Depth (cm)	Sand (%)	Silt (%)	Clay (%)	Bulk density (g cm^3)
A1	0–15	51.8	17.2	31.0	1.23
A1	15–30	54.5	14.5	31.0	1.22
A3	30–45	50.2	13.8	36.0	1.16
A3	45–60	50.0	14.0	36.0	1.17

al America have been reviewed by Lutz (1972). Oliveira (1968) surveyed the textural properties of the humid (Ultisol) and semi-arid (Alfisol) soils of Northeast Brazil. He analysed 1976 soil samples and noted that 85 per cent of the soils contained less than 20 per cent silt, and 20 per cent of the samples had clay content of between 0 and 20 per cent. The predominant textural classes of these soils were sand, loamy sand and sandy loam in the semi-arid regions and sandy clay and clay in the humid regions. Similar results of the analyses of physical properties of the soils of Cerrado region of Brazil have been obtained by Resck (1981). Stone and da Silveira (1978) reported the textural properties of an Ultisol from the Cerrado region of Goinia, Brazil (Tables 4.3 and 4.4). Although the soil contained 30–36 per cent clay, it behaved like sand. The data of Baena (1977) describing the textural properties of an Oxisol from the humid region of Para State, northern Brazil, shown in Table 4.5, indicate sand content of 70 to 76 per cent in the surface horizon. While the fine sand and silt content varies little with depth, coarse sand decreases and clay increases with depth.

The particle size distribution of some soils of the semi-arid and sub-humid regions of Venezuela has been described by Pla Sentis (1977). His data in Table 4.6 show that some Alfisols and Ultisols contain as much as 67 to 84 per cent sand. The silt content of Alfisols and Ultisols is low (5 to 14 per cent). It is the recently developed Inceptisols that contain high amounts of silt and clay. The predominant texture of uplands is sandy to sandy loam. Frómeta *et al.* (1979) analysed 1108 samples from the Los Lanos region of Venezuela. The majority of these soils were found to be of sandy loam texture with sand contents of 60 to 70 per cent. While silt and clay increased with depth, the sand progressively decreased.

The textural and physical properties of the forest zone soils of Peru are described by Meyer and Neumeyer (1980). González (1980) analysed physical properties of some Colombian soils. These soils had more clay (15–27 per cent) and silt (18–58) but less sand (8–48 per cent). Nevertheless, the available water holding capacity of these soils was low and ranged from 12.7 to 19.3 per cent. The soils were also either non-plastic or only slightly plastic. The textural properties of soils of Puerto Rico have been described by Philipson and Drosdoff (1972), Wolf and Drosdoff (1976a, b), and Agafonov *et al.* (1978). For some soils of Cuba, Belobrov (1978) reported that high rainfall in soils of

Table 4.4 Textural properties of an Oxisol from southern Bahia, Brazil (da Silva, 1981)

Horizon	Depth (cm)	Mechanical separates (%)				Natural clay (%)	Flocculation (%)
		Coarse sand	Fine sand	Silt	Clay		
A11	0–10	71.9	11.4	7.7	9.0	2.0	77
A12	10–25	47.9	13.9	13.0	25.2	12.1	52
A3	25–48	34.8	10.4	9.5	45.3	26.2	42
B1	48–68	30.8	9.0	12.8	47.4	28.0	40
B21	68–145	27.7	7.9	17.1	47.3	0.0	100
B22	145–200	26.9	7.1	14.5	51.5	0.0	100
B23	200–260	27.0	5.6	18.0	49.4	0.0	100

Table 4.5 Particle size distribution (%) of 175 samples of Oxisols from Para State, Brazil (Baena, 1977)

Depth (cm)	Coarse sand	Fine sand	Silt	Clay
0–10	30	46	10	14
10–20	27	44	11	18
20–30	25	44	11	20
30–40	24	44	11	21
40–60	24	43	11	22
60–80	23	43	11	23
80–100	24	42	10	24

Table 4.6 Physical properties of some upland soils from Venezuela (Pla Sentis, 1977)

Property	Soil series			
	Barinas (Alfisol)	Guanipa (Ultisol)	Fanfurria (Inceptisol)	Guanaguanare (Inceptisol)
Sand (%)	67	84	38	11
Silt (%)	14	5	37	47
Clay (%)	19	11	25	42
Upper plastic limit	17.4	21.0	30.0	36.0
Lower plastic limit	17.2	19.9	24.1	27.4
Plasticity index	0.2	1.1	5.9	8.7
Texture	Sandy loam	Sandy	Loam	Silty clay

the humid tropics results in the depletion of the clay from the upper soil horizons and their enrichment with sand.

(ii). Africa: The textural properties of African soils have been extensively studied (Vine, 1954, Vine et al., 1954; Greenland, 1981). The properties of the soils of the African savanna and semi-arid regions have been reported by Jones and Wild (1975) and Buanec (1974). Asamoa (1973) reported the particle size distribution of some Afilsols of Ghana. He observed low clay content in the surface but >40% in the B horizon. From the analyses of granite-gneiss derived soils in Togo, Lévêque (1977) noted that the surface horizon loses clay by eluviation, and that the eluviated clay is transported towards streams by selective surface erosion. This transported clay is rarely incorporated into the profiles over which it passes. The increase in sand fraction of the red ferrallitic soils of Guinea was attributed to the wetting and drying cycles and to the loss of clay by eluviation (Dimov, 1980).

Sedentary soils of West Africa have a well-defined gravelly horizon of varying thickness and a highly variable gravel content. In addition to the more resistant weathering material, i.e., quartz, localized concentrations of concretionary materials (Fe and Mn concretion) formed due to fluctuating water table

Table 4.7 Gravel content of different horizons for some soils of southwest Nigeria (IITA, 1975)

Series	Location coordinate	Gravels in different horizons (%)					
		1	2	3	4	5	6
1. Egbeda	8° 41'N, 3° 42'E	5	57	56	54	45	37
2. Sepeteri	8° 41'N, 3° 42'E	11	56	69	74	67	39
3. Iso	8° 6'N, 3° 20'E	5	34	58	29	18	—
4. Fashola	8° 6'N, 3° 20'E	5	39	65	69	72	—
5. Iwaji	7° 51'N, 4° 6'E	29	68	72	62	—	—
6. Erinoke	7° 51'N, 4° 6'E	0	43	73	63	39	—
7. Ekiti	7° 29'N, 3° 53'E	3	7	32	85	88	—
8. Apomu	8° 41'N, 3° 42'E	0	0	0	6	12	23

are also observed in soils located at the transition zone where the predominant slope changes from convex to concave. In West Africa, soils with Fe and Mn concretions are of inherently low fertility because of their shallow depth to *in situ* developed iron pan. Together these soils cover approximately 25 million hectares in West Africa out of which at least 130 million hectares are in the humid forest zone (Obeng, 1978). The data in Table 4.7 show a highly variable concentration of gravel in some soil series of western Nigeria. For most upland soils, the gravel content ranges from 30 to 80 per cent. The possible effects of soil fauna on the formation of these gravelly horizons is discussed in Chapters 7 to 10. Soils of West Africa are characterized by low silt content. Soils with silt content exceeding 10% are uncommon. Consequently, these soils occupy the extreme left-hand corner of the textural diagram (Fig. 4.1).

In contrast to the sedentary soils, the loess soils of the semi-arid and sahel zone contain high amounts of silt and fine sand fractions. These soils are low in

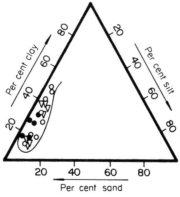

● Onne - Oxic Paleudult (Onne)
○ Alagba - Oxic Paleustalf (Ikenne)
▽ Egbeda - Oxic Paleustalf (Ilora)

Figure 4.1 The textural properties of some soils of West Africa with low silt contents (Mbagwu *et al.*, 1983). *Reproduced by permission of Williams & Wilkins Co.*

Table 4.8 Mechanical analysis and bulk density of surface soil horizons (0–10 cm) of some soils of the semi-arid regions of Kenya (Kilewe et al., 1983)

Location	Bulk density (g cm^{-3})	Mechanical analysis (%)			Texture
		Sand	Silt	Clay	
KARI Farm, Muguga	1.11	28.8	31.9	35.1	Clay loam
KNDRS Farm 1, Machakos	1.52	24.9	28.0	22.2	Sandy clay loam
KNDRS Farm 2, Machakos	1.61	13.7	23.7	25.7	Sandy clay loam
KNDRS Farm 3, Machakos	1.59	10.3	16.5	28.2	Sandy clay loam

Table 4.9 Mechanical analysis of the surface horizon of some Alfisols in the dry zone of Sri Lanka (Panabokke, 1958). *Reproduced by permission of* Tropical Agriculturist

Property	Soil A	Soil B	Soil D
Depth (cm)	0–22.5	0–22.5	0–17.5
Coarse sand (%)	28.6	33.6	47.5
Fine sand (%)	31.5	23.1	28.3
Silt %)	9.5	13.5	6.5
Clay (%)	31.5	30.1	17.8
Organic matter (%)	1.7	1.8	1.1

clay and coarse sand. Consequently, these soils are easily compacted. They develop crust and surface seal following heavy rains and subsequent drying. High silt content is also reported for soils of East Africa (Table 4.8). These soils are predominant in regions where more rain falls during the 3 to 4 summer months, and where there is a prolonged dry season.

(iii) Asia and Australia: Alfisols of south Asia (southern India and Sri Lanka) and northern Australia are similar in textural properties to that of West Africa especially in regions with similar rainfall characteristics and parent material. The data on mechanical analyses of the surface horizon of three Alfisols from the dry zone of Sri Lanka (Table 4.9) also show high sand (57 to 75 per cent) and relatively low silt contents. The gravelly horizon of soils in south Asia occurs relatively deeper in the profile than in soils of West Africa. The data of an Alfisol from Hyderabad, Central India (Table 4.10) show a gravelly horizon occurring at a depth of about 1 m.

Table 4.10 Gravel content and bulk density of two Alfisol profiles at Hyderabad, India (ICRISAT, 1978–79)

Depth (cm)	Gravel content (%)		Bulk density (g cm^{-3})	
	ST 2	RA 10	ST 2	RA 10
0–15	2.4± 0.8	10.9± 9.1	1.55±0.22	1.57±0.20
15–30	1.5± 0.8	7.7± 6.0	1.64±0.15	1.72±0.19
30–45	0.6± 0.8	2.9± 1.7	1.59±0.07	1.67±0.10
45–60	2.2± 4.8	1.4± 0.5	1.59±0.12	1.63±0.11
60–75	4.3±10.4	2.3± 1.3	1.64±0.04	1.71±0.10
75–90	5.0±13.0	11.1±19.7	1.60±0.04	1.79±0.12
90–105	6.6±10.1	16.2±20.2	1.68±0.10	1.83±0.15
105–120	11.2±12.6	30.3±27.8	1.65±0.09	1.85±0.14
120–135	12.4±14.5	21.7±18.4	1.75±0.10	1.81±0.16
135–150	18.1±17.8	28.3± 8.1	1.70±0.09	1.86±0.16
150–165	28.0±14.7	33.1±14.1	1.80±0.11	1.86±0.05
165–180	30.7±10.0	31.0±15.9	1.74±0.08	1.83±0.13
180–195	16.6±14.2	35.1±21.6	1.80±0.14	1.89±0.20

Table 4.11 Mechanical composition of some upland soils from Northeast Thailand (Takahashi et al., 1983)

Vegetation	Depth (cm)	Particle density (g cm^{-3})	Organic matter (%)	Gravel content (%)		Sand (%)		Silt (%)	Clay (%)
				>4.8 mm	2–4.8 mm	Coarse	Fine		
Cleared	0–5	2.51–2.65	4.87–8.49	0.1– 0.5	0.3–3.2	5.1– 9.7	30.3–36.9	29.3–38.7	17.2–30.9
	5–15	2.66–2.69	2.37–4.35	0.1–20.3	0.6–9.3	2.8–13.5	22.8–36.6	23.5–30.4	15.0–29.8
Forest	0–5	2.50–2.57	7.77–9.73	0.3– 0.6	0.9–2.9	5.3– 8.0	27.5–32.2	35.0–41.2	22.1–24.8
	5–15	2.65–2.76	2.54–4.49	0.0– 5.1	1.3–6.5	7.0– 7.4	29.7–32.0	27.0–36.0	23.1–26.6

Figure 4.2 Some Alfisols with low soil organic matter content develop slowly permeable crust and surface seal

Takahashi et al. (1983) analysed physical properties of upland soils of northeast Thailand. Their data (Table 4.11) show that the surface soils had relatively more silt content than soils in the humid zone of West and Central Africa. The clay content of the surface horizon is in the range of 15–25 per cent, and the gravel content in the subsoil is relatively less. Profile characteristics of upland soils of northern Australia are described by CSIRO (1965) and Northcote et al. (1975). The surface soils have less clay content, which gradually increases with depth. The predominant texture of the surface horizon is sand to sandy loam and very rarely clay loam. Mott et al. (1979) analysed the textural properties of some Alfisols in Northern Territory, Australia. These soils are susceptible to formation of hard, impermeable surface seal following heavy grazing and depletion of the vegetation cover (Fig. 4.2). Similar to the soils of West Africa under identical environment, these soils have a low silt content of 12 to 13 per cent and a high sand content, i.e. 55 to 60 per cent (Table 4.12).

2. Bulk density

In the humid and subhumid tropics, the bulk density of the surface horizon of most uplands is generally low especially if the soils are either cultivated by traditional methods or they have been under fallow for some time. The bulk density of cultivated soils is generally high especially when managed by mechanized farming operations. Some uncultivated soils of the savanna and semi-arid regions, however, are compacted even prior to their development for arable land-use. These soils have scarce vegetation cover and are exposed to

Table 4.12a Particle size analysis from surface 0.5 cm and bulk density of survey 2.0 cm of soil from sealed and grassed areas (Mott *et al.*, 1979). *Reproduced by permission of CSIRO*

Size fraction	Percentage		LSD (0.05)
	Grass	Seal	
Clay (<2 μm)	23.0	24.9	ns
Silt (2–20 μm)	12.2	13.0	ns
Fine sand (0.02–0.2 mm)	36.0	34.6	ns
Coarse sand (0.2–2 mm)	22.1	20.4	ns
Bulk density (g cm^{-3})	1.49	1.55	0.06

Table 4.12b Water transmission properties of surface seal developed on an Alfisol after structural collapse due to heavy grazing in Northern Territory, Australia (Mott *et al.*, 1979). *Reproduced by permission of CSIRO*

Property	Grassed soil	Seal	LSD
Soil-water sorptivity (cm min$^{-1/2}$)	2.14	0.35	1.18***
Hydraulic conductivity (cm min)	0.180	0.044	0.133**
Rainwater penetration depth (mm)	24.6	6.4	—
Rainwater penetration (%)	95	25	—

high intensity monsoons. The low ground cover is due to grazing and to accidental fire that destroys most of the leaf litter.

An example of low bulk density profile of a Cerrado Ultisol is shown in Table 4.3. The high bulk densities of intensely cultivated Alfisols in central India and in East Africa are shown in Tables 4.8 and 4.10, respectively. The high bulk density of some Alfisols (Fig. 4.3) is partly due to the high proportion of gravels and skeletal material. The data in Table 4.15 show that upland soils of northeast Thailand also have a low bulk density of 1.0 to 1.2 g cm^{-3} when under

Table 4.13 Range of the physical properties of soils of a catena developed on basement rocks in western Nigeria (Bonsu and Lal, 1983). *Reproduced by permission of the Nigerian Journal of Soil Science*

Property	Surface horizon	B2t horizon
Sand (%)	67.0 – 84.3	67.7 – 86.4
Silt (%)	4.8 – 12.8	1.5 – 13.8
Clay (%)	10.9 – 21.2	12.0 – 23.2
Gravels (%)	5.0 – 62.3	31.6 – 73.9
Bulk density of the fine earth (g cm^{-3})	1.00 – 1.48	1.17 – 1.49
Overall bulk density (g cm^{-3})	1.23 – 1.78	1.58 – 1.89
Total porosity (%)	33.0 – 53.0	29.0 – 42.0

Table 4.14 The bulk density of some western Nigerian uncultivated soils under natural vegetation cover (IITA, 1975)

Soils	Horizon 1		2		3		4		5		6	
	a	b	a	b	a	b	a	b	a	b	a	b
1. Egbeda	1.29	1.24	1.33	0.86	1.65	1.37	1.49	1.27	1.64	1.58	1.55	1.50
2. Sepeteri	1.55	1.43	1.37	1.17	1.77	0.86	1.70	0.92	1.52	0.65	1.70	1.26
3. Iwo*	1.35	1.32	1.61	1.61	1.64	1.49	1.39	1.27	1.49	1.45	1.43	1.42
4. Apomu	1.30	1.30	1.34	1.34	1.47	1.47	1.62	1.62	1.67	1.67	1.79	1.79
5. Ekiti	1.25	1.22	1.40	1.32	1.60	1.33	Bed Rock		—	—	—	—

a, overall bulk density. b, bulk density of the < 2 mm fraction. *Soil of different location than that in Table.

Table 4.15 Bulk density of some upland soils in Northeast Thailand (Takahashi et al., 1983)

Vegetation	Depth (cm)	Bulk density (g cm^{-3})	Moisture content (%)	Volume (%)		
				Solid	Water	Air
Cleared	0–5	1.001–1.247	8.2–14.0	39.5–49.5	8.6–13.5	39.5–50.8
	5–15	1.221–1.362	7.6–11.7	46.3–51.1	9.7–14.7	35.7–43.7
Forest	0–5	0.845–1.121	10.2–12.0	33.5–43.7	10.1–12.2	44.1–56.4
	5–15	1.236–1.263	9.9–11.5	43.7–46.3	12.4–14.2	39.6–41.3

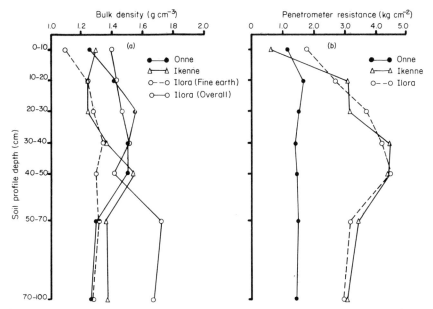

Figure 4.3 The profile of soil bulk density and penetrometer resistance of some Alfisols and Ultisols (Mbagwu et al., 1983). *Reproduced by permission of Williams & Wilkins Co.*

forest cover, which increases to 1.2 to 1.3 g cm^{-3} immediately after clearing.

Soils in regions with prolonged dry seasons are desiccated up to about 1 m depth. These soils are thus subjected to high temperature extremes when their moisture content is low. Under these conditions these soils acquire an extremely hard consistency. Extremes of temperature, low moisture, and inadequate vegetation cover are some of the factors responsible for these soils being compacted in their natural state.

Bertrand (1971) reported soil bulk density measurements of some arid zone soils in Sine-saloum, Senegal. His data (Table 4.21b) indicate the presence of some very dense soils even when the clay content exceeded 40 per cent. These soils have presumably been under fallow for sometime, because the cultivated soils in the Sahel are known to be even more dense. Williams (1979) measured physical properties of similar soils in adjacent regions of Gambia. His data (Fig. 4.4) indicate even higher bulk densities ranging from 1.6 to 1.9 g cm^{-3} These high bulk densities have important implications towards appropriate technology for soil surface management and water conservation in dryland farming. These aspects will be discussed in Chapter 15.

V. SOIL STRUCTURES

Soil structure is the most difficult parameter to assess and define, because it is a dynamic entity and its qualitative and quantitative aspects are in a metastable

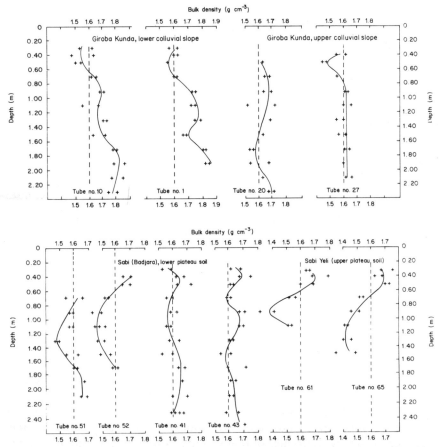

Figure 4.4 Bulk density profiles of some colluvial and plateau soils in the Gambian savanna. (Williams, 1979)

equilibrium with biotic environments and the prevailing microclimate. There is a no single, universally accepted definition of soil structure, and, ironically, the term has different meanings to scientists of different disciplines. To pedologists, soil structure means the shape and size of the peds and the orientation and nature of the clay film, if any, on them. To an agronomist soil structure may refer to 'tilth' and trafficability; to a soil conservationist it may imply percentage aggregation and stability against raindrop impact and shearing force of water run-off. To a hydrologist structure refers to water transmission and retention properties, and to a geomorphologist it may mean pore size distribution and pore channel continuity. A universal definition of soil structure must encompass all these aspects and also have a numerical criterion that lends itself to routine laboratory analysis. This 'structural index', so to say, must also reflect management potential and constraints for the intended land-use. It is important and relevant that scientists develop a definition encompassing all

Table 4.16a Erodibility of some LAC soils determined on field plots

	Country	Region	Erodibility	Reference
A.	*Alfisols*			
	Benin	Subhumid	0.10	Roose (1977)
	Ivory Coast	Subhumid	0.10	Roose (1977)
	Kenya	Subhumid	0.03–0.49	Barber *et al.* (1979)
	Nigeria	Subhumid	0.06–0.36	Lal (1976)
	Nigeria	Subhumid	0.058	Wilkinson (1975)
	Tanzania	Semi-arid	0.121–0.160	Ngatunga *et al.* (1984)
B.	*Ultisols*			
	Hawaii	Humid	0.09	Dangler and El-Swaify (1976)
	Nigeria	Humid	0.04	Vanelslance *et al.* (1984)
	Thailand	Subhumid	0.09–0.19	Tangtham (1983)
	Trinidad	Humid	0.03–0.06	Lindsay and Gumbs (1982)
C.	*Oxisols*			
	Costa Rica	Humid	0.103–0.155	Amezquita and Forsythe (1975)
	Hawaii	Humid	0.14–0.22	Dangler and El-Swaify (1976)
	Ivory Coast	Humid	0.10	Roose (1977)
	Puerto Rico	Humid	0.01	Barnett *et al.* (1971)

these factors related to characterization of structure and its implications towards soil management practices with intensive land-use.

For the purpose of this text, soil structure refers to 'arrangement and interaction of inorganic and organic components into aggregates that provide an array of voids and solids and that facilitate water retention and transmission, gaseous exchange, heat conductance, root proliferation, and biotic activity of soil fauna and flora'. Although this definition has shortcomings, it is an attempt to encompass the multi-faceted aspects. Obviously, it is difficult to develop a single quantitative parameter to express these multi-faceted and ever changing characteristics. From the point of view of the soil management constraints to crop production in the tropics, perhaps the pore size distribution and pore stability and continuity are the vital components that affect all other soil physical processes. It is the pore size distribution on which water retention and transmission, root system development, gaseous exchange and heat conductance depend. The alterations in pore space by management or through interaction with environments lead to alteration of the physical processes of edaphological importance. The pore size distribution is also dependent on the degree and stability of structural aggregates.

A. Factors affecting soil structure

Many researchers have indicated that the structural properties of soils in the tropics are profoundly influenced by the organic matter content (Godefroy *et al.*, 1977) and the biotic activity of soil fauna (Lal *et al.*, 1980). In the semi-arid regions, the composition of soil solution and the predominance of cations also influence the degree and stability of aggregates (El-Swaify, 1970). Management practices that affect organic matter content and biotic activity, therefore, affect soil structure. Among the cultural practices, the mode and intensity of tillage (Lal, 1983; Pagel, 1975) and the nature and frequency of cover crop and fallowing (Lal *et al.*, 1979) have significant effects on soil structure.

The structural properties of some South Asian soils were reported by Joachim and Pandilteskera (1948) and of soils of Malaysia by Soong (1980). These authors attributed the seasonal changes in structural properties to changes in soil organic matter content. In regions with prolonged dry seasons, i.e., semi-arid and subhumid zones, ultra-desiccation is considered by some to cause structural modification and changes in the hydrological characteristics of the soil (Chauvel and Pédro, 1978).

B. Micro-aggregation in tropical soils

Many tropical soils, especially Oxisols, exhibit a phenomenon of micro-aggregation. The clay sized particles are strongly aggregated to form silt- and fine sand-sized aggregates. For all practical purposes, even the soils with high clay content behave as sand (Table 4.2). These soils have high water intake rates and retain less water. The trafficability and tilth of these soils is also like those of coarse-textured sands. The micro-aggregation properties of some soils in Africa have been described by Ahn (1979) in relation to their importance to crop growth and management. Similar characteristics have been noted in some soils of Venezuela (Guedez and Langohr, 1978) and Brazil (Stone and da Silveira, 1978). These Oxisols and Ultisols were observed to contain considerable amounts of pseudo-silts with the highest percentage in the argillic horizon, less in the oxic horizon and much less in the coloured B horizon. Guedez and Langohr estimated 15-bar moisture content and the available water holding capacity of these soils and found these properties to be identical to soils containing high silt contents. The pseudo-silt behaves mainly as silt sand-sized particles.

C. Structural stability and soil erodibility

Surface horizons of most soils in the tropics, with relatively less clay and organic matter content, are less aggregated than soils of the temperate zone. Whatever structural aggregates exist are due mainly to the biotic activity of soil fauna. This is particularly true for soils of the humid and subhumid tropics. Very sandy horizons usually have loose single-grain structure. Subsoils containing high clay and Fe and Al oxides have sub-angular blocky or massive structure with thin clay-skin linings, depending on the rainfall characteristics and the degree of soil development.

Table 4.17 Aggregate analysis of some uplands from Thailand (Takahashi et al., 1983)

Vegetation	Depth (cm)	Aggregate analysis (%)				Aggregation (%)			Dispertion ratio (%)	Erosion ratio (%)
		2–0.2 mm	0.2–0.02 mm	0.2–0.002 mm	<0.2 mm	>0.2 mm	0.02–0.2 mm	0.002–0.02 mm		
Cleared	0–5	63.8–78.9	15.7–27.9	2.5– 6.7	0.9–2.2	55.0–70.6	47.6–55.2	16.6–29.8	6.5–14.5	7.4–22.7
	5–15	44.0–69.5	22.0–43.4	6.0–11.1	1.1–2.0	27.9–58.3	38.2–48.3	16.8–28.9	14.5–24.9	10.6–20.3
Forest	0–5	47.1–83.7	11.3–44.9	4.1– 6.7	0.9–1.4	39.0–76.2	52.8–59.9	21.7–23.7	8.5–13.2	9.2–15.1
	5–15	53.4–71.1	20.2–36.7	5.0– 9.9	1.1–2.0	45.9–63.2	50.1–50.8	22.0–25.3	10.7–19.0	7.6–15.7

Structural units and weakly developed aggregates are unstable to raindrop impact—they slake readily on quick wetting and form a slowly permeable surface seal or crust that reduces infiltration and inhibits seedling emergence. Consequently, soils with rolling to steep relief are severely prone to wind and water erosion. Susceptibility of soil to water erosion or the erodibility factor (K) of the Universal Soil Loss Equation (USLE), however, is often said to be low (Lal, 1984). This is because the K factor of the USLE is inversely proportional to the rainfall erosivity ($K = A/R$). Soils in regions of high rainfall erosivity, therefore, tend to have low erodibility and vice versa (Table 4.16). Soils with low organic matter content and high amounts of fine sand and silt are prone to developing surface seal (Fig. 4.2). The development of appropriate technology for soil surface management to enable intensive land-use for sustained and economic production from these soils is linked with structural characteristics of erodibility and surface seal. Mott et al. (1979) observed that the structure of heavily-grazed Alfisols in Northern Territory is easily destroyed. When devoid of vegetation cover, these soils develop a surface seal. The surface crust so formed has low sorptivity and hydraulic conductivity (Tables 4.12a, b) and most rains received are lost as surface run-off.

An example of aggregate analysis of some upland soils from Northeast Thailand is shown in Table 4.17 (Takahashi et al., 1983). Under forest cover or immediately after clearing a fallow vegetation, as much as 50 per cent of the soil particles are aggregated into water-stable aggregates exceeding 0.02 mm. Consequently, the erosion and dispersion ratios are low (10–20 per cent). With cultivation, however, the per cent of aggregated soil declines rapidly. The aggregate analysis of a Vertisol from Central India is compared with that of an alluvial soil from Delhi, Northern India (Table 4.18). Because of the high-activity clay, the mean weight diameter of a Vertisol is greater and the soil is more aggregated than coarse textured alluvial soil that contains low amounts of low-activity clay.

Table 4.18 Aggregate analysis of a Vertisol from Hyderabad, Central India and of an Alluvial soil from New Delhi, India (unpublished Lal, 1965)

Vertisol			Alluvial		
Depth (cm)	Mean weight diameter (MW) (mm)	Per cent Aggregation >0.25 mm	Depth (cm)	Mean weight diameter (MWD) (mm)	Per cent Aggregation >0.25 mm
0–7.5	0.452	55.1	0–25	0.275	29.4
7.5–30	0.015	82.4	25–60	0.135	19.4
30–60	0.554	67.6	60.90	0.108	14.3
60–100	0.546	67.0	80.125	0.075	13.3

VI. SWELL-SHRINK PROPERTIES

The coarse-textured soils in the humid and subhumid tropics containing predominantly low-activity clays exhibit little swell-shrink properties. Even some clayey subsoils behave as sand or silt because of highly stable microaggregation. The upper and lower plastic limits and the plasticity index of most uplands reflect little plasticity. This is one of the reasons that some of these soils can be cultivated and can support heavy machinery even a few hours after a heavy rain.

The consistency limits of yellow and red Latosols in the Central Amazonian region of Brazil have been studied by Corrêa (1982). The plasticity limits of cultivated soils were slightly lower than of those uncultivated. This is because the plasticity limits depend on the organic matter content, and the latter often declines with cultivation. Mott et al. (1979) observed (few or no) swell-shrink properties of an Alfisol from Northern Territory, Australia. The swell-shrink properties of some upland soils from western Nigeria are shown in Table 4.19. These soils are representative of similar soils elsewhere in the tropics for similar parent materials and ecologies. The surface layers have a zero plasticity index, and the shrinkage is also low.

VII. pF CURVES AND THE AVAILABLE WATER-HOLDING CAPACITY

The upland soils of the tropics have a low water holding capacity in the surface soil horizon. The plant available water in the rooting zone is generally less than 100 mm. The effective rooting depth for seasonal crops is shallow due either to adverse soil physical properties (Vine et al., 1981; Babalola and Lal, 1977) or to nutrient toxicity and imbalance (Kang and Juo, 1981, and Lal et al., 1983). Consequently, seasonal crops are prone to periodic drought stress throughout the growing season whenever rains fail for more than 7 to 10 days.

In northern Nigeria, Kowal (1970a, b) noted that the available water capacity for the loess soils was about 5 to 6 cm of water per 30 cm depth of soil (Fig. 4.5). For soils of southwest Nigeria, the available water capacity of the rooting depth is estimated to be 3 to 5 cm of water (Lal, 1979). Although some plants, e.g., sunflower and okra, can extract water even beyond the 1.5 MPa suction, the amount of water that can be extracted is rather small (Obi, 1974). Amezquita (1981) noted that some western Nigerian Alfisols attain equilibrium moisture content within 24 hours after irrigation and subsequently little change in water content occurs (Table 4.20). The equilibrium moisture content of the surface soil ranges from 12 to 29 per cent with the corresponding suction value of 25 to 50 cm of water (Fig. 4.6). Bonsu and Lal (1983) observed a field moisture capacity between 50 and 90 cm of water suction for some Alfisols in western Nigeria (Table 4.21). This rapid attainment of equilibrium is partly due to the presence of underlying gravel at 20 to 30 cm depth. Kowal and Kassam (1978) observed that the moisture potential at field capacity of some sedentary soils of West African savanna soils was in the range of 0.008 and 0.0125 MPa.

Table 4.19 Swell–shrink properties and plasticity indices of some soils of western Nigeria (Moormann et al., 1975)

Horizon	Soil series																					
	Ekiti			Iwo			Egbeda			Ibadan			Gambari			Apomu			Iregun			Matako		
	PI	SL	SR	PI	SL	SR	PI	SL	SR	PI	SL	SR	PI	SL	SR	PI	SL	SR	PI	SL	SR	PI	SL	SR
I	0	–	–	0	–	–	3	19.1	1.8	0	–	–	3	19.0	1.8	0	–	–	0	–	–	0	–	–
II	0	–	–	2	16.1	1.9	5	17.9	1.9	0	–	–	0	–	–	0	–	–	0	–	–	0	–	–
III	6	18.6	1.8	15	16.7	1.9	9	24.5	1.7	0	–	–	2	18.0	1.9	0	–	–	6	15.1	1.9	0	–	–
IV	–	–	–	6	20.1	1.8	10	29.3	1.6	0	–	–	8	18.5	1.9	0	–	–	7	15.9	1.9	–	–	–
V	–	–	–	7	23.4	1.7	6	26.9	1.6	5	14.9	2.0	9	24.6	1.7	0	–	–	5	14.6	2.0	11	14.9	2.0
VI	–	–	–	8	26.3	1.6	10	25.3	1.6	6	18.7	1.8	6	21.3	1.8	0	–	–	5	17.6	1.9	–	–	–

PI, Plasticity index; SI, shrinkage limit; SR, shrinkage ratio.

Table 4.20 The volumetric moisture content (%) and suction (MPa) of some soils in western Nigeria at different times after irrigation (Amezquita, 1981)

Depth (cm)	24 h θ_v	24 h ψ	48 h θ_v	48 h ψ	72 h θ_v	72 h ψ	96 h θ_v	96 h ψ	168 h θ_v	168 h ψ
				Ekiti (Alfisol)						
0–12	29.3	0.0038	28.5	0.0041	28.2	0.0041	28.0	0.0042	27.8	0.0042
12–25	16.2	0.0030	15.7	0.0034	15.4	0.0037	15.1	0.0038	14.9	0.0039
25–45	19.1	0.0030	19.0	0.0030	18.9	0.0031	18.8	0.0033	18.7	0.0034
				Apomu (Psammentic Ustorthent)						
0–13	16.3	0.0029	13.1	0.0043	12.9	0.0044	—	—	12.8	0.0045
13–32	17.9	0.0035	17.1	0.0040	16.9	0.0042	—	—	16.7	0.0043
32–72	21.2	0.0025	19.7	0.0035	19.2	0.0037	—	—	18.9	0.0040
				Adio						
0–10	13.0	0.0045	12.3	0.0050	12.1	0.0052	—	—	—	—
10–30	14.5	0.0052	14.1	0.0055	14.0	0.0057	—	—	—	—
30–45	16.0	0.0048	15.5	0.0050	15.3	0.0052	—	—	—	—
				Alagba						
0–15	12.5	0.0044	20.6	0.0045	20.2	0.0046	19.8	0.0048	19.5	0.0052
				Owode						
0–15	26.0	0.0027	25.5	0.0028	25.1	0.0029	24.7	0.0032	24.2	0.0034

θ_v = volumetric moisture content (%); ψ, moisture potential (MPa), (1 MPa = 10 bar).

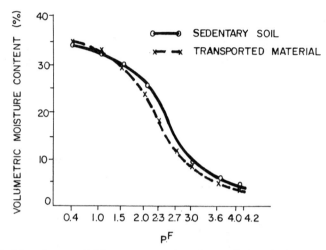

Figure 4.5 The plant available water capacity of some loes soils in northern Nigeria (Kowal, 1970)

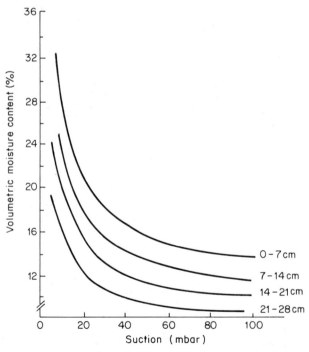

Figure 4.6 The equilibrium values of soil moisture content and soil moisture potential corresponding to the field moisture capacity after the free drainage has ceased (Amezquita, 1981)

Table 4.21 Measurement of field moisture capacity *in situ* and in the laboratory (Bonsu and Lal, 1983). *Reproduced by permission of the Nigerian Journal of Soil Sciences*

Horizon depth (cm)	In-situ measurement			Laboratory measurements at 0.03 MPa	
	Volumetric moisture content (%)	Tension (MPa)	Air capacity (%)	Volumetric moisture content %	Air capacity (%)
Profile I					
0–15	8.5	0.005	40.5	6.9	42.1
15–30	9.4	0.008	36.5	8.0	38.0
30–68	12.8	0.009	31.2	6.4	37.6
68–115	17.3	0.008	23.7	14.0	27.0
Profile II					
0–15	10.8	0.006	43.2	12.4	41.6
15–33	9.4	0.005	37.6	13.5	33.5
33–93	9.1	0.008	32.9	9.3	32.7
93–125	11.0	0.007	31.0	15.7	19.3

Table 4.21a Mean available water capacity over many (a) colluvial soils and (b) plateau soils determined by Bertrand (1971) working in Siné-Saloum, just north of The Gambia in Senegal

Horizon depth cm	Particle size distribution			Dry bulk density	Water characteristic, gravimetric percentage			Available water capacity pF 2.5–pF 4.2	
	Sand %	Silt %	Clay %		pF 2.5	pF 3.0	pF 4.2	% gravimetric	% volumetric
(a) Colluvial soils									
0–20	74	19	7	1.51	8	6	2.5	5.5	8
20–40	68	18	14	1.55	9	8	5	4	6
80–120 (maximum clay)	58	18	24	1.69	15	12	8	7	12
(b) Plateau soils									
0–20	62	28	10	1.6	12	6	3	9	14
20–40	54	23	23	1.5	12	9	6	6	9
(maximum	43	22	35	1.6	17	13	9	8	13

Table 4.21b Available water, pF values of soil water, bulk density and clay content with depth of three profiles studied by Bertrand (1971) in Senegal

Depth cm	Clay %	Bulk density	pF 2.5	pF 3	pF 4.2	Available water capacity % gravimetric	Available water capacity % volumetric
(a) Plateau Profile 1. No iron pan or concretions (similar to Sabi-Badjara soil)							
0–15	7	1.59	17.1	6.2	2.8	14	22
15–26	13	1.67	11.8	7.8	4.1	8	23
26–50	31	1.60	16.7	12.1	8.5	8	13
50–80	39	1.51	18.8	14.6	10.6	8	12
80–110	34	1.54	20.1	14.2	9.7	10	15
110–140	32	1.59	20.7	14.6	9.6	11	17
(b) Plateau Profile 2. No ironpan or concretions (similar to Sabi experimental site, valley head soil)							
0–12	5	1.60	5.7	4.0	1.9	5	8
12–35	9	1.46	7.0	5.2	3.1	4	6
35–100	22	1.55	11.9	9.4	6.7	5	8
100–160	23	1.50	11.9	9.5	6.7	5	8.5
160–180	23	1.77	14.5	11.3	7.6	7	12
180–220	25	1.70	16.1	11.9	8.2	8	14
220–260	25	1.76	17.5	13.7	8.7	9	16
(c) Plateau Profile 3. Iron pan within 1 m of surface (similar to Giroba plateau soil, upper plateau)							
0–20	18.4	1.54	19.7	18.8	5.7	14	22
20–50	46.2	1.87	26.0	22.2	13.0	13	24
50–80	44.3	1.77	30.4	22.5	13.9	16.5	29
80–100	37.8	ironpan	25.7	23.3	13.9	12	—

Table 4.21c Observed variation of available water within the top metre of soil at the lower colluvial slope site at Giroba Kunda, mm/m

Profile No.	Water content		Maximum available water
	Minimum	Maximum	
1	80	215	135
2	103	215	112
3	71	206	135
4	108	230	122
5	93	202	109
6	107	225	118
8	96	219	123
10	82	197	115
12	78	196	118
13	98	223	125
14	114	244	130
15	107	227	120
Mean	95	217	122
Standard deviation	14	14	8

For leached feruginous and ferrallitic soils, however, field moisture capacity was reported at pF 2.3 (Charreau, 1974).

The concept of field capacity is, however, ill defined and widely ranging values of tensions are used to compute available water capacity for soils of similar texture.

The available water holding capacity of soils of the semi-arid regions of West Africa have extensively been studied by scientists of IRAT. In Senegal, Bertrand (1980) computed available water as difference in moisture retained at pF 2.5 or pF 3 and pF 4.2 depending on soil texture. The data of Bertrand (1971) from soils at Sine-Saloum in Senegal (Tables 4.21a and 4.21b) indicate low available water capacity of these sandy soils. Although the permanent wilting point is widely presumed to be at pF 4.2 (Dancette, 1969, 1973a, 1973b), the amount of water released beyond tension of 3 bar is rather small.

Soil water investigations in Gambia by Williams (1979) support the findings of Bertrand and Dancette from Senegal. Williams data confirmed low available water capacity of these Aridisols of about 100 mm/m of soil (Table 4.21c and 4.21d).

The permanent wilting point, with no potential gradients in the soil or in the plant, of these soils is also attained very readily. In fact, for some soils with low activity clays there is often little difference in moisture content between 1 and 15 bar suctions. The wilting point also depends on the crops. Some crops can extract more water than others. For example, data in Table 4.22 show that cowpea wilted at a lower moisture potential than maize.

Table 4.21d Variation in observed soil water availability with crop and landscape position

Site	Landscape position	Crop in 1973	Mean observed maximum available water in top metre of soil, mm	Mapping association
Giroba Kunda	Lower colluvium	Groundnut	125	6
		Bare and fallow	120	
	Upper colluvium	Groundnut/sorghum	100	6
		Sorghum	110	
		Groundnut	105	
Mankamang	Lower colluvium	Groundnut	100	6
	Upper colluvium	Sorghum & millet	110	6
Sabi (Badjara)	Lower plateau	Cotton	90	12
		Sorghum	96	
		Groundnut	90	
Sabi (experiment)	Valley head (lowest plateau)	Sorghum	125	12
Sabi (Yeli)	Upper plateau	Cotton	85	11
		Sorghum	100	
Giroba plateau	Upper plateau	Sorghum	100	8
Mankamang plateau	Upper plateau	Bush and fallow	80	8
Nyambai	Shallow depression	Gmelina	140	5
		Bush	145	
		Groundnut	110	

Table 4.22 The volumetric moisture content θ and corresponding suction ψ at the permanent wilting point for some soil series in western Nigeria (Amezquita, 1981)

Series	Depth (cm)	Maize		Cowpea	
		θ_v (%)	ψ (MPa)	θ_v (%)	ψ (MPa)
Ekiti	0–12	7.5	3.2	3.9	5.8
Iwo	0–12	9.9	3.1	6.0	6.3
Apomu	0–13	3.9	2.8	3.1	4.1
Adio	0–10	4.2	3.0	3.2	5.6
Algaba	0–13	5.5	3.2	3.5	5.8
Owode	0–15	7.4	3.4	6.5	4.5

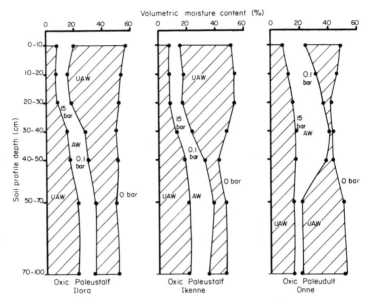

Figure 4.7 The water-holding capacity of some Alfisols and Ultisols profiles in southern Nigeria; AW, available water; UAW, unavailable water (Mbagwu *et al.*, 1983). *Reproduced by permission of Williams & Wilkins Co*

Mbagwu *et al.* (1983) evaluated the available water-holding capacity of some Alfisols and Ultisols of southern Nigeria. The available water-holding capacity ranged from 7.9 to 8.2 cm per 100 cm depth for Alfisols and up to 11.3 cm per 100 cm depth for an Ultisol (Fig 4.7 and 4.8). The available water reserves for the effective rooting depth are, of course, much lower because most seasonal crops rarely extend their root systems beyond 30 cm depth.

The available water holding capacity of uplands of Northern Australia has been reported by Arndt *et al.* (1963) and van de Graaff (1965). Their data showed a range of 4.3 to 6.8 per cent available water in various horizons of a Tippera family soil and 1.8 to 4.7 per cent in various horizons of a sandy member of Blain family.

In South Asia, soil moisture retention characteristics of uplands have been determined in the laboratory by Selvakumari *et al.* (1973), Banerjee and Chand (1981), and Oswall and Khanna (1981). There is a drastic difference, however, between the pF determination *in situ* and that measured from undisturbed samples in the laboratory. The upper and lower limits of available water of an Alfisol from Hyderabad, Central India, are shown in Table 4.23. The available water-holding capacity of the surface soil horizons (0–105 cm depth) ranged from 2.3 to 4.6 per cent. Kyuma and Pairintra (1983) reported the results of laboratory-measured pF curves of upland soils in northeast Thailand. The volumetric moisture content of the surface 0–5 cm layer range from 60 to 70 per cent at 0 suction to 15 to 20 per cent at pF 4.0 (Fig. 4.9). The moisture holding capacity, however, declined drastically after deforestation and burning.

McCown (1971) reported that three Alfisols in Northern Territory, Austra-

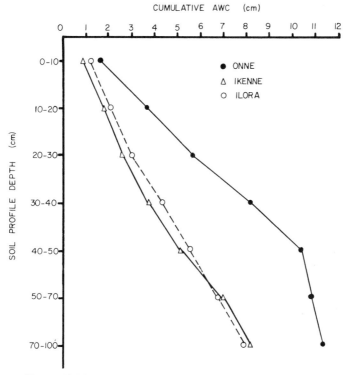

Figure 4.8 The available water holding capacity of some Ultisols in southern Nigeria (Mbagwu et al., 1983). *Reproduced by permission of Williams & Wilkins Co.*

lia, respectively retain 17, 24, and 33 per cent of gravimetric water content at field moisture capacity and 3, 10, and 11 per cent soil moisture at wilting point. The available water-holding capacity of these soils was calculated to be 3.5 cm, 4.9 cm and 6.2 cm for 18 cm, 25 cm and 20 cm depth of the surface soil horizon. Mott et al. (1979) analysed the moisture release characteristics of a similar soil prone to surface seal and crusting. Their data showed that these soils retain 10 to 12 per cent of the available water and most of it is released below 1 bar suction.

The upland soils of low activity clays in tropical America also have low available water holding capacity. However, soils of Central America, particularly those derived from volcanic ash and characterized by low bulk density, have high water retention capacity (Fig. 4.9). In Venezuela, Pla Sentis (1977) noted that soils of the Guanipa series retain 7.5 per cent moisture (on weight basis) at 0.01 bar suction, 5 per cent at 2 bar, and less than 2 per cent at 15 bar (Fig. 4.10). Most of the available water in Alfisols with low activity clays is released at suctions of less than 0.3 bar (Fig. 4.11). Zusevics (1980) reported seasonal variations in the wilting point of some soils from Miranda State, Venezuela, depending on the rainfall distribution and the organic matter content. For soils of Puerto Rico, Wolf and Drosdoff (1976a, b) noted the maximum available water storage of the upper 30 cm of some soils ranged from

Table 4.23 Some soil moisture constants of two Alfisols in Central India (ICRISAT, 1978–79)

Depth (cm)	Gravimetric moisture content (%)		
	1.5 MPa		0.03 MPa
	A	B	A
0– 15	6.0	8.6	8.3
15– 30	10.2	10.3	13.5
30– 45	14.8	10.4	18.7
45– 60	16.1	14.2	20.3
60– 75	17.1	15.7	21.1
75– 90	17.2	15.8	21.8
90–105	17.7	16.0	21.7
105–120	17.6	16.9	12.5
120–135	16.8	16.6	20.6
135–150	16.4	16.4	20.6
150–165	15.7	16.8	20.1
165–180	14.9	16.4	19.6
180–195	15.0	—	19.6

A and B refer to different sampling sites.

3.6 to 6.0 cm, but most of the water was released at suctions of less than 1 bar (Fig. 4.12). Similar levels of available moisture reserves for soils of Puerto Rico have been reported by Escolar and Lugo Lopez (1969).

Soil moisture retention properties of the soils of the Amazon region of Brazil have also been studied and are similar to those of the upland soils in West Africa. Roeder and Bornemisza (1968) reported that soils retained 20 to 30 per cent moisture at 0.3 MPa tension and 8 to 10 per cent at 0.5 MPa (Table 4.24). In comparison with soils of West Africa, these soils have a relatively higher silt

Table 4.24 Mechanical analysis and moisture retention characteristics of some soils from the Amazon region, Maranhao, Brasil (Roeder and Bornemisza, 1968)

Soil series	Mechanical analysis (%)			Moisture retention (% by weight) at different tensions (MPa)			
	Sand	Silt	Clay	0.033	0.10	0.5	1.2
Capinambi	34.9	45.3	19.7	20.4	15.7	9.4	8.6
Gurupi	63.1	21.1	15.7	26.4	10.6	7.0	6.6
Maracaja	44.9	40.3	14.8	29.4	10.6	7.9	7.2
Paxiuba	76.5	3.4	20.1	24.5	18.4	8.5	7.7
Saba	65.9	23.3	10.9	31.7	8.4	6.8	5.7
Toa	48.3	32.7	18.9	29.9	10.3	6.6	5.6
Turi	79.3	11.2	9.5	32.7	7.2	5.2	1.0
Ze Doca	46.8	34.8	18.4	34.9	12.3	10.8	8.2

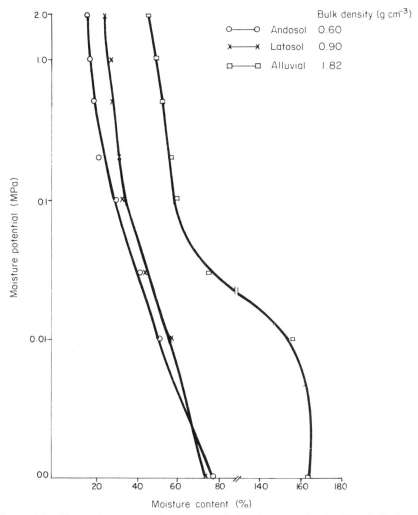

Figure 4.9 The moisture retention characteristics of some uplands of low bulk density in Costa Rica (Gonzalez and Gavande, 1969)

content. In Bahia, Brazil, Cadima and Elvim (1973) observed that cacao production was related to the available water holding capacity of soils studied. Soils with low yield had lower plant available water, and higher penetrometer resistance (Table 4.25). Reichart *et al.* (1980) observed that sedentary uplands have low water-holding capacity, measuring approximately 17.7 mm for 0–30 cm, 39.4 mm for 0–60 cm, and 65.7 mm for 0–100 cm depth.

One of the problems of assessing soil moisture constants, e.g., field moisture capacity and permanent wilting points, is that they are dynamic entities and vary with crops, season, evapotranspiration and the microclimate. The field

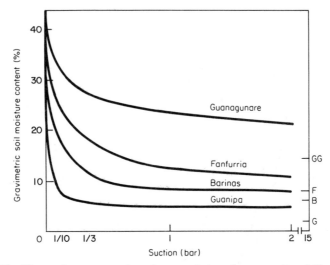

Figure 4.10 The moisture retention characteristics of some soils of Venezuela. X denotes the field moisture capacity (Pla Sentis, 1977)

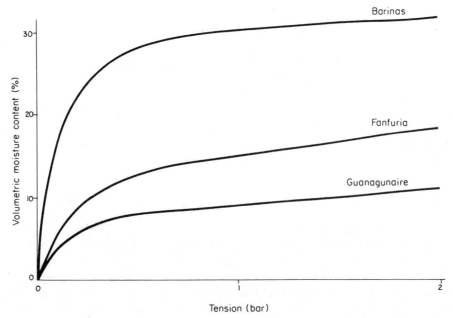

Figure 4.11 The available water in some soils of Venezuela is released at low suctions (Plat Sentis, 1977)

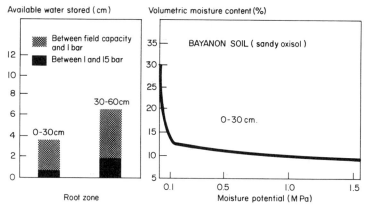

Figure 4.12 The low available water storage capacity of some soils of Puerto Rico (Wolf and Drosdoff, 1976). *Reproduced by permission of the Agricultural Experiment Station, University of Puerto Rico*

Table 4.25 Physical properties of some soils of Bahia, Brazil (Cadima and Elvim, 1973). *Reproduced by permission of CEPLAC*

Order	Region	Production potential*	Texture (%) Coarse sand	Fine sand	Silt	Clay	Penetrometer resistance (kg cm^{-2})	Available water (%)
Alfisol	Buerareme	A	21.0	18.4	33.0	27.6	0.20	6.64
		B	15.2	14.8	23.4	46.6	0.74	4.82
	Camacan	A	31.2	22.3	19.8	26.7	0.22	6.52
		B	29.2	10.3	18.7	41.8	0.34	7.0
	Coaraci	A	30.0	16.1	21.3	32.6	0.24	8.97
		B	31.4	13.1	17.6	37.9	0.32	6.73
	Gandu	A	30.9	13.9	17.1	38.1	0.22	10.50
		B	20.3	13.5	19.8	46.4	0.31	5.11
	Ibirapitanga	A	31.9	11.0	8.1	49.0	0.13	7.75
		B	25.9	10.1	12.4	51.6	0.35	7.28
	Ibirataia	A	22.2	12.1	16.3	49.4	0.11	7.64
		B	16.9	10.7	22.3	50.1	0.32	7.42
	Itajuipe	A	25.2	12.8	17.3	44.7	0.22	7.66
		B	17.8	10.7	15.7	55.8	0.32	6.23
	Rio Branco	A	15.2	14.9	23.0	46.9	0.21	9.04
		B	12.2	13.4	17.2	57.2	0.34	6.11
	Urucuca	A	13.9	10.1	22.1	53.9	0.22	8.95
		B	17.4	13.0	18.3	51.3	0.40	8.02
Ultisol	Ibicarai	A	21.2	18.8	20.4	39.6	0.22	10.50
		B	29.0	19.9	16.3	34.8	0.31	5.11
	Ipiaù	A	25.2	16.5	16.2	42.1	0.18	7.57
		B	33.1	11.7	3.0	52.2	0.32	4.72

* A, regions with high production of Cacao; B, regions with low production of Cacao.

moisture capacity, for example, is a tentative value at a given point in time. The profile drainage never ceases, contrary to what it may imply, and the equilibrium moisture content at which plants are no longer able to extract soil water is also influenced by soil and crop management, evaporative demand and the crop species. The assessment of the available water capacity of soils with restricted subsoil permeability is also difficult because there is a risk of over-assessment of the moisture content at both the upper and the lower limits (McCown et al., 1976).

VIII. WATER TRANSMISSION PROPERTIES

Upland soils with low activity clays and those that are under forest or the savanna vegetation or those that are being cropped with mulch farming and no-till systems generally have high infiltration rates and saturated hydraulic conductivity. In general, the soils in the savanna and semi-arid regions have lower infiltration rates than those within the forest ecology, all other factors being the same.

A. Water infiltration

Infiltration studies of the soils at Hyderabad, India (ICRISAT, 1975–76) indicate that these soils, which have been cultivated for many centuries, have low infiltration rates (Fig. 4.13). With a flat-planted treatment, the accumulative infiltration in 5 hours ranged between 5 and 12 cm for three sites. The equilibrium infiltration rate for these soils are no more than 1 to 2 cm/hr. The infiltration rate of these soils from a semi-arid ecology is also influenced by the nature of cations in the exchange complex, and decreases rapidly in the presence of sodium (Rengasamy et al., 1976; Saha et al., 1980). In the semi-arid regions of northeast Thailand, the uplands also are prone to crusting and have low infiltration rates (Bridge et al., 1975).

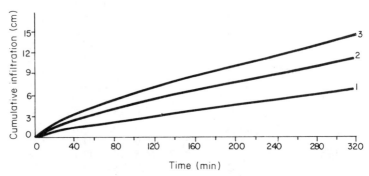

Figure 4.13 Infiltration characteristics of three Alfisols at Hyderabad, India (ICRISAT, 1975–76)

In Africa, the infiltration rates of Alfisols and Ultisols are higher than those in South and Southeast Asia. In general, the infiltration rates are higher for the soil in a forest ecology than in savanna and for uncultivated than cultivated soils. Water intake rates of semi-arid soils of Gambia measured by Williams (1979) indicated that infiltration rate greatly varied with landuse history and termite activity. The annually cultivated soils with low termite activity have lower infiltration rates than soils under undisturbed vegetation cover or perennial plantation. Williams (1979) observed that the rate of infiltration of soil measured in the gmelina plantation and bush was 'moderately rapid'. However, that of a similar soil in the groundnut field nearby was 'moderately slow'. The importance of faunal activity on infiltration rates has also been shown by studies in semi-arid soils of northern Nigeria. For the savanna ecology, infiltration rates have been studied by IAR (1975) at Samaru, Nigeria. Wilkinson and Aina (1976) studied the infiltration rate of forest zone soils in western Nigeria and reported it to be very high. Similar observations have been made by Moormann et al. (1975) and by Lal et al. (1980). High infiltration rates of Alfisol and Ultisols for southern Nigeria were reported by Mbagwu et al. (1983). Kelly and Walker (1976) compared the infiltration rates of four sites in the semi-arid regions of southeastern Rhodesia. The authors noted 9 times faster water intake rates in soil covered by litter than into bare soil.

Some infiltration data are also available from tropical regions of Central and South America and the Caribbean. The high water intake rates of some soils of Cuba have been related to their physical and mechanical properties (Siméon, 1979; Sagué Diaz et al., 1979). The allophone latosolic soils of Dominican Republic are also highly permeable and generate <20% of the rainfall as run-off. The water transmission properties of soils of Venezuela are related to their crusting properties and the nature of predominant cations on the exchange complex (Zusevics, 1981). In Sao Paulo, Brazil, Libera et al. (1977) observed that red Latosols have low permeability.

The infiltration characteristics of some Ultisols and Oxisols of Central America have also been studied. Wolf and Drosdoff (1976) observed very rapid (9 cm/hr after one hour of continuous flooding) water infiltration into clayey Ultisols, clayey Oxisol and a sandy Oxisol in Puerto Rico. Strong structural stability of clayey soils permitted infiltration rates in excess of that for the sandy soil. High rates of infiltration were partly attributed to lateral water movement downslope. Logu-López et al. (1968; 1970; 1981) also reported high infiltration rates (7.5 − 8.6 cm/hr) of Oxisols, and Ultisols in Puerto Rico. In some sandy soils the equilibrium infiltration rate was as high as 28.5 cm/hr but some clayey soils had an extremely low water intake rate, less than 0.2 cm/hr.

The rapid water infiltration rates reported for most uplands apparently contradict the widespread fears of accelerated erosion of the same soils. The anomaly lies in the method of measuring infiltration rate by the double-ring infiltrometer. This method overestimates the water acceptability of soils as observed under natural rainfall conditions. During natural rainstorms the creation of surface seal is a major factor responsible for generating the overland flow. That is why the prevention of raindrop impact by mulching is an effective measure to curtail run-off and erosion.

B. Hydraulic conductivity

Similar to infiltration, the saturated hydraulic conductivity of these soils is also high (Fig. 4.14). In uncultivated soils with high biotic activity there are also differences in the horizontal and vertical components of the hydraulic conductivity (Table 4.26). It is partly because of this variability that there are often considerable differences in field and laboratory measured values of saturated hydraulic conductivity.

Unsaturated hydraulic conductivity decreases drastically with a decrease in soil moisture potential (Fig. 4.15) (Amézquita, 1981; Bonsu and Lal, 1983). Stone and da Silveira (1978) developed empirical relations for a Cerrado Ultisol in Brazil relating unsaturated hydraulic conductivity with the soil moisture content (Fig. 4.16). These relations are numerically stated as follows:

$$K = 10^{-10.2+36.1} \; r = 0.94, \text{ for 0–30 cm depth}$$
$$K = 10^{-8.0+31.2} \; r = 0.87, \text{ for 30–60 cm depth}$$

where K is in cm day^{-1} and 0 in cm^3 cm^{-3}. Attempts have also been made to estimate unsaturated hydraulic conductivity from the pore size distribution or from the moisture retention characteristics. This is, however, feasible at low soil moisture suctions only (Bonsu and Lal, 1983) (Table 4.27).

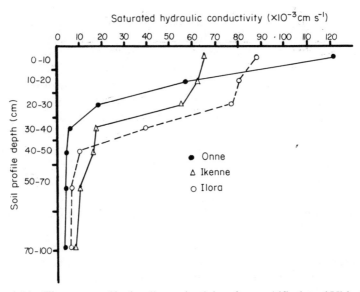

Figure 4.14 The saturated hydraulic conductivity of some Alfisols and Ultisols (Mbagwu et al., 1983). *Reproduced by permission of Williams & Wilkins Co.*

Table 4.26 Comparison of average *in situ* measured saturated hydraulic conductivity K_s (cm s^{-1}), with that determined in the laboratory and calculated from Marshall's model. Coefficients of variability, CV (%), for laboratory K_s are also included (Bonsu and Lal, 1983). *Reproduced by permission of the* Nigerian Journal of Soil Science

Horizon depth (cm)	Vertical sampling		Horizontal sampling		In situ	Calculated
	K_s	CV	K_s	CV		
Profile I						
0–15	6.1×10^{-2}	23	6.6×10^{-2}	60	2.5×10^{-3}	4.2×10^{-3}
15–30	6.6×10^{-2}	15	4.1×10^{-2}	7	1.8×10^{-3}	5.5×10^{-3}
30–68	1.9×10^{-2}	20	1.9×10^{-2}	36	5.0×10^{-4}	5.4×10^{-3}
68–115	4.8×10^{-4}	5	4.6×10^{-4}	11	2.2×10^{-5}	2.9×10^{-3}
Profile II						
0–15	1.7×10^{-2}	37	5.7×10^{-2}	32	1.7×10^{-3}	1.5×10^{-2}
15–33	7.3×10^{-3}	45	4.8×10^{-2}	58	1.4×10^{-3}	4.2×10^{-3}
33–93	5.5×10^{-3}	3	8.5×10^{-3}	25	3.2×10^{-3}	2.3×10^{-3}
93–125	4.4×10^{-3}	21	2.2×10^{-3}	34	3.1×10^{-3}	3.9×10^{-3}

Table 4.27 Comparison between field and laboratory measured unsaturated hydraulic conductivity (Bonsu and Lal, 1983). *Reproduced by permission of the* Nigerian Journal of Soil Sciences

Profile	Horizon depth (cm)	Moisture suction (bar)	Field $K(\theta)$ (cm s^{-1})	Lab. $K(\theta)$ (cm s^{-1})
3	33–73	−0.16	5.22×10^{-7}	1.12×10^{-8}
4	15–33	−0.08	3.47×10^{-6}	4.58×10^{-7}
5	33–93	−0.16	6.48×10^{-7}	3.22×10^{-9}
6	30–50	−0.16	6.00×10^{-7}	1.10×10^{-8}

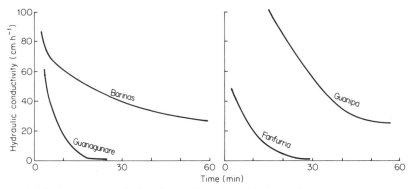

Figure 4.15   Decrease in hydraulic conductivity with time of four soils in Venezuela due to slaking of structural aggregates (Pla Sentis, 1977)

IX. AERATION

Most sedentary uplands are well drained and, even at low soil-water potential, have good air capacity. This is generally the case even for soils of heavy texture, because the aggregated clay behaves like sand or silt, depending upon its size. Some heavy textured Inceptisols, however, may have a high water table and low air capacity. These soils are found in flood plains and valley bottoms and are suitable for rice cultivation.

Soil air has not been extensively studied in the tropics. Some biologists have reported the evolution of CO_2 in different ecologies in relation to biotic activity and the decomposition rate of leaf litter and biomass (Crowther, 1983). Lal (1984) studied the oxygen diffusion rate in cultivated uplands and observed that gaseous exchange was not a constraint if the surface soil was not disturbed and was kept mulched (Table 4.28). The oxygen diffusion rates are generally low for soils and management systems that render surface soil prone to crusting. Soils high in silt content and low in organic matter and those that are exposed to raindrop impact have low oxygen diffusion rates. In general, sedentary uplands have well drained profiles and desirable levels of aeration capacity.

Table 4.28   The oxygen diffusion rate of an Alfisol with and without crop residue mulch (unpublished date of Lal, 1984)

Treatment	Oxygen diffusion rate ($\mu g\ cm^{-2}\ min^{-1}$)			
	I	II	III	II
No-till with mulch	0.516	0.652	0.722	0.682
No-till without mulch	0.441	0.630	0.720	0.665
Ploughed with mulch	0.568	0.629	0.703	0.666
Ploughed without mulch	0.458	0.604	0.637	0.653

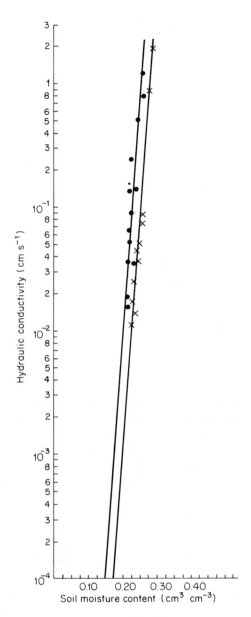

Figure 4.16 The relationship between soil moisture content and the unsaturated hydraulic conductivity of a Cerrado soil in Brazil X, O–30 cm; 0, 30–60 cm. (Stone and Silveira, 1978). *Reproduced by permission of Pesquisa Agropecuaria Brasileira*

X. THERMAL PROPERTIES

Although the temperature regime of some tropical soils has been reported for different ecologies (Lal and Greenland, 1979), thermal properties (heat capacity and thermal conductivity, etc.) have not been extensively studied. Banage and Visser (1967) reported the effects of rainfall on the temperature regime of some Ugandan soils. These authors observed that rainfall usually lowered soil temperatures but the effect lasted only during the actual rainfall. This was generally the case for well-drained coarse-textured soils. The moisture content of the surface horizons was readily depleted. On the contrary, Millington (1983) observed for some soils in Gambia that the major influence on soil temperature was climatological and not pedological. This conclusion of Millington seems to be an over-generalization because soil texture, organic matter content and mineralogical composition of the clay fraction, and soil colour affect heat capacity and thermal conductivity. For example, Ramamohan and Rao (1969) reported that specific heat of black soils of Mysore, India, was twice that of the coarse sand in its vicinity. The effects of texture on thermal capacity for some Indian soils were also reported by Yadav and Saxena (1973).

Ghuman and Lal (1985) studied the thermal properties of some Nigerian soils. Thermal capacity was different among soils of different mechanical composition (Table 4.29). In general, clayey soils had higher thermal capacity than sandy soils, probably due to the presence of a water flim around charged

Table 4.29   Thermal properties of some Nigerian soils (Guman and Lal, 1985). *Reproduced by permission of Williams & Wilkins Co.*

Soil texture	Organic carbon (%)	Thermal capacity (cal $g^{-1} °C^{-1}$)
Sandy clay loam	1.42	0.350±0.012
Clay loam	1.56	0.271±0.007
Loam	1.47	0.366±0.008
Loam	1.61	0.219±0.016
Sandy loam	0.39	0.322±0.005
Sandy loam	0.98	0.264±0.023
Sandy loam	0.63	0.217±0.010
Sandy clay loam	1.70	0.339±0.030
Sandy loam	1.79	0.269±0.009
Sanda loam	2.03	0.235±0.003
Sandy clay loam	1.11	0.342±0.016
Clay	2.19	0.248±0.013
Sandy clay loam	1.70	0.355±0.007
Clay loam	2.19	0.224±0.020
Loam	1.75	0.279±0.022
Sandy loam	0.89	0.175±0.006
Sandy clay loam	1.75	0.279±0.011
Sandy clay loam	1.70	0.367±0.010
Clay	2.40	– –
Clay	1.42	0.324±0.017

Table 4.30 Thermal conductivity of some Nigerian soils as a function of water content (Ghuman and Lal, 1985). *Reproduced by permission of Williams & Wilkins Co.*

Water content (cm^3 cm^{-3})	Thermal conductivity			
	Sandy loam	Loam	Sandy clay loam	Clay
0.02	0.37	0.37	0.35	—
0.04	0.45	0.51	0.43	—
0.06	0.62	0.83	0.47	—
0.08	0.87	1.24	0.51	—
0.10	1.02	1.59	0.62	0.39
0.12	1.10	1.68	0.74	0.40
0.14	1.19	1.78	0.95	0.41
0.16	1.42	1.90	1.17	—
0.18	—	1.93	1.35	—
0.20	—	2.09	1.52	—
0.22	—	2.33	1.70	—
0.24	—	2.57	1.95	—
0.46	—	—	3.34	—
0.52	—	—	—	1.15

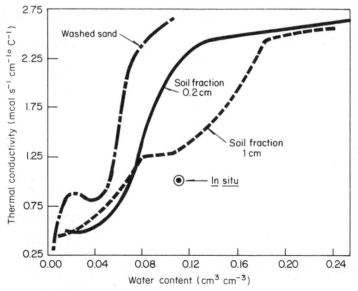

Figure 4.17 Comparison between the thermal conductivity of a gravelly and a gravel-free soil (Ghuman and Lal, 1983). *Reproduced by permission of Williams & Wilkins Co.*

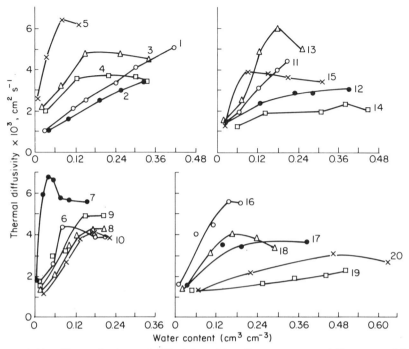

Figure 4.18 Change in thermal diffusivity with soil water content (Ghuman and Lal, 1983). *Reproduced by permission of Williams & Wilkins Co.*

clay particles. Thermal conductivity of some Nigerian soils in relation to soil moisture content (Table 4.30) indicates the maximum value of 3.34 mcal/s cm°C for a sandy clay loam at saturated moisture content. Clayey soils had lower conductivity than coarse-textured soils. Thermal conductivities of most soils studied lie within the range of 0.70–5.40 mcal/s cm°C for sand and 0.60–3.80 mcal/s cm°C for clay at soil water contents of 0 and 0.40 $cm^3 cm^{-3}$ respectively. In general, soils with low bulk density have low thermal conductivity. High gravel content, a common feature of many soils in West Africa and South Asia, has a profound effect on thermal conductivity. The data in Fig. 4.17 show that gravelly soil has lower thermal conductivity than gravel-free sand. Thermal diffusivity (ratio of thermal conductivity and volumetric heat capacity) of some Nigerian soils as a function of water content shown in Fig. 4.18 indicates that thermal diffusivity increased with water content to a maximum value and then decreased.

XI. SOIL PHYSICAL PROPERTIES OF TROPICAL VERSUS TEMPERATE ZONE ALFISOLS

Most Alfisols and Ultisols of the tropics are those containing predominantly low activity clays, and soil physical constraints are major limitations towards

Table 4.31 Profile characteristics of a temperate zone Alfisol (Aridic Paleustalf) (Dregue, 1983). *Reproduced by permission of Westview Press*

Depth (cm)	Particle size (%)			pH	CEC (meq/100 g)	Base saturation (%)
	Sand	Silt	Clay			
0–10	84	10	6	8.1	6.3	100
10–23	66	20	14	7.7	11.0	91
23–35	63	18	19	7.5	12.5	85
35–74	58	21	21	7.5	14.2	91
74–96	49	25	26	7.6	17.3	96
96–150	57	19	24	7.9	15.8	—

intensive arable land-use. Major soil physical constraints are low plant available water reserves, prone to crusting and compaction, and accelerated soil erosion. In general, however, the same soil orders of the temperate region have either mixed mineralogy or contain high activity clays, in addition to high soil organic matter content. Consequently, Alfisols and Ultisols of the temperate zone do not have as severe soil physical constraints towards an intensive land-use as those of the tropics. These differences are illustrated by the data in Tables 4.31 to 4.33. The data in Table 4.31 showing characteristics of a temperate zone Alfisol indicate relatively high silt content, pH, CEC and base saturation throughout the profile. In contrast Alfisols from India (Table 4.32) and West Africa (Table 4.33) contain lower silt, higher gravels, and lower effective cation exchange capacity (e.c.e.c.) and base saturation. The plant available water reserve is also low. The low organic matter content and low e.c.e.c. are responsible for rapid leaching losses of plant nutrients and widespread nutrient deficiencies. Their susceptibility to crusting and high run-off makes it necessary to develop soil water management systems to alleviate these constraints.

XII. PHYSICAL PROPERTIES OF HAC SOILS

The physical properties of soils with HAC are determined by the predominant cations on the exchange complex, soil organic matter content and moisture regime, governed by the rainfall characteristics and the position in the landscape. Two contrasting examples of physical behaviour are Vertisols of the semi-arid tropics and relatively scarce Mollisols.

A. Vertisols

Most Vertisols are soils with high amounts of montmorillonitic clays, which develop deep and wide cracks, are easily dispersed, and have some severe trafficability problems. Because of high clay content, soil bulk density is generally between 1.3 and 1.45 g cm^{-3} and rarely exceeds 1.5 gm cm^{-3}. The

Table 4.32 Major characteristics of Udic Rhodulstalf at Hyderabad, India (El-Swaify et al., 1984). Reproduced by permission of ICRISAT

Depth (cm)	Texture (%)			Gravels (%)	e.c.e.c. (meq/100 g)	Organic carbon (%)	Moisture retention (%)		Base saturation (%)
	Sand	Silt	Clay				0.3 bar	15 bar	
0–5	79.3	6.4	14.3	17	3.5	0.55	16.2	6.3	74
5–18	66.7	5.5	27.8	17	5.2	0.52	20.0	12.4	64
18–36	41.6	6.8	51.6	36	10.2	0.63	21.9	13.9	69
36–71	45.0	4.4	50.6	54	11.6	0.40	24.8	17.4	82
71–112	54.1	7.4	38.5	50	8.6	0.10	23.6	16.2	85
112–140	70.6	4.1	25.3	63	8.4	0.18	18.7	11.5	92

Table 4.33 Physical properties of an Alfisol* from western Nigeria (Moormann et al., 1975)

Depth (cm)	Texture (%)			Gravel (%)	C.E.C. (meq/100g)	Organic carbon (%)	Base saturation (%)
	Sand	Silt	Clay				
0–5	66.9	11.8	21.3	25	5.5	1.5	97
5–15	57.9	16.8	25.3	20	7.0	1.5	98
15–45	48.9	13.8	37.3	45	5.3	0.8	97
45–65	29.9	13.8	56.3	40	4.3	0.3	96
65–95	29.6	6.1	64.3	40	4.1	0.3	96
95–100	20.2	14.4	65.4	30	4.5	0.2	96

* Uncultivated soil of Edbeda series.

clay content generally exceeds 40 per cent and often increases with depth. Textural and mechanical properties of two Vertisols from Central India are shown in Table 4.34. The maximum clay content, of 66 to 69 per cent, is observed in the fifth layer of the soil profiles.

When measured under laboratory conditions, the mean weight, diameter and percentage of aggregation are often high. Under field conditions, however, soils slake readily and develop a surface seal of low permeability. Water permeability is particularly low for soils containing high exchangeable sodium percentage (ESP). Krantz et al. (1978) compared the water intake rate of an Alfisol with that of a Vertisol at Hyderabad, India. The infiltration rate in Vertisol declined from 76 mm h^{-1} at 0.5 h to 4 mm h^{-1} at 2h and to 0.21 mm h^{-1} at 144h after initiating the test. The corresponding rates for an Alfisol were 73, 15, 7.7 mm h^{-1}, respectively. Structural properties of a Vertisol are greatly influenced by the method of determinations and specifically by the soil moisture potential. Bruce-Okine and Lal (1975) observed that the detachability index of a Vertisol was high at low moisture potential (Table 4.35). Structural stability is generally higher when aggregates are equiliberated slowly to a high soil moisture potential.

In addition to structural stability and slaking behaviour, the infiltration rate of a Vertisol is also influenced by the soil moisture potential. The equilibrium infiltration rate is generally higher in soils at or above pF 4.44 than at high pF values. The low structural stability to quick wetting and a slow water infiltration rate in dry Vertisols is related to the effect of entrapped air and to the heat of wetting released (Collis-George and Lal, 1971; 1973). For a dry soil, there is an increase in soil temperature ahead of the visible wetting front. The increase in soil temperature for an oven dry Vertisol can be as much as 30°C to 40°C. The drier the soil and the more negative the soil water potential the higher the heat of wetting, the more the surface aggregates disintegrate and the lower the infiltration rate. In addition to the quantity of heat released, the rate of its release and dissipation are also important factors affecting slaking of the surface soils. If a mechanism exists in the soil body, so that the heat released at

Table 4.34 Texture, structure and mechanical properties of two Vertisols from Central India (Lal, 1965)

Depth (cm)	Bulk density (g cm^{-3})	Particle density (g cm^{-3})	Texture (%)			Mean weight diameter (mm)	Percent aggregates >0.25 mm	Permeability
			Sand	Silt	Clay			
(a) Hyderabad								
0–7.5	1.31	2.57	36.8	14.0	49.2	0.45	55	0.68
7.5–30	1.42	2.58	35.8	13.0	51.2	1.02	82	0.10
30–60	1.41	2.63	37.8	12.0	50.2	0.55	67	0.18
60–100	1.45	2.52	25.8	13.0	61.2	0.55	67	0.02
100–130	1.41	2.41	13.8	17.0	69.2	0.29	46	0.01
(b) Chhindwara								
0–10	1.40	2.68	25.8	18.0	56.2	0.55	62	0.16
10–50	1.45	2.73	22.8	18.0	59.2	0.45	67	0.19
50–75	1.44	2.69	20.8	17.0	62.2	—	—	0.07
75–127	1.40	2.72	15.8	18.0	66.2	0.53	75	0.08
127–155	1.44	2.70	16.8	17.0	66.2	0.92	82	0.07
155+	1.45	2.70	21.8	16.0	62.2	0.96	78	0.08

Table 4.35 Effect of soil moisture potential pF on the number of raindrops required to detach an aggregate of a Vertisol (Bruce–Okine and Lal, 1975). *Reproduced by permission of Williams & Wilkins Co.*

Horizon	pF	Average number of drops required at different temperatures		
		30 °C	40 °C	50 °C
A1	4.44	194	173	143
A2	4.44	86	65	34
A3	4.44	48	37	36
A1	7.0	23	23	17
A2	7.0	11	11	9
A3	7.0	21	19	11

Table 4.36 Moisture release characteristics of a Vertisol from Hyderabad, Central India (Lal, 1965)

Depth (cm)	Moisture retention (% by weight) at different suctions (bar)						
	0	0.1	1.0	3.0	5.0	10.0	15
0–7.5	54.1	35.9	25.1	20.7	18.6	16.6	16.1
7.5–30	48.1	40.5	25.7	21.8	19.2	17.1	16.6
30–60	47.7	38.1	25.0	20.6	18.8	16.7	16.2
60–100	57.4	49.6	33.3	27.3	25.2	22.2	21.8

the wetting front is dissipated as quickly as it is released, then the aggregates retain their stability.

Some Vertisols also exhibit self-mulching characteristics. Surface soil crumbles into a fine crumb structure due to alternate swelling and shrinkage caused by the changes in soil moisture content. The self-mulching characteristics have some important implications towards tillage requirements in relation to seedbed preparation.

Vertisols have high soil moisture retention capacity (Table 4.36). This does not imply, however, that most water retained is available. In contrast with the coarse-textured soils containing low activity clays, the moisture content at 15 bar in Vertisols is similar to that at 0.1 bar in Alfisols.

Draught characteristics and workability of clay soils are related to high cohesion and adhesion, plasticity, swell-shrink characteristics, the coefficient of linear expansion, the sticky point and other rheological properties of these soils. The range of moisture content at which Vertisols are workable for dry seedbed preparation is generally small. Vertisols and HAC soils containing high clay content are hard to puddle because of severe trafficability problems.

B. Mollisols

Mollisols have better structural stability and structural aggregates that are more resistant to dispersion than Vertisols. Some Mollisols contain as much clay as Vertisols (Fig. 4.19), but these soils are more productive because of their high organic matter content and more favourable soil moisture regime. Similar to Vertisols, Mollisols have high water retention capacities (Fig. 4.20) and high swell/shrink properties, and they develop large and deep cracks. In general, subsoil layers retain more water than the surface soil for all matric potentials (Hundal and De Datta, 1984).

The traffic problem in heavy-textured HAC soil is related to very hard consistency when the soil is dry and the very sticky consistency when it is wet. These soils with high shrinkage-swell capacity exhibit the lowest strength when they are at their maximum swelling capacity. The soil strength is related to the pore water pressure as is shown by the Terzaghi's effective stress equation:

$$\acute{\sigma} = \sigma - u$$

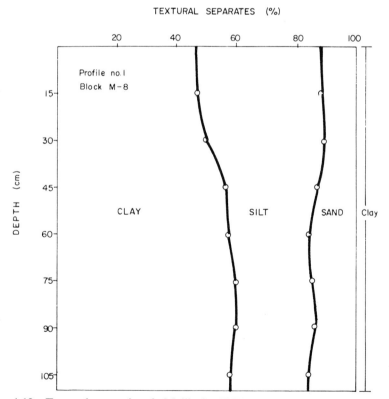

Figure 4.19 Textural properties of a Mollisol at IRRI, Los Banos, Philippines (Hundal and De Datta, 1984). *Reproduced by permission of IRRI*

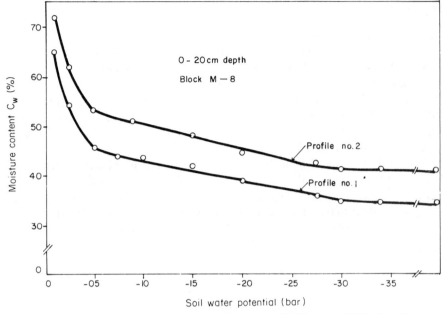

Figure 4.20 Moisture retention characteristics of a Mollisol at IRRI, Los Banos, Philippines (Hundal and De Datta, 1984). *Reproduced by permission of IRRI*

where $\bar{\sigma}$ = the effective stress, σ = the externally applied stress, and u = the pore water pressure.

This implies that the negative pore water pressure contributes positively to the effective stress. The mechanical strength of shearing resistance is also related to the soil strength. Ploughing a dry soil that has undergone a considerable shrinkage requires high-powered machinery because it is often beyond the capacity of draught animals. The strength during wet tillage, or puddling, as is traditionally used for rice cultivation, can be too low to bear the weight of the equipment. The traction performance is related to soil's resistance or its bearing strength. Kisu (1978) related trafficability of wet clay soils to its cone index (Table 4.37). The higher the cone index the better the trafficability during wet tillage.

Table 4.37 Relation between the cone index (kg cm^{-2}) and trafficability of a wet soil for different tillage operations (Kisu, 1978). *Reproduced by permission of IRRI*

Tillage operations	Trafficability		
	Easy	Possible	Impossible
Rotary tilling	>5	3–5	<3
Ploughing	>7	4–7	<4
Ploughing (with girdle)	>4	2–4	<2

XIII. PHYSICAL PROPERTIES OF TROPICAL ANDOSOLS

Soils developed on volcanic ash are extremely fertile and are intensively cultivated in Central America and parts of Southeast Asia. The physical and mechanical characteristics of Andosols are described in detail by Warkentin and Maeda (1980). These soils contain high amounts of amorphous clay, have low bulk density, and undergo irreversible changes in physical and mechanical properties on drying. Their physical properties are only slightly affected by the nature of exchangeable cations. The soil physical properties of Andosol are not controlled by the surface electrical properties.

The textural properties of Andosols are hard to characterize because the clay content changes with drying and the soils are hard to disperse. These soils have high liquid and plastic limits, and the plasticity is a useful property for characterizing these soils.

In comparison with high bulk density and low porosity in other soils, Andosols have low bulk density and high total porosity. The data in Table 4.38 by Gonźalez and Gavande (1969) indicate a low bulk density (0.6 to 0.7 g cm^{-3}) for Andosol from Costa Rica. Despite the large total porosity, however, poor aeration can be a problem in Andosols. The air permeability of these soils is generally low. Because of favourable soil structure and high total porosity, the infiltration rate of Andosols is generally high in comparison with that of some easily dispersed non-ash soils (Table 4.39).

Because of high total porosity Andosols have a high water-holding capacity and a large quantity of water is stored in the soil profile. The data in Table 4.40 and Fig. 4.9 compare moisture retention characteristics of Andosols with those of non-ash soils in Costa Rica. Gravimetric moisture content at low suctions exceeding 100 per cent is not uncommon. The field capacity of Andosols corresponds to a suction of 0.1 bar (Warkentin and Maeda, 1980). Although

Table 4.38 Mechanical properties of some soils from Costa Rice, Central America (Gonźalez and Gavande, 1969)

Soil series	Classification	Depth (cm)	Bulk density (g cm^{-3})	Particle density (g cm^{-3})	Total porosity (%)
Alajuda Plano		0–18	0.60	2.14	72.0
		18–33	0.61	2.20	72.3
		33–60	0.73	2.16	66.2
Grecia	Latasol	0–18	0.90	2.16	63.4
		18–33	1.04	2.13	52.6
		33–60	1.14	2.27	47.1
Instituto	Alluvial	0–30	1.82	2.49	29.9
		30–60	1.72	—	—
		>60	1.74	—	—

Table 4.39 Water infiltration into some soils of Costa Rica (Gonźalez and Gavande, 1969)

Series	Classification	Infiltration rate (cm h^{-1})	
		Accumulative	Instantaneous
Alajuela Plano	Andosol	15.4	11.2
Grecia	Latoso	9.5	8.0
Complejo Arcilloso Ciruelas	Fluvio lacustre	4.1	3.2
Alajuela Ondulado	Latasol	13.1	12.8

the water retention is also high at 15 bar suction, the available water holding capacity is usually high. Common methods of measuring the available water-holding capacity of non-ash soils may, however, not be applicable for Andosols.

Structurally most Andosols are relatively resistant to dispersion and to raindrop impact. The stability of these soils to water erosion is related to their high infiltration rate and high cohesion. Cohesion, however, decreases on drying although the infiltration rate of a dried Andosol is more than that of a wet soil.

XIV. PHYSICAL PROPERTIES AND SOIL CLASSIFICATION

Soil physical properties are not often used as diagnostic features for classifying soils. The soil temperature and moisture regimes considered to designate hydric and thermal regimes are for the subsoil horizons below 50 cm depth. This is done to minimize the effects of management and the interference created by agricultural activity. The physical characteristics at 50 cm depth, however, have only a little edaphological significance. Over and above the difficulties of using physical properties as diagnostic criteria, the problems of classifying tropical soils have been discussed by Aubert (1967), Babin and Maignien (1979) and Verheye (1974) for Africa; Jacomine (1969), Buol (1972) and Falesi et al. (1980) for South America; and Fridland (1969) for Southeast Asia.

Structural characteristics have been proposed by some as a diagnostic criterion for classifying soils of the tropics. Fridland et al. (1964) used stability of aggregates of 3-5 mm to differentiate some soils of Vietnam. Using this criterion, these authors differentiated the dark red ferrallitic soil developed on basalt, red–yellow ferrallitic soil on gneiss, mountain-humid-ferrallitic soil on rhyolite, and the ferrallitic margelitic soil on basic volanic tuff. Fridland (1969) observed that most ferrallitic soils had stable structural separates but those of the red–yellow and mountain humus ferrallitic soils were less stable. Moberg and Mmkonga (1977) suggested the use of stable micropeds as a diagnostic feature for classifying strongly weathered soils of Tanzania and the tropics. For

Table 4.40 Moisture release properties of some soils from Costa Rica, Central America (González and Gavande, 1969)

Soil series	Classification	Depth (cm)	Moisture retention (%, gravimetric) at different suctions (bar)							
			0	0.1	0.34	1.03	2.05	5.15	10.3	15.1
Alajuda Plano	Andosol	0–15	130.0	85.5	70.5	51.0	39.0	36.8	31.8	30.0
		18–33	115.5	84.4	71.6	49.0	40.0	38.4	33.9	31.5
		33–60	103.0	79.7	59.5	47.5	42.5	40.5	38.5	34.5
		60–75	93.3	63.0	50.0	43.3	41.5	39.1	37.5	33.0
Grecia	Latisol	0–18	82.0	64.0	50.0	37.5	36.0	33.0	32.1	29.5
		18–33	90.0	67.0	56.5	40.8	40.0	34.8	33.5	32.0
		33–60	62.0	61.0	55.5	39.5	38.5	37.5	36.0	35.0
		60–75	80.9	54.8	47.0	38.5	37.0	36.0	34.5	34.0
Instituo	Alluvial	0–30	90.0	86.0	42.0	33.0	32.0	30.0	29.0	27.0
		30–60	100.0	90.0	60.0	35.0	33.0	31.0	30.0	29.0
		>60	85.0	74.0	48.0	38.0	37.0	34.0	33.0	32.0

some soils of Central Africa and the Congo Basin, Frankart and Sys (1974) suggested characteristics that are useful to define 'pedoagronomic index'.

Most of the six major classification systems currently being used in the tropics have major limitations in that they were developed for specific objectives. Because the systems were developed in different regions for different objectives, there are few common bases for comparative evaluation. That has also been one of the constraints towards extrapolation of the results and transfer of technology from one soil to another.

From the edaphological point of view, the important soil physical properties suggested as diagnostic criteria are: (i) soil temperature and moisture regimes, (ii) structural stability, and (iii) effective rooting depth. Agropedologically important criteria are the maximum soil temperature of the surface horizon and the magnitude of its diurnal fluctuations, the available water-holding capacity of the root zone, and the ratio of rainfall acceptance (water intake rate under forest or uncultivated soil to that under bare ploughed soil surface). As with the soil erodibility index, these properties should also be evaluated under ploughed bare soil surface. It is the maximum hazard of supra-optimal soil temperature and the sub-optimal moisture conditions under inappropriate or undesirable management that determines the physical constraints to crop production. Considerable research has yet to be done to develop edaphologically appropriate criteria for classifying soils of the tropics.

XV. CONCLUSIONS

The physical properties of soils with low activity clays (e.g. tropical Alfisols, Ultisols and Oxisols) are crucial in maintaining high levels of agronomic productivity and in preventing degradation of soil characteristics with intensive and continuous cultivation. Alfisols and Ultisols have less stable structures than Oxisols. These soils develop surface seal and crust, are prone to erosion, have low available water-holding capacity and are easily compacted. Shallow-rooted seasonal crops suffer from periodic drought stress even in regions of high annual rainfall. Although the structural properties of Oxisols are somewhat better, these soils are also susceptible to erosion and drought, and are easily compacted.

Mollisols are relatively fertile soils with favourable structural properties. Vertisols are easily dispersed and have severe management problems. These soils are highly susceptible to water erosion even on gentle slopes. Andisols have high water holding capacity and low bulk density. These soils are resistant to water erosion.

REFERENCES

Agafonov, O., Delgado Diaz, M., Rivero Ramos, L., and Latevosian, G. 1978, Physical properties of Cuban Vertisols in relation to aspects of their formation, *Ciencias Agric.* **3**, 47–80.

Ahn, P.H., 1979, Micro-aggregation in tropical soils: its measurements and effects on the maintenance of soil productivity, in R. Lal and D.J. Greenland (eds.) *Soil Physical Properties and Crop Production in the Tropics*, John Wiley & Sons, Chichester pp 75–86.

Amezquita, E.C., 1981, A study of the water regime of a soil during approach to field capacity and wilting point, *Ph.D. (Thesis)*, University of Reading, U.K., 244 pp.

Amezquita, E.C., and Forsythe, W.F., 1975, Application de la ecuacion universal de perdade de suelo en Turrialba, Costa Rica. *V. Latin American congr. Soil Sci.*, Aug. 1975, Medellin, Colombia.

Arndt, W., Phillips, L.T., and Norman, M.J.J., 1963, Comparative performance of crops on three soils of the Tipperary region, *N.T. CSIRO, Aust. Div. Land Res. Reg. Surv. Tech.* Paper 23.

Asamoa, G.K., 1973, Particle size and free iron oxide distribution in some latosols and groundwater laterites of Ghana *Geoderma*, **10**, 285–297.

Aubret, G. 1967, *Tropical Soils, Trans. 8th Int. Congr. Soil Sci.*, **1**, 213–219.

Babalola, O., and Lal, R., 1977, Sub-soil gravel horizon and maize root growth, Parts I & II, *Plant Soil* **46**, 337–357.

Baena, A.R.C., 1977, The effects of pasture on the chemical composition of the soil after clearing up and burning up a typical highland rainforest. M.S. Disst., Agronomy Dept., Iowa State Univ., Ames, Iowa, USA, 172 pp.

Banage, W.B., and Visser, S.A., 1967, Soil moisture and temperature levels and fluctuations in one year in a Uganda soil catena, *E. Afri. Agric. For. J.*, **32**, 450–455.

Banerjee, S.P., and Chand, S., 1981, Physico-chemical properties and moisture characteristics of soils as influenced by forest fire, *Indian Forester*, **107**, 178–182.

Barber, R.G., Moore, T.R., and Thomas, D.B., 1979, The erodibility of two soils from Kenya, *J. Soil Sci.*, **30**, 579–591.

Barnett, A.R., Carreker, J.R., Abruna, F., and Dooley, A.E., 1971, Erodibility of selected tropical soils, *Trans. Amer. Soc. Agric. Engrs.* **14**, 496–199.

Belobrov, V.P., 1978., Lessivage and textural differentiation in some Cuban soils, *Soviet Soil Sci.*, **10**, 243–254.

Bertrand, R., 1971, Morpho-pedolgie et orientations culturales des regions soudaniennes du Siné Saloum-Senegal. IRAT, Paris, France.

Bertrand, R., 1970, Etude pédalogique de reconnaissance de quelques zones dans le department de Sedhiou (Senégal). En vue du développement de la riziculture. IRAT, Paris, France.

Bonsu, M., and Lal, R., 1983. Hydrological properties of soils in Western Nigeria: a comparison of field and laboratory methods, *Nigerian J. Soil Sci.*, **3**, 101–117.

Bowden, J.M., Posner, A.M., and Quirf, J.P., 1980, Adsorbing and charging phenomena in variable charge soils. p. 147–166. In B.K.G. Theng (ed) *Soils with Variable Charge*, New Zealand Soc. Soil Sci., Lower Hutt, New Zealand.

Bridge, B.T., Collis-George, N., and Lal, R., 1970, The effects of wall lubricants and column confinement on the infiltration behaviour of a swelling soil in the laboratory, *Aust, J. Soil Res.* **8**, 259–272

Bridge, B.T., Sarmnun, B., and Aromratria, U., 1975, Properties affecting water entry in Central Plain Soils, *Thai J. Agri. Sci.* **8**, 177–193.

Bruce-Okine, E., and Lal, R., 1975, Soil erodibility as determined by raindrop technique, *Soil Sci.*, **119**, 149–157.

Buanec, B. Le, 1974, Observations on soil profile loosening in ferrallitic soils. Effect on soil characteristics and growth of annual crops, *Agron, Trop.*, **29**, 1079–1099.

Buol, S.W., 1972, Soil genesis, morphology and classification, in *Review of Soils Research in Tropical Latin America*, North Carolina State University, pp. 1–51.

Buringh, P., 1979, *Introduction to Study of Soils in Tropical and Subtropical regions*, PUDOC, Wageningen, Holland.

Cadima, A.A., and Elvim, P. de T., 1973, Algunos factores del suelo asociados con la productividad del cacaotero en Bahia, Brazil. *Revista Theobroma* (Brazil), 13–26.

Charreau, C., 1974, Soils of tropical dry and dry-wet climatic areas and their use and management, Lecture Notes, Cornell Unviersity Ithaca, N.Y.

Chauvel, A., and Pédro, G., 1978, The importance of extreme soil dessication (Ultra-desiccation) in pedologic evolution in tropical zones with contrasting seasons, *Comptes Rendus Hebdomadaires des Séances de l' Académie des Sciences, Sao Paulo Brazil*, **D 286** (22), 1581–1584.

Collis-George, N., and Lal, R., 1971, Infiltration and structural changes as influenced by initial moisture content, *Aust, J. Soil Res.*, **9**, 107–116.

Collis-George, N., and Lal, R., 1973, The temperature profiles of soil columns during infiltration, *Aust, J. Soil Res.*, **11**, 93–105.

Commonwealth Bureau of Soils, 1978, *Biology of Tropical Soils, 1950—1964, Annotated Bibliography*, Commonwealth Bureau of Soils S932R, 21 pp.

CSIRO, 1965, *General Report on Lands of the Tipperary Area, Northern Territory, 1961*, Land Research Series, No. 13, CSIRO, Melbourne, 112 pp.

Corrêa, J.C., 1982, Consistency limits and their agricultural significance in soils of the central Amazônian region, *Pesquisa Agropecuaria Brasileiria*, **17**, 917–921.

Crowther, J., 1983, Carbon dioxide concentrations in some tropical karst soils, West Malaysia, *Catena*, **10**, 27–39.

Dabin, B., and Maignien, R., 1979, The main soils of West Africa and Thai agricultural potential, *Cahiers ORSTOM, Pédologie*, **17**, 235–257.

Dancette, C., 1969, Determination au champ de le capacité de retention aprés irrigation, dans un sol sableux du Sénegál. IRAT, Paris, France.

Dancette, C., 1973a, Les besoins en eau des plantes de grande culture au Senegal = Symposium on Isotopes and radiation techniques in studies of soil physics, irrigation and drainage in relation to crop production. FAO/IAEA, Vienna, Austria.

Dancette, C., 1973b, Mesures d'évapotranspiration potentielle et d'évaporation d' une nappe d'eau libre au senegal. Orientations des travaux portant sur les besoins en eau des cultures, IRAT, Paris, France.

Dangler, E.W., and El-Swaify, S.A., 1976, Erosion of selected Hawaii soils by simulated rainfall, *Soil Sci. Soc. Amer. J.*, **40**, 969–779.

Deshpande, T.L., Greenland, D.J., and Quirk, J.P., 1968, Changes in soil properties associated with the removal of iron and aluminium oxides in soils. *Trans. 8th Int. Congr. Soil Sci., Bucharest*, **3**, 1213–1235.

Diestel, H., 1979, *Quantitative Aspects of the Estimation of the Water Available to Plants in Tropical Soils: A Case Study*, Mitteilungen der Deutschen Bodenkunlichen Gesellschaft, **29**, pp 111–122.

Dimov, D.I., 1980, Some differences in the behaviour of yellow and red tropical forest soils during wetting and drying, *Grskostopanska Nauka*, **17**, 69–74.

Dregne, H.E., 1983, Soils of semi-arid regions, pp. 53–62, in Campos-Lopez, E. and Anderson, R.J. (eds), *Resources and Development in Arid Regions*, Westview Press, Boulder, Colorado.

El-Swaify, S.A., 1970, The stability of saturated soil aggregates in certain tropical soils as affected by solution composition, *Soil Sci.*, **109**, 197–202.

El-Swaify, S.A., Walker, T.S., and Virmani, S.M., 1984, Dryland management alternatives and research needs for Alfisols in the semi-arid tropics. *Consultant's Workshop 1–3 Dec., 1983*, ICRISAT, Hyderabad, India, 34 pp.

Escolar, R.P., and Lugo-Lopez, M.A., 1969, Availability of moisture in aggregates of various sizes in a tropical Ultisol and a typical Oxisol in Puerto Rico, *J. Agric. Univ. P. Rico*, **53**, 113–117.

Falesi, I.C., Baena, A.R.C., and Dutra, S., 1980, Effects of agricultural exploitation on the physical and chemical properties of the soils of the sub-regions of northeastern Pare, Brazil. *EMBRAPA—CPATU, Belem, Brazil.*

Frankart, R., and Sys, C., 1974, Significance of soil characteristics used for soil appraisal in humid tropical regions, *Trans. 10th Int. Cong. Soil Sci.*, **V**, 34–39.

Fridland, V.M., 1969, The differences between crusts of weathering and soils developed on acid and basic rocks in the humid tropics, in *Soils and Tropical Weathering; Proc. of the Bundung Symposium, Nov. 1969: Natural Resources Res.* **11**, 111–116.

Fridland, V.M., Dorokhova, K. Ya., and Zhitkova, A.I., 1964, Nature of the structure of humid tropical soils (North Vietnam), *Dokl. Akad. Nauk,* **154**, 707–709.

Frómeta, B.L., Rodriguez, M.M., Caballero, A., and Rangel, L., 1979, Interrelationships of some physical properties of soils from the Biolgogical Research Station of Los Llanos, Calabozo, Venezuela, *Agronomia Tropical,* **29**, 349–365.

Ghuman, B.S., and Lal, R., 1985, Thermal conductivity and diffusivity of some Nigerian soils, *Soil Sci.* (In Press).

Godefroy, J., Gaillard, J.P., and Hawry, A., 1977, Evolution of the chemical characteristics and structure of a brown eutrophic soil in Cameroon under pineapple culture, *Fruits,* **1977**, 591.

González, M.A., 1980, Physical properties of some Colombian soils, *Acta Agronomica (Colombia),* **30**, 19–48.

González, M.A., and Gavande, S.A., 1969, Physical properties of some sugarcane soils in Costa Rica, *Turrialba,* **19**, 235–245.

Graaff, R.H.M. van de, 1965, Soils of the Tipperary area, in *General Report on Lands of the Tipperary Area, Northern Territory, 1961,* Land Research Series No. 13, CSIRO, Melbourne, pp. 68–80.

Greenland, D.J. (ed.), 1981, *Characterization of Soils,* Clarendon Press, Oxford, 446 pp.

Guedez, J.E., and Langohr, R., 1978, Some characteristics of pseudo-silts in a soil-toposequence of the Llanos Orientales (Venezuela), *Pédologiea,* **28**, 118–131.

Hundal, S.D., and De Datta, S.K., 1984, *Soil & Tillage Res.,* **5** (In Press).

IAR, 1975, Institute of Agricultural Research, Technical *Bulletin,* ABU, Zaria.

ICRISAT, 1975–76, International Crops Research Institute for the Semi-Arid Tropics, *Annual Report,* pp.16–21.

ICRISAT, 1976–77, International Crops Research Institute for the Semi-Arid Tropics, *Annual Report,* pp.16–21.

ICRISAT, 1977–78, International Crops Research Institute for the Semi-Arid Tropics, *Annual Report,* pp.172–173.

ICRISAT, 1978–79, International Crops Research Institute for the Semi-Arid Tropics, *Annual Report,* pp.173–179.

ICRISAT, 1982, International Crops Research Institute for the Semi-Arid Tropics, *Annual Report,* pp.234.

Institute for Agricultural Research, Samaru, Nigeria, 1975, Observations on dry-season moisture profiles at Samaru, Nigeria. *Samaru Misc. Paper* 51, 9 pp.

I.I.T.A., 1975, International Institute of Tropical Agriculture, Report on Collaborative Research on Major Soils of West Africa, Mimeo.

Jacomine, P.K.T., 1969, Morphological, Physical, Chemical, and mineralogical, characteristics of some soil profiles under cerrado vegetation, *Boletim Técnico,* **11**, 126 pp.

Joachim, A.W.R., and Pandilteskera, D.G., 1948, Investigations on crumb structure and stability of local soils, *Trop. Agric.,* **104**, 119–129.

Jones, M.J., 1973, The organic matter content of the savanna soils of West Africa, *J. Soil Sci.,* **24**, 42–53.

Jones, M.J., and Wild, A., 1975, Soils of West African Savanna *Techn. Comm. No.* 55, CAB, Harpenden, U.K.

Kang, B.T. and Juo, A.S.R., 1981, Management of soils with low activity clays in tropical Africa. Proc. 4th Int. Soil Classification Workshop. June 2–12, 1981, Kigali, Rwanda.

Kelly, R.D., and Walker, B.H., 1976, The effects of different forms of landuse on the ecology of a semi-arid region in south-eastern Rhodesia, *J. Ecol.*, **64**, 553–576.

Khadr, M., and Abdel-Tawab, A.A., 1971, Contribution to the study of some soils of the Bani Basin in Mali, *Beitr. Trop. Landw. Vet. Med.,* **9**, 39–54.

Kilewe, A.M., Ulsaker, C.G., Kamu, F.N., Ngugi, F.K., Kaiyare, J.M., Njoroge, P.K., and Gathiaka, G.N., 1983, Soil Physics, *Annual Report*, KARI, Nairobi, Kenya.

Kisu, M., 1978, Tillage properties of wet soils, pp. 307–316. in *Soils and Rice*, IRRI, Los Banos, Philippines.

Koenigs, F.F.R, 1961, The mechanical stability of clay soils as influenced by the moisture conditions and some other factors. Versl. Landbouwk. Onderz. Nr. 67.7. p.171.

Kowal, J.M., 1970a, Soil physical properties of soil at Samaru, Zaira, Nigeria: storage water and its use by crops. I. Physical status of soils, *Res. Bull.,* Samaru 111, 20 pp.

Kowal, J.M. 1970b, Some physical properties of soil at Samaru, Zaria, Nigeria: Storage of water and its use by crops. II. Water storage characteristics. *Res. Bull.,* Samuru 118, 52 pp.

Kowal, J.M., and Kassam, A.H., 1978, *Agricultural ecology of Savanna*, Clarendon Press, Oxford, 403 pp.

Krantz, B.A., Kampen, J., and Russell, M.B., 1978, Soil management differences of Alfisols and Vertisols in the semi-arid tropics, pp. 77–95. in Drosdroff *et al.* (eds), *Diversity of Soils in the Tropics*, ASA Spec. Publ. 34, Madison, Wisconsin.

Kyuma, K., and Pairintra, C., 1983, *Shifting cultivation*, Ministry of Science, Technology and Energy, Bangkok, Thailand, 219 pp.

Lal, R., 1965, Physical properties of soils from major maize growing regions of India, *M.Sc. Thesis*, IARI, New Delhi, India, 145 pp.

Lal, R., 1976, Soil erosion problems on Alfisols in western Nigeria and their control, *IITA Monograph 1*, IITA, Ibadan, 208 pp.

Lal, R., 1979, Physical characteristics of soils of the tropics: determination and management, pp. 7–46. in R. Lal and D.J. Greenland (eds.). *Soil Physical Properties and Crop Production in the Tropics*, John Wiley and Sons, Chichester.

Lal, R. 1980, Physical and mechanical characteristics of Alfisols and Ultisols with particular reference to soils in the tropics, pp. 253–280, in B.K.G. Theng (ed.). *Soils With Variable Charge*, DSIR, Lower Hutt, New Zealand.

Lal, R., 1983, No-Till Farming, *IITA Monograph Series 2*, Ibadan, Nigeria, 68 pp.

Lal, R., 1984, Erosion on tropical lands and its control, *Adv. Agron.,* **37** (In Press).

Lal, R., Bridges, B.J., and Collis-George, N., 1970, The effect of column diameter on the infiltration behaviour of a swelling soil, *Aust. J. Soil Res.,* **8**, 185–193.

Lal, R., and Greenland, D.J. (eds), 1979, *Soil Physical Properties and Crop Production in the Tropics*. John Wiley & Sons, Chichester, 551 pp.

Lal, R., and Greeland, D.J., 1984, Physical properties of soils with low activity clays. *Annual Meetings, Amer. Soc. Agron., Las Vegas*, 26–30 Nov. 1984.

Lal, R., De Vleeschauwer, D., and Nganje, R.M., 1980, Changes in properties of a newly cleared Alfisol as affected by mulching, *Soil Sci. Soc. Amer. J.* **44**, 827–833.

Lal, R., Wilson, G.F., and Okigbo, B.N., 1979, Changes in properties of an Alfisol produced by various crops, *Soil Sci.,* **127**, 337–382.

Lal, R., Juo, A.S.R., and Kang, B.T., 1983, Chemical approaches towards increasing water availability to crops including minimum tillage systems. Proc. CHEM-RAWN II Conference, Manila, Philippines, 57–72.

Lévêque, A., 1977, The clay fraction. The essential characteristics of its distribution in the different soils of the granite-gneiss platform in Togo, *Cahiers, ORSTOM, Pédologie*, 15, 109–130.

Libera, C.L.F., Corsini, P.C., and Perecin, D., 1977, Variations in the permeability of three soil profiles, *Cientifica Numero especial*, pp.11–16, Sao Paulo.

Lindsay, J.I., and Gumbs, F., 1982, Erodibility indices compared to measured values of

selected Trinidad soils, *Soil Sci. Soc. Amer. J.*, **46**, 393–396.
Lugo-López, M.A.L., Juàrez, J. Jr., and Bonnet, J.A., 1968, Relative infiltration rate of Puerto Rican soils, *J. Agric. Univ. P. Rico*, **52**, 233–240.
Lugo-López, M.A., Juàrez, J., and Perez-Escolae, R., 1970, Correlation between the rate of water intake of tropical soils at hourly intervals to the 8th hour, *J. Agric. Univ. P. Rico*, **54**, 570–575.
Lugo-López, M.A., Wolf, J.M., and Perez-Escolar, R., 1981, Water loss, intake, movement, retention and availability in major soils of Puerto Rico, *Bull. Agric. Exp. Sta., Univ. P. Rico*, **264**, 24 pp.
Lutz, J.F., 1972, Soil physical properties, In *A Review of Soils Research in Latin America*, North Carolina State University, pp.52–61.
Mbagwu, J.S.C., Lal, R., and Scott, T.W., 1983, Physical properties of three soils in southern Nigeria, *Soil Sci.*, **136**, 48–55.
McCown, R.L., 1971, Available water storage in a range of soils in northeastern Queensland, *Aust. J. Agric. Anim. Husb.*, **11**, 343–348.
McCown, R.L., Murtha, G.G., and Smith, G.D., 1976, Assessment of available water storage capacity of soils with restricted subsoil permeability, *Water Resources Res.*, **12**, 1255–2159.
Medina, E., 1982, Physiological ecology of Neotropical Savanna plants, in Huntley, B.J. and Walker, B.H. (eds.), *Ecology of tropical savannas*, Springer Verlag, New York, pp.308–335.
Meyer, B., and Neumeyer, K., 1980, Holocene soil associations on young Pleistocene sediments under tropical rainforest in the transitional region of the Cordilleras – Amazon basin in Peru, *Gottinger Bodenkunliche Berichte*, **62**, 257 pp.
Millington, A.C., 1983, Soil temperature in the Gambia, *Weather*, **38**, 353–358.
Moberg, J.P., and Mmikonga, A.A., 1977, Content of stable micropeds in tropical soils, *CLAMATOPS, Conf. on Classification and management of tropical soils*, Kuala Lumpur, Malaysia, August, 1977.
Moormann, F.R., Lal, R., and Juo, A.S.R., 1975, *Soils of IITA*, Ibadan, 38 pp.
Mott, J., Bridge, B.J., and Arndt, W., 1979, Soil seals in tropical tall grass pastures of Northern Australia, *Aust. J. Soil Res.*, **30**, 483–494.
Ngatunga, E.L.N., Lal, R., and Uriyo, A.P., 1984, Effect of surface management on runoff and soil erosion from some plots at Mlingano, Tanzania, *Geoderma*, **33**, 1–12.
Northcote, K.H., Hubble, G.D., Isbell, R.F., Thompson, C.H., and Bettenay, E., 1975, A description of Australian soils, *CSIRO*, Canberra, 170 pp.
Obeng, H.B., 1978, Soil, water, management and mechanization. *Afr. J. Agric. Sci.* **5**, 71–83.
Ober, Zh, 1974, Problems of soil formation and characteristics of tropical soils, *Pochvovedenie*, **8**, 6–19.
Obi, A.O., 1974, The wilting point and available moisture in tropical forest soils of Nigeria, *Expl. Agric.*, **10**, 305–312.
O'Keefe, P., and Kristorferson, L., 1984, The uncertain energy path: Energy and third world development, *Ambio*, **13**, 168–170.
Oliveira, L.B. de, 1968, Observations on particle-size distribution in soils of the North-East Brazil, *Pesquisa Agropecuaria Brasileira*, **3**, 189–195.
Orvedal, A.C., 1975, Bibliography of soils of the tropics. Vol. I. Tropics in general & Africa, *Tech. Ser. Bull.*, Office of Agriculture, AID 17, 225 pp.
Oswall, N.C., and Khanna, S.S., 1981, Effect of soil physical and chemical properties on soil–water functional relationships, *J. Indian Soc. Soil Sci.*, **29**, 7–11.
Pagel, H., 1975, Soil science aspects of mechanised soil cultivation in the tropics, *Beiträge Zur Tropischen Landwirtschaft und Veterinärmedizin*, **13**, 165–172.
Panabokke, C.R., 1958, A pedologic study of dry zone soils, *Trop. Agric. (Ceylon)*, **114**, 151–174.
Philipson, W.R., and Drosdoff, M., 1972, Relationships among physical and chemical

properties of representative soils of the tropics from Puerto Rico, *Soil Sci.Soc. Amer. Proc.*, **36**, 815–819.

Pla Sentis, I., 1977, *Metodologia para la caracterizacion fisica con fines de diagnostico de problemas de manejo y conservacion de suelos en condiciones tropicales,* Instituto de Edfologia, Faculted de Agronomia, UCV, Marecay, Venezuela, 112 pp.

Ramamohan, R.V., and Rao, N.A.N., 1969, Specific heat of some Mysore soils and its significance, *Mysore J. Agric. Sci.*, **3**, 224–248.

Ramanathan, G., and Krishnamoorthy, K.K., 1977, Quality contributions of soil clays towards certain physical properties of soils of Tamil Nadu, *Madras Agric. J.*, **64**, 491–496.

Reichart, K., Ranzani, G., Freitas, E. De, Jr., and Libardi, P.L., 1980, Hydrology of some Amazonian soils–the lower Rio Negro Region, *Acta Amazonia*, **10**, 43–46.

Rengasamy, P., Murty, G.S.R.K., and Kathavate, Y.V., 1976, Cationic environment and hydrolophysical properties or tropical soils, *Zeitschrift fur Pflanzenernahrung und Bodenkunde*, **4**, 409–419.

Resck, D.V.S., 1981, Physical parameters of soils of the cerrado region, *Beletim Pesquiza*, EMBRAPA-CPAC 2, 17 pp.

Roeder, M., and Bornemisza, E., 1968, Some properties of soils from the Amazon region of the state of Maranhas, Brazil, *Turrialba*, **18**, 39–44.

Roose, E.J., 1977, Use of the Universal Soil Loss Equation to predict erosion in West Africa. SCSA Special Publication 21, pp.60–74.

Saha, A.K., Chauhan, C.P.S., and Yadav, D.S., 1980, Studies on soil–water behaviour and crop production under rainfed conditions (Agra), *J. Indian Soc. Soil Sci.* **28**, 277–285.

Sagué Diaz, H., Hernandez Martinez, L., Ortege Valdes, J., and Lastres L., 1979, Water balance erosion in Sierra del Rosario, *Voluntad Hidraulica*, **16**, (49/50) 28–38.

Selvakumari, G., Raj, D., and Krishnamoorthy, K.K., 1973, Moisture retention characteristics of various groups of soils of Tamil Nadu, *Madras Agric. J.* **60**, 849–853.

Siméon, F.R., 1979, Hydrophysical properties of the principal agricultural soils of Cuba, *Voluntad Hidraulica*, **16** (49/50) 16–23.

Soong, N.K., 1980, Influence of soil organic matter on aggregation of soils in Peninsular Malaysia, *J. Rubber Res. Inst. (Malaysia)*, **28**, 32–46.

Stone, L.F., and da Silveira, P.M., 1978, Conductividade hidraulica de un latossolo Vermelho-Amarelo, *Pesquisa Agropecuaria Brasileira* **4**, 63–71.

Takahashi, T., Nagahori, K., Mongkolasawat, C., and Losirikul, M., 1983, Runoff and soil loss in shifting cultivation. in Kyuma, K.and Pairintra, C. (eds.), *Shifting Cultivation*, Ministry of Science and Technology, Bangkok, Thailand, pp.84–109.

Tangtham, M., 1983, Estimating K- and C- factor in the USLE for Hill evergreen forest in northern Thailand, *Malama Aina Conf, 16—22 Jan., 1983, Honolulu,* Abstract. 59–60.

Vanelslande, A., Rousseau, P., Lal, R., Gabriels, D., and Ghuman, B.S., 1984, Testing the applicability of soil erodibility nomogram for some tropical soils, *IAHS Publ.*, **144**, 463–467.

Verheye, W., 1974, Nature and evolution of soils developed on the granite complex in the subhumid tropics (Ivory Coast). I. Morphology and Classification, *Pedologie*, **24**, 266–282.

Vine, H., 1954, Latosols of Nigeria and some related soils *Proc. Inter-African Soils Conf. Leopoldville*, pp. 295–308.

Vine, H., Esteron, V.J., Montgomery, R.F., Smyth, A.J., and Moss, R.P., 1954, Progress of soil survey in South Western Nigeria, *Proc. Inter-African Soils Conf., Leopoldville*, pp. 211–236.

Vine, P.N., Lal, R., and Payne, D., 1981, The influence of sands and gravels on root growth of maize seedlings, *Soil Sci.*, **131**, 124–129.

Walsh, R.P.D., 1980, Runoff processes and models in the humid tropics, *Zeitschrift fur Geomorphologie*, Supplementband **36**, 176–201.

Warkentin, B.P., and Maeda, T., 1980, Physical and mechanical characteristics of Andisols, p. 281–302. in B.K.G. Theng (ed.), *Soils with variable charge*, DSIR, Lower Hutt, New Zealand.

Wilkinson, G.E., 1975, Rainfall characteristics and soil erosion in the rainforest area of Western Nigeria, *Expl. Agric.* **11**, 247–255.

Wilkinson, G.E., and Aina, P.O., 1976, Infiltration of water into two Nigerian soils under secondary forest and subsequent arable cropping, *Geoderma*, **15**, 51–59.

Williams, J.B., 1979, Soil water investigations in the Gambia. Land Resource Development Centre, ODA, U.K. Tech.Bull 3, 183 pp.

Wolf, J.M., and Drosdoff, M., 1976a, Soil water studies in Oxisols and Ultisols of Puerto Rico: I. Water movement, *J. Agric. Univ. P. Rico*, **60**, 375–385.

Wolf, J.M., and Drosdoff, M., 1976b, Soil water studies in Oxisols and Ultisols of Puerto Rico: II. Moisture retention and availability, *J. Agric. Univ. P. Rico*, **60**, 386–394.

Yadav, M.R., and Saxena, G.S., 1973, Effect of compaction and moisture content on specific heat and thermal capacity of soils, *J. Indian Soc. Soil Sci.*, **21**, 129–132.

Zusevics, J.A., 1980, Seasonal changes of permanent wilting coefficient in some selected tropical soils, *Communications in Soil Science and Plant Analysis*, **II**, 843–850.

Zusevics, J.A., 1981, The potential of plant nutrient leaching from light tropical soil during the seasonal changes, *Soil Crop Sci. Soc. Florida*, **40**, 25–30.

Chapter 5
Variability In Soil Physical Properties

I	DEFINITIONS AND BASIC CONCEPTS	179
II	DIVERSITY OF TROPICAL VERSUS TEMPERATE REGION SOILS	182
III	VARIABILITY IN SOIL PHYSICAL PROPERTIES	182
	A. Particle Size Distribution and Mechanical Properties	182
	B. Soil Moisture	187
	1. Neutron Moisture Probe	188
	(i) Effects of soil texture	192
	(ii) Gravel content	193
	(iii) Bulk density	194
	2. Gravimetric Moisture Content Measurements	197
	C. Hydrological Properties	197
	D. Soil Temperature and Microclimate	201
	E. Run-off and Erosion Processes	204
	F. Factors Responsible for Stochastic or Random Variability	205
	(i) Vegetation cover	205
	(ii) Effects of soil animals	211
	(iii) Slope and micro-relief	212
	(iv) Man's activity	212
IV	MICROVARIABILITY IN PHYSICAL VERSUS CHEMICAL SOIL PROPERTIES	214
V	FIELD RESEARCH AND ANALYTICAL METHODS	215
	1. Replications and Design of Field Experiments	215
	2. Understanding Sources of Variability	216
	3. Crop Performances and Soil Properties	217
VI	SOIL SAMPLING AND STATISTICAL ANALYSES OF VARIABLE SOIL DATA	217
	A. Geostastical Techniques	219
	B. Kriging	219
VII	SUMMARY AND CONCLUSION	220

The diversity or lateral variability in soil physical properties is of common occurrence, and its magnitude, trends, and causes should be comprehended and evaluated for assessment of the productive potential and management constraints of soils in the tropics. Land assessment involves developing quantitative relationships between crop growth and soil properties. An understanding of soil heterogeneity and its effects on the precision of quantitative determinations of its properties is essential. Tremendous progress has recently been made in quantitative physical theory and analytical techniques for evaluating soil – water phenomena, and in basic and applied problems of soil physics, soil mechanics and ground-water hydrology. These techniques are, however, developed for controlled environments, and are based on mathematical solutions to deterministic models. An understanding of soil diversity is essential to apply these techniques to large areas in the field.

I. DEFINITIONS AND BASIC CONCEPTS

Many recent reviews have outlined the basic concepts of soil diversity (Beckett and Webster, 1971; Miehlich, 1976; Wright and Wilson, 1979; Philip, 1980). Miehlich (1976) defined a 'homogeneous' soil to be the one whose smallest identifiable fractions possess identical properties. By this definition, all natural soils are heterogenous. The degree of heterogeneity or diversity, however, varies among soils and the parameters being evaluated. An important question to be considered, therefore, is how much variation to expect within a 'pure' or a 'homogeneous' soil body or a mapping unit. At present, there is no criterion available on how much variability can be accepted within a mapping unit. The magnitude of this acceptable level of variability also depends on the scale of observations and the properties to be evaluated. Although Jansen and Arnold (1976) and other researchers have attempted to define ranges of soil characteristics, but found it difficult because some soil physical properties are more variable than others. For example, a variation of a few per cent in porosity and pore size distribution may lead to a variation in hydraulic conductivity by a few orders of magnitude. For precise measurement of these properties, the unit of observation is usually a ped or an undisturbed soil core. From the point of view of the assessment of soil physical properties in relation to crop growth, the scale of operations is the mapping unit or soil series within a toposequence.

There are various categories of soil diversity or field heterogeneity, as have been discussed by Freeze (1975), Webster (1976), Wilding and Drees (1978) and Philip (1980). There are two commonly used systems of classifying the variability: (a) the nature of variability, and (b) the scale of variability.

The nature of variability refers to whether or not the diversity in soil properties is explainable and predictable on the basis of landform, parent material and other factors. With this criteria in mind, the soil variability is commonly classified into the following two categories:

(i) Deterministic or systematic variability refers to both spatial and temporal diversity in relation to known variables, i.e., landform, hydrology, parent material, etc. For example, soil physical properties vary due simply to differences in the depth and duration of the perched water table. The hysteresis properties in pF curves and other characteristics are also influenced by the position in the landscape.

(ii) Stochastic or random variation is complex, irregular and unidentifiable diversity in soil properties. It is not easily comprehended. Freeze (1975) defined random variation as 'non-uniform homogeneity' because the randomness in soil property is considered to be stationary and independent of position in the landscape. Some of the random variation may be classified systematic when their causative factors are understood and predictable. Diversity in soil properties due to differential lithology, erosion, weathering and errors due to sampling is referred to as random variability.

The second criterion to classify soil variability is on the basis of the scale of observations. It is well established that the degree of systematic variation depends on the scale of observations, and it generally increases with an increasing scale factor (Fig. 5.1). The nature and magnitude of increase, however, differs among soils and depends on the property to be evaluated. Reynolds (1974) observed that variability in soil moisture content increased with increasing size of the area sampled (Fig. 5.2). The variability could, however, be grouped into three arbitrary groupings on the basis of the coefficient of variation and the sample size required to estimate true mean to ±5 at 0.05 probability level. These groupings were:

	Size (m²)	Coeff. of variation	Sample size
(i)	1–1000	≤10	<10
(ii)	1000–100 000	10–20	10–40
(iii)	100 000–approx. 10 million	>20	>40

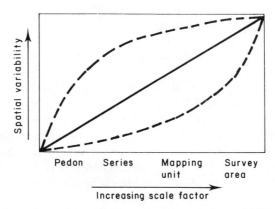

Figure 5.1 Schematic spatial variability reflecting scale factor (Wild and Drees, 1980)

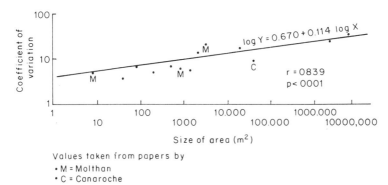

Figure 5.2 Effect of the size of area sampled on variability in soil moisture content (Reynolds, 1974). *Reproduced by permission of Elsevier Science Publishers*

Contrary to these, many authors have observed no effect, or a slight effect, of the scale on the magnitude of soil variability. Ferrari and Vermeulen (1956) reported that the size of sampling had no effect on the magnitude of the sampling error. Moltham (1966) also observed no effect of the sample size on soil moisture variability.

On the basis of scale, soil variability is generally classified into two categories:

(i) Macrovariability refers to the variability that persists even on a larger scale and results in the existence of distinct geographic patterns, or to large variations in soil properties over short distances resulting in different taxonomic groupings (Wambeke and Dudal, 1978).

(ii) Microvariability, as defined by Moormann and Kang (1978) refers to local and often recurrent variations at a small scale. These variations may be permanent or persist through several cropping cycles and affect crop growth. These effects on crop growth are generally more pronounced under low levels of management and their effects on crop growth decrease with increasing inputs.

In addition to the above mentioned systems of classifying soil variability, it is perhaps useful to classify the variability on the basis of the factors responsible and/or the properties affected. For example, the variability may be:

(i) genetic or lithological, caused by the variations in parent material;
(ii) anthropogenic, caused by man's intervention, i.e. cultivation and soil management (tillage methods, fertilizer used, rotation, etc);
(iii) biotic, introduced by the biological factors, e.g. termites, trees, etc;
(iv) hydrologic, caused by seasonal or spatial variations in moisture regime, e.g. perched water table;
(v) physical, variability in physical properties in contrast to that of the chemical characteristics.

Classification of soil variability on the basis of these functional characteristics may be useful in soil management. From the practical point of view, this classification system can be used to identify constraints and to develop appropriate systems of soil management to alleviate the constraints. It is the knowledge and understanding of the soil's microvariability that is important for developing suitable soil management practices for tropical environment. The knowledge of the soil's macrovariability is relevant for land-use planning.

II. DIVERSITY OF TROPICAL VERSUS TEMPERATE REGION SOILS

All soils are variable and the diversity of soils in the tropics is neither an exception nor a unique phenomenon. Soils of the temperate regions are equally diverse and variable although the relative magnitude and importance of different factors and processes may be different. The magnitude of variability within the predominant soils in the tropics (Oxisols, Ultisols, Alfisols and Vertisols) is as great as within the soils that occur predominantly within the temperate latitudes. Misconceptions and ambiguity arise due to the lack of understanding and the paucity of analytical data available for the soils of the tropics.

The variability of temperate zone soils has long been investigated and an impressive body of knowledge has already been accumulated. The bibliography prepared by the Commonwealth Bureau of Soils (1967) lists 57 references from 1943 to 1966 about variability of soils of the temperate region. The available literature on the subject for soils of the temperate zone indicates that the interest in this topic has not yet diminished (Walker *et al.*, 1968a, b; Adams and Wilde, 1976a, b; Banfield and Bascomb, 1976; Bascomb and Jarvis, 1976; Trudgill and Briggs, 1981). The importance of soil variability in the tropics is increasingly being realized, and the diversity of mappable pedological units is now considered a prerequisite in land evaluation and soil classification studies (Beaudou and Collinet, 1977; Oliveira, 1975).

III. VARIABILITY IN SOIL PHYSICAL PROPERTIES

A. Particle size distribution and mechanical properties

Variations in soil particle size distribution with depth is a natural genetic process of horizon differentiation. However, the lateral variations in particle size distribution for each horizon over short distances result in a random or systematic heterogeneity that causes considerable variation in soil physical characteristics. These variations in particle size distribution are by no means unique among soils of the tropics and have been well documented for soils of other regions as well (Collins, 1976; Gajem *et al.*, 1981).

Variability in physical and mechanical properties of soils in West Africa and elsewhere in the tropics is partly caused by the variable concentrations of gravel and skeletal materials. This variability is particularly pronounced in soils derived from parent materials that consist of pre-Cambrian granite and meta-

Table 5.1 Variations in gravel concentrations (%) among five sampling sites at 1.7 m interval on an Alfisol in western Nigeria (Moormann, 1978)

Depth (cm)	Concentration of gravels (2–50 mm) at five sampling locations				
	1	2	3	4	5
0–10	1	4	5	4	2
10–20	16	3	15	4	4
20–50	52	58	40	26	12
50–75	34	37	37	25	26
75–100	18	13	14	15	17

morphic rocks, mainly paragneiss, quartzite and quartz–schist (Smyth and Montgomery, 1962; Collinet, 1969; Leveque, 1969; Riquier, 1969; Segalen, 1969; Santamaria, 1965; Lutz, 1972; Thomas, 1974; Panabokke, 1967; Ahn, 1970). There is a considerable variation, even over short distances, in the concentration, size distribution and thickness of this gravelly horizon (Table 5.1). The presence or absence of a gravel horizon is partially responsible for the variability in soil physical and mechanical properties (Moormann, 1972; Moormann and Kang, 1978). There occur sharp lithological changes in parent materials causing measurable variations in soil mechanical properties.

The data in Table 5.2 represent variations in gravel content of the surface layer of two 5 ha watersheds adjacent to one another. There were 140 samples obtained from watershed A and 110 samples from watershed B. The data indicate a tremendous variation both between and within the watersheds. While watersheds A and B were managed by the no-till system, watershed C was conventionally ploughed and harrowed. Ploughing altered the gravel content by turning over the soil and by bringing the subsoil material to the

Table 5.2 Variations in gravel contents of the surface 0–5 cm layer of three 5 ha watersheds on Alfisols in western Nigeria

Range in gravel concentration (%)	Percentage of samples		
	Watershed A	Watershed B	Watershed C
<5	26.4	86.4	9
5–10	32.1	10.0	28
11–15	13.6	1.8	16
16–20	8.6	0	10
21–26	2.9	1.8	12
26–30	7.9	0	13
31–35	5.7	0	5
36–40	0.7	0	3
>40	2.1	0	4
Total samples	140	110	100

Table 5.3 Variations in profile gravel concentration and soil strength over four locations about 10 m apart (Block BN5, IITA, Ibadan)

Depth (cm)	Location 1				Location 2				Location 3				Location 4			
	Gravel (%)		D_b (C)	P_R (D)												
	Weight (A)	Volume (B)			A	B	C	D	A	B	C	D	A	B	C	D
0–10	2.2	1.5	1.60	3.9	0.8	0.7	1.55	3.0	4.2	2.3	1.47	2.7	8.2	5.9	1.55	2.0
10–20	1.8	0.7	1.65	>4.5	0.6	1.1	1.74	4.4	6.2	3.1	1.41	2.6	114.7	66.5	1.33	>4.5
20–30	3.7	2.7	1.70	>4.5	4.2	2.7	1.62	4.3	64.8	38.7	1.22	2.6	116.6	69.2	1.50	>4.5
30–40	7.6	5.0	1.65	3.8	13.3	8.5	1.53	4.0	135.2	54.6	0.83	3.6	84.3	56.4	1.55	>4.5
40–60	29.0	18.5	1.63	>4.5	7.9	1.5	1.59	3.5	87.5	49.4	1.36	>4.5	73.9	47.8	1.49	>4.5
60–80	9.4	2.3	1.60	4.4	10.3	6.7	1.68	3.5	52.8	26.9	1.22	>4.5	41.3	29.3	1.63	>4.5
80–100	50.5	26.1	1.60	>4.5	37.0	28.7	1.64	>4.5	50.2	32.5	1.59	>4.5	33.2	25.1	1.76	>4.5
100†	89.9	53.7	1.31	2.0	15.2	11.6	1.65	4.5	44.3	28.1	1.56	4.5	29.1	21.2	1.76	4.5

A, Gravel weight (%) = (gravel weight/<2 mm fraction weight) × 100.
B, Gravel volume (V) = (gravel volume/<2 mm fraction volume) × 100.
C, D_b = dry bulk density (g cm^{-3}) corrected for gravels.
D, P_R = penetrometer resistance (kg cm^{-2}).

surface. Similarly, even slight variations in clay content of a predominantly sandy soil results in large changes in moisture retention and transmission properties.

In addition to the variations in gravel content, there also occur considerable variations in the texture of the fine earth fraction. Babalola (1978) observed coefficients of variation in sand, silt and clay content on a 91.6 ha farm near Ibadan, Nigeria, of 3.2, 16.5 and 34.0 per cent for 0–15 cm depth; 5.1, 28.5 and 39.9 per cent for 15–45 cm depth; 17.6, 18.2 and 46.6 per cent for 45–75 cm depth; 21.5, 24.2 and 41.7 per cent for 75–105 cm depth and 20.5, 28.0 and 35.3 per cent for the 105–135 cm depth, respectively. The corresponding coefficients of variation for the gravel content were 10.0, 90.0, 77.3, 42.7 and 60.4 per cent, respectively. The coefficient of variation of all soil constituents was less for 0.34 ha than for the 91.6 ha farm size.

Variations in particle size distribution and gravel concentrations result in diversity in bulk density, penetrometer resistance and related mechanical properties of the soil profile (Table 5.3). Frequently analysis of bulk density on the three watersheds discussed earlier is shown in Table 5.4. As with the gravel content, the bulk density varies widely even over short distances, and quite often has a skewed distribution. The frequency distribution of bulk density of the ploughed watershed (B) is also different than that of the no-till (A) watershed. In addition to random variation, the management system adopted has also caused variation in bulk density of the surface layer. A comparison of the corrected and uncorrected (corrected for variations in gravel content) bulk density data shown in Table 5.4 indicates the effects of factors other than gravels on variation in bulk density of the surface horizon. As a matter of fact, the variability in bulk density of a non-gravelly soil is as much or more than a gravelly soil.

Table 5.4 Variations in bulk density of the surface 0–5 cm layer of three 5 ha watersheds in western Nigeria

Range in bulk density (g cm^{-3})	Percentage of samples					
	Watershed A		Watershed B		Watershed C	
	I*	II	I	II	I	II
<1.1	0.0	0.7	0.0	0.9	0	0
1.1–1.2	9.3	25.7	5.5	9.1	2	25
1.21–1.3	20.0	30.0	9.0	12.7	6	33
1.31–1.4	31.4	24.3	28.2	35.5	13	37
1.41–1.5	22.9	12.9	27.3	29.1	24	3
1.51–1.6	13.6	5.0	17.3	12.7	39	2
1.61–1.7	9.3	0.7	12.7	0.0	13	0
>1.7	2.1	0.7	0.0	0.0	3	0
Total samples	140	140	110	110	100	100

*I, Uncorrected for gravel contents; II, corrected for gravel contents.

Table 5.5 Effects of soil texture on density probe calibration (modified from Lal, 1974). *Reproduced by permission of Williams & Wilkins Co.*

Soils*	Regression equations
A	Density count ratio $(Y) = 0.620-0.130$ Dry Density (x)
B	$Y\ 0.769-0.194x$
C	$Y\ 1.115-0.484x$
D	$Y\ 1.127-0.487x$
E	$Y\ 0.665-0.154x$

* These soils are the same as those described in Table 5.4.

Variations in bulk density measurements are also caused by the methods of measurement adopted. In general, the larger the sample size the less the experimental error. The sampling error is often large for the 'core' or the 'excavation' methods. Measurements of bulk density by probes that use radioactive material are also influenced by the variability in soil constituents. Variations in texture and particle size distribution influence the calibration of the equipment using radioactive materials. Lal (1974) observed that the calibration of density probe, similar to that of the neutron moisture meter, is affected by the variability in soil texture. In general, soils of coarse texture have greater slope of the density–count ratio plot than fine-textured soils. Analyses of the regression equations developed for five tropical soils (Table 5.5) indicate that the count ratio for bulk density ranged from 0.345, 0.412 to 0.450 for soils C and D, A and E, and B, respectively for a bulk density of 1.6 g cm^{-3}

In addition to the texture of the fine fraction (> 2 mm), the presence of gravels also affects the calibration of the depth density probe (Lal, 1979). The negative slope of the calibration curve is generally more for soils with a high gravel content. For the 8–15 mm gravel size, Lal observed the slopes of

Table 5.6 Regression equations relating dry density with density count ratio for different soil–gravel mixtures of 4–8 mm and 8–15 mm gravel size fraction (After Lal, 1979). *Reproduced by permission of Williams & Wilkins Co*

Gravel content (%)	Regression equation	R^2
	(a) 4–8 mm size fraction	
0	DD = 2.022 − 0.859 DCR	0.55*
20	DD = 3.238 − 0.093 DCR	0.72**
40	DD = 2.177 − 1.303 DCR	0.54**
60	DD = 1.977 − 1.025 DCR	0.61*
	(b) 8–15 mm size fraction	
0	DD = 2.338 − 1.430 DCR	0.81**
40	DD = 2.073 − 1.067 DCR	0.73**
60	DD = 2.346 − 1.669 DCR	0.79**

DD, dry bulk density; DCR, density count ratio.

regression lines relating dry density with the density count ratio to be −0.86, −1.43. −1.07, and −1.67 for gravel concentrations of 0, 20, 40 and 60 per cent, respectively. Furthermore, calibration curves also vary widely among soils containing similar gravel content but containing gravels of different size fractions (Table 5.6).

Variability in bulk density and mechanical properties generally decreases with duration of cultivation. Variability is considerably more immediately after the land has been developed from a forest than when a soil has been under cultivation for, says, 10 years. Similar observations on variability and its relation to duration of cultivation are also reported for soils of the temperate region (Bätz et al., 1972; Northrup and Boyle, 1975; Hartge et al., 1978; Cassel and Nielson, 1979; Gajem et al., 1981).

B. Soil moisture

Soil moisture affects crop growth, and the variations in crop yield have often been related to variations in soil moisture content (Fig. 5.3) (Moormann and Kang, 1978). Variability in soil moisture content for temperate region soils has also been reported for the Soviet Union by Oreshkina (1976), Zablotskii and Karpacehevskii (1977), Varazashvili et al. (1976) and Polyakov (1976); for the

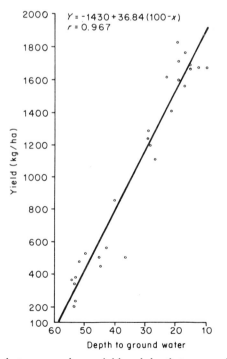

Figure 5.3 Relation between soybean yield and depth to ground water, 6 days after planting (Moormann and Kang, 1978)

United States by Sartz (1972), Nielsen et al. (1973), Reynolds (1974), Cassel and Bauer (1975) and Bell et al. (1980); and for Israel by Ben-Asher (1979).

There are a few studies from the tropics relevant to the variability in soil-water content, especially in relation to various methods of its determination. Over and above the variability caused by soil constituents and the micro-relief, the variation in soil moisture also depends on biotic and anthropogenic factors. Variation in soil moisture content is caused by variation in vegetation, mulch cover, water uptake and the root system distribution. This implies that in addition to the mean wetness of the soil, variability in soil moisture content depends on other factors. The soil moisture variability is an integrated effect of several factors and their spatial variability.

Babalola (1978) reported a mean coefficient of variability in volumetric moisture content on a 91 ha field near Ibadan to be 8.5, 14.3, 18.6, 19.8, 22.2, 20.7 and 27.1 per cent for corresponding soil moisture potential of 0, −20, −40, −60, −200 and −300 cm, respectively. The variability, as determined by the coefficient of variation, also of course depends on the degree of soil wetness. The coefficient of variation increases with decreasing soil moisture content. For the same moisture regime, the degree of variation is often more in soil moisture content than in soil moisture potential. It is advisable, therefore, to determine soil moisture potential, because it is influenced less by the variation in soil constituents than is the soil moisture content. In spite of that, some soils exhibit large variability in their pF curves (Table 5.7).

Similar to soil bulk density, the degree of variation in soil wetness is also caused by the method adopted for its determination. The most commonly used methods are:

1. Neutron moisture probe

In many soils of the tropics, the determination of soil moisture content by the neutron probe method gives erroneous results because of the difficulties in obtaining a reliable calibration curve. Variations in subsoil content of gravel, clay, and other constituents cause considerable variations in the calibration

Table 5.7 Spatial variations in moisture retention characteristics of surface (0–10) layer measured on a 360 m transect along a toposequence. Samples were taken every one meter apart

	Soil moisture retention (%) at different potentials (bars)		
	0.1	1	15
Minimum	5.1	2.7	2.1
Maximum	58.8	32.1	26.4
Mean	32.0	17.4	14.3
Standard deviation	26.9	14.7	12.2
cv (%)	84.1	84.5	85.3

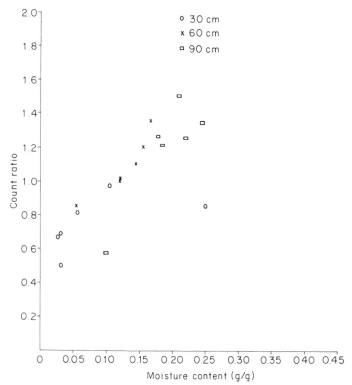

Figure 5.4 Neutron moisture probe calibration of a gravelly Alfisol for 30, 60, and 90 cm depths at 1-metre distance from the sampling pit (unpublished data, Lal, 1983)

curve even over short distances. Consequently, the use of a single calibration curve for these variable soils even over short distances within the same toposequence can lead to a considerable error.

The data in Figs. 5.4 to 5.8 show that the calibration curves are not only different for different horizons but also vary for each horizon even over a short, 1–metre distance. These calibration curves are obtained for one soil profile (cp 108) at the International Institute of Tropical Agriculture (IITA) research farm, Ibadan, Nigeria. The neutron moisture probe was calibrated for five locations in a line, exactly 1 m apart. At each of the five locations, calibration curves were obtained for three depths, 30, 60 and 90 cm below the soil surface. There was a separate calibration curve for each depth for each of the five locations, giving a total of 15 separate calibration curves.

Regression analyses relating moisture content with the count ratio are shown in Table 5.8. The regression equations were different for the 15 calibration points, and there was a large range in the value of correlation coefficients (from 0.01 to 0.94). There was a large variability in the gravel content and gravel size distribution among these 15 calibration points, that resulted in widely differing calibration curves.

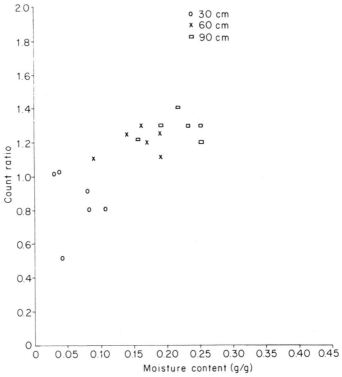

Figure 5.5 Neutron moisture probe calibration of a gravelly Alfisol for 30, 60 and 90 cm depths of 2-metre distance from the sampling pit (unpublished data, Lal, 1983)

Table 5.8a Regression equations relating moisture content with the neutron count ratio for five locations and three depths at each location (IITA date for CP 108)

Distance from CP 108 (m)	Depth (cm)	Slope (m)	Intercept (b)	Correlation coefficient (r)
1	30	0.008	0.733	0.55
	60	0.064	0.246	0.94
	90	0.014	1.033	0.35
2	30	0.041	0.533	0.85
	60	0.006	1.153	0.31
	90	0.001	1.308	0.01
3	30	0.010	0.862	−0.22
	60	0.014	0.976	0.42
	90	0.018	0.924	0.53
4	30	0.045	0.674	0.88
	60	0,009	1.044	0.43
	90	0.010	1.025	0.53
5	30	0.097	0.021	0.94
	60	0.013	0.855	0.21
	90	0.020	0.786	0.32

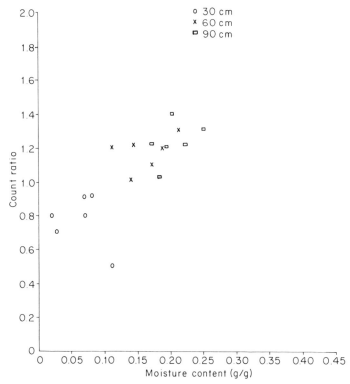

Figure 5.6 Neutron moisture probe calibration of a gravelly Alfisol 30, 60, and 90 cm depths at 3-metre distance from the sampling pit (unpublished data, Lal, 1983)

Table 5.8b Variation of neutron probe calibration with depth for two soils in the Philippines (Hundal and DeDatta, 1981). *Reproduced by permission of IRRI*

Soil depth (cm)	Regression equation	R^2
Profile 1		
15	$Y = 0.306 + 0.027\ C_v$	0.94**
30	$Y = 0.869 + 0.018\ C_v$	0.84**
45	$Y = 0.702 + 0.021\ C_v$	0.84**
60	$Y = 0.822 + 0.018\ C_v$	0.89**
75	$Y = 0.867 + 0.018\ C_v$	0.82**
90	$Y = 1.295 + 0.011\ C_v$	0.94**
105	$Y = 0.874 + 0.017\ C_v$	0.84**
Profile 2		
15	$Y = 0.616 + 0.021\ C_v$	0.95**
30	$Y = 1.258 + 0.013\ C_v$	0.70**
45	$Y = 1.248 + 0.011\ C_v$	0.65**
60	$Y = 1.395 + 0.008\ C_v$	0.59*
75	$Y = 1.314 + 0.012\ C_v$	0.56*
90	$Y = 1.271 + 0.009\ C_v$	0.94**
105	$Y = 1.346 + 0.006\ C_v$	0.48*

Y = Count ratio, C_v = Volumetric moisture content.

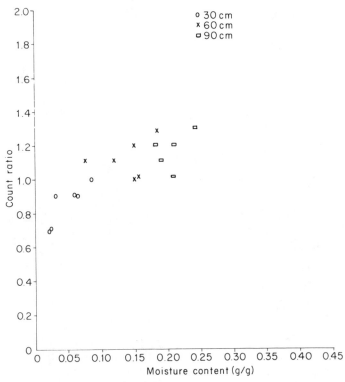

Figure 5.7 Neutron moisture probe calibration of a gravelly Alfisol for 30, 60, and 90 cm depths at 4-metre distance from the sampling pit (unpublished data, Lal, 1983)

(i) Effects of soil texture: In addition to soil moisture, neutron thermalization is also caused by bonded hydrogen and other neutron capturing elements in the soil. The calibration curve is drastically affected by differences in texture. For example, variations in calibration curves even over short horizontal and vertical distances reported above are due to differences in texture and density (Lal, 1974). The data in Fig. 5.9 show a highly significant effect of soil texture on neutron moisture probe calibration. The coarse textured soils (A, C and D) have greater slopes as compared to the fine textured soils (B and E) (Table 5.9). In addition to mechanical properties, differences in soil chemical constituents also affect the calibration and the precision of the determination of soil wetness. The steep slope is that of soil A, consisting of 71.5 per cent gravels and sand fraction, followed by that of soil D with 86.8 per cent gravels and sand fraction but containing a higher quantity of extractable Fe in the soil. The count ratio intercepts are higher for fine textured soils (B and E) because of their clay and extractable Fe content. In general, the higher the clay and Fe content the greater the intercept. For example, soil E, with a clay content of 63.9 per cent and extractable Fe content of 2.72 per cent, has an intercept of 0.372 in

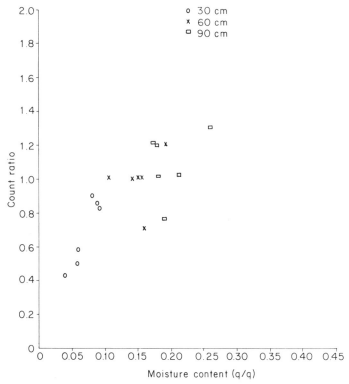

Figure 5.8 Neutron moisture probe calibration of a gravelly Alfisol for 30, 60, and 90 cm depths at 4-metre distance from the sampling pit (unpublished data, Lal, 1983)

comparison with 0.465 for soil B, with 50.2 per cent clay and 8.04 per cent extractable Fe. The magnitude of the intercept is determined by the quantity of bond water retained even after drying the sample at 105°C for 24 hours (Babalola, 1972; Lal, 1974). The relatively low organic carbon content in most cultivated soils of the tropics, and especially in the subsoil horizons, does not seem to have a marked effect on the neutron moisture probe calibration.

(ii) Gravel content: Variable concentrations of gravels have significant effects on alterations in neutron moisture probe calibrations (Lal, 1979). The data in Fig. 5.10 show a significant effect of gravel concentration on the moisture probe calibration of gravel–soil mixtures. The neutron count ratios for moisture content of 0.04 cm^3 cm^{-3} and for 4–8 mm gravel size are 0.19, 0.45, 0.48 and 0.42 for 0, 20, 40, and 60 per cent gravel, respectively. For a high moisture content of 0.20 cm^3 cm^{-3}, similar ratios are 0.71, 0.88, 0.97 and 1.07 for 0, 20, 40, and 60 per cent gravels. Similar variations in neutron count ratios are observed for other moisture contents.

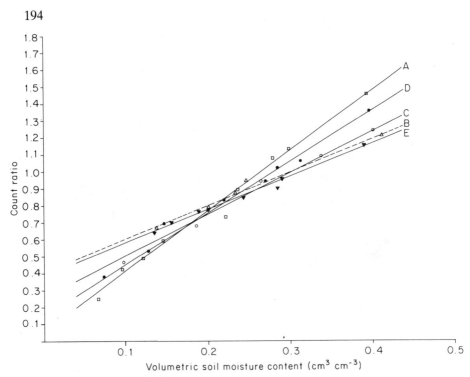

Figure 5.9 Effect of soil texture on neutron moisture probe calibration (Lal, 1974). *Reproduced by permission of Williams & Wilkins Co.*

Regression equations relating neutron count ratio to volumetric moisture content for different concentrations of 4–8 mm and 8–15 mm gravel size fractions are shown in Table 5.10. The data show a decrease in the slope of the regression line with increased gravel concentration. For example, the slope coeffficients for volumetric moisture content are 0.374, 0.331 and 0.252 for 4–8 mm gravel size compared with 0.306, 0.278, and 0.277 for 8–15 mm gravel size fraction, respectively, for 20, 40 and 60 per cent gravel. In addition to gravel concentration, the gravel size also affects neutron moisture probe calibration through its effect on thermalization of neutrons (Lal, 1979). The data in Fig. 5.10 show that for the 60 per cent gravel–sand mixture, the slope of the regression equations significantly increased with increased gravel size: 0.226, 0.239 and 0.256 for 4–8 mm, 8–15 mm, and 15–40 mm size fractions, respectively.

(iii) Bulk density: Neutron moisture probe calibration is also affected by the variation in bulk density among horizons (Lal, 1974). The data in Table 5.11 show that for 20 per cent volumetric moisture content, the count ratio ranges from 0.568, 0.643, 0.718, 0.793, 0.868 to 0.943 for bulk densities of 1.0, 1.2, 1.4, 1.6, 1.8 and 2.0 Mg m^{-3}, respectively. The effect of bulk density on neutron

Table 5.9 Textural properties of soils in relation to the neutron moisture probe calibration shown in Fig. 5.8 (Adapted from Lal, 1974). *Reproduced by permission of Williams & Wilkins Co.*

Soil	Mechanical analysis (%)			Organic Carbon (%)	Extractable Fe (%)	Soil moisture content at 105°C (gg^{-1})	Regression of count ratio with volumetric moisture content*		
	Gravels	Sand	Silt	Clay				r	m	b
A	30.5	41.0	13.4	15.2	1.88	4.64	0.0098	0.994	3.607	0.044
B	18.0	14.2	17.6	50.2	0.35	8.04	0.0670	0.984	1.788	0.465
C	0.8	73.1	14.0	12.1	1.30	8.80	0.0139	0.994	2.508	0.229
D	59.7	27.1	5.9	7.3	0.84	5.64	0.0533	0.988	2.839	0.200
E	8.2	16.3	11.6	63.9	0.16	2.72	0.0412	0.992	1.973	0.372

* r = correlation coefficient; m = slope coefficient; b = count ratio intercept.

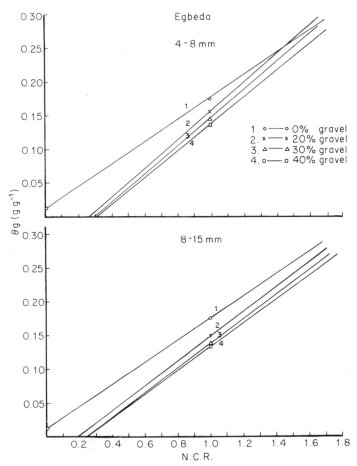

Figure 5.10 Effects of gravel concentration on neutron moisture probe calibration (Lal, 1979). *Reproduced by permission of Williams & Wilkins Co.*

Table 5.10 Regression equations relating volumetric moisture content with neutron count ratio for variable gravel concentrations of two size fractions (modified from Lal, 1979). *Reproduced by permission of Williams & Wilkins Co.*

Gravel content (%)	4–8 mm size fraction†			8–15 size fraction		
	r	m	b	r	m	b
0	0.81**	0.305	−0.019	0.81**	0.305	−0.019
20	0.97**	0.347	−0.128	0.99**	0.306	−0.084
40	0.93**	0.331	−0.119	0.99**	0.278	−0.072
60	0.98**	0.252	−0.065	0.97**	0.277	−0.079

† r = correlation coefficient; m = slope; b = intercept.

Table 5.11 Effect of soil bulk density on neutron moisture calibration (Adapted from Lal, 1974). *Reproduced by permission of Williams & Wilkins Co.*

Soil bulk density ($g\ cm^{-3}$)	Regression equation
1.0	Count ratio (Y) = 0.126 + 2.212 Volumetric moisture content (x)
1.2	$Y = 0.201 + 2.212x$
1.4	$Y = 0.276 + 2.212x$
1.6	$Y = 0.351 + 2.212x$
1.8	$Y = 0.426 + 2.212x$
2.0	$Y = 0.501 + 2.212x$

moisture meter calibration varies among soils, and is more pronounced in some than in others.

The determination of soil wetness by the neutron probe method is, therefore, influenced by the neutron moisture meter calibration. The latter is drastically influenced by the variations in texture, gravel content, gravel-size distribution, soil organic matter and variation in Fe and other chemical constituents. The moisture content determinations by this equipment for field scale hydrological investigations can, therefore, be highly variable unless a prior calibration is obtained for each soil and each horizon. The neutron moisture meter can, however, be useful in indicating the depth of wetting or drying fronts, and in determining the zone of water depletion.

2. *Gravimetric moisture content measurements*

Highly variable results are also obtained for the gravimetric determinations of soil moisture content. In addition to differences in vegetation cover and local relief, the presence of variable concentrations of gravels and coarse fractions also influence the degree of soil wetness as expressed on an oven-dry basis. Tensiometric measurements for soil moisture-potential determinations are subject to similar variations. The coefficient of variations in the gravimetric determinations of soil moisture content can be drastically reduced by making appropriate corrections for the gravel content.

Measurement of soil wetness by other indirect methods is equally subject to variations introduced by the technique used. For example, psychrometric measurements are influenced by variations in temperature. Measurements by gypsum blocks are affected by the contact, salt content, and the length of time the blocks have been left in the soil.

C. Hydrological properties

Spatial variability in hydrological properties is often more pronounced and striking than in textural and mechanical properties. A slight variation in textural and mechanical properties can result in a variability in hydrological properties by several orders of magnitude. Seemingly uniform land areas exhibit large variations in hydraulic conductivity even though variations in texture and mechanical properties are less. That is why an advance knowledge about the heteogeneity of soils on experimental areas with respect to soil water transmission and storage is desirable when planning studies of water use and water-balance (Sartz, 1972). Furthermore, high spatial variability in hydrological characteristics (e.g. water infiltration) makes it difficult to develop appropriate models of water-balance and run-off processes (Sharma et al., 1979).

The variability in water flux and hydrological properties of soils in the temperate zone and its consequences have been documented by many researchers (Carvallo et al., 1976; Rao et al., 1977; Peck et al., 1977; Springer and Gifford, 1980; Russo and Bresler, 1981a; b; Lascano and Van Bavel, 1982). Russo and Bresler (1982) reported from Israel that for a 0.8 ha field, performing 90 measurements to estimate mean values of hydraulic properties, the error was in the range of 7 to 46 per cent. Sisson and Wierenga (1981) observed that a large fraction of the water passes through only a small fraction of the plot area, because of the spatial variability of the steady-state infiltration rate.

In contrast to the soils of the temperate region, the spatial variability in hydrological properties of soils in the tropics has not been adequately described. Gurovich (1982) and Gurovich and Stern (1983) have evaluated field spatial variability in soil-water sorptivity in relation to the design of an irrigation system in Chile. Babalola (1978) reported spatial variability of soil-water properties in some soils of southern Nigeria. The coefficient of variability (cv) of the five tensiometers at each depth located on one single soil type ranged between 20 and 26 per cent. Although the cv in sand, silt and clay content did not exceed 42 per cent, that of the saturated hydraulic conductivity was as much as 178 per cent. Babalola reported that within each particular layer, the range in hydraulic conductivity values among the six cores taken from each layer was of the order of 10 to 150-fold. In comparison, Sartz (1972)

Table 5.12 Variability of soil-water content ($cm^3\ cm^{-3}$) and hydraulic conductivity ($cm\ day^{-1}$) measured on a 0.34 ha and a 91.6 ha size field on the 11th day after saturation (Modified from Babalola, 1978). *Reproduced by permission of Williams & Wilkins Co.*

Depth (cm)		91.6 ha		0.34 ha	
		Soil-water	Conductivity	Soil-water	Conductivity
45	x	0.155	0.081	0.159	0.116
	δ	0.036	0.093	0.011	0.086
	cv	23.23	114.81	6.92	74.14
75	x	0.209	0.194	0.216	0.246
	δ	0.023	0.110	0.010	0.339
	cv	11.00	56.70	5.63	137.8

Table 5.13 Coefficient of variability (cv) in accumulative water infiltration in two adjacent watershed of about 3 ha each (unpublished data, Lal, 1981)

Time (min)	Watershed No. 1			Watershed No. 3		
	Mean	δ	cv (%)	Mean	δ	cv (%)
10	7.1	2.8	40.1	13.9	10.4	74.8
20	11.7	4.3	36.8	26.1	20.8	79.7
30	17.3	7.1	41.0	32.9	24.6	74.8
45	21.7	8.2	37.8	44.7	33.7	75.4
60	25.1	9.3	37.1	56.0	44.2	78.9
90	31.7	12.5	39.4	77.2	66.9	86.7
120	36.7	14.8	40.3	95.0	83.1	87.5

reported a cv of 253 per cent for some soils in the United States. A comparison of the variability of soil-water content and hydraulic conductivity for a 0.34 ha and a 91.6 ha field is shown in Table 5.12. The variability in hydraulic conductivity was greater for the larger size.

There are also variations in water infiltration rate as monitored with the double ring infiltrometer. A very high cv is often observed by using this technique. The data in Table 5.13 show accumulative infiltration monitored on two adjacent watersheds of about 3 ha each. The cv on six separate measurements is about 40 per cent in one and about 80 per cent in the other. The infiltration rate over the 30 ha watershed of an Ultisol under a forest cover near Benin in southern Nigeria was found to be log-normally distributed (Fig. 5.11). The variability was random and was not related to soil bulk density (Fig. 5.12). The surface soil of the minimum and the maximum infiltration rate had the same bulk density.

The variability in infiltration determined on large watersheds generally changes with the duration of cultivation. This change may lead to an increase or a decrease in variability depending upon the change in relative magnitude of the mean infiltration and its standard deviation. The data in Table 5.14 show accumulative infiltration and its variability over a period of 5 years when the watershed was cultivated to maize with a no-till system. The accumulative infiltration decreased with the cultivation duration, and the mean accumulative infiltration decreased even more than the standard deviation. The cv, therefore, increased between 1976 and 1979. The cv for 1980, however, was less than that of 1976, and decreased even further in 1981. By using the cv, it is, therefore, difficult to conclude whether the variability decreased with the length of time the land had been under cultivation. What changed were the mean and the standard deviation.

Variability in water infiltration characteristics is also related to some biotic factors, e.g. activity of soil animals, including termites and earthworms. The infiltration rate in the vicinity of an abandoned termite mound is much lower than that of the surrounding soil. The burrowing activity of rodents and other

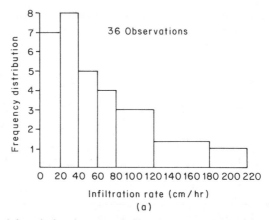

Figure 5.11a Spatial variation in water infiltration rate of an Ulfisol over a 30-ha watershed near Benin City, southern Nigeria (Ghuman and Lal, 1985)

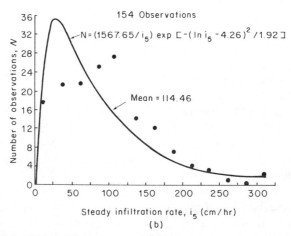

Figure 5.11b The infiltration rate over the forested watershed (Okomu) was lognormally distributed (Ghuman and Lal, 1985)

soil animals can have a positive or the negative effect on the infiltration rate depending whether the burrow opening is in contact with the positive water pressure or not. These aspects are discussed in more details in appropriate chapters dealing with earthworms, termites, and soil animals.

The effects of pre-clearing vegetation on infiltration rate are shown in Figs. 5.13, 5.14 and 5.15. The data indicate more favourable and rapid infiltration under secondary forest than under plantation or tropical root crops. The low infiltration under cocoa, cola etc is probably due to the compacting effect of farm labour during fruit picking and other management operations.

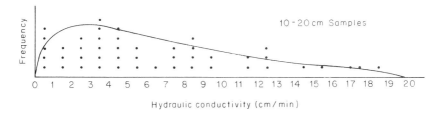

Figure 5.11c Spatial variation in saturated hydraulic conductivity of a newly cleared Alfisol at Ibadan (Unpublished data of Lal and Cummings)

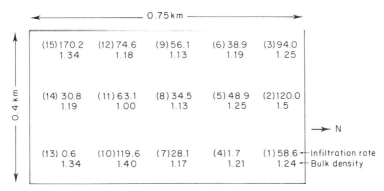

Figure 5.12 The infiltration rate over the watershed was not related to soil bulk density. The figure below the infiltration rate is that of dry soil bulk density (Ghuman and Lal, 1985)

D. Soil temperature and microclimate

Variability in hydrometeorological parameters is as pronounced as that of soil physical properties, and many researchers have emphasized the importance of the representative measurements of these parameters (Scharringa, 1976). As a matter of fact, the seasonal variability in many physical properties can partially be a response to corresponding variability in the magnitude and balance of the

Table 5.14 Change in variability in accumulative water infiltration (cm) on a 5 ha no-till watershed with the duration of cultivation (Calculated from Lal, 1984). *Reproduced by permission of Elsevier Science Publishers*

Time (min)	1976			1978			1979			1980		
	x†	δ	cv	x†	δ	cv	x†	δ	cv	x†	δ	cv
10	10	3.9	39.0	5.0	2.3	46.0	5.5	3.0	54.5	3.6	1.6	44.4
20	17.5	7.4	42.3	9.0	5.2	57.8	9.0	5.5	61.1	5.0	2.0	40.0
30	25.0	10.3	41.2	13.0	6.4	49.2	11.5	6.9	60.0	6.0	2.1	35.0
45	35.0	14.6	41.7	18.0	9.5	52.8	15.0	9.6	64.0	—	—	—
60	43.4	18.8	43.0	22.0	11.7	53.2	17.8	11.4	63.3	8.7	3.7	42.5
90	62.4	26.1	41.8	31.0	15.9	51.3	23.5	15.9	67.7	10.5	4.1	39.0
120	76.1	31.9	41.9	38.0	19.8	52.1	27.5	17.9	65.1	12.0	4.5	37.5
150	90.0	37.2	41.3	45.0	23.3	51.8	—	—	—	—	—	—
180	105.0	42.6	40.6	51.0	26.5	52.0	—	—	—	—	—	—

†x = mean; δ = standard deviation; cv = coefficient of variability (%).

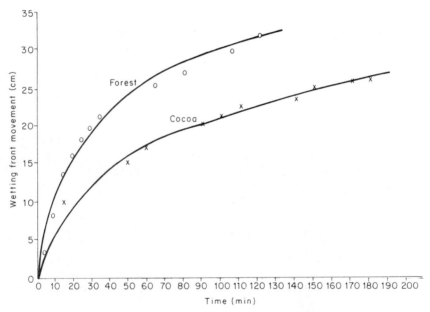

Figure 5.13 Effects of pre-clearing vegetation on infiltration into an Alfisol at IITA, Ibadan. The rate of wetting front movement from a semi-circular cavity was more under forest than under cocoa trees (unpublished data, Lal, 1970)

hydrometeorogical factors (Reid, 1977). Variability in soil temperature and micro-climate is also related to the shade, plant cover, micro-relief, slope aspect and other factors that influence the direct impact of solar radiation on the soil surface. Canopy characteristics under different types of vegetation

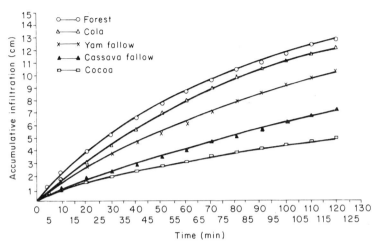

Figure 5.14 The rate of advance of wetting front from a semi-circular cavity was in the order of forest > cola > yam fallow > cassava fallow > cocoa (unpublished data, Lal, 1970)

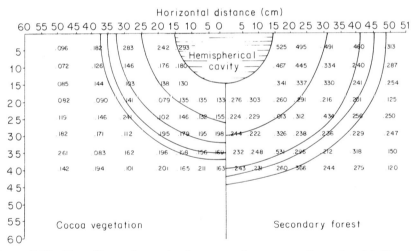

Figure 5.15 The effects of pre-clearing vegetation on two-dimensional infiltration rate. The curves indicate the rate of advance of the wetting front and the figures are the gravimetric moisture content measured 24 hours after (unpublished data, Lal, 1970)

cover have an obvious impact on the micro-climatic variables (Leger and Millette, 1975; Hatfield et al., 1982).

Although soil temperature is an important edaphological factor in the tropics, little is reported about the variability over a field plot and the number of measurements needed to obtain a representative mean. It is important to realize that monitoring equipment such as the mercury-in-glass thermometer or the thermister probe registers the temperature of the sensor and not that of soil. The sensor temperature, though an indication of the temperature of the surrounding medium, is significantly affected by the contact, local relief, exposure and many other uncontrolled variables.

E. Run-off and erosion processes

The processes of water run-off, soil splash, and sediment discharge from an agricultural land are affected by its soil physical properties and their interaction with the rainfall characteristics. There are three predominant processes that govern storm run-off: (i) Hortonian flow that occurs when rainfall intensity exceeds the surface infiltration capacity; (ii) saturated overland flow that occurs when the surface layer becomes saturated above the horizon of low permeability; and (iii) through-flow that occurs through the direct conduction of water into and through the subsoil through interconnected macropores. There is, however, a need to quantify the spatial variability of these processes on agricultural watersheds in the tropics. Even though the spatial variability in the water balance and hydrological processes can be quantified, as has for example been done for the U.S. and Australian watersheds (Vreeken, 1973; Rogowski, 1972; Anderson and Furley, 1975; Dunin and Aston, 1981), it is also important to develop analytical methods for developing models and for interpreting the results obtained. Spatial variation of rainwash, splash, and run-off occurs even on small plots of apparently homogeneous areas and on small samples subjected to simulated rains (Luk and Morgan, 1981; Bryan and Luk, 1981). While evaluating the surface and the subsurface run-off processes in an Australian watershed, Pilgrim *et al.* (1978) observed grossly non-uniform run-off processes, both spatially over the plot and laterally and vertically within the soil, on a plot that was superficially uniform. Analytical techniques available to cope with the variability are discussed by Bakr *et al.* (1978), Philip (1980), Anderson (1983) and others, and readers are referred these reviews for additional details.

There is a little research information available about the variability in run-off and erosion processes in tropical agricultural watersheds. The data in Table

Table 5.15 Variability in annual surface run-off and soil erosion over 5 ha agricultural watersheds in southwestern Nigeria (recalculated from Lal, 1984). *Reproduced by permission of Elsevier Science Publishers*

Season	No-till watershed						Ploughed watershed					
	Run-off			Erosion			Run-off			Erosion		
	x†	δ	cv	x	δ	cv	x	δ	cv	x	δ	cv
First season	48.8	2.9	5.9	1.3	0.2	23.1	194.4	51.0	26.2	5.2	1.5	28.8
Second season	32.5	5.2	16.0	0.6	0.1	16.7	51.7	20.9	40.4	1.3	0.9	69.2
Total	81.3	4.3	5.3	1.9	0.2	10.5	246.1	61.0	24.8	6.5	2.0	30.8

†x = mean; δ = standard deviation; cv = coefficient of variability (%).

5.15 compare variability in annual run-off and soil erosion monitored on eight field run-off plots established within each of two 5 ha no-tillage and ploughed watersheds. The cv of run-off ranged from 5 to 16 per cent in the no-tillage watershed and 25 to 40 per cent in the ploughed watershed. The soil erosion was even more variable than the surface run-off. For example, the cv in soil erosion ranged from 11 to 23 per cent for no-tillage and 29 to 69 per cent for the ploughed watersheds. The soil surface management, therefore, influences the variability in run-off and sediment discharge processes. More research information is needed to quantify the magnitude of variability of these processes in tropical soils, and the factors responsible for creating this diversity.

F. Factors responsible for stochastic or random variability

It is the microvariability that occurs over short distances within a seemingly homogeneous plot of land that is of major concern to this study. The macrovariability is also important but can be easily detected and mapped and is attributed to variations in climate, parent material and its stage of weathering, and the soil moisture regime (Wambeke and Dudal, 1978). Furthermore, the trends and magnitude of the deterministic macro-variability are predictable and related to known factors whose effect on soil physical properties is easily quantified. It is the random or the stochastic variability whose magnitude and trends are unpredictable and difficult to relate to known factors. The effects of pedological factors and lithological discontinuities are discussed by Moorman (1972) and Moormann and Kang (1978).

(i) Vegetation cover: Many researchers have attributed the microvariability in soils of the temperate regions to the effects of tree species or of the vegetation cover. In a forested land, the patterns in soil variability observed are often related to the redistribution in litter fall, stand density and alterations in population due to mortality or regrowth (Charley and West, 1975; Mollitor *et al.*, 1980). In fact, a pattern in soil properties is often observed around individual tree species, with marked and detectable variations out from the tree in all directions. Grieve (1977) concluded from a study of a forested soil in U.K., that there exists a detailed pattern of soil properties corresponding to the vegetation pattern. This implies that the greatest random variability in soil properties would be observed on a land with the greatest diversity in vegetation. Alban (1974) remarked from his studies in forested lands in Minnesota, U.S.A., that the number of samples needed to obtain 95 per cent confidence in soil analysis also varies among the tree species and the soil properties to be determined.

As in the soils of the temperate region, high random variability in properties of a land recently developed from a forest cover is attributed to diversity of tree species (Moormann, 1972; Kang and Moorman, 1977). Kang (1977) attributed variability in soil chemical properties to various tree species. Crops growing near oil palm (*Elaeis guineensis*) were more vigorous than those farther away.

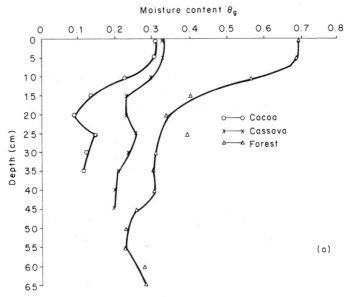

Figure 5.16a Soil moisture profile after free drainage had ceased in plots with different vegetation cover. The equilibrium moisture content was in the order of forest > cassava > cocoa

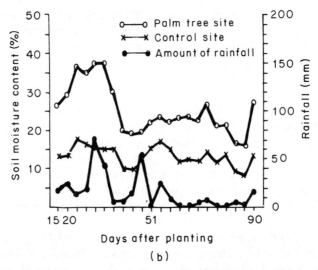

Figure 5.16b Soil moisture content is generally higher in the vicinity of an old palm tree stump (Kang, 1977). *Reproduced by permission of* Plant and Soil

Table 5.16 Maize grain yield as influenced by the vicinity of palm site (modified from Kang, 1977). *Reproduced by permission of* Plant and Soil

Yield per plant (g)	Control		Palm site	
	No fertilizer	NPK	No fertilizer	NPK
Grain	187 ± 46	253 ± 59	326 ± 43	371 ± 64
Stover	445 ± 122	634 ± 74	668 ± 94	653 ± 3

Maize grain yield near oil palm was 74 and 47 per cent more than away from the tree without and with fertilizer application, respectively (Kang, 1977). Superior crop growth near oil palm was due to more favourable soil organic matter content, low soil bulk density (Table 5.16) and consequently high available soil moisture reserves. The field capacity, or the equilibrium soil moisture content following saturation, also differs among different vegetation covers. The data in Fig. 5.16 show higher equilibrium moisture content or the field capacity of the surface 50 cm soil depth under forest than under cocoa or abandoned cassava plots. Crop growth following land development from these vegetations is obviously affected due to differences in available soil moisture reserves (Fig. 5.17).

There are also measurable differences in the available plant nutrient reserves. Similar effects of variability in soil properties caused by other tree species are reported for *Acacia albida* in Senegal (Charreau and Vidal, 1965; Dancette and Poulain, 1968) (Tables 5.17, 5.18, 5.19). Variations in soil moisture content due to vegetation cover also influence soil resistance to penetrometer and the root growth of the following seasonal crops (Fig. 5.18). Charreau and Vidal reported 2.5 times more millet grain yield in the vicinity of *Acacia* trees than farther away due to favourable soil physical and nutritional properties near the trees. The favourable effects of trees on soil productivity are also known for *Acio barteri* and *Chorophora excelsa* (Kellogg, 1975). Similar variability and its beneficial effects on crop growth are observed for some bamboo (*Phyllostachys* sp.) in Thailand (Moormann and Kang, 1978).

Table 5.17 Effects of *Acacia albida* on soil bulk density, total porosity and field moisture capacity in Senegal (Dancette and Poulain, 1968)

Depth (cm)	Field moisture capacity (%)		Bulk density (g/cm³)		Porosity (%)	
	A	B	A	B	A	B
0–10	7.54	8.62	1.51	1.51	43.4	43.2
10–20	7.83	7.91	1.55	1.54	42.2	41.7

A = away from *Acacia albida*; B = under the canopy of *Acacia albida*.

Figure 5.17 Better sorghum growth under this tree canopy in Nigerian savanna near Abuja is due to less drought stress under canopy than farther away

Table 5.18 Effect of *Acacia albida* on penetrometer resistance (Dancette and Poulain, 1968)

Depth of penetration (cm)	Control, away from tree		Under tree	
	Resistance (kg cm^{-2})	Moisture (%)	Resistance (kg cm^{-2})	Moisture (%)
4	7.6	0.47	7.2	0.59
20	32.3	1.45	39.0	1.13
35	45.8	3.28	57.6	2.97

Table 5.19 Mean soil moisture profile (average of four directions) as influenced by *Acacia albida* in Senegal (Dancette and Poulain, 1968)

Depth (cm)	2 m from tree trunk†			Limit of foliage canopy			Control, far away from tree		
	24/5/66	22/9/66	24/10/66	24/5/66	22/9/66	24/10/66	24/5/66	22/9/66	24/10/66
–20	0.30	6.72	4.05	0.36	6.10	3.20	0.15	4.40	2.67
20–40	0.58	5.13	4.41	0.60	5.74	3.83	0.86	6.10	4.05
40–60	0.85	6.38	4.69	0.92	5.97	4.46	0.93	6.09	4.32
60–80	0.88	6.33	4.70	1.05	5.86	4.77	1.08	6.30	4.29
80–100	0.99	6.82	5.16	1.22	6.50	5.09	1.03	6.58	4.57
100–120	0.97	7.52	5.92	1.06	6.04	5.16	1.06	7.39	5.14
140–160	1.31	7.05	5.46	0.95	6.63	6.01	0.95	6.94	5.52
160–180	1.62	7.13	6.13	1.96	6.41	5.71	1.24	6.37	5.89
180–200	2.48	6.95	6.62	2.13	6.87	5.91	1.10	7.04	5.75
240	2.57	6.40	6.26	2.69	3.91	6.28	1.77	6.42	6.96
280	2.83	6.31	6.27	2.52	4.22	6.49	2.59	4.95	6.54
320	2.81	5.26	6.02	3.05	4.43	6.57	2.32	5.42	6.38
360	2.58	3.45	5.41	2,83	4.40	6.12	2.49	3.02	6.21
400	4.28	5.76	6.53	4.24	4.80	6.11	2.36	4.08	5.16
	3.96	4.29	7.74	3.95	5.29	6.51	4.26	5.81	6.24

Moisture content at pF 3.0 = 2.9%. Moisture content at pF 4.2 = 1.4%.
†24/5/66 = before the onset of rains; 24/10/66 = one week after the last rain; 22/9/66 = during the rainy season.

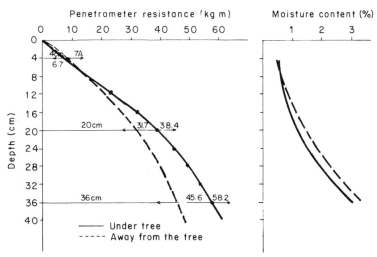

Figure 5.18 Effects of vegetation cover on soil moisture reserves and on the penetrometer resistance (Dancette and Poulain, 1968)

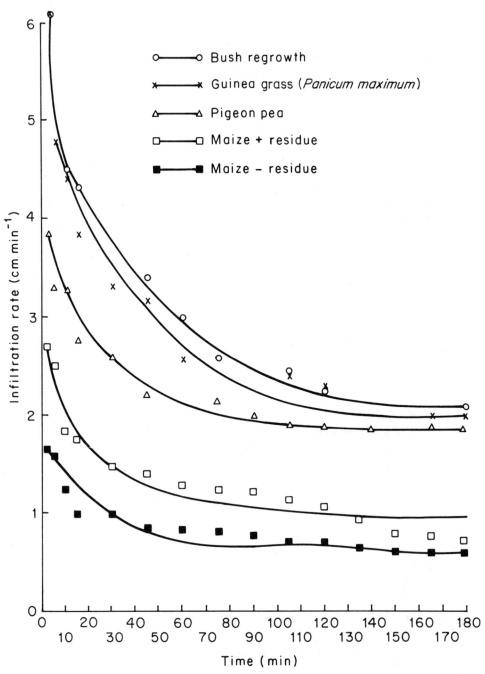

Figure 5.19 The effects of cropping systems and of *Panicum maximum* and pigeon pea on water infiltration rate (Juo and Lal, 1977). *Reproduced by permission of* Plant and Soil

Similarly to trees, various pasture species are also known to cause considerable alterations in properties of soils in their immediate vicinity. Variability in chemical properties may be related to differential ion uptake and nutrient recycling (Scott, 1975). In tropical Australia, McCown et al. (1977) related variations in subsoil permeability and water storage capacity to the patterns of perennial grass distribution. On an Alfisol in western Nigeria, Juo and Lal (1977) observed drastic differences in water intake rate in soils sown to perennial pigeon pea (*Cajanus cajan*) and *Panicum maximum* (Fig. 5.19).

(ii) Effects of soil animals: The biotic community in the soil influences its properties. Although separate chapter is devoted to earthworms and termites in relation to their effects on soil physical properties, it is important to realise that soil fauna is a major factor responsible for variations in soil physical and chemical properties. In Sierra Leone Miedema and Van Vuura (1977) examined the number of channels of 1–4 mm size at different distances from the termite mound. The number of channels was the highest near the mound and decreased with distance from the mound. (Fig. 5.20). The water transmission properties in the vicinity of a termite mound are, therefore, expected to be highly variable. The activity of animals in soil creates voids, alters soil structure, influences movement of water and air, affects soil erosion, changes soil microrelief, and regulates leaf litter decomposition and build-up. These animals alter soil fabric and influence shape, size, and the arrangement of soil particles. For additional details readers are referred to a comprehensive review article by Hole (1981).

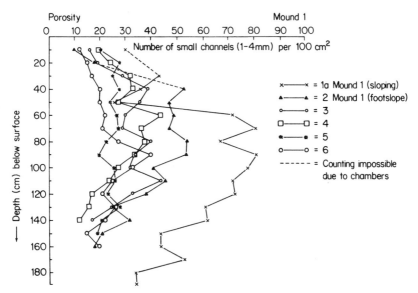

Figure 5.20 Number of open channels (φ 1–4mm) at different distances from termite mound (Miedema and Van Vuure, 1977)

(iii) Slope and micro-relief: Random microvariability in soil physical properties is also caused by microrelief and slope aspects that influence lateral and vertical water and sediment movements. In Bahia, Brazil, Leite and Ezeta (1982) attributed variability in particle size distribution and changes in the clay fraction to relief characteristics and to the drainage patterns influenced by relief.

(iv) Man's activity: In the tropics, the cultural practices associated with shifting cultivation also contribute to the random and unaccounted for microvariability. Soil physical properties are greatly influenced by the practice of burning (Fig. 5.21), mounding or making permanent ridges for cultivation of perennial crops. An uneven intensity of burning and ash distribution is an important factor that leads to soil variability. In poorly drained areas, farmers often make large mounds (Fig. 5.22), which are used for many years. These

Figure 5.21 Bush fire affects soil physical properties

213

Figure 5.22 Large mounds are made by farmers to grow upland crops on poorly drained soils

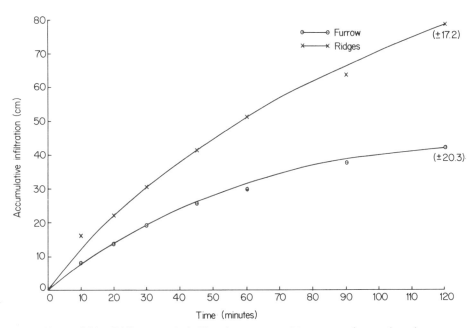

Figure 5.23 Differences in infiltration rate on ridge versus furrow locations

heaps are often as high as 150 cm and up to 6 metres in the base circumference. Shifts in the position of these mounds and ridges lead to variability in soil properties. Infiltration rate measured at different locations in relation to the ridge or mound position can be very different even over short lateral distances. The data in Fig. 5.23 show considerable differences in infiltration rates measured about 1 metre apart on the top of old abandoned ridges and in the furrows.

In those systems of traditional farming where homesteads are also rotated, a considerable variability in soil properties and its crop productivity is caused by the abandoned village sites. These differences are particularly noticeable on newly developed lands.

Soil erosion, caused by man's activities and diverse cultural practices, is another important factor responsible for soil microvariability. Differential removal and deposition of soil constituents by water and wind creates soil variability even over short distances. Pinho and Morais (1980) reported soil microvariability to erosion of the surface layers and due to intensive landuse.

Soils subjected to seasonal or periodic inundation are also subject to microvariability in their physical and hydrological properties. The depth of flooding in the presence or absence of straw is known to cause differences in microbial population and in soil aeration (Wakao and Furusaka, 1973).

IV. MICROVARIABILITY IN PHYSICAL VERSUS CHEMICAL SOIL PROPERTIES

Although this chapter is concerned primarily with the microvariability in soil physical properties, that is not to imply that chemical properties are either less or not at all variable. The microvariability of soil chemical and nutritional properties for the soils of the temperate zone has been documented for North America by Walker *et al.* (1968); Ball and Williams (1968); Drees and Wilding (1973); Lierop and Mackenzie (1977); Ricaud *et al.* (1974); Biggar (1978) and Robertson *et al.* (1980); for New Zealand by Lee *et al.* (1975) and Scott (1975); for Australia by Black and Waring (1977), and for Bulgaria by Slavov and Dinchev (1979). In contrast, there are only a few reports regarding the microvariability in chemical properties of soils in the tropics. Kang (1977), Kang and Moormann (1977) and Moormann and Kang (1978) related soil microvariability to biological factors, e.g., pre-clearing vegetation and activity of termites. In Cuba, Gonzàlez Abreau *et al.* (1976) observed a high cv in the mobile phosphate contents of upland soils. A high spatial variability in salt content of some Mexican soils was reported by Ramirez *et al.* (1981). In the grazed pastures of Queensland Australia, variability in N content of soil samples has been reported by Vallis (1973).

Comparing the coefficient of variability of soil physical with soil chemical properties of both temperate regions and soils of the tropics indicates that there is generally a much higher variability in physical than in chemical properties. This is particularly so for soils either under shifting cultivation or those that have just been cleared from a fallow vegetation. This differential variability may partly be due to the methodology used. Most soil physical properties are better determined *in situ* and the sample size is generally large. This is particularly the case for determination of infiltration rate, bulk density by the excavation method, and *in situ* measurements of pF curves. On the contrary, soil chemical properties are determined on relatively small amount of samples that have already been sieved through a 2 mm sieve. The variability among subsamples of sieved soil is generally less than that of *in situ* measured soil parameters on an undisturbed site.

V. FIELD RESEARCH AND ANALYTICAL METHODS

Variability in soil physical properties influences crop growth and development, as has been documented by the data of Dancette and Poulain (1968) in Senegal, by Gopalswamy and Raj (1973) in India, Hack (1976a,b) in Sudan, and Moormann and Kang (1978) in Nigeria. This heterogeneity has many agronomic implications in terms of: (i) interpreting field data and making valid recommendations; (ii) in designing field experiments; and (iii) in developing appropriate analytical methods of soil characterization. In addition to researchers, extension agents also face the problem of making valid recommendations for fertilizer use on a highly variable soil. Most of the land evaluation techniques do not consider variability, because it is an inherent soil property. And yet, an attempt to assess the land's potential on the basis of highly variable analytical data can be a frustrating experience. The basic principle is to understand the problem of soil variability, to come to terms and learn to live with it. Some suggestions on how to deal with the variability in soils of the tropics have been discussed by Hack (1976a, b), Moormann and Kang (1978), Buol and Couto (1978), and are briefly described below:

1. *Replications and design of field experiments*:

Smith (1970) remarked that variability and 'sampling errors are easily eliminated by the unthinking or the unscrupulous, in that only one sample is taken. The problem is then solved by being ignored'. In fact, the physical properties of soils in the tropics are so variable that not only it is hard to ignore this problem, but the conclusions based on one or few samples are worse than none at all.

It would be extremely useful if special experimental designs and analytical methods were available to cope with the variability (Hack, 1976a, b). The residual variation or the error term in the analysis of variance can be reduced by increasing the number of replicates. However, the more the replicates the more expensive and difficult to manage the experiment. Hack (1976a, b) estimated the effect on labour cost, in central Sudan, of having more replicates. His data show that the desirable level of cv is obtained from 8 replicates, but the cost of labour increases linearly with the increase in replicates between 2 and 40. Furthermore, increasing the number of replicates may also increase soil heterogeneity by enlarging the scale (Figs. 5.1 and 5.2). Uniformity trials should be conducted prior to initiating the experiment to estimate the magnitude of within-plot and within-block variability.

In addition to increasing the number of replicates, selecting appropriate plot shape to suit the variability trends is another possibility of decreasing the variability. In Taiwan, Wang (1971) reported that the optimum plot arrangement in the field was rectangular plots established parallel with the soil variability gradients. Similarly in Egypt, Kalla and Gomma (1977) observed that plot shape had an effect on plot-to-plot variability. Long narrow plots arranged with their greatest length in the direction of greatest variation (along rows) resulted in lower estimates of among-plot variance. Torres (1980) recommended random block experiments for computing the homogeneity index of some variable soils of Cuba. This method supposedly gives an unbiased estimate with a minimum variance for the homogeneity index and can be used as a guide for determining the optimum plot size for agricultural experiments.

2. *Understanding sources of variability*

Rather than increase replicates, another alternative is to identify the sources of variability. This may be easily done for the deterministic or known variability, but a detailed field survey would be needed to identify the sources of random or stochastic variability. If easily identified, these factors can be introduced as experimental treatments (Hack, 1976a, b). Statistical methods should be developed to express microscale heterogeneity (Flühler et al., 1976)

3. *Crop performance and soil properties*

In all agronomic experiments, crop yield of a seemingly homogeneous plot should be related to the properties of the soil for that location. Empirical relations should then be established between crop patterns during a growing season and the pattern of soil variation in the landscape. Because soil properties vary over short distances, crop yields may also differ from point to point. Establishing one-to-one relationships between soil properties and crop yields would facilitate understanding of the variability patterns. Across-gradient or sequential testing can be a useful system if the source of variability is known. For random variability, relating crop performance to soil properties is a useful approach (Milfred and Kiefer, 1976; Bresler *et al.*, 1981).

VI. SOIL SAMPLING AND STATISTICAL ANALYSES OF VARIABLE SOIL DATA

Obtaining representative soil samples is the first step in making reliable statistical analyses of variable soil data. If samples are not obtained from sites that represent patterns of soil variability, the results are likely to misrepresent the soil quality. The number of soil samples required for obtaining representative data also depends on the degree of soil variability. Burrough and Kool (1981) recommended the nested sampling technique to determine the optimum sampling spacing for a random variation in soil properties. A transact or grid sampling is useful for a systematic or known variability. Some transformation (log transformation for example) of data is necessary to fit these data into normal distribution (see Fig. 5.11b).

For soils with a high degree of random variability, it is advisable to obtain samples from the entire field. Decreasing the number of replications sampled, even if the number of samples obtained is kept the same, can be an inefficient sampling design. And yet, too large a number of samples is often uneconomic and impractical. Vintola and Canarche (1974) concluded that for 5 per cent allowable error, 5 to 10 samples are needed for a less variable soil and perhaps up to 15–20 for more variable soil properties. For 10 per cent allowable error, the number can be reduced to 3 – 5 and 8 – 10, respectively. Keisling et al. (1977) reported the need for as many as 56 sampling locations for evaluating physical parameters of some soils in Texas to achieve a standard deviation equal to 10 per cent of the mean. Alban (1974) related sampling intensity to soil variation under red pine and aspen in Minnesota, United States. In Canada, Patterson and Wall (1982) concluded that to assess the within-pedon variability in soil properties would require at least five samples. While considering the problem of soil sampling and analysis for some German soils, Thum (1974) and Geidel and Schafer (1976) observed that the number of samples required for a representative mechanical analysis was 30 to 60. For soil chemical analysis, on the other hand, the number of samples required may be 6 to 9 (Blyth and MacLeod, 1978). In Chile, Suarez (1979) estimated that for alluvial soils, at least 10 to 15 samples were required to obtain acceptable confidence limits. Yost and Fox (1981) and Yost et al. (1982) have analysed variability in chemical properties of some Hawaiian Andepts, using semi-variograms. They concluded that rainfall variability over the study area had imposed a degree of uniformity on surface soil properties not apparent in the subsoil. This is because the subsoil properties were observed to be highly variable, resulting from great variation over short distances in age and weathering of volcanic ash. On the contrary, some researchers have argued that sample size does not always have a significant effect on the magnitude of the sampling error (Ferrari and Vermeulen, 1956; Moltham, 1966).

The optimum number of samples required can be computed from statistical methods such as those described by Peterson and Calvin (1965). According to their method, the minimum number of soil samples required to obtain an acceptable estimate of the variation in question is computed according to the following equation:

$$N = t_\alpha^2 s^2 / D^2$$

where *t* is the Student's *t* with (*n*–1) DF at the α probability level, *s* the standard deviation and *D* the specified acceptable error of the variable in question. This equation may however be difficult to use for some related variables, e.g., bulk density and the volumetric moisture content (Cassel and Bauer, 1975). Reynolds (1974) computed the sample size required to estimate true mean (±5%) at 0.05 probability level by the following relation:

$$n = (\hat{\sigma})_2/\text{where } \hat{\sigma} \text{ is the standard error and } d$$

is the standard deviation.

Most statistical methods apply to those properties that are normally distributed. Nielsen *et al*. (1973) used the following normal frequency distribution function to compute the spatial variability of field-measured soil-water properties:

$$f = \frac{1}{\delta\sqrt{\pi}} \exp[-(x-m)^2/2\delta^2]$$

where *f* is frequency, *m* the mean, δ the standard deviation of the mean, and *x* the random variable. Some soil physical properties (e.g. hydraulic conductivity, permeability and infiltration rate) are generally non-normally distributed (Vintola and Canarche, 1974). Some transformations are necessary to fit these data into normal distribution. The use of other statistical techniques for such data have been described by Nagai and Menk (1976) and Rao *et al*. (1979).

Readers are referred to some appropriate and more comprehensive reviews of different statistical techniques available to analyse soil variability (Courtney and Nortcliff, 1977; Webster, 1978; De La Rosa, 1979). The use of multivariation analysis to quantify soil variability is proposed for Australian soils by Norris and his colleagues (Norris, 1970a, 1971; Norris and Loveday, 1971; Norris, 1972). Nortcliff (1978) recommended the Principal Component Analysis (PCA). A useful feature of the PCA solution is that the components, in turn, account for a maximum amount of the varian of the variables. In particular, the first component is that linear combination which contributes a maximum to the total variance, the second principal component, orthogonal to the first, contributes a maximum to the residual variances, and so on until the total variance, is analysed. Nortcliff recommended an analysis of variance technique which comprises breakdown of the total variation into parts which can be assigned to the separate contributing sources. The analysis of variance model recommended the nested or hierarchical design. The use of the range of semi-variogram was recommended by Yost *et al* (1982) to delineate areas of soils with similar properties.

Mathematical treatments for the analysis of variability of hydrological properties have been presented by many. Smith and Hebbert (1979) described the use of Monte Carlo analysis regarding the hydrological effects of spatial variability of infiltration. Stochastic analysis of spatial variability in subsurface flows has been used by Bakr *et al.* (1978). In addition, some non-experimental approaches to variability in soil erodibility assessment have also been proposed. Geissert et al. (1978) have recommended the use of cartography as a means of analysing spatial variability in soil erodibility.

A. Geostatistical techniques

Geostatistical techniques, used widely by mining engineers, are increasingly being used by soil physicists and hydrologists (Bregt, 1984). They are based on theory of regionalized variables (Matheron, 1971). The dependency between observation points can be geostatistically described by (a) autocorrelation, (b) semivariance, and (c) intrinsic random functions of order K. To allow application of geostatistical procedures some kind of stationarity of the data is required. Semi-variograms of soil physical and hydrological data may be parabolic, linear, or nugget (Huijbregts, 1975; Burrough, 1983a, b; Ten Berge *et al.*, 1983).

B. Kriging

The method of Kriging is widely recommended for spatial interpolation between point data. Kriging is also based on the theory of regionalized variables and produces estimates for unknown points which are optimal in the sense that they are unbiased and have minimum variance. An advantage of kriging over other interpolation methods is that it allows computation of the variance of the estimates, and the interpolation values can be used with known confidence. In addition the interpolated value for an unknown point is a weighted average of the observations in the neighbourhood. The value for the unknown points can be estimated by solving a set of linear equations (Journel and Huijbregts, 1978).

VII. SUMMARY AND CONCLUSION

Physical properties of soils in the tropics are highly variable, as are those of the soils of the temperate regions. The magnitude of variability in soil physical properties is partly due to the methods of their determination. It is the random or stochastic variability over short distances, i.e. microvariability, that is of relevance in terms of routine characterization of land assessment.

Microvariability in soil physical properties is introduced by a host of interacting factors. These factors include lithologically discontinuous parent material, effects of biotic factors, e.g. soil animals, vegetation cover, and tree species, microrelief and slope aspects, and the man-made factors including the history of land-use and methods of its management. The magnitude of variability also depends on plot size, the duration of cultivation, and the parameters to be considered.

Microvariability in soil physical properties has agronomic implications in terms of designing field experiments and in interpreting the analytical data for making valid recommendations. The problem can be tackled by understanding the sources of variability so that they can be introduced as treatments. The number of samples required, the experimental design, and the plot size and shape can be changed to cope with the variability.

A range of mathematical and statistical techniques have been proposed to analyse soil variability. The important concept is to understand this natural problem and to come to terms with it.

REFERENCES

Adams, J.A., and Wilde, R.H., 1976a, Variability within a soil mapping unit mapped at the soil type level in the Wanganui district. I. Morphological variation, *N. Z. J. Agric. Res.*, **19**, 165–176.

Adams, J.A., and Wilde, R.H., 1976b, Variability within a soil mapping unit mapped at the soil type level in the Wanganui district, II. Chemical variation, *N. Z. J. Agric. Res.*, **19**, 435–442.

Ahn, P.M., 1970, *West African Soils*, Oxford University Press, London.

Alban, D.H., 1974, Soil variation and sampling intensity under red pine and aspen in Minnesota. *Research Paper*, USDA Forest Service, NC–106, Washington, D.C.

Anderson, M.G., 1983, Selected effects of field variability on predictions from a soil water finite difference model, *Nordic Hydrology*, **14**, 1–18.

Anderson, K.E., and Furley, P.A., 1975, An assessment of the relationship between the surface properties of chalk soils and slope from using principal components analysis, *J. Soil Sci.*, **26**, 130–143.

Babalola, O., 1972, The influence of 'bound' water on the calibration of neutron moisture water, *Soil Sci.*, **114**, 323–324.

Babalola, O., 1978, Spatial variability of soil water properties in tropical soils of Nigeria, *Soil Sci.*, **126**, 269–279.

Babalola, O., and Lal, R., 1977a, Subsoil gravel horizon and maize root growth: I. Gravel concentrations and bulk density effects, *Plant Soil*, **46**, 337–346.

Babalola, O., and Lal, R., 1977b, Subsoil gravel horizon and maize root growth. 2: Effects of gravel size, inter-gravel texture and natural gravel horizon, *Plant Soil*, **46**, 347–357.

Bakr, A.A., Gelhar, L.W., Gutjahr, A.L., and MacMillan, J.R., 1978, Stochastic analysis of spatial variability in subsurface flows I. Comparison of one- and three-dimensional flows, *Water Resources Res.*, **14**, 263–268.

Ball, D.F., and Williams, W.M., 1968, Variability of soil chemical properties in two uncultivated brown earths, *J. Soil Sci.*, **19**, 379–391.

Banfield, C.F., and Bascomb, C.L., 1976, Variability in three areas of the Denchworth soil map unit. II. Relationship between soil properties and similarities between profiles using laboratory measurements and field observations, *J. Soil Sci.*, **27**, 438–480.

Bascomb, C.L., and Jarvis, M.G., 1976, Variability in three areas of the Denchworth soil map unit. I. Purity of the map unit and property variability within it, *J. Soil Sci.*, **27**, 420–437.

Bätz, G., Buhtz, E., and Heinze, G., 1972, Choice of sample size for determining physical properties of soil, *Tagungsbericht Deutsche Akademie der Landwirtschaftsswissenschaften Zu Berlin*, **116**, 97–109. GDR, Chernagen.

Beaudou, A.G., and Collinet, J., 1977, The diversity of mappable pedological units in the African ferrellitic region, *Cahiers ORSTOM Pedologie*, **15**, 19–34.

Beckett, P.H.T., and Webster, R., 1971, Soil variability: A review, *Soils Fert.*, **34**, 1–15.

Bell, K.R., Blanchard, B.J., Schmugge, T.J., and Witcqak, M.W., 1980, Analysis of surface moisture variations within large-fields sites, *Water Resources Res.*, **16**, 796–810.

Ben-Asher, J., 1979. Error in determination of the water content of a trickle irrigated soil volume. *Soil Sci. Soc. Amer. J.*, **43**, 665–668.

Biggar, J.W., 1978, Spatial variability of nitrogen in soils, in Nielson, D.R., and MacDonald, J.G. (eds.), *Nitrogen in the environment, Vol. 1. Nitrogen Behaviour in Field Soil*, Academic Press Inc., London, pp. 201–211.

Black, A.S., and Waring, S.A., 1977, The natural abundance of ^{15}N in the soil-water system of a small catchment area, *Aust. J. Soil Res.*, **15**, 51–77.

Blyth, J.F., and Macleod, D.A., 1978, The significance of soil variability for forest soil studies in northeast Scotland, *J. Soil Sci.*, **29**, 419–430.

Bregt, A.K., 1984, Geostatistical techniques and spatial variability of soil physical properties. *Proc. Int. Workshop on Physical Aspects of Soil Management in Rice-Based Cropping Systems, December 1984*, IRRI, Los Banos, Philippines.

Bresler, E., Dasberg, S., Russo, D., and Dagan, G., 1981, Spatial variability of crop yield as a stochastic soil process, *Soil Sci. Soc. Amer. J.*, **45**, 600–605.

Bryan, R.B., and Luk, S.H., 1981, Laboratory experiments on the variation of soil erosion under simulated rainfall, *Geoderma* **26**, 245–265.

Buol, S.W., and Couto, W., 1978. Fertility management interpretations and soil surveys of the tropics, in *Diversity of Soils in the Tropics*, ASA Special Publications, No. 34, pp. 65–75.

Burrough, P.A., 1983a, Multiscale sources of spatial variation in soil. I. The application of fractal concepts to nested levels of soil variation, *J. Soil Sci.*, **34**, 577–597.

Burrough, P.A., 1983b, Multiscale sources of spatial variation in soil. II. A non-Brownian fractal model and its application in soil survey, *J. Soil Sci.*, **34**, 599–520.

Burrough, P.A., and Kool, J.B., 1981, A comparison of statistical techniques for estimating the spatial variability of soil properties in trial fields, *Sol, France*, **4**, 29–37.

Carvallo, H.O., Cassel, D.K., Hammond, J. and Bauer, A., 1976, Spatial variability of *in situ* unsaturated hydraulic conductivity of Maddock sandy loam, *Soil Sci.*, **121**, 1–8.

Cassel, D.K., and Bauer, A., 1975, Spatial variability in soils below depth of tillage: bulk density and 15 atmosphere percentage, *Soil Sci. Soc. Amer. Proc.*, **39**, 247–250.

Cassel, D.K., and Nielson, D.A., 1979, Variability of mechanical impedance in a tilled one-hectare field of Norfolk sandy loam, *Soil Sci. Soc. Amer. J.*, **43**, 450–455.

Charley, J.L., and West, N.E., 1975, Plant-induced soil chemical patterns in some shrub-dominated semi-desert ecosystems of Utah, *J. Ecology*, **63**, 945–963.

Charreau, C., and Vidal, R., 1965. Influence de l' *Acacia* albida sur le sol, nutrition minerale et rendements des mils Pennisetum au Senegal, *Agric. Trop.*, **67**, 600–625.

Collinet, J., 1969, Contribution to the study of stonelines in the Middle-Ogoo's Region (Gabon), *Cah. ORSTOM Ser. Pedol.*, **7**, 1–42.

Collins, J.F., 1976, Soil heterogeneity and profile development in a stratified gravel deposit in Ireland, *Geoderma*, **15**, 143–156.

Commonwealth Bureau of Soils, 1967, Variability of soil samples taken for chemical analysis (1966–1973), *Bibliography No. 1080*, Rothamsted Exp. Station, Harpenden, Herts., U.K.

Courtney, F.M., and Nortcliff, S., 1977, Analysis of technique in the study of soil distribution, *Progress in Physical Geography*, **1**, 40–64.

Dancette, C., and Poulain, J.F., 1968, Influence de l'*Acacia albidia* sur les facteurs pedoclimatiques et les rendements des cultures, *Sols Afr.*, **13**, 197–239.

De Lal Rosa, D., 1979, Statistical analysis of soil properties, *Agrochimica* **23**, 72–83.

Dhanapalan, M.A., Soundarajan, S.K., and Lakshiminarayanan, S., 1975, Summary of soil test values of major soil series of Thanjavur Dist., *J. Indian Soc. Soil Sci.*, **23**, 371–379.

Drees, L.R., and Wilding, L.P., 1973, Elemental variability within a sampling unit, *Soil Sci. Soc. Amer. Proc.*, **37**, 82–87.

Dunin, F.X., and Aston, A.R., 1981, Spatial variability in the water balance of an experimental catchment, *Aust. J. Soil Res.*, **19**, 113–120.

Ferrari, T.J., and Vermeulen, F.H.B., 1956, Soil heterogeneity and soil testing, *O.E.E.C. Proj.*, **156**, 113–126.

Flühler, H., Stolzy, L.H., and Ardakani, M.S., 1976, A statistical approach to define soil aeration in respect to denitrification, *Soil Sci.*, **122**, 115–123.

Freeze, R.A., 1975, A stochastic conceptual analysis of one-dimensional ground water flow in nonuniform homogeneous media, *Water Resources Res.*, **11**, 725–741.

Gajem, Y.M., Warrick, A.W., and Myers, D.E., 1981, Spatial dependence of physical properties of a Typic Torrifluvent soil, *Soil Sci. Soc. Amer. J.*, **45**, 709–715.

Geidel, H., and Schäfer, P., 1976, Problems in soil sampling and analysis, *Landwirtschaftliche Forschung*, **29**, 149–160.

Geissert, D., Hoeblich, J.M., Messer, T., Mettauer, H., Monnon, J.M., El-Ghossain, T.S., Schwing, J.F., and Vogt, H., 1978, Non-experimental approaches to spatial variability of land erodibility on a large scale, *Recherches Géographiques a Strasbourg*, **9**, 53–55.

Gonzàlez Abreau, A., León, J., and Savich, V.I., 1976, Variation in the mobile phosphate content of soils, *Revista CENIC, Ciencias Biologicas*, **7**, 159–166.

Gopalswamy, A., and Raj, D., 1973, Influence of soil variability on available nutrients at different stages of plant growth, yield and uptake of nutrients, *Madras Agric. J.*, **60**, 308–312.

Grieve, I.C., 1977, Some relationship between vegetation patterns and soil variability in the Forest of Dean, *U.K. J. Biogeography*, **4**, 193–200.

Gurovich, L.A., 1982, Field spatial variability structure of soil hydrodynamic properties. *Ciencia Investigación Agraria*, **9**, 243–254.

Gurovich, L.A., and Stern, J., 1983. Field spatial variability of soil water infiltrability I. Data generation, *Ciencia Investigación Agraria*, **10**, 35–42.

Hack, H.R.B., 1976a, Components of error in field experiments with cotton, groundnuts, kenaf and sesame in the Central Sudan rainlands. I. Field and statistical methods, increasing precision by replication and its cost, *Expl. Agric.*, **12**, 209–242.

Hack, H.R., 1976b, Components of error in field experiments with cotton, groundnuts, kenaf and sesame in the central Sudan rainlands. II. Variation in plant measurements through the season and a discussion of surface drainage as a source of variation, *Expl. Agric.*, **12**, 225–240.

Hartge, K.H., Ellis, A., Nissen, J., and MacDonald, R.H., 1978, Anisotropy of penetration resistance in four soil profiles in Chile, *Geoderma*, **20**, 53–61.

Hatfield, J.L., Millard, J.P., and Goettelman, R.C., 1982, Variability of surface temperature in agricultural fields of central California, *Photogrammetric Engineering and Remote Sensing*, **48**(8), 1319–1325.

Hole, F.D., 1981, Effects of animals on soil, *Geoderma*, **25**, 75–112.

Hundal, S.S., and De Datta, S.K., 1981, IRRI Seminar, 10 Oct., 1981, IRRI, Los Banos, Phililippines.

Jansen, I.J., and Arnold, R.W., 1976, Defining ranges of soil characteristics, *Soil Sci. Soc. Amer. J.*, **40**, 89–92.

Journel, A.J., and Huijbegts, Ch. J., 1978, *Mining Geostatistics*, Academic Press, London.
Juo, A.S.R., and Lal, R. 1977, The effect of fallow and continuous cultivation on the chemical and physical properties of an Alfisol in western Nigeria, *Plant and Soil*, **47**, 567–584.
Kalla, S.E., El, and Gomma, A.A., 1977, Estimation of soil variability and optimum plot size and shape from wheat trials, *Agric. Res. Rev.* **55**, 81–88.
Kang, B.T., 1977, Effect of some biological factors on soil variability in the tropics. II. Effect of oil palm (*Elaeis guineensis*) tree, *Plant & Soil*, **47**, 451–462.
Kang, B.T., and Moorman, F.R., 1977, Effect of some biological factors on soil variability in the tropics, *Plant & Soil*, **47**, 441–449.
Keisling, T.C., Davidson, J.M., Weeks, D.L., and Morrison, R.D., 1977, Precision with which selected soil physical parameters can be estimated, *Soil Sci.*, **124**, 241–248.
Kellogg, C.E., 1975, *Agricultural Development, Soil, Food, People, Work*, Soil Sci. Soc. Amer., Madison, Wisconsin, 233 pp.
Klaassen, R., 1979, Aforestation on sites with very variable characteristics, *Bosbouw Tijdschrift (Nederland)*, **51**, 161–163.
Lal, R., 1974, The effect of soil texture and density on the neutron and density probe calibration for some tropical soils, *Soil Sci.*, **177**, 183–190.
Lal, R., 1979, Concentration and size of gravel in relation to neutron moisture and density probe calibration, *Soil Sci.*, **127**, 41–50.
Lal, R., 1984a, Mechanized tillage systems effects on soil erosion from an Alfisol in watersheds cropped to maize, *Soil & Tillage Res.* (In Press).
Lal, R., 1984b, Mechanized tillage systems effects on changes in soil properties of an Alfisol in watersheds cropped to maize, *Soil & Tillage Res.* (In Press).
Lascano, R.J., and Bavel, C.H.M. Van, 1982, Spatial variability of soil hydraulics and remotely sensed soil parameters, *Soil Sci Soc. Amer. J.*, **46**, 223–228.
Lee, R., Bailey, J.M., Northey, R.D., Barker, P.R., and Gibson, E.J., 1975, Variations in some chemical and physical properties of 3 related soil types: Dannevirke silt loam, Kiwitea silt loam, and Marton silt loam, *N.Z. J. Agric. Res.*, **18**, 29–36.
Leger, R.G., and Millette, G.J.F., 1975, Microvariations in soil climate under two kinds of forest cover, *Canad. J. Soil Sci.*, **55**, 447–456.
Leite, J. Del., and Ezeta, F.N., 1982, Soil variations in pastures of the Itapetinga agrosystem. I. The toposequence of Itaju do Colonia (Bahia, Brazil). *Boletim Tecnico*, CEPLAC 99, 20pp.
Leveque, A., 1969, The problem of stone lines: Preliminary observations on the gneiss–granite formation in Togo, *Cah. ORSTOM Ser. Pedol.*, **7**, 43–49.
Lierop, W. Van, and MecKenzie, A.F., 1977, Soil pH measurements and its application to organic soils. *Canad. J. Soil Sci.*, **57**, 55–64.
Luk, S.H., and Morgan, C., 1981, Spatial variations of rainwash and runoff within apparently homogeneous areas, *Catena*, **8**, 383–402.
Lutz, J.F., 1972, Soil physical properties, in *Review of Soils Research in Tropical America*, P.A. Sanchez (ed.). Soil Science Dept. North Carolina University, Raleigh, N.C., pp. 52–62.
Matheron, G., 1971, The theory of regionalized variables and its applications, *Cah. Centre Morphol. Math.* **5**.
McCown, R.L., Murtha, G.G., and Field, J.B.F., 1977, Pattern of distribution of Townsville stylo, annual grasses, and perennial grasses in relation to soil variation, *J. Applied Ecology*, **14**, 621–630,
Miedema, R., and Van Vuure, W., 1977, The morphological, physical, and chemical properties of two mounds of *Macrotermes Bellicosus* compared with surrounding soils in Sierra Leone. J. Soil Sci. 28, 112–124.

Miehlich, G., 1976, The homogeneity, heterogeneity and similarity of soil bodies, *Zeitschrift fur Pflanzenernahrung und Bodenkunde*, **5**, 597–609.

Milfred, C.J., and Kiefer, R.W., 1976, Analysis of soil variability with repetitive aerial photography, *Soil Sci. Soc. Amer. J.*, **40**, 553–557.

Mollitor, A.V., Leaf, A.L., and Morris, L.A., 1980, Forest soil variability on northeastern flood plains, *Soil Sci. Soc. Amer. J.*, **44**, 617–620.

Moltham, H.D., 1966, Influence of soil variability on soil moisture and strength predictions, USAE Waterways Exp. Station (Working draft).

Moormann, F.R., 1972, Soil microvariability, in *Soils of the Humid Tropics*, National Academy of Sciences, Washington, D.C., pp. 45–49.

Moormann, F.R., and Kang, B.T., 1978, Microvariability of soils in the tropics and its agronomic implications with special reference to West Africa, in *Diversity of Soils in the Tropics*, ASA Special Publication, No. 34, pp. 29–43.

Nagai, V., and Menk, J.R.F., 1976, Estimation of J and S in S_B curves of the distribution function of soil characteristics, *Bragantia*, **35**, 433–441.

Nielsen, D.R., Biggar, J.W., and Erh, K.T., 1973, Spatial variability of field-measured soil-water properties, *Hilgardia*, **42**, 215–259.

Norris, J.M., 1970a, Multivariate methods in study of soils, *Soils Fert.*, **33**, 313–318.

Norris, J.M., 1970b, Hypothesis generation for soil data using principal component analysis, in R.S. Andersen and M.R. Osborne (eds.), *Data Representation*, University of Queensland Press, Brisbane.

Norris, J.M., 1971, The application of multivariate analysis to soil studies. I. Grouping of soils using different properties, *J. Soil Sci.*, **22**, 69–80.

Norris, J.M., 1972, The application of multivariate analysis to soil studies. III. Soil variation, *J. Soil Sci.*, **23**, 62–75.

Norris, J.M., and Loveday, J., 1971, The application of multivariate analysis to soil studies. II. The allocation of soil profiles to established groups: a comparison of soil survey and computer method, *J. Soil Sci.*, **22**, 395–400.

Nortcliff, S., 1978, Soil variability and reconnaissance soil mapping: a statistical study in Norfolk, *J. Soil Sci.*, **29**, 403–418.

Northrup, M.L., and Boyle, J.R., 1975, Soil bulk densities after 30 years under different management regimes, *Soil Sci. Soc. Amer. Proc.*, **39**, 588.

Oliveira, J.B.De, 1975, Soil distribution and differentiation into various categories in two apparently homogeneous Oxisol areas, *Bragantia*, **34**, 309–348.

Oreshkina, N.S., 1976, Degree of variation of soil moisture as a function of moisture content, *Moscow Univ. Soil Sci. Bull.*, **31**, 7–9.

Panabokke, C.R., 1967, *Soils of Ceylon and Use of Fertilizer*, Metro Printers, Colombo.

Patterson, G.T., and Wall, G.J., 1982, Within-pedon variability in soil properties, *Canad. J. Soil Sci.*, **62**, 631–639.

Peck, A.J., Luxmoore, R.J., and Stolzy, J.L., 1977, Effects of spatial variability of soil hydraulic properties in water budget modeling, *Water Resources* Res., **13**, 348–354.

Peterson, R.G., and Calvin, L.D., 1965, Sampling, in C.A. Black (ed.) *Methods of Soil Analysis. Part I. Agronomy* 9, pp. 54–72, ASA, Madison, Wisconsin, USA.

Philip, J.R., 1980. Field heterogeneity: some basic issues, *Water Resources Res.*, **16**, 443–448.

Pilgrim, D.H., and Huff, D.D., 1978, A field evaluation of subsurface and surface runoff, *J. Hydrology*, **38**. 299–318.

Pilgrim, D.H., Huff, D.D., and Steel, T.D., 1978, A field evaluation of subsurface and surface runoff. II. Runoff processes, *J. Hydrology*, **38**, 319–341.

Pinho, A.F. Des, and Morais, F.I., 1980, Chemical properties related to microvariability in soils of the Recôncavo, Baiano, Brazil, *Revista Theobroma*, **1**, 41–46.

Polyakov, I.S., 1976, Spatial variation in the moisture content of Podzolic soils, *Pochvovedenie*, **3**, 65–76.
Ramirez, Ayala C., Palacios Velez, O., and Lara, G.P.Z. De, 1981, Spatial interpolation of data for salt content in soil, *Agrociencia, (Mexico)*, **45**, 89–103.
Rao, P.V., Rao, P.S.C., Davidson, J.M., and Hammand, L.C., 1979, Use of goodness-of-fit tests for characterizing the spatial variability of soil properties, *Soil Sci. Soc. Amer. J.*, **43**, 274–278.
Rao, P.S.C., Rao, P.V., and Davidson, J.M., 1977, Estimation of the spatial variability of the soil-water flux, *Soil Sci. Soc. Amer. J.*, **41**, 1208–1209.
Reid, I., 1977, Soil environment and the hydrometeorological mosaic, *Agric. Meteorology*, **18**, 425–433.
Reynolds, S.G., 1974, A note on the relationship between size of area and soil moisture availability, *J. Hydrology*, **22**, 71–76.
Reynolds, S.G., 1975, Soil property variability in slope studies: suggested sampling schemes and typical required samples sizes, *Zeitschfit fur Geomorphologie*, **19**, 191–205.
Ricaud, R., Golden, L.E., and Lytle, S.A., 1974, Physical and chemical properties of three groups of Mississippi River alluvial soils in the sugercane area of Louisiana, *Bulletin No. 683*, Agric. Expt. Station, Lousiana State University, 56 pp.
Riquier, J., 1969, Contribution to the study of stone lines in tropical and equatorial regions, *Cah. ORSTOM Ser. Pedol.*, **7**, 71–112.
Robertson, L.S., Warncke, D.D., and Baker, J.D., 1980, Chemical test variability within soil types, *Research Report No. 390*, Agric. Expt. Station, Michigan State University, 11 pp.
Rogowski, A.S., 1972, Watershed physics: soil variability criteria, *Water Resource Res.*, **8**, 1015–1025.
Russo, D., and Bresler, E., 1981a, Effect of field variability in soil hydraulic properties on solutions of unsaturated water and salt flows, *Soil Sci. Soc. Amer. J.*, **45**, 675–681.
Russo, D., and Bresler, E., 1981b, Soil hydraulic properties as a stochastic process: I. An analysis of field spatial variability, *Soil Sci. Soc. Amer. J.*, **45**, 682–687.
Russo, D., and Bresler, E., 1982, Soil hydraulic properties as a stochastic process. II. Errors of estimates in a heterogeneous field, *Soil Sci. Soc. Amer. J.*, **46**, 20–26.
Santamaria, F., 1965, Geographical distribution of the 'arrenife' horizon in Venezuela, *Bull. Soc. Venez. Cienc. Nat.*, **25** (108), 350–354.
Sartz, R.S., 1972, Anomalies and sampling variation in forest soil water measurement by the neutron method, *Soil Sci. Soc. Amer. Proc.*, **36**, 148–153.
Scharringa, M., 1976, On the representativeness of soil temperature measurements, *Agric. Met. (Nederland)*, **16**, 263–276.
Scott, D., 1975, Variation in soil pH under tussock grassland species, *N. Z. J. Expl. Agric.*, **3**, 143–145.
Segalen, P., 1969, Rearrangement of materials in soils and the formation of stone lines in Africa, *Cah. ORSTOM Ser. Pedol.*, **7**, 113–311.
Sharma, M.L., Gander, G.A., and Hunt, C.G., 1979, Spatial variability of infiltration in a watershed, *J. Hydrology*, **45**, 101–122.
Sisson, J.B., and Wierenga, P.J., 1981, Spatial variability of steady-state infiltration rates as a stochastic process, *Soil Sci. Soc. Amer. J.*, **45**, 699–704.
Slavov, D., and Dinchev, D., 1979, Variation in mineral nitrogen in soil under field conditions, *Pochvoznanie: Agrokhimiya*, **14**, 23–30.
Smith, L.P., 1970, The difficult art of measurement, *Agric. Meteorology*, **7**, 281–283.
Smith, R.E., and Hebbert, R.H.B., 1979, A Monte Carlo analysis of the hydrologic effects of spatial variability of infiltration, *Water Resource Res.*, **15**, 419–429.

Smyth, A.J., and Montgomery, R.J., 1962, *Soils and landuse in Central Western Nigeria*, Government Printer, Ibadan, Nigeria.
Springer, E.P., and Gifford, G.F., 1980, Spatial variability of rangeland infiltration rates, *Water Resource Bull.*, **16**, 550-552.
Suarez, F.D., 1979, Variability and soil sampling for fertility diagnosis, *Ciencia Investigacion Agraria*, **6**, 151-154.
Ten Berge, H.F.M., Stroornijder, L., Burrough, P.A., Bregt, A.K., and de Heus, M.J., 1983, Spatial variability of physical soil properties influencing the temperature of the soil surface. *Agric. Water Mgmt*, **6**, 213-226.
Thomas, M.F., 1974, *Tropical Geomorphology: A Study of Weathering and Land Form Development in Warm Climates*, MacMillan, London, 332 pp.
Thum, J., 1974, The variability of top soil characters and the sample size required for area-related mean values, *Archiv für Acker-und Pflanzenbau und Bodenkunde*, **18**, 909-915.
Torres, V., 1980, Estimation of the homogeneity index of the soil through random block experiments, *Cuban J. Agric Sci.*, **14**, 213-218.
Trudgill, S.T., and Briggs, D.J., 1981, Soil and land potential, *Progress in Physical Geography*, **5**, 274-285.
Vallis, I., 1973, Sampling for soil nitrogen changes in large areas of grazed pastures, *Comm. Soil Sci. and Plant Analysis (Australia)*, **4**, 163-170.
Varazashivili, L.I., Lytayev, I.A., and Petrova, M., 1976, Statistical soil moisture parameters as functions of the soil moisture potential and method for determining them, *Soviet Soil Sci.*, **8**, 98-104.
Vine, P.N., Lal, R., and Payne, D., 1981, The influence of sands and gravels on root growth of maize seedlings, *Soil Sci.*, **131**, 124-219.
Vintola, I., and Canarche, A., 1974, Some general features of the frequency distributions used in soil science, *Trans. 10th Int. Congr. Soil Sci.*, **VI**, pp. 676-683.
Vreeken, W.J., 1973, Soil variability in small loess watersheds, clay and organic carbon content, *Catena*, **1**, 181-196.
Wakao, N., and Furusaka, C., 1973, Distribution of sulfate-reducing bacteria in paddy field soil, *Soil Sci. Plant Nutr.*, **19**, 47-52.
Walker, P.H., Hall, G.F., and Protz, R., 1968a, Soil trend, and variability across selected landscapes in Iowa, *Soil Sci. Soc. Amer. Proc.*, **32**, 97-101.
Walker, P.H., Hall, G.F., and Protz, R., 1968b, Relation between landform parameters and soil properties, *Soil Sci. Soc. Amer. Proc.*, **32**, 101-104.
Wambeke, A. Van, and Dudal, R., 1978, Macrovariability of soils of the tropics. In *Diversity of Soils in the Tropics*, ASA Special Publication, No. 34, pp. 13-28.
Wang, C.M., 1971, What coefficient is the best indicator of the amount of soil heterogeneity. II. *Memoirs of the College of Agriculture*, National Taiwan University, 12, pp. 1-23.
Webster, R., 1976, The nature of soil variation, *Classification Soc. Bull.*, **3**, 43-55.
Webster, R., 1977, Spectral analysis of gilgai soil, *Aust. J. Soil Res.*, **15**, 191-204.
Webster, R., 1978, Mathematical treatment of soil information, *Trans. 11th Int. Congr. Soil Sci.*, **3**, 161-190.
Wilding, L.P., and Drees, L.R., 1978, Spatial variability: a pedologist's viewpoint, in *Diversity of Soils in the Tropics*. ASA Spec. Publ. No. 34, pp. 1-12.
Wright, R.L., and Wilson, S.R., 1979, On the analysis of soil variability, with an example from Spain, *Geoderma*, **22**, 297-313.
Yost, R.S., and Fox, R.L., 1981, Partitioning variation in soil chemical properties of some Andepts using soil taxonomy, *Soil Sci. Soc. Amer. J.*, **45**, 373-377.
Yost, R.S., Uehara, G., and Fox, R.L., 1982, Geostatistical analysis of soil chemical properties of large land areas I. Semi-variograms, *Soil Sci. Soc. Amer. J.*, **46**, 1028-1032.

Zablotskii, V.R., and Karpachevskii, L.O., 1977, Causes of spatial variation of soil moisture content in forests, *Moscow Univ. Soil Sci. Bull.*, **32**, 6–9.

PART II

Ecological Factors and Soil Physical Properties

Chapter 6
Vegetation and Soil

I	INTRODUCTION	231
II	FOREST RESOURCES AND DISTRIBUTION	233
III	MICROCLIMATE OF THE RAINFOREST	235
IV	SOIL PHYSICAL PROPERTIES	242
	A. Texture	244
	B. Bulk density and soil structure	246
	C. Water balance	248
	D. Water movement	249
	E. Soil moisture retention properties	250
	F. Run-off and soil erosion	251
V	SOIL–VEGETATION INTERACTION	252
VI	CHEMICAL AND NUTRITIONAL PROPERTIES OF SOILS	252
VII	CONCLUSIONS	254

I. INTRODUCTION

The tropical rainforest is a diverse and complex system that occupies approximately 10 per cent of the world's area and comprises some 40–50 per cent of the earth's 5 to 10 million species (Golley *et al.*, 1975; Myers, 1981). Even with the aid of the space age technology however, tropical forests are an intriguing challenge. Ecological characteristics of rainforests, described by many researchers (Sanchez, 1972, 1976; Farnworth and Golley, 1974; Klinge, 1977; UNESCO, 1978; Herrera, 1979; Moran, 1981), include high plant biomass, concentration of nutrients within the plant biomass, rapid rates of nutrient recycling, high annual rainfall with little seasonal variation in temperature and humidity, and a relatively closed system for nutrients and water. The rainfall in tropical forests is 3 times the world average; tropical forests contribute some 58 per cent of the earth's available water vapour (Moran, 1981), and receive 2.5 times more annual solar radiation than the poles. Sanchez (1972) estimated that rainforest biomass ranges from 200 to 400 t ha^{-1}, out of which 75% is trunk and branches, 15 to 20% roots, 4 to 6% leaves and only 1 to 2% litter. Most of the biomass in the rainforest is synthesized in a relatively short period of 8 to 10 years (Sanchez, 1976). It is widely believed that the ecosystem's nutrient capital is concentrated in the plant biomass. Some researchers argue that it is a misconception and that a considerable amount of N and P may be in the soil layers containing high concentrations of soil organic matter (Klinge, 1966, 1977; Anderson and Swift, 1983). Whatever nutrients exist in the soil, however, are stored in the upper few centimetres (Greenland and Kowal, 1960). Rainforest vegetation feeds on itself by the most rapid and efficient nutrient recycling system.

The number, size and height of tree species existing within a tropical rainforest are highly variable and diverse (Figs. 6.1 and 6.2). For example, Golley *et al.* (1975) counted 4800 stems of different plants in one-quarter hectare of a tropical moist forest. In a high forest near Benin City in southern Nigeria, Ghuman and Lal (unpublished data) estimated 1800 stems per hectare exceeding 15 cm in circumference at one metre above the ground surface. The number of trees exceeding 20 cm diameter one metre above ground can be 84 (Golley *et al.*, 1975) to 474 per hectare (Grubb *et al.*, 1963). A leaf area index of 7 to 28 is commonly observed for rainforests of different regions in the tropics (Odum, 1970; Ogawa *et al.*, 1961; Golley *et al.*, 1975). The canopy cover of a rainforest drastically reduces the insolation reaching the ground underneath. The microclimate within a forest cover is warm with high relative humidity.

The objective of this chapter is to describe the effects of forest on soil physical properties and on microclimate. The word 'forest' in this context is used to include rainforest, open woodland, and long forest fallows.

Figure 6.1 A high rainforest at Okomu Plantation, near Benin City, Nigeria. The forest has three well defined strata and exists in a region with mean annual rainfall of 2200mm

Figure 6.2 In the cloud forest along the Atlantic coast of Brazil, near Sao Paulo, continuous moisture supply from the Ocean supports luxurious growth of epiphytes

II. FOREST RESOURCES AND DISTRIBUTION

There are few reliable estimates of the actual area covered by tropical rainforest. The estimates vary widely because of the differences in criteria used to define this ecosystem, and because the statistics are often unreliable. Most estimates are 10 to 20 years old, and due to rapid rates of deforestation for diverse land-uses, it is difficult to assess reliably the exact area. The accuracy of most estimates is within 40 to 50 per cent at best (Myers, 1983), and most estimates are notably imprecise (Table 6.1a).

A report by UNESCO (1978) estimated the total area under tropical rainforest to be about 700 million hectares, with relative distribution of 557 million hectares in the Amazon basin, 125 million hectares in Asia and 83 million hectares in tropical Africa. According to this report, Brazilian Amazon occupies some 362 million hectares (Table 6.1).

In *State of the World, 1984*, Postel's (1984) estimates of tropical forest

Table 6.1a Extent of tropical forest (10^6 hectare)

Authors	Asia	Latin America	Tropical Africa	Total
Persson (1974)				
Closed Forest	294.0	576.8	195.9	1066.7
Total Forest	406.0	734.1	755.0	1895.1
Sommer (1976)				
Moist Forest	254.0	506.0	175.0	935.0
Total Forest	417.0	964.0	334.0	1715.0
UNESCO (1978)				
Moist Forest	124.9	557.0	83.25	765.15
Total Forest	128.2	725.0	138.75	1021.95
Myers (1981)				
Moist Forest	271.4	641.6	151.4	1064.4
FAO (1981)				
Moist Forest	357.0	943.1	669.2	1969.3
Lanly (1982)				
Closed Forest	305.5	678.7	216.6	1200.8
Total Forest	336.5	895.7	703.1	1935.3
Postel (1984)				
Moist Forest	305.0	629.0	217.0	1201.0
Total Forest	445.0	1212.0	1312.0	2969.0

resources are higher than that of UNESCO (1978). According to Postel (Table 6.2), the area of closed forest in the tropics is 1201 million hectares, open woodland 703 million hectares, and forest fallow and shrub land 1034 million hectares. Because shifting cultivation and related bush fallow systems are practiced in the tropics only, 98% of the world's forest fallow and shrubland lies within the tropics.

Estimates of forest resources within an ecology also vary widely. The estimates of forested land in Africa by UNESCO (Table 6.1) are similar to those of Phillips (1974) for Africa (Table 6.3). However, Postel's estimates (Table 6.2) are much higher. A recent survey by FAO (1981) observed that tropical Africa contains 216.5 million hectares of closed forest and 486.5 million hectares of mixed forest–grassland formations. The undisturbed productive closed forests cover about 118.5 million hectares, of which 94% are in the Cameroon–Congolese block and 67.3% in Zaire alone.

It is obvious from the various estimates presented that there is a need for reliable and accurate assessments of the actual forest resources of the tropics. This should be done by an international organization with a standard set of criteria, and with methodology that is frequently verified by ground truth.

Table 6.1 Estimates of forest in the tropics (millions of hectares) (UNESCO, 1978). *From* Tropical Forest Ecosystems. © *Unesco–Unep 1978. Reproduced by permission of Unesco*

Country		Evergreen rain forest	Semi-deciduous forest	Total area
(a)	*South East Asia*			
	India	4.5	1.8	6.3
	Srik Lanka	0.2	0.1	0.3
	Burma	—	—	
	Thailand	—	—	
	Indonesia	89.2	1.4	90.6
	Malaysia	21	—	21
	Philippines	10	—	10
	Cambodia	—	—	6
	Laos	—	—	5
	Vietnam	—	—	8
	Total	124.9	3.3	128.2
(b)	*Tropical Africa*			
	Cameroon	6.5	6.5	13.0
	Ivory Coast	4.5	4.5	9.0
	Congo	3	7	10
	Gabon	17	5	22
	Central African Empire	0.75	3	3.75
	Zaire	50	25	75
	Madagascar	1.5	4.5	6
	Total	83.25	55.5	138.75
(c)	*Tropical America*			
	Bolivia	25	10	35
	Brazil	362	101	463
	Colombia	49	22	71
	Ecuador	14	4	18
	Guyana	14	—	14
	French Guyana	7	—	7
	Peru	62	6	68
	Surinam	7	—	7
	Venezuela	17	25	42
	Total	557	168	725
	Grand Total	765.15	226.8	1021.95

III. MICROCLIMATE OF THE RAINFOREST

The approximate global estimate of water balance in the tropical region shown in Fig. 6.3 (UNESCO, 1978) indicates an annual water surplus in the equatorial belt. The drawback of the annual water balance of this nature

Table 6.2 Forest resources of the tropics (million hectares) (Modified from Postel, 1984)

Region	Closed forest	Open woodland	Forest fallow and shrubland	Total
Tropical America	679	217	316	1212
Tropical Africa	217	486	609	1312
Tropical Asia	305	31	109	445
Total in the tropics	1201	703	1034	2969
World Total	2676	1159	1055	4890
Percentage of the World	44.9	60.7	98.0	60.7

Table 6.3 Estimates of potential and remaining forest reserves in tropical and subtropical Africa in 1972 (Phillips, 1974). *Reproduced by permission of Academic Press Inc.*

Region/Zone	Potential forest area	Existing forest area
	(millions of hectare)	
A. *Forest zone*		
Guinea (Guinea, Sierra Leone, Ivory Coast, Ghana)	33	17
Nigerian (Togo, Benin, Nigeria)	16	4
Equatorial (Cameroon, Gabon, Congo, Central African Republic, Maiombe–Cabinda–Angola)	63	43
Zaire	105	74
Eastern montane (Zaire, Rwanda, Burundi)	1	1
Total	218	139
B. *Other Ecologies*		
Ethiopia and southern Sudan	9	4
East Africa (Uganda, Kenya, Tanzania, Mozambique, Malawi)	8	3
Southern Africa (Zimbabwe, South Africa)	0.6	0.2
Total	17.6	7.2
Grand total	235.6	146.2

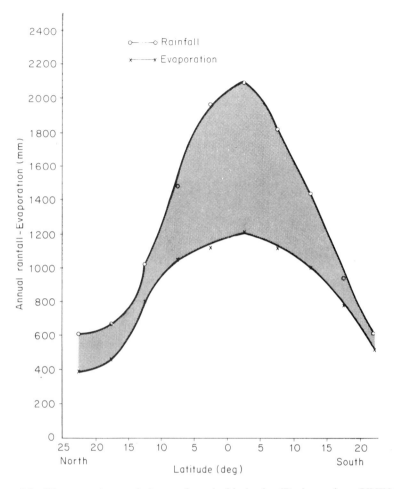

Figure 6.3 The annual water balance of tropical latitudes (Redrawn from UNESCO, 1978)

is that it does not reflect short-range (5 to 15 days) water deficits that have significant effects on plant-water status and growth. The existence of large rivers (Amazon, Congo, Mekong etc.) in tropical forest region is obviously an indication of the water surplus of this ecosystem. The region influenced by the Inter-Tropical Convergence Zone (ITCZ) supports these large river systems. Tropical forests also contribute greatly to the global water vapour pool, i.e. 58% of the global vapour pool is contributed by 40% of the earth's surface. The water surplus is more in the equatorial belt and decreases with increasing latitude. Medina (1983) observed in Venezuela that water surplus decreased from 1857 mm per annum for latitude 3°N (tropical rainforest) to 0 for 6°N (Fig. 6.4). In addition to the mean annual water balance, the length of dry season is important in determining predominant vegetation. For global energy balance,

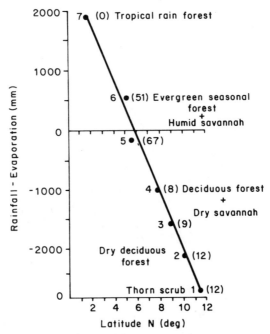

Figure 6.4 Change in water balance with change in vegetation from tropical rainforest at 3°N in Venezuela (Medina, 1983). *Reproduced by permission of Elsevier Science Publishers*

readers are referred to reports by UNESCO (1978) and other reviews on tropical climatology.

The microclimate within the forest canopy has not been extensively studied. Because the relative humidity is generally high, the use of ink-recorders within the humid forest environments is likely to cause an error of 2 to 5%. The lack of wind within the dense canopy is another factor responsible for high humidity. The measurements of temperatures and humidity for a primary forest in southwestern Nigeria are shown in Fig. 6.5. An example of the earlier data of 1939 is deliberately chosen because undisturbed primary forest existed in southwest Nigeria 50 years ago, but not any more. It is apparent from the data that the maximum relative humidity is about 95% regardless of the season. In the undisturbed primary forest, the minimum relative humidity is also not drastically lessened even in the dry season.

Maximum and the minimum temperatures are slightly lower in the rainy than in the dry season (Fig. 6.6a and 6.6b). The measurements of microclimate in the high forest near Benin City in southern Nigeria shown in Fig. 6.6 were made in April, 1984. The maximum temperature within the forest canopy on a sunny day was 24–26 °C with little diurnal fluctuation. The maximum relative humidity was between 87 and 90 per cent. Closed-forest canopy effectively intercepts the incoming radiation, which is why the undergrowth is minimum in a primary or a high forest. The increasing solar radiation being low under the forest canopy (Fig. 6.7a) also causes significantly lower soil temperatures

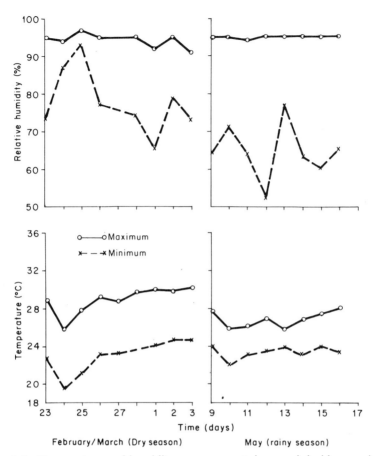

Figure 6.5 Temperature and humidity measurements in a semi-deciduous rainforest near Ibadan, Nigeria (After Evans, 1939). *Reproduced by permission of the British Ecological Society*

under forest than on cleared land (Fig. 6.7b). The temperature profile of the forest canopy near Benin City, from 50 cm below to 9m above ground surface for sunny and cloudy days were also compared. On a clear day, the soil temperature exceeded the air temperature at 0600 hours but was lower than the air temperature at 1500 hours. On a cloudy day, however, the soil temperature was slightly more than the air temperature regardless of the time of the day. The diurnal fluctuations in soil temperature within the forest canopy are compared in Fig. 6.7. There were little diurnal fluctuations (range 6.5 °C at 1 cm depth) in soil temperature under the forest cover.

The data presented are representative of other regions under high or primary rainforest. The forest cover has a strong buffering effect against sudden fluctuations in temperature, humidity and wind velocity. There are minimal diurnal fluctuations in microclimate under the forest cover. Similar observations on the microclimate of tropical forest have been reported by Aubert (1961) for Ivory Coast, Lawson *et al.* (1970) for Ghana, Whitmore (1975) for

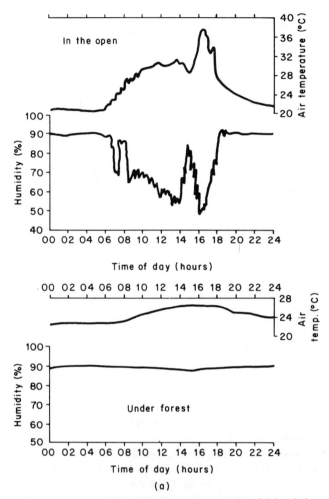

Figure 6.6a Temperature and humidity measurements in a high rainforest at Okomu, Benin in May, 1984 (Ghuman and Lal, 1985)

the Far East, Ghosh et al. (1982) in India, Tracey (1969) in Australia, and by Richards (1952, 1962) and Holdridge et al. (1971). Even in the semi-arid climates of Senegal, Dancette and Poulain (1968) observed that presence of established tree canopy increased relative humidity and soil moisture content.

Although seasonal and diurnal variations in microclimate within the forest canopy are less, the vertical and horizonal variations are not. Many researchers have reported considerable variations in air temperature and relative humidity at different heights under the forest canopy (Evans, 1939; Richards, 1952; Cachan, 1963; Cachan and Duval, 1963). In Ivory Coast Cachan and Duval (1963) observed that the diurnal range of air temperature increased with increase in height above ground (Fig. 6.8). The data indicate a linear increase in the temperature range up to 25 m above ground. There were also significant

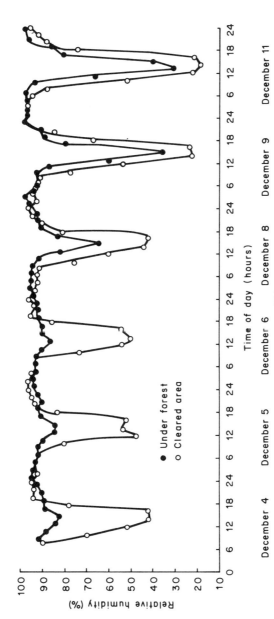

Figure 6.6b Relative humidity measurements in Okɔmu forest near Benin, Nigeria (Ghuman and Lal, 1985)

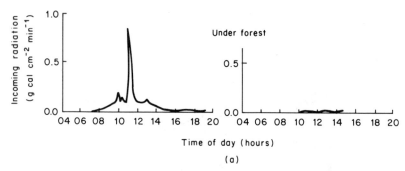

Figure 6.7a   In-coming solar radiation under forest cover on a sunny and cloudy day (Ghuman and Lal, 1985)

seasonal variations in the temperature range between the dry and the rainy season. The magnitude of the vertical variations depends on the tree species and the number of climbers and epiphytes. During day, the air temperature near the ground is lower than up in the canopy, and the reverse is the case at night. Nevertheless, dew formation does not occur under the forest cover.

The horizontal differences in microclimate under the forest canopy are due to spatial variations in the canopy cover and the leaf area index. The air temperature is higher under a canopy-gap than under full cover even a few meters apart. The radiation transmitted through the canopy-gap has an important effect on energy balance and the localized air temperature (Grubb and Whitmore, 1967; Bjorkman and Ludlow, 1972). Sunfleck light is a very important component (Longman and Jenik, 1974), which is particularly difficult to quantify. In southern Nigeria, Evans (1956) reported the average daily sunfleck light to be 500 cal dm^{-2} in comparison with less than 200 cal dm^{-2} under the shade. The relative light intensity in the rainforest is generally low, i.e., 2–5% in Surinam (Schulz, 1960), 0.2–0.7% in an Indonesian rainforest (Bunning, 1947), in Guyana (Carter, 1934) and in Ivory Coast (Cachan, 1963).

Tropical forests, therefore, create a unique under-canopy microclimate characterized by high humidity, low radiation and wind movement, relatively constant temperatures with little seasonal and only slight diurnal changes. The microclimate, however, has both horizontal and vertical variations due to differences in canopy cover and sunfleck light received through canopy-gaps. Although tropical rainforests have large annual water surpluses, short-period water deficits commonly observed influence plant-water status and microclimate.

IV.   SOIL PHYSICAL PROPERTIES

In contrast to the extensive literature available on chemical properties of soils under tropical rainforests, there is little information on soil physical properties. With some exceptions (Andosols and Inceptisols), most soils supporting tropical rain forests are very old, weathering has occurred to greater depths in

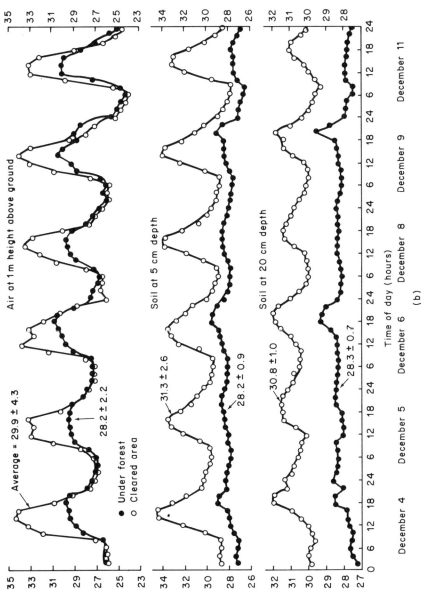

Figure 6.7b Diurnal fluctuations in soil temperature under the forest cover (Ghuman and Lal, 1985)

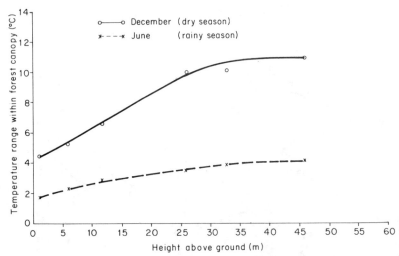

Figure 6.8 Diurnal range of ambient temperature at different heights within a forest canopy in Ivory Coast (Cachan and Duval, 1963)

Table 6.4 Textural properties of some soils of Cambodia (Tkatchenko, 1948). *Reproduced by permission of IRAT*

Property	Battambang	Kampot
Sand (%)	87.1	73.4
Silt (%)	11.3	12.0
Clay (%)	1.6	14.6

the bedrock, the clay has been eluviated to deeper horizons and is strongly aggregated. These are the reasons for the peculiar physical properties of the predominant uplands (Oxisols and Ultisols) supporting tropical forests.

A. Texture

The texture of the surface horizons of the strongly weathered Oxisols and Ultisols under rainforest is generally sandy to loamy sand, because the clay-sized particles have been eluviated to the deeper soil horizons. In Ivory Coast Roose and Godefroy (1977) observed a considerable loss of clay from the upper horizons. Examples of textural properties of some soils from Cambodia indicate sand contents of 73 to 87 per cent (Table 6.4). Mechanical analysis of some Oxisols from the Amazon region of Brazil indicate low silt and clay and high amounts of coarse and fine sands (Tables 6.5 and 6.6). Similar to the soils of West Africa, the silt content of sedentary soils of the Amazon region is low. Textural properties of an oxisol from Costa Rica, however, show a high silt content of 23% (Table 6.7).

This is not to say that all upland soils that support rainforests are coarse-

Table 6.5 Textural properties of soils from the Amazon basin in Brazil – Latosol Vermelho (EMBRAPA, 1976)

Depth (cm)	Particle size (%)			
	Coarse sand	Fine sand	Silt	Clay
0–7	28	53	6	13
7–21	26	49	14	11
21–44	27	51	6	16
44–80	23	49	8	20
80–113	23	49	10	18
113–150	22	50	9	19

Table 6.6 Textural properties of the surface horizons of some Oxisols from the Amazon region of Brazil (EMBRAPA, 1976)

Soil/Depth (cm)		Particle size distribution (%)			
		Coarse sand	Fine sand	Silt	Clay
A	0–5	35	46	11	8
B	0–20	25	54	7	14
C	0–8	37	48	7	8
D	0–7	34	49	7	10
E	0–11	18	56	12	14
F	0–2	33	41	17	9

Table 6.7 Textural properties of a surface soil (0–13cm) under primary forest in Costa Rica (Holdridge *et al.*, 1971)

Property	
Gravel (%)	8
Sand (%)	54
Silt (%)	23
Clay (%)	15
Liquid limit	71
Plastic limit	60
Plasticity index	11

textured. Heavy-textured soils (especially some Inceptisol, Entisols, and Andosols) also occur in the tropical forest zone, particularly in Central America along the flood plains of large river system and in the volcanic-ash-derived soils of Southeast Asia. High silt content is reported in soils derived from different parent material. For example, Tanaka *et al.* (1984) observed high silt content in soils developed on redeposited saprolite of Tertiary mudstones. The flood-plain soils generally have high silt and clay content.

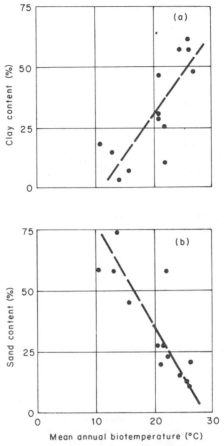

Figure 6.9 Relationship of (a) clay and (b) sand content in A horizon with mean annual biotemperature in a humid forest region in Costa Rica with annual rainfall of 3750±850 mm (Holdridge et al., 1971). *Reproduced by permission of Pergamon Press*

Holdridge et al. (1971) observed in soils of tropical rainforest in Costa Rica that in regions with the same rainfall, the clay content of the soil generally increased and sand content decreased with the increasing mean annual temperature (Fig. 6.9).

B. Bulk density and soil structure

With the exception of soil texture, soil physical properties (such as bulk density, porosity, infiltration, strength) are dynamic characteristics and are easily altered by the climate and man's intervention. In primary forest, however, these transient properties have attained a steady state level and experience only moderate seasonal fluctuations. The bulk density of soils supporting undisturbed forests is generally low in the surface horizon. Some

Table 6.8a Soil bulk density of some Peruvian Ultisols (Tanaka et al., 1984)

Horizon	Typic Paleudult (Yurimaguas)		Aquic Paleudult (Yurimaguas)	
	Depth (cm)	Bulk density (g/cm^3)	Depth (cm)	Bulk density (g/cm^3)
A	0–7	1.37	0–13	1.32
A_3	7–21	1.41	13–42	1.47
B	21–102	1.43	—	—
C	102–170	1.65	42–50	1.56

Table 6.8b Bulk density of some forest soils

Soil	Region	Bulk density (g/cm^3)	References
Oxisols	Hawaii	1.1–1.2	UNESCO (1978)
Inceptisols	Hawaii	0.8	UNESCO (1978)
Inceptisols	Hawaii	0.2–0.5	UNESCO (1978)
Ultisols	Hawaii	1.2	UNESCO (1978)
Ultisols	Thailand	0.48–0.67	UNESCO (1978)
Ultisol	Peru	1.32–1.37	Tanaka et al. (1984)
Alfisol	Nigeria	1.0–1.30	Lal (1979)
Ultisol	Nigeria	1.13–1.40	Ghuman and Lal (1984)
Oxisol/Ultisol	Puerto Rico	0.66–1.18	Edmisten (1970)
Oxisol/Ultisol	Puerto Rico	0.75–0.99	Jordan (1970)
Oxisol	Costa Rica	0.76	Holdridge et al. (1971)

data presented in chapter 4 indicate that soils in the forest zone of West Africa have a bulk density of less than one. The data in Table 6.8 show that in spite of the predominantly sandy texture, the bulk density of surface horizons of two paleudults from Yurimaguas, Peru, is low, i.e. 1.32 to 1.37 g cm^{-3}. The low bulk density of soils supporting tropical rainforests is mainly due to high activity of soil fauna. The compacting effect of the raindrop impact on soil is prevented by the leaf litter, and the high activity of termites, ants, earthworms and other soil animals keeps the soil highly porous, with a predominance of macropores and a loose and friable consistency.

Highly weathered and well-drained Oxisols and Ultisols, especially those that have not been disturbed by man's intervention due to logging activities, are often described to have good soil structure in comparison with, say the Alfisols and Quartzipsammentic subgroups. Good soil structure, as described for Oxisols by Buringh (1968), is caused by the flocculation/cementation of clay by organo-mineral complexes to form stable aggregates. The stable microaggregates are within the size range of silt to sand and develop weak or moderate granular or subangular blocky structure. If the soils are heavy textured, as in poorly drained flood plains, soil matrix lacks the stable micro-aggregates and thus develops into a compact subangular blocky structure of hard consistency.

The good soil structure of well-drained Oxisols and Ultisols should not be confused with their being easily degraded when forest is removed and the soil is cultivated by mechanized farm operations. Once cleared of its forest, soil structure can deteriorate rapidly, especially when exposed to high intensity rains and extremes of temperatures. Compaction is a common problem in soils managed with mechanized farm operations. Under cultivation these soils are also susceptible to severe erosion and the stable micro-aggregates are easily dislodged and transported downslope in water run-off.

C. Water balance

The water balance over a short time scale, i.e. 5 to 15 days or even a month, is a very important climatic parameter even in the per-humid and humid environments. In the humid ecology, short-term water deficit is frequently observed and it has drastic effects on vegetation, fauna, and on soils. Water deficit has been observed in the rainforest of Malaya (Nieuwolt, 1965), Amazon (Jordan, 1981), East Africa (McCulloch and Dagg, 1965), and West Africa (Laurie, 1957). In addition to rainfall amount and its distribution, the degree and duration of stress experienced by forest vegetation depends on the water-holding capacity of the soil and its rooting depth.

Soil protected by forest vegetation cover has less evaporation than bare soil or free water surface. The lower evaporation losses are due to low temperatures, low insolation, and less wind movement within the forest canopy. That is why the presence of trees even in the semi-arid savanna vegetation is known to reduce potential evapotranspiration (Schoch, 1968). In the humid tropics the forested land loses less water from the soil surfaces than land without forest cover.

The water budget of a tropical rainforest near San Carlos de Rio Negro, Venezuela, was computed by Jordan and Heuveldop (1981). From the annual rainfall of 3664 mm, throughfall was 87%, stemflow 85%, transpiration 47% and evaporation from the leaf surfaces was 5%. Many studies indicate that throughfall in tropical rainforest is about 70 to 80 per cent (Mohr and Van Baren, 1954; Malaisse, 1973; Dabral and Rao, 1968). Most researchers, however, have only indirectly estimated the throughfall and direct measurements are few. A survey of the literature of water balance of tropical rainforest is reviewed in a report by UNESCO (1978).

Soil moisture regime and available water reserves are important edaphological factors. Few studies have been made regarding the seasonal variations in soil moisture content in tropical forests of varying rainfall amounts. In the Yun-nan province of China, Zonn and Li (1961) observed considerable seasonal variation in soil moisture content under tropical forest. During the rainy season the soil was wet to field moisture capacity up to about 100 cm depth under *Geronniera plycnemia* and to 150 cm depth under bamboo forest. During the dry season, however, the soil moisture content declined to the wilting point. Perched water table was never observed even in the rainy season under an actively growing forest. In Zaire, Bronchart (1963) observed that soil moisture under forest often drops below wilting point during the dry season. Unless poorly drained, soils supporting an active forest undergo wide fluctuations in soil moisture. Dry soils can also develop locally because of physiographic conditions, due to actively growing trees, or on the upper part of the catena. Holdridge *et al.* (1971) observed in a Costa Rican rainforest that the percentage moisture content of the 15 to 30 cm layer increased exponentially with an increase in mean annual rainfall. The relative increase in soil moisture content with increasing rainfall, however, depends on soil texture, vegetation density, position within the catena and other ecological factors. In a forest soil at El Verde, Puerto Rio, Edmisten (1970) observed highly porous soils retaining as much as 170% soil moisture content. These soils have often high organic matter content (30%). Similar soils have been observed in those regions of Central America where soils are influenced by volcanic ash.

D. Water movement

The infiltration rate of uplands and strongly weathered Oxisols supporting undisturbed primary vegetation is generally high. This high rate is partly due to coarse texture of the surface horizon and partly to the well-developed friable structure with stable micro-aggregates. Infiltration rates of soils supporting rainforest are reported to be 14 cm hr^{-1} for tropical Africa (UNESCO, 1978); 13.3, 15.4 and 23.7 cm hr^{-1} for Inceptisols, Oxisols, and Ultisols, respectively, of Puerto Rico (Bonnet, 1968), and 60 to 70 cm hr^{-1} for forest soils at El Verde, Puerto Rico (Edmisten, 1970). In soils of low bulk density, water movement under small positive pressure can be easily 1 cm min^{-1}. High rates are often observed on the upper slopes of the catena where soils are loose, friable and deeply weathered. In western Nigeria, Moormann *et al.* (1975) observed extremely high infiltration rates of soils under semi-deciduous rainforest (Fig. 6.10). These high rates, however, decline rapidly after deforestation.

In Puerto Rico Kline and Jordan (1968) and Jordan (1970) conducted water movement studies using tagged water. One of the drawbacks of these lysimetric studies was the high rate of absorption $^{134}C_s$ and $^{85}S_r$) within the litter layer. Furthermore, determining the concentrations of the radioactive materials in different horizons is not necessarily an indication of the amount of water that has passed through the soil profile.

Some soils supporting tropical rainforests have low infiltration rates, most forested soils are freely drained and water transmission through the profile is very rapid. Some Inceptisols and Entisols along the flood plains are, however, poorly drained because of heavy texture and moderate to low structural development.

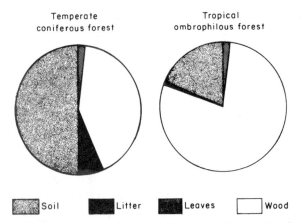

Figure 6.10 Distribution of organic carbon in the abiotic portion (soil litter) and biomass (wood, leaves) of tropical ombrophilous forest and temperate coniferous forest (After Kira and Shidei, 1967; Longman and Jenik, 1974). *Reproduced by permission of Longman Group Ltd*

E. Soil moisture retention properties

With good soil structure, high porosity, and friable consistency, the maximum water-holding capacity of these soils is often high. The moisture retention at the field moisture capacity, however, is usually low. These soils attain field moisture capacity at 30 to 50 cm of water suction. That is why laboratory-determined pF curves often give erroneous overestimated available water holding capacities. Few data are available regarding *in situ* measured pF characteristics. Tanaka *et al.* (1984) measured the water holding capacity of some Oxisols and Ultisols from the Amazon basin, and observed it to be 5 to 15 cm for the top 50 cm of the root zone (Table 6.9). A major portion of the available water, however, is released below pF 3. This is a peculiar characteristic of soils containing small amounts of low activity clays and in those where strongly aggregated clay behaves like sand. The low available water-holding capacity is partly responsible for frequent drought stress experienced by the forest vegetation during rainless periods.

Table 6.9 Available water-holding capacity of the surface horizons of some soils of the Amazon river system (Tanaka *et al.*, 1984)

Soil	Country	Depth (cm)	Available water-holding capacity (cm)		
			pF 1.7–3.0	pF 3.0–4.2	pF 1.7–4.2
Typic Haplustox	Brazil	0–45	6.1	4.2	10.3
Typic Acrustox	Brazil	0–36	5.9	2.6	8.5
Typic Acrustox	Brazil	0–53	11.5	3.9	15.4
Typic Paleudult	Peru	0–53	1.9	2.8	4.7

F. Run-off and soil erosion

Although the annual water surplus is high for the regions supporting tropical forest, the surface run-off under the forest vegetation is generally low. Run-off characteristics (amount, rate, velocity, etc) depend on type of vegetal cover, rainfall, soil, and the topography. Even though the decomposition rate is high (Wanner, 1970; Swift et al., 1981), there is adequate litter on the forest floor during the rainy season (Zonn and Li, 1960; Klinge, 1977) to provide protection against raindrop impact. That is probably the reason for the low sediment load of river systems that drain undisturbed tropical rainforests.

In per-humid and humid climates that support primary forests, there is often no shrub undergrowth and the surface soil is easily saturated due to frequent heavy rains. Large concentrated water drops from leaves and canopy attain a terminal velocity and cause considerable soil detachment. Under these conditions a large proportion of rainfall received is lost as surface run-off. Odum et al. (1970) observed that 45% of the annual rainfall of 3759 mm in a Puerto Rican forest was lost as run-off. In an Amazonian rainforest Jordan and Heuveldop (1981) reported that 48% of the annual rainfall of 3664 mm was lost as run-off. In northern Australia, 63% of the annual rainfall of 4175 mm was observed to be run-off (Gilmour and Bonell, 1977; Bonell and Gilmour, 1978; Bonell et al., 1979). These supposedly high run-off rates are likely to contain a considerable amount of interflow. It is often difficult to separate the direct run-off from the interflow. In low rainfall regimes of western Africa, however, the semi-deciduous or evergreen forest vegetations are virtually closed systems and little run-off is observed (Pereira, 1965; McCulloch and Dagg, 1965; Roose, 1977; Lal, 1981) (Table 6.10). In the savanna vegetation, however, run-off rates can be high (Rougerie, 1958).

Table 6.10   Run-off and soil erosion under undisturbed vegetation in different rainfall regimes of West Africa (Roose, 1977)

Vegetation	Country	Annual rainfall (mm)	Slope (%)	Run-off (%)	Erosion (t ha^{-1} y^{-1})
Evergreen secondary forest	Ivory Coast	2100	7	0.14	0.03
Evergreen secondary forest	Ivory Coast	2100	20	0.7	0.2
Evergreen secondary forest	Ivory Coast	2100	65	0.7	1.0
Semi-deciduous forest	Ivory coast	1750	9	—	0.5
Shrub savanna	Ivory Coast	1200	4	0.03	0.01
Thicket	Republic of Benin	1300	4	0.1–0.9	0.3–1.2

Regardless of the run-off rate and amount, erosion levels from a forested catchment are usually low. In Cambodia, Carbonnel (1964) observed that erosion rates under undisturbed forest have not changed over the last 5000 years. Most erosion under primary forest may occur as splash, slope wash, and soil creep. More erosion normally occurs in regions with wet/dry and in very humid climates than in regions of intermediate rainfalls.

V. SOIL–VEGETATION INTERACTION

Just as forests influence soil physical properties, some edaphological parameters (texture, water holding capacity, rooting depth etc.) have a strong effect on the forest and its species diversity. Soil moisture regime and the soil moisture depletion patterns vary widely under different vegetations depending on the tree density, rooting patterns, leaf area index, and the successional stage.

In southern Africa Strang (1969) observed that soil moisture in topsoil under *Hyparrhenia* grasses was depleted to wilting point shortly after the end of the rainy season. Moisture depletion continued until the next rains, and there was greater withdrawal by grass than by trees. Similar differences in moisture depletion patterns were observed between forest and savanna vegetations by Sarlin (1983). Sarlin observed different empirical relations between the available soil moisture and textural properties for different climax vegetations. Soils under live vegetation cover retained more water than those without good plant cover. The vegetal cover, in fact, is important in maintaining high levels of moisture reserves. The available moisture reserves were related to texture as follows:

$$W = 1.5\,S + 3C$$

where S and C are per cent sand and clays, respectively. Similar to moisture depletion, soil moisture recharge also differs among different vegetations (Glover and Gwynne, 1962).

Tree size and density and predominance of a specific climax vegetation often depend on soil properties (Tergas and Popenoe, 1971; Aweto, 1981). This general trend, however, is not easily observed in soils that have supported a forest vegetation for a long time. For example, Medina and Grohmann (1966) reported that the presence of Cerrado vegetation in São Paulo was not related to soil physical properties. In general, however, the presence of predominant species in evergreen forest is related to soil moisture regime and to soil texture. UNESCO (1978) reported soil texture–specie relationship for forests in Guyana, Amazonia, Borneo, Sarawak and Vietnam.

VI. CHEMICAL AND NUTRITIONAL PROPERTIES OF SOILS

Nutritional and chemical properties of soils supporting rainforests have been extensively described (Greenland and Kowal, 1960; Stark, 1969; 1970; 1971;

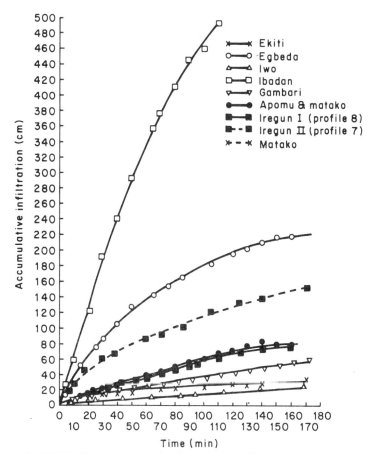

Figure 6.11 High infiltration rates of soils under semi-deciduous rainforest in Western Nigeria are due to high soil biological activity (Moormann et al., 1975). Extremely high rates are due to lateral flow under the ring

1972; Walter, 1973; Sanchez, 1976; Stark and Spratt, 1978; Jordan and Uhl, 1978; Herrera et al., 1978; 1981; Jordan, 1978; Moran, 1981). To a casual observer the lush green forest vegetation gives the deceptive and often misleading impression of high soil fertility. Most of the nutrient reserves of the forest ecosystem are tied in the vegetation and little nutrient is stored in the soil beneath (Fig. 6.11). Furthermore, most of the nutrients stored in the soil are in the top 30 cm, and a high proportion of active roots are located within the litter–soil interphase. The rainforest vegetation feeds on itself by a very efficient, rapid, and closed nutrient recycling system. Nutrients tied in the forest biomass are from an earlier time when soil was only partly weathered (Walter, 1973), and if these nutrients are removed from the ecosystem the nutrient-depleted soil cannot support another dense forest vegetation. There are some exceptions to these soils. Some forests are supported on very fertile soils of recent origin, i.e., volcanic-ash-derived soils of Java and Central America, and the soils developed on limestone (Terra Rossa).

Just like the interaction between soil physical properties and vegetation, the forest vegetation on nutrient-depleted soils behaves differently from that grown on fertile soils. In nutrient-depleted Oxisols and Ultisols forest vegetation acts as a very efficient filter through development of a thick above-ground root mat on top of the mineral soil. This root mat conserves mineralized nutrients through various adaptive mechanisms (Herrera, 1979; Moran, 1981). These nutrient-capturing and nutrient-conserving mechanisms are not developed in fertile soils. Research information on soil–vegetation interaction for both soil physical and nutritional properties is lacking. Disruption of these nutrient-cycling mechanisms by human activity would increase losses and enhance nutrient escape out of the ecosystem. For example, Herrara *et al* (1981) studied the status of soil fertility following clearing, burying and cultivation in the Amazonian forest of Venezuela (Fig. 6.12). Their data showed an initial increase in calcium and phosphorus level due to the nutrient release from decomposing biomass. However, the soluble nutrients, nitrogen and potassium, decreased rapidly because of the loss in drainage and run-off. The level of calcium and phosphorus would also eventually fall below that of the forest control, if cultivation continues.

VII. CONCLUSIONS

Undisturbed forest vegetation has profoundly favourable effects on soil physical properties, and has a moderating effect on its microclimate. Physical properties, i.e., structure, bulk density, porosity, water retention, and water

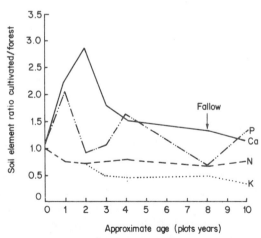

Figure 6.12 Changes in nutrient status of a soil under rainforest in Venezuela following clearing, burning, and cultivation (Herrera *et al.*, 1981). *Reproduced by permission of the Royal Swedish Academy of Sciences*

transmission, of soils supporting primary forests are better than soil in equivalent ecology but without a forest cover. In spite of predominantly coarse-textured soils, porosity and pore size distribution favour adequate water–air balance. Favourable soil physical properties are due to high soil fauna activity and the lack of human interference in the delicate soil–vegetation–climate balance.

In contrast, nutritional properties of the soil are poor. With few exceptions, soils under primary forests have been depleted of their nutrient reserves, and are excessively leached. If the present biomass is removed, these soils may not support another forest vegetation for a long time.

Forest vegetation, therefore, regulates physical (favourably) and nutritional and chemical (unfavourably) properties of soil that supports it. Soil properties in turn also affect the climax vegetation and favour the predominance of some species over the others. Important soil physical properties that influence the climax vegetation are soil moisture regime, soil texture and the depth of root penetration.

Tropical forest ecosystem is often referred to as 'fragile'. The term fragility is rather loosely used and has more emotional than rational basis. Nonetheless, the 'fragile' environments imply a quasi-equilibrium that is readily disturbed by man's intervention. For example, soil physical properties preserved at a favourable level under the forest cover are easily degraded by deforestation and subsequent cultivation. Drastic changes are observed in water balance, increasing surface run-off and accelerating soil erosion. The loss of soil fauna leads to compaction, and soft plinthite becomes irreversibly hardened laterite. Because most nutrients are tied in the biomass, this ecosystem is prone to rapid nutrient depletion out of the ecosystem.

Although the system is fragile, it is appropriate to develop a system of quantatively evaluating the rate and degree of degradation of soil physical and chemical properties by deforestation and change of land-use. It is advisable not to disturb the vegetation if the rate of decline in soil physical and nutritional properties is very rapid and drastic. Quantification and rationalization of the term 'fragile' is a prerequisite to developing criteria for choice of new land development for arable and other uses.

REFERENCES

Anderson, J.M., and Swift, M.J., 1983, Decomposition in tropical forests, in S.L. Sutton, T.C. Whitmore and L.C. Chadwick (eds), *The Tropical Rainforest*, Blackwell Scientific Publications, N.Y., pp. 287–309.

Aubert, G., 1961, The effects of various types of vegetation on the characteristics and evolution of soils in equatorial, sub-equatorial and bordering semi-humid tropical regions, *Tropical Soils and Vegetation, Proc. Abidjan Symp., 1959*, pp. 41–47.

Aweto, A.O., 1981, Secondary succession and soil fertility restoration in south-western Nigeria. III. Soil and vegetation interrelationships, *J. Ecol.*, **69**, 959–963.

Baumgartner, A., and Reichel, E., 1975, *The world water* balance. Oldenbourg, München, and Elsevier, Amsterdam, 179 pp.

Bjorkman, O., and Ludlow, M.M., 1972, Characterization of the light climate on the floor of a Queensland rain forest, *Carn. Inst. Yb. 1971-2*, **71**, 85-94.
Bonell, M., and Gilmour, D.A., 1981, The development of overland flow in a tropical rainforest catchment, *J. Hydrol.*, **39**, 365-382.
Bonell, M., Gilmour, D.A., and Sinclair, D.F., 1979, A statistical method for modelling the fate of rainfall in a tropical rainforest catchment, *J. Hydrol.*, **42**, 251-267.
Bonnet, J.A., 1968, Relative infiltration rates of Puerto Rican soils, *J. Agr. Univ. Puerto Rico*, **52**, 233-240.
Bronchart, R., 1963, Recherches sur le developpement de *Geophila renaris* De Wild. et Th. Dur. dans les conditions écologiques d'un sous-bois forestier équatorial. Influence sur la mise à fleurs d'une perte en eau disponible du soil, *Mém. Soc. Roy. Sci. Liège, Ser. 5*, **8**(2), 1-181.
Buringh, P., 1968, *Introduction to the Study of Soils in Tropical and Subtropical Regions*, Centre Agric. Publ. Docum., Wageningen.
Bunning, E., 1947, quoted by Longman, K.A., and Jenik, J. 1974.
Cachan, P., 1963, Signification écologique des variations micro-climatiques verticales dans la forest sempervirente de Basse Côte d'Ivoire, *Ann. Fac. Sci. Dakar*, **8**, 89-155.
Cachan, P., and Duval, J., 1963, Variations microclimatiques verticales et saisonnières dans la fôret sempervirente de basse, Cote D'Ivoire, *Ann. Fac. Sci. Dakar*, **8**, 5-87.
Carbonnel, J.P., 1964, Erosion in Cambodia, *C.R. Acad. Sci. Paris*, **259**, 3314-3319.
Carter, G.S., 1934, Reports of the Cambridge Expedition to British Guiana, 1933. Illumination in the rainforest at ground level, *J. Linn. Soc. (Zool)*, **38**, 579-589.
Dabral, B.G., and Rao, B.K.S., 1968, Intterception studies in cheri and teak plantations, *Indian Forester*, **94**, 541-551.
Dancette, C., and Poulain, J.F., 1968, Effect of *Acacia albida* on pedoclimatic factors and crop yield, *Afr. Soils*, **13**, 197-239.
Edmisten, J., 1970, Soil studies in the El Verde Rain Forest, in H.T. Odum (ed.), *A Tropical Rainforest*, U.S. Atomic Energy Commission, Washington, D.C., pp. H79-H37.
Embrapa, 1976, Ecossistema de pastagem cultivate na Amazonia, Brasileira. EMBRAPA-CPATU, Belem, Brazil, Boletim Técnico No. 1 193 pp.
Evans, G.C., 1939, Ecological studies on the rainforest of Southern Nigeria, II. The atmospheric environmental conditions, *J. Ecol.*, **27**, 436-482.
Evans, G.C., 1956, An area survey method of investigating the distribution of light intensity in woodlands, with particular reference to sunflecks, including an analysis of data from rainforest in southern Nigeria, *J. Ecol.*, **44**, 391-425.
FAO, 1981, *Tropical forest resources assessment project, forest resources of tropical Africa Part I; Regional analysis*, FAO, Rome, Italy.
Farnworth, E., and Golley, F., 1974, *Fragile Ecosystems: Evaluation of Research and Application in the Neotropics*, Springer-Verlag, New York.
Flenley, J., 1981, *The Equatorial Rain Forest: a Geological History*, Butterworths, London.
Ghuman, B.S., and Lal, R., 1984, A report to UNU, *Mimeo*, IITA, Ibadan, Nigeria.
Ghosh, R.C., Kaul, O.N., and Rao, B.K.S., 1982, Environmental effects of forests in India, *Ind. For. Bull.*, **1982**, 275.
Gilmour, D.A., and Bonell, M., 1977, Streamflow generation process in a tropical rainforest catchment, *Inst. Eng. Austr.* (Brisbane, Qld.), *Publ. No. 77/5*, pp. 178-179.
Glover, J., and Gwynne, M.D., 1962, Light rainfall and plant survival in East Africa. II, Dry grassland vegetation, *J. Ecol.*, **50**, 199-206.
Golley, F.B., McGinnis, J.T., Climents, R.G., Child, G.I., and Duever, M.J., 1975, *Mineral Cycling in a Tropical Moist Forest Ecosystem*, University of Georgia Press, Athens, USA.

Gómez-Pompa, A., 1972, The Tropical Rain Forest: A Non-Renewable Resource, *Science*, **177**, 762–765.
Greenland, D.J., and Kowal, J., 1960, Nutrient content of a moist tropical forest of Ghana, *Plant and Soil*, **12**, 154–174.
Grubb, P.J., and Whitmore, T.C., 1967, A comparison of montane and lowland forest in Ecuador, III, The light reaching the ground vegetation, *J. Ecol.*, **55**, 33–57.
Grubb, P.J., Lloyd, J.R., Pennington, T.D., and Whitmore, T.C., 1963, A comparison of montane and lowland rainforest in Ecuador, I, The forest structure, physiognomy and floristics, *J. Ecol.*, **51**, 567–601.
Herrera, R., 1979, Nutrient distribution and cycling in an Amazon Coatinga Forest on Spodosols in Southern Venezuela, *Ph. D. Thesis*, Univ. of Reading, UK.
Herrera, R., Jordan, C., Klinge, H., and Medina, E., 1978, Amazon Ecosystems: Their structure and functioning with particular emphasis on nutrients, *Interciencia*, **3**, 223–231.
Herrera, R., Jordan, C., Medina, E., and Klinge, H., 1981, How human activities disturb the nutrient cycles of a tropical rainforest in Amazonia. *Ambio* 10, 109–114.
Holdridge, L.R., Grenke, W.C., Hathaway, W.H., Liang, T., and Tosi Jr., J.A., 1971, *Forest Environments in Tropical Land Zones*, Pergamon Press, New York.
Jordan, C.F., 1968, Flow of soil water in the lower Montane tropical rain forest, in H.T. Odum (ed.), *A Tropical Rainforest*, U.S. Atomic Energy Commission, Washington, D.C., H199–H200.
Jordan, C.F., 1970, Movement of ^{85}Sr and ^{134}Cs by the soil water of a tropical rain forest, in H.T. Odum (ed.), *A Tropical Rainforest*, U.S. Atomic Energy Commission, Washington, D.C., H201–H204.
Jordan, C.F., 1978, Nutrient dynamics of a tropical rainforest ecosystem and changes in the nutrient cycle due to cutting and burning. *Annual Report, U.S. National Science Foundation, Institute of Ecology*, University of Georgia, Athens, GA, 207 pp.
Jordan, C.F., (ed.), 1981, *Tropical Ecology*, Stroudsberg, Pennyslvania, USA. Hutchinson Ross Publishing Co.
Jordan, C.F., and Heuveldop, J., 1981, The water budget of an amazonian rain forest, *Acta Amazonica*, **11**, 87–92.
Jordan, C.F., and Uhl, C., 1978, Biomass of a 'Tierra Firme' Forest of the Amazon Basin, *Oecologia Plantarum*, **13**, 387–400.
Kline, J.R., and Jordan, C.F., 1968, Tritium movement in soil of tropical rainforest, *Science*, **160**, 550–551.
Klinge, H., 1966, Verbreitung tropischer Tieflandspodsole *Naturwissenschaften* **17**, 442–443.
Klinge, H., 1977, Fine litter production and nutrient return to the soil in 3 natural forest stands of eastern Amazonia, *Geo-Eco-Trop*, **1**, 158–167.
Lal, R., 1979, Physical properties of soils of the tropics, in R. Lal and D.J. Greenland (eds.), *Soil Physical Properties and Crop Production in the Tropics*, J. Wiley & Sons, Chichester, UK.
Lal, R., 1981, Deforestation of tropical rainforest and hydrological problems, in R. Lal and E.W. Russell (eds.) *Tropical Agricultural Hydrology*, J. Wiley & Sons, Chichester, UK, pp. 138–140.
Lanly, J.P., 1982, Tropical Forest Resources, FAO Forestry Paper 30, FAO, Rome, 106.
Laurie, M.V., 1957, The effect of forests on water catchment areas on the water losses by evaporation and transpiration, *Emp. For. Rev.*, **36**, 55–58.
Lawson, G.W., Armstrong-Mensah, K.O., and Hall, J.B., 1970, A catena in tropical moisture semi-deciduous forest near Kade, Ghana, *J. Ecol.*, **58**, 371–398.
Lemée, G., 1961, Effets des caractèrer du sol sur la localisation de la végétation et zones equatoriale et tropicale humide, in *Sols et vegetation des Régions Tropicales*, UNESCO, Paris, pp. 25–39.

Longman, K.A., and Jenik, J., 1974, *Tropical forest and its environments*, Longman, London, 196 pp.
Malaisse, F., 1973, Contribution à l'etude de l'ecosysteme forêt claire, Note 8, Le Projet Miombo. *Ann. Univ. Abidjan, E.* **6**, 227–250.
McCulloch, J.S.G., and Dagg, M.,.1965, Hydrological aspects of protection forestry in East Africa, *E. Afr. Agric. For. J.*, **30**, 390–394.
Medina, E., 1968, Bodenatming und streuproduktion verschiedener tropischer Pflanzengemeinschaften, *Ber. Dtsch. Bot. Ges.*, **81**, 159–168.
Medina, E., 1973, Adaptations of tropical trees to moisture stress, in F.B. Golley (ed.), *Tropical Rainforest Ecosystems*, Elsevier, Amsterdam, pp. 225–238.
Medina, H.P., and Grohmann, F., 1966, Available water in some soils under 'cerrado' vegetation, *Bragantia*, **25**, 65–75.
Mohr, E.J.C., and Van Baren, F.A., 1954, *Tropical soils*, Interscience, London.
Moormann, F.R., Lal, R., and Juo, A.S.R., 1975, Soils of IITA, *IITA Tech. Bull. Bull. 3*, IITA, Ibadan, Nigeria, 48pp.
Moran, E., 1981, *Developing the Amazon*, Indiana Univ. Press, Bloomington, USA.
Myers, N., 1981, Conversion rates in tropical moist forests, in F. Mergen (ed.), *Tropical Forests Utilization and Conservation*, Yale Univ., New Haven, pp. 48–66.
Myers, N., 1983, Conversion rates in tropical moist forests. in F.B. Golley (ed.), *Tropical Rainforest Ecosystems*, Elsevier, Amsterdam, pp. 289–300.
Nieuwolt, S., 1965, Evaporation and water balances in Malaya, *J. Trop. Geogr.*, **20**, 34–53.
Odum, H.T., 1970, Rain forest structure and mineral cycling homeostasis, in H.T. Odum and R.F. Pigeon (eds.), *A Tropical Rain Forest*, U.S. Atomic Energy Commission, Washington, D.C., pp. H3–H52.
Ogawa, H., Yoda, K., and Kira, T., 1961, A preliminary survey of the vegetation of Thailand, *Nature Life S.E. Asia*, **1**, 21–157.
Pereira, H.C., 1965, Landuse and streamflow, *E. Afr. Agric. For J.*, **30**, 395–397.
Persson, R., 1974, *World Forest Resources*, Roy. College Forestry, Stockholm, No. 17, 261 pp.
Phillips, J., 1974, Effects of fire in forest and savanna ecosystems of sub-Saharan Africa, in T.T. Kozlowski, T.T. and C.E. Ahlgren (eds.), *Fire and Ecosystems*, Academic Press, New York, pp. 435–481.
Postel, S., 1984, Protecting forests, in *State of the World*, A Worldwatch Institute Report on Progress Toward a Sustainable Society, W.W. Norton and Co., N.Y., USA, pp. 74–94.
Reichle, D.E. (ed.), 1981, *Dynamic properties of forest ecosystems*, Cambridge Univ. Press, London.
Richards, P.W., 1952, *The Tropical Rain Forest*, Cambridge Univ. Press, London.
Richards, P.W., 1962, *Plant Life and Tropical Climate*, Pergamon Press, Oxford.
Roose, E.J., 1977, Application of the Universal Soil Loss Equation in West Africa, in D.J. Greenland, and R. Lal (eds.), *Soil Conservation and Management in the Humid Tropics*, J. Wiley & Sons, Chichester, UK, pp. 177–188.
Roose, E.J., and Godefroy, J., 1977, Pedogenesis of a reworked ferrallitic soil on schist, under forest and under a fertilized banana plantation in the lower Ivory Coast, Azaquie, 1969–1973, *Cahiers ORSTOM, Pedologie*, **15**, 409–436.
Rougerie, G., 1958, The existence and nature of runoff in the dense forest of the Ivory Coast, *C.R. Acad. Sci. Paris*, **246**, 290–292.
Sanchez, P. (ed.), 1972, A review of soils research in tropical Latin America, *Tech. Bull.* 219, *N.C. Ag. Exp. Sta.*, Raleigh, USA.
Sanchez, P.A., 1976, *Properties and Management of Soils in the Tropics*, Wiley-Interscience, New York.
Sarlin, P., 1963, Water and the soil, Soil moisture under forest, under savanna and under replanted forest, *Bois For. Trop.*, **89**, 11–29.

Schoch, P.G., 1968, Effect of the tree component on potential evapotranspiration in Senegal and its agronomic consequences, In *Proc. Symp. Agroclimatological Methods, Reading 1966*, pp. 313–319.

Schulz, J.P., 1960, *Ecological Studies on Rainforest in Northern Surinam*. North-Holland, Amsterdam.

Sommer, A., 1976, Attempt at an assessment of the world; moist forest. *Unasylva*, 28, 5–25.

Stark, N., 1969, Direct Nutrient Cycling in the Amazon Basin, in *II Symposio Foro de Biologic Tropical Amazónica*, Bogotá, Editorial Pax.

Stark, N., 1970, The nutrient contents of plants and soils from Brazil and Surinam, *Biotropica*, 2, 51–60.

Stark, N., 1971, Nutrient Cycling, I, Nutrient distribution in some Amazonian soils, *Trop. Ecol.*, 12, 24–50.

Stark, N., 1972, Nutrient cycling, Pathways and litter fungi, *Bioscience*, 22, 355–360.

Stark, N., and Spratt, M., 1978, Root biomass and nutrient storage in rain forest Oxisols near San Carlos de Rio Nere, *Trop. Ecol.*, 18, 1–19.

Strang, R.M., 1969, Soil moisture relations under grassland and under woodland in the Rhodesian highveld, *Commonw. For. Rev.*, 48, 26–40.

Swift, M.J., Russell-Smith, A., and Perfect, T.J., 1981, Decomposition and mineral-nutrient dynamics of plant litter in a regenerating bush-fallow in sub-humid tropical Nigeria, *J. Ecol.*, 69, 981–985.

Tanaka, A., Sakuma, T., Okagawa, N., Imai, H., and Ogata, S., 1984, *Agro-ecological condition of the Oxisol-Ultisol Area of the Amazon River System*, Faculty of Agric., Hokkaido Univ., Japan.

Tergas, L.E., and Popenoe, H.L., 1971, Young secondary vegetation and soil interactions in Izabal, Guatemala, *Plant and Soil*, 34, 675–690.

Tkatchenko, B., 1948, The sugar palm of Cambadi, *Agron. Trop.*, 3, 563–593.

Tracey, J.G., 1969, Edaphic differentiation of some forest types in eastern Australia, I, Soil Physical Factors, *J. Ecol.*, 57, 805–816.

UNESCO, 1978, *Tropical Forest Ecosystems: A state of Knowledge Report*, UNESCO, Paris.

Walter, H., 1973, *Vegetation of the Earth*, Springer-Verlag, New York.

Wanner, H., 1970, Soil respiration, litter fall and productivity of tropical rain forest, *J. Ecol.*, 58, 543–547.

Went, F.W., and Stark, N., 1968, Mycorrhiza, *Bio Sci.*, 1968, 1035–1039.

Whitmore, T.C., 1975, *Tropical Rain Forests of the Far East*, Clarendon Press, Oxford.

Zonn, S.V., and Li, C.K., 1960. Characteristics of the energy relations of biological processes in tropical forest soils, *Pochvovedenie*, 12, 1–15.

Zonn, S.V., and Li, C.K., 1961, Moisture regime of tropical-forest soils, *Pochvovedenie*, 3, 12–22.

Chapter 7
Soil Fauna and Flora

I	DIVERSITY OF SOIL ANIMALS IN THE TROPICS	265
II	SOIL PROPERTIES AND SOIL FAUNA	266
	A. Microclimate	267
	B. Soil moisture	267
	C. Soil temperature	267
	D. Voids	268
	E. Organic matter	268
III	CULTURAL PRACTICES AND SOIL FAUNA	269
	A. Deforestation	270
	B. Fire	271
	C. Cropping systems and agrochemicals	273
IV	FAUNAL ACTIVITY AND SOIL PROPERTIES	275
V	MICROFLORA AND SOIL PHYSICAL PROPERTIES	277
	1. Decomposition of organic matter	278
	2. Effects on soil physical properties	278
	3. Effects of cultural practices	280
VI	CONCLUSIONS	280

The role of soil animals in influencing properties and productivity of soils in the tropics has not been given the emphasis it deserves, particularly the emphasis on ecological research by pedologists and agronomists concerned with improving tropical agriculture. Consequently, we have little understanding of the interrelationship of soil fauna and agricultural practices. Vast areas of tropical land are being put to arable land-use without consideration for the ecological consequences of these practices. The result is a rapid degradation of soil and a decline of its productivity. The following quotation is very relevant in this connection:

'The plain truth is that this critically important subject of the ecology of soil has been largely neglected even by scientists and almost completely ignored by control men. Chemical control of insects seems to have proceeded on the assumption that the soil could and would sustain any amount of insult via the introduction of poisons without striking back. The very nature of the world of the soils has been largely ignored'.

— Rachel Carson in *'Silent Spring'*.

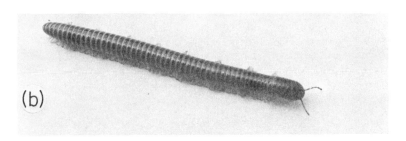

Figure 7.1 Some examples of soil macro-fauna commonly observed in the tropics

Table 7.1 An ecological classification of representative animals* that affect soil (After Hole, 1981) *Reproduced by permission of* Elsevier Science Publishers

1. Endo-pedonic animals	A. Litter dwellers	1. In litter	a. Rollers (Isopoda; Glomeridae)
			b. Crawlers and runners (Diplopoda)
			c. Leapers (Collembola)
		2. In refuges under stones, logs, near seeps, in cavities	
	B. Soil builders at the surface	1. Soil casters	a. Earthworms
			b. Millipedes
		2. Soil mounders	a. Termites
			b. Ants
			c. Crayfish
			d. Mammals 1. Moles
			2. Rodents
			3. Rabbits
			4. Badgers
	C. Soil burrowers	1. Peristaltic contractors	a. Earthworms
			b. Potworms
		2. Non-peristaltic movers	a. Large predatory nematodes
			b. Ants

c. Bumblebees, crickets, wasps, hornets, spiders, beetles
e. Millipedes
f. Mole crickets, moles
g. Skunks, lizards, snakes
h. Bears, badgers
i. Birds (desert owl, petrel)

D. Soil pore occupiers
   1. Non-aquatic mites, etc.
   2. Aquatic or semi-aquatic dwellers in water films in soil (protozoa, nematodes, tardigrades)

A. Ground walkers
   1. Ungulates
   2. Plantigrades

B. Tree dwellers
   1. Insects
   2. Amphibia
   3. Birds
   4. Mammals

2. Exo-pedonic animals
   C. Cliff, bluff and building dwellers
      1. Wasps
      2. Birds
   D. Cave dwellers
      1. Bats
   E. Water dwellers
      1. Amphibia
      2. Petrels

* Some animals belong to more than one category. The contrasting life forms of many insects during life history permit operation of one form in a different realm from that of another form of the same animal.

There is an urgent need to study tropical soil ecology. Some useful research, though, has been done by soil zoologists relating the activity and diversity of soil fauna to soil properties rather than on effects of fauna on soil. Little is known regarding the effects of soil fauna on properties and productivity of tropical soils. Since the science of 'Physical Edaphology' deals with the interaction of soil physical processes (i.e., the dynamics of water, heat and gases) with agricultural practices, it is relevant that the role of soil animals on physical processes be understood and used for advancement of agricultural production in the tropics. The invention of the plough as a technique for mechanical manipulation of soil and for alleviating constraints of soil physical and biotic factors has had mixed blessings. Its widespread use in advancing temperate zone agriculture has been partially responsible for de-emphasising, neglecting and even bypassing the important role of soil animals in tropical soil ecology. Whatever its use in temperate zone agriculture may be, the use of the plough as a soil manipulator can be harmful in harsh tropical environments. Although ploughing and mechanical soil manipulation are short cuts to a rapid seedbed preparation, they are not substitutes for the natural mixing and soil turnover achieved through the biotic activity of soil fauna.

It is, therefore, appropriate to define soil in 'edaphological' and 'ecological' terms. Soil is a living entity with its diverse soil fauna and flora (Fig. 7.1). When devoid of their integral fauna, the upper layers of the earth cease to be 'soil'. Soil is thus defined as a 'three-dimensional multi-layered upper part of the earth crust that comprises an inseparable mixture of solid, liquid and gaseous phases and intense and diverse faunal activity; is capable of supporting biological growth, and is in equilibrium with its environment' (Fig. 7.2). It is a

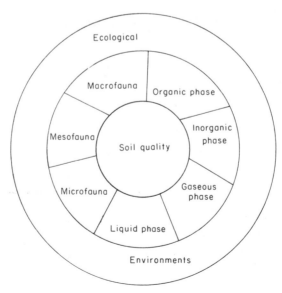

Figure 7.2 An ecological definition of soil

complex system. Laboratory analyses of the chemical and physical characteristics of a sample that disregard its faunal activity under natural conditions may lead to erroneous results and gross misinterpretations. Only an ecologically stable soil is edaphologically productive. Soil becomes unproductive when its ecological balance is disturbed, and it is then that the human inhabitants also lose their social and economic stability.

The effects of animals on soil have recently been reviewed by Hole (1981), and technical reports on soil zoology are presented by numerous authors (Murphy, 1958; Kevan, 1968; Bourliere and Hadley, 1971; Graff and Satchell, 1967; Madge and Sharma, 1969; Wallwork, 1970, 1976.) Most of these reports deal with fauna of temperate region soils. An attempt is made here to collate and analyse the existing information on the role of soil animals in regulating physical processes in tropical soils. This chapter deals with soil animals in general. Separate chapters are devoted to each of the important groups of soil fauna, e.g., earthworms, termites and ants.

I. DIVERSITY OF SOIL ANIMALS IN THE TROPICS

Soil is a habitat for a vast number of diverse species, some of which are yet to be identified and classified (Thorp, 1949). Soil fauna can be classified on the basis of: (i) taxonomy, (ii) body size, (iii) habitat preference (iv) feeding habits and (v) ecological distribution. For the purpose of physical edaphology, soil fauna are conveniently classified according to body size into five categories (Wallwork, 1970; Lavelle, 1983, Hole, 1981).

(a) Microflora: comprise bacteria, actinomycetes, fungi, and algae. Their biomass ranges from 1 to 100 g/m^2 and their number varies up to 1,000,000 millions per m^2.

(b) Microfauna: comprise mostly protozoa (*Flagellata, Rhizopoda, Ciliphora*, etc) with a biomass range of 1.5 to 6 g/m^2. Their number ranges from 1 million to 1,000,000 millions per m^2, and size is <0.2mm.

(c) Mesofauna: include *Rotifera, Nemotoda, Acari, Collembola, Protura,* and *Pauropoda*, etc. Their number ranges from a few hundreds to several millions per m^2, biomass from 0.01 to 10 g/m^2, and size from 0.2mm to 10 mm.

(d) Macrofauna: consists of *Enchytraeidae, Symphyla, Diptera* larvae, *Coleoptera, Diplopoda, Gastropoda, Chilopoda, Araneae* and other insecta. Their biomass ranges from 0.1 to 2.5 g/m^2, number from a few hundreds to several thousands, and size 10 to 20 mm.

(e) Megafauna: consists of earthworms (Lumbricidae) with a number from a few to several hundreds per m^2, biomass weight of about 40 g m^2, and size greater than 20 mm.

Equally important is the ecological classification of soil fauna as given by Hole (1981) (Table 7.1). The macrofauna and megafauna consist of endo-pedonic animals that live inside the soil in contrast to the exopedonic that live outside. Most soil animals have habitats both above and below ground, depending on soil, microclimate and stage in the life cycle. These broad categories are further subdivided according to their habitat and nesting.

The diversity of animals in tropical soils is large even within a small area. In the humid tropical forest region of French Guyana, Couteaux (1979) noted 27 species with a total population of 1616/mm^2. Lasebikan (1974, 1981) reported the population of microarthropods (termites, ants, mites, springtails, etc.) in a tropical rainforest of Nigeria. The population ranged from 3821 to 12 032/m^2 for Cryptostigmata, 122 to 6789/m^2 for Gasmasina, 203 to 3049/m^2 for Uropodina and 488 to 4888/m^2 for Prostigama. These animals are partly responsible for the decomposition of plant and animal remains. In Uganda, the population of microarthropods has been studied by Salt (1952) and Block (1970). Pomeroy and Rwakaikara (1975) reported from Muko in western Uganda that the Collembola population was about 2000/m^2 and that of mites about 10 000/m^2. In this grassland savanna, the biomass of termites and ants exceeded that of microarthropods, and earthworms were practically negligible. Athias (1971) has reported the microarthropod population in the Ivory Coast. In northern India, Singh and Pillai (1975) reported that the total microarthropod population ranged from 16 970/m^2 to 20 761/m^2. In the rainforest region of North Queensland, Australia, Holt (1981) observed the population of Cryptostigima in the surface 0–4 cm layer to range from 6937/m^2 to 8343/m^2. Serafino and Merino (1978) presented data on the populations of Collembola, Diplura, Protura, Homoptera, Acrina, Palpigrada, Pseudoccorpionida and Symphila etc. in different soils of Costa Rica.

Millipedes, an important constituent of tropical soil fauna, have also been studied only by zoologists and soil biologists. Bellairs *et al.* (1983) studied the life cycle and swarming behaviour of the larvae of an Indian polydesmoid millipede *Streptogonopus phipsoni*. Six principle millipede species in a cultivated soil of Senegal, West Africa, were described by Gillon and Gillon (1976, 1979a,b). In an open field there were an average of 12 Julids/m^2 with a fresh biomass of 3g/m^2, and around sprouting stumps densities were slightly higher, reaching 20/m^2.

Rodents are also an important component of tropical soil fauna. The African mole rat (*Tachyoryctes*), for example, has a wide range of habitat and creates a considerable soil disturbance. These animals excavate the soil and deposit it on the surface.

II. SOIL PROPERTIES AND SOIL FAUNA

For most groups the fauna are confined to the top few centimetres of tropical soil. The animal life is compressed into the shallow surface horizon where most of the profile's organic matter reserves are also concentrated. In Congo, Maldague (1959) observed the 80% of the fauna was confined to the top 2.5 cm layer of partially decomposed leaf litter and soil layer rich in organic matter content. Also in Uganda, most animals were found in the upper layer of the soil (0-3 cm) (Pomeroy and Rwakaikara, 1976). The layer contained 96% of the oribatid mites, 79% of the mesastigmatids, 64% of the Collembolans, 94% of the termites and 48% of the ants. In Senegal, Gillon and Gillon (1979b) observed that during the rainy season from July to October, over 50% of the millipede population was found in the upper 10 cm. In Northern Queensland, Holt (1981) noted that the cryptostigmatic population was concentrated in the upper 4 cm of the soil. In India Bhattachrya et al. (1981) noted that Cryptostigmata were most common in the upper 5 cm of the soil.

A. Microclimate

The most important soil factors affecting animal population are soil moisture, soil temperature, porosity and pore size distribution, and of course, the general fertility of the soil. Microclimate is a very important factor regulating the population of some groups of soil animals. Hot and dry climate generally suppresses soil fauna diversity and activity (McColl, 1975). Wood (1971) noted that in Southern Australia the densities of *Folsomides drerticola* and other microarthropods are low in hot deserts ($2100-3200/m^2$) and increase with increasing rainfall, decreasing aridity and increasing amount of plant litter. In the Amazon pasture, Dantas and Schubart (1980) observed that population density of Collembola was negatively related to air temperature, and that of Collembola and Acari were positively related to rainfall amount.

B. Soil moisture

Soil fauna are drastically affected by the fluctuations in soil moisture content. In West Bengal, India, the Collembolan population in a deciduous forest floor was found to be positively correlated with the soil moisture content (Hazra, 1978). In Assam, northeast India, Darlong and Alfred (1982) observed that a high arthropod population during the rainy season was highly correlated with the soil moisture content. The population density of *Hymenoptera* and *Acrina* were negatively correlated with the moisture content. In West Africa, Belfield (1970) observed a significant correlation between the rainfall of the previous month and the population of the soil fauna (Fig. 7.3). The correlation coefficients were higher with the rainfall of the previous month than with the same month: the correlations were 0.79 and 0.10 for total population, 0.82 and 0.09 for Collembola, 0.73 and −0.03 for acarina, 0.64 and 0.41 for pauropoda and 0.57 and 0.02 for symphyla, respectively. In Uganda, however, Pomeroy and Rwakaikara (1975) noted no correlation between soil moisture content and the population of Collembola, Oribatei, or Mesostigmata.

C. Soil temperature

Strongly interacting with the soil moisture content is the soil temperature. In the tropics, dry surface soil has higher soil temperature than moist soil. The amplitude of diurnal fluctuations in soil temperature is also the most in the

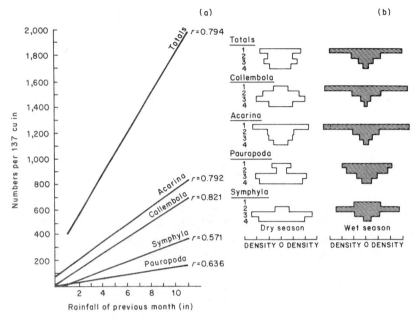

Figure 7.3 The correlation between rainfall of the previous month and the population of soil fauna (Belfield, 1970). (a) Numbers and (b) typical depth distributions. In (b) each depth density is expressed as a percentage of the total

surface layer where soil animals are mostly concentrated. Temperature is probably less important to soil animals than to surface-dwelling species. Lavelle and Meyer (1980) identified some factors responsible for 63.7% of the variance of soil fauna in the tropical savanna of the Ivory Coast. These variables were fire, soil moisture, soil temperature, soil structure and microclimate. Darlong and Alfred (1982) observed a positive correlation of the air and soil temperatures with the population of Collembola, Coleoptera and Symphyla.

D. Voids

The movement of non-burrowing soil animals is facilitated through existing soil pores. The preferred pore size range is generally the same as the body size range. A decrease in diameter of soil cavities often limits the penetration by large-size animals. In the rainforest of North Queensland, Holt (1981) noted a significant correlation between the percentage of Cryptostigmata with body size of 50–125 μm and the percentage of macropores having diameters in the 50–125 μm range ($r = 0.999$). The microarthropod population was however, not related to the total macroporosity.

E. Organic matter

The organic matter and its diversity and supply, are obviously important factors determining faunal activity and species diversity. In Varanasi, northern India, Singh and Pillai (1975) observed that the microarthropod population was related to the soil organic matter content. In Uganda, Pomeroy and Rwakaikara (1975) reported positive correlations between soil organic matter content and the populations of Oribatei and Mesostigmata but not with that of Collembola. Holt (1981) noted that the macro-organic matter content, the fraction of organic matter content which consists of fragments <0.25 mm, was positively and significantly correlated with the population of Cryptostigmatids ($r = 0.884$). Species diversity is also influenced by the soil reaction, the cation exchange capacity, and the predominance of different cation on the exchange complex. In the Australian subtropical forest, Plowman (1981) noted a distinct separation between three species groupings of Cryptostigmata and Mesostigmata present in different ecologies on the basis of soil pH.

III. CULTURAL PRACTICES AND SOIL FAUNA

Alterations in soil and microclimate by agricultural practices can have a profound effect on soil fauna. The practices detrimental to faunal activity are those that disturb their habitat, alter the microclimate, and reduce the food supply and its diversity. The detrimental practices are land clearing, mechanical tillage and use of heavy machines, and the indiscriminate use of agrochemicals, particularly pesticides. All ecologically sound agricultural practices and those that support good crop growth also favour soil fauna (Fig. 7.4).

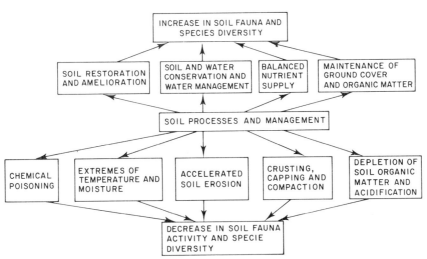

Figure 7.4 Effects of cultural practices on soil fauna and flora

Cultural practices that create extremes of temperature and moisture regimes, accelerate soil erosion and deplete organic matter content, have seriously adverse effects on soil faunal activity and diversity. Readers are referred to a review by Dindal (1979) for discussions on soil biology in relation to general land-use practices.

A. Deforestation

Vast areas of tropical humid forest are annually being cleared for agricultural production. The effects of clearing practices and of change in land-use on soil fauna have not been widely quantified. In South Africa, Van den Berg and Ryke (1967) observed slightly higher populations of acari under natural forest ($341\,032/m^2$) than under pine plantation ($307\,380/m^2$). The population of mesostigmates under pine plantation was 78% and that of the sarcoptiformes was 84% of that under natural forest. In contrast, the population of trombidiformes under pine plantation was about 3% more than under natural forest. In Senegal, Gillon and Gillon (1979b) surveyed the biomass of millipedes in cleared and forested plots and noted that the population of diplods was about four times more under trees than in the cleared field. The effects of bush clearing on the invertebrate fauna of the humid forest zone of western Nigeria are reported by Critchley et al. (1979). Clearing and cultivation resulted in a substantial reduction in activity of the majority of the active surface fauna, i.e., crickets, ants, spiders and millipedes. The catches in cultivated plots were 50–60% below those for uncleared plots. The mean population densities of subterranean fauna comprising major groups of Acari and Collembola in cleared and uncleared plots are shown in Table 7.2. In the 0–5 cm layer clearing resulted in a reduction in population densities of all groups of animals with the exception of the Prostigmata, which became the dominant microarthropod group in cleared land. Caveness (1982) reported from southwest Nigeria that population densities of plant-parasitic nematodes declined after land clearing to a mean level of 42% less than population densities in non-cleared land. In Shillong, India, Darlong and Alfred (1982) showed that clearing for shifting cultivation reduced soil fauna to half of the original forest stand. Watnabe et al. (1983) observed little effect of deforestation in Northeast Thailand on the total number of surface dwellers trapped before and after clearing. There were, however, differences in relative predominance of different species. The population of soil-dwellers was, however, drastically reduced by deforestation and burning. Densities of total meso-animals (including Collembola, Acari and other Arthropods) were $16\,122/m^2$ in the forest area and $8253/m^2$ in the cleared land.

Deforestation has both direct and indirect effects on soil fauna. Directly, mechanical disturbance and soil scraping damages habitats and results in the lowering of faunal populations. Among indirect effects are the changes in microclimate, decrease in food diversity, and exposure to predators and parasites. Prior to implementing large-scale deforestation schemes, it is, therefore, important to understand the consequences in terms of microclimate and shift in food supply for soil fauna. A knowledge of food preferences for

Table 7.2 Mean population densities of Acari and Collembola over the period May 1973 to March 1974 for the 0–50 mm depth (number m^{-2} ± standard error) (After Critchley et al., 1979). *Reproduced by permission of* Pedobiologia

Depth		Bush		Cultivated	
		n m^{-2}	SE±	n m^{-2}	SE±
0–50 mm	Acari				
	Cryptostigmata	17 403	749	1991	239
	Prostigmata	8477	358	9838	919
	Mesostigmata	3557	920	1121	246
	Astigmata	1372	359	708	216
	Collembola				
	Isotomidae	6963	900	1425	366
	Entomobryidae	2421	213	754	288
	Onychiuridae	1628	129	243	72
	Sminthuridae	1128	190	620	23
	Poduridae	665	142	54	11
50–100 mm	Acari				
	Cryptostigmata	7695	610	2230	358
	Prostigmata	6353	580	7862	2025
	Mesostigmata	1013	248	766	363
	Astigmata	237	188	17	10
	Collembola				
	Isotomidae	3644	1069	733	192
	Entomobryidae	355	68	425	168
	Onychiuridae	364	57	105	29
	Sminthuridae	280	21	322	16
	Poduridae	200	25	9	6

different groups of animals would be useful for developing land management and cropping systems that would support high faunal activity.

B. Fire

Fire is widely used in the tropics for land clearing and for pasture renovation. Burning alters soil moisture and temperature, food availability, soil pH and soil organic matter content and, therefore, has far reaching ecological effects (Ahlgren and Ahlgren, 1960). The effects of burning on soil fauna, however, depend on the intensity and frequency of burning and vary among species and ecological environments. The direct effect of high temperature at the fire front is the instant killing of many fauna. The soil, desiccated and exposed to solar radiation, attains a wilting point that is also fatal to many species. The indirect effects of fire include the depletion of litter and other food reserves. Recurring fire is known to increase the predominance of soil-eating over litter-eating animals (Lavelle, 1983).

In India, burning has been demonstrated to reduce soil fauna (Buffington, 1967; Darlong and Alfred, 1982). In Northeast Thailand, Watnabe *et al.* (1983)

Table 7.3 Changes in numbers and biomass of soil macro-animals due to clearing and burning in northeast Thailand (per 50 cm square) (Watnabe et al., 1983)

	Dry season				Rainy season			
	Before burning		After burning		Undisturbed forest		Maize field	
	Number	Biomass (mg)	Number	Biomass (mg)	Number	Biomass (mg)	Number	Biomass (mg)
Mollusca	0.4	7.9			0.3	48.9		
Oligochaeta	2.5	85.7	0.8	43.7	8.3	2067.7	4.0	*479.7
Diplopoda	1.0	256.7					0.6	6.9
Chilopoda	4.1	63.4	3.8	23.4	2.3	9.5	0.4	85.0
Isopoda	0.1	1.3	0.3	3.3				
Cheriferidea	0.1	0.2	0.2	0.2	0.1	0.6		
Araneida	3.1	16.3	1.2	13.8	2.6	39.3	0.4	13.3
Blattidae	5.4	204.2	1.0	36.0	4.2	401.8	0.2	16.3
Orthoptera	2.8	106.9	0.7	170.3	1.5	291.8		
Dermaptera	0.1		0.2	0.5				
Isoptera	*25.0	23.2	1.8	0.7	*16.7	48.6		
Coleoptera	7.5	*1169.0	5.5	*623.2	*9.7	*501.5	*24.4	63.2
Hemiptera	1.1	142.4	1.3	61.8	5.0	35.7	*7.6	*226.8
Lepidoptera	1.4	212.8	0.2	121.0	1.8	50.8	1.2	20.8
Hymenoptera	*21.5	104.8	*20.0	17.8	*34.8	113.2	0.4	31.0
Diptera	0.9	2.5	0.8	4.6	4.5	68.8	4.2	18.9
Others	0.6	2.4	1.0	20.2	3.5	44.3	3.0	21.8
							1.2	12.5

* Dominant groups.

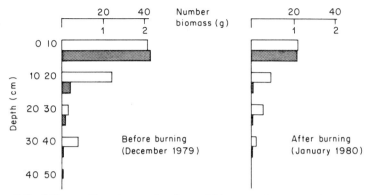

Figure 7.5 Effects of burning on density and biomass of macro-animals (Watnabe *et al.*, 1983)

observed a drastic decline in the population of soil macro-animals following burning. The density and biomass of macro-animals respectively was reduced from 310.4/m^2 and 9.6 g/m^2 before burning to 155.2/m^2 and 4.6 g/m^2 after (Table 7.3). Surface-dwelling macro-animals were particularly sensitive to direct heat. The density and biomass of macro-animals decreased with burning (Fig. 7.5). In New Zealand, Springett (1979) noted that burning reduced the number of arthropods.

In Uganda, Pomeroy and Rwakaikara (1975) reported that the population of mites tended to increase with the frequency of burning. The frequency of burning was positively correlated with the population of Oribatei ($r = 0.719*$) and Mesostigmata ($r = 0.577*$) but not with that of the Collembola ($r = 0.244$). Lavelle *et al.* (1981) observed in the savanna region of Mexico that the relative biomass percentage of different categories of soil animals was influenced by burning. The biomass percentage in the burned and unburned savanna was 77 and 52 for soil eaters, 0.3 and 0.3 for root eaters, 17.6 and 38.5 for detritus eaters and 4.6 and 8.9 for predators, respectively. Fire, therefore, seems to have different effect on different species.

In Bahia Blanca, Argentina, De Izara (1977) observed that the composition and population of microarthropod fauna was not affected by burning. In French Guiana, Couteaux (1979) noted that burned areas were devoid of any *Testacea* spp.

C. Cropping systems and agrochemicals

Land-use and cropping systems affect soil fauna through their effects on soil properties and the microclimate. Subsequent to land clearing, the introduction of a crop provides a food source for a wide range of insect pests. Therefore, different groups of soil fauna may predominate on cultivated as compared to forested land. However, the crops do not provide buffering effects comparable to the forest stand. In South Africa, Ryke and Loots (1967) observed

Table 7.4 Populations of the most abundant groups of microarthropods in the 0–50 mm horizon of plot soils treated with DDT (After Perfect et al., 1981) (numbers m^{-2} ± SE). *Reproduced by permission of* Pedobiologia

Group	Untreated		Treated with DDT		Contaminated with DDT	
Onychiuridae	1300± 640a	(3.5)	1370± 370a	(0.6)	1600± 730a	(1.4)
Isotomidae	8830±1720a	(1.4)	7190±1030a	(2.7)	18 820± 3230b	(2.5)
Entomobryidae	3830± 810a	(3.6)	6200±1400ab	(8.3)	11 410± 4660b	(7.4)
Sminthuridae	3620± 570a	(1.7)	18 420± 930b	(10.1)	25 150± 3590c	(9.5)
Total Collembola	19 060±1890a	(1.7)	33 310±2250b	(4.6)	61 300±10 590c	(4.4)
Prostigmata	20 540±5060a	(1.5)	14 640±8600a	(6.7)	11 680± 1180a	(1.9)
Mesostigmata	10 260±2450a	(2.5)	590± 80b	(5.4)	1130± 500b	(4.5)
Cryptostigmata	21 860±2620a	(2.8)	20 990±2110a	(4.4)	23 520± 960a	(5.9)
Total Acari	52 660± 180a	(2.0)	36 220±8580b	(4.7)	36 360± 2460b	(3.9)

Numbers within a row followed by the same letter are not significantly different ($p<0.05$). Figures in parentheses show the increase in population relative to the same soil pesticide treatments without added maize straw.

significant differences in the population of microarthropods under different cropping systems. For example, the minimum and the maximum populations of microarthropods were 6102 and 63 027/m^2 under pasture, 4498 to 117 095/m^2 under acacia, 21 637 and 69 846/m^2 under kikuyu, 2087 and 52 716/m^2 under eragrostis, and 28 287 and 57 126/m^2 under maize, respectively. In India, Singh and Pillai (1975) observed that the total microarthropod population was 20 761/m^2 in an uncultivated field, 20 119/m^2 in fodder, 19 098/m^2 in banana and 16 976/m^2 in citrus. The difference in faunal populations under different land-use was related to differences in organic matter and surface temperature. The relative proportion of different species was also found to be different under different crops. In Tamil Nadu, the population density of microflora of soil grown to cassava was observed to be higher with no-till and mulching than with clean seedbed preparation (Natarajan et al., 1980). In Northeast Thailand, Watnabe et al. (1983) observed that even after cultivation of maize following clearing and burning, the population of macro-animals did not recover.

The use of insecticides and herbicides on cropped land affects soil fauna (Mathys, 1975), although little research information is available on this subject from tropical environments. In western Nigeria Critchley et al. (1980) noted that crop protection with DDT reduced the activity of adult Grylline and immature crickets. None of the ant subfamilies showed significant differences in activity though Dorylinae were generally more active, and Formicinae were less active in DDT sprayed plots. The population of lycosid spiders was appreciably reduced. Among subterranean mesofauna, numbers of ants and Coleoptera were significantly reduced in treated soil. Millipedes were affected by ingestion of contaminated food. In a similar study, Perfect et al. (1981) reported the effects of DDT on the microarthropod population in western Nigeria. The populations of Collembola and Acari were doubled in untreated plots (Table 7.4). The effect of DDT was obviously different among different species. Subagja and Snider (1981) studied the side effects of the herbicides atrazine and paraquat upon *Folsomia candida* and *Tullbergia granulata* under laboratory conditions. Significantly high mortalities were observed at feeding rates of 5000 ppm of atrazine. In addition, repellent properties of atrazine accounted for the mortalities over an extended period of time. Egg viability in *T. granulata* was also affected by 5000 ppm paraquat.

IV. FAUNAL ACTIVITY AND SOIL PROPERTIES

Soil animal activities have a profound effect on soil properties and physical and chemical soil processes (Hole, 1981). Little research information exists relating properties of tropical soils to faunal activity. Faunal activities that affect soil physical processes are:

(i) Humification and formation of organo-mineral complexes by microfauna. Microorganisms (bacteria, fungi, algae, etc.) play an active

role in development of soil structure and promotion of soil aggregation.
 (ii) Soil mixing and turnover resulting in biological homogenization (Hoeksema and Edelman, 1960) and faunal pedoturbation (Hole, 1961). Earthworms play a significant role in mixing and soil turnover.
 (iii) Casting, mounding and nest building. Subsoil is brought to the surface in 'profile inversion'. Soil animals important in these activities are earthworms, termites and ants.
 (iv) Burrowing and tunnelling, leading to creation of channels and interconnected pores. In addition to mammals, rodents and reptiles vertebrates, earthworms, termites and ants influence soil through their burrowing activities. These activities create and back-fill voids that are preferentially occupied for root development especially in compacted and gravelly soils.

Soil edaphological processes affected by these faunal activities are:

 (i) *Aggregation*: Development of structural aggregates is facilitated by the activity of micro- and meso-fauna. Intra- and inter-aggregate porosity influences water and gaseous movement.
 (ii) *Water Movement*: Large channels and tunnels created by burrowing activity facilitate water movement. If the channels are connected to the surface and to the source of water, water is very rapidly transmitted to the layers beneath.
 (iii) *Gaseous Exchange*: Increase in porosity and pore size distribution facilitates gaseous exchange between the soil and the atmosphere. Soil fauna play an important role in regulating its environment. In the absence of external influences, soil animals create optimum environments that promote growth and enhance their activity. Aerobic conditions that support economic plants and crops are created and facilitated by faunal activity.
 Metabolic processes also regulate the microclimate (temperature and humidity) within the soil system.
 (iv) *Humification and Mineralization*: Soil fauna plays an important role in decomposition, mineralization and regulation of plant and animal litter and biomass, and in recycling plant nutrients and converting them into easily available nutrients and other soil constituents.

In addition to these physical processes, soil animals also influence pedogenetic and geological processes. Important processes involved are:

 (i) *Weathering and Soil Formation*: Mechanical action involving burrowing, mounding, casting and transporting results in mechanical breakdown of soil particles. Nye (1955) and others have recognized termites and earthworms as important biotic factors of soil formation in the tropics. Burrowing animals can bring enormous quantity of soil to the surface. In Ivory Coast Abaturov (1972) estimated the deposition of excavated soil to be 55t/ha/y. Similar magnitude of soil excavation has been reported for Kisii area of Kenya by Wielemaker (1984).

(ii) *Erosion*: Decomposition of plant and animal remains, soil turnover, aggregation and the effect of biotic activity on water transmission properties strongly influence soil erosion. Soil fauna are also influenced by run-off and soil erosion. Some soil animals are washed into lower soil horizons. The soil freshly activated by animal activity is easily displaced downslope by raindrop impact or water runoff. In Kenya Wielemaker estimated that soil displacement due to mole activity amounted to 7 kg air dry soil per 100 m^2 per 5 days or equivalent of 50t/ha/y. His data showed that rodent activity alone in uncultivated soil resulted in an annual displacement of a soil layer with an average thickness of 0.04 mm through a distance of 7 cm.

(iii) *Landform and Micro-relief*: Nesting activities of soil fauna (termites, earthworms, etc.) influence microrelief.

(iv) *Nutrient availability*: Soil fauna have a definite effect on the relative abundance of easily available plant nutrients.

V. MICROFLORA AND SOIL PHYSICAL PROPERTIES

Similarly to soil fauna, microflora also affect soil physical properties for tropical soils; however, little is known regarding the effects of microorganisms on soil physical processes in different ecologies and varied land-use systems. Soil microbiologists have been preoccupied with studies on the role of microorganisms in transforming organic and mineral matter, in mineralization or organic systems, in the nitrogen cycle and in nitrogen fixation. The study of the role of microorganisms in transforming the physical properties of soils in the tropics i.e., structure, water retention, soil air and gaseous exchange, has virtually been neglected, and has not been given the emphasis it deserves. The fact is that whatever benefits chemical and nutritional inputs might render towards boosting agricultural production in the tropics can be easily negated by deterioration in soil structure, low available soil water, inhibited gaseous exchange and supra-optimal soil temperatures. The potential benefits of microflora in ameliorating soil physical properties and in increasing agricultural production in the tropics have sadly not been realised because of ignoring this important aspect of tropical soil biology. The flora in tropical soils have been studied by some researchers (Stout, 1971; Balasubramanian *et al.*, 1972; Ayanaba and Omayuli, 1975; Martinez Cruz *et al.*, 1981; Katz, 1981; Kaiser, 1983), but few have related the microbial activity to soil physical properties. Nonetheless, the beneficial effects of microorganisms in increasing productivity of tropical soils are widely recognized (Lynch, 1977; Vincent *et al.*, 1977).

1. Decomposition of organic matter

Maintenance of crop residue mulch on the soil surface in the tropics can be an uphill task because the rate of decomposition is often higher in the tropics than in the temperate regions. Jenkinson and Ayanaba (1977) noted that the rate of mineralization of organic matter at Ibadan, Nigeria, was four times more than that at Rothamsted, UK. Kilbertus et al. (1980) evaluated the decomposition rate of sawdust in French Guiana and Ivory Coast. The decomposition rate was assessed at four sites in each country. Sterilized beech sawdust was buried at a depth of 5 cm in containers permeable to microorganisms. In French Guiana, 77 to 81 per cent of the sawdust decomposed after 7 months in containers only permeable to microorganisms, compared to 51 to 76 per cent in containers also permeable to soil animals. In the Ivory Coast, 73 to 83.6 per cent of the sawdust decomposed after 10 months.

The rate of decomposition depends on many factors including the climate (temperature and humidity), cultural practices (seedbed preparation, burning, etc.) and the quality of the crop residue. The C:N ratio in the plant is an important factor affecting the decomposition rate. In southern Nigeria, Lal et al. (1980) reported that rice straw decomposed faster than Guinea grass (Table 7.5). Sixty days after application, rice straw had decayed 49, 42, 39, and 34% at 2, 4, 6 and 12 tons ha^{-1} season^{-1} mulch rates, respectively. Leguminous mulch material with a narrow C:N ratio decomposes at even faster rates.

2. Effects on soil physical properties

Microbial by-products are known to affect soil aggregation and improve soil structure (Dabek-Szreniawska 1977). In Taiwan, Yang (1982) reported that

Table 7.5 Decomposition rate of rice straw and guinea grass for different mulch rates (Lal et al., 1980). *Reproduced from* Soil Science Society of America Journal, *Volume 44, 827–833, by permission of the Soil Science Society of America*

Mulch rate (tons/ha)	Regression equation	Correlation coefficient (r^2)
	Guinea grass	
2	$Y = -4.74 + 0.93t - 0.0022\ t^2$	0.97
4	$Y = -4.00 + 0.81t - 0.0020\ t^2$	0.97
6	$Y = -4.00 + 0.75t - 0.0015\ t^2$	0.98
12	$Y = -3.09 + 0.66t - 0.0011\ t^2$	0.99
	Rice straw	
2	$Y = -5.66 + 1.05t - 0.0024\ t^2$	0.95
4	$Y = -4.93 + 0.89t - 0.0017\ t^2$	0.95
6	$Y = -4.54 + 0.79t - 0.0012\ t^2$	0.95
12	$Y = -4.37 + 0.69t - 0.0009\ t^2$	0.95

Y = percent decomposition; t = time in days after mulch application.

Table 7.6 Relationship between percentage of actinomycetes antagonistic to Gram-negative bacteria and organic matter content of soils as well as maximum water-holding capacity of soil (% to all antinomycete isolates). Values as correlation r. (After Araragi, 1979) *Reproduced by permission of the author*

Test organisms	Soil properties	
	Organic matter content	Maximum water-holding capacity
Enterobacter aerogenes IAM 1063 (ATCC 8303)	0.67**	0.60**
Erwinia aroidea (Townsend) Holland	0.65**	0.51*
Pseudomonas marginalis (Brown) Steunes	0.64**	0.57*
Escherichia coli Najjar strain IAM 1239	0.57*	0.54*

*, ** significant at 1% and 0.1% levels, respectively.

amino sugar organic nitrogen compounds and other products of microbial decomposition all act as binding agents. Araragi (1978) assessed the effects of Actinomycete flora of tropical uplands from Trinidad, Thailand and Japan on soil physical properties. His data (Table 7.6) show that soils with high percentages of actinomycets antagonistic to Gram-positive bacteria were significantly high in organic matter content and maximum water-holding capacity.

Table 7.7 Mesofauna in 1000 cm³ soil from 0–5 cm layer of an eastern Amazon Oxisol as influenced by different mulches (8 months after mulch application) (from Schöningh and Alkämper, 1984)

Mesofauna (Taxonomic groups)	Mulches used*							
	1	2	3	4	5	6	7	8
Acari	306	169	422	397	264	1512	357	130
Collembola	67	31	40	27	28	62	30	11
Diplura			1	1	1			
Protura		4	12	3	7	2	7	
Coleoptera (adult)	11		1	3	1	5	2	3
Coleoptera (larva)	7	10	15	13	15		8	4
Corrodentia	9	4	9	15	8	13	2	2
Diptera (adult)	6	3	1	2		2		1
Diptera (larva)	1					1		
Homoptera	2	5	2	1	3		1	1
Hymenoptera (Formic.)	2		2	3	14	2	1	
Chilopoda	1				1			
Diplopoda	2	7	9	14	1	3	3	
Pauropoda			1	3	6			
Total	414	234	517	479	349	1602	411	152

1 = Elephand grass; 2 = Pueraria; 3 = weeds; 4 = sec. forest (2–3 yrs old); 5 = sec. forest 4–5 yrs old); 6 = rice husks; 7 = maize cobs + husks; 8 = bare soil.

3. Effects of cultural practices

Mulching, fallowing and fire affect the microbial population. Regular and frequent supply of organic matter to the soil improves the population of soil flora. In an eastern Amazon Oxisol, Schöningh and Alkämper (1984) observed larger population and diversity of mesofauna in mulched than in bare plots. Most mulches used had about 2–10 times higher numbers of mesofauna, except *Pueraria* which decomposed readily (Table 7.7). The highest number of mesofauna species were found in plots covered by rice husks. Cultural practices that increase erosion and raise the maximum soil temperature to supra-optimal levels decrease the floral activity. Burning has a detrimental effect on soil flora. In Ivory Coast, Schaefer (1974) observed a decrease in the fungal and bacterial population (following burning). Similar observations have been reported in India (Tiwari and Rai, 1977) and Malagasy, (Dommergues, 1954; Maureaux, 1959). Also in Malagasy, Bosser *et al.* (1956) studied the effects of erosion on soil flora and reported that the biological activity of the soil decreased with increase in erosion. Frequent and uncontrolled burning also increases the erosion hazard. The soil devoid of vegetation cover has usually less biological activity (Kaiser, 1983).

VI. CONCLUSIONS

Although the study of soil animals is at least 150 years old, little progress has been made in understanding their effects on the physical, mechanical and nutritional properties of soils in the tropics. Vast areas of humid tropical forest and savanna regions are rapidly being converted to arable land-use. The soil ecology and microclimate is drastically disturbed by mechanical systems of land clearing and development, and subsequent tillage operations. The use of fire as a clearing tool is widely practiced. Indiscriminate use of pesticides, herbicides and fertilizers is increasingly being made with the hope of boosting agricultural production. This is all being done with little knowledge regarding the ecological consequences of these practices, and their effects on soil fauna and soil properties. Tropical soils are often described as shallow with most of plant nutrient reserves being concentrated in the top few centimetres. Although soils are deeply weathered, the effective root zone is rather restricted. That is why the depletion of the topsoil by surface scraping and accelerated erosion causes a rapid decline in soil productivity. The soil fauna, like plant nutrients, are also concentrated in the top layer. Tropical soils devoid of meso- and macro-fauna are vulnerable to rapid soil degradation because they do not receive the beneficial effects of these soil animals. Lush green forests often make tropical soils appear deceptively fertile, but they should be managed with great caution because they are easily destablilized once the surface fauna are destroyed by land clearing, burning and other destructive agricultural practices.

REFERENCES

Abaturov, B.D., 1972, The role of burrowing animals in the transport of mineral substance in soil. *Pedobiologia* 12, 261–266.

Ahlgren, I.F., and Ahlgren, C.E., 1960, Ecological effects of forest fire, *Bot. Rev.*, **26**, 483–533.

Ajayi, S.S., and Tewe, O.O., 1978, Distribution of burrows of the African giant rat (*Cricetomys gabianus* Waterhouse) in relation to soil characteristics, *E. Afr. Wildl. J.*, **16**, 106–111.

Araragi, M., 1979, Actinomycete flora of tropical upland farm soils on the basis of genus composition and antagonistic property, *Soil Sci. & Plant Nutr.* **25**, 513–521.

Athias, F., 1971, Recherches écologiques dans la savane de Lamto (Côte-d'Ivoire): Etude quantitative préliminaire des Microarthropodes du Sol, *Terre et Vie*, **3**, 395–409.

Ayanaba, A., and Omayuli, A.P.O., 1975, Microbial ecology of acid tropical soils: A preliminary report, *Plant and Soil*, **43**, 519–522.

Balasubramanian, A., Shantaram, M.V., Sardeshpande, J.S., Siddaramappa, R., and Rangaswami, G., 1972, Studies on certain physico-chemical and microbiological properties of Tamil Nadu soil, *Madras Agric. J.* **59** (9/10), 496–502.

Belfield, W., 1970, The effect of shade on the arthropod population and NO_3 content of a West African soil, in *IVth Int. Colloquium for Soil Zoology, Dijon*, 14/19 IX – 1970, *Anales de Zoologie, Ecologie Animale (1971) Hors Sér.*, 557–567.

Bellairs, V., Bellairs, R., and Goel, S., 1983, Studies on an Indian polydesmoid millipede *Streptogonopus phipsoni* life cycle and swarming behaviour of the larvae, *J. Zool.*, **199**, 31–50.

Bhattacharya, T., Jos, S., and Joy, V.C., 1981, Community structure of soil cryptostigmata under different vegetational conditions at Santiniketan, *J. Soil Biol. Ecol.*, **1**, 27–41.

Block, W., 1970, Micro-arthropods in some Uganda soils, in J. Phillipson (ed). *Methods of Study in Soil Ecology, Proc. UNESCO/IBP Symp., Paris 1967*, pp. 195–202.

Bosser, J., Moureaux, C., and Pernet, R., 1956. Evolution biologique de deux sols à Madagascar, in *6 ème Congr. Int. Sci. Sol, Paris*, III, **67**, 399.

Bourliere, F., and Hadley, M., 1970, The ecology of tropical savannas, *Ann. Rev. Ecol. System*, **1**, 125–152.

Buffington, J.D., 1967, Soil arthropod population of the New Jersey Pine barrens as affected by cultivated fields in Nadia district (West Bengal) with a correlation between monthly population and individual soil factor, *Rev. Ecol. Biol. Sol.* **4**, 507–515.

Caveness, F.E., 1982, Deforestation induced changes in nematode soil population, *Int. Symp. Land Clearing and Development*, 23–26 Nov., 1982, IITA, Ibadan.

Couteaux, M., 1979, The effects of deforestation on Testacea populations in French Guiana: preliminary study, *Rev. Ecol. Biol. Sol*, **16**, 403–413.

Critchley, B.R., Cook, A.G., Critchley, U., Perfect, T.J., Russell-Smith, A., and Yeadon, R., 1979, Effects of bush clearing and soil cultivation on the invertebrate fauna of a forest soil in the humid tropics, *Pedobiologia*, **19**, 425–438.

Critchley, B.R., Cook, A.G., Critchley, U., Perfect, T.J., and Russell-Smith, A., 1980, The effects of crop protection with DDT on some elements of the subterranean and surface active arthropod fauna of a cultivated forest soil in the humid tropics, *Pedobiologia*, **20**, 31–38.

Dabek-Szreniawska, M., 1977, The influence of Arthrobacter sp. on soil aggregation, *Zeszyty Problemowe Postepow Nawk Rolniczych*, **197**, 319–328.

Dantas, M., and Schubart, H.O.R., 1980, Correlation between aggregation indices of

Acari and Collembola and environmental factors in an Amazonian dryland pasture, *Acta Amazonica*, **10**, 771–774.

Darlong, V.T., and Alfred, J.R.B., 1982, Differences in arthropod population structure in soils of forest and Jhum sites of North-East India, Shillong, *Pedobiologia*, **23**, 112–119.

Dindal, D.L., (ed.), 1979, Soil biology as related to landuse practices, *Proc. VII Int. Soil Zoology Colloquium of ISSS*, Syracuse, N.Y., 29 July – 3 August, 1979, Office of Pesticide & Toxic Substances, EPA, Washington, D.C., 880 pp.

Dommergues, Y., 1954, Action du feu sur la microflore des sols de prairie, *Mém. Inst. Sci. Madagascar*, DG, 149–158.

Gillon, Y., and Gillon, D., 1976, Comparison of millipede activity in an area of natural vegetation and a groundnut field using pitfall traps, *Cah. ORSTOM, Biologie*, **11**, 121–127.

Gillon, D., and Gillon, Y., 1979a, Spatial distribution of principal millipede species in a cultivated zone of Senegal, *Bull. Ecol.*, **10**, 83–93.

Gillon, D., and Gillon, Y., 1979b, Estimation of the number and the biomass of millipedes (Myriapoda, Diplopoda) in a cultivated zone of Senegal, *Bull. Ecol.*, **10**, 95–106.

Graff, O., and Satchell, J.E. (eds.), 1967, *Progress in Soil Biology*, North Holland Publishing Co., Amsterdam, 656 pp.

Harrington, G.N., 1974, Fire effects on a Ugandan grassland, *Trop. Grasslands*, **8**, 87–101.

Harrington, G.N., and Ross, I.C., 1974, The savanna ecology of Kidepo Valley National Park. I. The effects of burning and browsing on the vegetation, *E. Afr. Wildl. J.*, **12**, 93–105.

Hazra, A.K., 1978, Ecology of collembola in a deciduous forest floor of Birbhum district, West Bengal, in relation to soil moisture, *Oriental Insects*, **12**, 265–274.

Hoeksema, K.J., and Edelman, C.H., 1960, The role of biological homogenization in the formation of Gray-Brown Podzolic soils, *7th Int. Congr. Soil Sci. Trans.*, **4**, 402–405.

Hole, F.D., 1961, A classification of pedoturbations and some other processes and factors of soil formation in relation to isotropism and anisotropism, *Soil Sci.*, **91**, 375–377.

Hole, F.D., 1981, Effects of animals on soil, *Geoderma*, **25**, 75–112.

Holt, J.A., 1981, The vertical distribution of cryptostigmatic mites, soil organic matter and macroporosity in three North Queensland rainforest soils, *Pedobiologia*, **212**, 202–209.

Izara, D.C. De., 1977, The effects of burning on soil micro-arthropods in the semi-arid pampas zone, *Ecol. Bull. (Stockholm)*, **25**, 357–365.

Jenkinson, D.S., and Ayanaba, A., 1977, Decomposition of carbon[14] labelled plant material under tropical conditions, *Soil Sci. Soc. Amer. J.*, **41**, 912–915.

Kaiser, P., 1983, The role of soil micro-organisms in savanna ecosystems, in F. Bourliere (ed.) *Tropical Savannas*, Elsevier Scientific Publishing Co., Amsterdam, pp. 541–585.

Katz, B., 1981, Preliminary results of leaf litter decomposing microfungi survey, *Acta Amazonica*, **11**, 410–411.

Kevan, D.K. McE, 1968, *Soil Animals*, H.F. & G. Witherby Ltd., London, 244 pp.

Kilbertus, G., Kiffer, E., Mangenot, F., and Arnould, M.F., 1980, Biological activity in tropical soils (French Guiana and the Ivory Coast): Decomposition of lignified tissue, *Bois et Forets des Tropiques*, **190**, 3–15.

Lal, R., De Vleeschauwer, D., and Malafe Nganje, R., 1980, Changes in properties of a newly cleared tropical Alfisol as affected by mulching, *Soil Sci. Soc. Amer. J.*, **44**, 827–833.

Lasebikan, B.A., 1974, Preliminary communication on microarthropods from a tropical rainforest in Nigeria, *Pedobiologia*, **14**, 402–411.

Lasebikan, B.A., 1981, Comparative studies of the arthropod fauna of the soil and a decaying log of an oil palm tree in a tropical forest, *Pedobiologia*, **21**, 110–116.

Lavelle, P., 1983. The soil fauna of tropical savannas: II. The earthworms, in F. Bourliere (ed.) *Tropical Savannas*, Elsevier Scientific Publishing Co., Amsterdam, pp. 465–504.

Lavelle, P., and Meyer, J.A., 1980, Du tri des données à l'élaboration de modéles de simulation: exemple de l'étude écologiques des vers de terre de la savane de Lamto (Côte-d-Ivoire), in J.P. da Fonseca (ed.), *Colloque Informatique et Zoologie*, Informatique et Biosphère, Paris, pp. 311–326.

Lavelle, P., Maury, M.E., and Serrano, V., 1981, Estudio cuantitative de la fauna del suelo en la region de Laguna Verde (Vera Cruz, Mexico). Epoca de lluvias, in P. Reyes Castillo (ed.), *Estudios Ecologicas en el Tropico Mexicano*, Instituto do Ecologia, Mexico, pp. 65–100.

Lynch, J.M., 1977, *Soil Biotechnology; Microbial Factors in Crop Productivity*, Blackwell Scientific Publications, Oxford, U.K., 191 pp.

Madge, D.S., and Sharma, G.D., 1969, *Soil Zoology*, Ibadan Univ. Press, Ibadan, 54 pp.

Maldague, M., 1959, Importance et role de la microfaune du sol, *Bull, Agric. Congo Belge*, **50**, 5–34.

Martinez Cruz, A., Palenzuele, A., and Cahng, I., 1981, Biological characteristics of the main soils of Cuba I. Total Microflora, *Ciencias Agric.*, **9**, 91–102.

Mathys, G., 1975, Weed control and the environment, *Bull. OEPP*, **5**, 87–100.

McColl, H.P., 1975, The invertebrate fauna of the litter surface of a *Nothofagus truncata* forest floor, and the effect of micro-climate on activity, *N.Z.J. Zool.*, **2**, 15–34.

Moureaux, C.,1959, Fixation de gaz carbonique par le sol, *Mém. Inst. Sci. Madagascar*, D IX, 109–120.

Murphy, P.W. (ed.), 1958, *Progress in Soil Zoology*, Butterworths, London, 398 pp.

Nair, K.S., Krishna Menon, K.M., and Mariakulandai, A., 1957, Effect of 2,4–D on soil micro-organisms, *Madras Agric. J.*, **44**, 667.

Natarajan, T., Santhanakrishnan, P., Thamburaj, S., Shanmugavelu, K.G., and Oblisami, G., 1980, Effect of zero tillage and organic mulches on the microflora of soil grown to tapioca, *National Seminar on Tuber Crops Production Technology, 22–22 Nov., 1980*, Coimbetore, India (Tamil Nadu Agric. Univ.).

Nye, P.H., 1955, Some soil forming processes in the humid tropics, IV. The action of soil fauna. *J. Soil Sci.*, 73–83.

Perfect, T.J., Cook, A.G., Critchley, B.R., Critchley, U., Moore, R.L., Russel-Smith, A., Swift, M.J., and Yeadon, R., 1977, The effects of DDT on the populations of soil organisms and the processes of decomposition in a cultivated soil in Nigeria, *Ecol. Bull. (Stockholm)*, **25**, 565–568.

Perfect, T.J., Cook, A.G., Critchley, B.R., and Russell-Smith, A., 1981, The effect of crop protection with DDT on the microarthropod population of a cultivated forest soil in the sub-humid tropics, *Pedobiologia*, **21**, 7–18.

Plowman, K.P., 1981, Inter-relation between environmental factors and Cryptostigmata and Mesotimata (Acari) in the litter and soil of two Australian subtropical forest, *J. Anim. Ecol.*, **50**, 533–542.

Pomeroy, D.E., and Rwakaikara, D., 1975, Soil arthropods in relation to grassland burning, *E. Afr. Agri. For. J.*, **41**, 114–118.

Ryke, P.A.J., and Loots, G.C., 1967, The composition of the micro-arthropod fauna in South African soils, in O. Graff and J.E. Satchell (eds.), *Progress in Soil Biology*, North Holland Publishing Co., Amsterdam, pp. 538–546.

Salt, G. 1952, The arthropod population of the soil in some East African pastures, *Bull. Ent. Res.*, **43**, 203–220.

Schaefer, R., 1974, Le peuplement microbien due sol de la savane de Lamto, *Bull. Liaison Chercheurs Lamto, Numéro Spec.*, **5**, 39–44.

Schaller, F., 1968, *Soil Animals,* Univ. of Michigan Press, Ann Arbor, 144 pp.

Schöningh, E., and Alkämper, J., 1984, Effects of different mulch materials on soil properties and yield of maize and cowpea in an eastern Amazon Oxisol, *Proc. 1st Symp. in the Humid Tropics*, November 1984, Belem, Brazil.

Singh, J., and Pillai, K.S., 1975, A study of soil microarthropod communities in some fields, *Rev. Ecol. Biol. Sol.*, **12**, 579–590.

Serafino, A., and Merino, J.F., 1978, Microarthropod populations in different soils of Costa Rica, *Rev. Biol. Trop.*, **26**, 139–151.

Springett, J.A., 1979, The effect of a single hot summer fire on soil fauna and on litter decomposition in jarrah (*Eucalyptus marginata*) forest in western Australia, *Aust. J. Ecol.*, **4**, 279–291.

Stout, J.D., 1971, The distribution of soil bacteria in relation to biological activity and pedogenesis. Part 2, Soils of some Pacific Islands, *N.Z. J. Sci.*, **14**, 834–830.

Subagja, J., and Snider, R.J., 1981, The side effects of herbicides atrazine and paraquat upon *Folsomia candida* and *Tullbergia granulata*. Pedobiologia 22, 141–152.

Thorp, J., 1949, Effects of certain animals that live in the soil, *Sci. Mon.*, **68**, 180–191.

Tiwari, V.K., and Rai, B. 1977, Effect of soil burning on microfungi, *Plant Soil*, **47**, 693–697.

Van den Berg, R.A., and Ryke, P.A.J., 1967, The effect of vegetation change on soil acari, in O. Graff and J.E. Satchell (eds.), *Progress In Soil Biology*, North Holland Publishing Co., Amsterdam, pp. 267–274.

Vincent, J.M., Whitney, A.S., and Bose, J., 1977, Exploiting the legume–Rhizobium symbiosis in tropical agriculture, *Proc. Workshop Kahului, Maui, Hawaii, 23–28 Aug., 1976*, Univ. of Hawaii, 469 pp.

Wallwork, J.A., 1970, *Ecology of soil animals*, McGraw Hill, London, 283 pp.

Wallwork, J.A., 1976, *The Distribution and Diversity of Soil Fauna*, Academic Press, London, 355 pp.

Watnabe, H., Ruayssongnern, S., and Takeda, H., 1983, Soil Animals, in Kyuma, K., and Pairintra, C. (eds.), *Shifting Cultivation*, Ministry of Science, Technology and Energy, Bangkok, Thailand, pp. 110–126.

Wielemaker, W.G., 1984, *Soil formation by termites* a study in the *Kisii area*, Kenya. Agricultural University, Wageningen, The Netherlands.

Wood, T.G., 1971, The distribution and abundance of *Folsomides deserticola (Collembola: Isotomidae)* and other micro-arthropods in arid and semiarid soils in Southern Australia, with a note on nematode populations, *Pedobiologia*, **11**, 446–468.

Yang, T.W., 1982, Methods for determining factors contributing to soil aggregate formation, *Soils and Fertilizers in Taiwan*, 59–69.

Chapter 8
Earthworms

I	INTRODUCTION	286
II	HISTORICAL PRESPECTIVES	286
III	EARTHWORM SPECIES	287
IV	ECOLOGICAL FACTORS AND POPULATION OF EARTHWORMS	287
	1. Temperature	288
	2. Soil moisture and relative humidity	289
V	CASTING AND ITS IMPORTANCE IN SOIL TURNOVER	291
VI	PHYSICAL PROPERTIES OF WORM CASTINGS	296
	A. Texture	297
	B. Bulk density	299
	C. Structural stability	300
	D. Moisture retention characteristics	302
VII	CHEMICAL PROPERTIES OF WORM CASTINGS	305
VIII	EARTHWORM AND SOIL PROPERTIES	309
	A. Soil Formation	309
	B. Effects on Soil Physical Properties	311
	C. Decomposition of Crop Residue	314
IX	EARTHWORMS AND SOIL PRODUCTIVITY	315
X	INTERACTION WITH OTHER SOIL ANIMALS	316
XI	CULTURAL PRACTICES AND EARTHWORMS	317
	1. Deforestation	317
	2. Tillage methods	318
	3. Mulching	319
	4. Planted fallows	321
	5. Grazing	322
	6. Burning	322
	7. Soil compaction	323
	8. Erosion	324
	9. Agrochemicals	324
	(i) Fertilizers	324
	(ii) Herbicides	325
	(iii) Pesticides	325
XII	SOIL RESTORATION AND EARTHWORMS	326
XIII	SUMMARY	326

I. INTRODUCTION

Earthworms are an important invertebrate macrofauna in many tropical soils, especially when under the protective native vegetation cover. Their intense activity seemingly leads to ingestion of soil and organic matter, ejection of soil mixed with humified organic matter content and body secretions as worm castings, breakdown of crop residue and other biomass and their incorporation into the soil, creation of a network of stable channels into the soil through burrowing activity, and recycling of subsoil and leached-out nutrients to the soil surface. Earthworms are, in fact, the most natural ploughing agents. They turn over the soil and make it a dynamic living body without causing any of the adverse effects for which the man-invented mechanical plough is so notorious.

Because tropical environments are considered extremely aggressive and harsh, especially in relation to seasonal and grain crop production, the role of earthworms in improving soil structure and maintaining soil fertility deserves particular attention. Some important questions to be considered in evaluating their role in tropical soils are: (i) do earthworms improve soil physical and nutritional properties and/or do they merely thrive in a good soil; (ii) what are the relative merits of soil turnover by earthworms in organic farming in comparison with mechanical tillage; (ii) is there a relation between changes in soil physical and nutritional properties by earthworms to crop yields and land productivity; (iv) what is the interaction of earthworms with other soil fauna; and (v) what is the impact of agrochemicals and of farming systems on earthworms.

This chapter reviews the state-of-art of knowledge regarding the activity of earthworms in soils of different agro-ecological regions of the tropics, and attempts to answer the questions listed in the preceding paragraph.

II. HISTORICAL PERSPECTIVES

The literature regarding the role of the earthworm is fairly ancient. Aristotle, the Greek philosopher, referred to them as 'intestines of the earth' because of their habit of ingesting and ejecting the soil (Minnich, 1977). In the sub-tropical regions of Egypt and India, the success of the ancient civilization of the Nile and the Indus valley was partly due to the fertile soils created by the activity of the earthworms and by the continual renewal of the land by the alluvium process (Voisin, 1960). During the Cleopatra era (69–30 B.C.), the earthworm was declared a sacred animal in the ancient Egypt (Minnich, 1977). More recently, Darwin (1881) remarked that earthworms have played more roles throughout the history of the world than any other animal. This remark obviously referred not only to the fertility of the soil, but also to the rapid turnover of the soil and burying under of buildings and other archaeological evidence.

Table 8.1 Some earthworm species of the tropics

Specie	Region/Country	References
Hyperiodrilus africanus	Western Nigeria	Nye (1955)
Eudrilus eugenial	Nigeria	Madge (1969)
Hippopera nigeriae	Western Nigeria	Nye (1955)
Metapheretima jucchana	New Guinea	Lee (1967)
Agastrodilus dominicae	Ivory Coast	Lavelle (1981)
Millsonia anomala	Ivory Coast	Lavelle (1978)
Microchaetus sp.	South Africa	Reinecke (1983)
Glyphidrilus sp.	Northern Uganda	Harker (1955)
Alma stuhlmanni	Northern Uganda	Harker (1955)
A. emini	Northern Uganda	Wasawo and Visser (1959)
Pheretima alexandri	Nagaland, India	Reddy (1983)
Allolobophora sp.	India	Agarwal et al. (1958)
Eutyphoeus incommodus	Pakistan	Khan (1966)
Notoscolox sp.	Burma	Gates (1961)

III. EARTHWORM SPECIES

There are five large families and more than 1800 species of earthworms and many of them are reported from soils in the tropics (Lavelle, 1983a, b). This chapter merely explains the role of earthworms in properties and productivity of tropical soils, and readers are referred to the literature below for descriptions of various species of earthworms in different ecologies of the tropics. Some of the most common earthworm species reported in the tropics are listed in Table 8.1. The description and distribution of most predominant species in tropical Asia is given by Baweja (1939), Bahl (1950), Soota and Julka (1970), Gates (1972) and Kale and Krishnamoorthy (1978). A survey of predominant species of Brazil and Colombia is given by Arle (1981) and of the Venezuelan tropical rain forest by Nemeth and Herrera (1982). The description of the most predominant earthworm species found in tropical Africa is given for Ivory Coast by Athias *et al* (1975a, b); Lavelle and Meyer (1977), and Couteaux (1978), for Nigeria by Madge and Sharma (1969) and Segun (1977); for New Guinea by Lee (1967a, b); for Uganda by Ljungstron (1974), for Madagascar by Betsch (1974) and for Lesotho in Southern Africa by Reinecke and Ryke (1969). The description of predominant species in Australia, New Zealand and the Pacific is given by Lee (1969), Martin (1978), Easton (1982) and Springett (1983). A general description of most useful species for temperate regions and of their behaviour, habits and role in agriculture is found in Edwards and Lofty (1972), Minnich (1977), and Satchell (1983).

IV. ECOLOGICAL FACTORS AND POPULATION OF EARTHWORMS

Bouché (1977) and Lavelle (1983a, b) recognized three major ecological categories of earthworms: (i) *epigeic*, the litter-dwelling, living in the soil litter;

Table 8.2 Distribution of earthworms biomass in tropical grassland and forest ecosystems (Adapted from Nemeth and Herrera, 1982). *Reproduced by permission of Pedobilogia*

Ecology	Location	Number/m^2	Biomass (g/m^2)	Reference
Savanna	South Africa	74	96	Ljungstrom and Reinecke (1969)
Grassland	Nigeria	33	10	Madge (1969)
Grassland	Lamto, Ivory Coast	91–400	13.4–54.4	Lavelle (1978)
Gallery forest	Lamto, Ivory Coast	74.7	3.4	Lavelle (1978)
Swamp (forest)	Uganda	7.4	0.23–3.64	Block and Banage (1968)
Rain forest	Amazonas, Venezuela	32.7–68.4	8.7–16.6	Nemeth and Herrera (1982)

(ii) *hypogeic*, the soil-dwelling, living within the soil and feeding upon its humic substances; and (iii) *anecic* earthworms, large-sized, living in soil but also feeding on litter. The population of these categories of earthworms in soil varies from a few to several hundreds or even thousands per m^2 depending on soil moisture and temperature regimes, availability of food supply, and competition with other soil fauna. There is a little information on population density of earthworms in tropical soils. In Uganda, East Africa, Block and Banage (1968) observed populations of between 7.4 and 101.8 per m^2 (0.5–300g/m^2). Mean biomass ranged from 0.06 g/m^2 in arable soil to 4.55 g/m^2 under banana. In Egypt, Ghabbour (1966) observed populations to range from 8 to 788 per m^2 (1.96 to 578.2g/m^2). In Western Nigeria, Madge (1969) reported earthworm populations of 33 per m^2 (10g/m^2). Dash and Patra (1977) reported the population density of a tropical earthworm specie *Lampito mauritii* to range from 64 to 800 per m^2 in a grassland site in Orissa, India. The mean biomass corresponding to this population was about 30.3 g/m^2.

The adult weight of different earthworm species also varies widely. Nemeth and Herrera (1982) reported the adult weight of some Venezuelan species to vary from 10 mg to 20 g. They also observed that tropical grassland and savanna ecologies are more favourable and support higher earthworm biomass than tropical forests (Table 8.2).

The population density, biomass, and the burrowing and casting activity of different species depend on soil temperature and moisture regimes which are the most important ecological factors responsible for seasonal changes in earthworm population.

1. Temperature

Most earthworms have a definite temperature range for their optimum activity, growth, reproduction, metabolism and respiration. The optimum temperature range for most tropical species seems to be between 20 and 35°C. In this respect, the soil temperature has more effect on activity of earthworms than air or ambient temperature. A few earthworms survive very high soil temperatures especially when it exceeds 40°C in the root zone. The optimum temperature range for earthworm species from the tropics is more than that of the temperate zone species.

In Egypt, Duweini and Ghabbour (1965) reported that the optimum temperature range for *Pheretima* and *Alma* species was between 26 to 30°C and 24 to 26°C, respectively. The upper lethal temperatures were 37 and 38°C, respectively. In contrast the activity of *A. calignosa* was spread equally over the range from 10 to 28°C. In Western Nigeria, Madge (1969) reported the optimum range of temperature for *H. africanus* to be 23.9 to 31.5°C, and temperatures exceeding 34°C were avoided. Also in western Nigeria, Critchley *et al.* (1979) reported that although air temperatures were the same, earthworms avoided the extremes of temperatures observed in cultivated (23–41°C) soil than that in forested plots (23–36°C).

2. Soil moisture and relative humidity

Earthworms cannot tolerate as high temperatures in dry environments as in moist environments because of the adverse effects of desiccation when the relative humidity is low. Prevention of water loss from the body is an important factor in survival and activity of earthworms, and therefore, soil wetness and the overall moisture regime (relative humidity) of the environment are very important ecological factors. Lavalle (1983) observed a linear increase in earthworm biomass with increase in rainfall amount in different regions of the tropics (Fig. 8.1).

For coarse textured Egyptian desert soils of high gravel content, *A. calignosa* prefers the soil moisture range of 15 to 34% by weight (Duweini and Ghabbour, 1965). The number of earthworms in this desert soil was associated with the water-content/(gravel + sand) ratio of the soil. This ratio is obviously an index of soil moisture availability. The scarcity of earthworms in Egyptian soils was attributed to the dry climate and to the fact that most land is disturbed by agricultural activity. In Western Nigeria the activity of *H. africanus* was found to be the best for soil moisture range of 12.5 to 17.5%, i.e. equivalent to the field capacity. The casting rate, however, was the most at soil moisture content of 23.3% (Madge, 1969). The *H. africanus* can also survive a period of up to 9 weeks when submerged in water. Critchley *et al.* (1979) related the seasonal fluctuations in activity of *H. africanus* to changes in soil moisture and temperature regimes. The casting activity virtually ceased during the dry season and commenced again with the onset of rain.

Similar to the temperature regime, the preferred range of soil wetness also varies among species, because some species can tolerate extremes of soil moisture conditions better than others. The optimum range of soil moisture

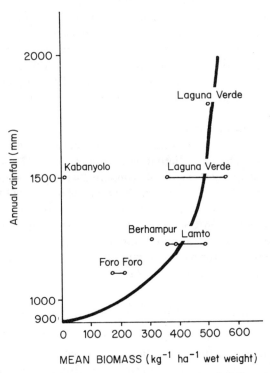

Figure 8.1 The relationship between average annual rainfall and the earthworm biomass in various tropical savannah study sites (Lavelle, 1983)

also differs among soils and with the degree of soil compaction. Springett *et al.* (1970) observed that low value of the humidity index was significantly correlated with low frequency of earthworms in an organic soil but not in a mineral soil. Abrahemsen (1971) studied the influence of temperature and soil moisture on population density of *Cognettia* sp. in culture with homogenized raw humus. The threshold value of soil moisture below which the worms did not survive corresponded to about pF 4.0 or about 10% of the water-holding capacity corresponding to 0.5 pF. The maximum abundance of worm activity occurred at a pF range of 0.6 to 1.5 or 50 – 95% of water-holding capacity.

In Santa Fe Province of Argentine, Ljungstrom *et al.* (1973) observed that the soil moisture regime was the most important edaphic factor influencing the population distribution and activity of earthworms. In Central Amazonia, the effects of soil moisture on the activity of earthworms has been reported by Irmler (1976) and Ayres and Guerra (1981). Of the 40 species of earthworms from Central Amazonia, 33 species occurred only in humid environments. Four species were eurihydric, occurring in diverse habitats with a wide range of water content, while the remaining three species had a very narrow humidity range. Bulk density interacts strongly with moisture content for mineral soils in determining the optimum environment. Thomson and Davies (1974) observed that high bulk density could compensate for low moisture levels in supporting a high level of activity.

Additional information regarding the effects of other ecological factors on earthworms is obtainable from Duweini and Ghabbour (1965), Satchell (1966), Wood (1974), Laird and Kroger (1981), Southwell and Majer (1982) and Lavelle (1979; 1983a; 1983b).

V. CASTING AND ITS IMPORTANCE IN SOIL TURNOVER

The usefulness of the earthworm in improving water transmission properties of the soil profile and in facilitating root growth and development is attributed to the channels created by its burrowing activity. Earthworms create channels either by pushing aside the soil particles or by ingesting them. Burrowing and casting habits differ greatly among different species, and deep vertical channels are created only by a few species. These channels are often stabilised through lining with body fluids. The extent of channels also depends on soil texture,

Figure 8.2 Turret-shaped casts of *Hyperiodrilus africanus*

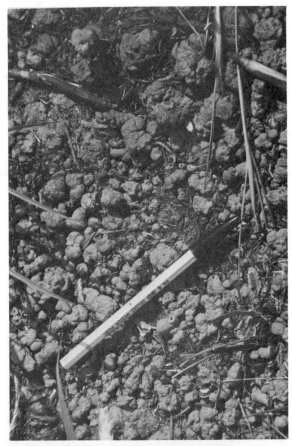

Figure 8.3 Casts of *Hyperiodrilus africanus* in a recently burnt savannah in western Nigeria

organic matter content and the drainage conditions. The availability of food supply, the soil moisture and temperature regimes, and ecological factors discussed before also affect the burrowing activity.

Earthworm castings are the by-product of burrowing activity, and are ejected either in the voids within the soil or on the soil surface. The shape, size and mode of castings differ among species, and the casting activity is also influenced by soil texture, compactness and the organic matter content. The casts of some African earthworms (i.e., *H. africanus* and *D. jaculatrix*) are generally 2–12 cm long and from 1 to 4 cm in diameter (Baylis, 1915; Madge, 1969) (Figs. 8.2 to 8.4). The *H. africanus* produces funnel-shaped casts in shade and *Eudrilus eugeniae* produces mound-shaped (spherical) casts in the open (Madge, 1969) (Figs. 8.5 to 8.7). A Burmese earthworm, *Notoscolex* sp., is known to produce giant tower-shaped casts 20–25 cm long and up to 4 cm in diameter (Gates, 1961). The shape, size, consistency and weight of castings

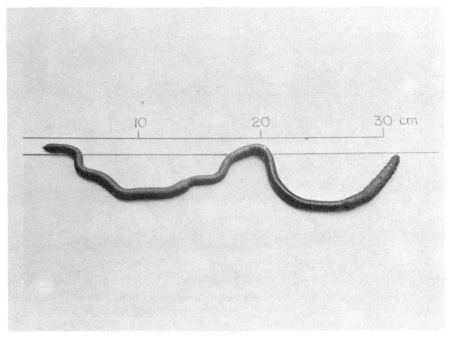

Figure 8.4 A common earthworm observed at IITA, Ibadan, Nigeria

Figure 8.5 Spherical casts of *Eudrilus eugeniae*

Figure 8.6 A single large surface casts about 20 cm long (range 6 cm to 20 cm) and weighing as much as 1000g (200 to 1000g) of an unidentified specie in Savanna near Abuja, Nigeria. These casts are similar to that of *Microchaetus* sp. reported by Reinecke (1983) in Southern Africa

Figure 8.7 Large casts create highly irregular microrelief in the undisturbed savanna landscape

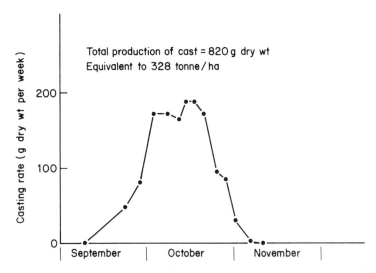

Figure 8.8 Casting activity of earthworms (*Hyperiodrilus* sp.) under forest cover ceases during the dry season beginning in November (Lal and Cummings, 1979). *Reproduced by permission of Elsevier Science Publishers*

differ so much among species as to afford species identification from examining their casts (Bhatti, 1962). The casting activity of most tropical species, however, is limited to the wet season only (Fig. 8.8).

Castings, being rich in plant nutrients and of stable structure, are desirable for agricultural soils. On average, earthworms can produce their own weight daily in castings. A study conducted by United States Department of Agriculture (1949) in Sudan (reported by Minnich, 1977) indicated the importance of worm castings on the fertility of soil in the valley of the White Nile. The annual castings recorded during a 6-month rainy season amounted to about 250 t/ha^{-1}yr^{-1} In Nigeria, Madge (1969) observed that about 175 t/ha of casts are produced by each of the two species (*H. africanus* and *E. eugenia*) during the wet season of 5 to 6 months. At Lamto in the savanna zone of Ivory Coast, the casting activity of *Millsonia anomale* was related to seasonal fluctuations in soil moisture content (Lavelle, 1971). Earthworms were found to be quiescent for 8 months on the plateau and slopes and for 7 months in the hydromorphic soils and swamps. The diurnal pattern of casting showed two peaks, at midnight and 0900 hours, during a wet period and one peak only, at 0500 hours, during a dry period. Lavelle (1975a, b) estimated that in one year a field population of 215 000 earthworms/ha ingested 507 t/ha of oven-dry soil. The forest floor in some parts of West Africa is usually covered with a 2–3 cm layer of worm casts (Nye, 1955; Madge, 1969).

In Pakistan, Khan (1966) observed that the population of *Eutyphoeus incommodus* fluctuated from nil in dry weather to approximately 150 000/ha with abundant soil moisture and adequate organic matter supply. The mean casting rate was observed to be 4.1 g per worm per day.

Table 8.3 The casting rate (A) in relation to the availability of grass meal (g day^{-1} g^{-1} live worm) (Martin, 1982). *Reproduced by permission of* Pedobiologia

Grass meal (g kg^{-1} dry soil)	A. caliginosa		A. trapezoides		L. rubellus	
	A	D	A	D	A	D
17.6	3.84a	0.45a	2.63a	0.17a	2.59bc	0.16a
4.4	4.09a	0.46a	3.65bc	0.26b	3.01c	0.18a
1.1	4.73a	0.60a	4.19c	0.30b	2.28ab	0.15a
0.0	3.75a	0.48a	3.21ab	0.25b	1.92a	0.14a
Standard error	0.91	0.015	0.22	0.014	0.13	0.022

Table 8.4 Rate of soil turnover by earthworms

Region	Soil	Earthworm species	Equivalent depth (mm h^{-1} y^{-1})	Reference
Nigeria	Alfisol	*Hyperiodrilus africanus*	3.46	Nye (1955)
Nigeria	Alfisol	"	2.77	Madge (1965)
Nigeria	Alfisol	"	3.7	Lal (1976)
Ivory Coast	Alfisol	*Millsonia anomala*	39.0	Lavelle (1975)
Adelaide	—	—	0.18	Barley (1959)

Nye (1955) and Lal (1976) observed casting rates in forested land in the semi-deciduous rainforest zone of western Nigeria. The casting rate ranged from 18 to 90 casts m^{-2} day^{-1} from *H. africanus* and from 1 to 10 casts m^{-2} day^{-1} from *E. eugeniae*. The equivalent weight in casts can be as much as 300 t ha^{-1} yr^{-1} under ideal conditions.

The casting rate not only depends on the species of earthworm but also on the organic matter available as food. Martin (1982a, b) observed that earthworms construct few burrows in the soil while there is ample organic matter available on the surface. The soil egested, or the casting rate, therefore, also depends on the food availability (Table 8.3) and distance burrowed (D) in metres. With high casting activity, the rate of annual soil turnover can be substantiated (Table 8.4).

VI. PHYSICAL PROPERTIES OF WORM CASTINGS

Worm castings, or egesta, are influenced by digestive mechanisms involving physical and biochemical processes. The physical, chemical and biological properties of egesta and their energy equivalents are thus likely to be different from the soil ingested. The egesta is a thoroughly mixed blend of ingested soil and organic matter, body wastes, and other microorganisms. Soil particles are cemented together by the metabolic by-products and are, therefore, converted into stable aggregates. The physical properties of worm castings are generally more favourable to crop production than the surrounding soil. The surface soil characteristics and micro-relief patterns are often influenced by the quantity and stability of worm castings (Haantjens, 1965).

A. Texture

The mean particle size of worm castings is generally different from the surrounding soil. In India, Joshi and Kelkar (1952) reported that worm castings contained less sand than the surrounding soil. In contrast, Nijhawan and Kanwar (1952) observed more sand in cast than in adjacent soil (Table 8.5). In Nigeria, Nye (1955) reported that casts contained virtually no grains exceeding 0.5 mm and contained a low proportion between 0.2 and 0.5 mm. The ratio of clay: silt + fine sand in cast was found to be similar to that of the top 30 cm of the soil (Table 8.6). In Zaire, Tran-Vinh-An (1973) also observed a higher content of fine fraction in casts than in soil (Table 8.7). In Ivory Coast, Lavelle (1983 a, b) observed higher contents of clay, fine and coarse silt and fine sand in casts of *Eudrilidae* and *M. anomala* than in adjacent soil (Table 8.8). In

Table 8.5 Physical properties of soil and of castings of *Pheretima* sp. and of *Euthyphoeus Waltoni* earthworms in northern India (After Nijhawan and Kanwar, 1952)

Property	E. Waltoni	Parent soil (0–15 cm)	Pheretima sp.	Parent soil (0–15 cm)
Coarse sand (%)	39.8	32.0	45.2	38.7
Fine sand (%)	32.4	37.6	38.0	39.6
Silt (%)	14.1	11.7	7.5	8.0
Clay (%)	8.9	15.8	7.9	10.0
Dispersion coefficient (15 mm)	5.9	9.8	1.8	5.0

Table 8.6 Mechanical composition of worm casts (Nye, 1955). *Reproduced by permission of the Journal of Soil Science*

Sample	Percent size fraction (mm)								Clay: silt + fine sand
	7–4	4–2	2–1	1–0.5	0.5–0.2	0.2–0.02	0.02–0.002	<0.002	
Worm casts	—	—	—	1	15	45	14	25	0.42
Soil (0–2.5 cm)	1	3	12	12	29	26	7	13	0.39
Soil (2.5–15 cm)	2	7	23	19	26	13	5	7	0.39
Soil (15–30 cm)	10	16	19	14	22	14	6	9	0.45
Soil (30–45 cm)	13	17	21	13	17	10	5	17	1.13
Soil (45–60 cm)	11	19	29	8	12	8	6	28	2.00

Table 8.7 Mechanical analysis (%) of worm casts and soil for Zaire (After Tran-Vinh-An, 1973). *Reproduced by permission of* Cah. ORSTOM, Serie Pedol

Particle size (μm)	Casts		Soil		Coefficient of variation (%)		t test
	Mean	Sx	Mean	Sx	Cast	Soil	
0–2	10.75	0.310	6.97	0.284	16.3	23.0	8.97***
2–20	2.33	0.122	1.80	0.099	5.1	31.3	3.37***
20–50	2.36	0.059	1.77	0.088	14.1	28.3	5.58***
50–100	12.18	0.068	9.18	0.277	3.1	17.1	8.52***
100–250	52.53	0.278	47.25	0.460	3.0	5.5	9.44***
250–500	20.70	0.473	33.19	0.703	13.3	12.0	12.97***

*** Significant at 1% level of probability.

Table 8.8 Textural composition of the casts of two earthworms from Ivory coast and of the uncontaminated adjacent soil (Lavelle, 1983)

Particle size (μm)	Casts		Soil (0–40 cm)
	Eudrilidae	*M. anomala*	
Clay (<2)	5.0	5.2	3.1
Fine silt (2–20)	7.3	8.6	6.1
Coarse silt (20–50)	12.0	11.5	8.0
Fine sand (50–200)	40.0	27.0	27.3
Coarse sand (200–2000)	32.0	46.5	53.2

comparison, the casts contained a lower content of coarse sand. Lavelle (1971) observed that casts of *M. anomala* contained higher proportions of silt and clay fractions than the adjacent soil. Sharpley and Syers (1976) found that surface casts of earthworms contained a higher proportion of finer particles than the underlying surface soil.

The data in Table 8.9 for six soil series in western Nigeria show that casts contain less sand and a higher quantity of silt and clay than the adjacent soil. It is also apparent from the data that high sand content in the surrounding soil is reflected in the relatively high sand content in the cast. For comparison, 59.9% sand was observed in casts from Iwo soil containing 73.6% sand; and 67.5% sand was recorded in casts from Matako soil containing 86.1% sand. The ratio of sand content in cast to sand content of soil was 0.74, 0.81, 0.71, 0.77, 0.79, and 0.78 for the Ekiti, Iwo, Ibadan, Apomu, Adio and Matako soil series, respectively. Most of the decrease in the sand content of casts was compensated for by a corresponding increase in the silt content (Table 8.9). The ratio of silt content in cast to silt content of soil was 2.39, 2.27, 2.56, 3.23, 2.04, 2.45 for different soils in that order. The sand + silt content in cast and soil, respectively, was 87.3 and 89.7 for Ekiti, 82.6 and 83.6 for Iwo, 89.4 and 95.6 for

Ibadan, 95.2 and 95.7 for Apomu, 92.8 and 94.8 for Adio, and 95.7 and 97.6 per cent for Matako soils. Furthermore, casts had a higher silt: clay ratio than any of the original soil.

This change in particle size distribution may be due to one or a combination of the following factors. While passing through the digestive system, soil particles are mechanically broken down to smaller size. Many authors support this argument (Joshi and Kelkar, 1952; Teotia *et al.*, 1950). But the decrease in sand particle size that would occur during single passage through the gut would be relatively small and of little significance. The second possibility, therefore, is that worms preferentially absorb silt-sized particles (De Vleeschauwer and Lal, 1981). On the contrary, Nijhawan and Kanwar (1952) observed that earthworm castings contained more coarse fractions than the parent soil and suggested that worms preferentially feed on coarse particles. If in fact preferential feeding is involved, the preferred particle size obviously differs among different species.

The data in Table 8.9 indicate that the silt content of casts in some soils (Iwo series) was equivalent to that in their subsoil (70–150 cm) horizons. The third possibility, therefore, is of bringing the soil from deep horizons and voiding it on the soil surface. Few experimental data exist to substantiate this hypothesis. If the subsoil material were in fact being brought to surface, it would have tremendous implications in terms of nutrient recycling. This hypothesis of soil turnover was also put forth by Edwards and Lofty (1972). In that case, turnover of 200 to 500 t ha^{-1}yr^{-1} of worm castings would imply bringing 15 to 35 mm of subsoil to the surface everywhere and leaving coarse sand and gravels beneath This biological turnover has significant pedological significance in the development of tropical soils.

B. Bulk density

The available experimental data indicate differences in bulk density of cast and the surrounding soil. The data in Table 8.10 are from soils that had been under

Table 8.9 Textural compositions of casts and soils for some soil series in western Nigeria (Adapted from De Vleeschauwer and Lal, 1981). *Reproduced by permission of Williams & Wilkins Co.*

Soil series	Sand (%)			Silt (%)			Clay (%)		
	Cast	Soil	t	Cast	Soil	t	Cast	Soil	t
Ekiti	57.0	77.0	**	30.3	12.7	**	12.7	10.3	**
Iwo	59.9	73.6	**	22.7	10.0	**	17.4	16.4	ns
Ibadan	59.2	83.8	**	30.2	11.8	**	10.6	4.4	**
Apomu	66.8	86.9	**	28.4	8.8	**	4.8	4.3	**
Adio	64.2	80.8	**	28.6	14.0	**	7.2	5.2	**
Matako	67.5	86.1	**	28.2	11.5	**	4.3	2.4	**

Table 8.10 Comparison of bulk density of cast and surface soil in a forest fallow for different soil series (De Vleeschauwer and Lal, 1981). *Reproduced by permission of Williams & Wilkins Co.*

Soil series	Bulk density (g cm^{-3})	
	Cast	Soil
Ekiti	1.33	1.01
Iwo	1.40	1.17
Ibadan	1.15	0.98
Apomu	1.21	1.20
Adio	1.29	—
Matako	1.32	1.21

Table 8.11 Comparison of bulk density, particle density and total porosity of cast and surrounding soil from a cultivated soil (recalculated from Lal and Oluwole, 1983)

Property	Cast	Soil	LSD (0.05)
Bulk density (g cm^{-3})	1.48	1.68	0.09
Particle density (g cm^{-3})	2.43	2.46	0.15
Porosity (%)	38.6	32.7	5.7

forest fallow and thus uncultivated for at least 15 years at the time of sampling. Under these conditions, the bulk density of surface soil (including casts, leaf litter, and partially decomposed organic matter content, etc) is generally 1.0 or even less. In comparison, casts consist of cemented particles with fewer macropores than the soil, and therefore, have a higher bulk density than the soil. The data in Table 8.11 compare bulk density of cast with soil for Iwo soil that had been continuously cultivated for at least 12 years. The bulk density of cultivated soil is generally higher than fallowed soil (comparison of Tables 8.10 and 8.11). However, the bulk density of casts from cultivated Iwo soil is not drastically different from that of fallowed land. The mean bulk density of cast was about 1.48 g cm^{-3} in cultivated and 1.40 g cm^{-3} in uncultivated soil. The particle density was not affected and was the same for the cast and the soil. The total porosity, however, was significantly more in cast than in cultivated soil (Table 8.11). The total porosity in uncultivated soil was more in soil than in cast.

C. Structural stability

Worm casts are structurally more stable than the surrounding soil because they are formed by the cementing action of body fluids and other microbial by-products. Not only do casts have higher mean weight diameter but they also have more resistance to raindrop impact than the surrounding soil. Casts have a

Table 8.12 Dispersion coefficient of earthworm castings and adjacent soil for some soils of Ludhiana, northern India (Nijhawan and Kanwar, 1952)

Sample	Species	Dispersion coefficient at different times (hours)				
		0.25	1	4	12	24
Cast	E. Waltoni	5.9	11.8	2.9	1.6	1.1
Soil	(0–22 cm)	9.8	18.9	9.3	9.2	6.4
Cast	Pheretima sp.	1.8	5.0	3.4	1.5	1.3
Soil	(0–22 cm)	11.5	16.5	7.1	2.7	1.2

Table 8.13 Structural stability of casts and soils (De Vleeschauwer and Lal, 1981). *Reproduced by permission of Williams and Wilkins Co.*

Soil series	Kinetic energy (J) required to disrupt	
	Cast	Soil
Ekiti	1.12 ± 0.24	0.06 ± 0.02
Iwo	1.05 ± 0.19	0.14 ± 0.02
Ibadan	1.04 ± 0.14	0.22 ± 0.01
Apomu	1.09 ± 0.02	0.02 ± 0.01
Adio	1.07 ± 0.21	0.03 ± 0.01
Matako	1.08 ± 0.21	0.02 ± 0.01

lower dispersion coefficient and a higher aggregation percentage than the parent soil. The data from soils of northern India indicate that castings had higher amounts of water stable aggregates and that they were bigger and more stable than the soil. The dispersion coefficient of the castings was less than the corresponding parent soil (Table 8.12). Similar evidence is available by comparing the energy required to disrupt a cast and a soil aggregate (Table 8.13). In general, casts required 19 to 55 times more energy for complete dispersion by raindrop impact than soil aggregates. In another study, Lal and Oluwole (1983) reported the mean weight diameter of the cast to be 6.7 mm in comparison with that of 2.4 mm and 1.0 mm for surrounding soil from 0–5 cm depth, respectively. Worm castings left on the soil surface can stay stable for many months and resist the forces of soil erosion, especially if not destroyed by the rainfall when wet.

The mechanisms responsible for aggregate stabilization have been put forth by many authors (Edwards and Lofty, 1972). Murillo (1966) reported from Spain that earthworms markedly improved aggregate stability in soils receiving farmyard manure. In China, Huang (1979) used electron microscopy to demonstrate that organo-mineral complexes in worm castings promoted soil aggregation, and that aggregates formed by earthworms had high water stability. Pierce (1981) supported the argument that body fluids promote aggregation and that liquid lost from burrowing earthworms contributes to soil aggregation. The microbial by-products, the high amount of humified organic matter content, the higher silt and clay content, the presence of fungal hyphae, and many other factors are responsible for better structural stability of worm castings.

D. Moisture retention characteristics

Differences in texture, structural stability, total porosity and pore size distribution are responsible for differences in the moisture retention characteristics of casts and soil. Lal and Oluwole (1983) reported significantly more moisture retention at all suctions in casts than in the surrounding soil. The data in Fig. 8.9 indicate significantly higher moisture retention in casts than in soil, particularly at suctions exceeding 0.1 bar. The available water holding capacity in casts is often more than in soils. The mean available water-holding capacity, calculated from the data in Fig. 8.9, was 23.4 and 10.7 per cent for cast and the soil,

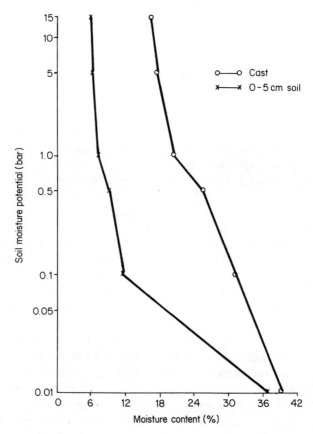

Figure 8.9 Soil moisture retention characteristics of earthworm cast compared with that of an adjacent soil (Lal and Oluwole, 1983)

Table 8.14 Comparison of pore size distribution of cast and the surrounding soil for the cast of *H. africanus*

Pore size (μm)	Cast	Soil	LSD (0.05)
>15	20.3	69.0	3.5
3–15	15.5	6.5	2.2
1.5–3	13.5	5.1	1.8
0.3–1.5	7.3	3.8	2.3
0.1–0.3	4.6	1.7	1.2
<0.1	38.8	13.9	1.4

respectively. Differences in pF curves of casts and soil samples are due to differential pore size distribution. It is evident from the data in Table 8.14 that worm castings have more retention pores (<3.0 μm) and less transmission pores (>3.0μm) than the surrounding soil. Casts, therefore, retain more water but do not conduct water through the profile. The water transmission through the soil profile is obviously influenced by channels created by their burrowing activity.

The heterogeneity of soil physical and chemical properties, discussed in Chapter 5, is partly due to the variable activity of earthworms in the soil. The presence or absence of casts on the soil surface (their concentration, frequency, and variable chemical and physical properties) is definitely an important factor that contributes towards soil's heterogeneity. Furthermore, a decrease in soil variability due to a prolonged period of cultivation may partly be attributed to a uniform decline in earthworm activity.

Table 8.15 Comparison of the chemical charactristics of soil with that of earthworm castings of *Pheretima* sp. and of *E. Waltoni* in northern India (After Nijhawan and Kanwar, 1952)

Property	E. Waltoni	Parent soil (0–15 cm)	Pheretima sp.	Parent soil (0–15 cm)
Organic matter (%)	1.72	1.49	0.74	0.60
Total nitrogen (%)	0.12	0.09	0.05	0.04
NO_3-nitrogen (ppm)	4.34	1.22	0.37	0.33
C/N ratio	8.3	10.8	8.8	8.7
Ca^{2+} (meq/100g)	6.5	6.7	2.6	9.8
Mg^{2+} (meq/100g)	3.4	3.1	2.5	4.0
K^+ (meq/100g)	0.92	1.13	0.59	0.85
Na^+ (meq/100g)	0.42	0.03	0.30	0.24
CEC (meq/100g)	11.24	10.96	5.99	14.89
Available P (ppm)	9.3	6.2	4.5	7.1
pH	7.2	7.7	8.0	8.0

Table 8.16 Chemical characteristics of earthworm casts and adjacent soil for samples from northern India (Gupta and Sakal, 1967). Reproduced by permission of the Indian Society of Soil Science

Soil	Depth (cm)	CaO (%)	K_2O (%)	Total P_2O_5 (%)	Available P (ppm)	Organic carbon (%)	Total N (%)	NO_3–N (ppm)	C/N ratio
Garden soil									
I Earthworm casts	Surface	1.02	0.896	0.29	87	0.579	0.070	9.7	8.3
Soil	0–15	0.85	0.712	0.20	74	0.380	0.035	7.0	10.8
II Earthworm casts	Surface	1.10	0.791	0.25	84	0.572	0.067	8.5	8.5
Soil	0–15	0.80	0.662	0.21	66	0.303	0.030	4.6	10.1
III Earthworm casts	Surface	0.98	0.620	0.22	72	0.480	0.059	7.7	8.1
Soil	0–15	0.75	0.517	0.17	49	0.291	0.026	6.2	11.2
Cultivated soil									
I Earthworm casts	Surface	1.35	0.525	0.15	77	0.485	0.055	11.6	8.8
Soil	0–15	1.09	0.480	0.10	62	0.242	0.024	7.7	10.8
II Earthworm casts	Surface	1.20	0.385	0.15	82	0.478	0.051	9.9	9.2
Soil	0–15	1.00	0.290	0.12	59	0.235	0.022	6.9	10.3
III Earthworm casts	Surface	1.10	0.415	0.19	75	0.502	0.053	12.5	9.4
Soil	0–15	0.92	0.317	0.17	52	0.335	0.030	8.9	11.2

VII. CHEMICAL PROPERTIES OF WORM CASTINGS

Similar to physical properties, there are measurable differences in chemical properties of worm castings in comparison with the adjacent soil. In India Nijhawan and Kanwar (1952) observed that worm casts contained more organic matter, total nitrogen and available P than parent soil (Table 8.15). Gupta and Sakal (1967) reported that worm casts contained more available P, organic matter, total N and NO_3-N than soil and that the C:N ratio in casts was also narrower (Table 8.16). Analyses of worm castings made for some soils in the semi-arid region near Hissar, India also indicated higher amounts of organic carbon and nitrogen in casts than in the soil from which the casts were collected (Dhawan et al., 1955) The organic carbon in the casts contained a higher percentage of polysaccharides than the soil.

Chemical analyses of worm casts from African soils have extensively been reported. In Nigeria, Nye (1955) reported higher amounts of exchangeable cations and organic carbon in casts than in soil (Table 8.17). Bates (1960) analysed the organic carbon (Table 8.18) and other chemical constituents (Table 8.19) on different particle size fractions and compared them with that of

Table 8.17 Chemical characteristics of worm casts (Nye, 1955). *Reproduced by permission of the* Journal of Soil Science

Sample	pH	Exchangeable cations and acidity (meq/100g)						Sat. (%)	Organic carbon (%)
		Ca	Mg	K	Na	Mn	Acidity		
Worm casts	6.9	7.0	2.5	0.34	0.38	0.11	2.4	81	2.96
Soil 0–2.5 cm	6.2	nd	nd	nd	nd	nd	nd	nd	0.96
Soil 2.5–15 cm	6.4	2.5	1.3	0.43	0.45	—	2.0	70	0.53
Soil 15–30 cm	6.7	1.9	0.8	0.34	0.50	—	1.6	69	0.32
		Total cations (meq/100g)							
Leaves	—	84	26	5.3	2.6	0.77	—	—	—

The proportions of exchangeable cations in the casts agree well with the relative amounts in the leaves overlying them.

Table 8.18 Per cent organic carbon content of aggregate fractions in soil profile and worm casts (Bates, 1960). *Reproduced by permission of the* Journal of Soil Science

	Worm cast	Depth (cm)				
		0–5	5–17.5	17.5–30	30–50	50–75
Coarse silt	17.8	12.1	1.6	0.6	0.6	0.4
Silt	12.0	10.4	1.7	0.9	0.5	0.5
Clay	12.4	9.7	1.3	1.0	0.7	0.9

Table 8.19 Distribution of organic matter in the profile and in various size fractions in comparison with worm casts (Bates, 1960). *Reproduced by permission of the* Journal of Soil Science

Soil Fraction/Property	Worm casts	Soil samples from different depths (cm)		
		0–5	5–17.5	17.5–30
Total stones over/mm (%)	nil	10	33	63
Concretions over/mm (%)	nil	1.4	16	40
Quartz ground over/mm (%)	nil	8.6	17	23
Mechanical analysis (%)				
1.0–0.2 mm	17	29	33	18
0.02–0.02 mm	38	39	26	11
0.2–0.02 mm	32	12	5.4	4.4
<0.002 mm	13	10	2.6	3.6
pH	7.1	7.7	6.4	6.
Free Fe_2O_3 (%)	1.22	1.15	1.05	1.47
Carbon (%)	6.9	3.95	0.38	0.30
N (%)	0.44	0.29	0.047	0.043
C/N ratio	15.7	14.6	8.1	7.0

Table 8.20 Phosphorus content in fine earth fraction and in worm casts (Bates, 1960). *Reproduced by permission of the* Journal of Soil Science

	Worm casts	Profile depth (cm)		
		0–5	5–17.5	17.5–30
Total P in fine earth (ppm)	603	553	210	283
Organic P in fine earth (ppm)	450	405	95	70
Extractable P (ppm)				
(i) Acetic acid	32	24	2.1	0.9
(ii) Truog's reagent	32	29	1.5	1.7
(iii) NH_4F, pH 7.0	46	33	1.8	3.6
(iv) 0.1N NaOH	74	34	19	33

Table 8.21 Chemical properties of worm casts and surface soil from Zaire (Tran-Vinh-An, 1973). *Reproduced by permission of* Cah, ORSTROM, Serie Pedol

Chemical property	Cast		Soil		cv (%)		t test
	Mean	Sx	Mean	Sx	Cast	Soil	
Organic carbon (%)	2.24	0.050	1.12	0.054	17.1	31.3	14.08***
Total nitrogen (%)	0.133	0.003	0.07	0.004	17.0	36.9	11.88***
pH	6.43	0.024	6.22	0.072	2.2	7.0	2.68**
Exchangeable Ca^{2+} (%)	4.54	0.305	3.27	0.364	31.5	52.3	2.67**
Fe_2O_3 (%)	6.94	0.326	10.26	1.157	22.5	54.0	2.99***
CEC (meq/100g)	64.31	0.816	55.08	1.165	5.8	9.7	2.05**

the worm casts. Once again worm casts were found to be richer in chemical fertility than the parent soil. There were more total and organic P in casts than in soil (Table 8.20). Similar results of high chemical fertility of casts are reported from Zaire (Table 8.21).

Duweini and Ghabbour (1971) observed a high ratio of ammonia:urea in casts collected from some Egyptian soils. In New Zealand, Sharpley and Syers (1976) reported that the release of inorganic P to solution from casts was approximately four times greater than that from surface soil. These authors observed that casts contained a pool of loosely bound inorganic P which was readily released to solution. Furthermore, the amount of exchangeable P was found to be three times greater in casts than in surface soil.

Although worm casts contain more organic matter and nutrients than soil, the relative magnitude of nutrient content in casts varies among soils, land-use, and management systems. In general, worm casts have lower pH than adjacent soils by as much as a half unit (Tables 8.17, 8.19, 8.21), and thus, the total acidity is also higher. On the contrary, Nye (1955) reported higher pH and lower total acidity of worm casts than the surface soil.

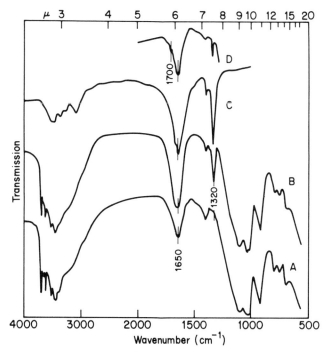

A : untreated soil
B : untreated cast
C : calcium oxalate
D : cast after HCl treatment

Figure 8.10 Infrared spectograph of >2μm fraction showing the calcium oxalate contents of earthworm cast (Tran-Vinh-An, 1973)

Table 8.22 Chemical characteristics of worm casts and surface soil for some soil series under forest cover in western Nigeria (De Vleeschauwer and Lal, 1981). *Reproduced by permission of Williams & Wilkins Co.*

Soil characteristics	Ekiti			Iwo			Ibadan			Apomu			Adio			Matako		
	C	S	t	C	S	t	C	S	t	C	S	t	C	S	t	C	S	t
pH (1:1 H_2O)	5.4	5.4	ns	5.3	5.7	**	5.7	6.4	**	5.5	6.3	**	5.5	5.9	**	6.0	6.4	**
CEC (meq/100g)	8.9	2.9	**	17.7	4.5	**	16.1	3.1	**	12.2	1.9	**	12.6	4.9	***	15.5	3.4	**
Ca^{2+} (meq/100g)	5.0	1.5	**	12.2	2.7	**	10.6	2.0	**	8.0	1.1	**	7.5	2.9	**	10.2	1.8	**
Mg^{2+} (meq/100g)	3.2	0.8	**	4.3	1.3	**	4.1	0.7	**	2.8	0.4	**	4.3	1.6	**	4.6	1.2	**
K^+ (meq/100g)	0.4	0.2	**	0.7	0.2	**	0.7	0.2	**	1.0	0.2	***	0.3	0.15	**	0.2	0.1	**
Na^+ (meq/100g)	0.10	0.06	**	0.16	0.07	**	0.17	0.06	**	0.14	0.05	**	0.13	0.08	**	0.14	0.06	**
Bray P (ppm)	11.2	6.4	**	12.6	4.5	**	20.1	9.3	**	32.1	4.0	**	15.8	5.7	**	14.7	6.9	**
Total N (%)	0.36	0.09	**	0.38	0.15	**	0.36	0.14	**	0.34	0.11	**	0.30	0.14	**	0.26	0.11	**
Organic C (%)	3.05	1.43	**	3.10	1.08	**	3.12	0.80	**	2.83	0.50	**	2.90	0.98	**	3.15	0.70	**

ns = not significant; * = significant at 5% level of probability; ** = significant at 1% level of probability; C = cast; S = soil.

Chemical analyses of worm casts of *H. africanus* are compared with adjacent soil for six soil series along a toposequence in western Nigeria (Table 8.22). Worm casts contain 2 to 5 times as much organic carbon and total nitrogen, and 3 to 6 times as much CEC and exchangeable cations as the adjacent soil. The C:N ratio of casts is narrower than that of the soil. The available P content of the worm casts is almost double the amount in adjacent soil. Worm casts have been reported to contain calcium oxalate while no such compounds were found in the soil (Fig. 8.10) (Tran-Vinh-An, 1973). The analyses presented in Tables 8.15 and 8.22 indicate that castings are in fact a high quality compost. An annual addition of 100 to 500 t/ha of this high grade compost should improve soil productivity.

VIII. EARTHWORM AND SOIL PROPERTIES

Not all species of earthworms have burrows and produce casts. It is, therefore, difficult to evaluate the effects of earthworms only on the basis of the physical and chemical properties of worm castings. Of the four predominant species in western Nigeria, only *H. africanus* produces casts that can be easily identified and analysed. The direct effects of non-casting species of earthworms on physical and chemical properties of soils in the tropics have not been quantified. There are some reports indicating the indirect effects of earthworms on soil and crops.

A. Soil formation

In Africa, the measured annual rate of new soil formation varies from 0.011 to 0.045 mm (Leneuf and Aubert, 1960; Owens and Watson, 1979). The annual rate at which Alfisols are formed is estimated to be 0.0013 mm in the semi-arid and arid regions (Nahon and Lappartient, 1979) and 0.07 mm in the humid region (Boulad et al., 1977). In comparison, Andisols are formed at a higher rate i.e. 0.06–0.7 mm/annum (Lal, 1983). The rate of soil formation is very slow indeed.

Soil fauna play a very important role in the rate of soil formation and in influencing the microrelief. Earthworms turn over the soil at the rate of 50 to 600 t ha^{-1} y^{-1}. The soil particles passing through their systems are physically reduced in size and are thoroughly mixed with soil organic matter. They are a very important soil forming factor (Anon, 1950; Nye, 1955; Heath, 1965; and Nakamura, 1980).

In South Africa Pickford (1926) (quoted by Story, 1952) reported that earthworm activity was responsible for the formation of a mound and hollow microrelief with an amplitude of 60–90 cm. This microrelief is caused by a giant (little-finger thick) earthworm, and is supposedly initiated by the tendency of the worms to throw their castings in groups, near solid objects and away from puddles. Similarly, some swamp worms construct small mounds. Wasawo and Visser (1959) reported that a vast area of grass swamps in the Teso region of

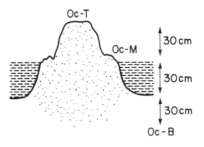

Oc-T—Top of mound
Oc-M—Fresh worm casts
Oc-B—30 cm. below the surface of the mud
  between two neighbouring mounds

Figure 8.11 Seasonal activity of earthworms in relation to the fluctuations in water table in Ochuloi swamp, Uganda (Wasawo and Visser, 1959). *Reproduced by permission of East African Agricultural and Forestry Journal*

Uganda is covered with mounds believed to be constructed by swamp worms. Three earthworms found in these mounds are *Alma stuhlmanni*, *Alma emini* and *Glyphidrilus* sp. The size and shape of these mounds vary within wide limits. The mound height ranges from 25 cm to 60 cm. The mounds are either cylindrical or rectangular. Of the cylindrical mounds, the diameter of the top varies between 20 and 100 cm, whereas the rectangular ones can be as long as 4 m with a breadth varying between 40 and 60 cm. The distance between two adjacent mounds ranges from 10 to 44 cm. In an area of 25 m^2, 80 such mounds were counted by these authors. It is obvious that even cattle would find it difficult to walk about on such mounds. As this swamp dries out, the worms bury deep into the mound and resume their activities when the swamps are flooded in the rainy season (Fig. 8.11). The casts produced during the next rainy season are piled up on to the surface of the already existing mounds. The properties of these mounds in relation to earthworm population are shown in

Table 8.23 Earthworm count in Teso mounds in Uganda (Wasawo and Visser, 1959)

Mound	Diameter (cm)	Height (cm)	Volume (cm^3)	Number of worms	Volume per worm (cm^3)	Depth of (cm)
1	20	30	9 420	74	127	20
2	32	31	24 919	67	372	20
3	31	25	20 069	80	251	20
4	31	33	26 527	169	157	20
5	25	28	14 858	118	126	20
6	28	33	20 310	126	161	21
7	28	33	20 310	69	294	25
8	25	35	18,573	59	315	22

Table 8.23. These worms are present in this region in large numbers and the annual deposition rate is estimated to be 50 mm.

In New Guinea, Haantjens (1965) reported that two types of patterned ground occur on rolling to undulating plains in the humid tropical (1500–2000 mm annual rainfall) lowland regions. These pitted soils are found on gentle slopes and are characterized by sharply defined holes or unconnected trenches developed in coarser textured upper soil horizons, above slowly permeable finer textured subsoil. This pitted soil microrelief is due to the transportation of soil from depressions on to rises by earthworms. During the initial stages, relatively shallow depressions are formed; but prolonged intensive activity leads to the formation of trenches of greater depth as the result of the linking of individual holes by the gradual removal of the saddles between them. In West Africa, Nye (1955) designated a shallow layer (2–5 cm thick) immediately below the leaf litter solely due to the activity of earthworms.

B. Effects on soil physical properties

The burrowing activities of earthworms create channels that improve the water transmission properties of the soil. The networks of interconnected channels improve the ability of plant roots to penetrate soil horizons. In northern Nigeria, Wilkinson (1975) observed that the improvements in equilibrium infiltration rate brought about by fallowing are primarily due to the channels created by earthworms. This argument was strengthened by the fact that fallowing apparently caused no durable aggregation in the sandy surface soil but produced easily identifiable water conducting pores and channels. The equilibrium infiltration rate (EIR) increased with the duration of fallowing (F) according to the following relationship:

$$EIR = 0.760 + 1.982\ F^{\frac{1}{2}}$$

where EIR is in cm h^{-1}, and F is in years. The data in Table 8.24 show that the improvements brought about by fallowing were negated by a single cultivation operation. Whatever infiltration capacity was retained by the end of the first

Table 8.24 Effects of grass fallow rotation and of earthworm channels on the infiltration of water into into a savanna zone soil of northern Nigeria (Wilkinson, 1975). *Reproduced by permission of* Tropical Agriculture

Event	Equilibrium infiltration rate (cm h^{-1})	Month
End of 6-year fallow	5.84	November
End of seedbed preparation for cotton	2.34	June
Peak of rainy season	1.30	August
End of cotton season	0.97	November

Table 8.25 Effects of earthworm species on soil properties as influenced by mulch type (Mba, 1982). *Reproduced by permission of Pedobiologia*

Mulch		Aggregation status (GMD, mm)		Structural stability		pH		Soil structure	Total carbon (%)
		Absolute	Relative	GMD	(%)	H_2O	$CaCl_2$		
Paspalum dilitatum –	Pc	19.4	185	37.3	(113)	6.4	5.7	Very spongy	1.20
	Ee	16.4	156	40.2	(122)	6.3	5.6	Spongy	1.11
Penisetum purpuisum	Pc	18.2	173	33.2	(101)	6.2	6.0	Spongy	1.08
	Ee	14.7	140	29.4	(89)	6.4	6.2	± Spongy	0.75
Gmelina arborea	Pc	18.4	175	73.2	(222)	7.2	7.1	Spongy	1.74
	Ee	14.2	135	48.3	(146)	7.5	7.1	Compact	1.42
Largo stroemia	Pc	14.8	141	45.4	(137)	7.6	7.5	Very spongy	0.60
	Ee	11.0	105	33.1	(100)	7.6	7.4	Compact	0.42
E. Hamufu soil		10.5	100	33.0	(100)	4.5	3.5	Compact	0.33
lsd (0.05)		0.8		2.3		0.2	0.1		
		1.2		3.3		0.3	0.2		

Pc = *Pontoscolex corethrurus*   Ee = *Eudrilus eugeniae*

Aggregation status according to De Leenheer and Wageman (1948) followed by graphical determination of GMD by Hartge (1971). Aggregate water stability was by method of Hartge (1971).

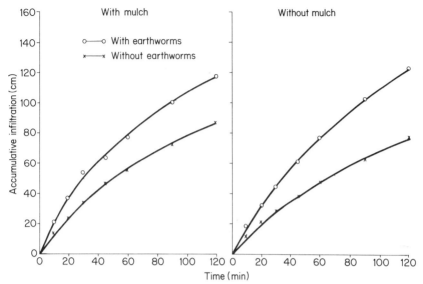

Figure 8.12 Effects of earthworm activity on water infiltration rate of an Alfisol at IITA, Ibadan, Nigeria, (unpublished data of Lal, 1984)

cropping cycle was due to the stabilized worm channels (Kowal, 1968). Similar improvements in infiltration rate have been reported under various planted fallows in western Nigeria by Lal et al. (1978). Data in Fig. 8.12 show significant differences in the infiltration rate of soils with and without earthworms. The earthworms were eliminated from the control plots. Improvements in the water infiltration rate of soil due to earthworm activity are also extensively reported for soils of the temperate region (Ehlers, 1975; Abbott and Parker, 1981; Carter et al., 1982). Earthworms improve total and macroporosity, decrease soil bulk density and improve soil aggregation.

Mba (1982) reported significant improvements in soil physical properties by earthworms during a short period of 3 weeks (Table 8.25). The improvement in soil physical properties was greater by *P. corethrurus* than *E. eugeniae*. Relative to the original soil, aggregation status was 169% on the average with *P. corethrurus* compared to 134% with *E. eugeniae*. Soil structural stability was also influenced by worm species and the mulch type. Within 3 weeks *P. corethrurus* transformed the compact soil structure to a spongy, porous structure. Similar observations regarding the beneficial effects of earthworms on soil physical properties were observed for an Alfisol near Ibadan, Nigeria. Field experiments conducted at IITA showed that plots from where earthworms had been eliminated had high soil bulk density, low water retention capacity and massive structure (Table 8.26).

Some workers have reported adverse effects of earthworms on soil physical properties. Agarwal et al. (1958) reported that activity of *Allolobophora* results in cloddy and massive structure. Similar observations regarding the adverse

Table 8.26 Effect of soil compaction on casting rate of *H. africanus* under field conditions (unpublished data, Lal, 1984)

Plot	Bulk density (g/cm^3)		Penetrometer resistance (kg/cm^2)		Casting rate (number/m^2/day)	
	Compacted	Control	Compacted	Control	Compacted	Control
1	1.51	1.45	2.0	1.0	—	20
2	1.56	1.41	2.25	1.0	—	—
3	1.55	1.34	2.50	0.75	4	4
4	1.68	1.43	2.50	1.25	4	—
5	1.49	1.54	2.75	0.75	—	9
6	1.62	1.42	2.25	1.25	—	—
7	1.65	1.46	2.50	1.25	—	3
8	1.59	1.50	2.75	1.25	—	—
9	1.41	1.40	2.0	1.0	5	—
10	1.68	1.54	2.5	0.75	—	—
11	1.59	1.55	2.5	1.0	—	8
12	1.60	1.48	2.5	0.75	—	40
Mean	1.58	1.46	2.37	1.0	1	7

effects on soil structure and crop yield were reported for *Pontoscolex* sp. and *Pheretima* sp. by Puttarudriah and Sastry (1961).

C. Decomposition of crop residue

Tropical lowland rain forest and grassland savanna are characterized by high annual litter production, but have a low litter accumulation on the ground. Therefore, the decomposition rate of leaf litter or crop residue must be extremely high in the tropics. Madge (1965, 1966) and Irmler and Furch (1980) reported that litter decomposition in tropical rain forest is 3–10 times faster than in the temperate woodlands. Many authors (Jenny *et al.*, 1949; Nye, 1969; Olson, 1963; Cornforth, 1970) have observed that in most tropical forests the leaf litter decomposes in about 6 months with a range of 2.5–11 months. In India, Bharat and Srivastava (1982) reported that about 23% of the litter was left undecomposed after a year and it required about 16 months for complete decomposition.

The high rate of decomposition is attributed to high temperature and moisture, and to high activity of soil fauna, including that of earthworms. That soil fauna play an important role in disappearance of leaf litter and other organic residues is supported by a marked decline in decomposition rate when the soil is treated with herbicides and other chemicals (Malkomes, 1980). Senapati *et al.* (1980) investigated the effect of *Octochaetona surensis* on the decomposition of leaf litter under laboratory conditions in India. In the presence of earthworms, leaf litter decreased by 13% and 31% in 7 and 25 days of incubation at 25 ± 3°C ambient temperature and 15 ± 3% soil moisture content. In comparison, only 4% of the litter decomposed in 25 days when earthworms were eliminated. The rate of litter disappearance also depends on earthworm species (Mba, 1982), nature of the litter material (Mba, 1978), and the humus content in the soil (Plowman, 1979).

IX. EARTHWORMS AND SOIL PRODUCTIVITY

Improvements in soil physical and nutritional properties brought about by earthworms have beneficial effects on crop growth and yield. Earthworm casts contain 2 to 5 times more organic matter, total nitrogen and exchangeable cations than the soil. The high organic matter content in casts and egesta renders phosphorus and other nutrient elements into readily available form. Rapid mineralization of organic matter and the possibility of association of some N-fixing bacteria (Day, 1950; Khambata and Bhatt, 1957; Bouché, 1969) makes N more readily available (Lal and Vleeschauwer, 1982). The contribution to soil N by the dead bodies of earthworms can also be substantial. Earthworms add calcite spheroids to soil (Wiecek and Messenger, 1972; Kale et al., 1977). Castings have high silt and clay content and high effective cation exchange capacity. Earthworms definitely improve soil physical properties, especially the water retention and transmission characteristics, aeration, porosity and pore size distribution, and structural stability (Primavesi and Covolo, 1968). Although earthworms thrive in better soil, they also render soil a better habitat for plant growth.

In the temperate zone, the beneficial effects of earthworms on crop growth are well documented (Barley, 1961). Atlavinyte et al. (1968) observed in a pot experiment with different soils that the yield of barley increased with increasing numbers of earthworms. In Germany, Graff (1971) demonstrated that rape and oats planted in vertical earthworm burrows produced higher yields and absorbed more N and P than control plants.

A few experiments have been conducted in the tropics relating earthworm activity to crop growth. In India Nijhawan and Kanwar (1952) observed more significant improvement in growth and yield of wheat grown in worm casts than in soil (Table 8.27). Mixing 10 to 20% worm casts in the upper 30 cm of a sandy soil also resulted in significant improvements in wheat growth and yield (Table 8.28). In the Philippines, Modena (1978) reported that fertilizing soil with

Table 8.27 Growth of wheat in castings of *E. Waltoni* and *Pheretima* sp. and in parent soil (Nijhawan and Kanwar, 1952)

Growth	*Pheretima* sp.	*E. Waltoni*	Soil
Total dry matter (g/pot)	4.5	3.6	1.6
Grain yield (g/pot)	2.0	1.0	0.4
No. of grains (grain/pot)	54	45	20

Table 8.28 Effects of rates and mode of cast application on wheat yield (Nijhawan and Kanwar 1952)

Growth parameter	10% mixed in 30 cm soil (A)	20% casts mixed in 30 cm soil (B)	10% cast mixed in upper – 10 cm soil (C)	20% cast mixed in upper – 10 cm soil (D)	Control
Total dry matter (g/pot)	18.9	27.2	15.7	22.4	12.8
Grain yield (g/pot)	5.4	6.8	4.1	5.7	3.5
No. of grains (g/pot)	216	265	207	220	132
Shoot length (cm)	61	64	59	62	57
Tielers	3	3	2	3	2

worm casts increased crop yields by more than 40%. Inoculation of about 1000 tons of municipal waste with earthworms converted it into a high grade manure of worm casts at the rate of 350 tons per month. In Palmerston, Otago, New Zealand, Stockdill (1982) observed that pasture yield was 12 100 kg/ha with earthworms established, 11 180 kg/ha with earthworms just introduced, and 9400 kg/ha without earthworms. Also in New Zealand, McColl et al. (1982) reported that inoculation of a silt loam with earthworms greatly increased early growth of perennial ryegrass and increased dry matter production throughout the trial. Nutrient uptake was also improved by the presence of earthworms.

It is also possible that earthworms compete with crops for nitrogen and temporarily immobilize plant nutrients. The adverse effects of earthworms if any, would also depend on soil properties, residue and leaf litter composition, and crops grown. In Pakistan Khan (1966) reported that washings (solution) from casts of *Pheretima posthuma* retarded the early growth of wheat seedlings, although later growth was better than in control plots.

X. INTERACTION WITH OTHER SOIL ANIMALS

The effects of earthworms on crop yield may be partly due to their interaction with other animals, and due to competition for food and nutrients among different species of earthworms and between earthworms and plants. Abbot (1980) reported a stable coexistence among different species, although a possibility exists that earthworms do compete for food especially when the preferred food is scarce. Earthworms are also known to interact with parasitic and non-parasitic nematodes, although controversy exists regarding the effects of earthworms on nematodes. Dash et al. (1980) reported that nematodes decreased the earthworm population. Yates (1981) remarked that soil nematode populations are depressed in the presence of earthworms. In contrast, Caveness, (1984 personal communication) observed that some nematodes can survive the digestive mechanism and pass unharmed in egesta of *H. africanus*. Plant-parasitic and non-parasitic soil nematodes were recovered from casts of

H. africanus collected from cultivated and secondary forest soils at the end of a 4 month dry season. Using the modified Baermann tray method, 17 plant-parasitic nematodes (*Helicotylenchus pseudorobustus, Pratylenchus sefaensis, Criconemoides* sp.) were recovered from 2–100 g samples of casts from cultivated soil over a 39-day period, and 7144 non-parasitic nematodes were recovered. From secondary forest casts, 93 plant-parasitic nematodes (*H. pseudorobustus, P. sefaensis, Meloidogyne* sp., *Xiphinema nigeriense, X. brasiliense*) were recovered, while 12 259 non-parasitic nematodes were recovered.

To ascertain whether nematodes passed through the earthworm gut alive, the wet tips of the casts were collected within seconds of deposition. From 40 g of fresh cast tips, 226 non-parasitic nematodes were recovered from a 2-day extraction.

Some earthworms are reported to be carnivorous. Lavelle (1981) observed an individual of *Agastrodilus dominicae* to ingest small *Stuhlmannia poriferia* and *A. dominicae* as well as two other *Agastrodilus* species. This earthworm may have adaptation to predation.

XI. CULTURAL PRACTICES AND EARTHWORMS

Land-use, soil management, use of agro-chemicals, farming systems and cultural practices have significant effects on earthworm population and activity. In view of the beneficial effects on soils and crops, it is important and relevant that those cultural practices be adopted that promote earthworm activity. The effects of cultural practices, however, may be soil and site specific. It is perhaps difficult to generalize the effects of farm operations on earthworms for a wide range of soils and ecologies in the tropics.

1. Deforestation

The removal of vegetation for agricultural purposes often leads to drastic alterations in microclimate and hydrological balance (Lal and Cummings, 1979). Deforestation also leads to a reduction in the diversity of food sources and nesting and there is no buffering against sudden fluctuations in the microclimate. These changes in the environment affect soil faunal population and activity (Lasebikan, 1975a, b). At Ibadan, Nigeria, Critchley *et al.* (1979) reported that deforestation resulted in a reduction in activity of *H. africanus* (Fig. 8.13) initially due to physical disturbance and later due to the harsher environment caused by reduced shading and food resources, extremes in soil moisture and temperature and compaction. The data in Fig. 8.13 show that deforestation had less drastic effects on the activity of *E. eugeniae* than on *H. africanus*. Lal and Cummings (1979) reported that casting rate of *H. africanus* was approximately 1 t ha^{-1} per 10 weeks under forest compared to negligible activity on adjacent plots cleared mechanically. If the soil is sown to grass or leguminous cover crops immediately after forest removal, recolonization by

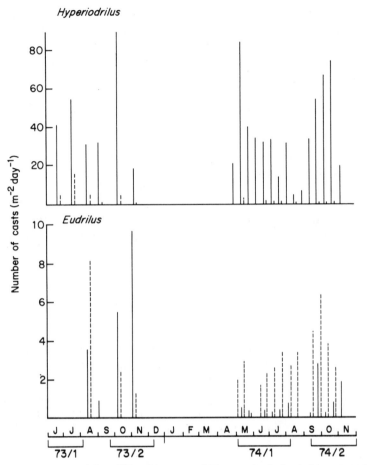

Figure 8.13 Casting activity of H. *africanus* and E. *eugeniae* in bush (-) and cultivated (---) soils (Critchley *et al.*, 1979). *Reproduced by permission of Pedobiologia*

some elements of the fauna occurs. It is the presence of the humus layer, a continuous supply of organic matter, and favourable temperature and moisture regimes in the forest cover that encourage earthworm activity.

2. Tillage methods

Methods of seedbed preparation that cause soil physical disturbance and result in extremes of soil temperature and moisture conditions affect the earthworm population. Ploughing and mechanical tillage often lead to a reduction in earthworm activity because they increase the maximum soil temperature and decrease soil moisture reserves. Those methods of seedbed preparation that retain crop residue mulch on the soil surface are known to encourage earthworm activity. Lal (1976) and Lal and Oluwole (1983) reported 2 to 5 times

Table 8.29 Effects of tillage methods on the casting rate of *H. africanus* for different cropping sequences

Cropping sequence	Number of casts/m²		Equivalent weight (t/ha)	
	No-till	Ploughed	No-till	Ploughed
Maize – Maize	1060	90	41	4
Maize – Cowpea	1220	372	48	15
Pigeonpea – Maize	464	100	18	4
Soybean – Soybean	42	3	2	0
Cowpea – Cowpea	28	36	1	1
Mean	563	120	22	5

Table 8.30 Effects of tillage methods on casting rate (kg ha^{-1} day^{-1}) (Lal and Oluwole, 1983)

Mulch	No-till	Ploughed
With	1052	357
Without	744	236
LSD (0.05)		
(i) Mulch	376	
(ii) Tillage	505	

higher earthworm activity in no-till than in ploughed soil (Tables 8.29 and 8.30). The lack of crop residue mulch also decreases earthworm activity even in no-till soil. The favourable faunal activity in no-till soil is attributed to lower soil temperature, higher soil moisture, and more organic matter than in ploughed land. Mechanical disturbance caused by soil turnover in ploughing destroys habitat and nests and exposes earthworms to predators and harsh climate.

Similar observations have been made for the soils of temperate zones where direct drilling and the no-till system have also been demonstrated to support a higher earthworm population than ploughing and mechanical tillage (Schwerdtle, 1969; Ehlers, 1975; Barnes and Ellis, 1979; Edwards and Lofty, 1977, 1982; Loring et al., 1981).

3. Mulching

The use of stubble mulch lowers the maximum soil temperature, improves soil moisture conditions, provides a buffer against sudden fluctuations in microclimatic environments and provides food for earthworms. Mulching and the use of farmyard manure reportedly promote casting activity on the surface of arable soils in the temperate zone (Graff, 1969; Ritter, 1976 Gerard and Hay, 1979).

Table 8.31 Influence of mulching treatment on earthworm activity (cumulative casts/m²) in 1973, measured about 100 days after seeding (Lal, 1978)

Treatment	Casts/m²				Equivalent weight (t/ha per 100 days)			
	First season		Second season		First season		Second season	
	Row zone	Inter-row zone	Inter-row zone		Row zone	Inter-row zone		Inter-row zone
Complete mulch	816	318	641		18.0	7.0		14.1
Inter-row mulch	30	224	611		0.7	4.9		13.4
Row mulch	192	38	180		4.2	0.8		4.0
Control	12	18	120		0.3	0.4		2.6
LSD (0.05)	44	32	87		1.0	0.7		1.9

Table 8.32 Casting activity of *H. africanus* as affected by the mulch rate (Lal et al., 1980). *Reproduced from* Soil Science of America Journal *Volume 44, 827–833, by permission of the Soil Science Society of America*

Month	Regression equation	Correlation coefficient (r)
April	$Y = 1.15X + 3.30$	0.96**
May	$Y = 0.94X + 2.44$	0.96**
June	$Y = 0.90X + 4.33$	0.93**
July	$Y = 1.81X + 0.45$	0.99**
August	$Y = 1.45X + 2.84$	0.97**
September	$Y = 1.62X + 1.93$	0.97**
October	$Y = 1.02X + 3.29$	0.97**
Annual	$Y = 1.41X + 2.66$	0.98**

** = significant at the 0.01 level.
Y = worm casts m^{-2} month^{-1}; X = mulch rate, t/ha.

Mulching is perhaps more beneficial to enhance faunal activity in the tropics than in the temperate regions. Lal (1978) observed that earthworm activity was significantly influenced by mulching treatments (Table 8.31), and that the casting activity was distinctly confined to the mulched zone. The mulched strip had 5 to 7 times more casting activity than the unmulched zone. The dry weight of worm casts produced under complete mulch treatment was about 25 t ha^{-1} per 100 days. In another study, Lal et al. (1980) attributed the beneficial effects of mulching on soil structure, water retention and transmission properties and soil bulk density to increased earthworm activity. The casting activity of *H. africanus* was linearly related to the mulch rate (Table 8.32). The worm population also varied with the quality of mulch material and its durability.

4. Planted fallows

Periodic fallowing with grass or leguminous cover crops provides a continuous ground cover, maintains a regular supply of organic matter, and through its shading effect lowers the soil temperature. Under these conditions earthworm activity in planted fallows is greater than under cultivation. Differences among cover crops regarding the nature of crop residue (C:N ratio, etc), growth habit, biomass production and other factors influence the population of different species of earthworms and their activity. The data in Table 8.33 show that the activity of *H. africanus* was high under *Brachiaria*, *Melinis* and *Stylosanthes* and was related to the quantity of dead sod and its persistence. In addition to the effect of sod species, there were differences in earthworm activity due to the crops grown. Casting activity was higher under maize than cowpea. Earthworm activity is also more under perennials than under seasonal crops. Akobundu and Okigbo (1984) also observed significantly higher casting rates under *Paspalum notatum* in comparison with maize stover mulch or a green

Table 8.33 Worm activity under different cover crops (casts m^{-2} week) (after Lal et al., 1978). *Reproduced by permission of Elsevier Science Publishers*

Cover crops	Weeks after planting					
	2 – 4		4 – 7		7 – 10	
	Maize	Cowpeas	Maize	Cowpeas	Maize	Cowpeas
Panicum	1680	1708	75	35	157	27
Setaria	1531	1477	109	64	192	56
Brachiaria	1879	1454	431	253	340	122
Melinis	1257	1280	463	343	419	153
Centrosema	1340	956	100	59	131	33
Pueraria	1233	1139	101	96	45	29
Glycine	1131	833	25	33	59	13
Stylosanthes	1011	976	227	38	189	9
Control	37	23	8	0	36	0
LSD (0.05)	1200	1066	165	81	193	20

cover of *A. repens*, *D. triflorum*, and *I. spicata*. The casting activity was also higher under *D. triflorum* than under other covers. The method of cover management, i.e., mechanical or chemical, frequency of mowing, use of fertilizer, etc., all influence earthworm activity. Alley farming is obviously a useful practice to maintain a high earthworm population on tropical arable lands.

5. Grazing

The trampling effect of animals and the depletion of biomass by grazing influence earthworm activity by mechanical disturbance of the habitat, compaction and depletion of food supply. The stocking rate, controlled or uncontrolled grazing and the pasture species grown all affect faunal activity. In Ghana, West Africa, Gerard (1967) studied the effects of variable soil moisture and of shading on earthworm population in pastures. In plots kept at field moisture capacity, earthworms were active even during the dry season. Only a few earthworms survived in the shaded treatment. In Orissa, India, Senapati and Dash (1981) studied the effects of grazing on earthworm population and net primary production. The average live biomass of *Oligochaeta* was 56 g (11 g dry weight) and 41 g (8 g dry weight) per m^2 per year in ungrazed and grazed pastures, respectively. The amount of secondary production of *Oligochaeta* amounted to 130 g live weight (26 g dry weight) and 155 g live weight (31 g dry weight) per m^2 per year in ungrazed and grazed plots. Grazing encouraged earthworm mortality.

6. Burning

Fire is a widely used principal clearing tool in the humid and subhumid tropics, and is therefore an important environmental factor. Traditionally burning is used to save labour. It is a cheap and ready-to-use herbicide, and adds some acidity-neutralizing ash to soils. Burning is also believed to improve grasslands and pastures. Uncontrolled burning, however, depletes the soil of its vegetation cover and leaves it bare. The soil is thus susceptible to erosion and compaction and undergoes extremes of moisture and temperature.

The effects of burning on faunal population and particularly on earthworms are difficult to generalize. The effects obviously depend on the itensity and frequency of burning, soil type, land-use, soil temperature during and after burning, the quantity of mulch left unburnt, and the agro-ecological environment. The effects of burning on earthworm population in tropical soils have not been widely studied. In Ivory Coast, Athias *et al.* (1975a, b) and Couteaux (1978) reported a decrease in earthworm population by annual burning of the savanna vegetation.

7. Soil compaction

Cultural practices influence the earthworm population through their effect on soil compaction. Since burrowing earthworms ingest the soil ahead or move it aside, there may be a critical strength beyond which earthworms find it difficult to manoeuvre in the soil. This critical bulk density, applying in a similar way to root penetration, differs among soils, moisture regimes, earthworm species, organic matter content and a multitude of other interacting factors. Aritajat *et al.* (1977a) reported from a laboratory investigation that in a silt loam grassland the body weight of earthworms significantly decreased in ten-times compacted soil following a compaction stress of 5.6 kg/cm^2 four months after treatment. Changes in body weight, however, were not observed in one-time compacted soil following a compaction stress of 2.8 kg/cm^2 or in loosened soil (Table 8.34). No significant differences in body weight were recorded among treatments 10 months after compaction. In a clay soil, the body weight of earthworms

Table 8.34 Effect of soil compaction on number, weight and recolonization of earthworms in silt loam grassland soil (Aritajat *et al.*, 1977b)

Treatment	Number of earthworms		Weight of earthworms (g)	
	4 months	10 months	4 months	10 months
Control	4.0	35.8	7.5	12.7
1× compaction	10.0 n.s.	33.5 n.s.	2.8 n.s.	12.1 n.s.
10× compaction	2.0 n.s.	20.0***	0.6*	6.6 n.s.
Loosened soil	1.8 n.s.	12.3***	2.5 n.s.	10.1 n.s.
SE of mean	2.1	1.1	2.1	3.1
F	3.30 n.s.	114.23***	1.64 n.s.	0.79 n.s.
LSD (.05)	6.5	6.5	6.4	9.5

n.s. = not significant; * $P = 0.05$; *** $P = 0.001$.

remained unaltered even 6 months after it was compacted. Thomson and Davies (1974) found the optimum bulk density for casting to be 1.75 g/cm^3 for a sandy and a silt loam soil. Dexter (1978) observed that tunnelling was independent of soil strength over the range of penetrometer resistance from 0.3 to 3 MPa.

Soil compaction has a drastic effect on casting rate. The rate of surface casting is often more on compacted than on uncompacted soil. Lal (unpublished) conducted a study on an Alfisol to relate the effects of soil density on casting rate. The mean bulk density was 1.58 and 1.46 g/cm^3 on compacted and uncompacted plots, respectively, corresponding to a penetrometer resistance of 2.37 and 1.0 kg/cm^2. The mean casting rate on uncompacted plots was 7 times more than on compacted plots. Soil compaction is a severe hazard in continuous crop production in tropical soils. It seems that even a slight level of soil compaction can adversely affect population and casting activity of earthworms.

8. Erosion

Accelerated soil erosion decreases faunal population both directly and indirectly. Soils devoid of earthworms have poor structure and are susceptible to erosion. Sharpley *et al.* (1979) observed that chemical elimination of earthworms doubled annual run-off from a 13° slope, from 1650 m^3/ha to 3210 m^3/ha. On the other hand, severely eroded soils have low organic matter content, low water retention and transmission properties, are easily compacted and are less productive. Accelerated erosion, through its adverse effect on soil organic matter content, decreases soil faunal activity. Many earthworms and other soil animals are washed off in water run-off. Atlavinyte (1965) and Atlavinyte *et al.* (1974) reported carrying away of earthworms in water runoff.

9. Agrochemicals

The increase in food demand in the tropics has necessitated the widespread use of a range of chemicals to promote crop growth through intensive and continuous soil use. An increase in the use of agrochemicals in the tropics, however, has come about without understanding their environmental impact and the fate of the by-products of their biochemical disintegration.

(i) Fertilizers: Organic manures have positive effects on earthworms and other soil animals, although few research data are available from the tropics in support of this argument. Those inorganic fertilizers that do not have drastic effects on soil pH and that promote crop growth and increase supply of organic matter also generally promote earthworm population. The application of fertilizer to provide balanced nutrients for crop growth should normally promote earthworm activity. On the other hand, use of acidifying fertilizers (such as $(NH_4)_2SO_4$) adversely affect earthworm population and casting activity.

(ii) Herbicides: Herbicides are increasingly being used in the tropics for weed control and for vegetation management. Whatever else the environmental consequences of the herbicides may be, they affect soil fauna. There is little research information regarding the effects of herbicides on earthworm activity in tropical soils. The most commonly used herbicides are paraquat (1,1 dimethyl 4,4 bipyridilium ion) and atrazine. Preliminary experiments conducted at IITA showed no consistent effect of paraquat on the casting rate of *H. africanus*. From some research conducted in the temperate region soils it is evident that different earthworm species respond differently to different herbicides (Edwards and Stafford, 1979). Caseley and Eno (1966) reported that *Eudrilus eugeniae* is more vulnerable than *Lumbricus terrestris* to 11 herbicides investigated. Furthermore, herbicides are more toxic in coarse-textured soils of low organic matter content than in clayey soils or those with high organic matter. Some herbicides have side effects (Subagja and Snider, 1981) and affect earthworms by eliminating some algal food supply. Martin (1982b) observed that trifluralin killed all earthworms at 100 ppm. Earthworms survived, but did not grow, in the same concentration of Oxadiazon, 2,4-D, prometryne, MCPB, metribuzin, terbacil, linuron and bromacil. Earthworms were only slightly affected by aziprotryne, diuron, methabezthiazuron, asulam and hexazinone. The herbicides Anitrole, ammonium thiocynate, 2,2-DPA, glyphosate, propazine and simazine had no consistent effect on earthworm growth.

Some herbicides are reported to have a beneficial effect on earthworms and other soil fauna. Leibundgut (1981) studied the effect of Deserpan (35% Simazine, 15% triazine derivatives, and 10% 2,4-D Na salt) applied 1, 2 or 4 g/m^2 to leaf compost/forest garden soil/rotted manure mixture stored out of doors. Total earthworm populations were significantly greater in the treated mixture than in the control.

With only a few relevant studies reported from the tropics, it is urgent that field and laboratory research be conducted to investigate the effects of variable concentrations of a range of herbicides on earthworm population for soils of varied physical and chemical properties.

(iii) Pesticides: Among commonly used chemicals, carbamate insecticides (carbaryl and carbofuran) are the most toxic and are deadly to earthworms. Even small concentrations at recommended rates of application can severely reduce earthworm population (Finlayson *et al.*, 1975; Martin, 1976; Lebrun *et al.*, 1981; Medts, 1981). Although some species are more sensitive than others (Stenersen, 1979; Edwards and Brown, 1982), the LC$_{50}$ for carbofuran is 10 to 12 ppm over a 5-day period (Stenersen *et al.*, 1973).

The commonly used fungicide Benomyl is also very toxic to erthworms (Stringer and Lyons, 1974; Keogh and Whitehead, 1975). Benomyl is recommended for control of earthworms around airports and recreational facilities

Table 8.35 Effects of planted fallows on eroded soil on weight of castings of H. africanus accumulated in 3 years

Slope (%)	Weight of earthworm castings (t/ha)				
	Pueraria	Stylosanthes	Centrosema	Mucuna	Panicum
1	16.1	35.7	16.1	12.5	16.1
5	5.4	17.9	9.0	12.6	9.0
10	7.2	9.0	10.8	23.2	10.8
15	50.0	28.6	7.2	12.5	17.9
Mean	19.7	22.8	10.8	15.2	13.5

(Tomlin, 1981; Tomlin et al., 1981). Many soil pollutants and other chemicals considered to be non-toxic also adversely affect earthworms (Zachariae and Ebert, 1970; Thompson and Sans, 1974; Rhee, 1977).

XII. SOIL RESTORATION AND EARTHWORMS

Vast areas of tropical soils are being rendered unproductive by accelerated erosion and by the degradation of their physical, chemical and biological qualities. The need to develop new land for arable use can be drastically reduced by restoring the productivity of degraded lands. Amelioration in physical properties can be readily brought about if the degraded soil can be colonized by earthworms. Recolonization of earthworms in degraded soils may be difficult if the soil is too compact, pH is too low, and a food supply comprising a regular and diverse type of organic matter is not available. Ease of recolonization also depends on the plant cover. Data in Table 8.35 for eroded soils at Ibadan, Nigeria, show that casting activity was greater under *Stylosanthes* and *Pueraria* than other plant species. Woody perennials may have different effects on earthworm colonization than seasonal grass or legume species. Some earthworm species may recolonize more readily than others. Research is needed to develop practical methods of re-establishing earthworms and other soil fauna on degraded lands to facilitate their restoration.

XIII. SUMMARY

The review presented indicates that earthworms substantially alter soil physical, chemical and biological properties. Soil physical properties are improved by earthworms, and their beneficial effects on soil porosity, aggregation, and water transmission properties are significant. Worm channels created through their burrowing activity conduct water, facilitate gaseous exchange between soil and the atmosphere, and promote root growth and development into the otherwise hard-to-penetrate subsoil horizons. Worm casts are structurally stable and reduce soil splash. Under favourable conditions the annual casting rate in many tropical soils may be 50 to 500 t/ha, equivalent to an annual turnover of 4 to 40 mm of soil depth. Worm casts are finer in texture than adjacent soil, and the soil in the cast may be brought from a depth of 15 to 30 cm. Earthworms are, therefore, a natural ploughing agency. They over turn the soil without expensive fuel consumption and without any adverse effects on soil and environments.

Although earthworms thrive better in a fertile soil, they also render nutrients more readily available by assisting in decomposition and mineralization of organic matter content. Earthworm casts are a high grade farmyard manure, and contain 2 to 5 times as much N, organic matter and exchangeable cations, and have higher CEC than adjacent soil. Castings of some earthworm species also have a high calcium content and are associated with symbiotic N-fixing bacteria. Earthworms also add directly to the N content of the soil through decay of their dead bodies. Crop growth and economic yield on worm-infested soil is generally better than on soils that do not support earthworm activity.

Modern methods of mechanized intensive agriculture based on frequent and heavy doses of diverse agrochemicals have adverse effects on earthworms. Cultural practices that disturb the habitat, decrease buffering against extremes of temperature and moisture regimes, disrupt food supply, and increase chemical inputs and soil pollutants adversely affect earthworm population and activity. Deforestation, moldboard ploughing, and use of fungicides, insecticides, and ammonium sulphate decrease earthworm activity. In general, fallowing with grass or legume cover crops, use of mulch and farmyard manure, balanced fertilizer application, no-till and organic farming, encourage earthworm activity and ameliorate soil properties.

Although the effects of earthworms on physical soil manipulation have long been recognized, there is little research information available from tropical soils regarding the effects of: (i) earthworms on soil properties and crop productivity; (ii) cultural practices on earthworms; (iii) agro-chemicals and soil pollutants on earthworm ecology and survival; and (iv) interaction among different earthworm species and with other animals. Cultural practices should be developed to stimulate and enhance earthworm activity. The importance of earthworms in restoration of eroded and degraded lands cannot be overemphasized. The productivity of vast tracts of abandoned lands in the tropics can be restored by their amelioration through earthworms. Soil biologists will find the earthworm a very challenging and a professionally rewarding research topic for generations to come.

REFERENCES

Abbott, I., 1980, Do earthworms compete for food? *Soil Biology and Biochemistry*, **12**, 523–530.

Abbot, I., and Parker, C.A., 1980, The occurrence of earthworms in the wheat-belt of Western Australia in relation to landuse and rainfall, *Aust. J. Soil Res.*, **18**, 343–352.

Abbott, I., and Parker, C.A., 1981, Interactions between earthworms and their soil environment, *Soil Biology and Biochemistry*, **13**, 191–197.

Abrahamsen, G., 1971, The influence of temperature and soil moisture on the population density of *Cognettia sphagnetorum* in cultures with homogenised raw humus, *Pedobiolgia*, **11**, 417–424.

Agarwal, G.W., Rao, K.S.K., and Negi, L.S., 1958, Influence of certain species of earthworms on the structure of some hill soils, *Curr, Sci.*, 213.

Akobundu, I.O., and Okigbo, B.N., 1984, Preliminary evaluation of ground covers for use as live mulch in maize production, *Field Crops Res.*, **8**, 177–186.

Anon, 1950, The role of earthworms and insects in soil formation, *Soils and Fert.*, **13**, 157–160.

Aritajat, U., Madge, D.S., and Gooderham, P.T., 1977a, The effects of compaction of agricultural soils on soil fauna. I Field investigations, *Pedobiologia*, **17**, 262–282.

Aritajat, U., Madge, D.S., and Gooderham, P.T., 1977b, The effects of compaction of agricultural soils on soil fauna. II Laboratory investigations, *Pedobiologia*, **17**, 283–291.

Arle, R., 1981, Survey of the Earthworm. Brazilian species of *Pseudochorutinae*, with a description of new species from Colombia, *Acta Amazonica*, **11**, 583–593.

Athias, F., Joseus, G., and Lavelle, P., 1975a, The effect of annual bush fires on the indigeneous population of the Lamto savanna (Ivory Coast), in *Soil Zoology. Proc. Vth Int. Coll. Soil Zoology, Prague, Sept. 1972*, W. Junk, The Hague, Netherlands, pp.389–396.

Athias, F., Joseus, G., and Lavelle, P., 1975b, General characteristics of the indigenous animal population of the Lamto savanna (Ivory Coast), in *Progress in Soil Zoology, Proc. 5th Int. Coll. Soil Zoology, Prague, 1972*, W. Junk, The Hague, Netherlands, pp.375–387.

Atlavinyte, O., 1965, The effect of erosion on the population of earthworms in the soils under different crops, *Pedobiologia*, **5**, 178–188.

Atlavinyte, O., Bagdonavicene, Z., and Budavieience, I., 1968, The effect of Lumbricidae on the barley crops in various soils, *Pedobiologia*, **8**, 435–423.

Atlavinyte, O., Kuginyle, Z., and Pileckis, S., 1974, Erosion effect on soil fauna under different crops, *Pedobiologia*, **14**, 35–40.

Atlavinyte, O., and Vanagas, J., 1973, Mobility of nutritive substances in relation to earthworm numbers in the soil, *Pedobiologia*, **13**, 344–352.

Ayres, I., and Guerra, R.A.T., 1981, Water as a limiting factor in the distribution of earthworms (Annelide, Oligochaeta) in central Amazonia, *Acta Amazonica*, **11**, 77–86.

Bahl, K.N., 1919, On a new type of nephridia found in Indian earthworms of the genus Pheretiuva, *Q. Fl. Micros. Sci.*, 64.

Bahl, K.N., 1922, On the development of the 'enthrouephric' type of nephridial system found in Indian earthworms of the genus Pheretiuva, *Q. Fl. Micros. Sci.*, **66**, 49–103.

Bahl, K.N., 1927, On the reproductive processes of earthworms: Pt I. the process of copulation and exchange of sperm in Eutyphoeus waltoin, *Q. Fl. Micros. Sci.*, **71**, 479–502.

Bahl, K.N., 1947, Excretion in the Oligochaeta, *Biol. Rev.*, **22**, 109–147.

Bahl, K.N., 1950, *The Indian Zoological Memoirs. I. Pheretima*, 4th ed, Lucknow Pub. House, Lucknow.

Barley, K.P., 1959, The influence of earthworms on soil fertility. II: consumption of soil and organic matter by the earthworm (Savigny), *Aust. J. Agric. Res.*, **10**, 179–185.

Barley, K.P., 1961, The abundance of earthworms in agricultural lands and their possible significance in agriculture, *Adv. Agron.*, **13**, 249–268.

Barnes, B.T., and Ellis, F.B., 1979, Effects of different methods of cultivation and direct drilling, and disposal of straw residues, on populations of earthworms, *J. Soil Sci.*, **30**, 669–679.

Bates, J.A.R., 1960, Studies on Nigerian forest soils: I. The distribution of organic matter in the profile and in various soil fractions, *J. Soil Sci.*, **11**, 246–256.

Baweja, K.D., 1939, Studies of the soil fauna with special reference to the recolonisation of sterilised soil, *J. Anim. Ecol.*, **8**, 120–161.
Baylis, H.A., 1915, A new African earthworm collected by Dr. C. Christy, *Ann. Mag. Nat. Hist.*, **8**, 16.
Betsch, J.M., 1974, Study of the Collembola of Madagascar III. A new genus of Bourlatiellidae, *Revue Ecol. Biol. Sol*, **11**, 561–567.
Bhandari, G.S., Randhawa, N.S., and Maskina, M.S., 1967, On the polyscaccharide content of earthworm casts, *Curr. Sci.*, **36**, 519–520.
Bharat, R., and Srivastara, A.K., 1982, Decomposition of leaf litter in relation to microbial populations and their activity in a tropical dry mixed deciduous forest, *Pedobiologia*, **24**, 151–159.
Bhatti, H.K., 1962, Some studies on castings of earthworms on the physical and biological properties of soil, *Z. Pflauz. Düng. Bodeukunde*, **3(B)**, 198–210.
Block, W., and Banage, W.B., 1968, Population density and biomass of earthworms in some Uganda soils, *Revue Ecol. Biol. Sol* **5**, 515–521.
Bolton, P.J., and Phillipson, J., 1976, Energy equivalents of earthworms, their egesta and a mineral soil, *Pedobiologia*, **16**, 443–450.
Bouché, M.B., 1969, (*Ailoscolax lacteospumosus*): An earthworm with remarkable biological and morphological characteristics, *Revue Ecol. Biol. Sol* **6**, 525–531.
Bouché, M.B., 1977, Strategies lombriciennes, *Ecol. Bull. (Stockholm)*, **23**, 122–132.
Boulad, A.O., Muller, J.P., and Bocquire, G., 1977, Determination of the age and rate of alteration of a ferrolitic soil from Cameroon, *Sci. Geo. Bull.*, **30**, 175–178.
Carter, A., Heinonen, J., and De Vries, J., 1982, Earthworms and water movement, *Pedobiologia*, **23**, 395–397.
Caseley, J.C., and Eno, C.F., 1966, Survival and reproduction of two species of earthworm and a rotifer following herbicide treatments, *Proc. Soil Sci. Soc. Amer.*, **30**, 346–350.
Chawla, O.P., and Nijhawan, S.D., 1972, Effects of aggregate and burrows made by earthworms on the yield of wheat, *Agric. Agro-Ind. J.*, **5**, 12.
Cook, A.G., Critchley, B.R., Critchley, U., Perfect, T.J., and Yeadon, R., 1980, Effects of cultivation and DDT on earthworm activity in a forest soil in the sub-humid tropics, *J. Appl. Ecol.*, **17**, 21–29.
Cornforth, I.S., 1970, Leaf fall in a tropical rainforest, *J. Appl. Ecol.*, **7**, 603–608.
Couteaux, M.M., 1978, Quantitative study of soil Testacea in a Loudetia savannah at Lamto, Ivory Coast, *Revue Ecol. Biol. Sol*, **15**, 401–412.
Critchley, B.R., Cook, A.G., Critchley, U., Perfect, T.S., Russell-Smith, A., and Yeadon, R., 1979, Effects of bush clearing and soil cultivation on the intervertebrate fauna of a forest soil in the humid tropics, *Pedobiologia*, **19**, 425–438.
Darwin, C., 1881, *The Formation of Vegetal Mould Through the Action of Worms, with Observations of their Habits*, Murray, London, 326 pp.
Dash, M.C., and Patra, U.C., 1977, Density, biomass and energy budget of a tropical earthworm population from a grassland site in Orissa, India, *Revue Ecol. Biol. Sol*, **14**, 461–471.
Dash, M.C., Senapati, B.K., and Mishra, C.C., 1980, Nematode feeding by tropical earthworms, *Oikos*, **34**, 322–325.
Day, G.M., 1950, The influence of earthworms on soil microorganisms, *Soil Sci.*, **69**, 175–184.
De Leenheer, L., and Wageman, G., 1948, *La Science du Soil*, Bruxelles, 210 pp.
De Vleeschauwer, D., and Lal, R., 1981. Properties of worm casts in some tropical soils, *Soil Sci.*, **132**, 175–181.
Dexter, A.R., 1978, Tunnelling in soil by earthworms, *Soil Biology and Biochemistry*, **10**, 447–449.
Dhawan, C.L., Sharma, R.L., Singh, A., and Handa, B.K., 1955, Preliminary investigations on the reclamation of saline soils by earthworms, *Proc. Natn. Inst. Sci. India*, **24**, 631–636.

Duweini, A.K.El., and Ghabbour, S.I., 1965, Population density and biomass of earthworms in different types of Egyptian soils, *J. Appl. Ecol.*, **2**, 271–287.

Duweini, A.K. El., and Ghabbour, S.I., 1971, Nitrogen contribution by live earthworm to the soil. *Annales de Zoology: Ecologie Animale*, Hors Ser, 449–501.

Easton, E.G., 1982, Australian Pheretimoid earthworms: a synopsis with the description of a new genus and five new genera, *Aust.J. Zoology*, **30**, 711–735.

Eaton, T.H., and Chandler, R.F., 1942, *The fauna of forest-humus layers in New York*, Cornell Univ. Agric. Exp. Sta. Mem. 247, 26 pp.

Edwards, C.A., and Lofty, J.R., 1972, *Biology of Earthworms*, Chapman and Hall, London, 283 pp.

Edwards, C.A., and Lofty, J.R., 1977, The influence of invertebrates on root growth of crops with minimum or zero cultivation, *Ecol. Bull. (Stockholm)*, **25**, 348–356.

Edwards, C.A., and Lofty, J.R., 1982, The effect of direct drilling and minimal cultivation on earthworm populations, *J. Appl. Ecol.*, **19**, 723–734.

Edwards, C.A., and Stafford, C.J., 1979, A review of long-term effects of herbicides and the soil fauna, *Annals Appl. Biol.* **91**, 132–137.

Edwards, P.J., and Brown, S.M., 1982, Use of grassland plots to study the effect of pesticides on earthworms, *Pedobiologia*, **24**, 145–150.

Ehlers, W., 1975, Observations on earthworm channels and infiltration on tilled and untilled loess soil, *Soil Sci.*, **119**, 242–249.

El-Duweini, A.K., and Ghabbour, S.I., 1965, Temperature relations of three Egyptian oligochaete species, *Oikos*, **16**, 9–15.

El-Duweini, A.K., and Ghabbour, S.I., 1965. Population density and biomass of earthworms in different types of Egyptian soils, *J. Appl. Ecol.*, **2**, 271–287.

Finlayson, D.G., Campbell, C.J., and Roberts, H.A., 1975, Herbicides and insecticides, their compactability and effects on weeds, insects and earthworms in the minicauliflower crop, *Annals Appl. Biol.*, **79**, 95–108.

Gates, G.E., 1961, Ecology of some earthworms with special reference to seasonal activity, *Am. Midl. Nat.*, **66**, 61–86.

Gates, G.E., 1972, Burmese earthworms. An introduction to the systematics and biology of megarile Oligochaetes with special reference to southeast Asia, *Trans. Am. Philosophical Soc.*, **62**, 326 pp.

Gerard, B.M., 1967, Factors affecting earthworms in pastures, *J. Anim. Ecol.*, **36**, 235–252.

Gerard, B.M., and Hay, R.K.M., 1979, The effect on earthworms of ploughing, tined cultivation, direct drilling and nitrogen in a barley monoculture system, *J. Agric. Sci.*, **93**, 147–155.

Ghabbour, S.I., 1966, Earthworms in agriculture: a modern evaluation, *Rev. Ecol. Biol. Soc.*, **111** (2), 259–271.

Graff, O., 1967, Translocation of nutrients into the subsoil through earthworm activity, *Landw. Forsch.*, **20**, 117–127.

Graff, 1969, Earthworm activity in arable soil under different mulches, measured by means of casts, *Pedobiologia*, **9**, 120–127.

Graff, 1970, The phosphorus content of earthworm casts, *Landw. Forsch. Volkenrole*, **20**, 33–36.

Graff, O., 1970, Nitrogen, phosphorus, and potassium in earthworm casts on grassland plots of the Solling project, in IVth Colloquim for Soil Zoology, *Annales Zoologie: Ecologie Animale, Hons Series*, DJ, 503–511.

Graff, O., 1971, Is the nutrition of plants affected by earthworm channels? First communication, *Landbauferschung Volkenrode*, **21**, 103–108.

Guild, W.J. McL., 1955, Earthworms and soil structure, in D.K. Mc. Kevan, (ed.), *Soil Zoology*, Butterworth, London, pp.83–98.

Gupta, M.L., and Sakal, R., 1967, The role of earthworms on the availability of nutrients in garden and cultivated soils, *J. Indian Soc. Soil Sci.*, **15**, 149–151.

Haantjens, H.A., 1965, Morphology and origin of patterned ground in a humid tropical lowland area, New Guinea, *Aust.J. Soil Res.*, **3**, 111–129.

Harker, W., 1955, Quoted by Lee, K.E., 1983, Earthworms of tropical regions – some aspects of their ecology and relationships with soils. In Satchell, J.E. (ed) *Earthworm Ecology*, Chapman and Hall Ltd, London, 179–193.

Hartge, K.N., 1971, *Die Physikalische Unterschung von Boden*, Stuttgart. Ferd. Enke Verlag.

Heath, G.W., 1965, The part played by animals in soil formation, *Experimental Pedology, Proc. 11th Easter School Agric. Sci.* Univ. Nottingham 1964, pp.236–247.

Huang, F.Z., 1979, Effect of earthworms on soil structure, *Acta Pedologica Sinica*, **16**, 211–217.

Irmler, U., 1976, Composition, population density and biomass of the macrofauna of soil in the unsubmerged phase of the Central Amazonian inundation forest, *Biogeographics*, **7**, 79–99.

Irmler, U., and Furch, K., 1980, Weight, energy, and nutrient changes during the decomposition of leaves in the immersion phase of Central-Amazonian inundation forests, *Pedobiologia*, **20**, 118–130.

James, S.W., 1982, Effects of fire and soil type on earthworm populations in a tallgrass prairie, *Pedobiologia*, **24**, 37–40.

Jenny, H., Gessel, S.P., and Bingham, F.T., 1949, Comparative study of decomposition rates of organic matter in temperate and tropical regions, *Soil Sci*, **68**, 419–432.

Joachim, A.W.R., and Pauditesekera, D.G., 1948, Soil fertility studies, IV. Investigations on crumb structure on stability of local soils, *Trop. Agric.*, **104**, 119–139.

Joshi, N.V., and Kelkar, B.V., 1952, The role of earthworms in soil fertility, *Indian J. Agric. Sci.*, **22**, 189–196.

Kale, R.D., Bano, K., and Krishnamoorthy, R.V., 1977, Feeding zones and interspecific zonation in earthworms, *Current Sci.*, **46**, 79.

Kale, R.D., and Krishnamoorthy, R.V., 1978, Distribution and abundance of earthworms in Bangalore, *Proc. Indian Acad. Sci.*, **37**, 23–25.

Keogh, R.G., and Whitehead, P.H., 1975, Observations on some effects of pasture spraying with benomyl and carbendazim on earthworm activity and litter removal from pasture, *N. Z. J. Exp. Agric.*, **3**, 103–104.

Khalaf El-Duweini, A., and Ghabbour, S.I., 1964, Effect of pH and of electrolytes on earthworms, *Repr. Bull. Zool. Soc. Egypt*, **5**, No. 19, 89–100.

Khambata, S.R., and Bhatt, J.V., 1957, A contribution to the study of intestinal microflora of Indian earthworms, *Arch. Microbiol.*, **28**, 69–80.

Khan, A.W., 1966, Earthworms of West Pakistan and their utility in soil improvement, *Agric. Pakist.*, **17**, 415–434.

Kowal, J., 1978, Some physical properties of soil at Samaru, Zaira, Nigeria: Storage of water and its use by crops. I. Physical status of soils, *Niger. Agric. J.*, **5**, 13–20.

Laird, J.M., and Kroger, M., 1981, Earthworms, *Critical Reviews in Environmental Control*, **11**, 189–218.

Lal, R., 1976, No-tillage effects on soil properties under different crops in Western Nigeria. *Soil Sci. Soc. Amer. J.*, **40**, 762–768.

Lal, R. 1978, Influence of within and between row mulching on soil temperature, soil moisture, root development and yield of maize in a tropical soil, *Field Crops Res.*, **1**, 127–139.

Lal, R., 1983, Soil erosion in the humid tropics with particular reference to agricultural land development and soil management, *IAHS Publ.* No. 140, pp.221–239.

Lal, R., and Cummings, D.J., 1979, Clearing a tropical forest I. Effects on soil and microclimate, *Field Crops Res.*, **2**, 91–108.

Lal, L., Mukherji, S.P., and Katiyar, O.P., 1971, The role of earthworms on soil fertility and yield of crops, *Farm Journal (Calcutta)*, **13**, 2.

Lal, R., and Oluwole, A., 1983, Physical properties of earthworm casts as influenced by management, *Soil Sci.*, **135**, 114–122.
Lal, R., and Vleeschauwer, D. De, 1982, Influence of tillage methods and fertilizer application on chemical properties of worm castings in a tropical soil, *Soil and Tillage Res.*, **2**, 37–52.
Lal, R., Vleeschauwer, D. De, and Malafa, R., 1980, Changes in properties of a newly cleared tropical Alfisol as affected by mulching, *Soil Sci. Soc. Amer. J.*, **44**, 827–833.
Lal, R., Wilson, G.F., and Okigbo, B.N., 1978, No-till farming after various grasses and leguminous cover crops in tropical Alfisol I. Crop performances, *Field Crops Res.*, **1**, 71–84.
Lasebikan, B.A., 1975a, The effect of clearing on the soil anthropods of a Nigerian rainforest, *Biotropica*, **7**, 84–89.
Lasebikan, B.A., 1975b, The effect of clearing on the soil anthropods of a Nigerian rainforest. in *Progress in Soil Zoology: Proc. 5th Int. Coll. on Soil Zoology, Prague, 1972*, W. Junk, The Hague, Netherlands, pp. 533–544.
Lavelle, P., 1971, Preliminary study of the nutrition of an African earthworm, *Millsonia anomale, Annales de Zoologie: Ecologie Animale (1971), Hors Series*, 133–145.
Lavelle, P., 1975a, Annual consumption of soil by a natural population of earthworms (*Millsonia anomala* Omodeo; Acanthodrilida: Oligochetes) in the Lamto savanna (Ivory Coast), in *Progress in Soil Zoology: Proc. 5th Int. Coll. on Soil Zoology, Prague, 1972*, W. Junk, The Hague, Netherlands, pp.299–304.
Lavelle, P., 1975b, Annual consumption of soil by a natural population of earthworms (*Millsonia anomola* omodes) in the Lamto savanna (Ivory Coast), *Revue Ecol. Biol. Sol*, **12**, 11–24.
Lavelle, P., 1978, Les Vers de terre della savane de Lamto (Côte d'Ivoire;, peuplements, populations et fonctions dans l' écosystem, *Publ. Labor. Zool. ENS.*, **12**, 301 pp.
Lavelle, P., 1979, Relationships between ecological types and demographic profiles among earthworms of the Lamto savannah (Ivory Coast), *Revue d' Ecol. Biol. Sol*, **16**, 85–107.
Lavelle, P., 1981, A carnivorous earthworm from the savannahs of central Ivory Coast, *Agastrodrilus dominical* nov. sp., *Revue Ecol. Biol. Sol*, **18**, 253–258.
Lavelle, P., 1983a, The soil fauna of tropical savannas, I. The community structure, in F. Bourliere, (ed.) *Tropical Savannas*, Elsevier, Amsterdam, pp.447–484.
Lavelle, P., 1983b, The soil fauna of tropical savannas, II. The earthworms, in F. Bourliere, (ed), *Tropical Savannas*, Elsevier, Amsterdam, pp.485–504.
Lavelle, P., and Meyer, J.A., 1977, The modelling and simulation of the dynamics, formation and feeding of the geophagous earthworm (*Millsonia anomala*) in the Lamto savanna (Ivory Coast), *Ecol. Bull. (Stockholm)*, **25**, 420–430.
Lebrun, P., Medts, A. De, and Wauthy, G., 1981, Comparative ecotoxicology and bioactivity of three carbamate insecticides on an experimental population of the earthworm *Lumbricus herculeus*, *Pedobiologia*, **21**, 225–235.
Lee, K.E., 1967a, Microrelief features in a humid tropical lowland area, New Guinea, and their relation to earthworm activity, *Aust. J. Soil Res.*, **5**, 263–274.
Lee, K.E., 1967b, A new species of earthworm from the Sepik district, New Guinea, *Trans. R. Soc. S. Aust.*, **91**, 59–63.
Lee, K.E., 1969, Earthworms of the British Solomon Islands Protectorate, *Phil. Trans. R. Soc. B*, **255**, 345–354.
Leibundgut, H., 1981, The effect of a herbicide on earthworm, *Schweizerische Zeitschrift fur Forttwesen*, **132**, 977–979.
Leneuf, B., and Aubert, G., 1960, Attempts to measure the rate of ferrallisation, *Trans. 7th Int. Cong. Soil Sci.*, **4**, 225–228.
Lidgate, H.J., 1966, Earthworm control with chlorolane, *J. Sports Turf Res. Inst.*, **42**, 5–8.

Ljungstrom, P.O., and Reinicke, A.J., 1969, Ecology and natural history of microchaetid earthworms of South Africa, *Pedobiologia*, **9**, 152–157.

Ljungstrom, P.O., De Orellana, J.A., and Prieno, L.J.J., 1973, Influence of some edaphic factors on earthworm distribution in Santa Fe Province, Argentina, *Pedobiologia*, **13**, 169–185.

Ljungstrom, P.O., Orellana, J.A. De, and Priano, L.J.J., 1973, Influence of some edaphic factors on earthworm distribution in Santa Fe Province, Argentina, *Pedobiologia*, **13**, 236–247.

Ljungstrom, P.O., 1974, On two species of *Eudrilidae* from Malawi and Uganda and resurrection of the family *Almidae* (Oligochaeta), *Revue Ecol. Biol. Sol*, **11**, 141–148.

Loring, S.J., Snider, R.J., and Robertson, L.S., 1981, The effects of three tillage practices on *Collembola* and *Acarina* populations, *Pedobiologia*, **22**, 172–184.

Madge, D.S., 1965, Leaf fall and litter disappearance in a tropical forest, *Pedobiologia*, **5**, 273–288.

Madge, D.S., 1966, How leaf litter disappears, *New Sci.*, **20**, 113–115.

Madge, D.S., 1969, Field and laboratory studies on the activities of two species of tropical earthworms, *Pedobiologia*, **9**, 188–214.

Madge, D.S., and Sharma, G.D., 1969, *Soil Zoology*, Ibadan University Press, Ibadan, 54 pp.

Malkomes, H.P., 1980, Straw decomposition as a test to determine the side effects of herbicides on the activity of soil organisms, *Pedobiologia*, **20**, 417–427.

Martin, N.A., 1976, Effect of four insecticides on the pasture ecosystem. V. Earthworms. (*Oligochaeta: Lumbricidal*) and Arthropode extracted by wet sieving and salt plotation, *N. Z. J. Agric. Res.*, **19**, 111–115.

Martin, N.A., 1978, Earthworms in New Zealand agriculture, in *Proc. 31st New Zealand Weed and Pest Control Conference*, M.J. Hartley, (ed.), Palmerston North, New Zealand, New Zealand Weed and Pest Control Society, pp.176–180.

Martin, N.A., 1982a, The interaction between organic matter in soil and the burrowing activity of three species of earthworms, *Pedobiologia*, **24**, 185–190.

Martin, N.A., 1982b, The effects of herbicides used on asparagus on the growth rate of the earthworm *Allolobophora caliginosa*, in *Proc. 35th New Zealand Weed and Pest Control Conference*. M.F., Hartley, (ed.) Palmerston North, New Zealand, New Zealand Weed and Pest Control Society, pp.328–331.

Mba, C.C., 1978, Influence of different mulch treatments on growth rate and activity of earthworm *E. eugeniae*, *Z. Pflanzeneriahr. Bodenk*, **141**, 453–458.

Mba, C.C., 1982, Dietary influence on the activity of two Nigerian *Oligochaetes*: *Eudrilus eugeniae* and *Pontoscolex corethulus*, *Pedobiologia*, **23**, 244–250.

McColl, H.P., Hart, P.B.S., and Cook, F.J., 1982, Influence of earthworms on some soil chemical and physical properties, and growth of ryegrass on soil after top soil stripping—a pot experiment, *N. Z. J. Agric. Res.*, **25**, 239–243.

Medts, A.D., 1981, Effects de residus de pesticides sur les lombriciens en terre de culture, *Pedobiologia*, **21**, 439–445.

Minnich, J., 1977, *The Earthworm Book*, Rodale Press, Emmaus, PA, 372 pp.

Modena, A.C., 1978, Earthworm makes fertilizer, *Sugar News (Philippines)*, **54**, 228, 241.

Murillo, B., 1966, Effect of earthworms on aggregate stability in soils under different organic matter treatments, *An. Edafol. Agrobiol.*, **25**, 91–99.

Nahon, D. and Lappartient, J.R., 1979, Time factor and geochemistry in iron crusts genesis, *Catena*, **4**, 249–254.

Nakamura, Y., 1980, Earthworms and soil formation, *Pedologist*, **24**, 43–50.

Nemeth, A., and Herrera, R., 1982, Earthworm populations in a Venezuelan tropical rain forest, *Pedobiologia*, **23**, 437–443.

Nijhawan, S.D., and Kanwar, J.S., 1952, Physicochemical properties of earthworm castings and their effects on the productivity of soil, *Indian J. Agric. Sci.*, **22**, 357–373.

Nye, P.H., 1955, Some soil-forming processes in the humid tropics. IV. The action of soil fauna, *J. Soil Sci.*, **6**, 73–83.

Nye, P.H., 1969, Organic matter plant and nutrient cycles under moist tropical forest, *Pl. Soil*, **13**, 333–345.

Olson, J.S., 1963, Energy storage and balance of producers and decomposers in ecological systems, *Ecology*, **44**, 322–331.

Owens, L.B., Watson, J.P., 1979, Rates of weathering and soil formation on granite in Rhodesia, *Soil Sci. Soc. Amer. J.*, **43**, 160–166

Patel, H.K., 1960, Earthworms in tobacco nurseries and their control, *Indian Tobacco*, **10**(1) 56.

Patel, H.K., and Patel, R.M., 1959, Preliminary observations on the control of earthworms by soapdust (*Sapindus Laurifolius* Vahl) extract, *Indian J. Ent.*, **21**, 251–255.

Pierce, T.G., 1981, Losses of surface fluids from lumbricid earthworm, *Pedobiologia*, **21**, 417–426.

Plowman, K.P., 1979, Litter and soil fauna of two Australian subtropical forests, *Aust. J. Ecology*, **4**, 87-104.

Prabho, N.R., 1960, Studies on Indian *Enchytraeida* (Oligochaeta: Annelide). Description of three new species, *J. Zool. Soc. India*, **12**(2), 125–132.

Primavesi, A.M., and Covolo, G., 1968, Comparison between the activity of termites and of earthworms in relation to nutrients and soil structure, in Primavesi, A.M. (ed.), *Progress in Soil Biodynamics and Soil Productivity*, Pallotti, Santa Mana, Brazil, pp.149–154.

Puh, P.C., 1941, Beneficial influence of earthworms on some chemical properties of the soil, *Contr. Biol. Lab. Sci. Soc. China*, **15**, 147–155.

Puttarudriah, M. and Sastry, K.S.S., 1961, A preliminary study of earthworm damage to crop growth, *Mysore Agric. F.*, **36**, 2–11.

Reddy, M.V., 1983, Effects of fire on the nutrient content and microflora of casts of *Pheretima alexandri*. In Satchell, J.E. (ed), *Earthworm Ecology*, Chapman and Hall Ltd, London, 209–213.

Reinecke, A.J., 1983, The ecology of earthworms in southern Africa. In Satchell, J.R. (ed), *Earthworm Ecology*, Chapman and Hall Ltd, London, 195–207.

Reinecke, A.J., and Ryke, P.A.J., 1969, A new species of the genus *Geogenia* from Lesotho, with notes on two exotic earthworms, *Revue Ecol. Biol. Sol*, **6**, 515–523.

Reinecke, A.J., and Kriel, J.R., 1981, Influence of temperature on the reproduction of the earthworm *Eisenia foetida*, *S. Afr.J. Zoology*, **16**, 96–100.

Rhee, J.A. Van, 1977, Effects of soil pollution on earthworms, *Pedobiologia*, **17**, 201–208.

Ritter, M., 1976, Effects of cultural practices on soil organisms, *Bulletin SROP (France)*, **3**, 7–19.

Satchell, J.E., 1958, Earthworm biology and soil fertility, *Soils and Fert.*, **21**, 209–219.

Satchell, J.E., 1960, Earthworms and soil fertility, *New Sci.*, **7**, 79–81.

Satchell, J.E., 1966, Lumbricidae, in *Soil Biology*, Academic Press, New York, pp.259–322.

Schwerdtle, F., 1969, Investigations on the population density of earthworms in relation to conventional tillage and direct drilling, *Z. Pflkrankh. PflPath. PflSchutz.*, **76**, 635–641.

Segun, A.O., 1977, The genus Iridodrilus (Eurdrilidae Oligochaeta) in Nigeria, *J. Nat. Hist.*, **11**, 579–585.

Senapati, B.K., and Dash, M.C., 1981, Effect of grazing on the elements of production in vegetation and *Oligochaete* components of a tropical pasture, *Revue Ecol. Biol. Sol*, **18**, 487–505.

Senapati, B.K., Dash, M.C., Rana, A.K., and Panda, B.K., 1980, Observation on the effect of earthworm in the decomposition process in soil under laboratory conditions, *Comparative Physiology and Ecology*, **5**,(3), 140–142.

Sharpley, A.N., and Syers, J.K., 1976, Potential role of earthworm casts for the phosphorous enrichment of runoff waters, *Soil Biology and Biochemistry*, **8**, 341–346.

Sharpley, A.N., Syers, J.K., and Springett, J.A., 1979, Effect of surface casting earthworms on the transport of phosphorus and nitrogen in surface runoff from pastures, *Soil Biology and Biochemistry*, **11**, 459–462.

Shrikhande, J.G., and Pathak, A.K., 1948, Earthworms and insects in relation to soil fertility, *Current Sci.*, **17**, 327–328.

Smyth, A.J., and Montgomery, R.F., 1982, *Soils and Land Use in Central Western Nigeria*, Gov. Printer, Ibadan, Nigeria.

Soota, T.D., and Julka, J.M., 1970, Notes on earthworms of the Andeman and Nicobar Islands, India, *Proc. Zoological Society (Calcutta)*, **23**, 201–206.

Southwell, L.T., and Majer, J.D., 1982, The survival and the growth of earthworm (*Eisenia foetida*) in alkaline residues associated with the bauxite refining process, *Pedobiologia*, **23**, 42–52.

Springett, J.A., 1983, Effects of 5 species of earthworm on some soil properties, *J. Appl. Ecol.*, **3**, 865–872.

Springett, J.A., Brittain, J.E., and Springett, B.P., 1970, Vertical movement of Enchytraeidae in moorland soils, *Oikos*, **21**, 16–21.

Stenersen, J., 1979, Action of pesticides on earthworms Part I: the toxicity of chlinesterase—inhibiting insecticides to earthworms as evaluated by laboratory tests, *Pesticide Sci.*, **10**, 66–74.

Stenersen, J., Gilman, A., and Vardanis, A., 1973, Carbofuran: its toxicity to and metabolism by earthworm, *J. Agric. & Food Chem.*, **21** 166–171.

Stockdill, S.M.J., 1982, Effects of introduced earthworms on the productivity of New Zealand pastures, *Pedobiologia*, **24**, 29–35.

Story, R., 1952, A botanical survey of the Keis Kammanhock district, union of South Africa, *Dept. Agric. Bot. Survey Mem.*, N. 27.

Stringer, A., and Lyons, C.H., 1974, The effect of benomyl and thiophanatemethyl on earthworm populations in apple orchards, *Pesticide Science*, **5**, 189–196.

Subagja, J. and Snider, R.J., 1981, The side effects of the herbicides atrazine and paraquat upon *Folsomia candida and Tullbergia granulate, Pedobiologia*, **22**, 141–152.

Teotia, S.P., Duley, F.L., and McCalla, T.M., 1950, Effect of stubble mulching on number and activity of earthworms, *Neb. Agric. Exp. Sta. Res. Bull.*, **165**, 20.

Thompson, A.R., and Sans, W.W., 1974, Effects of soil insecticides in Southwestern Ontario on non-target invertebrates: earthworms in pasture, *Environmental Entomology*, **3**, 305–308.

Thomson, A.J., and Davies, D.M., 1974, Mapping methods for studying soil factors and earthworm distribution, *Oikos*, **25**, 199–203.

Thomson, A.J., and Davies, D.M., 1974, Production of surface casts by earthworm *Eisenia rosea, Can. J. Zool.*, **526**, 659.

Tomlin, A.D., 1981. Effect on soil fauna of the fungicide, benomyl, used to control earthworm populations around an airport, *Protection Ecol.*, **2**, 325–330.

Tomlin, A.D., Tolman, J.H., and Thorn, G.D., 1981, Suppression of earthworm population around an airport by soil application of the fungicide benomyl, *Protection Ecol.*, **2**, 319–323.

Tran-Vinh-An, 1973, The effect of *Hyperiodrilus africanus* earthworms on some pedological properties of a sandy soil in the Kinshasa region (Zaire), *Cahiers ORSTOM, Pedologie*, **11**, (3,4), 249–256.

Vimmerstedt, J.P., and Finney, J.H., 1973, Impact of earthworm introduction on litter burial and nutrient distribution in Ohio strip-mine spoil banks, *Soil Sci. Soc. Amer. Proc.*, **37**, 388–389.

Voisin, A., 1960, *Better Grassland Sward*, Crosby Lockwood and Sons, Ltd, London.

Wasawo, D.P.S., and Visser, S.A., 1959, Swamp worms and tussock mounds in the swamp of Teso, Uganda, *E. Afr. Agric. J.*, **25**, 86–90.

Wiecek, C.S., and Messenger, A.S., 1972, Calcite contributions by earthworms to forest soils in northern Illinois, *Soil Sci. Soc. Amer. Proc.*, **36**, 478–480.

Wilkinson, G.E., 1975, Effect of grass fallow rotations on the infiltration of water into a savanna zone soil of Northern Nigeria, *Trop. Agric.*, **52**, 97–103.

Wood, T.G., 1974, The distribution of earthworms in relation to soils, vegetation and altitude on the slopes of Mt Kosciusko, Australia, *J. Animal Ecol.*, **43**, 87–106.

Yates, G.W., 1981, Soil nematode populations depressed in the presence of earthworms, *Pedobiologia*, **22**, 191–195.

Zachariae, G., and Ebert, H., 1970, Does pest control in forests constitute a hazard to earthworms? *Pedobiologia*, **10**, 407–433.

Chapter 9
Termites

I	INTRODUCTION	340
II	TERMITARIA	341
III	EFFECTS ON SOIL PHYSICAL PROPERTIES	352
	A. Mechanical composition	352
	B. Bulk density	364
	C. Soil structure	367
	D. Water transmission properties	370
	E. Soil–water retention	374
IV	MICROCLIMATE WITHIN THE TERMITE MOUND	377
	A. Air movement	377
	B. Relative humidity	378
	C. Temperature	381
V	PEDOLOGICAL IMPLICATIONS OF ALTERATIONS IN SOIL PHYSICAL PROPERTIES	383
	A. Soil formation	383
	B. Alterations in soil profile	384
	C. Gravel horizon	385
	D. Termites and soil erosion	386
	E. Laterization	389
VI	EFFECTS ON SOIL CHEMICAL PROPERTIES	392
	A. Organic matter and total nitrogen	392
	B. pH and Exchangeable Bases (Ca^{2+}, Mg^{2+}, K^+ and Na^+)	395
VII	TERMITES AND VEGETATION	402
VIII	TERMITE ACTIVITY AND AIR POLLUTION	405
IX	TERMITES AND CULTURAL PRACTICES	407
	A. Deforestation	407
	B. Shifting cultivation and termites	408
X	TERMITES AND CROP GROWTH	409
XI	CONCLUSIONS	415

Table 9.1 The most common termite species observed in Africa, Asia and Australia

Specie	Region	Habit	Reference
Pseudacanthotermes spinger	East Africa	Mound builder	Wilkinson (1965)
P. militaris	East Africa	Ground builder	Wilkinson (1965)
Macrotermes goliath	East Africa	Mound builder	Hesse (1955)
M. bellicosus	East Africa	Mound builder	Hesse (1955)
M. natalensis	East Africa	Mound builder	Hesse (1955)
Odontotermes bodius	Kenya	Ground dweller	Robinson (1958)
Odontotermes sp.	Kenya	Mound builder	Glover et al. (1964)
M. subhyalinus	Kenya	Mound builder	Arshad et al. (1982)
Cubitermes	Kenya	Mound builder	Wielemaker (1984)
Odontotermes	Kenya	Subterranean nests	Wielemaker (1984)
Bellicositermes natalensis	Uganda	Mound builder	Harris (1949)
M. nigeriensis	Nigeria	Mound builder	Nye (1955)
A. evuncifer	Nigeria	Mound builder	Omo Malaka (1967)
O. letericius	Zimbabwe	Mound builder	Watson (1967)
M. subhyalinus	Zimbabwe	Mound builder	Watson (1974)
M. flacifer	Zimbabwe	Mound builder	Watson (1976)

M. bellicosus	Sierra Leone	Mound builder	Miedema and Van Vuune (1979)
Trinervitermes geminatus	Senegal	Mound builder	Leprun (1976)
Gnathamitermes tubiformans	Texas, USA	Ground dweller	Ueckert et al. (1965)
O. obesus	Varanasi, India	Mound builder	Singh and Singh (1981)
O. redemanni	Varanasi, India	Mound builder	Singh and Singh (1981)
O. gurdaspurensis	Northern India	Mound builder	Gupta et al. (1981)
Cyclotermes redemanni	Sri Lanka	Mound builder	Joachim and Kandiah (1940)
Nasutitermes exitiosus	South Australia	Mound builder	Lee and Wood (1968)
Amitermes vitiosus	North Qld, Australia	Mound builder	Holt et al. (1980)
Drepanotermes rubriceps	North Qld, Australia	Mound builder	Holt et al. (1980)
Tumulitermes pastinator	North Qld, Australia	Mound builder	Holt et al. (1980)
Amitermes herbertensis	North Qld, Australia	Ground dweller	Holt et al. (1980)
A. meridionalis	Darwin, Australia	Mound builder (magnetic ant hills)	Lee and Wood (1971)
A. latidens	North Qld, Australia	Ground dweller	Holt et al. (1980)
Termes harrisi	North Qld, Australia	Ground dweller	Holt et al. (1980)
Macrotermes malaccensis M. carbonarius	Malaysia	Ground dweller	Abe and Matsumoto (1979)
Dicuspiditermes nemorosus	Trinidad	Mound builder	Griffith (1953)
Nasutitermes ephratae	Trinidad	Mound builder	Griffith (1953)

I. INTRODUCTION

Soil fauna, through their diversity and activity, influence the soil's physical and nutritional properties, and its ability to sustain crop growth. Ignoring the effects of fauna on soil properties can lead to gross misinterpretations of the analytical data and can have far reaching consequences towards maintenance of soil productivity. A casual observer passing through the Guinean or the Sudanian ecological zones of tropical Africa is impressed by the frequency and diversity of termitaria (termite mounds). The termite (an insect belonging to order Isoptera) is, in fact, a very important constituent of tropical soil fauna. These insects move the soil from deep horizons, create galleries and channels and improve pore space, decompose crop residue and leaf litter, and have a significant role in socio-economic aspects of many tropical cultures. Following Darwin's (1881) report, 'Formation of Vegetal Mould', indicating the importance of earthworms in soil and agriculture, Drummond (1887, 1895) concluded that termites are the tropical analogue of the earthworm, and play as vital a role as earthworms in: (i) turning the soil over, (ii) altering mechanical properties of the soil, and (iii) mixing organic matter with the soil and facilitating its decomposition. Termites play an important role in flow of energy and cycling of nutrients, and are economically and ecologically an important aspect of tropical ecosystem. However, their damage and control costs, worldwide, are estimated to be billions of dollars per year. This tiny insect, with far reaching consequences, can hardly be ignored.

There are approximately 2000 termite species of which at least 260 are from India (Singh and Singh, 1981a) and 600 or more from Africa (Harris, 1954). Many species live together and compete for food and living space. In Malaysia, Abe and Masumoto (1979) reported 52 species in a 1 ha plot, 14 species in a 16 m^2 area, and 10 species in a 2 m^2 area. In this forest ecology, termites were observed to dwell at heights of more than 30 m on and in the standing trees, and down to a depth of 30 cm in the soil. Some common species in Asia, Africa and Australia are listed in Table 9.1. The dominant species and their habits are described for Zaire by Maldague (1964), Ethiopia by Bouillon (1969) and Ruelle (1969), Cameroon by Collins (1977a), and South Africa by Ferrar (1982a). For detailed description of biology, classification and interaction with other soil fauna, readers are referred to the excellent reviews by Krishna and Weesner (1969; 1970), Lee and Wood (1971), Wilson (1971), Brian (1978), and Hermann (1979, 1981).

This chapter describes the role of termites in altering soil physical environment and its implications for soil and crop management strategies in the tropics. The important questions to be considered are: (i) the quantity of soil being turned over and its effect on soil physical properties; (ii) the decomposition of leaf litter and crop residue and its consequences for soil management; (iii) the association between termites and vegetation; (iv) the influence of land-use and cultural practices under intensive agriculture on termite population; and (v) the effect of termites on crop growth and other economic consequences.

II. TERMITARIA

The size, shape and architecture of termite nests vary with species. Some species build galleries in wood and soil, and others construct complex and above-below-ground structures called termitaria. The termitarium is a system of inter-connected cavities. The central nest is often connected with feeding galleries from where termites go out to forage for organic matter and other foodstuffs. Termite species that construct above-ground mounds have a marked effect on soil physical properties. The characteristics and evolution of termite hills have been described by Noirot (1970) and by Pullan (1979).

Mounds and tunnels partly serve the function of regulating temperature and humidity conditions and partly of protecting the colony from rainfall and predators. The activity of mound dwellers can be evaluated from the number of mounds and their relative shape, size, volume and weight. The mound architecture, however, varies widely among species, soils, vegetation, and many other interacting factors. Some commonly observed mounds are shown in Figs. 9.1–9.6. Mound architecture has been described by Lee and Wood (1971) for Australian species, by Roonwal (1973) and Singh and Singh (1981a, b) for Indian termites and Ohiague (1979) for Nigerian species. The mounds range in size from a few centimetres to 8 or 9 m in height and 20–30 m in base circumference (Howse, 1970). Mounds of Macrotermes in southern Zaire (Fig. 9.7) and in Ethiopia are huge, indeed. The weight of soil in the mounds also varies, from a few kilograms to several thousand tonnes (Meyer, 1960). The

Figure 9.1 Some commonly observed shapes and sizes of termite mounds in West Africa (A, B, C)

B

C

343

A

B

Figure 9.2,3,4,5 Some commonly observed shapes and sizes of termite mounds in Northern Territory, Australia (A, B)

Figure 9.6 Large termite mound (5–7 m high) in Nigerian Savanna, near Mokwa

Figure 9.7 Large *Macrotermes* mounds in southern Zaire

surface area of the soil covered by mounds on newly developed land may range from >0.01 per cent (Kang, 1978) to as much as 30% (Meyer, 1960).

The weight of soil in a termitarium is related to its dimensions, and empirical relations have been developed relating soil weight to the above-ground geometry. The volume of the mound is computed assuming the best fitted geometric shape (e.g., cone, dome, cylinder or combination of these shapes). In Australia Holt et al. (1980) developed the following relationships between mass and volume of *A. vitiosus* mounds on red and yellow earths of northern Queensland:

Red earth: Mass (kg) = 1.48 Volume $(L)-1.518$ $r = 0.9967$
Yellow earth: Mass (kg) = 1.413 Volume $(L)+0.001$ $r = 0.9988$
Both soils: Mass (kg) = 1.446 (Volume $(L)+0.001$ $r = 0.997$

In north India, Singh and Singh (1981) noted linear relationships between the volume and the mass and the volume and the termite population for two species. The following regression equations were developed:

$Y_1 = -0.389 + 0.00005X$ (*O. obesus*) $r = 0.852$
$Y_2 = 0.4858 + 0.0004X$ (*O. redemanni*) $r = 0.997$

where Y_1 and Y_2 are the volumes in m³ and X is the population (number per mound) of termites (soldiers, workers, nymphs and imagoes). The volume and weight of the mounds of both species were also linearly related as follows:

$Y = 3.32 + 2041.87(X)$ (*O. obesus*) $r = 0.998$
$Y = 27.58 + 2146.11(X)$ (*O. redemanni*) $r = 0.999$

where Y is the volume (m³) and X is the weight of the mound (tonnes).

Although the maximum life-span of some colonies (*M. bellicosus*) may be 15–20 years (Collins, 1981) the abandoned and fossil mounds may take several decades to be completely weathered. Watson (1967) reported a termite mound that was obviously several hundred years old. In any ecology, therefore, there are both active and abandoned termitaria. The ratio of active to fossil mounds varies widely among species and ecologies. Valiachmedov (1981) observed from Central Asia that only 5 out of 181 termitaria in a 4900 m² area were active. Singh and Singh (1981) found that the density of unoccupied mounds of *O. obesus* and *O. redemanni* was low (20–30%). The abundance of termite mounds may vary from a few to as many as 1000/ha depending on the species, mound size, soil properties, available food, and other environmental factors (Table 9.2).

Termite density and the density of termitaria depend on soil, moisture regime and climatic variables (Table 9.3). In Uganda Pomeroy (1978) studied the effects of 14 environmental variables on the activity of mound-building *Macrotermes bellicosus* and *M. subhyalinus* and developed regression equations on the basis of a multivariate analysis. The density of large Macrotermes mounds was related to the mean annual rainfall and the mean annual temperature. The parabolic transformations were found to be the best related to the

Table 9.2 Termitaria density in different ecologies (modified and upgraded from Lee and Wood, 1971)

Species	No/ha	Habitat and locality	Reference
Macrotermes spp.	2–4	Savanna, East Africa	Hesse (1955)
Odontotermes sp	5–7	Savanna, Kenya	Glover et al. (1964)
Cubitermes	250	Savanna, Kenya	Wielemaker (1984)
Pseudocanthotermes	50	Savanna, Kenya	Wielemaker (1984)
Macrotermes	5–8	Savanna, Kenya	Wielemaker (1984)
M. bellicosus	2–3	Savanna, Congo	Bouillon and Kidieri (1964)
Cubitermes fungigaber	875	Tropical rainforest, Congo	Maldague (1964)
C. exiguus	0–652	Steppe savanna, Congo	Bouillon and Mahlhot (1964)
C. sankurensis	8–550	Steppe savanna, Congo	Bouillon and Mahlhot (1964)
—	3–5	Savanna, S. Zaire	Sys (1955)
—	4500–6000	Savanna, Shaba, Zaire	Soyer (1983)
Cubitermes fungifaber	875	Humid forest, Zaire	Maldague (1964)
Cubitermes servus	85–131	Savanna, Ivory Coast	Bodot (1964)
Amitermes evuncifer	3–17	Savanna, Ivory Coast	Bodot (1964)
Trinervitermes occidentalis	6–31	Savanna, Ivory Coast	Bodot (1964)
T. trinervius	3–9	Savanna, Ivory Coast	Bodot (1964)
M. bellicosus	2–68	Savanna, Ivory Coast	Bodot (1964)
Trinervitermes geminatus	63–753	Savanna, northern Nigeria	Sands (1965)
T. trinervius	1–2	Savanna, northern Nigeria	Sands (1965)
T. togoensis	1–2	Savanna, northern Nigeria	Sands (1965)

Species	Value	Location	Reference
M. bellicosus	2–25	Savanna, northern Nigeria	Sands (1965)
O. sudanensis	3–15	Savanna, northern Nigeria	Sands (1965)
T. germinatus	175–232	Savanna, Nigeria	Ohiagu (1979)
C. Fungifaber	150–370	Forest, South Nigeria	Ghuman and Lal (1984, unpublished)
M. bellicosus	17	Semi-deciduous forest, S. Nigeria	Kang (1978)
M. subhyalinus			
A. ahngerians	500	Central Asia, USSR	Kozlova (1951)
Anacanthotermes ahngerianus	162	Steppe. Central Asia	
A. ahngerianus	180–370	Central Asia, USSR	Valiachmedov (1981)
O. obesus	5	Shores forest, Varanasi, India	Singh and Singh (1981)
O. redemanni	4	Shores forest, Varanasi, India	Singh and Singh (1981)
N. exitiosus	4–9	Sclerophyll forest, S. Australia	Wood and Lee (1971)
N. triodiae	61	Pasture, E. Australia	Wood and Lee (1971)
Trinervitermes trinavoides	534	Savanna, S. Africa	Murray (1938)
Cubitermes pretorianus	385–496	Savanna, S. Africa	Ferrar (1982)
Drepanotermes spp.	upt to 350	Semi-arid woodland, Australia	Gay and Calaby (1970)
A. leurensis	28–210	Savanna woodland, N. Australia	Wood and Lee (1971)
Tumulitermes hastilis	180–500	Savanna, NT, Australia	Lee and Wood (1971)
Amitermes vitiosus	60–268	Savanna, NT, Australia	Lee and Wood (1971)
Drepanotermes spp.	354	Savanna, NT, Australia	Lee and Wood (1971)
Coptotermes acinaciformis	5	Savanna, NT, Australia	Lee and Wood (1971)
A. vitiosus	225–450	Savanna, north Qld, Australia	Holt et al. (1980)

Table 9.3 Factors affecting mound density, termite population and its activity

Climate	Soil	Vegetation	Land-use
Rainfall (amount distribution)	Texture	Succession	Arable
Temperature	Water availability	Diversity	Shifting cultivator
Humidity	Depth	Root and above-ground biomass	Agrochemicals
Radiation and shade	Nutrient status	Chemical characteristics	Plantation
	Horizonation		Pastures
	Drainage		Seedbed preparation
	Mineralogy		
	Soil temperature		
	Swell–shrink and cracking behaviour		

Table 9.4 The correlation coefficient of some environmental variables on the mound density of two termite species in Uganda (Pomeroy, 1978). *Reproduced by permission of Blackwell Scientific Publications Ltd.*

Variable	Correlation coefficient (r)	
	M. subhyalinus	*M. bellicosus*
Rainfall	0.128**	0.317***
P:E	0.162***	0.226***
Maximum temperature	0.123**	0.311***
Minimum temperature	0.196***	0.385***
Sand content (10–30 cm depth)	0.103*	0.211***
Sand content (1-m depth)	0.106**	0.199***
Clay content (10–30 cm depth)	0.173***	0.169***
Clay content (1-m depth)	0.098*	0.212***
Exchangeable bases	0.106**	0.246***
pH	0.238***	0.218***
Organic carbon (%)	0.18***	0.259***
Nitrogen (%)	0.086*	0.226***
C:N ratio	0.191***	0.186***
Human population	0.083*	0.056 NS

NS = not significant; * = significant at 0.05 level; ** = significant at 0.01 level; *** = significant at 0.001 level.

data (Table 9.4) indicating both the limiting and the optimal values of these variables. The optimum values for *M. bellicosus* and *M. subhyalinus* were 1250 and 1150 mm per annum for rainfall, 18 and 17°C for the mean maximum temperature, respectively. The lower limit for the same variables were 700 mm yr^{-1}, 12°C and 23°C for *M. bellicosus* and 300 mm yr^{-1}, 9°C and 21°C for *M. subhyalinus*. The upper limits were 1900 mm yr^{-1} 24°C and 37°C for *M. bellicosus* and 2000 mm yr^{-1}, 23°C and 37°C for *M. subhyalinus*. The optimum

Table 9.5 Regression equations relating activity (%) of *O. gurdaspurensis* with environmental variables (Gupta *et al.*, 1981). *Reproduced by permission of* Pedobiologia

Independent variable	Intercept	Slope	Correlation coefficient (r)
Mean monthly air temperature (°C) (X_1)	91.81	−1.51	−0.31
Cumulative rainfall (mm) (X_2)	73.43	−0.17	−0.81
Rainy days (%) (X_3)	83.96	−1.37	−0.93
Litter biomass (g.m^{-2}) (X_4)	21.99	+0.25	0.57

Multiple regression equations were:
$Y = 71.77 - 0.16 X_2 + 0.009 X_4$ $R^2 = 0.65$
$Y = 100.22 - 1.59 X_3 - 0.087 X_4$ $R^2 = 0.88$

values for pH and C:N ratio were 5–6 and 7 for *M. bellicosus* and 6 and 12.5 for *M. subhyalinus*, respectively. The optimum and upper limit of the clay content for *M. subhyalinus* was 20 and 60%, respectively. The clay content in the subsoil was also found to be a very important factor by Leprun and Roy-Nöel (1976).

In northern India, Gupta *et al.* (1981) developed empirical relations between the activity of *O. gurdaspurensis* and the environment variables, i.e., air temperature, per cent number of rainy days in a month and cumulative rainfall (mm). Regression equations shown in Table 9.5 indicate that the termite activity was not related to temperature. Cumulative rainfall and per cent number of rainy days had an inverse relationship with termite activity. These researchers also observed a positive correlation of termite activity with the litter biomass. On the contrary, Ueckert *et al.* (1976) reported a significant decline in population of *G. tubiformans* in the upper 30 cm of the soil in Texas when the soil temperature was less than 9°C at 15 cm (Fig. 9.8). Regression analysis of population density with environmental variables on the basis of the Ueckert *et al.* (1976) data is shown in Table 9.6. Ueckert *et al.*, however,

Table 9.6 Multiple regression equations relating population (Y) of *G. tubiformans* in Texas, USA, to environmental variables.
X_1 = air temperature (°C), X_2 = soil temperature at 15 cm (°C), X_3 = soil moisture at 45–60 cm (%) and X_4 = rainfall (cm) (Ueckert *et al.*, 1976)

Time period	Regression equation	Correlation coefficient (r)
2 years	$Y = -4032.63 - 5.02 X_1 + 155.53 X_2 + 395.69 X_3 + 64.17 X_4$	0.66
Year 1	$Y = -3963.21 + 336.02 X_1 + 152.35 X_4$	0.93
Year 2	$Y = -4637.11 + 66.03 X_2 + 623.05 X_4$	0.90

Y = number of termites per m^2 in the upper 30 cm of soil.

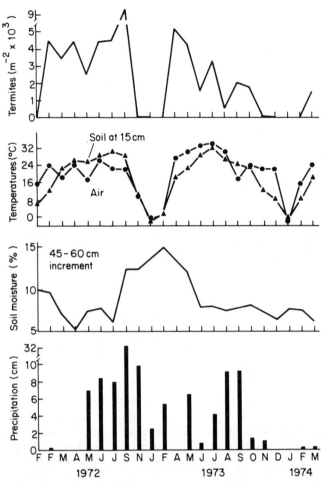

Figure 9.8 Relationship between soil temperature and the population of *G. tubiformans* (Uechkert et al., 1976). *Reproduced by permission of the Ecological Society of America*

observed that environmental variables important in regulating activity over long intervals of time (2 years) were not important over shorter time interval (1 year). Rainfall accounted for 28% of the variation in a 2-year period and 41% in year 1 but for only 1% during year 2. Similar differences were observed for soil moisture. The influence of environmental variables thus depends on the length of the period considered.

In an equatorial rainforest region of Yangambi, Zaire, Maldague (1964) observed a termite population of $3376/m^2$ averaging 11.5 mg live weight per individual. In the Southern Guinea Savanna zone near Mokwa, Nigeria, Ohiagu (1979) reported 222 termites/m^2 in primary savanna woodland corresponding to a fresh-weight biomass of 1.09 g/m^{-2}. The changes in abundance of mound population of *T. geminatus* were related to rainfall distribution and to the forage cycle. The maximum number 399/m^2, 2.03 g/m^2 biomass) was recorded during the period of peak rainfall in August – September. The minimum population (126/m^2, 0.57 g/m^2) was recorded in April – May during the dry season when the available forage was scarce. In Malaysia Abe and Matsumoto (1979) estimated the density of the entire population at 3000–4000/m^2, respectively.

In addition to the forage cycle, termitaria density and termite population also depend on the land-use. Mound density generally increases with advancement of plant succession from cultivation to varying lengths of natural fallow. This increase is partly due to change of microclimate under fallow vegetation compared with extremes of moisture and temperature regime on cultivated lands, and partly due to more quantity and diversity of the food supply with increase in plant succession. Those soils, through their favourable nutrient and moisture regime, that favour luxuriant vegetation growth (roots and the above ground biomass) also support high termite populations. Infertile soils (too dry or too wet, shallow and stony, eroded and compacted) with poor vegetation growth have low termite population.

For mound dwellers, an important soil property is the relative proportion and distribution of sand, silt and clay-sized particles. Coarse textured soils with little silt and clay are not suitable for mound construction. Soils that develop deep and wide cracks (e.g., Vertisols and other soils with high swell: shrink properties) are also not suitable for mound construction because of their structural instability. In soils of low activity clays (LAC), however, the preference for building mounds generally increases with increasing clay content. In southern Zaire, for example, Sys (1955) and Meyer (1960) observed 4.9, 4.0, 3.7 and 2.7 mounds/ha of *Macrotermitinae* on soils with clay contents of 60, 50, 40 and 30–35%, respectively. The corresponding surface area occupied by termite mounds was 7.8, 6.4, 5.9, and 4.3%, respectively.

Different species are also adapted to different climatic regions. For example, fungus-feeding termites are rare in the semi-arid and arid regions and are more abundant in humid and subhumid West Africa (Grassé, 1950). The effects of environmental variables on the activity of ground-dweller species are somewhat different than on mound-builders. Soil textural and mineralogical composition, for example, are less important to ground-dwellers. On the contrary, soil temperature, soil moisture and precipitation are more important because ground-dwellers are not protected by the environmental regulatory influence of the termitaria (Ueckert *et al.*, 1976).

The population density, as affected by environmental variables, influences soil properties through foraging activities and litter consumption. The litter consumption also varies among species and is influenced by environmental factors. Many authors have observed a relationship between litter consumption and the mean annual rainfall:

$$C = 0.14 \text{ AR} - 37.7 \quad \text{(Buxton, 1981)}$$
$$C = 0.16 \text{ AR} - 28 \quad r = 0.92 \text{ (Josens, 1983)}$$

where C is the consumption in $g/m^{-2}\ yr^{-1}$ and AR is the annual rainfall (mm). Consumption also depends on litter production, that in turn depends on the rainfall. For example, Collins (1977b) observed a linear relationship betwen litter production and the rainfall amount in the Southern Guinea savanna region of Nigeria:

$$LP = 0.59\ AR - 112$$

where LP is the annual litter production in $g/m^{-2}\ yr^{-1}$.

III. EFFECTS ON SOIL PHYSICAL PROPERTIES

Being earth dwellers, termites influence soil physical properties through construction of mounds, nests, forage galleries and surface runways. They bring a considerable quantity of fine earth material to the surface for building termitaria and burrow deep and extensive channels in search of food and water. Considerable research has been done in Africa and elsewhere in the tropics in regard to the effects of termite activity on soil properties. Most of the available information relates properties of termitaria to that of the surrounding soil and is briefly reviewed below:

A. Mechanical composition

Termites carry soil particles used for the construction of mounds in their mandibles. This implies an upper limit (in dimensions and weight) of the particle size that a termite can transport against gravity from beneath the soil surface. The size of portable material depends on the size of the worker and may thus differ among various species. Those termite species whose workers are comparatively large can carry larger-sized particles than small-sized workers. Small-sized (clay fraction) particles are ingested and later either regurgitated or excreted. The preference for regurgitation depends on the feeding habit of the species.

Boyer (1958a) reported that *Bellicositermes rex* utilizes clay for the cementing of the mounds and, therefore, does not build its nest in a very sandy soil. Sometimes the clay is replaced by other materials such as micas (muscovite) which becomes plastic and adhesive by trituration in water. The construction materials, containing about 70% clay and the remainders as inert matter mainly quartz, are taken not only from the horizons of clay accumulation but also from the partially decomposed parent rock. In Kenya Wielemaker (1984) observed that the grain size distribution of a termite mound depends on that of the corresponding substratum. His data showed that termites prefer finer sand ($<100\ \mu m$) than coarser sand ($>100\ \mu m$) so that continued termite activity will shift medium diameter of sand in top soil to lower values (Fig. 9.9a, b, c). Maldague (1959) and Stoops (1964) reported differences in textural composition due to species. Maldague reported that the texture of the mound-material is finer than adjacent soil for *Bellicositermes bellicosus* and *B. natalensis*, while (a)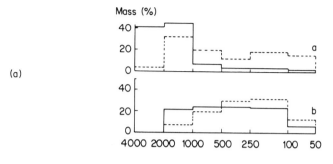

(b)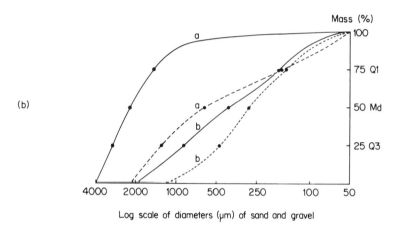

(c)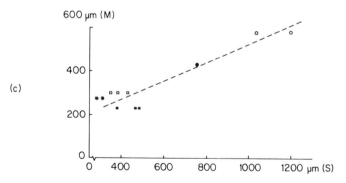

Figure 9.9a (a) Relation between mass percentage of different size fractions in sand and gravel of termite mound (– – –) and the substraturn (—). (b) Cumulative curves of mass percentages of sand and gravel for 14–40 cm (a) and 95–110 cm (b) depths for substraturn (—) and termite mound (– – –). (c) Relation between the median diameter of sand and gravel in mound (M) and in substraturn (S) (Wielemaker, 1984). *Reproduced by permission of the author*

Table 9.7 Effects of termite species on mechanical composition of termite mound soils (Adapted from Omo Malaka, 1977). *Reproduced by permission of CSAIRO*

Soil separate (%)	A. evuncifer		Cubitermes sp		M. bellicosus		T. geminatus	
	TM	S	TM	S	TM	S	TM	S
Sand	61.7	62.5	41.4	75.3	53.5	65.8	64.5	72.0
Silt	9.5	7.7	23.3	8.2	9.9	12.8	12.3	9.8
Clay	28.8	24.9	35.2	17.2	36.6	22.5	23.2	18.2

TM, termite mound; S, adjacent soil.

there was no difference in the texture of the mound material of *Amitermes unidentatus*. The data in Table 9.7 show that mounds of *A. evuncifer* and *T. geminottus* contained more sand and less clay than those of *Cubitermes* sp. and *M. bellicosus*. Pomeroy (1976a, b) noted in Uganda that the range of sand content in the mound of *M. subhaylinus* was 41 to 73% (compared with 38–94% in the subsoil) and that of *M. bellicosus* was 27 to 78% (subsoil 27–90%). The equivalent range for the clay content was 15 to 49% (for subsoil 0 to 52%) for *M. subhyalinus* and 17 to 63% (for subsoil 9 to 63%) for *M. bellicosus*, respectively. In spite of the textural differences observed in the mound material of different species, Lee and Wood (1971) concluded that different termite species do not have precise requirements of particle size for building their mound structures.

In addition to the effect of species, the particle size distribution of the mounds is also affected by the properties of the soil on which the mounds are constructed. For example, Kang (1978) reported that the clay content in the outer casing of *Marcrotermes* mounds was 35% for soil that contained 15% clay in the surface layer and 23% clay in the subsoil. In contrast the clay content in the mound casing was only 23% in soil that contained 9% clay in the surface and 23% clay in the subsoil. There is probably some selectivity involved in choice of the particle size used for mound construction.

Comparison of particle size distribution of mound with the surrounding soil indicates a marked change in the sand: clay ratio (Table 9.8). Most termitaria of different species from Africa and Australia contain lower sand:clay ratio than the surrounding surface soil. For some Nigerian soils the sand:clay ratio changes from 11:1 in the surface soil to 1.4:1 in the mound casing. Various researchers have shown that particle size distribution of mounds approaches but is not quite equivalent to that of the subsoil (Nye, 1955; Harris, 1954; Watson, 1976; Kang, 1978). Lee and Wood (1971) observed that the choice of fine material from the surface or subsoil horizon depends on their relative textural composition. The mound soil contains considerably more clay than the surrounding soil. Watson (1976) reported for some termite mounds in Zimbabwe that, irrespective of the total rainfall amount, mounds contained less fine and coarse sand and more silt and clay than the adjacent soil (Fig. 9.10). Similar

Table 9.8 Particle size distribution of the surface soil and the termitaria outer casing for some soils of the tropics

Soil	Location	Sand (%)		Silt (%)		Clay (%)		Sand : Clay ratio		Organic matter (%)		Reference
		TM	S	TM	S	TM	S	TM	S	TM	S	
Alluvial	India	6.2	60.1	15.1	13.2	15.6	21.0	1:2.5	2.9:1	62.8	0.8	Pathak and Lehri (1959)
Alluvial/Colluvial	Nigeria	70.0	80.0	7.0	11.0	23.0	9.0	3.0:1	8.9:1	0.5	1.6	Kang (1978)
Alfisol	Nigeria	56.0	76.0	9.0	9.0	35.0	15.0	1.6:1	5.1:1	0.65	1.4	Kang (1978)
Alfisol	Nigeria	46.0	78.0	19.0	18.0	33.0	7.0	1.4:1	11.1:1	0.8	0.5	Nye (1955)
Alfisol	Kenya	46.0	58.0	18.0	20.0	36.0	22.0	1.3:1	2.6:1	0.8	1.5	Arshad et al. (1982)
Ultisol	Zimbabwe	60.0	84.0	18.0	8.0	22.0	8.0	2.7:1	10.5:1	0.9	0.5	Watson (1967)
Alfisol	Australia	64.7	74.7	7.8	7.7	27.5	17.6	2.4:1	4.2:1	—	—	Holt et al. (1980)
Alfisol	Upper Volta											Leprun (1976)
Oxisol	Zaire											Sys (1955)

TM, termite mound;   S, adjacent soil.

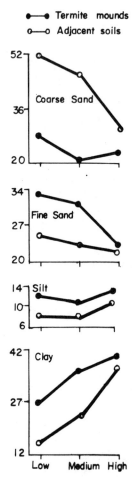

Figure 9.10 Textural composition of termite mound casing from Zimbabwe (Watson, 1976)

observations were made for soils of Central Africa by Sys (1955) (Table 9.9).

The textural composition of *Macrotermes* mound casing and the subsoil material for an occupied and an abandoned termite mound on Alfisol in western Nigeria are shown in Figs. 9.12 and 9.13, respectively. For the occupied mounds, the sand and clay profiles are the mirror images of one another (Fig. 9.11). The sand and clay content of the above-ground mound casing are about equal (45% each). The below-ground nest has about 30% sand and 60% clay and is similar in textural composition to the subsoil at 90 to 100 cm depth. The above-ground mound casing contained a slight amount or none of the material exceeding 2 mm particle size. The termite nest below the soil surface, however, contained increasing amounts of gravel, with the maximum amount of 80% observed at 75 cm depth. In contrast, the maximum gravel

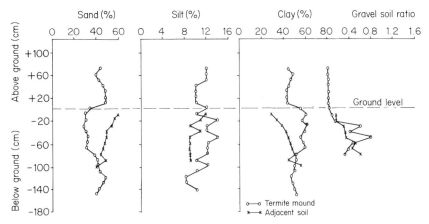

Figure 9.11 Textural composition of the occupied termite mound at IITA, Ibadan, Nigeria.

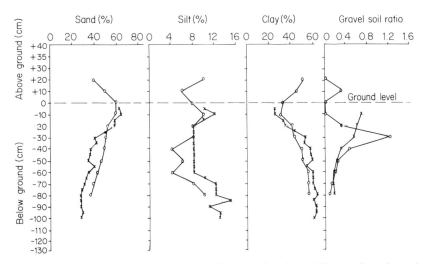

Figure 9.12 Textural composition of the abandoned and partially weathered termite mound at IITA, Ibadan, Nigeria (unpublished data, Lal, 1983)

content in adjacent soil was 49% and occurred at a depth of 60 cm (Fig. 9.11). The sand:clay ratio of the above-ground material was about one (Table 9.10) and ranged between 0.45 and 0.77 for the nest material beneath the soil surface. The sand:clay ratio of the adjacent soil, however, was generally higher and decreased from 2.02 at the ground surface to about 1.0 at 90 cm depth. This implies that the clay content of the underground nest was enriched up to about 90 cm depth. Lee and Wood (1971) also reported a higher clay content in the termitaria. They concluded that all termite species select fine particles in preference to coarse ones for building their mounds, and that the soil is being

Table 9.9 Comparison of the mechanical analysis of a normal soil profile with that of a termitaria in southern Zaire (Sys, 1955)

Profile	Horizon	Depth (cm)	Particulate size range (µm)							
			0–2	2–20	20–50	50–100	100–250	250–500	500–1000	1000–2000
Normal	Ap	0–15	53.4	11.8	13.6	5.7	5.0	2.3	2.5	5.7
	B_1	15–27	59.0	8.7	14.5	6.0	4.4	1.7	2.0	3.7
	B_2	27–66	57.7	8.9	15.6	6.0	5.0	1.8	2.5	2.5
	B_3	66–85	53.2	9.1	16.1	6.5	4.2	1.6	2.5	6.8
	C	85–180	51.8	8.8	16.3	6.0	4.7	1.6	2.5	8.2
Termitaria	Ap	0–11	63.8	11.2	13.1	4.7	3.7	1.2	0.8	1.5
	B_1	11–35	64.4	11.4	12.6	5.3	3.5	1.3	0.8	0.7
	C_1	35–75	63.8	11.9	12.1	4.9	3.7	1.3	1.0	1.3
	C_2	75–110	60.6	11.5	12.8	6.5	4.0	1.7	1.6	1.3
	C_3	110–140	56.9	11.5	14.0	6.1	5.0	1.8	1.9	2.8
	C_4	140–180	44.5	8.9	13.3	5.2	6.3	4.0	6.4	11.4

High clay content in termitaria indicates a selective transportation of fine elements.

Table 9.10 Sand: Clay ratio for mound and parent soils for two sites in western Nigeria (unpublished data, Lal, 1983)

Height (+) or Depth (−) (cm)	Plot 4 (Occupied)		Height (+) or Depth (−) (cm)	CP 108 (Abandoned)	
	TM	S		TM	S
(+) 60–71	0.96	—	(+) 10–20	0.77	—
(+) 50–60	0.86	—	(+) 0–10	1.10	—
(+) 40–50	0.88	—	(−) 0–10	1.80	2.32
(+) 30–40	1.01	—	(−) 10–20	1.92	1.69
(+) 20–30	1.10	—	(−) 20–30	1.25	0.92
(+) 10–20	1.11	—	(−) 30–40	1.15	0.71
(+) 0–10	1.11	—	(−) 40–50	0.97	0.59
(−) 0–10	0.61	2.02	(−) 50–60	0.93	0.65
(−) 10–20	0.48	1.69	(−) 60–70	0.75	0.51
(−) 20–30	0.51	1.39	(−) 70–80	0.68	0.42
(−) 30–40	0.45	1.17	(−) 80–90	0.64	0.45
(−) 40–50	0.50	1.11	(−) 90–100	—	0.45
(−) 50–60	0.57	1.02			
(−) 60–70	0.55	0.97			
(−) 70–80	0.50	0.85			
(−) 80–90	0.73	1.06			
(−) 90–100	0.77	0.52			

TM, termite mound; S, adjacent soil.

brought from the B horizon. Similar conclusions were made by Laker et al. (1982) who observed that particle size distribution of the mound crust resembled the subsoil adjacent to each mound. The occupied or recently constructed termite mound, therefore, resembles an inverted soil profile. Termite workers collect fine clay particles from considerable distances from their nests or bring them from depth within the profile. Some termites have been demonstrated to even grind up coarser material into clay-sized particles (Boyer, 1966).

The textural profile of the abandoned termitaria (Fig. 9.12) was considerably different from that of the occupied. The above-ground mound casing contained less sand and more clay than the soil surface. The textural composition of the underground nest, however, was similar to that of the adjacent soil (Fig. 9.12). The sand:clay ratio of the underground nest was generally higher than the adjacent soil. This trend in sand:clay ratio is exactly the opposite to that of the occupied termite mound.

The relatively low clay content of the above-ground mound material of the abandoned colony is attributed to soil erosion. The mound casing is subject to high intensity rains and the clay particles are gradually washed away from the upper layers of the mound crust. In fact the sand:clay ratio of the above-ground material of the occupied mound is almost twice that of the underground nest. The argument of the preferential removal of the clay-sized particles by erosion is also supported by the data of the abandoned mound (Table 9.11). Although

Table 9.11 The particle size distribution of the sand fraction of an abandoned Macrotermes mound and the parent soil (unpublished data, Lal, 1983)

Height (+) or Depth (−) (cm)	1–2 mm		0.5–1.0 mm		0.25–0.50 mm		0.125–0.25 mm		0.053–0.125 mm		<0.053 mm		Total sand	
	TM	S	TM	S	TM	S	TM	S	TM	S	TM	S	TM	S
(+) 20–10	20.14	—	1.97	—	5.15	—	6.18	—	4.03	—	1.73	—	39.2	—
(+) 10–0	24.61	—	3.67	—	10.96	—	4.96	—	3.53	—	1.47	—	49.2	—
(−) 0–10	28.97	28.55	3.45	7.43	4.76	12.01	9.78	8.77	4.80	3.69	1.94	1.75	59.2	57.2
(−) 10–20	29.13	26.42	3.83	6.31	10.20	10.91	8.67	8.29	5.78	3.40	1.59	1.87	59.2	44.2
(−) 20–30	24.20	19.45	5.09	4.98	10.77	8.53	6.17	6.36	3.18	3.12	1.79	1.76	51.2	38.2
(−) 30–40	22.01	16.56	5.12	4.91	11.30	6.79	5.79	5.60	3.17	2.79	1.82	1.55	49.2	34.2
(−) 40–50	23.05	14.92	4.65	4.69	8.20	6.65	6.38	4.40	3.36	2.19	1.56	1.35	47.2	34.2
(−) 50–60	21.23	18.06	6.16	4.26	8.76	6.53	4.93	4.18	2.81	2.41	1.31	0.76	45.2	36.2
(−) 60–70	20.96	13.41	3.79	3.29	7.52	5.77	5.12	4.36	2.61	2.18	1.20	1.19	41.2	30.2
(−) 70–80	18.40	10.93	3.17	2.49	6.41	5.07	4.90	3.79	2.81	2.34	1.51	1.58	37.2	26.2
(−) 80–90	17.90	11.26	2.89	2.83	6.81	4.95	4.36	3.96	2.21	2.53	1.03	1.47	35.2	27.0

TM, termite mound; S, adjacent soil.

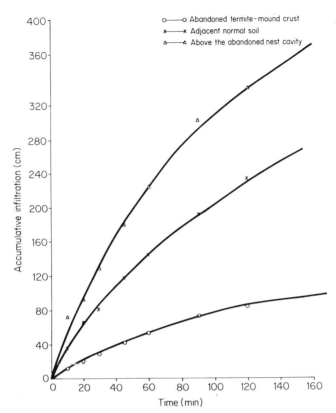

Figure 9.13a Water infiltration characteristics of an abandoned termite mound in a forest fallow (unpublished data, Lal, 1983)

the exact age of the abandoned mound was hard to judge, it was obvious that most of the clay-sized particles had been eroded out of the mound crust. The below-ground abandoned nest also had a higher sand:clay ratio than the adjacent soil. The logical explanation is the subsidence of the above-ground mound casing. Progressive weathering of the abandoned mound resulted in the collapse of the nest cavity followed by caving-in of the above ground material. In Australia, Ettershank (1968) attributed the high proportion of coarse sand in the reworked material to the blowing away of the friable fine fraction by wind erosion.

The particle size distribution of the sand fraction of the occupied-termite mound (Fig. 9.11) is shown in Table 9.12. About 50% of the sand fraction of the outer crust of the occupied mound lies within the 1 and 2 mm size range. Furthermore, the proportion of 1–2 mm sand size fraction in the above-ground mound crust exceeds that in the below-ground nest, further supporting the erosion hypothesis. In the below-ground nest, the 1–2 mm sand fraction is less

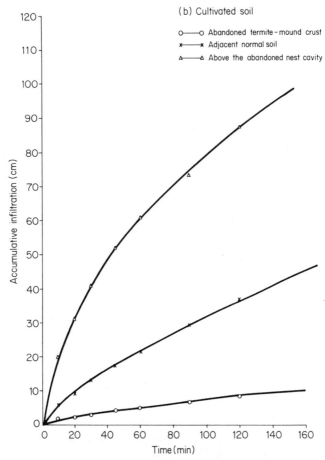

Figure 9.13b Water infiltration characteristics of an abandoned termite mound in a cultivated soil (unpublished data, Lal, 1983)

than 50% of the total sand in the mound and is also less than the similar fraction in adjacent soil. The 1–2 mm sand fraction in the adjacent soil is equivalent to that in the above-ground mound crust. In the abandoned mound (Table 9.11) the 1–2 mm sand fraction is about 50% of the total sand in both the above- and below-ground mound material. In addition, the 1–2 mm sand fraction in the mound and nest is equal to or more than the similar fraction in the adjacent soil. The differences in sand fraction of the occupied and abandoned mound in comparison with the soil adjacent to them also indicates the caving-in of the abandoned mound following weathering.

The gravel profile of the termite mound and of the adjacent soil shown in Figs. 9.11 and 9.12 indicates an increase in gravel concentration below the termite mound between the depths of 20 and 60 cm. For this depth, the gravel content of the occupied termite mound is 1.3 to 2.4 times more than that of the

Table 9.12 Particle size distribution of the sand fraction of the occupied Macrotermes mound crust and parent soil

Height (+) or Depth (−) (cm)	1–2 mm		0.5–1.0 mm		0.25–0.50 mm		0.125–0.25 mm		0.053–0.125 mm		<0.053 mm		Total sand	
	TM	S	TM	S	TM	S	TM	S	TM	S	TM	S	TM	S
(+) 60–70	21.69	—	3.59	—	6.09	—	5.36	—	4.41	—	1.79	—	43.2	—
(+) 50–60	20.03	—	2.11	—	4.92	—	5.75	—	4.55	—	1.84	—	39.2	—
(+) 40–50	21.12	—	2.71	—	6.08	—	5.43	—	4.12	—	1.74	—	41.2	—
(+) 30–40	24.73	—	2.43	—	6.04	—	6.06	—	4.71	—	1.23	—	45.2	—
(+) 20–30	24.24	—	2.22	—	6.08	—	6.56	—	5.63	—	2.47	—	47.2	—
(+) 10–20	24.98	—	2.59	—	5.85	—	6.03	—	5.60	—	2.15	—	47.2	—
(+) 0–10	23.93	—	2.68	—	5.88	—	6.41	—	5.59	—	2.71	—	47.2	—
(−) 0–10	17.06	27.94	1.27	4.20	3.69	7.80	5.44	9.10	3.86	6.61	1.88	3.37	33.2	59.0
(−) 10–20	13.18	29.08	1.27	4.04	3.94	8.53	5.31	7.71	3.48	6.03	2.02	2.95	29.2	56.0
(−) 20–30	13.10	25.23	1.09	3.69	3.33	7.51	5.28	7.48	4.65	5.96	1.75	3.15	29.2	53.0
(−) 30–40	11.21	22.41	1.07	3.74	3.55	7.00	5.28	6.70	3.95	5.46	2.14	2.70	27.2	48.0
(−) 40–50	14.08	23.66	0.90	3.95	3.25	6.78	5.50	6.40	3.69	4.99	1.78	2.23	29.2	48.0
(−) 50–60	12.91	22.61	0.91	3.84	4.04	6.26	6.47	6.44	4.79	4.30	2.08	2.56	31.2	46.0
(−) 60–70	15.36	21.71	1.10	4.55	3.02	6.65	5.81	5.99	4.14	3.88	1.77	2.22	31.2	45.0
(−) 70–80	14.87	19.19	0.56	6.30	3.03	6.82	5.57	4.65	3.43	3.20	1.74	1.85	29.2	42.0
(−) 80–90	15.79	21.81	2.81	7.55	5.56	7.03	5.90	5.29	5.07	3.41	2.07	1.93	37.2	47.0
(−) 90–100	18.72	10.19	2.59	3.51	5.72	5.71	5.86	4.03	4.47	3.33	1.84	2.24	39.2	29.0

TM, termite mound; S, adjacent soil.

adjacent soil. Similar differences exist in the gravel content of the unoccupied mound. It is also this zone of the termite mound that contains more clay and less sand than the adjacent soil. It is likely, therefore, that clay has been preferentially removed from this zone for mound construction, leaving gravels and sand behind.

Analysis of the data of particle size distribution of the termitaria with adjacent soil indicates:

(i) The clay size fraction is used preferentially for mound construction. The material used often resembles the subsoil. Some evidence suggests even mechanical grinding of coarse particles into finer fraction by some species.

(ii) The concentration of gravels below the mound cavity between 20–60 cm depth is an indication of the depth from which the fine particles have been transported for mound construction.

(iii) The relative sand:clay ratio depends on the soil and the age of the termitaria. The clay fraction from the above-ground mound is washed away by erosion, leaving relatively more sand in the above-ground crust than in the underground nest.

(iv) The weathering of abandoned termite mound leads to caving-in and burying of the above-ground mound crust.

(v) The spatial variability in textural composition of soils in the tropics is partly due to the termite activity. This is a random variation and depends on the density of active, abandoned, and fossil termitaria present.

B. Bulk density

During mound construction, single grained (or small) aggregates are cemented together with saliva and other body wastes. The mound crust is carefully constructed to prevent water seepage through it. The macropores are therefore eliminated during the packing process and subsequently due to the raindrop impact. The termite mounds, therefore, may have higher bulk density in comparison with a normal soil of a similar particle size distribution. There is, however, little quantitative data on bulk density of termitaria or of termite-worked soil in comparison with termite-free soil of similar textural composition.

Since bulk density is affected by the particle size distribution and its packing arrangement, the density of mound material depends on parent soil, termite species, age of the colony, and whether the mound is abandoned or active. A few researchers have compared the bulk density of termitaria with that of the parent soil. In Nigeria, Omo Malaka (1977) observed that the bulk density of the unvegetated mound crust was about 1.8 g/cm^3, with corresponding porosity of less than 33% (Table 9.13). There was little difference in bulk density of termite mounds of the same species in savanna and the rainforest zone. The

Table 9.13 Bulk density and porosity of the termitanic material for various termite species in Nigeria (Omo Malaka, 1977). *Reproduced by permission of CSIRO*

Termite species	Location	Bulk density (g/cm³)		Porosity (%)	
		Range	Mean	Range	Mean
A. evuncifer	Ile-Ife	1.24–1.54	1.33	40–51	48
Cubitermes sp.	Ile-Ife	1.61–1.82	1.74	31–39	34
	Mokwa	1.85–1.91	1.88	28–30	29
	Average		1.81		32
M. bellicorum	Ile-Ife	1.80–2.07	1.87	25–35	32
	Mokwa	1.83–1.99	1.91	30–36	33
	Average		1.89		33
T. geminatus	Ile-Ife	1.69–1.91	1.82	28–37	31
	Mokwa	1.63–1.85	1.78	31–41	35
	Average		1.80		33

bulk density of the vegetated mound of *A. evuncifer* species was low, i.e., 1.33 g/cm³ with corresponding porosity of 48%.

The low bulk density may be due to the presence of roots and other biomass of the associated vegetation, and also due to differences in textural composition. In northern Australia, Holt *et al.* (1980) compared the bulk density of termite mounds with that of the individual horizons of each soil (Table 9.14). In red earth, the soil bulk density of various horizons was lower than that of the mound crust. The overall bulk density of the mound was 1.48 g/cm³, giving a mean gallery space of about 12.9%. The differences between the bulk density of individual horizons of yellow earth and termite mounds were, however, not significant. The overall bulk density of the mound was 1.41 g/cm³, giving a mean gallery space of 13.5%.

The bulk density of an occupied *Macrotermes* mound, fungal comb, and of the adjacent uncultivated Alfisol in southwestern Nigeria is shown in Table

Table 9.14 Bulk density (g/cm³) of the mound of *A. vitiorus* and of adjacent red and yellow earths (Adapted from Holt *et al.*, 1980). *Reproduced by permission of CSIRO*

	Red earth			Yellow earth	
Horizon (cm)	Mean	Range	Horizon (cm)	Mean	Range
0–20	1.61	1.58–1.69	0–20	1.73	1.62–1.78
20–40	1.58	1.54–1.63	20–30	1.60	1.56–1.64
40–120	1.58	1.46–1.76	30–56	1.53	1.49–1.61
Total B Horizon	1.58	1.46–1.76	56–69	1.58	1.53–1.62
A. vitiorus mound	1.70	1.60–1.80	Total B Horizon	1.56	1.49–1.64
			69–120(c)	1.89	1.81–1.91
			A. vitiorus mound	1.63	1.43–1.82

Table 9.15 Bulk density and penetrometer resistance of an occupied Macrometes mound (unpublished data, Lal, 1983)

Height (+) or Depth (−)	Bulk density (g/cm³)		Porosity (%)	Bulk density of adjacent soil (g/cm³)	Penetrometer resistance (kg/cm²)	Field moisture (%)	Bulk density of fungal comb (g/cm³)
	D′b	Db					
(+) 71–60	1.75	1.69	36.2	—	>5.0	9.4	—
(+) 60–50	1.77	1.66	37.4	—	>5.0	7.8	—
(+) 50–40	1.79	1.66	37.4	—	4.97	7.2	—
(+) 40–30	1.77	1.69	36.2	—	>5.0	8.7	—
(+) 30–20	1.82	1.71	35.5	—	>5.0	4.9	—
(+) 20–10	1.83	1.74	34.3	—	4.88	5.0	—
(+) 10–0	1.80	1.69	36.2	—	4.92	9.3	—
(−) 0–10	1.66	1.63	38.5	1.20	4.83	6.9	0.37
(−) 10–20	1.65	1.62	38.9	1.20	4.88	5.6	0.36
(−) 20–30	1.63	1.60	39.6	1.36	>5.0	4.3	0.38
(−) 30–40	1.68	1.65	37.7	1.52	>5.0	8.6	0.39
(−) 40–50	1.63	1.60	39.6	1.50	4.92	8.5	—
(−) 50–60	1.59	1.56	41.1	1.48	>5.0	10.2	—
(−) 60–70	1.68	1.64	38.1	1.48	>5.0	12.2	—
(−) 70–80	1.64	1.61	39.2	1.49	>5.0	10.0	—
(−) 80–90	1.69	1.67	37.0	1.49	>5.0	10.1	—
(−) 90–100	1.65	1.62	38.9	1.49	—	—	—
(−) 100–110	1.65	1.63	38.5	1.49	—	—	—
(−) 110–120	1.70	1.67	37.0	1.49	—	—	—
(−) 120–130	1.73	1.68	36.6	1.49	—	—	—
(−) 130–140	1.74	1.71	35.5	1.49	—	—	—
(−) 140–150	1.67	1.63	38.5	1.49	—	—	—
(−) 150–160	1.62	1.59	40.0	1.49	—	—	—

9.15. The average bulk density of the above-ground mound material was 1.69 g/cm^3. The bulk density of the mound material below-ground up to 1 metre depth was 1.62 g/cm^3. In contrast the bulk density of adjacent soil was 1.20 g/cm^3 for the surface horizon and 1.50 g/cm^3 for the subsoil material including gravels. The termite mound material is, therefore, rather densely packed, with low total porosity of about 36% for the above-ground and 39% for the below-ground material. In comparison the total porosity of uncultivated soil was 55% and 43%, respectively, for the surface and the subsoil horizons. These observations were made during the dry season and the moisture content of the mound material ranged between 5 and 10%. In spite of no gravels, the mound material had a high strength and the penetrometer resistance often exceeded 5 kg/cm^2 (0.5 MPa). The bulk density of the fungal comb was less than 0.4 g/cm^3.

The bulk density of the abandoned mound was even higher than that of the active mound (Table 9.16). The mean bulk density was 1.73 g/cm^3 for the above-ground and 1.69 g/cm^3 for the below-ground abandoned mound. The bulk density of some layers was as high as 1.92 g/cm^3. Once again, the uncultivated soil in the vicinity had low bulk densities of 1.15 to 1.56 g/cm^3 for different horizons. In spite of high bulk density and low field moisture content, the penetrometer resistance of the weathered mound was low (Table 9.16). The bulk density of the fungal comb was higher than that of the occupied mound, i.e., 0.53–0.56 g/cm^3.

From the scanty literature and the few field observations made, it is apparent that the bulk density of the termite-infested soil may be lower than a soil free of termite. The low soil bulk density of uncultivated in comparison with cultivated soils is partly attributed to the channels created and soil turnover by termites (Ghilarov, 1962; Maldague, 1964). Under natural conditions, however, a range of soil animals may be responsible for soil loosening and it is difficult to isolate the effect of termites. The other soil factors that affect comparative effects of termite activity on soil bulk density are the texture, organic matter content, and other soil fauna. It is, therefore, difficult to quantify the effects of termite activity on soil bulk density under uncontrolled conditions in the field.

The bulk density of termite mounds, however, is more than the adjacent soil. The mound material, being densely packed, has low porosity and high strength. The bulk density of mounds associated with vegetation growth may be lower than those without it.

C. Soil structure

Termites affect soil structure in many ways. The direct effects consist of construction of feeding galleries and the resulting channels and burrows that alter water transmission properties. The mound building species drastically alter the textural and structural properties of the termitaria. This reworked material covers a considerable proportion of the soil surface. The subterranean galleries may radiate as far as 40 to 50 m from the mound (Greaves, 1962) and may extend deep into the profile to a depth of 10 to 15 m (Ghilarov, 1962). The

Table 9.16 Bulk density and penetrometer resistance of an unoccupied Macrotermes mound (unpublished data, Lal, 1983)

Height (+) or Depth (−) (cm)		Bulk density of mound material (g/cm³)		Porosity (%)	Bulk density of adjacent soil (g/cm³)	Penetrometer resistance (kg/cm²)	Moisture content (%)
		D'b	Db				
(+)	20–10	1.77	1.74	34.3	—	0.63±0.24	3.3
(+)	10–0	1.74	1.72	35.1	—	0.46±0.17	2.2
(−)	0–10	1.72	1.68	36.6	1.15	—	—
(−)	10–20	1.49	1.45	45.2	1.15	—	—
(−)	20–30	1.94	1.92	27.5	1.40	0.58±0.19	2.1
(−)	30–40	1.85	1.83	30.9	1.40	0.54±0.17	2.0
(−)	40–50	1.77	1.74	34.3	1.40	1.17±0.49	4.9
(−)	50–60	1.69	1.66	37.4	1.56	0.79±0.44	3.9
(−)	60–70	1.72	1.69	36.2	1.56	0.50±0.20	2.6
(−)	70–80	1.61	1.58	40.4	—	1.71±0.70	6.1
(−)	80–90	1.65	1.62	38.9	—	2.33±0.42	7.3

Bulk density of the fungal comb = 0.56–0.53 g/cm³.

indirect effects of termite activity on soil structure are through the rapid mineralization of crop residue and organic wastes.

The material used for mound construction is densely packed with low porosity and has a 'massive' structure. The massive structure is created by cementing together of particles by saliva, by plastering of different layers by regurgitated material or by other body excrements (Noirot, 1970). The net result is a densely packed, cemented, hardened mass of high strength. The mound material is thus resistant to raindrop impact and has a high degree of stability in running water.

A little research has been done to study the effects of termite activities on structural properties of soil. In Sri Lanka Joachim and Pandittisekera (1948) reported higher aggregation percentage in termite mounds than in adjacent soils. In India Pathak and Lehri (1959) observed that the soil from the mound of *Odontotermes* sp. contained 68% clay aggregates exceeding 0.002 mm compared with 48.1% in control. The dispersible clay was 5 and 10% in mound and adjacent soil, respectively. Lee and Wood (1971) reported significant flocculating effect of organic matter present in the mound material of some Australian termites.

In Texas, Spears *et al.* (1975) observed that the percentge of aggregates exceeding 2 mm was greater in soils whose populations of *Gnathamitermes tubiformans* had been destroyed by spraying with chlordane than they were in an infested soil. However, the percentage of aggregates <0.5 mm in size were greater in termite-infested soil (Table 9.17). The increase of soil aggregates in the smaller size classes on termite-infested soils was attributed to the bringing of fine soil particles to the surface for construction of mud casts and sheetings. During the 1973 growing season, *G. tubiformans* deposited a total of 403 g/m^2 of soil on the surface providing a 0.9% ground cover by mud casts. Contrary to the observations of Pathak and Lehri and of Lee and Wood, Spears *et al.* observed that aggregates from termite-infested soils were easily dispersed in water. The presence of large aggregates in termite-free soil was attributed to the increase of litter on the soil surface.

Rajagopal *et al.* (1982) observed higher aggregate stability and large aggregates in termite mound soils of Karnataka, India. This apparent contradiction

Table 9.17 Mean percentage of soil aggregates on termite-infested and termite-free soil (Adapted from Spears *et al.*, 1975)

Size class (mm)	Termite-infested	Termite-free	P-value
>3	22.6	23.7	<0.20
2–3	4.5	4.9	<0.05
1–2	6.3	6.5	—
0.5–1	4.9	4.8	—
0.25–0.5	5.3	4.8	<0.05
>0.25	56.5	55.3	<0.30

Table 9.18 Pore size distribution and water transmission properties as affected by *G. tubiformans* (Spears *et al.*, 1975)

Property	Termite-infested	Termite-free	*P*-value
Non-capillary pore space (%)			
(i) Summer and winter	10.5	9.8	<0.01
(ii) Spring	9.7	10.4	<0.20
Capillary pore space (%)			
(i) Summer and winter	32.1	32.8	<0.01
(ii) Spring	32.6	32.0	<0.05
Total pore space (%)			
(i) Summer and winter	42.6	42.6	
(ii) Spring	42.3	42.4	
Bulk density (g/cm^3)	1.55	1.55	
Hydraulic conductivity (ml min^{-1} cm^{-2})	0.075	0.072	
Evapotranspiration (%)	4.4	4.9	<0.001
Antecedent soil moisture (%)	4.6	5.0	<0.001
Sediment load (kg/ha)	161	122	<0.05

Sediment load measured with a rainfall simulator.

may be due to difference in the mound-dweller versus ground-dweller termites. Soils of the termite mounds predominate in micropores and have, therefore, slow infiltration.

Spears *et al.* (1975) compared the total porosity in a soil infested with *G. tubiformans* with that from which termites had been controlled. Their data (Table 9.18) show that non-capillary pore space was significantly greater on termite-infested soils than on termite-free soils for soil samples obtained during summer and winter. For spring sampling, however, non-capillary pore space was significantly greater on termite-free soils. In contrast, the capillary pore space was significantly greater on termite-free soils in summer and winter sampling and in termite-infested soil for spring sampling. The increase in non-capillary pore space during summer and spring was attributed to the construction of underground galleries and tunnels.

D. Water transmission properties

Little is known about the effect of termite activity on water transmission properties of the soil. Termite runways and feeding galleries are an extensive network of often interconnected tunnels that extend to some distance away from the nest. A colony of many thousands or millions of insects makes an impact on the soil's porosity and pore size distribution. Termite galleries are often back-filled and later occupied by plant roots that enrich them with organic matter and stabilize them (Lee and Wood, 1971). These channels and feeding galleries created in a termite-infested soil may increase infiltration and water transmission (Harris, 1965). The effect of subterranean tunnels and

Table 9.19 Hydraulic conductivity and infiltration rate of the crusts of mounds of *A. evuncifer* in Nigeria (Omo Malaka, 1977). *Reproduced by permission of CSIRO*

Samples	Thickness of crust (cm)	Infiltration index (min)	Hydraulic conductivity (cm/min)
A	0.8	35	0.0356
B	0.7	40	0.0291
C	0.8	32.5	0.0383
D	0.5	26	0.0374
Average	0.75	42.5	0.0305
Normal rainforest soil in arable crops			0.0624
Normal rainforest soil in bush			0.3125

Infiltration index is the time for the 3 cm depth of water to pass through the crust.

chambers on water infiltration, however, depends on whether these channels are connected to the soil surface. If they are not connected to the source of water, they serve as physical barriers to water infiltration until a positive water head is developed. The infiltration rate is improved only if these channels are connected to the source of water supply. Many researchers have observed that the subterranean channels are rarely flooded (Adamson, 1943) implying thereby that they may decrease water infiltration.

The water transmission properties are influenced by the pore size distribution and stability of soil aggregates. The infiltration rate is improved if termite activity increases macro-pores (non-capillary) and encourages formation of large-size water stable aggregates. Increase in capillary or micro-porosity and in small-size, easily dispersed aggregate results in a decrease in water infiltration. In Texas, Spears *et al.* (1975) reported lower infiltration on termite-infested than on termite-free soil because of high organic matter content in the latter. Similar observations on low infiltration rates of termite-infested soils have been reported by Ghilarov (1962) and Boyer (1969).

In Nigeria Omo Malaka (1977) studied the crust thickness and water transmission properties of the mound crust of *A. evuncifer*. The mound crust had low hydraulic conductivity and low infiltration rate in comparison with cultivated or fallowed soil (Table 9.19). The data of field measurements of water infiltration using a double-ring infiltrometer on an abandoned *Macrotermes* mounds are shown in Figs. 9.13 and 9.14. In cultivated soil (Fig. 9.13), the accumulative infiltration at 120 minutes was 8.5 cm on a mound-crust compared with 36.5 cm on an adjoining normal soil. When the test was performed directly above the nest cavity (in the centre of a levelled-abandoned mound), the infiltration rate was higher than even the adjoining soil. The accumulative infiltration in 120 minutes was 87.5 cm, 2.4 times more than the adjacent soil. The mound cavity was about 5–10 cm below the crust, and was directly connected with a positive water head when the double-ring was pushed 15 cm below the ground surface. The data of similar tests conducted on a

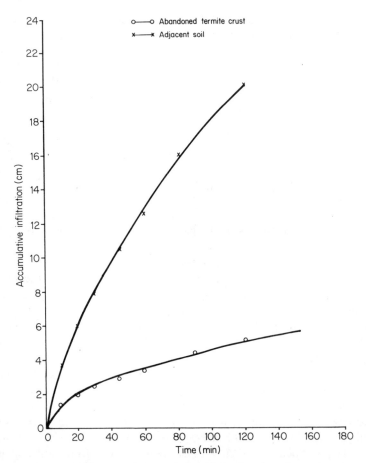

Figure 9.14 Water infiltration characteristics of an abandoned termite mound with voids on the surface connected to the central weathered cavity (unpublished data, Lal, 1983)

forested soil are shown in Fig. 9.13a. Although the trends in relation to the effects of mound crust and of nest cavity on infiltration were similar to that of cultivated soil, the infiltration rate was generally higher on forested land. The accumulative infiltration at 120 minutes was 84 cm, 232 cm and 341 cm for mound crust, normal soil and nest cavity site, respectively.

The mound crust has a generally low total porosity and contains predominantly micropores (capillary or retention pores). Termite mounds, therefore, shed most of the rainwater received as surface run-off and little water percolates through an occupied or an abandoned mound. The mound soils are thus less leached and generally contain more soluble plant nutrients than the parent soil. Watson (1969) studied the water movement pattern of termite mounds of *Odontotermes bodius* and *Macrotermes bellicosus* in Zimbabwe using radioactive chromium-51. The distribution of radioactivity measured after 773 mm

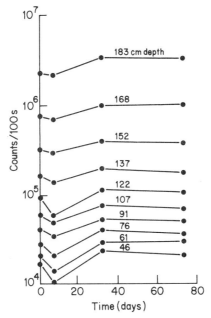

Figure 9.15 Water movement through a termite mound using the radioactive chromium-51: field count of radioactivity inside the mound. Injection depth 226 cm (Watson, 1969). *Reproduced by permission of Blackwell Scientific Publications Ltd.*

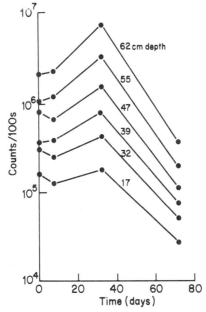

Figure 9.16 Water movement through a termite mound using the radioactive chromium-51: field count of radioactivity outside the mound, Injection depth 116 cm (Watson, 1969). *Reproduced by permission of Blackwell Scientific Publications Ltd.*

of rain was received following the chromium injection showed a slight spread to about 66 cm upward and over 75 cm downward from the injection depth within the *O. bodius* mound. No ^{51}Cr was observed outside the mound. In contrast, no ^{51}Cr activity was found at the injection depth in the parent soil. In the soil adjacent to the *M. bellicosus* mound, while no activity was found in the parent soil 257 days after injection (676 mm rainfall), the ^{51}Cr had moved slightly to a distance of 30 cm away from the injection depth. Field measurements also showed that the radioisotope remained within the termite mound but was completely removed from the adjacent soil (Figs 9.15 and 9.16). Although the author concluded that the use of radioactive tracer can give erroneous results regarding the amount or rate of water movement, the data showed that termite mounds are subject to less leaching than the surrounding soil. The reduced leaching may partly be due to higher run-off from the steeply sloping termitaria and partly to low water conductance of the mound crust.

The effects of termite activity on water movement in soils are difficult to generalize and depend on soil texture, termite species and their foraging habits, presence or absence of the water table, quantity of galleries and tunnels and their continuation to the soil surface or to the source of water supply, stability of aggregates, presence of litter and so on. The scanty data available show that ground-dwellers and mound-dwellers both affect water movement through the soil. Mound crust inhibits the amount and rate of water infiltration. The abandoned termite mounds subjected to different degrees of weathering may improve infiltration if the nest galleries are connected directly to the source of water supply.

E. Soil-water retention

The soil-water retention properties of termite-infested soil also depend on the alterations created by their activity in soil's pore-size distribution. A literature survey regarding the effects of termite activity on water-holding capacity leads to contradictions and anomalies. Some researchers report an increase in water retention capacity of termite-infested soils, whereas others have shown a decrease. Systematic investigations relating the water retention capacity of termite-infested soils to their pore size distribution, textural and structural properties, and humified organic matter content are few.

In Thailand Pendleton (1941) reported about 50% increase in water retention capacity of termite mounds relative to adjacent soil. Kemp (1955) also reported a slight increase in the water-holding capacity of the termite-mound soils of northeastern Tanzania. In East Africa Hesse (1955) reported that inhabited mounds of *M. goliath*, *M. bellicosus* and *M. natalensis* contained significantly more moisture than those of nearby uninhabited mounds. This high moisture was obviously due to high humidity and due to the activity of termites (see section on humidity). In India, Pathak and Lehri (1959) observed significantly more soil moisture equivalent, the maximum water-holding capacity and sticky point of the samples of termitarium built by *Hypotermes*

Table 9.20 Moisture retention properties of samples of termitarium of *H. obscuripes* and the adjacent soil in U.P., India (Pathak and Lehri, 1959). *Reproduced by permission of the Indian Society of Soil Science*

Property		Termite mound	Adjacent soil
Organic matter	(%)	62.8	0.80
Sand	(%)	6.2	60.1
Silt	(%)	15.1	18.1
Clay	(%)	15.6	21.0
Oven dry moisture	(%)	5.3	1.6
Moisture equivalent	(%)	64.2	17.9
Water-holding capacity	(%)	239.5	46.0
Sticky point		4.5	8.1

Table 9.21 Moisture retention properties of termite mound soils of *Odontotermes* sp. in Upper Volta (Leprun, 1976). *Reproduced by permission of* Pedobiologia

Property		Surrounding soil	Termitaria-pocket walls
Clay	(%)	13.2	26.7
Silt	(%)	45.9	31.1
Sand	(%)	39.3	25.1
Organic matter	(%)	0.8	3.6
Antecedent moisture	(%)	0.66	0.93
pF 3.0	(%)	23.9	25.3
pF 4.2	(%)	8.2	16.5
Porosity	(%)	24.5	24.8
Bulk density (g/cm^3)		1.49	1.66

obscuripes as compared to those of adjacent soils (Table 9.20). The increase in water retention was obviously due to high organic matter content. In Upper Volta Leprun (1976) observed higher moisture retention of termite pockets at pF 3.0 and 4.2 than the soil, although the total porosity was similar (Table 9.21). The higher clay and lower sand and silt content of the termite pockets was responsible for higher proportion of water retention pores.

Many researchers have reported lower water-holding capacity in termite-affected than adjacent soils. In Central Asia, Ghilarov (1962) observed about 8% less water-holding capacity in termite-affected soil than in parent soil. Similar findings have been reported from the Rupununi savanna region of Guyana by Goodland (1965).

The moisture retention characteristics of an occupied Macrotermes mound crust in western Nigeria are compared in Fig. 9.17 with those of an adjacent soil sampled from different depths. The moisture retention properties of the mound crust are similar to the subsoil obtained from 80–85 cm depth. The moisture retained in mound-crust samples varies from 44% at zero suction to

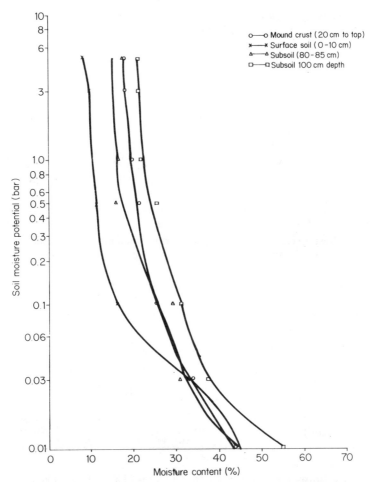

Figure 9.17 Soil moisture retention characteristics of an occupied termite-mound soil. (Unpublished data, Lal, 1983)

18% at 5 bar, in comparison with 44.5% and 15% respectively for the subsoil. In contrast, the surface soil retained 45% moisture at zero and 8.5% at 5 bar suction. The surface soil has obviously a broader range of pore size distribution and more available water-holding capacity than a mound-crust sample, which has massive structure and a narrow range of pore size distribution (predominantly micropores). The moisture retention characteristics of a mound-crust sample of an abandoned *Macrotermes* mound (Fig. 9.18) show an extremely narrow range of pore size. Both mound crust and surface soil retained about 34% moisture at zero suction. At 5 bar suction, however, the moisture retained in surface soil was only 7.5% in comparison with 23.5% in the mound crust. Although at 5 bar suction the subsoils sampled from 50 and 100 cm depths retained moisture similar to that of the mound crust, their moisture retention at zero suction was considerably greater—55 and 60%, respectively. This data

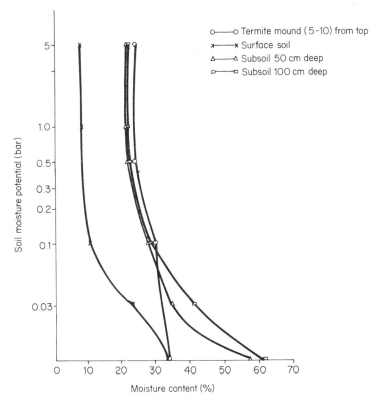

Figure 9.18 Soil moisture retention characteristics of an abandoned termite mound soil (unpublished data, Lal, 1983)

implies that the mound-crust sample had virtually no macropores whereas samples of well-structured surface and subsoil had a wide range of pore size distribution.

IV. MICROCLIMATE WITHIN THE TERMITE MOUND

Termitaria, being enclosed chambers, maintain a different microclimate within compared with their surroundings. They buffer the nest cavity against sudden fluctuations in temperature and relative humidity, and protect the colony from air movement and from inundation by rains. The microclimate of the nest is reviewed in detail by Noirot (1970).

A. Air movement

With a large termite population within the mound, uninterrupted gaseous exchange between the nest cavity and the atmosphere is essential. An adequate gaseous exchange should be facilitated without exposing the colony to direct

Table 9.22 Composition of air at different depths within the nest of *H. mossambicus* (Nel and Hewitt, 1970)

Depth (cm)	Nitrogen (%)		Oxygen (%)		Carbon dioxide (%)	
	Mean	Range	Mean	Range	Mean	Range
60	80.0	79.5–81.1	19.7	18.7–20.2	0.27	0.21–0.32
300	79.8	79.2–80.1	19.8	19.5–20.4	0.44	0.26–0.63
540	79.2	78.9–79.7	20.2	19.7–20.6	0.55	0.48–0.61

wind currents. Most termite mounds are described as having an adequate ventilation system through a series of interconnected galleries and canals (Lüscher, 1961). The chamber walls are often very thin and porous and permit gaseous exchange (Grassé and Noirot, 1958).

A little information is available describing the composition of the air within the nest. The concentration of CO_2 in the nest air has been described to vary from 3% (Grassé and Noirot, 1958; Ruelle, 1962, 1964) to as much as 15% (Skaife, 1955). Nel and Hewitt (1970) analysed the composition of air within the nest of *Hodotermes mossambicus* in South Africa (Table 9.22). Air samples within the nest were collected from 60, 300 and 540 cm depths. With the exception of CO_2 the composition of nest air was similar to the atmospheric air. There was only a slight increase in the CO_2 concentration with depth and the CO_2-tension in the nest was relatively low. This favourable composition was obviously maintained through free gaseous exchange between the soil and the nest.

Some researchers believe that there exists a 'thermosiphon system' to facilitate air circulation within the nest of *B. natalensis* (Lüscher, 1961). The temperature gradient within the nest may facilitate air circulation, and the predominant current direction may be variable depending on the nest architecture and the external air movement. Loos (1964) used a sensitive anemometer to measure air currents within the nests of *M. natalensis*. He observed irregular air morements depending on the outside wind and the wall and tower porosity. The current velocity was usually found to be less than 1 m min^{-1} and rarely exceeded 2 m min$^-$.

B. Relative humidity

Being poorly protected against dehydration, termites require an environment of high relative humidity. In addition to the effects of enclosed environments and reduced evaporation from within the nest, the metabolic activity of a large termite population is partly responsible for maintaining a high relative humidity. The finer particle size preferentially used for construction of mounds and feeding galleries is also hygroscopic and often retains more water than adjacent soil.

Data available on relative humidity measurements within the nest of various

Table 9.23 Relative humidity of the fungus garden of *O. obesus* mound in India (Cheema *et al.*, 1960)

Date	Time (hours)	Relative humidity (%) of different chambers		
14.4.1959	1145	89	97	97
	1545	89	97	96
15.4.1959	1000	88	97	91
	1600	87	97	91
16.4.1959	0930	86	100	95
	1600	86	98	93
17.4.1959	1000	89	95	93
	1600	90	94	92
18.4.1959	0830	91	88	93

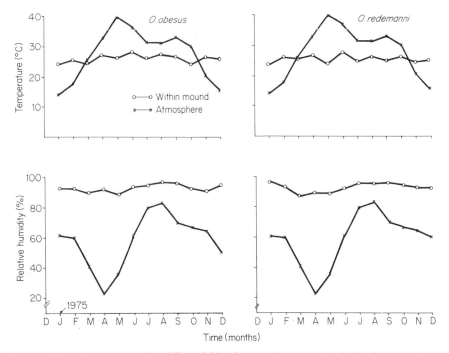

Figure 9.19 The relative humidity within the termite mound of *O. obesus* and *O. redemanni* (Singh and Singh, 1981)

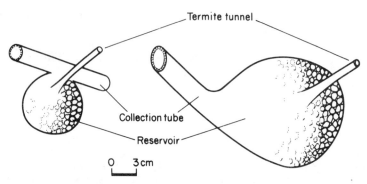

Figure 9.20 Special water receptacles constructed within the colony to maintain high relative humidity (Leprun, 1976). *Reproduced by permission of* Pedobiologia

species indicate a value ranging between 90 and 100% (Lüscher, 1961; Noirot, 1970). Cheema *et al.* (1960) reported the relative humidity of different chambers within the occupied mound of *O. obesus* to range from 90 to 100% irrespective of the atmospheric vapour pressure (Table 9.23). Singh and Singh (1981b) investigated the relative humidity of mounds of *O. obesus* and *O. redemanni*. Irrespective of the seasonal fluctuations, the relative humidity within the mound was always 90% or above (Fig. 9.19).

In the dry Sahelsudanian climate, termites are known to construct special water receptacles to meet the colony's water requirements and maintain high relative humidity during prolonged dry season (Leprun, 1976). These pockets are specially constructed by *T. geminatus* and *Odontotermes* sp. (Fig. 9.20). The walls of these receptacles are made of clay fraction and are hygroscopic (Table 9.24). It is also due to the high relative humidity within an occupied termitarium that the mound soils often have a far higher moisture content than

Table 9.24 Moisture retention properties of the soil from walls of the 'water receptacles' made by termites (Leprun, 1976). *Reproduced by permission of* Pedobiologia

Property	Soil	Receptacle
Clay (%)	13.2	26.7
Fine Silt (%)	36.6	20.1
Coarse silt (%)	9.3	11.0
Fine sand (%)	12.7	15.4
Coarse sand (%)	26.6	25.1
Organic matter (%)	0.8	3.6
Bulk density (g/cm^3)	1.49	1.66
Porosity (%)	24.5	24.8
Air dry moisture content (%)	0.66	0.93
Moisture content at pF 3.0 (%)	23.9	25.3
Moisture content at pF 4.2 (%)	8.2	16.5

Table 9.25 Moisture content of inhabited and nearby uninhabited termite mounds in East Africa (Hesse, 1955)

Mound	Characteristics	Location	Sampling depth (cm)	Field moisture content (%)
A	Inhabited	Ngomeni	Centre	18.7
	Uninhabited	Ngomeni	90 cm	4.9
B	Inhabited	Magadi	Centre	26.9
	Uninhabited	Madadi	60 cm	8.0
C	Inhabited	Kampala	150 cm	12.0
	Inhabited	Kampala	Near queen's cell	15.1
	Uninhabited	Kampala	150 cm	5.6

those of nearby uninhabited mounds (Hesse, 1955). The data in Table 9.25 show that in spite of similar textural make-up, the higher moisture content of the inhabited compared with uninhabited mounds is definitely due to the activity of the termites and due to their ability to regulate the internal relative humidity. When the colony dies, the metabolic source of water is lost and the moisture regime is then influenced by environmental factors rather than by termite activity.

C. Temperature

Temperature fluctuations within the nest have lower amplitude and higher damping than its surroundings because of the insulating effect of the termitarium. Temperature measurements made within the nests of *B. natalensis* indicate a relatively constant temperature of about 30°C with little diurnal or seasonal fluctuations (Grassé and Noirot, 1958; Cheema *et al.*, 1960; Lüscher,

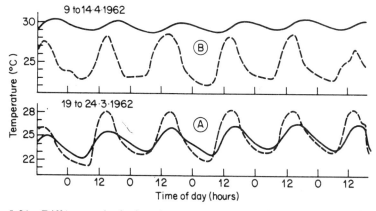

Figure 9.21 Differences in the interior nest temperature (solid line) between the two species (Noirot, 1970): A *Thoracotermes brevinotus*; B *Cephalotermes rectangularis*. The dotted line is the exterior temperature

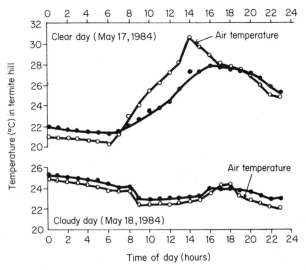

Figure 9.22 Diurnal fluctuations in the interior nest temperature of *Cubitermes fungifaber* (Ghuman and Lal, 1985)

1961; Ruelle, 1962, 1964). Noirot (1970) recorded the temperatures of the interior of the nests of *Thoracotermes brevinotus* and *Cephalotermes rectangularis* in the Ivory Coast. There were differences in the nest temperature between the two species (Fig. 9.21). The temperature within the nest of *C. rectangularis* was relatively constant and higher than the ambient temperature. Singh and Singh (1981a) observed a stable temperature at about 25°C within the nests of *O. obesus* and of *O. redemanni* with little or no seasonal fluctuation (Fig. 9.19). Mound temperature also depends on size, and there exists a temperature gradient from the top to the underground chamber. Small mounds have a large surface area and get overheated during the day and lose heat more rapidly during the night (Josens, 1971). In a forested area with less diurnal fluctuation, the mounds are generally smaller than in open savanna and woodlands. The diurnal fluctuations in air temperature with that of the mound temperature of *Cubitermes fungifaber* in the forested area near Benin, Nigeria are shown in Fig. 9.22. The maximum temperature within the mound was considerably less than the ambient.

In addition to the insulating effect of the termitaria, some biologists believe that termites regulate nest temperature through their metabolic activity. Holdaway and Gay (1948) compared the temperatures of 'dead' and inhabited termitaria (Fig. 9.23) of *Nausitermes exitiosus* and *B. natalensis*. It is apparent from the data that the metabolism of the insects plays an important role in maintaining the temperature within the nest, because killing the colony by application of arsenic resulted in the same temperature within and outside the termitaria.

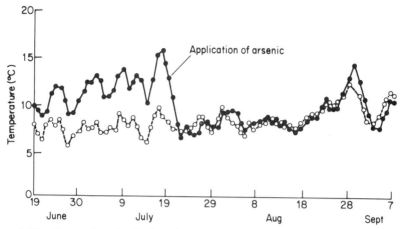

Figure 9.23 Comparison in the interior nest temperature (●) of the alive and dead termitaria with soil temperature (○) (Holdaway and Gay, 1948)

V. PEDOLOGICAL IMPLICATIONS OF ALTERATIONS IN SOIL PHYSICAL PROPERTIES

Alterations in physical properties greatly influence the processes of soil formation in the tropics (Nye, 1955; Sys, 1955). Some of the major processes attributed to the termite activity are briefly described below:

A. Soil formation

Some soil biologists have suggested that Egypt is not only the 'gift of the Nile', but also of the diligent termites that work on the soils between lakes Nyasa and Tanganyika (Lee and Wood, 1971). Whether this is an exaggeration or not, the fact remains that termites play a very significant role in soil formation in the tropics through: (i) soil turnover and physical disturbance; (ii) creation of deep galleries and burrows; and (iii) decomposition of soil organic matter content. The fine earth fraction brought to surface by mound-builder species can be substantial. In North Cameroon Mire (1975) put forth an hypothesis that the sterile 'harde' soils result, at least in some cases, from termite-induced cementation, aggravated by human activity.

Nye (1955) observed that a mound of *Macrotermes nigeriensis* contained about 2.5 tons of fine earth fraction. Nye defined two new soils horizons, the creep-termite layer and the underlying creep-gravel layer. Formation of these horizons is attributed to the work of termites in bringing fine material up, leaving the gravels and the coarse fraction at the bottom of the worked zone. Nye described a termite-formed gravel-free horizon just above the gravelly horizon. He described the termite-formed horizon as having (a) a considerable eluviation of clay, (b) a loss of fine sand and silt due to weathering, and (c) a progressive decrease in the 0.5–1, 1–2 and 2–4 mm fractions possibly because

Table 9.26 Rate of soil turnover by termites

Region	Soils	Rate (mm ha^{-1} yr^{-1})	Reference
Senegal	Aridisol	0.13	Lepage (1974)
Upper Volta	Aridisol	0.086	Roose (1976)
Zaire	Oxisols	0.05–0.5	de Heinzelin (1955)
Zaire	Oxisols	0.1–0.2	De Ploey (1964)
Nigeria	Alfisols	0.025	Nye (1955)
Uganda	Ultisol	0.1	Pomeroy (1976)
NT Australia	Alfisols	0.0125	Williams (1968)
Northern Australia	Alfisols	0.02–0.1	Lee and Wood (1971)
North Qld, Australia	Alfisols	0.025–0.05	Holt et al. (1980)

termites are unable to bring up these fractions and possibly due to some physical comminution of the grains. The termite influenced horizonation has also been described in South Africa by Watson (1974b) and in northern Australia by Williams (1968). Nye estimated that the rate of soil being transferred to the creep horizon every year by termite activity was about 0.025 mm. In northern Senegal, Lepage (1974) estimated the annual soil turnover by *M. subhylinus* to be about 1800 kg ha^{-1} yr^{-1}. In Upper Volta, Roose (1976) estimated the amount of soil brought up by *Trinervitermes* to be about 1200 kg ha^{-1} yr^{-1}. Some gigantic mounds in southern Zaire contain as much as 2400 tons of soil (Meyer, 1960). In Uganda Pomeroy (1976) estimated that *M. bellicosus* mounds attained a volume of about 2.4 m^3 in 3 years, after which they grew more slowly to a maximum of 4–6 m^3. He estimated that the fresh earth was brought to the surface at a minimum rate of about 1.0 m^3 ha^{-1} yr^{-1}. Holt *et al.* (1980) reported from Australia that mounds of *A. vitiosus* have a persistence of 25–50 years. In one generation, termite mounds reworked up to 12.5 m^3/ha of soil, i.e., 20 t/ha. This implied that the 20 cm thick A-horizon may have taken about 8000 years to accumulate. Similar estimates for other regions range from 0.01 mm/yr for an Alfisol in northern Australia to 0.5 mm/yr for an Oxisol in Zaire (Table 9.26). These estimates are based on the assumption that the amount of soil packed by termites in feeding galleries and in above-ground runways is often very large and may exceed that in mounds, and that the soil being brought up is continuously being subjected to erosion. Termites are, therefore, an important soil forming factor in the tropics.

B. Alterations in soil profile

The role of termites in the morpho-genetic processes or profile development of tropical soils has long been recognized. In tropical Africa, the action of termites in altering soil profile characteristics has been described by Chevalier (1949); Nye (1955); Stoops (1964); Heath (1965); Léveque (1969) and Boyer (1973). Similar effects have been described for Australian soils by Derbyshire (1968), Williams (1968) and Lee and Wood (1971). In central Asia Valiachmedov (1981) reported that the activity of a mound builder specie *A. ahngerians* has resulted in a complete mixing up of the surface layer 25 to 35 cm deep. Even the deep horizons were found to be pierced by tunnels and chambers up to 25 m from the centre of a termitarium.

A few interesting alterations attributed to termite activity need special mention. One is the observation made by Stoops (1968) in soils of lower Zaire. The author describes the presence in some soils of 'Striotubules', with banded aspect possibly representing alteration of different plasma types. The author attributed the presence of skeletal grains to the same origin and formation as the 'bow structure' observed in the interior of termite mounds. In southern Zimbabwe, Watson (1962) noted concentric cones of whitened soil extending as much as 6.5 m below a termite mound. The discolouration was attributed to the accumulation of finely divided carbonates (4%). Another observation is from the Sudan zone of Upper Volta. Leprun (1976) reported the presence of clayey underground pockets. These pockets, presumably constructed by *T. geminatus*, are situated deep down in weathered material below the lateritic crust. The water stored in these pockets is partly from the rainwater filtering through the permeable worked material. In northern Brazil Taltasse (1957) reported the occurrence of 'cabecas de yacare' or the 'alligator's head'. These are rather dense schist-like coloured formations that occur to several metres depth. The top horizon (3 m deep) is dark-coloured, sandy and compact, perforated by interconnecting tubes of 7–12 cm diameter and covered with a red deposit. The middle-horizon (2 m) is loose, containing channels of about 5 cm diameter. The bottom horizon, a sandy marl, is 3 m deep. These formations are also attributed to termite activity.

C. Gravel horizon

The role of earthworms in the development of the gravel horizon has already been described. Many soils in Africa and elsewhere in the tropics are characterized by a conspicuous gravelly horizon and a sharply defined boundary with the horizon of finer material above (Figs. 9.7 and 9.24). The gravelly horizons have a high proportion of gravel and increasing proportion of clay. The three-layered soils commonly observed in Africa are believed to be formed mainly by transportation of the fine materials upward by termites for mound construction (Charter, 1949; Watson, 1974a, b; Wielemaker, 1984). The mounds are eventually flattened by weathering and erosion giving rise to the upper stone-free layer. Nye (1955) observed that the average proportion of coarse quartz sand and gravels >1 cm in the gravel horizon was about double that in the sedentary horizon. The coarse sand and the gravel was presumably the residue from a thickness of rock equal to about twice the thickness of the gravel horizon. The finer material in the earthworm- and termite-influenced horizons is, therefore, transported above from this gravelly horizon. In southern Sudan the widespread occurrence of the gravel horizon is attributed to the activity of termites by Fólster (1964). He considered gravels as a degradation

Figure 9.24 A cross section of the *Macrotermes* mound. Termite, bring fine material above leaving coarse sand and gravel beneath the mound. The main cavity is used to stone organic matter harvested from the surroundings

residue deposited during pedimentation, and the inter-gravel loam partly as the product of erosion of a younger surface and partly as a pedogenetic horizon built by termites carrying material to the surface. In Australia, Williams (1968) estimated that the gravelly horizon and the finer layer above would have taken about 12 000 years to develop. Similar mechanisms of gravelly horizon development have been described for Zaire by Stoop (1968) and De Ploey (1964). De Ploey attributed the development of gravel horizon to intense sheet erosion of fine material continuously being brought to the surface by termites and faunal activity. Léveque (1969) supported these concepts and attributed the development of a fine texture horizon above the stonelines to termite activity.

D. Termites and soil erosion

There is a little research data available from the tropics quantifying the effects of termites on erosion. Termite activity can affect sheet erosion both directly and indirectly. In semi-arid regions, termites forage on grass and other herbage during the dry season. In conjunction with accidental fire and other factors, the vegetation cover is gradually dwindled, exposing the soil to high intensity rains at the onset of monsoons. The most severe erosion hazard in the semi-arid tropics is attributed to scanty vegetation cover early in the season.

Termitaria crusts contain high clay and fine soil particles cemented together. These crusts have low water infiltration. Most of the rain water received on termite mounds becomes surface run-off. In some regions of tropical Africa as much as 30% of the surface area is covered by active and abandoned termitaria (Meyer, 1960). Termitaria of some species do not support any vegetation cover, which further accentuates the surface run-off and soil erosion. Furthermore, the surface soil of the termite mounds often has a low organic matter content and is thus more erodible than soil aggregated by the action of higher organic matter content.

The effects of termites on soil erosion in Africa were first described by Drummond (1887) and later supported by the observations of Harris (1949). The positive effects of termite activity in reducing erosion are due to high infiltration of water through the channels and feeding galleries. The anti-erosion effect of vacuolated crust structures formed by termite activity was reported from French Sudan by Erhart (1951a, b). These crusts occur extensively and reportedly protect the soil beneath them from erosion. If these crusts occur beneath the soil surface, they also conserve soil-water in the root zone above the crusted horizon.

Erosion from termite mounds means additions of nutrient-rich clayey material to the surrounding soil. At Gouse, Upper Volta, Roose (1976) estimated the erosion from abandoned mounds to be at 400 kg ha^{-1} yr^{-1} and that from the occupied mounds at 800 kg ha^{-1} yr^{-1}. The clay and fine silt fractions are selectively removed by erosion leaving a sandy top soil in the upper crust of the mound. Out of the gross soil turnover of 0.086 mm yr^{-1}, the net amount of soil accumulated by termite activity was 0.06 mm yr^{-1}. The eroded material, equivalent to 0.022 mm yr^{-1}, adds a considerable amount of plant nutrients to the adjacent soil. Pomeroy (1976) estimated that at Natete and Naluvule in Uganda one-tenth of the above-ground volume of the mound is annually eroded. His data (Table 9.27) show that the estimated rate of addition of calcium to the soil by erosion was comparable to the rate of removal by two crops. This much calcium addition to the soil (3 to 5 kg ha^{-1} yr^{-1}) would also meet the needs for the optimal stocking rate of cattle—about 2.4 ha per 200 kg animal. Wielemaker (1984) reported from his study in Kiisi region of Kenya that soil with intense activity of termite was so porous and stable that no surface runoff and erosion was observed on this site even on intensively cultivated slopes of up to 20 percent. The exposed surface of mounds, however, was readily washed by water runoff. Wielemaker estimated soil erosion from the termite mounds to be as much as 1163 kg ha^{-1} yr^{-1} or equivalent of 6.3 kg of soil per meter contourline per year. In addition to soil creep, the amount of soil transported in water runoff downstream was estimated to be 27.3 to 54.5 kg yr^{-1} per meter of contourline.

In Australia, Watson and Gay (1970) reported that termites were mainly responsible for grass removal, and sheet erosion observed in denuded areas was due to the lack of protective vegetation cover. Surface crust of termitaria exposed by sheet erosion are slowly permeable, do not support vegetation, and further accelerate the degradative process through enhanced soil erosion.

The anti-erosion effects of termite mounds have been demonstrated by

Table 9.27 Estimated quantities of nutrients added to soil by erosion of termite mounds in Uganda (Pomeroy, 1976a)

Locality	Termite	Estimated soil erosion from mound (mm ha^{-1} yr^{-1})	Quantities of nutrients (kg ha^{-1} yr^{-1})			
			P	Ca	K	N
Natete	M. bellicosus	0.115	0.014	1.22	0.56	1.70
	Pseudocanthotermes spp.	0.0112	0.001	0.15	0.05	0.15
	Total	—	0.015	1.37	0.61	1.85
Naluvule	M. bellicosus	0.039	0.001	0.45	0.09	0.66
	Pseodcanthotermes spp.	0.10	0.003	1.36	0.22	1.20
	Total	—	0.004	1.81	0.31	1.86
Muko	M. subhyalinus	0.437	0.149	8.05	2.54	7.86
Mweya	M. subhyalinus	0.026	0.031	1.69	0.47	0.39
Chobe	Macrotermes spp.	0.0218	0.007	4.10	0.67	0.22
Annual crop removal						
Coffee (beans)			2.2	1.2	16	25
Maize (ears)			16	1.4	23	52

Erhart (1951). He observed that fossil or sub-fossil crusts formed from termite nests occur among still inhabited nests in forests and clearings of equatorial Africa. In desert and semi-desert regions these crusts protect from erosion the soil layers beneath them. They also conserve moisture, especially when the crusts are buried.

E. Laterization

The role of termites in the formation and decomposition of laterites and lateritic soils is a controversial issue and is reviewed by Lee and Wood (1971). Many authors have attributed the formation of laterites to the ferruginization of termite nests and gallery systems (Erhart, 1951; Tessier, 1959; Harris, 1965; Yakushev, 1968; Raunet, 1979). Other researchers attribute laterization to natural physico-chemical processes, and the development of vesiculae to the removal of soft material by percolating water and not by burrowing termites (Griffith, 1953; Grassé and Noirot, 1959; Boyer, 1959). Similar observations were made in 1979 by the author during the experiment conducted on watershed development at Ibadan, Nigeria. The exposed plinthite (soft inititally) hardened within 1 year and developed vesiculated structures because of the removal of soft material by run-off and percolating water (Fig. 9.25). There was no termite activity in the vicinity during the period when the laterite shown in Fig. 9.25 was formed.

Some authors believe that termite activity may lead to eventual decomposition or burial of lateritic crusts (Grassé, 1950; Tricart, 1957; Boyer, 1959).

Figure 9.25 The vesiculated structure of hardened plinthite

Table 9.28 Soil organic matter contents in termite mounds and adjacent soil

Region	Termite species	Organic carbon content (%)			References
		Termi-taria	Surface soil	Sub-soil	
(a) *Termitaria with low organic matter content*					
Sierra Loane	M. bellicosus	0.2–1.5	2.5	0.90	Miedeme and Van Vuune (1977)
Zimbabwe	M. bellicosus	0.81	0.88	0.14	Watson (1969)
Nigeria	Macrotermes sp.	0.65	1.42	0.55	Kang (1978)
Nigeria	Macrotermes sp.	0.48	1.60	—	Kang (1978)
Nigeria	Macrotermes sp.	0.58	1.70	0.70	Nye (1955)
Kenya	O. badius	2.28	3.01	1.18	Robinson (1958)
Kenya	O. badius	1.32	1.51	0.87	Robinson (1958)
Kenya	O. badius	1.70	2.37	1.21	Robinson (1958)
Kenya	Macrotermes sp.	0.2	0.5	0.2	Hesse (1955)
Kenya	Macrotermes sp.	0.3	0.5	0.2	Hesse (1955)

Kenya	*Macrotermes* sp.	0.7	1.20	0.7	Hesse (1955)
Kenya	*Macrotermes* sp.	0.5	0.9	0.5	Hesse (1955)
Kenya	*Macrotermes*	0.7	1.3	0.7	Hesse (1955)
Kenya	*Macrotermes*	0.4	0.7	0.4	Hesse (1955)

(b) Termitaria with high organic matter content

India	*O. badius*	36.5	0.5	—	Pathak and Lehri (1959)
Zimbabwe	*O. badius*	1.92	0.88	0.30	Watson (1969)
Zimbabwe	*Macrotermes falciger*	0.7	0.2	—	Watson (1976)
Kenya	*O. badius*	2.27	2.12	—	Robinson (1958)
Kenya	*O. badius*	2.91	2.80	—	Robinson (1958)
Kenya	*Macrotermes*	2.4–5.1	1.0–1.5	0.8	Wielemaker (1984)
Upper Volta	*T. germinatus*	3.62	0.78	—	Leprun (1976)
Nigeria	*Macrotermes* sp.	0.79	0.53	0.32	Nye (1955)
Australia	*Coptotermes* sp.	2.7	0.2	—	Lee and Wood (1971)
Australia	*Amitermes* sp.	2.9	0.8	—	Lee and Wood (1971)
Australia	*Drepanotermes* sp.	1.4	0.2	—	Lee and Wood (1971)
Australia	*Nasutitermes* sp.	2.8	0.3	—	Lee and Wood (1971)
Australia	*Tumulitermes* sp.	4.4	0.4	—	Lee and Wood (1971)

Turnover of fine soil from beneath the crust to the surface for nest and mound building and alterations of crust strength by chemical and physical changes caused by the high humidity environments of the mound lead to eventual degradation of the hardened laterites. The disintegration process is accelerated once the vegetation is re-established on hardened laterites.

VI. EFFECTS ON SOIL CHEMICAL PROPERTIES

Through their influence on texture, structure, water retention and transmission properties, termites also alter chemical and nutritional properties of the soil they inhabit. In contrast to scanty research data on soil physical properties, a considerable amount of data exists on soil chemical properties.

A. Organic matter and total nitrogen

There are conflicting reports about the organic matter content of termite-infested soils in comparison with termite-free soils. Some researchers have observed more organic matter in termitaria than in surrounding soils, which is probably due to the use of body fluids and excrements in mound construction, and to the accumulation of biomass as food reserves. There are also reports of lower organic matter content in soils infested with termites and in termitaria than in surrounding soils. The data in Table 9.28 compare the organic matter content of termitaria and adjacent soils for different termite species in a range of ecological zones. In Belgian Congo, Maldague (1959) observed that mounds of *Bellicositermes bellicosus* and *B. natalensis* contained less organic matter than the adjacent soils, and the opposite was the case in the mounds of *Amitermes unidentatus*, *Nausitermes ueleensis* and *Cubitermes fungifaber*. On the contrary, no such clear demarcations among species are shown by the data in table 9.28. Neither can the differences in organic matter content be attributed to soils or rainfall patterns. Mounds of most Australian termite species seemingly contain more organic matter content than adjacent soils (Lee and Wood, 1971).

One of the reasons for this discrepancy may be the portion of termitaria sampled. Similar to a normal soil profile, there are differences in chemical properties among different termitaria components, e.g., mound crust, whether above or below ground, fungal comb, nest, galleries, internal chambers, etc. The above-ground mound crusts of most *Macrotermes* sp. generally contain less organic matter than adjacent soils. The organic matter content, however, increases from outer casing to the inner royal chamber (Stoops, 1964).

The organic matter content of various depths for an occupied and an abandoned *Macrotermes* mound is compared with soils in their vicinity in Fig. 9.26a. The above-ground mound consisted of a hollow cavity immediately below the crust. No samples were thus available for analysis. Samples for analyses were obtained from the centre of the mound and from adjacent normal soil for 0 to 100 cm depth. With the exception of the surface 0 to 10 cm

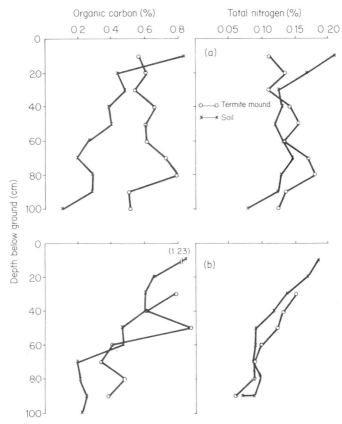

Figure 9.26a The organic matter content distribution within the mound of an occupied *Macrotermes* mound in a cultivated soil (unpublished data, Lal, 1983)
Figure 9.26b The organic matter content distribution within a *Macrotermes* mound in a forest (unpublished data, Lal, 1983)

layer, the mound contained more organic matter content than the soil. The organic matter content of the termite mound was relatively uniform at about 0.6% between 0 and 60 cm depth and then increased in the royal chamber between 60 and 80 cm depth. Between 20 and 100 cm depth the organic matter content of the mound was more than adjacent soil by 11 to 265 per cent.

The hollow cavity in the abandoned mound existed from 0 to 20 cm depths below the soil surface. The maximum organic carbon was present in the section containing the royal chamber, i.e. 40 to 60 cm depth (Fig 9.26b). The royal chamber contained 2.4 times more organic matter than the adjacent soil. The soil profile had more organic carbon content than termitaria samples for 0 to 40 cm and 60 to 100 cm depths. With the exception of the surface 0 to 30 cm depth, termitaria contained more nitrogen than the adjacent profile of the cultivated soil. Samples from the unoccupied termitaria also contained more organic carbon between 30 and 70 cm depths than adjacent fallowed soil.

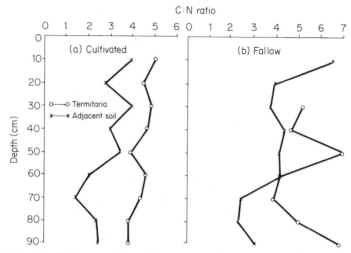

Figure 9.27a   The C:N ratio of the cultivated soil and the termite mound (unpublished data, Lal, 1983)
Figure 9.27b   The C:N ratio of the fallowed soil and the abandoned termite mound (unpublished data, Lal, 1983)

The C:N ratio of the cultivated soil was narrower than that of the termitaria (Fig. 9.27a) and ranged between 1.35 and 4.0. In comparison, the C:N of the occupied termitaria ranged between 3.8 and 4.0. The narrow C:N ratio of the cultivated soil is probably due to the application of nitrogenous fertilizers at 100 kg N/ha twice a year. The fertilizer nitrogen has leached out up to 1 m depth. The C:N ratio of the unoccupied termitaria is also more than that of the uncultivated soil (Fig. 9.27b), whilst the C:N ratio of the uncultivated soil is also more than that of the cultivated soil. Some researchers have reported little difference between the C:N ratio of the mound and the subsoil (Hesse, 1955; Maldague, 1959), others have observed higher C:N ratio in soil than in mound (Joachim and Kandiah, 1940; Boyer, 1956a, b), and still others found a higher C:N ratio in mound than in soil (Pathak and Lehri, 1959). Lee and Wood (1971) observed 46 termitaria and only two contained a lower C:N ratio than the adjacent soil; 26 had significantly higher C:N ratios. The C:N ratio of termitaria depends on soil properties and the feeding habits of the termites. The wood-feeders generally have a higher C:N ratio than grass feeders (Lee and Wood, 1971).

Some researchers consider that because termites can live on a diet of cellulose, there may be a possibility of nitrogen fixation in termites. In Turkenia, Central Asia, Kozlova (1951) estimated that the total amount of NO_3 in the 500 or more mounds/ha was about 208 kg or 47 kg of readily available nitrogen. Pathak and Lehri (1959) observed that nitrification and nitrogen fixation in the soil of a termite nest was higher than that in the control. They observed an overall higher level of microbiological activity in the termite nest than in the normal soil (Table 9.29). From East Africa Meiklejohn (1965)

Table 9.29 Biological activity of termite (*Hypotermes obscuripes*) nest soils (After Pathak and Lehri, 1959)

Biological activity	Termite nest	Control
Nitrogen fixation (g)	0.350	0.011
Nitrification (ppm)	1064	140
CO_2 evolution (mg % after 24 hours)	31.7	10.1
Total N after incubation (g)	1.52	0.061
N fixed per g of C oxidized	0.14	0.013

reported that termite mounds of *Macrotermes* sp. contain more fertile materials than surrounding soil because they contain more cellulose decomposers, denitrifiers, ammonifiers and nitrifiers than the surrounding soil. Goodland (1965) observed that the fixed nitrogen in the termitaria heaps was triple and the organic carbon was double that of the adjacent soil. Arshad et al. (1982) observed that the bacterial population in the mounds of *M. subhyalinus* and *M. michaelseni* in Kenya was maximum in the termite-active nursery section. Whereas the population of Actinimycetes, bacteria and fungi generally declined with increasing depth in the adjacent soil, it increased with depth in the open mound up to the section of the fungal comb.

The quantitative data on nitrogen fixation by different termites is reported by Breznak et al. (1973). They compared the C_2H_2 reducing ability of 17 genera of other insects with four termite spp. (Table 9.30). Of all the insects tested only termites reduced C_2H_2 to C_2H_4, and all castes of termites tested reduced C_2H_2 to C_2H_4, but soldiers possessed the lowest activity. There were also differences among species regarding their ability to reduce C_2H_2. If their nitrogen fixation ability is high, as it seems to be, termites are obviously important both economically as well as ecologically.

B. pH and exchangeable bases (Ca^{2+}, Mg^{2+}, K^+ and Na^+)

Although many workers have observed higher pH levels in termitaria than in the surrounding soil (Table 9.31), there are also examples of the pH decline (Maldague, 1959; Lee and Wood, 1971; Laker et al., 1982). Stoops (1964) and Goodland (1965) observed similar pH of the termite mound and the adjacent soil. Lal (unpublished data) observed that the pH of the mound ranged from 6.2 to 7.4 and that of the uncultivated adjacent soil from 6.3 to 6.5. In general, the pH of the termite mounds depends on the pH of the surrounding soil. The pH of the termite mound soil is usually higher if the adjacent soil is highly acidic. The pH of the termite mound soil is unaffected if that of the adjacent soil is neutral.

Many researchers have observed an accumulation of exchangeable bases in the termitaria. Accumulation of calcium as concretions of $CaCO_3$ is often observed at the base of *Macrotermes* mounds (Boyer, 1956; Hesse, 1955; Robinson, 1958; Watson, 1962; Laker et al., 1982). In India, Gokhale et al.

Table 9.30 N_2 (C_2H_2) fixation in termites (Adapted from Breznak et al., 1973)

Termite	Caste	Diet	C_2H_4 (nmol formed per h)	
			per 50 termites	per g termites
Coptotermes formosanus	Worker	Wood	0.122	0.695
	Soldier	Wood	0.027	0.139
Reticulitermes flavipes	Worker	Wood	0.043	0.204
	Soldier	Wood	0.014	0.073
Zootermopsis sp.	Reproductive nymphs & workers	Wood	0.517	0.272
Cryptotermes brevis	Reproductive nymphs	Moist filter paper	0.622	1.705

Table 9.31 pH of termitaria and the surrounding soil

Region	Specie	pH			
		Termitaria	Surface soil	Subsoil	
Nigeria	T. geminatus	5.1	4.7	—	Omo Malaka (1977)
Nigeria	Macrotermes sp.	5.8	6.5	5.6	Kang (1978)
Nigeria	Macrotermes sp.	6.0	5.8	—	Kang (1978)
Nigeria	Macrotermes sp.	5.4–7.4	6.4	6.7	Nye (1955)
Nigeria	Macrotermes sp.	5.6–6.8	6.5	5.2	Nye (1955)
Upper Volta	T. geminatus sp.	8.7	8.7	—	Leprun (1976)
Zaire	—	6.4–7.9	5.5	5.5–5.7	Sys (1955)
Savanna	—	5.0	5.0	—	Goodland (1965)
Kenya	O. badius	5.9	5.6	5.5	Robinson (1958)
Kenya	M. subhyalinus and M. michaelseni	5.2–6.7	6.0	5.7	Arshad et al. (1982)
Kenya	Macrotermes	6.7–7.6	5.8	5.5	Wielemaker (1984)
Zimbabwe	Macrotermes sp.	6.0–7.8	4.8	—	Watson (1976)
Zimbabwe	M. goliath	6.2–7.5	4.8	5.4	Watson (1967)
India	H. obscuripes	4.5	8.1	—	Pathak and Lehri (1959)

Table 9.32 Exchangeable bases and the available phosphorous of an occupied *Macrotermes* mound and adjacent soil (unpublished data, Lal, 1983)

	Bray-1 P (ppm)		Exchangeable cations (meq/100g)										Total acidity		CEC	
			Ca^{2+}		Mg^{2+}		K^+		Mn^{3+}		Na^+					
	TM	S	TM	S	TM	S	TM	S	TM	S	TM	S	TM	S	TM	S
0–10	0.6	6.1	1.57	2.13	1.13	0.68	0.42	0.30	0.327	0.199	0.103	0.067	0.10	0.18	3.65	3.55
10–20	0.6	1.0	1.50	2.13	1.15	0.68	0.38	0.20	0.228	0.098	0.106	0.063	0.33	0.26	3.69	3.43
20–30	0.6	2.3	1.50	2.50	1.11	0.71	0.31	0.16	2.34	0.060	0.108	0.072	0.33	0.22	3.59	3.72
30–40	0.6	1.5	1.20	2.95	1.42	0.80	0.61	0.12	0.017	0.024	0.077	0.067	0.17	0.24	3.49	4.20
40–50	0.6	1.0	1.64	3.00	1.27	0.75	0.38	0.15	0.245	0.019	0.108	0.155	0.55	0.38	4.19	4.45
50–60	0.6	0.9	1.64	2.95	1.38	0.73	0.31	0.11	0.212	0.037	0.111	0.125	0.52	0.22	4.17	4.17
60–70	0.6	0.7	1.87	3.10	1.42	0.82	0.46	0.10	0.272	0.018	0.116	0.074	0.67	0.24	4.81	4.35
70–80	2.04	0.6	1.94	3.25	1.44	0.77	0.57	0.10	0.207	0.030	0.134	0.057	1.07	0.33	5.36	4.54
80–90	1.20	0.6	1.72	2.65	1.28	0.91	0.19	0.29	0.239	0.011	0.098	0.062	0.38	0.37	3.91	4.29
90–100	0.72	0.8	1.42	3.29	1.08	0.93	0.15	0.29	0.174	0.007	0.191	0.072	0.32	0.28	3.34	4.87

TM, termite mound;   S, adjacent soil

(1958) and Pathak and Lehri (1959) observed 5.4 times more calcium in mounds than in the parent soil. In Nigeria Nye (1955) reported a higher content of exchangeable calcium in one mound and lower in another in comparison with the adjacent soil. In Guyana Goodland (1965) reported 3 to 7 times more calcium in termite mounds than in adjacent soil. In Australia, Lee and Wood (1971) reported 2 to 5 times more calcium in mounds of Australian termites than in adjacent soil. In Indian desert, Roonwal (1975) observed that earth mounds had a higher percentage of water soluble Na than the surrounding soil. In Kenya Robinson (1958) reported a higher percentage of exchangeable bases in the termite soil samples than in either the topsoil or the subsoil samples. Arshad et al. (1982) observed only a slight increase in the calcium content of the termite mound soil in comparison with the normal subsoil, and a decrease in calcium content in comparison with the surface soil. In Kiisi region of Kenya Wielemaker (1984) reported 3 to 5 times more Ca and Mg in termite mounds than in uncontaminated adjacent soil. Consequently the base saturation of mound soil was also high.

In western Nigeria Kang (1978) reported lower calcium content in mounds than in adjacent Alfisol in western Nigeria. The exchangeable bases of a *Macrotermes* mound in western Nigeria are compared with adjacent soil in Table 9.32. In general, mound soil contained less calcium but higher amounts of Mg, K and Mn in comparison with the normal soil. There were no significant differences between the total CEC of the mound soil and the normal soil.

The literature reported indicates higher calcium and exchangeable bases in mound soil of some species, lower in others, and no differences in still others. It seems that the base status of mound soil is influenced by the termite species, rainfall regime, and other environmental factors. In Nigeria Omo Malaka (1977) observed significant differences in the calcium and magnesium status of mound soils of different termite species (Table 9.33). The calcium and magnesium content of mound soil from *A. evuncifer* and *Cubitermes* was more and that of *M. bellicosus* was less than adjacent soil. The mound soil from *T. geminatus* contained more magnesium than the parent soil and similar calcium. It is apparent, therefore, that some termite species accumulate more bases in their nests than others. Some species are known to build their mound on soils of specific mineralogical composition. For example, Leprun and Roy-Noël (1976) observed from West Senegal that mounds of *M. bellicosus* occur only on ferruginous and ferrallitic soils with kaolinitic clay. On the contrary, mounds of *M. subhyalinus* occur in soils with montmorillonite or attapulgite clay and more rarely on kaolinitic soils.

The differences in nutrient status between the mound soil and the adjacent soil are partly due to differences in their texture and organic matter content. A soil with more colloids would generally contain more nutrient elements. Differences in textural and structural properties also result in differential leaching. Mound crust, being slowly permeable, loses most of its rain as surface run-off with little or no seepage through it. The leaching losses of basic cations and of soluble elements are, therefore, less. The bases are believed to be

Table 9.33 The calcium and magnesium content of mound soil of different termite species in Nigeria (Omo Malaka, 1977)

Nutrient element (kg/ha)	Amitermes evuncifer		Cubitermes sp.		M. bellicosus		T. geminatus	
	TM	S	TM	S	TM	S	TM	S
Phosphorus	17	4.7	25	11	4	4	17	10
Potassium	305	78	172	33	57	45	210	52
Calcium	691	136	253	139	109	329	243	232
Magnesium	225	46	91	56	55	119	116	54

TM, termite mound;   S, adjacent soil.

Table 9.34 The effects of the rains regime on the concentrations of calcium, magnesium and potassium in mound and adjacent soil (Watson, 1976)

Rainfall zone	Termite mounds (meq/100 g) A	Adjacent soil (meq/100 g) B	A − B Absolute concentration (meq/100 g)	A/B Relative concentration
Extractable calcium				
Low	43.4	5.4	38.0	8.0
Medium	40.5	1.7	38.8	23.8
High	5.9	1.1	4.8	5.4
Extractable Magnesium				
Low	3.5	1.9	1.6	1.8
Medium	3.1	0.8	2.3	3.9
High	1.1	0.6	0.5	1.8
Extractable Potassium				
Low	0.69	0.23	0.46	3.0
Medium	1.11	0.30	0.81	3.7
High	0.64	0.20	0.44	3.2

accumulated in mounds from vegetation material brought in as food. The amount of leaching of the accumulated bases depends on the rainfall and the permeability of the mound material. Watson (1976) compared the base status of termite mounds in different rainfall zones, and observed that calcium had been weakly leached in low and medium rainfall zones and strongly leached in the high rainfall zone (Table 9.34).

Watson (1974) advanced many biological reasons for accumulation of calcium in the mound soil of some species: (i) exchangeable calcium in soil collected incidentally by termites with their food, (ii) exchangeable calcium in soil collected purposely by termites requiring an alkaline environment, (iii) calcareous material collected from below the depth of soil said to be non-calcareous, and (iv) calcium-containing ground-water brought up by termites. Although the source and mode of entry of calcium-rich water into the mound may differ, it is its evaporation that results in accumulation of Ca. Watson (1974) advocated a two-stage chemical process of accumulation of calcium: (i) elevation of the pH status of the mound above that of the pH of the ground-water, and (ii) saturation of the base of the termite mound by ground-water containing calcium bicarbonate which eventually precipitates as $CaCO_3$. It is apparent from the review presented that termite mounds comprise complex physical, chemical and biological processes and considerably more research is needed to understand them.

Whatever the mechanism of the accumulation in termite mounds of these basic cations may be, it has far reaching practical implications. Mound soils are often used as a source of lime and fertilizer in East Africa (Mielke, 1978), and in Southeast Asia (Pendelton, 1941; Kozlova, 1951). This aspect will be discussed in more detail in the section dealing with crop growth. These calcium-rich mounds also have some archaeological significance. Some 700 to 800 year-old skeletal remains in a burial ground have been discovered in alkaline soil only within the termite mound (Watson, 1967). The skeletal remains have presumably long been dissolved in the surrounding acidic soil. Some researchers had believed that termites carry Au-containing materials to their mounds from the Au-bearing fissures of the basement complex. Watson (1972) observed that kaolinitic soils contained 0.15–0.20 ppm of gold in comparison with 0.13–0.52 ppm in termite mounds, and the difference between the mean concentrations of Au in termite mounds and in soils was not significant.

VII. TERMITES AND VEGETATION

Vegatation, being the source of food and an important ecological factor, affects termite population and activity and is in turn affected by it. The diversity of food supply and the different microclimatic environments created are important factors. Some termites are associated with definite types of vegetation (Wild, 1952). Sands (1965a, b) reported the effects of vegetation type on termite density in nothern Nigeria.

Figure 9.28 Association of termites with a particular tree species is a common phenomenon

Some termites (e.g. *T. geminatus*) build nests in the open, unshaded areas and others under the shade (e.g. *T. oeconomus* and *T. occidentalis*). Some termite species build their nests only in association with particular trees (Harris, 1966) (Fig. 9.28). In the Loita plains of Kenya, Glover *et al.* (1964) identified different vegetation patterns around the termite mounds. They observed three concentric vegetation zones: (i) a tough low shrub around the central mound; (ii) short grass in the intermediate zone; and (iii) a ring of tall grass in the outer zone. The top of the mound was found to be bare, devoid of any vegetation. In the Shaba province of Zaire, Goffinet (1976) distinguished a gradual transition of dense dry forest to open forest to savanna vegetation accompanied by the gradual replacement of the fungus-feeder *Macrotermes falciger* (which consumes large amounts of litter and builds enormous mounds) by the humus-feeder *Cubitermes* spp (which builds relatively small mounds). In Pakistan Kayani and Sheikh (1981) described the distribution of 13 termite species in relation to various plant communities and soil characteristics.

Termites play a significant role in the decomposition of plant residue and vegetation biomass, and readers are referred to detailed reviews on this subject elsewhere (1971 Krishna and Weesner, 1970; Lee and Wood,). For the purpose of this volume two points deserve a special mention. Firstly, the termites usually change the lignin: cellulose ratio in the plant debris. For example, Lee and Wood (1968) reported that *Nasutitermes exitiosus* changed the lignin to cellulose ratio from 1:5 in their food (dead wood) to 5:1 in their mounds. The organic matter stored in mounds has a high C:N ratio (often as much as 70) with little N and P. That is why the organic matter in the abandoned mounds decomposes very slowly. Secondly, the termites tend to favour the mineralization of plant debris rather than its humification (Baschelier, 1977).

The data available from a few studies on the role of termites on vegetation decomposition indicate that termites are voracious feeders and consume a great amount of plant debris. In South Africa, termites are known to compete with stock for grazing crops, and termites destroy what is not eaten by stock (Farming in South Afica, 1960). Collins (1981a) reported that in the southern Guinea Savanna region of Nigeria, termites removed 836 kg ha^{-1} yr^{-1} of leaf litter, which amounted to 24% of the total annual litter production. In comparison, the annual bush fire was estimated to remove 1.17 t ha^{-1} yr^{-1} (49%) of the annual leaf fall but only 3 kg ha^{-1} yr^{-1} (0.2%) of annual wood fall. Fungus-growing *Macrotermitinae* consumed 95% of the litter removed by the termites or 23% of the annual litter production. In Uganda, Okwakol (1980) estimated soil and organic matter consumption by termites of the genus *Cubitermes*. He observed that consumption of the soil varied between between 0.72 and 0.91mg per termite per day of which organic matter constituted between 0.65 mg per and 0.84 mg termite per day. The field estimates showed a consumption of between 69 and 88 mg m^{-2} yr^{-1}. In the lowland humid forest region of Yangambi, Zaire, Maldague (1964) estimated that termites consume about 570g plant material m^{-2} yr^{-1}, which is about 50% of the total litter fall of

Table 9.35 Estimated consumption of surface wood litter by various termite species in northern Nigeria (Wood et al., 1977)

Species	Wood consumption (kg ha^{-1} yr^{-1})
Ancistrotermes	157.1
Macrotermes bellicosus	177.1
M. subhyalinus	36.8
Microtermes	300.8
Microcerotermes	40.1
Odontotermes	121.1
Other species	2.5
Total	835.5

the region. In West Malaysia Abe (1980) placed bole and branch samples of *Shorea parvifolia* and *Ixonanthes icosandre* on the lowland rainforest floor and determined their weight loss due to termites within a period of 1.5 yr. Small branches (3–6 cm diameter) lost 80% of their initial dry weight, medium-size (6–13 cm) 60% and large (13–30 cm) about 50%. Small branches of *Oxonanthes* lost only 37% of their initial weight.

Differences in food preferences by termites are probably due to differences in the physical and chemical nature of the plant material. Wood et al. (1977) investigated the feeding habits of some termite species in northern Nigeria, and observed that woody litter was avidly consumed. *Macrotermes bellicosus* was the only species which consumed significant quantities of tree leaf litter. The rate of wood consumption by all termite species was 835 kg ha^{-1} yr^{-1} (Table 9.35). In addition to wood, most species also consumed other sources of food, i.e., tree bark, grass, leaf litter and woody roots. There were differences in food preferences, and seasonal variations in surface foraging intensity among different species. Gupta et al (1981) observed marked differences in the probability of attack by *Odontotermes* sp. on shoot and root material of mixed grasses in northern India. The range of probabilities of attack was 1 and 0.67 for shoot and roots of *Chenopodium*, 0.67–0.444 for mixed grass shoots, 0.44–0.0 for *Desmostachya* shoots, 0.67–0.083 for mixed grass roots, and 0.05–0.0 for *Desmostachya* roots. Termites depend on a range of microflora in their guts to facilitate digestion of the plant material. Furthermore, the nitrogen fixation by gut symbionts may also play a significant role in termite nutrition.

VIII. TERMITE ACTIVITY AND AIR POLLUTION

Some researchers believe that methane content of the atmosphere, a greenhouse gas, has been increasing steadily by as much as 2% annually. Termites and many other fauna are believed to contribute to this atmospheric methane. The termites are particularly voracious eaters and occupy about two-thirds of the earth's land surface. Zimmerman et al. (1982), based on extrapolation of

Table 9.36 Estimates of methane production by termites on the basis of global biomass consumption in different ecologies (Zimmerman et al., 1982)

Ecological region	Area (10^{12} m^2)	NPP (g/m^2 year, dry weight)	Total biomass produced 10^{15} g/year, metre	Termites per square dry weight)	Total termite consumption (10^{15} g, calculated)	Percentage biomass by termites	Annual CH$_4$ production (10^{12} g)
Tropical wet forest	4.6	1200	5.5	1000	0.6	12	2.9
Tropical moist forest	6.1	1500	9.2	4450	3.8	41(30)	17.3
Tropical dry forest	7.8	1200	9.4	3163	3.4	36	15.7
Temperate	12.0	1250	15.0	600	1.0	7	4.6
Wood/shrub land	8.5	700	6.0	431	0.5	9	2.3
Wet savanna	14.2	1200	17.0	4402	8.7	51(31–47)	39.9
Dry savanna	4.3	900	3.9	861	0.5	13(10)	2.3
Temperate grass land	9.0	600	5.4	2139	2.7	50(47)	12.4
Cultivated land	11.9	650	7.7	2813	4.7	60	21.6
Desert scrub	18.0	90	1.6	229	0.6	38	2.8
Clearing burning	6.8		9.6	6825	6.5	68	29.8
Total	103.2		90.3	2.4×10^{12}	33.0	37	151.6
Percentage of total terrestrial	68		77		28		

Table 9.37 Estimates of global CH_4 emissions by termites (Seiler et al., 1984)

Sub-family	$^FCH_4/BC^a$ (gC/gC)	Consumption rate (g dry matter/yr)	CH_4 emission (Tg Ch_4/yr)
Macrotermitinae	6.0–6.3 × 10^{-5}	2.2 × 10^{15}	0.08–0.83
Termitinae	0.6–1.2 × 10^{-3}	4.0 × 10^{15}	1.44–2.83
Other termites	0.9–2.6 × 10^{-4}	1.1 × 10^{15}	0.59–1.72
Total		7.3 × 10^{15}	2.11–5.43

laboratory results, estimated the potential global production of CH_4 by termites to be 1.5 × 10^{14}g or the equivalent of 1.1 × 10^{14}g of carbon. In addition, termites are also estimated to release 4.6 × 10^{16}g of CO_2 (1.3 × 10^{16}g of carbon), 10^{13}g of CO (0.4 × 10^{13}g of carbon) and 7 × 10^{11}g of dimethyl disulphide (DMDS) (Table 9.36). Some recent studies have shown that Zimmerman et al. (1982) had overestimated the methane flux due to termites. This is expected since the extrapolation was based on limited laboratory studies. Data in Table 9.37 by Seiler et al. (1984) show more conservative flux of methane attributable to termites. A much larger contribution is from rapidly expanding rice paddies (see Chapter 15).

IX. TERMITES AND CULTURAL PRACTICES

The encroachment upon tropical forest and savanna vegetation by commerical agriculture has affected habitat, diversity of food supply and ecological factors that influence termite population and its activity. Among cultural practices that have drastic effects on termites are deforestation and land development, monocropping, grazing and, above all, the indiscriminate use of insecticides and other farm chemicals. Some quantitative data are, however, available relating the effect of these practices on the activity of termites in different ecologies.

A. Deforestation

In Nigeria, Sands (1965a) investigated termite distribution in man-modified habitats, and observed that the mound population density of *T. ebenerianus* was higher in a cleared land dominated by grass than in a scrub plagioclimax or an open woodland. In addition to depleting the source of the diverse food supply, deforestation also influences termite population through its effect on pedological, hydrological and climatic conditions (Bodot, 1967; Wood, 1975). Critchley et al. (1979) studied the effect of bush clearing on termite population in western Nigeria. There were differences in species composition in cleared and uncleared land. The tree-nesting and decomposing-wood-feeding species were understandably eliminated by deforestation. For example, the population densities of *Microceratotermes* sp., *Nasutitermes* sp. were low in cleared land. Also low in cleared land were the densities of soil-nesting but wood or litter-feeding species e.g. *Macrotermes bellicosus*, *M. subhyalinus*, *Microtermes* sp., *Ancistrotermes* sp. and *Pseudocanthotermes* sp. Some subterranean soil-feeding species (e.g. *Pericapritermes* sp., *Adaiphrotermes*, *Anenteotermes* sp, *Astalotermes*, sp, and *Basidentitermes* sp.) were eliminated due to the soil distrubance during land clearing. Out of 13 species in the forest only three species were recorded on cleared land, i.e., *Ancisotermes* sp., *Acanthotermes acanthothorax* and *Amitermes* sp.

B. Shifting cultivation and termites

Shifting cultivators of the tropics have long recognized the importance of termites as an ecological factor and have learned to live with them and often use them to their advantage. Pendleton (1941, 1942) observed that farmers in Thailand use the abandoned mounds (2–3 m high and from 5 to 7 m in diameter at the base) to build the straw stack or to place the rice milling mortar and walking-beam pestle. Mounds are also planted to important crops that require high-fertility soils.

During cultivation, shifting cultivators rarely destory these mounds (Fig. 9.29). Pendleton (1941) observed that spreading of mound soil in the field is not advisable because it creates uneven crop growth. The importance of termites in tropical agriculture has also been indicated by Mielke (1978). In southeastern

Figure 9.29 Termite mounds are left undisturbed by the shifting cultivators. Yam mounds are constructed around the termite mound

Tanzania and northern Malawi, termite mounds are traditionally used by farmers to enhance soil fertility, provide food stores during critical periods, and yield constuction material. As a source of human diet, 100 g of some termites contains 561 calories.

X. TERMITES AND CROP GROWTH

Termites affect crop growth both directly and indirectly. Termites are known to damage many crops. Exotic and introduced crops are particularly vulnerable to termite attacks. Table 9.38 lists trees of economic importance attacked by different termite species. Wood *et al.* (1977) observed a positive linear relationship between the abundance (x) of *Mictotermes* in northern Nigeria and the incidence of maize lodging and loss in yield.

$$Y_1 = 0.86 + 0.004211 x \qquad r = 0.59$$
$$Y_2 = 0.68 + 0.001179 x \qquad r = 0.86$$

where Y_1 is % lodging Y_2 is % loss of grain yield and x is the abundance of *Microtermes* (thousands/m^2). In pastures, these authors observed that *T. geminatus* consumed 3.2% of the annual grass production.

Indirectly termites affect crop production through soil turnover, decomposition of mulch material, and alterations of soil physical, chemical and biological properties. The role of termites in altering the fertility of tropical soils has long been recognized (Adamson, 1943; Harris, 1949; Hesse, 1955). In addition to their destructive powers with regard to their attack on timbers and other crops of economic importance (Pullan, 1974), soil nesting members have a particularly useful role in influencing soil productivity.

In most regions soils of the mound are found to be more fertile and more productive than adjacent soils (Pendleton, 1942; Wild, 1975; Mielke, 1978). Mound soils are traditionally used as fertilizer and a source of lime (Rounce,

Table 9.38 Tree species attacked by termites (Wilkinson, 1965)

Termite	Tree species
(a) Limited range	
Neotermes tectonae	*Tecona grandis* (Java)
Protermers adamsoni	*Eucalyptus* sp. (Australia)
Neotermes aburiensis	Cocoa (W. Africa)
Kolatermes durbanensis	Cashew (E. Africa)
Neotermes cytops	*Olea welwitschii*
(b) Free range	
Coptotermes niger	*Swietenia macrophylla* (B. Honduras)
Coptotermes niger	*Pinus caribea* (B. Honduras)
Coptotermes niger	*Araucaria cunninghami* (Kenya)
Ancistrotermes amphidon	*Callitris grandis* (W. Africa)
Pseudocanthotermes militaris	Large trees (E. Africa)

Table 9.39 Effects of *Macrotermes* mound on grass production and soil properties (Arshad, 1982). *Reproduced by permission of Elsevier Science Publishers*

Distance from mound (m)	Production (g m^{-2} yr^{-1})			Distance from mud (m)	Bulk density (g/cm^3)	Infiltration rate (cm h^{-1})	Available water (%)	Organic matter (%)	CEC mEq/100 g
	1978	1979	1980						
Upslope									
6–8	750c	486c	517bc	0 (top)	1.76	1.3	6.1	0.91	26.2
3–6	840bc	558bc	492bcd	Base	1.25	2.8	5.9	—	—
1–3	1020a	600b	653ab	1	1.28	12.0	6.4	1.32	24.0
0 (mound)	—	—	—	2	1.31	16.3	7.2	1.40	23.1
Downslope				3	1.30	22.3	7.9	—	—
1–3	983a	810a	728a	10	1.33	15.2	7.8	1.95	20.9
3–6	1031a	825a	722a	15	1.37	8.7	6.0	1.60	20.0
6–10	885b	490c	468cde	20	1.37	8.8	5.8	1.59	17.4
10–14	545d	320d	307ef	25	1.39	8.0	6.0		
14–18	375e	270d	332def						
18–22	395e	310d	325def						
22–26	335e	330d	265f						
26–28	350e	300d	280f						
SE (±)	21.4	19.6	39.2						

Figures followed by same letters are not significantly different.

Figure 9.30 The growth of grain crops around termite mounds is superior than on adjacent soils because of natural water runoff from crusted soil alleviates drought stress

1949). In Tanzania, some termite-mound soils contain as much as 3.5% $CaCO_3$. When the mound soil from an acre ground is spread evenly, it adds about 4.5 t/acre of lime (Rounce, 1949). The growth of grain crops is often superior on these mounds (Fig 9.30). Farmers in East Africa use these mounds as a source of fertilizer and for improving the texture of sandy soils. The data of Arshad (1982) from Kenya (Table 9.39) show significantly higher grass production in the vicinity of *Macrotermes* mounds than further away from it. High grass production was due to more favourable soil physical and chemical properties.

Not all termite soils, however, provide a favourable environment for crop growth. Erhart (1951) reported poor or no crop growth on outcrops of the fossil-termite crusts. Goodland (1956) noted that termites impoverish the surrounding soil by concentrating soil nutrients in their termitaries. By decreasing plant cover, termites decrease the water-holding capacity. In Western

Table 9.40 Maize growth on mound and adjacent soil in western Nigeria (Kang, 1978)

	Mound soil	Egbeda soil
Plant height (cm)	295 ± 26.0	314 ± 11.3
Stover yield (g/plant)	358 ± 89.5	418 ± 88.5
Grain yield (g/plant)	132 ± 70.1	215 ± 40.8

Figure 9.31 In West Africa poor physical properties of mound soils restrict crop growth

Nigeria and northern Australia, the physical properties of mound soil restrict seedling emergence and subsequent crop growth (Fig. 9.31). The hardened caps resist water penetration. Williams (1979) observed from Gambia, for example, that water infiltration into the soil of an old mound surface was very slow compared to the adjacent soil. His data (Fig. 9.32) indicate that out of 603 mm rain received only 20 mm penetrated through the hardened mound surface in comparison with 90 mm gained in the natural soil. Restricted water acceptance and plant availability, therefore, are major problems in crop growth on abandoned termite mounds. Seed lodgement and seedling establishment are apparently for long periods. In spite of favourable nutrient status, poor crop growth on mound soils is often due to low water availability and restricted root growth and development (Table 9.40).

In southern Zaire and the adjoining regions of Zambia, rather large (4–6 m high) mounds are found in clayey soils (Fig. 9.7). These mounds have a low level of soil fertility and are an obstacle towards performing mechanized farm operations. Meyer (1960) observed that levelling these mounds for crop

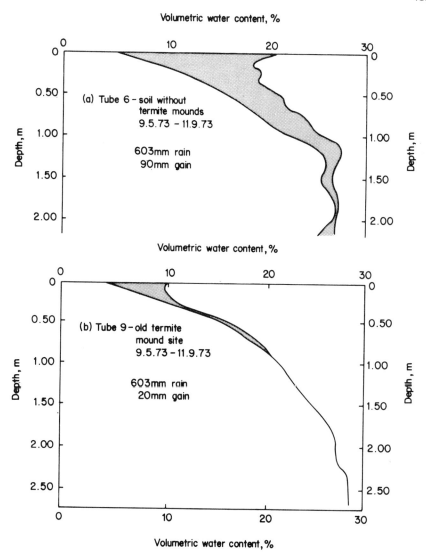

Figure 9.32 Wet and dry profiles of soil water content near Bureng, showing the change in content in (a) 'normal' soil and (b) an old termite mound site (Williams, 1979). *Reproduced by permission of the Land Resources Development Centre*

cultivation (maize-rice-ground rotation) requires the removal of the sterile or infertile soil after cutting-off the top of the mound. Alternatively, the sterile soil can be covered with the topsoil from the adjoining areas. The author observed that simple levelling without removal of the sterile soil would necessitate massive doses of chemical fertilizer i.e., 150 kg/ha N, 130 kg/ha P, 60 kg/ha K and 24 kg/ha of Mg.

Table 9.41 Effects of termite activity on soil formation, on soil properties and on suitability of soils for plant production (Wielemaker, 1984). *Reproduced by permission of the author*

Effects on soil formation	Effects on soil properties	Effects on plant production
Formation of pores (diameter >1 mm), lining and backfilling of channels and chambers with salivated pellets of earth	Increase in permeability for water and air and in structural stability. Decrease in bulk density	A. Beneficial effects: — Increase in rootability and potential for moisture uptake by plants — Increase in oxygen availability
Mixing of soil materials	Formation of homogeneous soil horizons without gravel and with diffuse transitions	— Decrease in risk of water stagnation, acid formation and shortage of oxygen
Transformation of saprolite into soil materials	Increase in soil depth	— Decrease in risk of erosion
Rapid decomposition of dead organic material including wood. Concentration of bases in organic matter. Formation of humus, rich in bases	If supply of litter is *small* termites may cause depletion of soil organic matter so that structural stability declines and pore systems become less heterogeneous (see text)	— More available nutrients through: decomposition of material rich in lignin and cellulose; upward transport of less weathered soil materials; creation of appropriate physical conditions for soil flora and fauna B. Adverse effects: — If litter is in short supply, termites may cause a rapid decline in functions of soil flora and fauna (see text)

XI. CONCLUSIONS

Termites constitute an important group of soil animals and play a major role in tropical ecosystems. Effects of termites on soils and crop production are best summarized by Wielemaker (1984) (Table 9.41). They affect soil properties through constuction of nest and gallery systems and by soil turnover. The physical and chemical properties of mound soil are different than the adjacent soil. The magnitude of the alterations in soil, however, depends on termite species and on surface and subsoil properties. Termites contribute toward soil development and are an important biotic factor in soil formation.

Termites have often been viewed as harmful insects because of the damage caused to crops, although termite do not attack native crops and tree species. Termites have both beneficial and harmful effects on agriculture. Among the beneficial effects are: (i) soil turnover; (ii) addition of plant nutrients and nutrient recycling; (iii) alterations in solid physical properties, i.e. texture; (iv) improvement in soil porosity; and (v) decomposition of plant litter. There are, of course, some harmful effects as well. They are: (i) denudation by depletion of vegetation cover; (ii) nutrient immobilization in the mound and stored organic matter; (iii) accelerating soil erosion; and (iv) damage to crops and property.

Contrary to the general conception, termites should not be viewed as a pest and as a barrier towards development of modern agriculture. The development of agriculture in the tropics should consider integrating soil fauna by adapting compatible crops and agricultural practices in recognition of their significant role in soil and ecology.

Soil scientists and agronomists involved in tropical agriculture have just begun to understand the important and delicate equilibrium that exists between soil, vegetation and termites. Many important aspects of this interaction are only vaguely understood. In view of the high energy and material demands on tropical soils, soil scientists should rather explore the potential of biotic factors in improving soil physical and nutritional properties rather than bypassing this ecological link in favour of expensive machines and agrochemicals that are neither economically feasible nor socially and ecologically compatible or acceptable.

REFERENCES

Abe, T., 1980 Studies on the distribution and ecological role of termites in a lowland rain forest of West Malaysia (4). The role of termites in the process of wood decomposition in the Pasoli Forest Reserve, *Revue Ecol. Biol. Sol*, **17**, 23–40.

Abe, T., and Matsumoto, T., 1979, Studies on the distribution and ecological role of termites in a lowland rain forest of West Malaysia (3). Distrubution and abundance of termites in Pasoli. Forest Reserve. *Japan. J. Ecol.* **29**, 337–351.

Adamson, A.M., 1943, Termites and the fertility of soils, *Trop. Agric. (Trinidad)*, **20** (6), 107–112.

Arshad, M.A., 1981, Physical and chemical properties of termite mounds of two species of Mactotermes (*Isoptera termitidae*) and the surrounding soils of the semi-arid savanna of Kenya, *Soil Sci.*, **132**, 161–174.

Arshad, M.A., 1982, Influence of the termite*Macrotermes michaelseni* on soil fertility and vegetation in a semi arid savannah ecosystem. *Agro-ecosystems*, **8**, 47-58.

Arshad, M.A., Mureria, N.K., and Keya, S.O., 1982, Effect of termite activities on the soil microflora, *Pedobiologia*, **24**, 161–167.

Bachelier, G., 1977, Review of the action of termites in soil, *Science du sol*, **1**, 3–12.

Badawi, A., Faragalla, A.A., and Dabbour, A., 1982, The role of termites in changing certain chemical characteristics of the soil, *Sociobiology*, **7**, 135–144.

Bernard, J., 1964, Termites and agriculture, *Bull. Ecole Nat. Super. Agric.* (*Tunis*), **3**, 83–95.

Bodot, P., 1964, Etudes ecologiques et biologiques des termites dans les savanes de Basse Cote d' Ivoire, in *Etudes sur les Termites Africaines* A. Bouillon, (ed.), Leopoldville Univ., Leopoldville, pp. 251–262.

Bodot, P., 1967, Ecological study of termites in savannas of the lower Ivory Coast, *Insects Soc.*, **14**, 229–258.

Bouillon, A., 1969. Studies on Ethiopian termite populations, *Revue Ecol. Biol, Sol.*, **6**, 469–482.

Bouillon, A., and Kidieri, S., 1964, Répartition des termitières de *Bellicositermes belicosus* rex Grassé et Noirot dans l' Ubangi, d' aprés les photos aériennes. Corrélations écologiques quélle révèle, In *Etudes sur les Termites Africains* A. Bouillon (ed.), Leopoldville Univ., Leopoldville, pp 373–376.

Bouillon, A., and Mahlhot, G., 1964, Observations sur l' écologie et le nid de *Cubitermes exiguus* Mathot. Description de nymphes-soldats et d' un pseudimago . In *Etudes sur les Termites Africains* A. Bouillon, (ed.), Leopoldville Univ., Leopoldville,
pp 215–230.

Boyer, P., 1956a, Action des termites constructeurs sur certains sols à Afrique Tropicale, *Proc. 6th Int. Congr. Soil Sci.*, *Paris*, **3**, 95–103.

Boyer, P., 1956b, Relations entre la flore intestinale de *Bellicositermes natalensis* et celle du sol. *Proc. 6th Int. Congr. Soil Sci.*, *Paris*, **3**, 111–113.

Boyer, P., 1958a, The material in the giant termite mounds of *Bellicositermes rex*. *C.R. Acad. Sci.* (*Paris*), **247**, 488–490.

Boyer, P., 1958b, Effect of reworking by termites and of erosion on the pedogenetic development of the mounds of *Bellicositermes rex.*, C.R.Acad. Sci. (*Paris*), **247**, 749–751.

Boyer, P., 1959, De l' influence des termites de la zone intertropicale sur la configuration de certains sols, *Rev. Geomorph. dyn.*, **10**, 41–44.

Boyer, P., 1966, Action de certains termites constructeurs sur l' evolution des sols tropicaux, *Ph.D. Thesis*, Univ. Paris.

Boyer, P. 11969, The effects of implantation of termitaries of *Bellicositermes* on the structure of savannah soils in the Central African Republic, *Bull. Mus. Nat. Hist. Natur.* (*Paris*), **41**, 789–800.

Boyer, P., 1973, Effect of some mound-building termites on the evolution of tropical soils, *Ann. Sci. Nat. Zool. Biol., Animale*, **15**, 329–498.

Breznak, J.A., Brill, W.J., Mertins, J.W., and Coppel, H.C., 1973, Nitrogen fixation in termites, *Nature*, **244** (5418), 577–580.

Brian, M.V. (ed.), 1978, *Production Ecology of Ants and Termites. The International Biological Program N. 13*. Cambridge Univ. Press, Cambridge, 409 pp.

Buxton, R.D., 1981. Changes in the composition and activities of termite communities in relation to changing rainfall, *Oecologia*, **51**, 371–378.

Charter, C.F., 1949, The characteristics of the principal cocoa soils, *Proc. Cocoa Conf.*, London, pp. 105–112.

Cheema, P.S., Das, S.R., Dayal, H.M., Koshi, T., Maheshwar, K.L., Nigam, S.S. and

Rangantham, S.K., 1960, Temperature and humidity in the fungus garden of the mound-building termite *O. obesus*, in *Termites in the Humid Tropics, New Delhi*, UNESCO, Paris, pp. 145–149.

Chevalier, A., 1949, Points de vue nouveaux sur les soils d' Afrique tropicale, sur leur dégradation et leur conservation, *Bull. Agric. Congo Belge*, **40**, 1057–1092.

Collins, N.M., 1977a, Two new termites (*Isoptera*) from the United Republic of Cameroon, *Systematic Entomology*, **21**, 95–104.

Collins, N.M., 1977b, Vegetation and litter production in Southern Guinea savanna, Nigeria, *Oecologia*, **28**, 163–175.

Collins, N.M., 1981a, The role of termites in the decomposition of wood and leaf litter in the Southern Guinea savanna in Nigeria, *Oecologia*, **51**, 389–399.

Collins, N.M., 1981b, Populations, age structure and survivorship of colonies of *Macrotermes bellicosus*, *J. Animal Ecology*, **50**, 293–311.

Critchley, B.R., Cook, A.G., Critchley, U., Perfect, J.S., Russell-Smith, A., and Yeadon, R., 1979, Effects of bush clearing and soil cultivation on the intervertebrate fauna of a forest soil in the humid tropics, *Pedobiologia*, **19**, 425–

Darwin, C.F., 1881, *The Formation of Vegetal Mould Through the Action of Worms with Observations on their Habits*, J. Murray, London.

Davies, O., 1959, Termites and soil stratification in equatorial Africa, *Antiquity*, **33**, 290–291.

De Ploey, J., 1964, Nappes de gravats et couvertures argilo-sableuses au Bas-Congo; leur génèse et l' action des termites, in *Etudes sur les Termites Africains* A. Bouillon, (ed.) Leopoldville Univ., Leopoldville, pp. 399–414.

De Heinzelin, J., 1955, Observations on the formation of stone lines in tropical soils. Publ. INEAC sén Sci. No 64, Bruxelles pp 37.

Derbyshire, E., 1968, Termites and soil development near Brocks Creek, Northern territoy, *Aust. J. Sci.*, **31**, 153–154.

Drummond, H., 1887, On the termite as the tropical analogue of the earthworm, *Proc. R. Soc. (Edinb.)*, **13**, 137–146.

Drummond, H., 1895, *Tropical Africa*, Hodder and Stoughton, London, 6th Edn.

Erhart, H., 1951a, The importance of biological phenomena in the formation of ferruginous crusts in the tropics, *C.R. Acad. Sci, (Paris)*, **233**, 804–806.

Erhart, H., 1951b, The role of termite crusts in the geography of tropical regions, *C.R.Acad. Sci. (Paris)*, **233**, 966–968.

Ettershank, G., 1968, The three-dimensional gallery structure of the nest of the meat ant *Irdomyrmex prupureus*, *Aust. J. Zool.*, **161**, 715–723.

Farming in South Africa, 1960, Destruction of grazing by harvester termites, *Fmg. S. Afr.*, **35**, 6–9.

Ferrar, P., 1982a, Termites of a South African savanna, I. List of species and subhabitat preferences, *Oecologia*, **52**, 125–132.

Ferrar, P., 1982b, Termites of a South African Savanna II. Densities and populations of smaller mounds, and seasonality of breeding, *Oecologia*, **52**, 133–138.

Folster, H., 1964, The pediseliments of the southern Sudanese pediplane. Origin and soil formation, *Pedologie*, **14**, 64–84.

Gay, F.J., and Calaby, J.H., 1970, termites from the Australian region, in K. Krishna and F.M. Weesner (eds) *Biology of Termites* vol 2, Academic Press, New York, pp. 393–448.

Ghilarov, M.S., 1962, Termites of the USSR, their distribution and importance, in *Termites in the Humid Tropics, New Delhi*, UNESCO, Paris, pp 131–135.

Glover, P.E., Trump, E.C., and Wateridge, L.E.D., 1964, Termitaria and vegetation patterns on the Loita plains of Kenya, *J. Ecol.*, **52**, 367–377.

Goffinet, G. 1976, The soil ecology of the natural ecosystem of the Haut-Shaba (Zaire). III. Populations of surface-living termites on latosols, *Revue Ecol. Biol. Sol*, **13**, 459–475.

Gokhale, N.G., Sarma, S.N., and Bhattacharya, N.G., 1958, Effect of termite activity on the chemical properites of tea soils, *Sci. Cult.*, **24**, 229.

Goodland, R.J.A., 1965, Termitaria in savanna ecosystem (Effects on soil chemistry), *Can. J. Zool.*, **53**, 641–650.

Grassé, P.P., 1950, Termites et sols tropicaux, *Revue Int. Bot. Appl. Agric. Trop.*, **30**, 549–554.

Grassé, P.P., and Noirot, C., 1958, Le comportement des termites a l' égard de l' air libre. L' atmosphere des termitieres et son renouvellement, *Ann. Sci. Nat. Zool. Biol. Animale*, **20**, 1–28.

Grassé, P.P., and Noirot, C., 1959, Rapports des termites avec les sols tropicaux, *Rev. Geomorph. dyn.*, **10**, 35–40.

Greaves, T., 1962, Studies on the foraging galleries and the invasion of living trees, by *C. acinaciformis* and *C. brunneus*, *Aust. J. Zool.*, **10**, 630–651.

Griffith, G., 1938, A note of termite hills, *E. Afr. Agric. J.*, **4**, 70–71.

Griffith, G., 1953, Vesicular laterite, *Nature*, **171**, 530.

Gupta, S.R., Rajvanshi, R., and Singh, J.S., 1981, The role of the termite *Odontotermes gurdespurensis* in plant decomposition in a tropical grassland, *Pedobiologia*, **22**, 254–262.

Harris, W.V., 1949, Some aspects of the termite problem, *E.Afric. Agric. J.*, 151–155.

Harris, W.V., 1954, Termites in tropical agriculture, *Trop. Agric.*, **31**, 11–18.

Harris, W.V., 1965, Termites and the soil, *Soil Zool. Proc. Nottingham Sch. Agric. Sci.*, pp. 62–72.

Harris, W.V., 1966, Termites and trees. A review of recent literature, *For. Abstr.*, **27**, 173–178.

Heath, G.W., 1965, The part played by animals in soil formation, *Expl. Pedology, Proc. 11th Easter School Agric. Sci.*, Univ. Nottingham, pp 236–243.

Hermann, H.R., 1979, *Social Insects, I*, Academic Press, New York, 437 pp.

Hermann, H.R., 1981, *Social Insects. II.*, Academic Press, New York, 491 pp.

Hesse, P.R., 1953, Study of the soils of termite mounds, *E. Afr. Agric. For. Res. Org. Rep. 1952*, pp. 73–75.

Hesse, P.H., 1955, A chemical and physical study of the soils of termite mounds in East Africa, *J. Ecol.*, **43**, 449–461.

Holdaway, F.G., and Gay, F.J., 1948, Temperature studies of the habitat of *Eutermes exitiosus* with special reference to the temperatures within the mound, *Aust. J. Sci. Res.*, **B1**, 464–493.

Holt, J.A., Coventry, R.J., and Sinclair, D.F., 1980, Some aspects of the biology and pedological significance of mound-building termites in red and yellow earth landscape near Charters Towers, north Queensland, *Aust. J. Soil Res.*, **18**, 97–109.

Howse, P.E., 1970, *Termites*. Hutchinson, London, 150 pp.

Joachim, A.W.R., and Kandiah, S., 1940, Studies on Ceylon soils XIV: A comparison of soils from termite mounds and adjacent land, *Trop. Agric. Mag. Ceylon Agric. Soc.*, **95**, 333–338.

Joachim, A.W.R., and Pandittesekera, D.G., 1948, Soil fertility studies IV, Investigations on crumb structure and stability of local soils, *Trop. Agric. Mag. Ceylon Agric. Soc.*, **104**, 119–129.

Josens, G., 1971, Variations thermiques dans les nids de *Trinervitermes geminatus* en relation avec le milieu exterieur dans la savanne da Lamto (Cote d'Ivoire), *Insects Soc.*, **18**, 1–14.

Josens, G., 1983, The soil fauna of tropical savannas. III. The termites, in *Tropical savannas* F. Bourliere (ed.), Elsevier, Amsterdam, pp. 505–524.

Kang, B.T., 1978, Effect of some biological factors on soil variability. III. Effect of Macrotermes mounds, *Plant Soil*, **50**, 241–251.

Kayani, S.A., and Sheikh, K.H., 1981, Inter-relationships of vegetation, soils and termites in Pakistan. I. Arid marine tropical coastlands, *Pakistan Journal of Botany*, **13**(2), 165–188.

Kemp, P.B., 1955, The termites of north-eastern Tanganyika: their disturibution and biology, *Bull. Ent. Res.*, **46**, 113–135.

Kozlova, A.V., 1951, Accumulation of nitrates in termite mounds in Tukmenia, *Pochvovedenie*, **10**, 626–631.

Krishna, K. and Weesner, F.M. (eds.), 1969, *Biology of termites*, Vol. I., Academic Press, New York.

Krishna, K. and Weesner, F.M. (eds), 1970, *Biology of Termites* Vol. 2, Academic Press, New York.

Laker, M.C., Hewitt, P.H., Nel, A. and Hunt, R.P., 1982, Effects of the termite *Trenervitermes trinervoides* sjuostedt on the pH, electrical conductivities, CEC and extractable base contents of soils, *Fort Hare Papers*, **7**, 275–286.

Lee, K.E., Wood, T.G., 1968, Preliminary studies of the role of *Nasutitiermes exitiosus* in the cycling of organic matter in a yellow podzolic soil under dry sclerophyll forest in South Australia, *Trans. 9th Int. Cong. Soil Sci.*, **2**, 11–18.

Lee, K.E. and Wood, T.G. 1971, Physical and chemical effects on soils of some Australian termites, and their pedological significance, *Pedobiologia*, **II**, 376–409.

Lepage, M.G., 1974, Les termites d'une savane sahélienne (Ferlo Septentrional, Sénégal): peuplement, populations consommation, role dans l' écosystem, *Thesis*, University of Dijon, Dijon, 344 pp.

Leprun, L.C., 1976, An original underground structure for the storage of water by termites in the Sahelian region of the Sudan zone of Upper Volta, *Pedobiologia*, **16**, 451–456.

Leprun, J.C., and Roy-Nöel, J., 1976, Clay mineralogy and the distribution of the above-ground nests of two species of the genus *Macrotermes* in West Senegal (near Cape Verde peninsula), *Insectes Sociaux*, **23**, 535–547.

Léveque, A., 1969, The problem of stone lines. Preliminary observations on the gneiss-granite formation in Togo, *Cah. Pedologie. ORSTOM* 7, 43–69.

Loos, R., 1964, A sensitive anemometer and its use for the measurement of air currents in the nest of *Macrotermes natalensis*, *Colloque Int. Termites Afr.*, UNESCO, Leopoldville, 1964, pp 363–372.

Lüscher, M., 1961, Air-conditioned termite nests, *Scient. Am.*, **205**, 138–145.

Maldague, M., 1959, Analysis of soils and materials from termite mounds of Belgian Congo, *Insectes Sociaux*, **6**, 343–359.

Maldague, M.E., 1964, Importance des populations de termite dans les sols équatoriaux, *Trans. 8th Int. Congr. Soil Sci.*, Bucharest, 1964 (3), 743–751.

Miedema, R., and Van Vuure, W., 1977, Termite mounds in Sierra Leone. *J. Soil Sci.*, **28**, 112–124.

Mielke, H.W., 1978, Termitaria and shifting cultivation: The dynamic role of the termite in soils of tropical wet-dry Africa, *Trop. Ecol.*, **19**, 117–122.

Meiklejohn, J., 1965, Microbiological studies on large termite mounds, *Rhod. Zambia Malawi J. Agric. Res.*, **3**, 67–79.

Meyer, J.A., 1960, Resultats agronomiques d' un essaie de nivellement des termitieres réalise dans la Cuvette centrale Congolaise, *Bull. Agric. Congo Belge*, **51**, 1047–1057.

Mire, P.B. De., 1975, The genesis of 'harde' soils in North Cameroon, *Agron. Tropicale*, **30**, 271–275.

Murray, J.M., 1938, An investigation of the interrelationships of the vegetation, soil and termites, *S. Afr. J. Sci.*, **35**, 288–297.

Nel, J.J., and Hewitt, P.H., 1970, The air supply available to the subterranean termite, *Hodotermes mossambics* (Hagen) in the Central Orange Free State, *Phytophylactica*, **2**, 227–299.

Noirot, C.H., 1970, The nests of termites, in K. Krishna and F.M. Weesner (eds.), *Biology of Termites*, Vol. II, Academic Press, New York, pp. 73–125.

Nye, P.H., 1955, Some soil forming processes in the humid tropics, IV. The action of the soil fauna, *J. Soil Sci.*, **6**, 73–83.

Ohiagu, C.E., 1979, Nest and soil populations of *Trinervitermes* spp. with particular reference to *T. geminatus* in Southern Guinea savanna near Mokwa, Nigeria, *Oecologia*, **40**, 167–178.

Okwakol, M.J.N., 1980, Estimation of soil and organic matter consumption by termites of the genus Cubitermes, *Afr. J. Ecol.* **18**(1), 127–131.

Omo Malaka, S.M. 1977a, A study of the chemistry and hydraulic conductivity of mound materials and soils from different habitats of some Nigerian termites, *Aust. J. Soil Res.*, **15**, 87–91.

Omo Malaka, S.l., 1977b, A note on the bulk density of termite mounds, *Aust. J. Soil Res.*, **15**, 92–94.

Parihar, D.R., 1980, Termite problem in desert plantations, *Ann. Arid Zone*, **19**(3), 329–334.

Pathak, A.N., and Lehri, L.K., 1959, studies on termite nests I. Chemical, physical, and biological characteristics of a termitarium in relation to its surroundings, *J. Indian Soc. Soil Sci.*, **7**, 87–90.

Pendleton, R.L., 1941, Some results of termite activity in Thailand soils, *Thai Sci. Bull.*, **3**, 29–53.

Pendleton, R.L., 1942, Importance of termites in modifying certain Thailand soils, *J. Am. Soc. Agron.*, **34**, 340–344.

Pomeroy, D.E., 1976a, Some effects of mound-building termites on soils in Uganda, *J. Soil Sci.*, **27**, 377–394.

Pomeroy, D.E., 1976b, Studies on population of large termite mounds in Uganda, *Ecol. Entom.*, **1**, 49–61.

Pomeroy, D.E., 1977, The distribution and abundance of large termite mounds in Uganda, *J. Appl. Ecol.*, **14**, 465–475.

Pomeroy, D.E., 1978, The abundance of large termite mounds in Uganda in relation to their environment, *J. Appl. Ecol.*, **15**, 51–63.

Portères, R., 1952, Linear cultural sequences in primitive systems of agriculture in Africa and their significance, *African soils*, **2**, 15–29.

Powers, W.L., and Bollen, W.B., 1935, The chemical and biological nature of certain forest soils, *Soil Sci.*, **40**, 321–329.

Prestiwich, G.D., Bentley, B.L., and Carpenter, E.J., 1980, Nitrogen sources for neotropical nasute termites: fixation and selective foraging, *Oecologia*, **46**, 397–401.

Primavesi, A.M., and Covolo, G., 1978, Comparison between the activity of termites and of earthworms in relation to nutrients and soil structure, in *Progress in Soil Biodynamics and Soil Productivity*, A. Primavesi (ed), Palloti, Santa Maria, Brazil, pp. 149–154.

Pullan, R.A., 1974, Biogeographical studies and agricultural development in Zambia, *Geography*, **59**, 309–320.

Pullan, R.A., 1979, Termite hills in Africa: their characteristics and evolution, *Catena*, **6**(3/4), 267–291.

Rajagopal, D., Sathyanarayana, T., and Veeresh, G.K., 1982, Physical and chemical properties of termite mound and surrounding soils of Karanataka, *J. Soil Biol. Ecol.*, **2**,(1), 18–31.

Raunet, M., 1979, The importance and interactions of geochemical, hydrological and biological (termites) processes on tropical granito-gneisic peneplains in Western Kenya, *Agron. Trop.*, **34**(1), 40–53.

Robinson, J.B.D., 1958, Some chemical characteristics of 'termite soils' in Kenya coffee fields, *J. Soil Sci.*, **9**, 58–65.

Roonwal, M.L., 1973, Mound-structure, fungus combs and primary reproductives (king and queen) in the termite *Odontotermes brunnaeus* in India, *Proc. Ind. Nat. Sci. Acad., Part B*, **39**, 63–76.

Roonwal, M.L., 1975, Field and other observations on the harvester termite, *Acacanthotermes macrocephalus* (Deoneux) (Hodotermitidae) from the Indian desert, *Zeitsch. Angewandte Entomologie*, **78**, 424–440.

Roonwal, M.L., 1978, Bioecological and economical observation on termites of Peninsular India, *Zeitsch. Angewandte Entomologie*, **85**, 17–30.

Roose, E.J., 1976, Contribution à l' étude de l'influence de la mésofaune sur la pédogénèse actuelle en milieu tropical, *Rapport ORSTOM*, Centre d' Adiopoudoumé, Ivory Coast, 56 pp.

Rounce, N.V., 1949, Manuring in its relation to land rehabilitation, in N.V. Rounce (ed.), *the Agriculture of the Cultivation Steppe*, Longmans, Green & Co., London, pp. 25–35.

Ruelle, J.E., 1962, Etude de quelques variables du microclimat du nid de *Macrotermes natalensis* en rapport avec le declenchement de l' essaimage, Ph.D. Thesis Lovanium Univ., Kinshasha.

Ruelle, J.E., 1964, L' architecture du nid de *Macrotermes natalensis* et son sens fonctionnel, in A. Bonillon, ed, *Etudes sur les Termites Africans*, Masson, Paris, pp. 327–362.

Ruelle, J.E., 1969, Distribution of the principal species of *Macrotermes* in the Ethiopian fauna, *Proc. VIth Cong. Int. Stud. Soc. Insects*, pp. 249–253, Univ. Lovanium, Kinshasa.

Sands, W.A., 1965a, Termite distribution in man-modified habitats in West Africa, with special reference to species segregation in the genus *Trinervitermes*, *J. Anim. Ecol.*, **34**, 557–571.

Sands, W.A., 1965b, Mound population movements and fluctuations in *Trinervitermes ebenerianus*, *Insects Soc.*, **12**, 49–58.

Seiler, W., Conrad, R., and Scharffe, D., 1984, Field studies of methane emission from termite nests into the atmosphere and measurements of methane uptake by tropical soils. *J. Atmospheric Chemistry*, **1**, 171–186.

Singh, U.R., and Singh, J.S., 1981a, Population structure and mound architecture of the termites of a tropical deciduous forest of Varanasi, India *Pedobiologia*, **22**, 213–223.

Singh, U.R., and Singh, J.S., 1981b, Temperature and humidity relations of termites, *Pedobiologia*, **21**, 211–216.

Skaife, S.H., 1955, *Dwellers in Darkness*, Longmans, Green and Co., New York.

Snymam, A., 1970, Havester termites: new control methods, *Fmg. Soc. Afr.*, **4**, 46.

Soyer, J., 1983, Microrelief of low mounds on seasonally flooded soils in southern Shaba (Zaire), *Catena*, **10**, 253–265.

Spears, B.M., Ueckert, D.N., and Whigham, T.L., 1975, Desert termite control in a short grass prairie: effect on soil physical properties, *Envir. Ent.*, **4**, 899–904.

Stoops, G., 1964, Application of some pedological methods to the analysis of termite mounds, in *Etudes sur les Termites Africains* A. Bouillon, (ed.), pp. 379–398, Leopoldville Univ., Leopoldville.

Stoops, G., 1964, Application of some pedological methods to the analysis of termite mounds, *Coll. Int. Termites Afr.*, UNESCO, 1964, pp 379–398.

Stoops, G., 1968, Micromorphology of some characteristic soils of the Lower Congo (Kinshasa), *Pedologie*, **18**, 110–149.

Strickland, A.H., 1944, The arthropod fauna of some tropical soils, *Trop. Agric.*, **21**, 107–114.

Sys, C., 1955, The importance of termites in the formation of latosols, *Sols. Afri.*, **3**, 392–395.

Taltasse, P., 1957, The 'Cabecas de yacare' (alligators' heads) and the role of termites, *Rev. Geomorph. dynam.*, **8**(11/12), 166–170.

Tessier, F., 1959, The laterite of the Manual Cape at Dakar and its fossil termitaries, *C.R. Acad. Sci. (Paris)*, **248**, 3320–3322.

Tricart, J., 1957, Observation sur le role ameubliss eur des termites, *Rev. Geomorph dynam.*, **8**, 170–172.

Ueckert, D.N., Bodine, M.C. and Spears, B.M., 1976, Population density and biomass of the desert termite *Gnathamitermes tubiformans* in a short grass prairie: relationship to temperature and moisture, *Ecology*, **57**, 1273–1280.

Valiachmedov, B.V., 1981, Termites (*Anacanthotermes ahngerianus*) and their influence on takyr formation in south-western Tadjikistan (Central Asia), *Pedobiologia*, **21**, 242–256.

Watson, J.A.L., and Gay, F.J., 1970, Role of grass-eating termites in the degradation of a mulga ecosystem, *Search*, **1**, No. 1, 43.

Watson, J.P., 1962, The soil below a termite mound, *J. Soil Sci.*, **13**, 46–51.

Watson, J.P., 1967., 1967, A termite mound in an Iron Age burial ground in Rhodesia, *J. Ecol.*, **55**, 663–669.

Watson, J.P., 1969, Water movement in two termite mounds in Rhodesia, *J. Ecol.*, **57**, 441–451.

Watson, J.P., 1972, The distribution of gold in termite mounds and soils at a gold anomaly in Kalahari sand, *Soil Sci.*, **113**, 317–321.

Watson, J.P., 1974a, Calcium carbonate in termite mounds, *Nature*, **247** (5435), 74.

Watson, J.P., 1974b, Termites in relation to soil formation, groundwater, and geochemical prospecting, *Soils Fert.*, **37**, 111–114.

Watson, J.P., 1976, The composition of mounds of the termite *Macrotermes falciger* on soil derived from granite in three rainfall zones of Rhodesia, *J. Soil Sci.*, **27**, 395–503.

Wielemaker, W.G., 1984, Soil formation by termites: a study in the Kiisi area, Kenya. Dept. of Soil Science and Geology, Agricultural University, Wageningen, T W Netherland, 132 pp.

Wild, H., 1952, The vegetation of southern Rhodesian termitaria, *Rhodesia Agric. J.*, **49**, 280–292.

Wild, H., 1975, Termites and the serpentines of the Great Dyke of Rhodesia, *Trans. Rhodesia Scient. Assoc.*, **57**, 1–11.

Wilkinson, W., 1965, The principles of termite control in forestry, *E. Afr. Agric For. J.*, **31**, 212–217.

Williams, J.B., 1979, Soil water investigation in the Gambia Land Resources Development Centre, ODA, UK, Tech. Bull 3, 183 pp.

Williams, M.A.J., 1968, Termites and soil development near Brock's Creek, Northern Territory, *Aust. J. Sci.*, **31**, 153–154.

Wilson, E.O., 1971, *The Insect Societies*, Belknap Press of Harvard University Press, Cambridge, Mass., 548 pp.

Wood, T.G., 1975, The effects of clearing and grazing on termite fauna (Isoptera) of tropical savannas and woodlands, in *Progress in Soil Zoology, Proc. 5th Int. Coll. on Soil Zoology*, Prague, 1972 W. Junk, The Hague, Netherlands.

Wood, T.G., and Lee, K.E., 1971, Abundance of mounds and competition among colonies of some Australian termite species, *Pedobiologia*, **1**, 341–366.

Wood, T.G., Johnson, R.A., Ohiagu, C.I., Collins, N.M. and Longhurst, C., 1977, Ecology and importance of termites in crops and pastures in northern Nigeria. COPR, London, *ODM Research Scheme R. 2709*, 131 pp.

Yakushev, V.M., 1968, Influence of termite activity on the development of laterite soil, *Soviet Soil Sci.*, **1**, 109–111.

Zimmerman, P.R., Greenberg, J.P., Wandiga, S.O., and Crutzen, P.J., 1982, Termites: a potentially large source of atmospheric methane, carbon dioxide and molecular hydrogen. *Science*, **218**, 563–565.

Chapter 10
Ants

I	INTRODUCTION	423
II	NESTS	427
III	SOIL TURNOVER	427
IV	SOIL PROPERTIES	429
	A. Texture	429
	B. Penetrometer resistance and bulk density	432
	C. Moisture retention characteristics	433
	D. Chemical properties	435
V	EFFECTS OF AGRICULTURAL PRACTICES	437
VI	CONCLUSIONS	437

I. INTRODUCTION

Little information exists regarding the effects of ants on tropical soils. In fact, ants are often confused with termites and the latter are misnomered as 'white ants'. These are different insects: termites belong to the group 'Isoptera' and the ants to the 'Hymenoptera' (Wallwork, 1970). Ants are also the most important predators of termites (Wheeler, 1936; Lee and Wood, 1971). In the southern Guinea savanna zone of Nigeria, Longhurst *et al.* (1979) reported that a predatory specie *Megaponera foetens* can kill as many as 142 Macrotermes m^{-2} yr^{-1}. Ants also attack earthworms (Fig. 10.1) and other soil fauna (Salt, 1950). Some scientists have even considered ant communities for biological control of tropical pests (Risch and Carroll, 1982).

Most ant taxa are primarily tropical insects and are believed to have originated in tropical rainforests (Lévieux, 1983). The species diversity decreases drastically from equator to temperate latitudes. Although many ant species construct nests in trees within tropical forest and savanna ecosystems, it is the soil-dweller ants that are of immediate interest to soil scientists. Mound building and excavation are common to some ants (Figs. 10.2 and 10.3). Most soil-dwelling ants belong to *Ponerinae*, *Myrmicinae*, and *Formicinae* sub-families.

In all there are 240 genera and 6000 species of ants (Bachelier, 1983). There are about 72 genera and 357 species in the Amazon (Kusnezov, 1963). The wet tropical savannas in northern Australia harbour 120 species (Greenslade and

Figure 10.1   Ants predate earthworms

Mott, 1978) and the Lamto savanna in Ivory Coast is a habitat for about 90 species (Lévieux, 1983). The specie diversity in Ivory Coast is representative of many similar ecologies in tropical Africa.

Ant population varies among different ecological zones and decreases from a high in the humid forest region to a relatively low diversity and low population density in the savanna. The population density of the most common species in African savanna is estimated at 350 adult nests ha^{-1}, i.e., 2.2×10^6 workers ha^{-1}. All species combined add up to a population density of about 20 million ha^{-1} (Lévieux, 1983), a value comparable to that of 15.5 million ha^{-1} observed in the rainforest regions of Panama in Central America (Williams, 1941). Lavelle et al. (1981) reported a population density of 1.5 to 5.8 million ha^{-1} in the savanna and from 7.8 to 14.0 million ha^{-1} in the forest zone of Mexico. Greenslade (1976) noted a density of 0.37 million ha^{-1} in the dry Australian savanna. Critchley et al. (1979) identified six sub-families in the forest region of western Nigeria: *Dorylinae, Ponerinae, Dolichoderinae, Fermicinae, Myrmicinae* and *Pseudomyrmecinae*. The latter are exclusively confined to the forested land.

Food habits differ widely among species and ecologies. In the humid and subhumid regions, most ant species (65 to 70%) are ominivorous and 30 to 33% are carnivorous (Deiomandé, 1981; Lévieux, 1983). As the rainfall decreases in the savanna and desert regions the proportion of 'seed harvesting' species increases: from 0 in the humid to 40% in the desert (Delye, 1968; Lévieux, 1983). The granivores (seed harvesters) are common in regions with a long dry season because seeds are stored for a long time. The same species are often observed to have different food preferences depending on the ecology, season,

Figure 10.2a,b Nest building by ants

Figure 10.3 Excavation and burrowing activities by ants

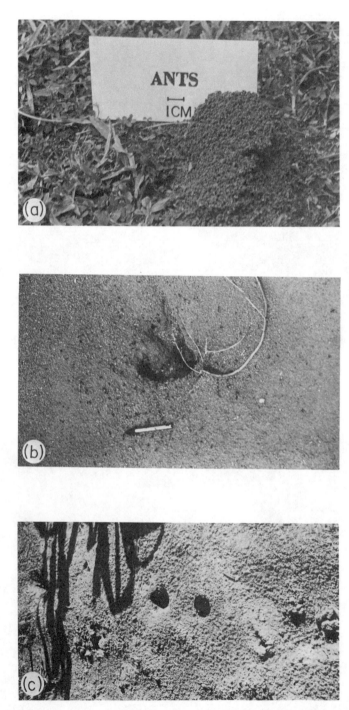

Figure 10.4a (a) A small earth crater around the nest. (b) Nest entrance is often protected by a branch on a twig. (c) Contrasting cavities made by termite activity

and the competition for the preferred food. This implies that most species are omnivorous.

II. NESTS

Some ants live in the deep subsoil layers and construct a network of interconnected tunnels and chambers. Although ants may turn over soil, they do not incoroporate organic matter into the soil they excavate as earthworms do. The ants living in the subsoil have a small earth crater around the nest entrance (Fig. 10.4). Some examples of the soil-dwelling species are *Amblyopone pluto, Apomyrma stygia, Asphinctopone silvestrii, Hypoponera gr., Megaponeaa foetens*. The soil-dwelling ant *Myrmicaria eumenoides* described in Kisii region of Kenya by Wielemaker (1984) is known to construct 15–40 cm high nests with a basal diameter of 60 to 100 cm. The surface area of ant nest varies among species and ranges from 50 to 600 m^2 (Weber, 1966; Alvarado et al., 1981). The density of ant nests in the Caribbean is estimated to range from 3 per hectare (Cherrett, 1968) to 153 per hectare (Lewis, 1975). There are differences in nest construction of ants and termites. In the tropics, an objective of the termite nest is to decrease the temperature, but it is believed that ants construct nests to increase the temperature, by intercepting the largest possible amount of solar radiation (Josens, 1983). That is one of the reasons for the abundance of ant nests in the forest and their scarcity in the savanna and open woodlands.

Ants escape the adverse microclimate by reducing activity during the hot period of the day, and by migrating to the deeper horizons during dry, hot summer (Lévieux, 1983). Some ants are nomadic and carve out small channels in the soil surface when the colony migrates from one site to another (Fig. 10.5). Some examples of the ground living species are *Acantholpis arenaria, A. canescens, Aneleus* sp., *Anochetus* sp. etc. For details on biology, feeding and nesting habits, readers are referred to more authoritative reviews by Brown (1972, 1973), Levings and Frank (1982) and Levieux (1983).

III. SOIL TURNOVER

The influence of leaf-cutting ants on properties of tropical soils have been studied by Weber (1966), Cherrett (1968), Arya *et al.* (1978), Glavis *et al.* (1979), Goosen (1979) and Alvarado *et al.* (1981). Some researchers estimate that about 1% of tropical land surface is affected by the activity of ants (Buol, 1973; Alvarado *et al*; 1981). Farmers and agricultural scientists in the tropics are aware of the presence of ants in forested and arable plants and of their possible effects on soil physical properties. Ants transfer large amounts of subsoil to the surface, and alter soil structure. In Brazil Weber (1966) estimated that *Attasexdens* turned over 40,000 kg of subsoil to the surface (Hölldobler, 1984) They are also one of the first animals to colonize recent soil deposits, i.e., flood plains, lacustrine deposits and other inceptisols (Kevan, 1968). These

Figure 10.5 Channels carved on the soil surface to facilitate colony migration

animals dwell in coarse-textured sandy soils. When present in large numbers, as in a humid forest environment, they influence soil properties.

The effect of ants on soils is considerably less than that of earthworms and termites. Nye (1955) quotes a study by Shaler (1891) near Cambridge, Mass., where ants transferred about 5 mm of soil to the surface each year during the 2-year observation period. Shaler also referred to a field in New England that was covered by a layer of fine sand a few centimetres thick supposedly deposited by ants. Raignier (1952) estimated that tree-dwelling ant species can transport about 100 kg soil ha for nest construction in the trees. In the western United States, Thorp (1949) estimated 49 nests/ha, representing about 1.5 t of the soil material. Madge and Sharma (1969) estimated in western Nigeria that ants turn over about 1 t ha^{-1} yr^{-1} of the soil through their nesting activity. In the United States, Baxter and Hole (1967) and Salem and Hole (1968) estimated that ants (*F. exsectoides*) had moved to the surface an amount of soil equivalent to 1.25% of a plough layer during a period of about 2 years. They noted that pedoturbation by ants could affect the soil of an entire profile in a period of 600 to 1200 years.

IV. SOIL PROPERTIES

A. Texture

Similarly to termites, some ants burrow deep passages into the soil and carry subsoil material to the surface. Mound building is also common with some ants

Table 10.1 Particle size analysis of material piled by ants (Nye, 1955). *Reproduced by permission of the* Journal of Soil Science

Sample	Particle size (mm)	Gravel (%)			Coarse sand (%)		Fine sand (%)	Silt (%)	Clay (%)
		7–4	4–2	2–1	1–0.5	0.5–0.2	0.2–0.02	0.02–0.002	<0.002
Large ant		0	5	22	18	32	14	4	5
Small ant		0	0	10	7	18	42	9	13
Parent soil									
0 – 2.5 cm		1	3	12	10	29	26	7	13
2.5 – 15 cm		2	7	23	19	26	13	5	7
15 – 30 cm		10	16	19	14	22	14	6	9

Table 10.2 Mechanical analysis (%) of soil from an ant nest in western Nigeria (Apomu soil series) (unpublished data, Lal, 1984)

Depth (cm)	Sand fraction (mm)														Total sand		Silt		Clay	
	0.053		0.053–0.125		0.125–0.25		0.25–0.5		0.5–1		1–2									
	A	B	A	B	A	B	A	B	A	B	A	B			A	B	A	B	A	B
0–10	2.6	1.8	5.5	7.6	10.4	7.8	16.4	16.0	7.9	4.1	37.2	37.5			80.0	74.8	8.8	12.0	11.2	13.2
10–20	3.7	V	6.7	V	11.4	V	14.9	V	6.2	V	27.1	V			80.0	V	8.8	V	11.2	V
20–30	3.1	2.4	7.1	7.6	10.8	9.4	15.3	12.7	5.9	5.5	37.8	37.2			80.0	74.8	8.8	12.0	11.2	13.2
30–40	3.9	2.9	6.5	7.3	9.9	10.7	15.6	13.6	6.6	6.1	37.5	38.2			80.0	78.8	8.8	8.0	11.2	13.2
40–50	3.0	3.4	6.6	9.8	10.7	10.5	15.7	10.8	7.0	5.2	36.0	37.1			79.0	76.8	8.8	10.0	12.2	13.2
50–60	3.7	3.8	6.8	12.5	11.0	10.6	14.8	10.8	6.5	4.5	38.2	34.5			81.0	76.8	8.8	10.0	10.2	13.2
60–70	4.0	4.4	7.1	9.1	10.7	10.8	15.3	13.4	8.2	4.4	36.7	40.7			82.0	82.8	8.8	6.0	9.2	11.2
70–80	3.4	4.3	7.4	8.1	11.0	11.8	16.3	13.1	7.1	4.4	36.8	37.1			82.0	78.8	8.8	8.0	9.2	13.2
80–90	4.0	3.5	8.1	7.1	10.9	10.7	15.4	13.4	6.0	5.3	39.6	40.0			84.0	74.8	7.8	10.0	8.2	15.2
90–100	3.6	—	9.6	—	13.3	—	13.8	—	5.0	—	40.7	—			86.0	—	5.8	—	8.2	—

A, parent soil; B, ant nest; V, void space.

(Fig. 10.2). While excavating, ants transport loose single-grained material through a central hole and deposit a shallow pile. The central hole is often covered with a leaf (Fig. 10.3). The excavated material is loosely piled in a crater-shaped heap. The crater is about 10 to 15 cm across the and consists of the soil-material of the same colour as the subsoil (Fig 10.4). The ant workers carry the soil particles in their mandibles, and the size of the grain varies with the size of the ant. Unlike termites, these ants do not cement together the excavated subsoil material nor mould it into a cohesive mound.

Ettershank (1968) described the three-dimensional gallery structure of a nest of the meat-ant and *Irdomyrmex purpureus* sp. of Australia. He observed that galleries and tunnels are lined with a cement of saliva and silt. In the friable sands of the South Ita Sandhills, the sand can be blown away leaving the fragile galleries intact.

Nye (1955) in western Nigeria and Watson (1960) in Rhodesia observed that ants generally use material derived from the top 30 cm of soil for their nest construction. Nye observed that in areas with fewer earthworms, the topmost layer of the soil is often formed through the activity of ants. The data of Nye (Table 10.1) and of Watson indicate that ants use higher proportions of coarse material in nest construction than termites. The nest soil of the large ant contained 86% sand, 4% slit and 5% clay. The particle size distribution of the material brought up by small ants, in comparison, contained 77% sand, 9% silt and 13% clay. The sand in the small-ant nest contained less coarse fraction than that of the large-ant nest. In Australia, Ettershank also observed higher proportions of coarse sand and silt in the reworked nest material of the meat-ant than in adjacent soil.

Table 10.2 gives the particle size distribution of ant nest material collected from different depths around the nest cavity and compares it with that of the uncontaminated adjacent soil material. The nest had a large cavity at 10–20 cm depth (Fig. 10.6). For all other depths, the nest material contained 2 to 5% less sand, I to 3% more silt and 1 to 2% more clay than adjacent soil. The difference in particle size distribution between the nest soil and adjacent soil were slight in comparison with that of the termite mound. Termites preferentially transport fine particles, while ants do not show such a degree of preference. Alvarado et al (1981) studied the effects of leaf-cutter ants on properties of a Costa Rican Andept. Their data in Table 10.3 show alterations in sand and clay contents of the active ant nest. These authors conlcuded that leaf-cutter ants are of major importance in soil development and genesis.

B. Penetrometer resistance and bulk density

Material excavated by ants and brought above ground is loosely piled and is eventually compacted only by the raindrop impact and other environmental factors. The bulk density of the nest, depending on the particle size distribution and the state of compaction, is generally low. The data in Table 10.4 show low

Figure 10.6 A large soil cavity created by the nest-building activity

bulk density and penetrometer resistance of the ant-worked soil up to about 40 cm depth. Similar observations are made by Alvarado *et al.* (1981).

C. Moisture retention characteristics

Loosely packed uncemented soil material has favourable moisture retention and transmission properties. Although it is difficult to conduct an infiltration test above or in the immediate vicinity of an ant nest, the implications of large channels and feeding galleries on facilitating rapid water movement through the profile are obvious.

The data in Fig. 10.7 compare the moisture retention characterisitics of ant material with the adjacent soil. Within the suction range of 0.1 to 15 bar, the ant material retained more water than the soil did. For this sandy soil in western Nigeria, the surface 0–10 cm layer of the ant nest contained 18.3, 79.6, 90.6, 62.5, and 75.0 per cent more water than the adjacent soil at soil water potentials of 0.1, 0.5, 1, 3, 6 and 15 bars, respectively. Similar differences in moisture

Table 10.3 Effects of leaf-cutter ant on properties of a Costa Rican Andept (Alvarado et al., 1981)

Horizon	Depth (cm)	Sand (%)	Silt (%)	Clay (%)	Particle density (g/cm³)	Bulk density (g/cm³)	Organic matter (%)
Undisturbed site							
A_1	0–8	40	16	44	2.21	0.69	7.17
AB	8–35	16	14	70	2.63	0.68	3.52
B_2	35–60	14	6	80	2.67	0.70	1.07
B_3	60–110+	14	6	80	2.69	0.90	0.69
Active mound							
A_i	13–0	16	14	70	2.60	0.62	3.05
A_{lli}	0–32	28	18	54	2.37	0.66	6.90
AB_i	32–40	16	24	60	2.59	0.70	4.46
B_2	40–103	36	24	40	2.61	0.72	2.75
B_3	103+	20	10	70	2.68	0.77	2.01
Abandoned nest sites							
A_{lli}	0–9	46	16	38	2.47	0.73	8.45
A_{12}	9–19	44	14	42	2.53	0.66	6.20
A_{13}	19–33	36	12	52	2.58	0.72	5.59
AB_i	33–58	14	14	72	2.62	0.79	3.05
A_{li}	58–73	24	12	64	2.63	0.88	1.91
B	73+	18	18	64	2.72	0.89	1.04

Table 10.4 Bulk density and penetrometer resistance of the ant nest soil (unpublished data, Lal, 1984)

Depth (cm)	Penetrometer resistance (kg/cm²) B	Bulk density (g/cm³) B	Gravels (%) B
0–10	1.04±0.42	1.32	0.0
10–20	V	V	V
20–30	1.67±1.12	1.38	0.0
30–40	1.04±0.67	1.54	0.0
40–50	1.88±0.99	1.62	4.7
50–60	2.21±0.24	1.60	3.0
60–70	3.21±0.53	1.50	2.5
70–80	2.88±0.80	1.53	2.6
80–90	3.33±0.67	1.69	4.2
90–100	—	—	—

B, ant nest; V, void space.

retention at all suctions existed until about 40 cm depth (Table 10.5). From 40 to 100 cm depth, however, the ant material retained more water than the soil for low suctions (>1 bar) only. Differences in moisture retention properties, however, do not imply more water available for plant growth. In fact, the ant

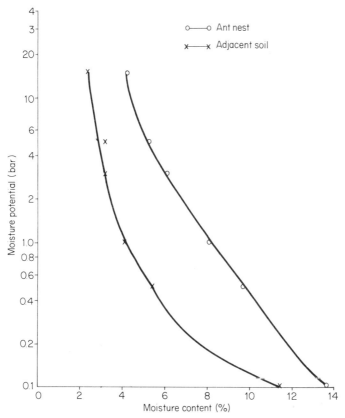

Figure 10.7 Soil moisture retention properties of the ant-nest material and of the adjacent soil (unpublished data, Lal, 1984)

material releases less water than the soil within the range of 0.1 and 15 bar suctions, respectively.

D. Chemical properties

As with the physical properties, little research information exists regarding the chemical properties of ant nest materials. In Kanpur, northern India, Shrikhande and Pathak (1948) compared the chemical properties of ant hill material with adjacent soil (Table 10.6). The ant hill material contained more organic carbon, total nitrogen, total and exchangeable Ca^{2+} and exchangeable K^+ than adjacent soil. The data in Table 10.7 compare the chemical properties of ant hill material with adjacent Apomu soil in western Nigeria. Up to 40 cm depth, the ant hill material contained more organic carbon, and more exchangeable cations (except Ca^{2+}) than adjacent soil. The differences were particularly pronounced in terms of organic carbon and exchangeable Mn^{3+}.

Table 10.5 Moisture retention characteristics (pF curves) of ant nest and an adjacent Apomu soil in western Nigeria (unpublished data, Lal, 1984)

Depth (cm)	Moisture potential (bars)											
	0.1		0.5		1		3		5		15	
	A	B	A	B	A	B	A	B	A	B	A	B
0–10	11.5	13.6	5.4	9.7	4.1	8.1	3.2	6.1	3.2	5.2	2.4	4.2
10–20	9.9	V	4.7	V	3.9	V	3.1	V	3.0	V	2.3	V
20–30	9.4	10.7	4.4	7.4	3.8	6.1	3.1	5.5	3.1	4.6	2.2	4.3
30–40	7.8	10.1	4.0	4.5	3.6	4.0	2.7	3.3	2.7	3.3	1.8	3.1
40–50	8.2	9.4	4.0	4.5	3.2	4.0	2.5	3.5	2.4	3.2	1.8	3.6
50–60	6.6	9.9	3.5	4.6	3.1	3.7	2.1	2.3	2.3	2.2	1.5	2.2
60–70	5.7	8.8	3.2	3.3	2.7	2.7	2.1	2.0	1.7	2.0	1.3	1.7
70–80	5.3	7.7	3.2	3.8	2.4	2.9	1.9	2.0	1.4	2.2	1.3	1.7
80–90	5.5	11.1	3.1	6.5	2.4	5.7	2.0	4.8	1.5	4.3	1.3	4.0
90–100	4.3	—	2.4	—	1.8	—	1.5	—	1.2	—	1.2	—

A, parent soil; B, ant nest; V, void space.

Table 10.6 Chemical analyses of ant hills and soils from northern India (Srikhande and Pathak, 1948). *Reproduced by permission of the Current Science Association, Bangalore*

Property	Ant hills	Soil
pH	7.5	7.3
Organic carbon (%)	0.88	0.54
Total nitrogen (%)	0.126	0.087
Total Ca (%)	1.03	0.83
Total Mg (%)	0.26	0.35
Total K (%)	0.37	0.48
Total P (%)	0.058	0.041
Total exchange capacity (meq/100 g)	16.83	14.81
Exchangeable Ca^{2+} (meq/100 g)	10.56	9.36
Exchangeable Mg^{2+} (meq/100 g)	0.009	0.015
Exchangeable K^+ (meq/100 g)	1.62	1.39

V. EFFECTS OF AGRICULTURAL PRACTICES

Little is known regarding the effects of land-use and of soil and crop management practices on ant population. Critchley *et al.* (1979) reported that the sub-family *Pseudomyrmecinae* dwell mostly within forested land. Deforestation, therefore, would drastically decrease its population and activity. Their data (Fig. 10.8) did not show marked seasonal trends in the activity of ants, and the total population on cultivated plots did not show any decline in comparison with the forested control. More quantitative data of this nature are needed to draw definite conclusions regarding the effects of bush clearing of cultural practices on ant population.

VI. CONCLUSIONS

Among soil fauna, the effect of ants on soil physical processes is the least understood. There are not enough research data available to draw any valid conclusions regarding the amount of soil turnover, the depth from which the material is being brought, the nesting characteristics, and the physical and nutritional properties of the nest material compared with adjacent soil.

There are also many gaps in our knowledge regarding the effects of agricultural practices on these animals. The effects of cultural practices, e.g., deforestation and land development, tillage methods and cropping systems, use of chemicals, e.g., pesticides, on species diversity and population density is least understood. Little is also known regarding the effects of environment and microclimatic factors on these animals.

Whatever limited information exists, it is apparent that ants are an important aspect of tropical soil fauna. In the tropical forest excosystem, the texture of the surface horizon is greatly influenced by ants and the material brought above from the subsoil layers. The water transmission properties are greatly affected by the channels and the excavation cavities.

Table 10.7 Chemical analyses of ant nest soil and the adjacent Apomu soil series (unpublished data, Lal, 1984)

Depth (cm)	Organic carbon (%)		Ca^{2+}		Mg^{2+}		Mn^{3+}		K$^+$		Na$^+$	
	A	B	A	B	A	B	A	B	A	B	A	B
0–10	0.91	1.29	2.60	1.87	0.95	1.41	0.04	0.09	0.14	0.14	0.10	0.11
10–20	0.32	V	1.10	V	0.46	V	0.05	V	0.04	V	0.03	V
20–30	0.33	1.17	1.40	1.64	0.54	1.28	0.04	0.12	0.05	0.15	0.05	0.13
30–40	0.31	0.45	1.10	0.97	0.36	0.69	0.02	0.11	0.03	0.12	0.04	0.12
40–50	0.08	0.10	1.00	0.60	0.32	0.41	0.02	0.04	0.03	0.06	0.04	0.10
50–60	0.17	0.17	1.20	0.60	0.27	0.31	0.01	0.03	0.03	0.04	0.03	0.10
60–70	0.26	0.17	0.90	0.37	0.25	0.22	0.01	0.02	0.02	0.05	0.04	0.11
70–80	0.23	0.20	0.90	0.37	0.27	0.25	0.09	0.01	0.02	0.03	0.04	0.08
80–90	0.10	0.18	0.90	0.67	0.35	0.48	0.007	0.01	0.02	0.06	0.05	0.13

A, adjacent soil; B, ant nest.

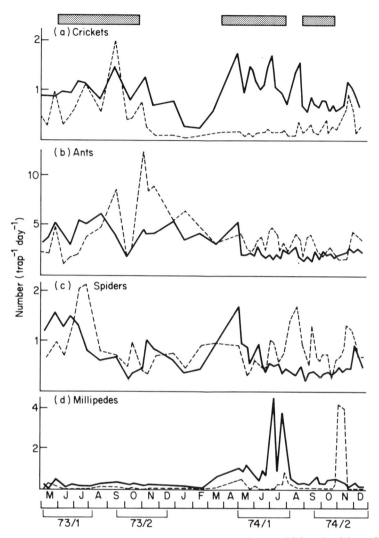

Figure 10.8 Seasonal variations in activity of ants in a forested (–) and cultivated (----) soil (Critchley *et al.*, 1979). ▨ , rainfall over 100 mm per month. Marked below the horizontal axis are the growing seasons

Many pedologists recognize the importance of ants in altering soil profile characteristics. It is appropriate to use a subindex (i) to designate insect-influenced soil, and to indicate the importance of insects as agents of soil formation.

REFERENCES

Alvarado, A., Berish, C.W., and Peralta, F., 1981. Leaf-cutter ant (*Atta cephalotes*) influence on the morphology of Andepts in Costa Rica. *Soil Sci. Soc. Am. J.*, **45**, 790–794.
Araya, L. and Alvarado, A., 1978. Influencia de la hormiga *Atta* spp. en la genesis de suolos. In III Congreso Agronómico Nacional, Resúmenes. Universidad de San José, San José, Costa Rica I: 86.
Bachelier, G., 1983, *La vie animale dans les sols*, ORSTOM, Paris, 279 pp.
Baxter, F.P., and hole, F.D., 1967, Ant (*Formica cinerea*) pedoturbation in a prairie soil, *Proc Soil Sci. Soc. Amer.*, **31**, 425–428.
Brown, W.L., Jr., 1972, The geographical distribution of ants, past and presnt, in *Proc. 14th Int. Congr. Entomol.*, *Canberra*, 189 pp.
Brown, W.L., Jr., 1973, A comparison of the hylean and Congo West African rainforest, in B.J. Meggers, Ayensu, E.S., and W.D. Duckworth, (eds), *Tropical Forest Ecosystems in Africa and South America: a Comparative Review*, Smithsonian Institution Press, Washington, D.C., pp 161–185.
Buol, S.W., Hole, F.D., and McCracken, R.J., 1973, *Soil genesis and classification*. Iowa State University Press, Ames, Iowa.
Cherrett, J.M., 1968. Some aspects of the distribution of pest species of leaf-cutting ants in the Caribbean. *Proc. Trop. Reg. Am. Soc. Hortic. Sci.*, **XV**, 12, 296–310.
Critchley, B.R., Cook, A.G., Critchley, U., Perfect, T.J., Russell-Smith, A., and Yeadon, R., 1979, Effects of bush clearing and soil cultivation on the invertebrate fauna of a forest soil in the humid tropics, *Pedobiologia*, **19**, 425–438.
Delye, G., 1968, Recherche su l' ecologie, la physiolgie et l' éthologie des fourmis du sahara, *Thesis*, Marseille University, Marseille, 115 pp.
Deiomandé, T., 1981, Etude du peuplement en fourmis terricoles des forests ombrophiles et des zones anthropiseés de la Côte d' Ivoire méridionale, *Thesis*, Université d' Abidjan, 254 pp.
Ettershank, G., 1968, The three-dimensional gallery structure of the nest of the meat ant *Irdomyrmex prupureus*, *Aust. J. Zool.*, **16**, 715–723.
Glavis, C., Valencia, H., Chamorro, C. and Cortes, A.H., 1977, Efecto edáfico de la hormiga arriera *Atta lacrigate* en algunos suelos de la orinoquio suelos de la orinoquio Colombiana. *Suelos Ecutat.*, **8**, 416–422.
Goosen, D., 1979. Management of acid soil savannas: the Llanos Orientales of Colombia. p 125–140. In H.O. Mongi and P.A. Huxley (eds.) *Soil Research in Agroforestry*, ICRAF, Nairobi, Kenya.
Greenslade, P.J.M., 1976, The meat ant *Irodomyrmex purpureus* (Hymenoptera formicidae) as a dominant member of ant community, *J. Aust. Entomol. Soc.*, **15**, 237–240.
Greenslade, P.J.M., and Mott, J.J., 1978, Ants (Hymenoptera formicidae) of native and sown pastures in Katherine area, N.T., Australia, in *Proc. 2nd Aust. Cong. Grassland Invertebrate Ecology*, Massey Univ., Palmerston North.
Hölldobler, B., 1984, The wonderfully diverse ways of the ant. *National Geographic*, **165** (6), 779–813.
Josens, G., 1983, The soil fauna of tropical savannas. III. The termites, in F. Bourliere, (ed.), *Ecosystems of the World: Tropical Savannas*, Elsevier, Amsterdam, pp. 505–524.
Kevan, K. McE, 1968, *Soil Animals*, H.F. & G. Witherby Ltd., London, 244 pp.
Kusnezov, N., 1963, Zoogeogragia de les hormigas en Sud America, *Act. Zool. Lilloana*, **19**, 25–186.
Lavelle, P., Maury, M.E., and Serrano, V., 1981, Las comunidades animales en suelos de pastizales selves pertenecientes a le region de laguna verde (Vera Cruz, Mexico)

durante la temporada de leuvas in R. Reyes Castilto (ed.), *Estudious Ecologicos en el Tropico Mexicano*, Instituto de Ecologia, Mexico, Publ. N. 6, 1–187.

Lee, K.E., and Wood, T.G., 1971, *Termites and Soils*, Academic Press, London, 251 pp.

Lévieux, J., 1973, Study of the ant population in soils of a pre-forest savanna of the Ivory Coast, *Revue Ecol. Biol. Sol.*, **10**, 379–428

Lévieux, J., 1983, The soil fauna of tropical savannas. IV. The ants, in F. Bourliere, (ed.), *Ecosytems of the World: Tropical Savannas*, Elsevier, Amsterdam, pp 525–540.

Levings, S.C., and Frank, N.R., 1982, Patterns of nest dispersion in a tropical ground ant community, *Ecology*, **63**, 338–344.

Lewis, T., 1975, Colony size, density, and distribution of the leaf-cutting ant *Acromyrmex octospinosus* in cultivated fields. *Trans. Real Ent. Soc. Long.*, **127**, 51–64.

Longhurst, C., Johnson, R.A. and Wood, T.G., 1979, Foraging recruitment and predation by *Decamrium uelense* (Santschi) (Formicidae Myrmicinae) on termites in Southern Guinea Savanna, Nigeria, *Oecologia*, **38**, 83–91.

Madge, D.S., and Sharma, G.D., 1969, *Soil Zoology*, Ibadan University Press, 54 pp.

Nye, P.H., 1955, Some soil-forming processes in the humid tropics, IV. The action of the soil fauna, *J. Soil Sci.*, **6**, 73–83.

Raignier, A., 1952, *Vie et Meurs des Fournis*, Edns Payot Paris, 224 pp.

Risch, S.J., and Carroll, C.R., 1982, The ecological role of ants in two Mexican agroecosystems, Oecologia, **55**, 114–119.

Salem, M.Z., and Hole, F.D., 1968, Ant (*Formica exsectioides*) pedoturbation in a forest soil, *Proc. Soil Sci. Soc. Amer.* **32**. 563–567.

Salt, G., 1950, Investigations of the arthropod fauna of the soil in East Africa, *E. Afr. Agric. For. Res. Org. Rep.*, **1949**, 42–43.

Shaler, N.S., 1891, The origin and nature of soils, *U.S., Geol. Survey 12th Ann. Rep. (1890–1891)*, 213–345.

Shrikhande, J.G., and Pathak, A.N., 1948, Earthworms and insects in relation to soil fertility, *Curr. Sci.*, **17**, 327–328

Thorp, J. 1949, *Sci. Month.*, **68**, 180–191.

Wallwork, J.A., 1970, *Ecology of Soil Animals*, McGraw–Hill, London, 283 pp.

Watson, J.P., 1960, Some observations on soil horizons and insect activity in granite soils, *Proc. Int. Fed. Sci. Cong., Rhodesia and Nyasaland*, pp. 271–276.

Weber, N.A., 1966. Fungus growing ants. *Science*, **153**, 587–604.

Wheeler, W.M., 1936, Ecological relations of ponerine and other ants to termites, *Proc. Am. Acad. Arts Sci.*, **71**, 159–243.

Williams, E.C., 1941, An ecological study of the floor fauna of the Panama rain forest, *Bull. Chicago Acad. Sci.*, **6**, 63–124.

PART III

Man as an Ecological Factor

Chapter 11
Man and Soil

I	MAN AND ECOLOGY	445
II	MAN AND SOIL	446
III	ANTHROPIC ACTIVITIES AND SOIL PROPERTIES	447
	1. Agriculture	450
	2. Urbanization	450
IV	CONCLUSION	450

I. MAN AND ECOLOGY

In the recent past of man's tortuous and long evolutionary history, he has become the most important ecological factor and a manipulator of the environment. He drastically modifies soil properties through his quest for 'mastering nature', and he alters the course of the normal soil-forming processes. Man-induced ecological instability for short-term economic gains leads to chain reactions that have far reaching effects on major biomes.

The ecological stability in the context of physical edaphology refers to the ability of an ecosystem to resist drastic changes in soil physical and microclimatic processes caused by man's perturbations. A stable ecosystem is dynamic and undergoes slow and gradual alterations. The community in a stable ecosystem endeavours to return the biophysical processes to the state and level that existed prior to man-induced distrubances. A stable ecosystem is very complex and diverse, with many alternative energy and nutrient pathways. It has a relatively large community population and little or no net community productivity (Holdren and Ehrlich, 1981). Odum (1971) reported that net primary productivity (NPP) and the net community productivity (NCP), respectively, are 15 200 and 14 000 cal m^{-2} yr^{-1} for alfalfa, 5 000 and 2 000 cal m^{-2} yr^{-1} for pine forest, and 13 000 and little or no cal m^{-2} yr for tropical rainforest. Even though the net community productivity of the tropical forest is not zero, it is certainly less than the agricultural ecosystems. Undistrubed, diverse and stable ecosystems often home less net community productivity than agricultural ecosystems. It is the NCC that is of economic importance to man, and he increases it at the cost of stability. In this connection, man's objectives are at cross purposes with nature. Whereas man tries to maximize productivity, nature tries to maximize stability. In order, therefore, to avoid major ecocatastrophe, man must strike a harmonious or ecologically compatible compromise.

The literature is full of emotional cries of pending ecological disasters and catastrophe set in motion by man's interference with nature. These emotional rhetorics, popular and mass appealing as they may be, are not objective and do not serve any useful purpose. For any suggestion to be useful, it must be rational and based on scientifically verifiable facts. Rather than be conservative, it is time to be positive and think about the possibilities of environmental manipulation and creating conditions optimum for raising the net economic productivity of the artificially created ecosystems. This positive approach implies creation of new environments to suit the dynamic needs of man, and does not necessarily imply working against nature but living in symbiosis with it. After all, vast areas of unproductive wilderness have been converted into very productive agricultural and urban ecosystems. There is no denying the fact that this conversion has created some adverse effects, but better understanding of man-ecological interaction should reduce the risks of these new environmental creations.

II. MAN AND SOIL

The impact of man as a soil manipulator, widely recognized by soil scientists, is increasing with increase in antecedent number and diversity of activities. The agricultural revolution, the rapidly developing modern agricultural technology, and the increase in man's demand for food and raw materials have drastically increased man's impact on soil properties. Some general aspects of man-made alterations to soil properties are reviewed by Tavernier and Pecrot (1957), Bidwell and Hole (1965), and Yaalon and Yaron (1966). The last authors termed man-induced processes and changes as 'metapedogenesis', and developed a process-response model to assess the impact of these changes on natural soil-forming factors.

Man-induced alterations in soil properites can be beneficial or detrimental depending on the specific property altered and its interaction with the soil-climate-vegetation system. There are examples of soils that have been improved by man, i.e., restoration of degraded soils, improved surface and subsurface drainage conditions, application of soil ameliorants to improve soil structure, development of elaborate terrace systems (such as in Southeast Asia) to cultivate steeplands, removing the unfavourable topsoil and replacing it with fertile soil, as is a common practice in China, and artificial soil created by refuse-fills in large urban centres. On the other hand, accelerated erosion, soil compaction and increase in salinity due to irrigation are examples of the detrimental effects of man's intervention on soil properties. Pedologists have defined soil horizons created by man's activities: (i) the anthropic epipedon; (ii) the plaggen horizon; and (iii) the agric horizon (USDA, 1975).

III. ANTHROPIC ACTIVITIES AND SOIL PROPERTIES

The range of activities of contemporary man that influence soil properties is diverse. Some important and relevant activities are:

Figure 11.1   Topsoil from large areas of arable lands is used annually for brick-making in China

Table 11.1 Man-induced alterations in soil-forming factors and ecological environments

Man's activity	Ecological effects	Effects on soil-forming factors
1. *Deforestation*	Denudation of vegetation cover leads to alterations in hydrological balance, energy balance, nutrient recycling, and biotic activity of soil fauna, a decrease in specie diversity and loss of nutrients.	Changes in some major soil-forming factors, e.g., climate, vegetation and organisms, alter the relative predominance of these factors and their interaction.
2. *Agriculture*		
(i) Plowing and tillage	Alterations in soil cover and biotic activity, accelerated soil erosion, high insolation, soil compaction.	Changes in microclimate, water transmission and biotic activity of soil fauna.
(ii) Monoculture	Reduction in species diversity, increase in potential epidemics from insects, diseases and weeds.	Decrease in root mass and depth distribution, shift in predominant flora and alteration in water and nutrient regime.
(iii) Chemicals	Agriculture increases both qualitative and quantitative and quantitative pollution, which alters natural pathways and cycles, and decreases community structure and complexity.	Soil-forming factors are altered by decrease in complexity and diversity. Decrease in biotic activity alters the rate of soil formation.

(iv) Grazing	Denudation of vegetal cover, decrease in specie diversity, soil compaction and accelerated erosion.	Changes in microclimate, reduction in vegetal cover and decrease in specie diversity slow down the rate of soil turnover and new soil formation.
(v) Fire	Alterations in microclimate, vegetal succession, loss of plant nutrients pollution, change in soil fauna, change in water balance.	Formation of plinthite. Alterations in vegetal cover influence the normal soil development processes.
(vi) Irrigation	Changes in water balance, growth cycle, crop cover, microclimate.	Changes in ground-water table, salinization, alkalization are direct consequences of faulty irrigation practices.
(vii) Drainage	Alteration of soil air–water balance, oxidation environment and soil microclimate.	Shifts in soil–water balance and nutrient availability.
(viii) Terracing or land levelling	Change in water regime, exposure of subsoil.	Exposed subsoil is subject to weathering processes.
3. *Urban development*	Alteration in microclimate, pollution, waste disposal, decrease in vegetation shift in hydrological cycle, exposure of subsoil.	Alternate uses of soil resource, disruption of natural soil-forming factors.

1. Agriculture

Man simplifies the diverse natural ecosystem into a simpler, unstable, but more productive agricultural ecosystem. The expansion of agriculture to new regions—forest or savanna biomes—brings about drastic alterations in soil properties and microclimates. The magnitude of alterations thus created depends on the specific activity, the intended land-use or farming system, soil suitability in relation to the land-use proposed, and the environment.

The agricultural activities of relevance to physical edaphology are deforestation, fire, tillage, alternate land-uses, and the intensive use of agrochemicals. The effects of each of these activities on soil properties are briefly outlined in Table 11.1 and some activities (fire, deforestation, tillage and farming systems) are discussed at length in the following chapters. In addition to the effects of these specific activities on soils and environment, expansion of agriculture has other ecological consequences, i.e., reduction in species diversity, environmental pollution and eutrophication by a wide range of chemicals, disturbance of natural elemental cycles. The important nutrient cycles that affect productivity of agricultural ecosystems are C, N, S and P. These elements have to be subsidized to enhance the net community productivity of agricultural ecosystems.

2. Urbanization

Expanding population centres are rapidly swallowing prime agricultural land in the tropics. The use of soil as a construction material is a serious threat to agriculture. In populous countries such as China and India fertile agricultural lands are being used for brick making (Fig. 11.1). The subsoils thus exposed are not always a favourable medium for crop growth. Prime agricultural lands are used for making roads and other infrastructures. Land is taken up for these activities which could have otherwise been used for agriculture. Nontheless, infrastructure is also essential for boosting agricultural production.

IV. CONCLUSION

Man's activities have drastic ecological impacts, and some effects are long-lasting and even irreversible. The impacts of man on soil and environments are both direct and indirect. In his attempt to increase net community productivity, man introduces ecological instability. Acceptable levels of instability differ among ecosystems and have to be defined in quantitative terms. To create excessive instability to accrue short-term economic gains is to be short-sighted, but favourable ecological conditions can be created, the new ecosystems to increase productivity and yet to cause acceptable levels of 'negative' effects.

REFERENCES

Bidwell, O.W., and Hole, F.D., 1965, Man as a factor of soil formation, *Soil Sci.*, **99**, 65–72.

Holdren, J.P., and Ehrlich, P.R., 1981, Human population and the global environment, in B.J. Skinner, (ed.), *Use and Misuse of Earth's Surface*, William Kaufmann Inc., Los Altos, California, pp. 6–16.

Kassas, M., 1970, Desertification versus potential for recovery in circum-saharan territories, in *Arid Lands in Transition*, Am. Assoc. Adv. Sci., Washington, D.C.

McNeil, M., 1972, Lateritic soils in distinct tropical environments: southern Sudan and Brazil, In H. Brown, and A. Sweezy, (eds.), *the Careless Technology, 1971*, Freeman-Cooper, San Francisco, pp. 591–608.

Odum, E.P., 1971, *Fundamentals of Ecology*, Saunders, Philadelphia, 46 pp.

Ravenholt, A., 1971, The Philippines, in H. Brown, and A. Sweezy, (eds.), *Population Perspective, 1971* Freeman-Cooper, San Francisco, pp 247–266.

Ravenholt, A., 1974, Man-land-productivity microdynamics in rural Bali, in H. Brown, J. Holdren, A. Sweezy, and B. West, (eds.), *Population Perspective, 1973* Freeman-Cooper, San Francisco.

Sabloff, J.A., 1971, The collapse of classic Maya civilization. in J. Harte, and R. Socolow, (eds.), *Patient Earth*, Holt, Rinehart and Winston, New York.

Tavernier, R., and Pecrot, A., 1957, L' homme et l' evolution due sol en Belgique, *Pedologie*, **1957**, 226–231.

Thomas, Jr., W.L. (ed.), 1956, *Man's Role in changing the Face of the Earth*, University of Chicago Press, Chicgo.

Thorkild, J., and Adams, R.M., 1958, Salt and silt in ancient Mesopotamian agriculture, *Science*, **128**, 1251–2158.

USDA, 1975, *Soil Taxonomy*, Agricultural Handbook, USDA, Washington, D.C.

Yaalon, D.H., and Yaron, B., 1966, Framework for man-made soil changes—an outline of metapedogenesis, *Soil Sci.*, **102**, 272–277.

Chapter 12
Fire

I	INTRODUCTION	453
II	FIRE-DEPENDENT AND FIRE-FREE ECOSYSTEMS	453
III	MAN-INDUCED FIRE AND TRADITIONAL FARMING	455
	A. Temperature During Fire	456
	B. Effects on Soil	457
	(i) Soil temperature during fire	457
	(ii) Soil temperature after the fire	459
	C. Effects on Vegetation	460
	D. Consequences on Soil Physical Properties	461
	(i) Changes in soil structure and wettability	461
	(ii) Soil texture	463
	E. Burning and Soil Fauna	463
IV	SOIL–VEGETATION INTERACTION	464
V	PRESCRIBED FIRE IN MODERN AGRICULTURE AND LAND MANAGEMENT	465
	A. Land Clearing	466
	B. Forest and Savanna Management	467
	C. Pasture Management	467
	D. Crop Residue Management	468
VI	EFFECTS OF PRESCRIBED FIRE ON SOIL AND ENVIRONMENT	468
	A. Vegetation	468
	B. Soil Physical Properties	470
	1. Particle size distribution	470
	2. Structural stability	472
	3. Wettability	473
	4. Moisture retention properties	474
	5. Soil moisture balance	475
	6. Water transmission properties	479
	7. Water run-off and soil erosion	480
	8. Soil compaction	485
	C. Nutrient Losses and Changes in Soil Fertility	485
	D. Fire and Soil Fauna and Flora	487
	E. Effects on Atmospheric Chemistry	489

VII PRESCRIBED BURNING AND CROP YIELD	489
VIII CONCLUSIONS	493

I. INTRODUCTION

Fire is an important ecological factor in the tropics. It has created lasting effects on vegetation, animals, landscape and soil fauna. Fire has been shaping the ecological environments for millennia before man's intervention by domesticating plants and animals. The causes of this natural fire were lightning and volcanoes (Komarek, 1964; 1965). The use of fire by primitive man in Africa dates back to 50 000 to 60 000 years ago (Clark, 1960; Howell and Clark, 1963; Gillon, 1983). Lacey *et al.* (1982) reported that aboriginal man in Australia has used fire since 40 000 years B.C. The effects of man-induced fire in Brazilian Cerrado are reportedly at least 10 000 years old (Coutinho, 1979). In Belize and other sites of Maya settlements in Central America, the occurrence of large amounts of carbonized plant fragments in lake cores at 95 to 280 cm depth, indicate that fire was commonly used by civilizations in the year 300 B.C. (Lambert and Arnason, 1978). The use of fire as a tool is probably even older in tropical Asia than in Africa and Central America. The effects of fire on terrestrial ocosystems have been extensively reported (Garren, 1943; Hodgkins, 1958; Shantz and Turner, 1958; Boughey, 1963, 1965; Phillips, 1965, 1974; Trollope, 1982; Vogl, 1977; Raison, 1979; Gillon, 1983; Lacey *et al.*, 1982; Coutinho, 1982), and readers are referred to these reveiws for more detail. The findings vary widely depending on the climate, soil, vegetation, duration of the use of repeated fire and whether the fire is natural or man-induced. The objective of this chapter is to review those effects of man-induced fire on soil physical properties and microclimate that have lasting effects on vegetation, soil productivity, and soil degradation.

II. FIRE-DEPENDENT AND FIRE-FREE ECOSYSTEMS

Fire may be a natural process in some ecosystems and a perturbation in others. The commonly observed fire-free ecosystems in the tropics are rainforest, cloud forests, wetlands, swamps and gallery forests. These systems have high water retaining capacities and are normally not affected by natural fires. In addition, there are fire-dependent ecosystems for which fire is a natural process. A community in a fire-dependent ecosystem often requires fire for its survival and continuation (Vogl, 1977). Fire-dependent ecosystems promote fire because of their low water-retaining capacities. These ecosystems are perturbed if fire is eliminated or its frequency is decreased. Tropical savanna and grasslands are examples of fire-dependent ecosystems.

The magnitude of the effect of fire on soil and vegetation depends on the kind, intensity, and frequency of fire. Phillips (1974) describes three kinds of fire—ground fire, crown fire, and the man-induced fire. The ground fire is common in the fire-dependent ecosystems, i.e., the subhumid and savanna regions where the dry season is long enough to enable the leaf litter to be readily flammable. The effects of ground fire are generally superficial and mostly confined to the thin surface layer, and are essential for the stability of the ecosystem. The crown fire, on the other hand, is more severe because the quantity of flammable material within the canopy is often greater. Generally, ground fire develops into crown fire if flammable material is present within the canopy. The crown fire has more drastic effects on vegetation than ground fire. The crown fire is often man-induced and occurs in a fire-free ecosystem with tall vegetation growth. In fire-free ecosystems man-induced fires are often the most severe. The fire is intentionally started for land preparation, hunting, habitat management or for access and mobility. By these operations, the quasi-equilibrium is disturbed and the soil properties are more drastically altered.

Man-induced fire can easily become catastrophic in the transition zones between fire-free and fire-dependent ecosystems, i.e., semi-deciduous forest, a transitional vegetation between rainforest and the grassland.

The effects of fire on soil and ecology are likely to be more drastic in a fire-free ecosystem because the community is unadapted and cannot survive the severe change. On the other hand, fires are a natural process in tropical savannas which have both rainy and dry seasons long enough to encourage luxuriant growth and then render it sufficiently dry for the biomass to become easily inflammable. Tropoical savannas, characterized by the prolonged dry season and sufficient rains to encourage growth, are affected by annual fire, whereas the wetter humid forest or the drier sahel and arid climates are not (Batchelder, 1967; Phillips, 1965, 1974). Fire is a natural process in all tropical savannas. For example, Myers (1936) observed that all savannas in tropical America are annually burnt. Top layers of the forest vegetation take longer to dry out. The dry season in the humid forest region is not long enough to create sufficient inflammable material for kindling natural fire. Neither is the rainy season long enough in the sahel and arid climates to generate enough biomass. The savanna regions most subject to frequent natural fire are the derived savanna (or forest/savanna mosaic), the moist wooded savannas (or Guinean savannas) and, to a lesser extent, the dry wooded savanna (or Sudanian savannas) (Gillon, 1983). In these savannas, the perennial grass species grow rapidly and provide the ground vegetation for the annual fire cycle during the dry season.

In addition to the rainfall, soil factors also play an important role in rendering an ecosystem fire-free or fire-dependent. Tropical soils (including Alfisols, Ultisols, Oxisols, and Aridisols), characterized by low available water-holding capacity and good internal drainage, strong to moderate leaching with low fertility, structural unstability, high susceptibility to erosion, and shallow rooting depth with impermeable and compacted subsoil, make the vegetation easily inflammable. Thus there exists a strong interaction among climate, soil and vegetation that renders the 'tropical savanna' ecosystem more prone to annual fire than tropical forest or the arid lands. In these areas the

Table 12.1 Differences in the soil, climate and vegetation of fire-free and fire-dependent ecosystems

Parameter	Fire-free (Forest)	Fire-dependent (Savanna)
(A) *Microclimate*		
(i) Insolation	Low	High
(ii) Relative humidity	High	Low
(iii) Ambient temperature (mean and range)	Low	High
(iv) Soil temperature (mean and range)	Low	High
(v) Rainfall effectiveness	High	Low
(vi) Potential evaporation	Low	High
(vii) Climatic aridity	Low	High
(B) *Vegetation*		
(i) Biomass	High	Low
(ii) Inflammable material (grasses)	Low	High
(iii) Relative nutrient concentration in vegetation	High	Low
(iv) Presence of large roots	High	Low
(C) *Soil*		
(i) Moisture depletion depth	Deep	Shallow
(ii) Surface moisture depletion	Less	More
(iii) Physiography	Rolling to undulating	Flat to gentle slopes
(iv) Nutrient leaching	Excess	Moderate to low
(v) Subsoil compaction	Less	More
(vi) Macropores in soil	High	Low
(vii) Permeability to water	High	Low
(viii) Water run-off	High	Low

insolation directly reaching the soil surface is also more than that in the forest zone. The comparative evaluation of factors that render savanna ecology more prone to natural fire are listed in Table 12.1.

III. MAN-INDUCED FIRE AND TRADITIONAL FARMING

Fire is a principal land management tool in traditional farming in the tropics. It is used for land clearing, weed control, disposal of unwanted crop residue, production of good pasture and for hunting. Uncontrolled fire, used traditionally for many reasons, is often responsible for annual burning of savannas and other regions of the tropics (Fig. 12.1). Man-induced fires have resulted in drastic environmental effects on vegetation, soil, etc.

Fire has both short-term (immediate) and long-term effects. The short-term effects are easy to monitor and quantify, but there is little scientific data about the long-term ecological effects. It is only recently that attempts have been made to standardize the terminology used so that results of different

Figure 12.1 Uncontrolled fire used traditionally can seriously injure tree regrowth and accelerates conversion of forests into savannahs

researchers are comparable (Byram, 1958; McArthur and Cheney, 1966). The environmental effects of fire depend on fire intensity, available fuel energy, combustion rate, and the rate of forward spread (Raison, 1979). These factors influence the volatilization of plant nutrients and changes in soil properties.

A. Temperature during fire

High temperatures developed during fire have direct effects on soil and vegetation. Intensity or severity of the burn depends on the season. Burning early in the dry season is generally light because the moisture content of the vegetation is still high. Burning late in the dry season often produces severe effects because most of the grass and vegetation is now dried and produces intense heat. The effects of late burning on soil are also more drastic because the exposed soil is subjected to torrential rains soon after.

The temperature during fire depends on the quantity of inflammable material, degree of dryness, wind velocity, etc. Fire temperatures vary over short distances depending on the variation in plant biomass. Many researchers have investigated the effects of fire on canopy temperature. The temperature within the canopy depends on the vegetation, i.e., grass or wooded savanna. The maximum temperature within a dense grass canopy has been observed to exceed 600°C (Pitot and Masson, 1951; Rains, 1963; Hopkins, 1965). In Sudan, Guilloteau (1956) reported temperatures of 715°C and 850°C recorded just above the soil surface. The quantity of combustible material exceeding 450

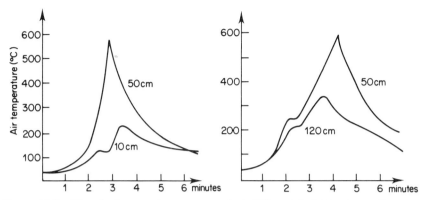

Figure 12.2 Rise in temperature measured at different heights above the ground surface during grass fire in an Ivorian savannah at Lamto, Ivory Coast. The average height of grass cover was 70 cm (Viani et al., 1973; Gillon, 1983)

g/cm^2 can increase the scorch height of the flame from 1 to 6 m (Rains, 1963). In Ivory Coast Viani et al. (1973) observed peak temperatures of 200°C, 600°C and 300°C at 10, 50, and 120 cm above the ground surface for the average grass height of 70 cm (Fig. 12.2). The effect of fire in increasing temperature above the burning grass is often observed 3 to 4 m above the ground surface. Hopkins (1965) studied the blaze temperature in the Olokemoji forest reserve in Nigeria. He reported considerable place-to-place variations in fire temperatures at different heights above the ground surface. The maximum temperature of 533°C was observed in the late-burnt plots from soil surface to the height of 3 m. All surface measurements and to the height of 0.1 m experienced temperatures exceeding 533°C while the upper limit of temperature at 3 m was variable at 101°C, 149°C and 316°C. Temperatures are generally lower on early burnt plots and rarely exceed 66°C at 3 m above the ground surface. In the late-burnt plots, however, temperatures exceeding 100°C are commonly observed even at 6 to 7 m above the ground surface.

B. Effects on soil

The effects of fire on soil are short-term, intermediate, and long-term. The short-term effects include the soil temperature during fire and its direct consequences.

(i) Soil temperature during fire: The soil temperature profile during the fire has been reviewed for various vegetation types by Raison (1979). The soil temperature profile is different for different vegetations. For example, grass vegetation produces a rapid and complete burn as the fire front passes. Fire temperatures rise rapidly and fall gradually over several minutes and their effects are usually limited to the top 3 or 4 cm of the soil surface. The duration and intensity of heat increases with increases in tree density. Forest fires create

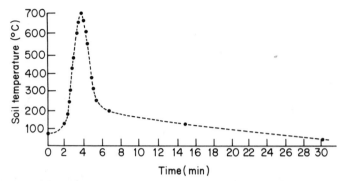

Figure 12.3 The maximum surface soil temperatures during burning observed in Senegal lasted only 30 seconds (Masson, 1948). *Reproduced by permission of IRAT*

higher soil temperatures to deeper soil depths and for longer durations than grass fires.

The maximum temperatures are observed at the soil surface. In Senegal Masson (1948) observed surface soil temperatures attain a level of 720 and 850°C and then decrease to less than 100°C within 8 to 12 minutes (Fig. 12.3). Very high temperatures lasted only for a fraction of a minute. No temperature over 200°C was maintained for longer than 30 seconds. Only in one instance did the temperature remain over 100°C for more than I minute, and temperatures over 50°C were maintained between 1 and 3 minutes. In another experiment, however, the maximum temperature recorded on the surface was only 200°C but it lasted for a period of 40 minutes and decreased to 35°C at 70 minutes after the fire. The duration of the fire itself was 1 minute and 10 seconds only (Fig. 12.4).

In spite of the very high surface temperatures, the heating front does not penetrate deep into the subsoil. Masson (1948) and Pitot and Masson (1951) observed only an increase of 3–14°C in soil temperature at 2 cm depth. Gillon

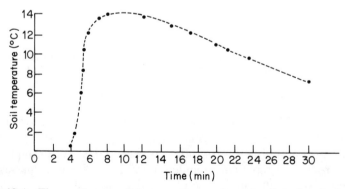

Figure 12.4 The maximum soil temperatures at 2 cm depth during the low-intensity fire are generally lower but last longer (Masson, 1948). *Reproduced by permission of IRAT*

and Pernés (1968) observed that soil temperature just beneath the surface did not exceed 60°C. Experiments conducted in Venezuela registered an increase in soil temperature of only 3°C at 3 cm depth (Vareschi, 1962). When savannas are burned under windy conditions temperatures are often not very high (Vareschi, 1962; 1969).

For Brazilian Cerrado the soil temperatures during fire in a red-yellow Latosol are reported by Coutinho (1976, 1978). The data (Fig. 12.5) showed only a slight increase in temperature during fire with a maximum of 74°C at the surface. Soil temperatures did not exceed 40°C at 1, 2, and 5 cm depths.

(ii) Soil temperatures after the fire: The ground is devoid of low-growing vegetation cover immediately after the fire, and the soil is covered by gray and black ash. Under these circumstances soil surface temperatures are higher than in unburnt savanna. Soil temperatures are highest immediately after the fire and prior to regrowth of the vegetation. Moureaux (1959) reported soil temperatures of burnt and unburnt grassland in Madagascar. His data (Table 12.2) show that at 11 a.m. the surface temperature of the bare soil under burnt savanna was 56°C in comparison with 40°C in soil under unburnt savanna. The increase in soil temperatures at 2.5 and 5 cm depths was 5 and 3°C, respectively. Athias et al. (1975) observed soil temperatures of the burnt and unburnt savanna in Lamto, Ivory Coast. They observed that the effect of burning on soil temperature was observable at 10 cm depth until 6 months after burning, after which the dense grass regrowth provided a protective vegetation cover against the intense insolation (Fig. 12.6).

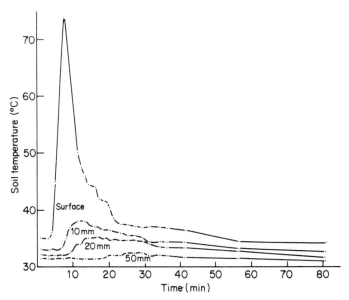

Figure 12.5 Soil temperatures in a red-yellow Latosol during Cerrado fire in Brazil the curves depict various depths in soil (Coutinho, 1976; 1978). *Reproduced by permission of Springer-Verlag*

Table 12.2 Surface temperature (°C) of soil in burnt and unburnt grasslands in Madagascar; temperatures measured in November at 11 a.m. (Moureaux, 1959)

Depth (cm)	Bare soil under burnt savanna	Soil under unburnt savanna
Surface	56	40
2.5	35	30
5	29.5	26.5

The changes in soil temperature and the increase in radiation levels reaching the soil also influence the microclimate near the ground. The air temperature is higher and the humidity is often lower in burnt than in unburnt savanna.

C. Effects on vegetation

The change in climax vegetation is an example of the long-term effect of fire on

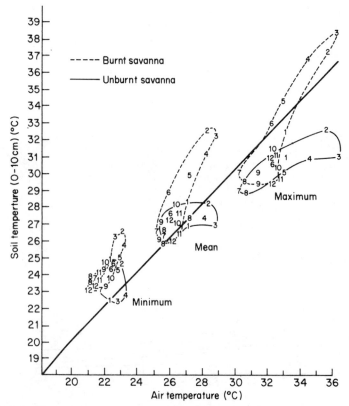

Figure 12.6 The minimum, average and maximum soil air temperatures during successive months of the year in burnt and unburnt savannah at Lamto, Ivory Coast (Athias *et al.*, 1975; Gillon, 1983)

an ecosystem. The high air temperatures within the grass/tree canopy during fire repeated over centuries have evolved a fire-adapted vegetation. Species adapt to these dangerously high temperatures through various modifications, and the fire-tender species are eventually eliminated. A community in the fire-dependent ecosystem often contains organisms that even depend on fire for their survival and continuance because fire has become an essential part of the environment (Vogl, 1977). Plants have evolved fire-dependent structures, mechanisms and functions. For example, plants have adapted to minerals or pH substrate changes for seedling establishment, growth-inhibiting chemicals, heat-treated semi-sterilized soil, altered soil–water relation, etc. In addition, some organisms have developed survivial mechanisms such as have been described by Coutinho (1982) for Brazilian Cerrado, and Lacey *et al.* (1982) for Australian tropical savanna. Some of these survival mechanisms are: (i) development of a thick and insulating corky bark around the trunks and branches; (ii) capacity to sprout from underground organs (i.e. gemmiparous roots); (iii) protection of apical buds by dense hairy cataphylls; (iv) protection of apical buds by densely imbricated sheaths as in some grasses; (v) presence of underground dormant buds; (vi) very extensive underground system penetrating deep into the subsoil; (vii) elevated and well separated crowns; and (viii) development of rhizomes and root suckers. Lacey *et al.* (1982) observed that the grass species in Australia have evolved two survivial mechanisms: (i) short growth cycle with seed reproduction; and (ii) protection of bud primordia by deeply buried rhizomes (*Imperata cylindrica, Coelorachis rottboellioides*), woody cataphylls around primordia, and the protection of growing shoots by a sheath of cataphylls. These particular morphological adaptations make a plant resistant to environmental stresses, i.e., drought, fire, grazing. These traits are termed by Grubb (1977) as 'regeneration niche'.

Quick growing trees and shrubs are suppressed by fire in favour of the herbaceous and undershrub species and perennial grasses. Trollope (1982) observed the effects of annual fire on vegetation in South Africa. Tree mortality was found to be the highest when the fuel load exceeded 3t/ha, fuel moisture content was less than 40%, air temperatures were greater than 25°C, and relative humidity was less than 30%. These conditions increased the above ground temperatures to lethal levels.

Fire also affects vegetation through its effect on primary productivity. The nutrients contributed by the ash have a dramatic effect on total biomass as has been observed in the Brazillian Cerrado by Cavalcanti (1978). The plots receiving ash produced significantly more biomass than those from which ash was removed. The time when ash is contributed in relation to the rainfall distribution also has an important effect on the biomass produced (Coutinho, 1982).

D. Consequences on soil physical properties

(i) Changes in soil structure and wettability: Repeated cycles of fire for

Table 12.3 Wettability of soil layers (given as mean wetting angle in degrees) before and the change resulting from fires of various intensities (Scholl, 1975). *Reproduced from Soil Science Society of America Journal, Volume 39, 356–361, by permission of the Soil Science Society of America*

Layer	Depth (cm)	High (331c) Lab.		270C Field		Medium 260 Lab.		Low 70C Field	
		Before	Change	Before	Change	Before	Change	Before	Change
1	0–1	74.9	−17.8*	74.6	+10.2*	70.3	+11.5*	65.9	+5.0
2	1–3	67.9	+10.0*	69.8	+3.4	66.2	+4.3*	63.7	+1.9
3	3–5	62.2	+7.4*	—	—	60.4	+0.4	—	—
4	5–7	60.5	+3.4	—	—	60.4	+0.4	—	—

centuries have changed the soil structure and wettability of many savanna soils. Water repellency is caused by coatings of organic materials on the soil surface. These materials are mainly aliphatic hydrocarbons (Scholl, 1975), which cause severe repellency problems in coarse-textured soils with low surface area. Intense heat during annual burning causes downward diffusion of volatile organic materials (De Ban et al., 1970). The materials condense immediately beneath the surface because of the steep thermal gradients. Coating of these condensed organic materials on the soil surface is the principal cause of water repellency of the burnt soils in tropical savannas. Scholl (1975) observed a significant increase in the wetting angle of the soil following burning. The wetting angle of the soil was increased up to a depth of 3 cm even at low field temperatures of 70°C (Table 12.3). The effect of soil heating on wetting angle decreases with increase in soil depth. The low-temperature fire increases the water repellence of the soil surface and the high-temperature burns water repellence in the subsoil layers up to a depth of 5 cm.

The water repellence thus created increases water run off, and with low vegetation cover following burning accelerates soil erosion. The loss of vegetation cover and large roots in the subsoil decreases total porosity and the proportion of macropores. The raindrop impact on the bare unprotected soil further compacts the soil surface. Soil compaction is aggravated by the diminished activity of soil fauna. Savanna fires, therefore, have severe effects on soil physical properties.

(ii) Soil texture: Fire-heating of soil is popularly practiced by subsistence farmers of Ethiopian highlands supposedly to improve soil fertility. According to this practice, called 'guie', the grass-fallowed land is repeatedly ploughed and then made into heaps (Tessema and Yirgou, 1967; Wehrmann and Johannes, 1965; Donahue, 1972). These heaps are about 60 cm in diameter, 80 cm high and number about 400/hectare. The heap is ignited by a small amount of cowdung placed in its centre and burned slowly for about 10 days. The soil temperature during ignition is variable and may be as high as 400°C (Sertsu and Sanchez, 1978). When cool, the soil from burnt heaps is spread over the entire field. This practice reportedly alters soil texture by fusion of clay-sized particles into sand. The mineralization of organic matter also improves soil fertility (Wehrmann and Johannes, 1965).

E. Burning and soil fauna

Physical properties of soils in the tropics are influenced by soil fauna, i.e., eartheworms, termites and other surface and subsoil-dwelling animals. Although little is known about the effects of man-induced forest and savanna fires on soil animals, one would expect direct adverse effects of high temperatures on surface dwellers. Indirect effects include a decrease in food quantity and diversity, increased exposure to predators, decreased soil moisture and increased environmental aridity. Similar to the effect on vegetation, fire creates a shift in fauna specie domination (Rambelli *et al.*, 1973). The effects of fire on soil fauna are reviewed by Meiklejohn (1955) and Gillon (1983), and have been discussed in chapters dealing with soil fauna.

IV. SOIL–VEGETATION INTERACTION

The area covered by savanna and grasslands within the tropics exceeds that of the rainforest. For examples, Shantz (1954) and Shantz and Turner (1958) estimated that tropical savanna occupies 6.7 million square miles compared with 5.8 million square miles covered by tropical forest vegetation. Some soil and climatic environments that would normally support evergreen rainforests are now under grassland and wooded savanna vegetation. Does the savanna ecosystem depend on fire or has fire transformed the fire-free ecosystem into savanna? This is a debatable issue indeed. There are examples of the man-made or derived savanna where man's perturbations have been so frequent and drastic that ecosystem instability has resulted. This instability has established, over centuries, the degraded vegetation, derived savanna, in place of the rainforest. If these savannas are protected from fire, the natural forest vegetation would be eventually restored.

However, there are two schools regarding the predominance of savanna vegetation, one believes in the relative importance of edaphic factors, and the other in the biotic factors. Some researchers argue that the characteristic soils of the savanna are the result of biotic factors, i.e., man's intervention in removing the climax vegetation. Sarmiento and Monasterio (1965) discussed the role of fire in relation to the origin and maintenance of natural savanna. The replacement of forest by man's intervention has led to the degradation of the exposed soil and the repeated burning has rendered it difficult for the perennial vegetation to re-establish. Similarly, Budowski (1956), while explaining the origin of savanna in tropical America, argued that repeated burning by man-induced fires created the initial dominance of fire-resistant species, which were eventually eliminated by the constant killing of regeneration until replaced by the fire resistant grasses and isolated tree species that now constitute the predominant vegetation. One of the most fire-resistant grasses is *Imperata* spp., which survives because of its rhizomes. *Imperata* savannas are widely observed in Asia, Africa and tropical America and have been created due to the degradation of forest vegetation by repeated fire (Fig. 12.7) (Chevalier, 1928; Waltson *et al.*, 1953). Man-induced fire in a fire-free ecosystem has disastrous effects on the community, which set back the vegetational development to some earlier successional stage (Lemon, 1968; Vogl, 1977). If fire is natural and mild, as it occurs in a fire-dependent ecosystem, a fire-conditioned community is maintained.

Instability created by repeated cycles of deforestation and fire in a fire-free ecosystem results in accelerated soil erosion, compaction due to the lack of large roots and decreased soil fauna activity, leaching and depletion of fertility due to larger amounts of rain water passing over and through the soil, and

Figure 12.7 Imperata savannah in Sumatra

formation of hard pans at shallow depths due to alterations in hydro-chemical processes. It is, therefore, likely that the depleted and infertile soils of the savanna have been created both by forest removal and man-induced repeated cycles of fire in an originally fire-free ecosystem. The two factors, however, interact and it is often difficult to differentiate the cause–effect relationship. Medina et al. (1978) have supported the argument that tropical savannas are characterized by soils containing low nutrient status.

Whatever the cause, the soils of savanna subjected to frequent fires have noticeably different physical, and biological properties. Soils of the savanna are more prone to erosion than those of the forest, are more compacted, and undergo higher extremes of temperature and moisture regimes than soils protected from annual fire by the perennial vegetation.

V. PRESCRIBED FIRE IN MODERN AGRICULTURE AND LAND MANAGEMENT

Although ecological effects of man-induced fire described in previous sections are rather adverse (Ahlgren and Ahlgren, 1960; US Forest Service, 1977; Gerakis et al., 1978; Rundel, 1982), yet fire is often prescribed as a technique/tool for land clearing, habitat management, pasture management and other purposes to restore soil productivity. This seems contradictory. In fact the use of prescribed fire in tropical agriculture is a debatable issue, and the answer to whether 'prescribed fire' can be used as a tool to our advantage depends on the purpose, the conditions in which this tool is used and its interaction with social and biophysical environments. The use of 'prescribed fire' in modern agriculture for specific objectives is different from the use of fire in 'traditional' subsistence agriculture because in the latter case the environmental consequences are often ignored.

In modern agriculture fire can be used as a tool for three broad objectives: (i) land clearing, (ii) vegetation management for improved pasture, and (iii) plant nutrient improvement.

A. Land clearing

It is generally agreed that deforestation for arable land-use is better ecologically if it is done by manual rather than mechanized methods. Manual felling followed by *in situ* burning of the felled vegetation is often prescribed for acid and highly leached soils. In addition to benefits on soil chemical and nutritional properties (Nye and Greenland, 1964; Seubert, 1975), soil physical properties are also better maintained by manual-cum-fire land clearing than by mechanized methods (Lal and Cummings, 1979).

The soil temperature during *in situ* burning of felled vegetation depends on the tree density and fuel available. In the sub-humid region of southwest Nigeria, Lal and Cummings (1979) observed soil temperatures exceeding 70 °C at 1 cm depth and exceeding 40°C at 10 cm depth (Fig. 12.8). Burning in windrows often results in drastic alterations in soil physical properties because the intensity and duration of burning is much higher than with *in situ* burning

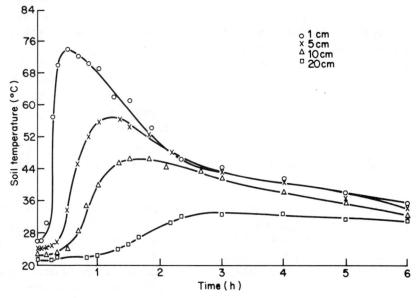

Figure 12.8 Soil temperatures during burning in windrows following deforestation (Lal and Cummings, 1979)

following manual clearing. The addition of a large quantity of nutrient-rich ash in the windrows can be advantageous for the succeeding crops (Ling and Mainstone, 1983).

B. Forest and Savanna management

Prescribed controlled fire is also used for management of forest reserves and savanna lands and to create fire breaks to prevent wildfire hazard. This is achieved by the controlled burning of a broad strip of land, thereby reducing the amount of flammable material and lessening the chances of intense wildfire during the dry season.

C. Pasture management

The use of prescribed fire is widely recommended for pasture improvement and rangeland management (Vincent, 1936; Cook, 1939; Masefield, 1948; Daubenmire, 1968a, b; Kayll, 1974; San José and Medina, 1975). Fire is recommended for pasture improvement in Nigeria (Moore, 1960) and in tropical Australia (Stocker and Sturtz, 1966; Norman, 1963, 1969; Tothill, 1971a, b). In an 11 year experiment conducted at Katherine, Northern Territory, Australia, Norman (1969) observed that the annual dry matter production and total N output of control plots (protected from fire for 5 to 10 years) was low in comparison with plots burned biennially. On the other hand, production was high in the first 5 years after burning on plots that had been protected for 5 years and then burned. Norman recommended burning once every 5 years or so rather than annual or biennial burning or no burning at all.

The use of fire for pasture improvement in Africa is advocated by West (1965, 1971). In forest ecology, West recommends the repeated use of controlled burning to eliminate trees and shrubs and to establish grass. Some researchers argue that complete protection from burning and grazing for a long period can render the pasture most unattractive from a grazier's point of view Van Rensburg (1952) recommended that pastures should be burnt periodically to maintain palatability and nutritive properties of the herbage. He recommended that the best time to burn to promote vigorous growth is the end of the dry season, i.e., late burn. Burning early or mid-dry season often results in brush encroachment. Scott (1970) observed that some adverse effects of intense burning are minimized by occasionally substituting mechanical mowing for burning.

The following recommendations are made for controlled burning for pasture improvement (Kayll, 1974):

    (i) burn when plants are best able to recover quickly;
    (ii) burn only when soil and plant crowns are damp after a rain to minimize heat penetration to growing points;
    (iii) burn when there is a breeze to move the fire quickly; and
    (iv) avoid close and early grazing after burning.

Kayll concluded that where grazing is not intensive, fire seems to be an adequate and cheap management system for tropical grasslands to prevent the undesirable accumulation of organic matter. It is important, however, to establish precise methodology and conditions under which fire can be most beneficially used.

D. Crop residue management

Although it is a controversial issue and will be discussed at length later in Chapter 15, some researchers recommend burning the crop residue for disease control and for adding, through ash, some readily available plant nutrients. This practice, however, has adverse effects on soil and is not widely recommended for tropical environments.

VI. EFFECTS OF PRESCRIBED FIRE ON SOIL AND ENVIRONMENT

The use of controlled fire is a controversial issue, and it is difficult to discuss its pros and cons without reviewing the effects of fire on vegetation, soil, microclimate, and hydrological conditions. Although the uses and abuses of fire in the tropics have extensively been reported, it is difficult to separate the emotion from facts without sufficient scientific data. The following is a review of the effects of prescribed fire as a management tool on vegetation, soil, and ecology of the tropics:

A. Vegetation

Even controlled fire has significant effects on the predominant vegetation. That is the principle underlying its use for improving the pasture (Garren, 1943; Shantz, 1947; Hodgkins, 1958). In Venezuela Medina et al. (1977) observed that burning eliminated shrub and tree regrowth in savanna dominated by *Trachypogon plumosus*. In East Africa Bogdan (1954) observed that burning suppressed regeneration of thorn trees and shrubs. Consequently the proportion of couch grass in the pasture was increased by burning. The effects of fire on an African Plateau grassland in Malawi are reported by Lemon (1968). These experiments were conducted in a region 10° 30'S latitude at altitude 2000 to 3000 m asl with annual rainfall of about 1150 mm received in a period of 6 months. Lemon observed that the number of legume plants in an annually burned pasture was about double that on the unburned control.

Hopkins (1965) studied the vegetation shift in the Olokemeji Forest Reserve in Nigeria by annual burning from 1959 to 1964. The treatment plots were burnt during the dry season between late January and late February each ysar. All the trees and shrubs 2 m or more high on research plots were enumerated in February 1959 and again in February 1964. Of the 139 individuals present in 1959, only 92 were alive 5 years later, representing a net loss of 32% and an average loss of 7.9% per year. The survival rate increased with increasing height and basal area. Similar adverse effects of burning on young regeneration were reported by Trapnell (1959) from Zambia. The most fire-resistant species were *Burkea africana* and *Lophira lanceolata*. Less resistant species included *Butyrosperumum paradoxum* and *Peterocarpus erinaceous*. The four species that failed to survive 5 years of annual burn were *Bridelia ferruginea, Parkis clappertoniana, Parinari curatellifolia* and *Syzygium guineense*. The species that increased in number due to annual burn were *Burkea africana* and *Lophira lanceolata*.

The effects of fire on tropical vegetation have been reviewed by Phillips (1974). Trapnell (1959) studied shift in vegetation composition as a result of woodland burning in northern Rhodesia. He observed that late burning was more severe than early burning. The fire-sensitive species easily eliminated by annual fire were *Brachystegia specififormis, Julbernaridia paniculata*, and *Isoberlinia angolensis*. In Natal, South Africa, Killick (1963) reported that frequent burning not only maintained the grassland community but also favoured *Themeda triandra* over rank and coarse grass species such as *Hyparrhenia*. Tinley (1966) proposed the use of fire to control wood species in Botswana. Trollope (1974) observed that the use of a single intense head fire in a *Themeda triandra* dominated grassland infested by *Acacia karroo* and other bush species resulted in 81% topkill of trees and shrubs. Trollope (1982) reviewed the problem of bush encroachment in relation to veld management in South Africa, and concluded that fire *per se* favours the development and maintenance of a predominantly grassland vegetation by destroying the juvenile trees and shrubs and preventing the development of more mature plants to a taller, fire-resistant stage. The South African experience indicates, however, that once the bush has become dominant and is suppressing the grass, fire is no longer effective because the grass fuel is not enough to support an intense fire. The fire effectiveness in bush control depends on the season, frequency, type, and intensity of burning. There is also a strong interaction between grazing and fire as a determinant for predominant vegetation. For example, Boughey (1963) argued that in Rhodesia large animals and fire have prevented the development of the fairly dense woodland the climate and ecology would normally support.

In Venezuela San José and Medina (1975) observed that burning changed the competition relation among three predominant grass species of the llanos. Protection against fire increased the proportion of *Trachypogon montufari* against *T. plumosus* and *A. canescens*. In fact increased biomass production of the burned plots was due to higher productivity of subordinate and not of the dominant species. Repeated use of fire eliminates perennial grasses, leading to a rapid successional change in the savanna with high initial proportion of the dicot (Aristeguieta and Medina, 1965). Coutinho (1982) compared alterations in fire-protected vegetation of Brazillian Cerrado from 1944 to 1977. He observed change in the physiognomy of Campo Cerrado with low dwarf trees or shrubs. Herbaceous and undershrub species lost their vigour by fire protection, and the quick growing trees and shrubs became much more dense and shaded the undergrowth layer. The shrub was infested with tall molasses grass *Melinis minutiflora*, which was eventually eliminated and replaced by trees.

The effects of fire on vegetation succession for Australian tropics are described by Lacey *et al.* (1982). The seedling establishment of woody perennials is restricted by frequent burning and severe drought. Norman (1963, 1969) compared the effects of annual and biennial burning of perennial pasture on shifts in species composition. Norman reported that burning during dry intervals of the wet season can cause a dramatic change in perennial species composition. Also in northern Australia, Smith (1960) observed that mid-wet-season fire suppressed the population of *Sorghum plumosum* but that of *Chrusopogon fallax* was not affected.

B. Soil physical properties

The direct short-term effects of prescribed fire on soil physical properties are negligible, unless it is a very intense burning. Repeated cycles of frequent man-induced burning, as has been the case in most of the tropics, have created adverse effects on soil physical properties (Raison, 1979). The negative effects of burning on soil physical properties are aggravated by drastic decrease in activity of soil fauna.

1. Particle size distribution

Some researchers have observed a decrease in the clay content of soil that has been subjected to frequent burning over a long period of time. Sreenivasan and Aurangebadkar (1940) reported the effects of fire-heating on the physical and mechanical properties of a Vertisol at Indore, Central India. Their data of mechanical analysis (Table 12.4) show a marked decrease in clay content and a corresponding increase in silt and fine sand as a result of heating. The magnitude of the difference observed was less with the fire-heated as compared with the oven-heated soil. In the compost-treated soil, however, the decrease in clay content by heating was relatively small. The overall effect of fire heating was in creating a 'light-textured' soil. The authors concluded that fire-heating

Table 12.4 Effects of heating on texture of some soils from Central India (Sreenivasan and Aurangebadkar, 1940). *Reproduced by permission of Williams & Wilkins Co.*

Treatment	Percentage					Index of texture
	Coarse sand	Fine sand	Silt	Fine silt	Clay	
Untreated control	1.3	7.0	13.5	13.8	48.3	40.6
Lightly heated	1.1	9.7	22.1	23.3	19.7	42.0
Strongly heated	0.8	9.8	21.5	23.1	21.9	41.8
Fire heated	5.4	11.3	18.8	22.8	29.5	29.1

Table 12.5 Effect of soil heating on particle size distribution of some soils of Ethiopia (Sertsu and Sanchez, 1978)

Soil	Treatment (°C)	Sand (mm)					Total sand	Silt	Clay	Textural class
		2–1	1–0.5	0.5–0.25	0.25–0.10	0.1–0.05				
Ethiopian	26	2.2	2.1	1.2	2.3	2.5	12.0	37.4	48.8	Clay
	100	1.8	2.2	2.2	4.1	3.2	14.5	37.2	48.8	Clay
	200	2.7	3.2	1.4	3.7	3.3	15.4	41.4	44.5	Clay
	400	19.6	17.3	9.2	13.9	9.7	72.1	25.6	5.1	Sandy loam
	600	33.7	22.2	8.9	11.9	7.2	88.6	10.1	1.4	Loamy sand
Vertisol	26	2.1	6.5	4.9	7.0	3.7	24.8	12.8	55.2	Clay
	100	2.4	7.7	5.6	7.4	2.8	26.2	12.5	52.8	Clay
	200	3.0	6.1	5.3	7.4	2.6	24.8	16.5	52.6	Clay
	400	2.4	7.2	7.0	8.7	3.4	29.6	33.5	28.2	Clay loam
	600	26.3	30.1	14.7	16.0	3.9	91.3	2.6	0.4	Sand
LSD (0.05)		6.1	3.1	1.8	3.3	4.0	5.1	5.4	3.7	
cc (%)		52	24	23	23	25	9	9	7	

Table 12.6 Effect of burning on texture of Llanos soil at Calabozo, Venezuela (San José and Medina, 1975). *Reproduced by permission of Springer-Verlag*

Treatment	Sand (%)	Silt (%)	Clay (%)
Unburnt control	55.5	19.0	25.9
Burned	57.6	16.2	26.3

Each mean represents an average of 22 samples.

of Vertisol reduced its colloidability by increasing the relative proportion of the coarse fraction.

The effects of heating on change in soil texture have also been reported by Nishita and Haug (1972). These researchers observed change in soil texture only by temperatures exceeding 400°C. At higher temperatures the sand fraction increased with increase in temperature while silt and clay content decreased. Sertsu and Sanchez (1978) conducted laboratory experiments to simulate the effects of Ethiopian 'guie' practice on soil texture. They observed that heating up to 200°C had no effect on particle size distribution but at 400 and 600°C the sand content increased and the clay content decreased (Table 12.5). These textural changes were more pronounced in Vertisol with higher initial clay content than in a coarse-textured soil. The texture of the Vertisol with 55% clay changed to sand. These textural changes were attributed to fusion of clay particles into sand-sized particles.

Very high temperatures of the soil surface during natural fire may also alter soil texture. For example, under field conditions, San José and Medina (1975) observed in llanos in South America an increase in sand content of the soils from burned plots in comparison with the protected plot (Table 12.6). The effect may partly be due to fusion of clay into sand, and partly due to preferential removal of clay in fire-induced accelerated run-off and soil erosion.

2. *Structural stability*

The effects of natural and man-induced fire on structural stability depend on the soil and its mineralogical composition. Sreenivasan and Aurangebadkar (1940) observed an improvement in structural stability of the heated Vertisol. Their data (Table 12.7) showed an increase in the ratio of pore space to clay. The percentage of aggregated soil increased gradually with increase in heating intensity as is indicated by decrease in the dispersion coefficient. In contrast to their findings, Fogliata *et al.* (1967) reported decreases in porosity and structural stability of soil subjected to repeated cycles of burning of sugarcane trash. Similarly Giovannini and Lucchesi (1983) observed a decrease in aggregate stability of fire-heated soil. These seemingly contradictory results are explainable on the basis of intensity of burn and the temperatures attained. During

Table 12.7 Effects of heating on structural properties of some soils from central India (Sreenivasan and Aurangebadkar, 1940). *Reproduced by permission of Williams & Wilkins Co.*

Treatment	Total clay (D) (%)	Aggregated clay (S) (%)	Dispersion coefficient $\frac{D-S}{D} \times 100$
Untreated control	47.5	22.6	52.4
Lightly heated (90–100°C)	20.1	9.0	55.2
Strongly heated (140–150°C)	22.4	8.4	62.5
Fire heated	30.1	5.1	83.1

very intense burning e.g. in case of heated soils, structural stability may increase. For light burning, on the other hand, as in burning of sugarcane trash, destruction of crop residue may decrease structural stability. The effects of temperature on soil structure also depend on clay mineralogy. For example, Bruce-Okine and Lal (1975) observed that oven-drying at 105°C decreased the structural stability of a Vertisol but increased that of an Alfisol (Table 12.8). The number of drops required to break the structural ped of a Vertisol decreased from 194 for the air-dry sample at pF 4.4 to 21 for the oven-dry sample. In contrast, the drops required to disrupt the structural ped of the Alfisol increased slightly by oven-drying (Table 12.8).

3. Wettability

The effects of fire-induced water-repellence of soil have been widely recognized. For coarse-textured surface soil horizons, even low temperatures of 70°C, commonly observed during natural and man-induced fire, can impart water-repellence properties (Scholl, 1975). Hydrophobic characteristics of heated soils have also been reported by Kelly and McGeorge (1913), Rao and Madhawan (1955) and Rotini *et al.* (1963). On the contrary, Giovanni and Lucchesi (1983) did not observe any effect of fire on water repellence. The

Table 12.8 Effect of oven-drying on structural stability of a Alfisol and a Vertisol (Bruce-Okine and Lal, 1975). *Reproduced by permission of Williams & Wilkins Co.*

Soil	Treatment	Number of water drops (30°C) required for structural breakdown
Vertisol	Air-dry (pF 4.4)	194
Alfisol	Air-dry (pF 4.4)	23
Vertisol	Oven-dry (pF 7.0)	21
Alfisol	Oven-dry (pF 7.0)	25

magnitude of the effect of fire on water repellence is obviously more for coarse textured soils containing predominantly low activity clays than for fine textured soils containing expanding-lattice clay minerals.

4. Moisture retention properties

High temperature burns influence soil organic matter content, texture, pore size distribution and soil structure. It is apparent, therefore, that moisture retention properties would also be affected by burning that increases soil temperature sufficiently high to affect soil matrix and its arrangement. Edwards (1942) studied the effects of frequency and intensity of grazing and of fire on physical properties of an Alfisol in Kenya. His data (Table 12.9) show that burning decreased the hygroscopic soil moisture content, the moisture content at the sticky point, and the cation exchange capacity. The decrease in the hygroscopic coefficient and the sticky point are in fact due to loss of colloidal properties, i.e., decrease in soil organic matter and clay content. The worst treatment regarding the effects on soil physical properties was the combination of grazing for 5 years, fire control for 3 years followed by severe burning, and then protection for 2 years. Similar results of fire-induced decrease in the moisture retention properties of a Vertisol from Central India were reported by Sreenivasan and Aurangebadkar (1940). They observed that fire-heating decreased the hygroscopic coefficient, wilting coefficient, and the moisture content at the sticky point of both Vertisol and a Gray soil (Table 12.10). There was also a decrease in the maximum water-holding capacity and the total porosity. San José and Medina (1975) observed for Ilanos at Calabozo, Venezuela, that soils from plots receiving regular burning treatments retained less water at 15 bar suction, maximum water-holding capacity, and at field capacity than the unburnt control (Table 12.11). Significant differences were observed in *in situ* measured field moisture capacity –25.8 versus 31.2 per cent for the burnt and control plots, respectively.

The available literature indicates that repeated occurrence of intense fire decreases soil moisture retention properties, especially the moisture retention at field moisture capacity. This would imply a decrease in the available water-holding capacity of the regularly-burnt soil and an increase in its droughtiness for shallow-rooted crops.

5. Soil moisture balance

A few studies have been conducted regarding the effects of fire on soil water balance. Athias *et al.* (1975) observed in the derived savanna at Lamto, Ivory Coast, that the soil moisture of the burnt plots was depleted rapidly immediately after the burning. High soil-water loss was due to an increase in surface soil temperature and high soil evaporation. Burning increased the effects of the dry season by rapidly depleting the moisture reserves of the surface horizon.

Effects of fire on soil water balance have been investigated for tropical

Table 12.9 Effect of burning treatments on soil (0–7.5 cm depth) physical properties for some soils in Kenya (Edwards, 1942). *Reproduced by permission of Cambridge University Press*

Plot	Treatment	Hydroscopic moisture (%)	Loss on ignition (%)	Sticky point (%)	CEC (meq/100 g)	Mean placing
1	Complete protection from grazing and fire for 10 years	8.11 (1)	11.68 (1)	37.3 (1)	11.4 (1)	(1)
2	Ditto for 8 years, severe fire, and protection for 2 years	6.57 (6)	11.41 (3)	35.1 (6)	10.8 (4)	(5)
3	Periodical fire for 5 years protection for 3 years, severe fire and protection for 2 years					
4	Periodical fire with light grazing for 5 years, protection for 3 years, severe fire and protection for 2 years	6.51 (6)	9.38 (8)	33.4 (7)	10.2 (8)	(7)
5	Over grazing for 5 years and complete protection for 5 years	7.80 (2)	11.61 (1)	37.5 (1)	11.1 (2)	(1–2)
6	Light grazing for 5 years and complete protection for 5 years	7.48 (3)	11.11 (4)	37.4 (1)	11.0 (2)	(2–3)
7	Light grazing for 5 years, protection for 3 years, severe fire and complete protection for 2 years	7.31 (3)	10.74 (5)	35.7 (4)	10.7 (4)	(4)
8	Medium protection for 5 years, protection for 3 years, severe fire and complete protection for 2 years	6.34 (8)	9.82 (7)	32.9 (8)	10.4 (6)	(7)

The values in parentheses refer to the relative magnitude of value in decreasing order.

Table 12.10 Effects of fire-heating of some soils from Central India on soil moisture constants (%) (Sreenivasan and Aurangebadkar, (1940). *Reproduced by permission of Williams & Wilkins Co.*

Treatment	Hygroscopic moisture	Wilting coefficient	Sticky point	Saturation capacity	Pore space
Untreated control	9.5	14.0	42.4	64.3	59.8
Lightly heated (90–100°C)	8.0	11.8	44.3	72.7	60.0
Strongly heated (140–150°C)	7.8	11.5	44.1	80.0	60.0
Fire heated	5.5	8.0	32.6	57.6	57.6

Table 12.11 Effect of burning of Llanos in Venezuela on moisture retention properties (San José and Medina, 1975). *Reproduced by permission of Springer-Verlag*

Soil moisture constant	Control	Burned
Moisture content at -15 bar (%)	7.1	6.5
Maximum water-holding capacity $-$lab. (%)	32.8	29.1
Minimum soil–water content (field) (%)	5.1	5.0
Field capacity (*in situ*) (%)	31.2	25.8

savannas in the Estación Biólogica de los Llanos at Calabozo, Venezuela, by San José and Medina (1975). Following burning, soil water content in the burnt plot was higher than in the unburnt area due to reduced transpiration (Figs. 12.9 and 12.10). This difference disappeared after the first rain due to vigorous regrowth on the burnt plot. Subsequently, the soil moisture content of the burnt plot remained lower than the protected control. Furthermore, all soil layers in protected plots were moistened homogeneously, with the subsoil layers attaining the saturation level. The absence of complete soil cover in the burnt plot increased losses due to water run off and evaporation and thus decreased the amount of soil water recharge. As a result the soil of the burnt plot never attained the saturation point. The water balance for the burned and protected plots is shown in Table 12.12. The data show that unburnt protected plots had more percolation and soil water recharge especially since the vegetation cover caused little run off. The vigorous growth and high leaf area index of the burned plot, however, increased transpiration, i.e., 1440 versus 1105 mm for burned and protected treatments, respectively.

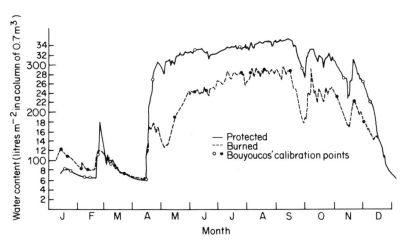

Figure 12.9 The effect of burning on soil moisture reserves in Llanos, Venezuela using field lysimeters (San José and Medina, 1975). *Reproduced by permission of Springer-Verlag*

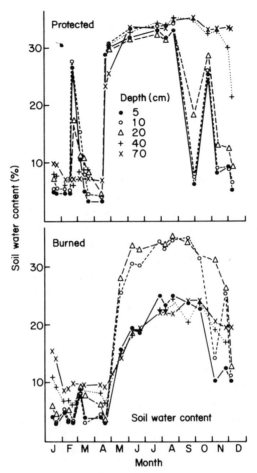

Figure 12.10 The effect of burning on soil moisture reserves in Llanos, Venezuela under natural field conditions (San José and Medina, 1975). *Reproduced by permission of Springer-Verlag*

Table 12.12 Soil–water balance for burned and protected plots at Calabozo, Venezuela (San José and Medina, 1975). *Reproduced by permission of Springer-Verlag*

Parameter	Burned	Control
Rainfall (mm)	1839	1839
Pan evaporation (mm)	2406	2406
Available soil–water (mm)	1327	1850
Evapotranspiration (mm)	1440	1105

Fire used for pasture regeneration also influences the plant-water relations of the regrowth. In Australia Fisher (1978) investigated the recovery of leaf water potential following burning of two tropical pasture species that had suffered from severe drought. The pasture species studied were (a) green panic cv *Petrie* and (b) *Macroptilium atropurpureum* cv *Siratro*. Measurements were made of the changes in leaf water potential (ψ_s and stomatal conductance (g_s) of the burned and unburned pasture plots. The ψ_s of the burned plots was higher (−14 to −18 bar) than that of the unburned control (−23 to −45 bar). The stomata of unburned plots were closed and no fresh growth was observed. The leaf water potential of both species 12 days after burning was −9 to −11 bars; the stomata were open with vigorous fresh growth. Fisher argued that complete defoliation allowed the plants to make use of a limited store of soil water. High rate of new growth in burned plots was maintained even 28 days after burning (CSIRO, 1979).

6. *Water transmission properties*

Changes in wettability, structural stability, pore size distribution and moisture retention properties due to repeated use of fire over a long period of time can also influence water infiltration rate and rainwater acceptance. The latter is also influenced by the lack of protective vegetation cover on burned plots especially for late burning. Impact of high intensity rains on unprotected soil results in slaking, crusting, and decrease in water infiltration. Burning also hastens the drying process, and the surface layer of the burnt plot attains higher pF than the protected soil of the unburnt plots. The water infiltration rate of a dry soil near wilting point is less than a soil at higher soil-water potential.

Lal (1981) studied the effects of oven drying of an Alfisol on its infiltration rate (Figs 12.11 and 12.12). The oven-drying of surface soil significantly decreased water penetrability, i.e., in 2 minutes the wetting front advanced 34 cm in u column containing oven-dried soil compared to 40 cm in the air-dried soil. The oven-drying of the Alfisol subsoil, containing low organic matter content but more iron and aluminium oxides, increased the rate of advance of the wetting front. In the oven-dry subsoil the wetting front penetrated to a depth of 43.5 cm in 10 minutes, about 4 times increase over that of air-dry soil. Collis-George and Lal (1971) also reported lower infiltration into oven-dried Kraznozem and Vertisol compared with soils at higher initial soil moisture potential (Fig. 12.13). Both the soil water sorptivity (S) and the transmissivity of the oven-dried Kraznozem were drastically reduced (Table 12.13). The decrease in infiltration into dry heated soil is partly due to entrapped air and partly, for soils of high surface area, due to heat of wetting suddenly released on quick wetting (Collis-George and Lal, 1973). The drier the soil the more the heat of wetting and the less the infiltration rate. The effect of heat of wetting on decreasing the infiltration rate is more for unstable slaking Vertisols and similar heavy-textured soils than for coarse-textured Alfisols, Oxisols or Ultisols.

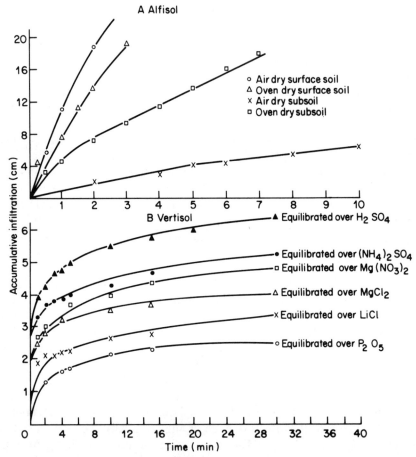

Figure 12.11 The effect of soil heating on water infiltration rate of an Alfisol and Vertisol (Lal, 1983)

Lal and Cummings (1979) observed a greater decrease in the infiltration rate of an Alfisol following *in situ* burning than in the unburnt control (Fig. 12.14). The cumulative infiltration at 2 hours was 140 and 175 cm for burned and control plots, respectively. The equilibrium infiltration rate was 44 and 63 cm/hr for burned and unburned plots.

7. Water run-off and soil erosion

Some ecologists argue that soil erosion by fire is increased more by man-induced fires in non-fire ecosystems than in fire-dependent savanna ecosystems (Viro, 1974; Vogl, 1974, 1977). In the fire-free ecosystem, the increased run-off and high risk of splash erosion are direct consequences of decreased

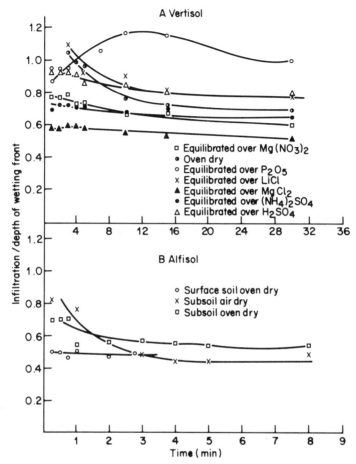

Figure 12.12 The effect of soil heating on water infiltration rate of an Alfisol and a Vertisol (Lal, 1983)

Table 12.13 The effects of oven-drying of an Alfisol on parameters S and A of the Phillip approximate infiltration equation (Collis-George and Lal, 1971). *Reproduced by permission of CSIRO*

pF	4.4	5.60	5.98	6.18	6.48	7.0 (oven-dry)
S (cm s$^{-1/2}$)	—	0.78	1.45	1.45	1.0	0.70
A (cm s^{-1})	0.333	0.156	0.113	0.066	0.037	0.020

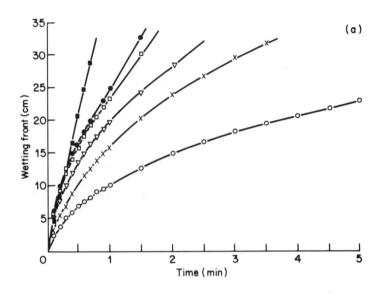

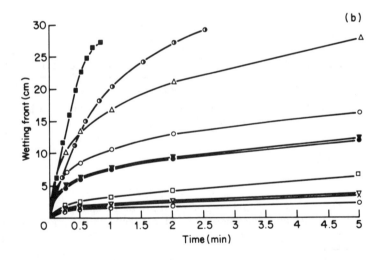

Figure 12.13 The effects of initial soil moisture potential on water infiltration into (a) Kraznozem and (b) Vertisol (Callis-George and Lal, 1983). ■ 98% RH, pF 4.44; ⊙ 95% RH, pF 4.84; △ 90.2% RH, pF 5.14; ○ 84.3% RH, pF 5.37; ▲ 80% RH, pF 5.49; ● 75.4% RH, pF 5.60; □ 50.9% RH, pF 5.98; ▽ 33% RH, pF 6.18; X 11.1% RH, pF 6.48; ○ over-dry, Pf 7.00

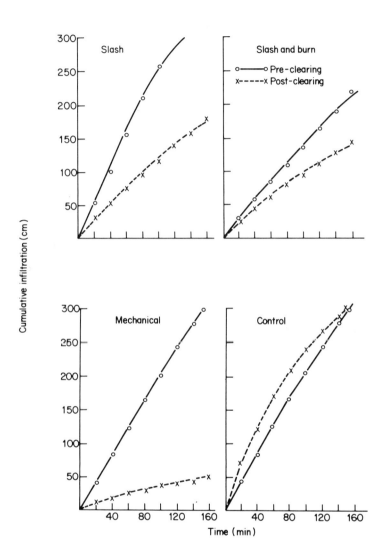

Figure 12.14 The effects of *in situ* burning following land clearing on water infiltration rate of an Alfisol (Lal and Cummings, 1979)

wettability and reduced infiltration rate of an unprotected and denuded burned soil surface. Qualitative observations in subtropical Africa by West (1965) and Du Pleiss and Mostert (1965) indicate an increase in erosion hazard on steep slopes following burning. In Ivory Coast Roose (1971) studied the effects of savanna burning on soil erosion. He estimated that annual soil erosion from burnt savanna plots was 3 times that of the unburnt control—150 and 50 kg ha^{-1} yr^{-1}, respectively. Similarly Granier and Cabanis (1976) measured annual soil erosion of 610 kg ha^{-1} from burnt grassland savanna in comparison with only 13 kg ha^{-1} from unburnt control. In Zaire, Central Africa, Soyer *et al.* (1982) observed that rainsplash in wooded savanna was lower than in cleared woodland, except after late savanna burning. Soil splash was less than 10 t ha^{-1} yr^{-1} if the annual late fires were adequately controlled.

The magnitudes of soil erosion even on burnt savanna plots reported above are low and within the acceptable limits of tolerable soil loss. In the forest ecology, however, burning following deforestation results in very severe soil erosion. For an example, in a prescribed burning experiment for pineapple plantation in the forest region of southern Ivory Coast, Roose and Asseline (1978) observed that burning of crop residue resulted in a significant increase in soil erosion in comparison with unburnt treatment where the residue was left on the soil surface (Table 12.14). In Papua New Guinea Klaer and Loffler (1980) observed severe soil erosion in wet/dry marginal tropical regions caused by desiccation of the soil surface due to prolonged dry season and uncontrolled fire.

The available literature indicates that use of fire accelerates soil erosion in the forest ecology more than in savanna ecology. Regardless of the ecosystem, late burning increases the erosion hazard more than early burning or rainy-season burning. Increase in soil erosion following burning is due to the lack of protective vegetation cover on the one hand and the decline in soil structure and infiltration rate on the other. Relating frequency, time, and season of burning to erosion hazard for different soils and ecology should be given a research priority.

Table 12.14 Effect of residue management on soil erosion (t/ha) in pineapples in Ivory Coast (Adapted from Roose and Asseline, 1978). *Reproduced by permission of ORSTOM*

Slope (%)	Bare soil	Soil erosion under different treatments		
		Residue burnt	Incorporated	Mulch
4	15	0.2	0.3	0.001
7	102	3.8	0.06	0.0001
20	253	16.7	9.7	0.007

8. Soil compaction

Fire-induced changes, e.g., the removal of the protective vegetation cover, the exposure of structurally unstable soils to high intensity rains, a decline in faunal activity and a decrease in root biomass are likely to cause compaction of the soil surface. A few experiments conducted on deforestation in the tropics have indicated that forest removal *per se*, by whatever means, results in soil compaction.

In Sri Lanka, Joachim and Kandiah (1942) observed that removal of forest cover resulted in subsoil compaction and slow internal drainage. Studies conducted in Brazil (Freise, 1934, 1939; Budowski, 1956) have shown that total porosity is greater in soils of the fire-free ecosystem (forest) than in those under fire-prone i.e. derived savanna. Proection of savanna vegetation from fire and re-establishment of forest cover increases total porosity and soil permeability. Deforestation-induced soil compaction in Brazilian tropics was attributed to the absence of large tree roots, and shrinkage of soil due to ultra-desiccation (Budowski, 1956). The exposure of soil to extremes of temperature, impedance of internal drainage, and alterations in the water balance resulting in greater fluctuations in water table are all related to the formation of iron pans in tropical soils. Occurrence of iron pans in the derived and guinea savanna zone of Africa is a common phenomenon (Aubert, 1950; Auberville, 1947; Kellogg and Davol, 1949). Sudres (1947) and Mohr and Van Baren (1954) ascribed the deforestation-induced iron pan formation to the following fire-caused alterations: (i) decrease in soil organic matter content, (ii) leaching out of the soluble bases and the resultant increase in the concentration of insoluble Fe, Al, and Mn, (iii) leaching of silica, (iv) deposition of insoluble bases as concretions, i.e., formation of plinthite, and (v) hardening of plinthite on exposure and desiccation. The fluctuating water table and the profile drainage conditions obviously play an important role in this process. The formation of iron pan and plinthite is an example of a long-term effect of fire-induced deforestation and changes in soil physico-chemical and hydrothermal regimes.

Soil compaction is also caused immediately after a fire as a direct short-term effect. In Zambian savannas, compaction of the surface soil was more in burnt than in fire-protected unburnt plots (Brockington, 1961; Trapnell *et al.*, 1976). In a forest ecology in southern Nigeria, Lal and Cummings (1979) observed higher penetrometer resistance in burnt than unburnt plots (Table 12.15).

Fire, therefore, causes compaction of surface soil immediately after the event and of the subsoil by repeated incidents over a long period of time. Subsoil compaction is a long-term effect of man-induced fires, the result of drastic alterations of the hydro-thermal regime and the soil physico-chemical and biological properties.

C. Nutrient losses and changes in soil fertility

The effects of fire in tropical ecosystems on nutrient cycling and modifications

Table 12.15 Effects of *in-situ* burning of manually cleared vegetation in western Nigeria on penetrometer resistance (kg cm/$^{-2}$) (Lal and Cummings, 1978)

Treatment	0–4 cm	5–9 cm
Forest control	0.59	1.60
Slash	1.86	3.31
Slash and burn	2.89	3.55
LSD (0.05)	0.35	

of soil fertility have been discussed in many recent reviews (Wells *et al.*, 1979; Raison, 1979, 1980; Lal and Kang, 1982). Although changes in nutrient cycling and soil fertility are important, only a general reference will be made here and readers are referred to the literature cited for more details. Two aspects that need special mention are (i) losses of nutrients during burning, and (ii) the addition of nutrients to the soil through the ash.

The losses of N due to volatilization and run-off have been discussed for Northern Territory, Australia, by Norman and Wetselaar (1960) and for West Africa by Rosswall (1980). Nutrient losses by natural fires are less in the fire-dependent savanna than in the forest ecosystem. For an example, Villecourt *et al.* (1979) estimated nutrient losses by annual fire in Lamto savanna, Ivory Coast. These authors observed that nutrient content of the above-ground plant material just before the annual fire is generally low (0.28% N, 0.06% P, and 0.28% K). Furthermore, the below-ground biomass is usually higher in the Lamto savanna at the time of burn than the above-ground biomass (7–12 t ha^{-1} versus 6 t ha^{-1}). The nutrient losses by annual fire, are, therefore, limited. In another study Villecourt *et al.* (1980) estimated the nutrient losses by annual savanna burning to be 2 t ha^{-1} C, 10 kg ha^{-1} N, 0.4 kg ha^{-1} P, 5.3 kg ha^{-1} K, 3.9 kg ha^{-1} Ca and 0.3 kg ha^{-1} Na.

In contrast, some researchers argue that the low nutrient losses in savanna fire are due to the fact that nutrient reserves have already been depleted by centuries of repeated fire, and that fire-dependent ecosystems are progressively being rendered infertile. For an example, the poor nitrogen status of the savanna soils has been attributed by some to the perpetual loss of N by the use of fire over centuries (De Rham, 1973). In addition to direct loss of N during fire, subsequent changes in moisture and temperature regimes are also responsible for N depletion in the savanna ecosystem.

The magnitude of nutrient loss depends on the amount of nutrients stored in the biomass and in the top soil layers. The higher the nutrient reserves, the more the loss. Okoro (1981) studied the effect of burning fallow vegetation on nutrient addition to an Alfisol and an Ultisol in Nigeria. The quantity of ash after burning was small, ranging from 310 kg/ha in grass fallow to 675 kg/ha in a forest-fallow. The nutrient recovery in ash was merely 41% of the amount stored in the above-ground biomass. Low recovery implies a considerable loss of nutrients during burning. The magnitude of nutrient loss, therefore, depends on the biomass, length of the fallow, vegetation, and the intensity and duration of the burn.

One of the often mentioned reasons for the use of fire in traditional farming is the addition of ash to enrich plant nutrients and supply acidity-neutralizing bases (Stent, 1933; Laudelout, 1954; Moore, 1960; Uribe et al., 1967; Brinkman and De Nascimento, 1973; Boyle, 1973; Fassbender, 1975; Afolayan and Ajayi, 1979; Coutinho, 1979). Wood ash formed by burning of trees and shrubs supplies Ca, P, K, Mg and S (Young and Golledge, 1948), and is generally very rich in K (Corbet, 1934). Some researchers have also observed an increase in N content of the surface layer of the burnt plot, probably due to fire-induced changes in microflora (Meiklejohn, 1955). In western Nigeria Lal and Cummings (1979) compared the effects of burning on chemical properties of soil sampled a week after the event. The data shown in Table 12.16 indicate that burning significantly increased soil N, Ca, Mg, K, Na, Mn and soil pH. The organic matter content was, however, not affected. The nutrient addition to the soil through ash is an example of the short-term effects of burning.

The cumulative effects of periodic fire on soil chemical properties can only be surmised from comparative analyses of soils under different vegetation. The results of such comparison are often difficult, if not impossible, to interpret because of the confounding effects of many interacting factors. Trapnell et al. (1976) studied the effect on soil properties of 23 years of annual burning of a Zambian woodland. The effects of early and late burning on soil chemical properties and on activity of termites (mounds of *Cubitermes*) were compared with a fire-protected plot. Burning improved soil pH by 0.3 to 0.5 units. There was a marked increase in exchangeable Ca (0.21 to 0.73 meq/100g) and Mg (0.28 to 0.55 meq/100g) of the burnt plots. The most remarkable feature of the data was the lack of any significant effect of 23 years of annual burning on soil organic matter and total nitrogen content. This lack of effect on the organic matter content was attributed to the complementary effect of the termite activity. In fire-protected plots, the nutrients (bases) are absorbed by plant roots and are thus immobilized in the woodland canopy. The fallen litter is consumed by termites and the nutrients (bases) are stored in the termitaria at cost of the nutrient supply to the surrounding soil. With complete late burning most of the nutrients are returned to the soil in a readily available form. In partial early burning termites return the nutrients to the mounds by consuming the unburnt litter. With the interacting effects of different ecological factors, it is difficult to isolate the long-term cumulative effects of centuries of natural or man-made fires on nutrient cycling of soils in the tropics.

D. Fire and soil fauna and flora

The effects of fire on soil chemical and physical properties are strongly linked with changes in faunal and floral composition and diversity. Soil physical properties are particularly influenced by soil fauna. The shift in flora by fire has a profound effect on nitrogen fixation and nutrient cycling. The effects of fire

Table 12.16 Soil chemical analysis before and after *in situ* burning of slashed vegetation and that of plot where vegetation was removed rather than burnt (Lal and Cummings, 1979)

Treatment	pH		Organic carbon (%)		Total nitrogen (%)		Calcium (meq/100 g)		Magnesium (meq/100 g)		Potassium (meq/100 g)		Manganese (meq/100 g)		Sodium (meq/100 g)	
	I*	F†	I	F	I	F	I	F	I	F	I	F	I	F	I	F
Slash and burn	6.6	9.0	2.05	2.34	0.353	0.412	9.35	33.5	2.85	14.41	0.55	11.81	0.06	0.17	0.11	0.40
Slash and remove	6.6	6.4	2.18	2.14	0.356	0.361	9.55	11.0	3.01	3.17	0.55	0.50	0.07	0.05	0.11	0.12
LSD (0.05)	0.2	1.4	1.09	0.65	0.068	0.077	3.12	9.5	0.92	2.50	0.13	2.49	0.01	0.03	0.03	0.10

*I = before; †F = after.

on soil fauna have been reviewed by Gillon (1983) and have been separately described in this text in different chapters dealing with earthworms, termites and soil fauna. Survey of the literature regarding the effects of fire on fauna often gives a conflicting view. For example, Gillon (1983) (Fig. 13) and Springett (1979) observed in Western Australia that burning reduced the number of arthropods whereas Izara (1977) reported that burning of pampas in South America had no effect on microarthropod fauna. Also in Australia Malajczuk and Hingston (1981) observed negative effects of burning on the mean number of ectomycorrhizae. The effects of burning on microbial population are often estimated in terms of the rate of litter decomposition (Santos and Grisi, 1979) or in terms of soil respiration (Grisi and Santos, 1978).

Fire effects on soil fauna and flora are both immediate and the cumulative long-term. While it is relatively easy to quantify the immediate effects of fire, the long-term effects are difficult to surmise.

E. Effects on atmospheric chemistry

Biomass burning, in addition, affects gaseous composition in the upper atmosphere. Burning leads to a substantial emission of air pollutants such as CO, NO, N_2O, CH_4 and other hydrocarbons. The amount of various gases ejected into the atmosphere, although debatable and highly speculative, is believed to be as large as the world-wide industrial input. Crutzen *et al.* (1985) measured CO volume mixing ratios in boundary layers of forest and Cerrado regions of Brazil. In comparison with marine and near coastal areas with a commonly observed range of 50–100 ppbv, the mixing ratios in the boundary layer of the Amazonian forest exceeded 300 ppbv (Fig. 12.15). Crutzen (1985) estimated $5-10 \times 10^{14}$g per year of CO supplied by the oxidation of reactive hydrocarbons emitted from tropical forests. The comparison of data in Fig. 12.15 a–A and Fig. 12.15d indicate that the CO values were even more in the Cerrado boundary layer where widespread burning is a common phenomenon during the dry season of August and September. This author estimated that the annual production of CO from biomass burning in the tropics is about $5-10 \times 10^{14}$ g CO.

In addition to CO, biomass burning is also a major source of methane into the atmosphere. It is estimated that the biomass burning adds about 2.7×10^{13} g CH_4 annually into the atmosphere. These greenhouse gases have implications on global climate.

VII. PRESCRIBED BURNING AND CROP YIELD

The effects of burning on growth and vigour of seasonal crops are also difficult to generalize. Crop yield is an integrated response to many interacting factors. That is why it is often difficult to relate yeild response to a variable. Many studies relating the effects of prescribed fire on pasture, discussed in a previous section, have shown that a periodic burn is often useful in improving the quality

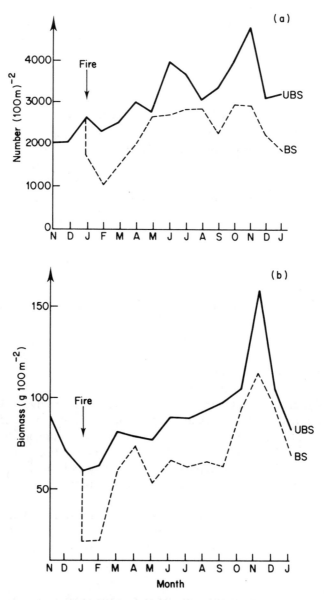

Figure 12.15 Seasonal variations in arthropod population in the grass layer of annually burnt savannah at Lamto, Ivory Coast: (a) number of arthropods; (b) biomass of arthropods (Gillon, 1970, 1983). BS, burned savannah, UBS, unburned savannah; (c) The average profile of CO volume mixing ratios and standard deviation (dotted curve) over Brazil in 1980 over the humid forest area near Manaus and (d) over the Cerrado region (Crutzen et al., 1985). *Reproduced by permission of D. Reidel Publishing Co.*

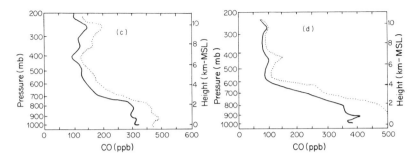

and quantity of herbage (Fig. 12.16). For example, Norman (1969) recommended a burn every 5 years for pasture improvement in Northern Territory, Australia, rather than an annual or biennial burning or no burning at all.

The effects of burning on growth and yield of seasonal food crops are difficult to generalize. Yield response to burning depends on inherent soil fertility and its base status, vegetation, length of fallow, crop grown, and on the level of fertilizers applied. It is generally believed that in traditional farming with little reliance on chemical fertilizers, burning produces positive yield response by adding the much needed plant nutrients and bases for neutralizing soil acidity. The yields are often low, but they are even lower when an acidic infertile soil does not receive even the minimum of readily available nutrients supplied through the ash. Burning may also initiate or increase soil nitrification, particularly on acidic soils (Raison, 1979). The meagre amount of nutrients in ash supports a modest growth for one or two consecutive crops, but the crop

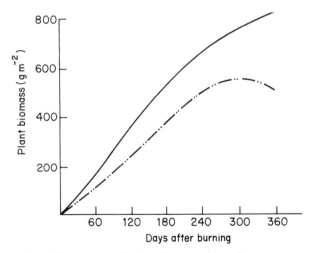

Figure 12.16 Plant biomass production in two plots of campo cerrado at Emas, Pirassununga, Estado de Sao Paulo, after a July burn. (—) control plot, (-.-.-.) plot from which ash had been removed (Cavalcanti, 1978; Coutinho, 1982). *Reproduced by permission of Springer-Verlag*

yields are limited by the nutrient elements not adequately supplied in the ash, e.g. nitrogen and some micronutrients. Although fire has the potential to change the ecosystem nutrient balance and mobility, its use to man's advantage in traditional farming depends on many factors.

The effects of fire and of the addition of ash on plant growth in intensive commercial agriculture are contradictory. Under some conditions the response is positive, in others negative, and in still others fire has been shown to have no substantial effect on crop growth. For example, while studying the effects of burning on productivity of a Lithosol in 'zona do Agreste' of Pernambuco, Brazil, Morgolis (1977) reported a decrease in maize yield on burned-over land. In Cameroon, Lyonga (1980) observed that burning increased maize yield by increasing soil pH and available P content. This variable response in plant producitivity for different soils is often difficult to relate to specific nutrient deficiencies or surplus.

In Western Nigeria Lal (1981) observed only slight improvement in maize grain yield by burning (Table 12.17). The increase in mean grain yield of maize in slash and burn compared with slash and removal treatment was 0.44 t ha^{-1} in the first season and 0.4 t ha^{-1} in the second. The effect of burning on improvement in maize grain yield was statistically insignificant for this Alfisol of nearly neutral pH and medium fertility. The low yield in the second crop was not due to nutrient deficiency but due to severe drought stress. The drought stress is better alleviated by mulching with crop residue than by burning or incorporating it into the soil. That is why Okoro (1981) observed that the maximum grain yield was obtained from mulched rather than from burned-over plots. Ajuwon *et al.* (1978) observed a slight decrease in maize yield from a burned-over Alfisol at Ilora compared with mulched treatments (Table 12.18).

Table 12.17 Effects of burning a fallow vegetation on the yield of the following no-till maize crop with and without chemical fertilizer (Lal, 1981)

Treatment	Fertilizer*	Maize grain yield (t/ha)	
		First crop	Second crop
Mechanical and removal	F_0	4.77	1.36
	F_1	4.96	2.20
Mean		4.86	1.78
Slash and burn	F_0	4.63	2.00
	F_1	5.75	2.40
Mean		5.19	2.20
Slash and removal	F_0	4.21	1.59
	F_1	5.29	2.01
Mean		4.75	1.80
LSD (0.05)			
(i) Vegetation removal method		0.69	0.71
(ii) Fertilizer		0.49	0.81

* F_0 = without fertilizer; F_1 = with fertilizer.

Table 12.18 Effects of residue management and tillage methods on maize grain yield at Ikenne and Ilora in western Nigeria (Ajuwon et al., 1978)

Residue management treatment	Grain yield in 1974 (t/ha)	
	Ikenne	Ilora
1. Ploughed-under	3.6a	5.1b
2. Ploughed after burning	3.8a	5.0b
3. No-till without herbicide	3.8a	4.3a
4. No-till with herbicides	3.7a	4.5ab
5. No-till, burning and with herbicides	3.5a	4.7ab

At Indore India Sreenivasan and Aurangebadkar (1940) observed significant improvements in cotton yield on burnt soil in a greenhouse study.

Other than the effects of fire on pasture improvement, there is no evidence that conclusively show beneficial effects of fire on crop yields in commercial-level agriculture. The fire-induced benefits to crop growth and yield in traditional subsistence-level agriculture are related to many factors including the addition of some easily available plant nutrients.

VIII. CONCLUSIONS

Vogl (1977) and other ecologists have argued whether fire is a destructive menace or a natural process. The research results presented in this chapter can be used to provide a rational answer to this highly debatable issue. Undoubtedly, fire has been a powerful manipulator of tropical ecosystems. Over the millenia, it has had lasting effects on vegetation, soils and fauna. Although fire can be used to man's advantage in terms of its immediate impact as an environmental manipulator, its long-term cumulative effects are mixed blessings if not totally negative. The adverse effects of fire are definitely enhanced by man-induced fires in a fire-free ecosystem.

One of the most spectacular long-term effects of fire is its role in transformation of the vegetation. Specific fire-adapted vegetation associations have developed through gradual evolution where periodic fire has been a normal feature of the environment. Large trees the 'forest vegetation' have given way to fire-resistant shrubs and grasses—'the savanna'. In addition to the natural fire-dependent savanna ecosystems, man's intervention has also created vast tracts of derived savanna in Africa, Asia and tropical America where rainfall and climatic factors would normally support a forest vegetation. Deliberately created monocultures by fire-induced vegetation control have also resulted in severe ecological problems. In contrast, attempts to eliminate or control fire form fire-dependent ecosystems have led to unpredictably adverse effects on some communities and environments (Vogl, 1977).

The properties of soils in derived savanna have also undergone drastic changes due to annual cycles of man-induced fire. High temperatures during

Table 12.19 Cotton yield on fired and unfired soils (see cotton in g/100 sq. ft) (Sreenivasan and Aurangebadkar, 1940). *Reproduced by permission of Williams & Wilkins Co.*

Variety	Unmanured			Manured		
	Untreated	Heated soil		Untreated	Heated soil	
		6" surface layer	50% in 6" surface layer		6" surface layer	50% in 6" surface layer
Indore 1	221	456	608	277	762	955
Malvi 9	658	1164	801	910	1399	1174

FYM at 7.4 t/acre.

and after burning, denudation of the vegetation cover with soil exposure to high intensity rains, and decline in activity and diversity of soil fauna have resulted in alterations of hydro-thermal regime, nutrient balance and recycling, and loss of plant nutrients from the ecosystem through volatilization, leaching or run-off. Changes in water balance and loss of soluble bases have been responsible for widespread occurrence of plinthite and hardened iron pans. Widespread occurrence of compacted soils with degraded physical properties in derived savanna are due to direct and indirect effects of the now fire-mastery of a once fire-free ecology.

Although the cumulative effects of fire on soil, vegetation and ecology in general have been adverse and even devastating, the short-term effects of fire can be used to man's advantage. Fire as a tool has been useful more to primitive man and to the subsistence agriculturist than in modern commercial agriculture. One of the limitations in using fire is the lack of our understanding of the fire ecology and its complex ramifications. That is why the opinions differ widely and the results of even carefully conducted experiments are often contradictory.

The ability of fire to transform vegetation has been effectively used in improving pastures. If used at the right time and with the right frequency, prescribed fire can be used to eliminate undesirable vegetation from pasture. But the overall ecological effects of this prescription-caused monoculture may be as harmful as the indiscriminate use of herbicides. Regrettably, the available research information does not provide any answer.

The use of fire as a short cut to nutrient recycling has often resulted in loss of plant nutrients out of the ecosystem. Although some nutrients become readily available to crops, the efficiency of nutrients released is often less than 50%. Furthermore, the released nutrients are often lost in surface run-off or with percolating water because of the nutrient imbalance created by the excess supply of some and deficiency of others. The magnitude of nutrient loss is more from the forest than from the savanna ecosystem.

The use of fire as a tool for residue management in commercial and intensive tropical agriculture is questionable at its best and disastrous at its worst. Erosion is less if the soil is covered, the efficiency of water utilization is more when run-off and evaporation losses are reduced by the ground cover, and soil stays porous and receptive to high intensity rains when worked over by fauna stimulated through the mulch cover. The use of fire to eliminate this protective cover is an ecologically destructive method of obtaining a few kilograms of plant nutrients. Once again, the fire-induced erosion hazard is more in the forest than in savanna ecology.

Many authors (Phillips, 1950, 1965; Daubenmire, 1968) consider that fire is a useful and cheap tool for habitat management. The review presented indicates that it is 'cheap' but not always useful. In fact, its overall effect seems to be ecologically negative at least for the fire-free ecosystems. Even if man can 'master' fire as a prescribed tool, its ecological effects are far from being convincingly useful.

No doubt well planned research experiments designed to understand fire ecology and its long-term and cumulative effects on soil and environments will dispel some myths. It is obvious that if used properly it has some short-term beneficial effects on habitat management. Fire can be used as a tool in regions where it does not disrupt natural processes. But in view of its harmful and lasting effects, let us use it carefully and sparingly until we understand better its ecology and how to manipulate it.

REFERENCES

Afolayan, T.A., and Ajayi, S.S., 1979, Reasons for further burning experiments in West African Savanna woodland, *Commonwealth Forestry Review*, **58**, 253–265, U.I., Ibadan.

Ahlgren, I.F. and Ahlgren, C.E., 1960, Ecological effects of forest fire, *Bot. Rev.*, **26**, 483–533.

Ajuwon, S., Olungua, O., and Lal, R., 1978, Suitability of no-tillage technique in the forest and savannah regions of western Nigeria, *Nigeria J. Sci.*, **12**, 247–266.

Alvim, P.T., and Araujo, W.A., 1952, el suelo como factor ecologico en el desarrollo de la vegetacion en centro-oeste del Brasil, *Turrialba*, **2**, 153–160.

Andrews, F.W., and Clouston, T.W., 1937, The effect on shoot and root development of 'firing' the surface soil, *Sudan Govt. Dept. Agr. For. Ann. Rpt.*, **193**, 39.

Aristeguieta, L., and Medina, E., 1965, Protection y quema de la sabana Llanera, *Bol. Soc. Ven. Cienc. Nat.*, **109**, 129–139.

Athias, F., Josens, G., and Lavelle, P., 1975, Influence due feu de brousse annuel sur le peuplement endogé de la savane de Lamto (Côte d'Ivoire), in *Proc. 5th Int. Coll. Soil Zoology, Prague, 1973*, pp. 389–397.

Aubert, G., 1950, Observations sur la dégradation des sols et le formation de la cuirasse latéritique dans le nord-ouest due Dahomey, *Trans. IVth Int. Congr. Soil Sci.*, pp. 127–128.

Auberville, A., 1947, Les brousses secondairres en Afrique équatoriale, *Bois et Forêts des Tropiques*, **2**, 24–49.

Batchelder, R.B., 1967, Spatial and temporal patterns of fire in the tropical world, *Proc. 6th Ann. Tall Timbers Fire Ecol. Conf.*, pp. 171–208.

Bogdan, A.V., 1954, Bush clearing and grazing trial at Kisokon, Kenya, *East Afr. Agric. J.*, **19**, 253–259.

Boughey, A.S., 1963, Interaction between animals, vegetation, and fire in Southern Rhodesia, *Ohio J. Sci.*, **63**, 193–209.

Boyle, J.R., 1973, Forest soil chemical changes following burning, *Comm. Soil Sci. Plant Anal.*, **4**, 369–374.

Brinkman, W.L.F., and De Nascimento, J.C., 1973, The effect of slash and burn agriculture on plant nutrients in the tertiary region of Central Amazonia, *Turrialba*, **23**, 284–290.

Brockington, N.R., 1961, Studies on the growth of Hyparrhenia-dominant grassland in Northern Rhodesia. I. Fertilizer response. II. The effect of fire, *J. Br. Grassl. Soc.*, **16**, 54–64.

Bruce-Okine, E., and Lal, R., 1975, Soil erodibility as determined by raindrop technique, *Soil Sci.*, **119**, 149–157.

Budowksi, G., 1956, Tropical savannas, a sequence of forest felling and repeated burnings, *Turrialba*, **6**, 22–33.

Byram, G.M., 1958, Combustion of forest fuels, in K.P. Davis, (ed.), *Forest Fire: Control and Use*, McGraw Hill Co., New York.

Cavalcanti, L.H., 1978, Efeito das cizas resultantes da queimada sobre a productividade do estrato herbaceo subarbustivo do cerrado de Emas, *Dsc. Thesis*, Universidade de Sao Paulo, Sao Paulo.

Chevalier, A., 1928, Sur l'origine des campos brésiliens et sur le rôle des *Imperata* dans la substitution des savanes aux forêts tropicales, *C.R. Acad. Sci, Paris*, **187**, 997–999.

Clark, J.D., 1960, Human ecology during the pleistocene and later times in Africa south of the Sahara, *Current Anthrop.*, **1**, 307–324.

Collis-George, N. and Lal, R., 1971, Infiltration and structural changes as influenced by initial moisture content, *Aust. J. Soil Res.*, **9**, 107–116.

Collis-George, N., and Lal, R., 1973, The temperature profiles of soil columns during infiltration, *Aust. J. Soil Res.*, **11**, 93–105.

Cook, L., 1939, A contribution to our information on grass-burning, *S Afr. J. Sci.*, **36**, 270–282.

Corbet, A.S., 1934, The bacterial numbers in the soil of the Malay Peninsula, *Soil Sci.*, **38**, 407.

Coutinho, L.M., 1976, Contribuiscao ao conhecimento do papel ecologico das queimadas na floracao de especies do cerrado, *Livre-Docencia Thesis*, Universidade de Sao Paulo.

Coutinho, L.M., 1978, Aspectors ecologicos do fogo no cerrado. I-A temperatura do solo durante as queimados, *Revista Brasileira de Botanica*, **1**, 93–97.

Coutinho, L.M., 1979, The ecological effects of fire in Brazilian cerrado, *Workshop Dynamic Changes Savanna Ecosystems*, Kruger National Park, May 1979.

Coutinho, L.M., 1982, Ecological effects of fire in Brazilian Cerrado, in B.T. Huntley, and B.H. Walker (eds), *Ecology of Tropical Savannas*, Springer-Verlag, New York, pp. 273–291.

Crutzen, P.J., 1985, The role of the tropics in atmospheric chemistry, in B. Dickinson (ed.), *Geophysiology*, UNU Japan, Tokyo (In Press).

Crutzen, P.J., Delany, A.C., Greenberg, J. Haagenson, P., Heidt, L., Lueb, R., Pollock, W., Seiler, W., Wartburg, A., and Zimmerman, P., 1985, Tropospheric chemical composition measurements in Brazil during the dry season, *J. Atmospheric Chem.* **3**, (In Press).

CSIRO, 1979, Burning for that green pick, *Rural Res.*, **103**, 29–30.

Daubenmire, R., 1968a, Ecology of fire in grassland, *Adv. Ecol Res.*, **5**, 209–283.

Daubenmire, R., 1968b, Ecology of fire in grasslands, in *Advances in Ecological Research V*, Academic Press, London, pp. 209–266.

De Bano, L.F., Mann, L.D., and Hamilton, D.A., 1970, Translocation of hydrophobic substances into soil by burning organic litter, *Soil Sci. Soc. Amer. Proc.*, **34**, 130–133.

De Rham, P., 1973, Recherches sur la minéralisation de l'azote dans les sols des savannes de Lamto (Côte d'Ivoire), *Rev. Ecol. Biol. Sol.*, **1973**, 169–196.

Donahue, R.L., 1972, *Ethiopia, taxonomy, cartography and ecology of soils*. African Studies Centre, Michigan State Univ., East Lansing, pp. 34–38.

Du Pleiss, M.C.F., and Mostert, J.W.C., 1965, Runoff and soil losses at the Agricultural Research Institute Glen, *S. Afr. J. Sci.*, **8**, 1051–1060.

Edwards, D.C., 1942, Grass burning, *Emp. J. Exper. Agric.*, **10**, 219–231.

Fassbender, H.W., 1975, The effect of burning plant residues on some soil properties, *Turrialba*, **25**, 249–254.

Fisher, M.J., 1978, The recovery of leaf water potential following burning of two droughted tropical pasture species, *Aust. J. Exp. Agric. Anim. Husb.*, **18**, 423–425,

Fogliata, F.A., Leiderman, J., and Matiussi, R.E., 1967, Effect of trash burning on soil temperature and microbial population, *Revita Ind. Agric. Tucumàn*, **45**, 27–53.

Freise, F.W., 1934, Beobachtungen uber den Verbleib von Niederschlagen im Urwald ünd der Einfluss von Waldbestand auf den Wasserhaushalt der Umgebung, *Forstwissent schaftliches Centralblatt*, **56**, 231–245.

Freise, F.W., 1939, Untersuchungen uber die Folgen der Brandwirtschaft aus tropischen Boden, *Tropenpflanzer*, **42**, 1–22.
Garren, K.H., 1943, Effects of fire on vegetation of the southeastern United States, *Bot. Rev.*, **9**, 617–654.
Gerakis, P.A., Veresoglou, D.S., and Sfakiotakis, E., 1978, Ethylene pollution from wheat stubble burning, *Bull. Env. Contam. Tox.*, **20**, 657–661.
Gillon, D., 1983, The fire problem in tropical savannas, in F. Bourlier (ed.), *Tropical Savannas: Ecosystems of the World 13*, Elsevier, Amsterdam, pp. 617–641.
Gillon, D., and Pernés, J., 1968, Etude de l'effet du feu de brousse sur certains groupes d'Arthropodes dans une savane prédofrestière de Côte d'Ivoire, *Ann. Univ. Abidjan*, **E1**, 113–197.
Giovannini, G., and Lucchesi, S., 1983, Effect of fire on hydrophobic and cementing substances of soil aggregates, *Soil Sci.*, **136**, 231–236.
Granier, P., and Cabanis, Y., 1976, Les feux courants et l'elevage en savane soudanienne, *Rev. Elev. Méd. Vét. Pays Trop.*, **29**, 267–272.
Grisi, B.M., and Santos, O.M., 1978, Soil respiration in undisturbed and burned areas, in a tropical rain forest ecosystem in Southern Bahia, Brazil, *Revista Brasileira de Biologia*, **38**, 579–586.
Grubb, P.J., 1977, The maintenance of species richness in plant communities: the importance of the regeneration niche, *Biol. Rev.*, **52**, 107–145.
Guilloteau, J., 1956, The problem of burns in land development and soil conservation in Africa south of the Sahara, *Sols Afr.*, **4**, 64–102.
Hodgkins, E.J., 1958, Effects of fire on undergrowth vegetation in upland southern pine forests, *Ecology*, **39**, 36–46.
Hopkins, B., 1965, Observations on savanna burnings in the Olokemeji Forest Reserve, Nigeria, *J. Appl. Ecol.*, **2**, 367–381.
Howell, F.C., and Clark, J.D., 1963, Acheulian hunter-gatherers of sub-Saharan Africa, in F.C. Howell and F. Bourliere, (eds.) *African Ecology and Human Evolution*, Aldine Publ. Co., Chicago, pp. 458–533.
Izara, D.C., De, 1977, The effects of burning on soil micro-arthropods in the semi-arid pampas zone, *Ecol. Bull.*, **25**, 357–365.
Joachim, A.W.R., and Kandiah, S., 1942, Studies on Ceylon soils. XVI. The chemical and physical characteristics of soils of adjacent contrasting vegetation formations, *Trop. Agriculturist (Ceylon)*, **98**, 15–30.
Kayll, A.J., 1974, Use of fire in land management, In T.T. Kozlowski and C.E. Ahlgren (eds.), *Fire and Ecosystems*, Academic Press, New York, pp. 483–511.
Kellogg C.E., and Davol, F.D., 1949, An exploratory study of soil groups in the Belgian Congo, *Institut National pour l'Etude Agronomique due Congo Belge, Ser. Sci.* **46**, 73 pp.
Kelly, W.P., and McGeorge, W., 1913, The effect of heat on Hawaiian soils, *Hawaii Agr. Exp. Sta. Bull.*, **30**, 5–38.
Killick, D.J.B., 1963, An account of the plant ecology of the Cathedral Peak area of the Natal Drakensberg, *Bot. Surv. S. Africa*, Mem. 34, Govt. Printer, Pretoria, South Africa.
Klaer, W., and Loffler, E., 1980, Effect of traditional agriculture on soil erosion processes in the tropical wood- and grasslands in Papua New Guinea, *Erde*, **111**, 73–83.
Komarek, E.V., 1965, Fire ecology—grasslands and man, *Proc. 4th Ann. Tall Timbers Fire Ecology Conf.*, pp. 169–220.
Lacey, C.J., Walker, J., and Noble, I.R., 1982, Fire in Australian Tropical Savannas, in B.J. Huntley and B.H. Walker, (eds.), *Ecology of Tropical Savannas*, Springer-Verlag, New York, pp. 246–272.

Lal, R., 1981, Clearing a tropical forest. II. Effects on crop performance, *Field Crop Res.*, **4**, 345–354.
Lal, R., and Cummings, D.J., 1979, Clearing a tropical forest I. Effects on soil and micro-climate, *Field Crops Res.*, **2**, 91–107.
Lal, R., and Kang, B.T., 1982, Management of organic matter in soils of the tropics and subtropics, *Trans. 12th Int. Congr. Soil Sci.*, **IV**, 152–178.
Lambert, J.D.H., and Arnason, T., 1978, Distribution of vegetation on Maya ruins and its relationship to ancient landuse at Lamana, Belize, *Turrialba*, **28**, 33–41.
Laudelout, H., 1954, Étude sur l'apport d'éléments minéraux résultant de l'incinération de la jachère forestiere, *Second Inter-African Soils Conf.*, **1**, 383–388.
Lemon, P.C., 1968, Effects of fire on an African Plateau grassland, *Ecology*, **49**, 316–322.
Ling, A.H., and Mainstone, B.J., 1983, Effects of burning of rubber timber during land preparation on soil fertility and growth of *Theobroma cacao* and *Gliricidia maculata*, *Planter*, **59**, 52–59.
Lyonga, S.N., 1980, Some common farming systems in Cameroon, their influence on the use of organic matter and the effects of soil burning on maize yields and on soil properties, *FAO Soils Bull.*, **43**, 79–86.
Malajczuk, M., and Hingston, F.J., 1981, Ectomycoarhizae association with jarrah, *Aust. J. Bot.*, **29**, 453–462.
Masefield, G.B., 1948, Grass burning: some Uganda experience, *E. Afr. Agric. J.*, **13**, 135–138.
Masson, H., 1948, La temperature du sol au cours d'un feu de brousse au Senegal, *Agron. Trop.*, **3**, 174–179.
McArthur, A.G., and Cheney, N.P., 1966, The characterization of fires in relation to ecological studies, *Aust. For. Res.*, **2**, 36–45.
Medina, E., Mendoza, A., and Montes, R., 1977, Nutritional balance and organic matter yield of *Trachypogon* savannas at Calabozo, Venezuela, *Boletin Sociedad Venezolana de Ciencias Naturales*, **33**, 101–120.
Medina, E., Mendoza, A., and Montes, R., 1978, Nutrient balance and organic matter production in Trachypogon savannas of Venezuela, *Trop. Agric.*, **55**, 243–253.
Meiklejohn, J., 1955, The effect of bush burning on the microflora of a Kenya upland soil, *J. Soil Sci.*, **6**, 111–118.
Menaut, J.C., 1971, Etude de quelques peuplements ligneux d'une savane guineenue de Cote d'Ivoire, *Thesis*, University of Paris, pp. 414.
Mohr, E.C.J., and Van Baren, F.A., 1954, *Tropical soils*, Mouton, The Hague, Netherlands.
Moore, A.W., 1960, The influence of annual burning on a soil in the derived savanna zone of Nigeria, *Trans. 7th Int. Congr. Soil Sci.*, **4**, 257–264.
Morgolis, E., 1977, Effect of burning on lithosols of the Zona do Agreste of Pernambulo, *Pesquisa Agropecuaria Pernambucana*, **1**, 81–88.
Moureaux, C., 1959, Fixation de gaz carbonique par le sol, *Mém. Inst. Sci. Madagascar*, D **IX**, 109–120.
Myers, J.G., 1936, Savannah and forest vegetation of the interior Guinea Plateau, *J. Ecol.*, **24**, 162–184.
Nishita, H., and Haug, R.M., 1972, Soil physical and chemical characteristics of heated soil, *Soil Sci.*, **113**, 422–430.
Norman, M.J.T., 1963, The short term effects of time and frequency of burning on native pastures at Katherine, N.T., *Aust. J. Exp. Agric. Anim. Husb.* **3**, 26–29.
Norman, M.J.T., 1969, The effect of burning and seasonal rainfall on native pasture at Katherine, N.T., *Aust. J. Exp. Agric. Anim. Hus.*, **9**, 295–298.

Norman, M.J.T., and Wetselaar, R., 1960, Losses of nitrogen on burning native pasture at Katherine, N.T., *J. Aust. Inst. Agric. Sci.*, **26**, 272–273.

Nye, P.H., and Greenland, D.J., 1964, Changes in soil after clearing tropical forest, *Plant Soils*, **21**, 101–112.

Okoro, G.E., 1981. The effects of burning and fertilizer application on crop yields and soil chemical properties under shifting cultivation in Southern Nigeria, *Ph.D. Thesis*, Dept. Soil Sci., University of Manitoba, Canada.

Phillips, J.F.V., 1959, *Agriculture and Ecology in Africa*, Faber and Faber, London.

Phillips, J.F.V., 1965, Fire—as master and servant: its influence in bioclimatic regions of trans-saharan Africa. *Proc. 4th Ann. Tall Timbers Fire Ecology Conf.* pp. 7–109.

Phillips, J.F.V., 1974, Effects of fire in forest and savanna ecosystems of sub-saharan Africa, In T.T. Kozlowski and C.H. Ahlgren, (eds.), *Fire and Ecosystems*, Academic Press, New York, pp. 435–481.

Pitot, A., and Masson, H., 1951, Quelques données sur la temperature au cours des feux de brousse aux environs de Dakar, *Bull. IFAN*, **A 13**, 711–732.

Rains, A.B., 1963, Grassland research in Northern Nigeria (1952–62), *Misc. Paper, Inst. Agric. Res. Samaru*, **11**, 1–67.

Raison, R.J., 1979, Modification of the soil environment by vegetation fires, with particular reference to nitrogen transformations: a review (NSW). *Plant Soil*, **51**, 73–108.

Raison, R.J., 1980. A review of the role of fire in nutrient cycling in Australian native forests, and of methodology for studying the fire-nutrient interaction, *Aust. J. Ecol.*, **5**, 15–21.

Rambelli, A., Puppi, G., Bartoli, A., and Aebonetti, S.G., 1973, Second contribution to the study of the fungal *Rev. Ecol. Biol. Sol.*, **10**, 13–18.

Rao, K.S., and Madhawan, S.K., 1955, Plasticity of heated soils, *Irrig. Pur.*, **12**, 294–299.

Roose, E., 1971, Influence de la modification du milieu sur l'erosion, le ruissellement, le bilan hydrique et chimique. Suite à la mise en culture. Résultats sous pluie naturelle. *Comité Technique* ORSTOM, Abidjan, pp. 20.

Roose, J., and Asseline, J., 1978, Measurements of erosion phenomena under simulated rainfall at Adiopodoume, solid and soluble loads in runoff water from bare soil and pineapple plantations, *Cahiers ORSTOM, Pedologie*, **16**, 43–72.

Rosswall, T., (ed.), 1980, *Nitrogen cycling in West African ecosystems*, SCOPE/UNEP Int. N Unit, Royal Swedish Academy of Sciences, pp. 45.

Rotini, O.T., Lotti, G., and Baldacci, P.V., 1963, Studies on the changes occurring in soils and clays heated at 50 and 100°C, *Agrochimica*, **7**, 289–295.

Rundel, P.W., 1982, Fire as an ecological factor, in O.L. Lauge, P.S. Nobel, C.B. Osmond, and H. Ziegler (eds.), *Physiological Plant Ecology: I. Response to the physical environment*, Springer-Verlag, Bolin, 625 1, pp. 501–538.

Rycroft, H.B., 1947, A note on the immediate effects of veld burning on stream flow in a Jonkerschoek stream catchment, *J.S. Afr. For. Assoc.*, **15**, 80–88.

San José, J.J., and Medina, E., 1975, Effect of fire on organic matter production in a tropical savanna, In F.B. Golley and E. Medina (eds.), *Tropical Ecological Systems*, Spinger-Verlag, Berlin, pp. 251–264.

San José, J.J., and Medina, E., 1977, Effect of fire on organic matter production and water balance in a tropical savanna, in F.B. Golley and E. Medina, (eds.), *Tropical Ecological Systems*, Springer-Verlag New York, 251–263.

Santos, O.M., and Grisi, B.M., 1979, Cellulose and leaf litter decomposition in a tropical rainforest soil in southern Bahia, Brazil: comparative study in undisturbed and burned areas, *Rev. Brasileria Ciencia Solo*, **3**, 149–153.

Sarmiento, G., and Monasterio, M., 1975, A critical consideration of the environmental conditions associated with the occurrence of savanna ecosystems in tropical America,

in F.B. Golley and E. Medina, (eds.), *Tropical Ecological Systems*, Springer-Verlag, New York, pp. 223–250.
Schmidt, D.R., 1976, Effects of burning upon soil fertility, *J. AAASA*, **3**, 40–44.
Scholl, D.G., 1975, Soil wettability and fire in Arizona chaparral, *Soil Sci. Soc. Amer. Proc.*, **39**, 356–361.
Scott, J.D., 1970, Pros and Cons of eliminating veld burning, *Proc. Grassl. Soc. S. Afr.*, **5**, 23–26.
Sertsu, S.M., and Sanchez, P.A., 1978, Effects of heating on changes in soil properties in relation to an Ethiopian land management practice, *Soil Sci. Soc. Amer. J.*, **42**, 940–944.
Seubert, C.A., 1975, Effects of land clearing methods on crop peformance and changes in soil properties in an Ultisol of the Amazon jungle in Peru, *M.Sc. Thesis*, N. Carolina State Univ., Raleigh, N.C.
Shantz, H.L. 1947, *Fire as a tool in management of brush ranges*, Calif. St. Bd. Forest, Sacramento, Calif.
Schantz, H.L., 1954, The place of grasslands in earth's cover of vegetation, *Ecology*, **35**, 143–145.
Shantz, H.L., and Turner, B.L. 1958, *Photographic documentation of vegetational changes in Africa over a third of a century*, Univ. Ariz. Coll. Agr. Ref. 169.
Smith, E.L., 1960, Effects of burning and clipping at various times during the wet season on tropical tall grass range in Northern Australia, *J. Rangelands Management*, **13**, 197–203.
Soyer, J., Miti, T., and Aloni, K., 1982, Comparative effects of rainfall erosion under suburban tropical conditions (Lubumbashi, Shaba, Zaire), *Revue de Géomorphologia Dynamique*, **31**, 71–80.
Springett, J.A., 1979, The effects of a single hot summer fire on soil fauna and on litter decomposition in Jarrah forest in Western Australia, *Aust. J. Ecology*, **4**, 279–291.
Sreenivasan, A., and Aurangebadkar, R.K., 1940, Effect of fire-heating on the properites of black cotton soil in comparison with those of gray and humus-treated soils, *Soil Sci.*, **50**, 449–462.
Stent, H.B., 1933, Observations on the fertilizer effect of wood-burning in the 'chitemense' system, *A. Bull. Dep. Agric., North Rhod.*, **2**, 48–49.
Stocker, G.C., and Sturtz, J.D., 1966, The use of fire to establish Townsville lucerne in the Northern Territory, *Aust. J. Exp. Agric. Anim. Husb.*, **6**, 277–279.
Sudres, R., 1947, La degradation des sols au Foutah Djalon, *Agron, Trop.*, **2**, 227–246.
Tessema, T., and Yirgou, D., *Soil burning (guie): its problems and possible solutions*, Debrezeit Agric. Exp. Stn, Haile Selassie I. Univ. Dire Dawa, Ethiopia.
Tinley, K.L., 1966, *An ecological reconnaissance of Moremi wildlife reserve—Botswana*, Okoungo Wildl. Soc., Johannesburg.
Tothill, J.C., 1969, Soil temperatures and seed burial in relation to the performance of *Heteropogon contortus* and *Themeda australis* in burnt native woodland pastures in eastern Queensland, *Aust.* J. Bot., **17**, 269–275.
Tothill, J.C., 1971a, Grazing, burning and fertilizing effects on the regrowth of some *Eucalyptus* and *Acacia* species in cleared open forest in southeast Queensland, *Trop. Grasslands J.*, **5**, 31–34.
Tothill, J.C., 1971b, A review of fire in management of native pasture with particular reference to northeast Australia, *Trop. Grasslands J.*, **5**, 1–10.
Tothill, J.C., and Shaw, N.W., 1968, Temperatures under fires in bunch spear grass of south-east Queensland, *J. Aust. Inst. Agric. Sci.*, **34**, 94–97.
Trapnell, C.G., 1959, Ecological results in woodland burning experiments in Northern Rhodesia, *J. Ecol.*, **47**, 129–168.
Trapnell, C.G., Friend, M.T., Chamberlain, G.T., and Birch, H. F., 1976, The effects of fire and termites on a Zambian woodland soil, *J. Ecol.*, **64**, 577–588.

Trollope, W.S.W., 1974, Role of fire in preventing bush encroachment in the eastern Cape, *Proc. Grassland Soc. S. Afr.* **9**, 67–72.

Trollope, W.S.W., 1982, Ecological effects of fire in South African Savannas, in B.J. Huntley, and B.H. Walker, (eds.) *Ecology of Tropical Savannas*, Springer-Verlag, New York, pp. 292–306.

Uribe, A.H.F., Suarez de Castro, and Alvar, R.G., 1967, Effets de les quemas sobre la productivardad de los suelos, *Cenicafe*, **18**, 116–134.

U.S. Forest Service, 1977, *Proc. Symp. Environmental Fuel Management in Mediterranean Ecosystems*, August 1–5, 1977, Palo Alto, California, U.S. Dept. Agric.

Van Rensburg, H.J., 1952, Grass burning experiments on the Msima River Stock Farm, Southern Highlands Tanganyika, *E. Afr. Agric. J.*, **17**, 119–129.

Vareschi, V., 1962, Le quema como factor ecologico en los Llanos, *Bol. Soc. Venez. Cienc. Nat.*, **23**, 9–26.

Vareschi, V., 1969, Las sabanas del Valle de Caracas. *Acta Bot. Venez*, **4**, 427–522.

Viani, R., Baudet, J., and Marchant, J., 1973, Réalisation diun appareil potaif d' enrogistrement magnétique de mesures. Application à l'étude de la température lors du passage d'un feu de brousse, *Ann. Univ.*, *Abidjan*, **E6**, 295–304.

Villecourt, P., Schmidt, W., and César, J., 1979, The chemical composition of the herbaceous layer of the Lamto savannah (Ivory Coast), *Rev. Ecol. Biol. Sol*, **16**, 9–15.

Villecourt, P., Schmidt, W., and César, J., 1980, Losses from an ecosystem during a bush fire in tropical savannah, Lamto, Ivory Coast, *Rev. Ecol. Biol. Sol*, **17**, 7–12.

Vincent, C., 1936, The burning of grassland, *Rev. Industr. Anim.* (From *Herbage Abstracts*, 1935, 6, 50).

Viro, P.J., 1974, Effects of forest fire on soils, in T.T. Kozlowski and C. E. Ahlgren, (eds.), *Fire and Ecosystems*, Academic Press, New York.

Vogl, R.J., 1974, Effects of fire on grassland, In T.T. Kozlowski and C.E. Ahlgren, (eds.), *Fire and Ecosystems*, Academic Press, New York.

Vogl, R.J., 1977, Fire: a destructive menace or a natural process, in J. Cairns, K.L. Dickson, and E.E. Herricks, (eds.), *Recovery and Restoration of Damaged Ecosystems*, Univ. Press of Virgina, Charlottesville, pp. 261–289.

Waltson, A.B., Barnard, R.C., and Wyatt-Smith, J., 1953, Silviculture of lowland dipterocarp forest in Malaya, *Usylva*, **7**, 19–23.

Wehrmann, J., and Johannes, W., 1965, Effects of 'guie' on soil conditions and plant nutritions, *Afr. Soils*, **10**, 129–145.

Wells, C.G., Campbell, R.E., De Bano, L.O. Lewis, C.E., Fredrickson, R.L., Franklin, E.C., Forelich, R.C., and Dunne, P.H., 1979, Effects of fire on soil. A state-of-knowledge review, *General Technical* Report, USDA Forest Service, Washington, DC (1979) No. WO-7.

West, O., 1965, Fire in vegetation and its use in pasture management with species reference to tropical and subtropical Africa, Commonw. Bur. Pastures and Crops, *Field Crops*, Mimeo Publ., **1**, pp. 1–53.

West, O., 1971, Fire, man and wildlife as interacting factors, limiting the development of vegetation in Rhodesia, *Proc. 11th Ann. Tall Timbers Fire Ecol. Conf.*, pp 121–146.

Wright, H.A., Churchill, F.M., and Stevens, W.C., 1982, Soil loss, runoff and water quality of seeded and unseeded steep watersheds following prescribed burning, *J. Range*, **35**, 382–285.

Young, R.S., and Golledge, A., 1948, Composition of woodland soils and wood ash in Northern Rhosdesia. *Emp. J. Exp. Agric.*, **16**, 76–78.

Chapter 13
Conversion of Tropical Rainforests

I	REASONS FOR DEFORESTATION	504
II	RATE OF DEFORESTATION	506
III	RELIABILITY OF CONVERSION RATES	511
IV	METHODS OF DEFORESTATION	513
V	EFFECTS OF DEFORESTATION	516
	A. Regional and Local Effects	516
	1. Micro- and Meso-climate	516
	(i) Temperature and relative humidity	518
	(ii) Evaporation	519
	(iii) Rainfall	519
	2. Soil Physical Properties	522
	(i) Soil temperature	522
	(ii) Soil texture and structure	524
	(iii) Soil compaction	527
	(iv) Water transmission properties	531
	(v) Soil-water retention properties	532
	3. Hydrology and Soil-water Balance	533
	4. Run-off and Soil Erosion	536
	(i) Logging and erosion	536
	(ii) Perennial crops and erosion	537
	(iii) Grain crops	539
	(iv) Grazed pastures	540
	5. Soil Chemical Properties	541
	6. Soil Fauna	543
	7. Crop Yields for Different Land-Uses Following Deforestation	545
	(i) Short-duration fallow under shifting cultivation	545
	(ii) Antecedent soil fertility	547
	(iii) Crop species	549
	(iv) Soil management	549
	B. Global Effects	550
	1. The 'Greenhouse Gas' Syndrome	550
	2. Climate	554
	3. World Water Balance	556
VI	CONCLUSIONS	558

An ecologically important activity of man that has attracted world-wide attention in the recent past is mass scale deforestation, especially the reportedly rapid rate of depletion of tropical rainforests. Deforestation is a debated issue because it is done to increase agricultural production and to meet other human needs but it also has far-reaching ecological consequences on the local, regional and global scale. There are, however, few quantitative data from long-term studies or complete ecological surveys of adequately equipped small- and medium-sized watersheds. Consequently, it is often difficult to isolate factual effects of deforestation on environments from arguments tangled with human emotions. To describe quantitatively and objectively the effects of deforestation on soils and environments is definitely a challenge in light of the scanty research information available from the tropics. This chapter is a state-of-knowledge review of the available research information on the ecological consequences of forest alterations, and indicates knowledge gaps and research and development priorities. Readers are also referred to a recent review by Hamilton and King (1983).

I. REASONS FOR DEFORESTATION

Most of the present agricultural and urban land was once under forest cover. The need to increase food production and to industrialize led to forest removal and the development of land resources for economic benefit. Deforestation, therefore, has been a necessary evil. An important issue thus to be considered is, whether and which additional deforestation is necessary. It is the reckless and unthinking destruction of the forest reserves that must be avoided. The principal causes of deforestation may be grouped as historic, economic, and socio-political (Table 13.1).

Historically, forest dwellers have sought their livelihood and basic necessities of life by harvesting forest products for various uses. Lea (1975) listed 32 uses of various forest products used traditionally by the inhabitants of Papua New Guinea. In addition to harvesting minor forest products, the farmers of the humid forest zone cultivate the land by shifting cultivation and bush fallow

Table 13.1 Principal causes of forest conversion

Causes of deforestation

Tradition/Historic	Economic	Socio-political
1. Shifting cultivation	1. Agriculture	1. Strategic
2. Fuel-wood	2. Ranching	2. Population migration
3. Harvesting forest products	3. Plantation crops	
	4. Timber and commercial wood	
	5. Urbanization	
	6. Infra-structure	

rotation, which involves incomplete clearing to encourage rapid regeneration during the fallow phase. In the tropics, forest clearance for shifting cultivation is believed to have been practiced for at least 3000 years in Africa, 7000 years in Central or Southern America, and for about 9000 years in India (Anonymous, 1982). Some reports indicate that perhaps as much as 50 per cent of the deforestation in the tropical rainforest zone is done by shifting cultivation (Postel, 1984). The forest removed by subsistence cultivators eventually regenerates if the fallow period is long enough to allow forest re-establishment; but this is not often the case, particularly in densely populated areas, e.g., southeast Nigeria. Jackson (1983) estimated that there are some 150 million forest farmers involved in traditional agriculture in the forest zone who depend solely on forest lands for their meagre living.

Deforestation in developing tropical countries is also done for procuring fuel-wood to meet household energy requirements. FAO (1981) reported that the fuel consumption in developing tropical countries (except China) in 1979 was 1300 million m^3, about 100 million m^3 short of the requirements. It is estimated that some 250 million people live in regions of fuel-wood shortage, and that the fuel-wood demand by the year 2000 will be 2600 million m^3. If the rate of deforestation is to be controlled, alternative economic sources of fuel-wood, or another fuel, will have to be developed.

Planned development of natural resources to improve gross economic production in the tropics involves deforestation for alternative land-uses including agriculture, ranching, plantation crops, timber, urbanization and infrastructure development. Deforestation for agriculture is often based on the common misconception that soils supporting lush green forest ought to be fertile enough to grow seasonal crops. The subsequent low and uneconomic returns result in more deforestation to meet the rising food demands. Deforestation and agriculture have thus become a vicious cycle from which planners find it hard to locate a safe escape.

Postel (1984) reported that some 4.4 million hectares per year of tropical moist forest are being logged over lightly for selected trees and 13.25 per cent of the tropical forests have so far been logged. Large areas of forested lands have been developed for oil palm plantations in West Africa and Malaysia, and for sugar-cane plantations in Cuba, Barbados and in the Caribbean. Ranching is rapidly encroaching upon the remaining forest lands in South and Central America. Myers (1981) estimated that 33 per cent of the land area in Costa Rica is devoted to ranching.

The transmigration of population from densely populated areas to forested land is a major socio-political reason for large-scale deforestation in some countries, e.g., Sumatra, Indonesia, and Brazil. Often the forest land is opened up for strategic reasons to claim ownership.

The net result of these inter-related socio-political and economic causes is the rapid depletion of the existing forest reserves in favour of new land development and colonization schemes. The effects of these rapid transformations will be felt during decades to come.

The traditional use of forest by shifting cultivators and those that harvest minor products does not have any major ecological effects even on a local scale. It is the intensive land use either by traditional or modern methods that leads to drastic detrimental ecological effects.

II. RATE OF DEFORESTATION

The actual rate of deforestation is hard to estimate and, as the literature presented indicates, the current rate is anyone's guess. Some reports estimate that about 11 million hectares of tropical forest cover are being cleared annually for miscellaneous uses (FAO, 1981) (Table 13, 2a,b,c,d,e). If these estimates and those of the remaining forest reserves are correct, only 12 per cent of the tropical forests will be depleted between 1980 and the year 2000. Other reports estimate annual conversion rates at only 5 to 6 million bectares (Seiler and Crutzen, 1980; Dickinson, 1982). Jackson (1983) reported that out of the 11 million hectares, 7 million hectares comprise tropical rainforest and 3.8 million hectares are deciduous forest and the wooded savanna.

Table 13.2a Variable rates of deforestation in the tropics reported in the literature

Author	Rate (million hectares)
Sommer (1976)	2.15
Myers (1980)	6.59
Seiler and Crutzen (1980)	5–6
Dickinson (1982)	6
Salati and Vose (1983)	9
FAO (1981)	11
Jackson (1983)	11
Grainger (1980)	15.8
Rubinoff (1983)	5.8–7.4
O'Keefe and Kristofferson (1984)	0.6%

Table 13.2b Rate of deforestation in the tropics (modified from FAO, 1981). *Reproduced by permission of the Food and Agriculture Organization of the United Nations*

Vegetation	Existing area (10^6 ha)	Annual change (10^6 ha)
Woody vegetation	2900	—
Forest land	1875	−11.0
Closed forest	1200	−7.5
Wooded savanna	675	−3.5
Former forest under shifting cultivation	400	+5.3
Shrubland	625	—
Miscellaneous land	1915	+5.7

Table 13.2c Estimated rates of forest conversion between 1980 and 1985 in different regions of the tropics (O'Keefe and Kristofferson, 1984). *Reproduced by permission of the Royal Swedish Academy of Sciences*

Vegetation	Area deforested (10^6 ha) in different regions of the tropics			Total
	Tropical America	Tropical Africa	Tropical Asia	
Closed forest	4339 (0.64%)	1331 (0.62%)	1826 (0.60%)	7496 (0.62%)
Open forests	1272 (0.59%)	2345 (0.48%)	190 (0.61%)	3807 (0.52%)
All forests	5611 (0.63%)	3676 (0.52%)	2016 (0.60%)	11303 (0.58%)

Numbers in parentheses represent deforestation rate as percentage of the remaining forest area on basis of a survey conducted in 76 countries.

The estimates of the annual rate of forest clearing vary widely among countries and depend on the area remaining under forest (Jackson, 1983). In tropical Africa, the annual rate of deforestation is 6.5 per cent in Ivory Coast, 2.77 per cent in Rwanda and 2.3 per cent in Zaire. In tropical America, the yearly deforestation rate is estimated at 3.8 per cent for Nicaragua, 2.59 per cent for Honduras, and 2.38 per cent for Ecuador. The forest is being depleted at 3 per cent per year in Thailand. It is estimated that Madagascar, Uganda and Malaysia lose their forest at more than twice the world average rate of 0.6 per cent per year. Presenting the loss of forests as per cent per year can be misleading, however. The absolute areas cleared can be very different, but still represent the same percentage of total forested land. For example, 0.38 per cent per year loss of forest in Brazil amounts to 13 600 km^2 whereas 0.53 per cent in Indonesia is approximately equal to 6 000 km^2.

Rubinoff (1983) estimated that 42 per cent of the tropics, or 1.9 billion hectares, is under forest, with 1 billion hectares under closed moist forest and 0.9 billion hectares under open woodland savanna. These authors estimated the area under forest to be 42 per cent in tropical America, 21 per cent in Asia, Australia and Oceania, and 37 per cent in Africa. The area that has already been lost to deforestation is estimated at 37 per cent in tropical America, 42 per cent in Asia and Australia, and 52 per cent in Africa.

Salati and Vose (1983) reported that about 6 million hectares are annually being cleared by shifting cultivators and loggers, and an additional 3 million hectares for pastures and for cropping. Some high estimates indicate an annual forest loss to pastures of about 10 to 30 million hectares. These authors observed that some 20 million hectares of secondary forest decline annually to more degraded secondary forest or are converted to agricultural land-use. In addition, about 1 million hectares of cropped land originally derived from forest are believed to be abandoned due to soil degradation. Salati and Vose estimated the remaining tropical forest reserves to be at about 75 million hectares, and the area lost to deforestation at about 10 million hectares in

Table 13.2d Estimates of rates of forest conversion by Myers (1980) and Sommer (1976). *Reproduced by permission of National Academy Press*

Country	Time interval	Current forest area	Myers (1980)		Sommer (1976)	
			Actual conversion (10⁶ ha)	Actual %/yr	conversion (10⁶ ha)	%/yr
Asia						
Bangladesh	—	—	—	—	0.01	0.8
Burma	—	36.5	0.142	0.39	—	—
Laos	—	14.0	0.30	2.14	0.30	5.17
Malaysia (Peninsular)	1972–79	7.2	0.22	2.65	—	—
Malaysia (Total)	—	40.0	—	—	0.15	0.64
Papua-New Guinea	—	11.5	0.025	0.06	0.02	0.05
Philippines	1971–76	13.18	0.30	2.62	0.26	2.05
Thailand	1972–78	—	1.15	5.73	0.30	1.05
Vietnam	—	—	—	—	10.0	—

Latin America						
Brazil	Last 20 yrs	286.0	1.3	0.42	—	—
	1976–75	—	1.28	0.43	—	—
Costa Rica	1967–77	1.6	0.04	2.0	0.06	2.73
Colombia	—	—	—	—	0.25	0.50
Nicaragua	1970–79	0.35	0.04	1.1	—	—
Guyana	—	18.7	0.01	0.05	—	—
Peru	1945–75	65.0	0.17	0.24	—	—
Venezuela	1950–75	34.2	0.07	0.20	0.05	0.29
Africa						
Gabon	—	20.5	0.03	1.46	—	—
Ghana	Last 25 yrs	2.0	0.16	2.67	0.05	2.50
Ivory Coast	1976–74	5.4	0.45	5.0	0.40	4.44
Liberia	recent	2.5	0.23	9.2	—	—
Madagascar	—	—	—	—	0.30	4.0
Nigeria	1970–79	2.6	0.28	6.2	—	—
Sierra Leone	1944–79	0.29	0.134	2.69	—	—
Total/Mean		565.6	6.59	0.75	2.15	1.09

Table 13.2e Forest reserves and the rate of deforestation (million hectares) (Sommer, 1976)

Region	Moist forest		Decrease	Percentage decrease
	Potential area	Actual area		
Central Africa	269	149	120	44.6
West Africa	68	19	49	72.0
Total (Africa)	362	175	187	51.6
South America	750	472	278	37.1
Central America	53	34	19	35.8
Total (Latin America)	803	506	297	37.0
Pacific Region	48	36	12	25.0
South East Asia	302	187	115	38.1
South Asia	85	31	54	63.5
Total (Asia)	435	254	181	41.6
Total (tropics)	1600	935	665	41.6

South America, 15 million hectares in Southeast Asia and more than 100 million hectares in tropical Africa. About 60 000 hectares of forest are annually being lost to ranching in Costa Rica, and only one-third of the original forest remains in Central America.

Yet another report observed that West Africa has already lost 72 per cent of its rainforest and Asia has lost 63 per cent (Anonymous, 1982). This report warns that half of the Amazon forest and 28 per cent of that remaining in Africa may be destroyed by the year 2000. An FAO-sponsored study reported the annual forest loss at about 15 million hectares of which 2, 5, and 5 to 10 million hectares are annually lost by Africa, Asia and Latin America, respectively (Sommer, 1976; Grainger, 1980). The Eighth World Forestry Congress stated that forests are being destroyed at 30 hectares/minute, and that the remaining forest reserves will disappear by the year 2040 (Fig. 13.1) (Grainger, 1980).

How accurate and reliable is this information? That is an important issue. How can scientists and planners develop a conservation and management policy without reliable data on the existing forest reserves and on their conversion rates? A comparison of the estimates of Myers (1980) and Sommer (1976) (Table 13.2d) indicate the magnitude of differences both in the area under the forest cover and in its conversion rate. Some other reports indicate the conversion rates to be twice or one-half as much as those shown in Table 13.2a to 13.2e depending upon whether the reporter is a conservationist or an exploiter (Anonymous, 1982). Whereas many conservationists feel that most forest will disappear within the next 50 years, others argue that this popular concern about the loss of tropical forests may be excessive and misdirected. For example, satellite pictures show that only 2 per cent of the forest reserves in the Brazilian Amazon have so far been cleared, and by the year 2000 the additional

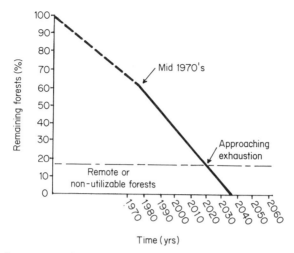

Figure 13.1 Some researchers estimate that forest reserves will disappear by the year 2040 (Grainger, 1980). *Reproduced by permission of The Ecologist*

area of the tropical forests that will have disappeared will be only 9 per cent in Latin America, 6 per cent in tropical Africa and 13 per cent in Asia. On the contrary, another report indicates conversion rates of about 24.5 million hectares per year, which mean that the entire forest reserve will disappear within 30 to 40 years (Myers, 1980).

III. RELIABILITY OF CONVERSION RATES

The current status of the 'tropical jungle' is as confusing and complex as is the jungle ecology itself. Reliable survey data are required to resolve this emotional issue, as it has implications in planning for development of natural resources and implementing conservation policies, if necessary. Most of the available reports are 'guestimates' based on a few facts and little validation. These estimates vary by several orders of magnitude (Tables 13.3 and 13.4). Myers (1980) observed that most reports on the current status of the forest are obsolete, irrelevant, and often of doubtful utility. In fact there is no systematic study of the current global forest reserves and their conversion rates for miscellaneous land-uses. Thorough and systematic regional and global surveys are thus needed. These surveys should use modern techniques including satellite pictures and the data should be locally validated.

Another important source of confusion is the lack of standard terminology and definition. Terms 'forest' and its 'conversion' should be standardized with universally accepted criteria. It is also relevant to standardize the classification of different types of forests. For example, the interchangeable use of terms like tropical moist forest, rainforest, lowland rainforest, closed forest, primary and secondary forests, deciduous forests, wooded savanna, etc., is very confusing

Table 13.3 Tropical forest resources in 1973 (Persson, 1974)

Vegetation	Area (10^6 ha)
Evergreen	560
Moist deciduous	308
Dry deciduous	588
Total	1456

Table 13.4 Total forest reserves of 76 countries of the tropics (Lanly, 1982)

Sub-region/Region	Area in 10^6 ha (including inland waters)	
Tropical America	1679.5	
Central America and Mexico		247.2
CARICOM		25.4
Other Caribbean		44.5
Tropical South/Latin America		1362.4
Tropical Africa	2189.4	
Northern savanna region		423.6
West Africa		212.1
Central Africa		532.7
East Africa and Madagascar		881.1
Tropical South Africa		139.9
Tropical Asia	944.9	
South Asia		448.8
Continental Southeast Asia		119.2
Insular Southeast Asia		255.5
Centrally planned tropical Asia		75.2
Papua New Guinea		46.2
Grand total for 76 countries	4813.8	

and leads to erroneous interpretations. Different methods of forest classification have been reviewed by Ross (1984). Having no data at all is better than reporting unreliable or obsolete data.

Another source of confusion is in the interpretation of the term 'deforestation'. Once again the interchangeable use of commonly used terms—deforestation, clearing, conversion, logging, forest regression, etc.—increases the confusion. The magnitude of alterations in different ecological parameters is different for clearing for different land-uses. For example, logging for timber and incomplete clearing by shifting cultivators do not produce the same effects as complete clearing for arable land-use. Similarly, clearing for seasonal crops produces grossly different ecological effects than clearing for plantation developments, improved forestry, or for ranching.

The criteria used for assessing conversion rates for different land-uses must also be standardized. Some land-uses involve complete removal of the forest vegetation whereas others cause disruption and depletion rather than complete transformation. Different forms of conversion ranging from slight alterations to complete transformation (Myers, 1983) should be separately assessed because they produce variable cological effects.

Lack of infrastructure and accessibility, and scarcity of trained manpower in many countries in the tropics, contribute to the scarcity of reliable surveys of the existing forest reserves and their conversion rates. Inadequate communication is also a major hindrance in updating the field records, particularly in countries where large areas are being cleared for agricultural settlements. Not only are some records 10 to 15 years out-of-date, their accuracy level is within plus or minus 40 per cent (Persson, 1977).

The presently available estimates of the forest reserves and their conversion rates are therefore unreliable and obsolete. An internationally coordinated effort should be made to develop a data bank on forest reserves and their conversion rates. This information should be collated by precise, reliable and easily verifiable methods.

IV. METHODS OF DEFORESTATION

A wide range of methods is used for forest removal depending on purpose, infrastructure and socio-economic factors (Table 13.5). Some of these methods have been discussed at length by Lal *et al.* (1985) and are only briefly outlined below. A discussion of methods is relevant because the effects of forest removal on soil, hydrology and micro-climate depend on the method used.

In traditional farming and shifting cultivation, native tools and fire are widely used as means of clearing. Traditionally, vegetation is cut by manually operated tools (axe, machette, etc), felled and then burned in place when suitably dried. Whereas shifting cultivation involves incomplete clearing, labour-intensive manual clearing methods are preferred for complete and thorough clearing for more stable and modern agriculture. Improved and semi-mechanized methods (e.g., chain-saw and winch) are increasingly being used. Some equipment, including root ploughs, can be drawn by animals. This technique may be useful to eradicate rhizomatous vegetation such as *Imperata cylindrica*.

Capital-intensive mechanical methods are commonly used for large scale clearing. These methods often cause drastic changes in upper soil horizons, and the magnitude of the alterations depends on the attachment used. Attachments that can be used with motorized equipment include tree pusher, tree extractors, tree crusher, shear blade, root plough, root rake and dozer blade. Mechanized clearing results in high degrees of soil disturbance, scraping and removal of the topsoil, and concentration of biomass and fertile topsoil in the windrows.

Tree poisoning by appropriate chemicals is used to kill isolated trees in wooded savanna and in deciduous rainforest with relatively low tree density.

Table 13.5 Methods of forest cultivation

Methods of deforestation

Traditional	Manual	Animal-drawn for rhizomatous grass, e.g. *Imperata cylindrica*	Mechanical, tractor-mounted/pulled attachments	Chemical	Combination of these methods
(i) Fire (ii) Native tools	(i) Native (ii) Semi-mechanized, e.g. chain saw (iii) Winch		(i) Chain (ii) Tree pusher (iii) Tree extractor (iv) Tree crusher (v) Shearblade (vi) Root plough (vii) Root rake (viii) Dozer blade	(i) Defoliant (ii) Systemic	

Table 13.6 Factors affecting the choice of land clearing methods used

Choice of land clearing method

Vegetation	Slope	Crops	Season	Conservative measure
(i) Rhizomatous grass (ii) Savanna (iii) Secondary forest (iv) Closed forest	(i) Degree (ii) Length (iii) Aspect	(i) Seasonal (ii) Perennial	(i) Dry (ii) Transition (iii) Wet	(i) Mechanical (ii) Biological

Chemical land-clearing methods are often used for pasture development and for eradicating some rhizomatous perennial grasses.

Numerous combinations of these methods are employed to eradicate the existing vegetation efficiently and rapidly. The ecological effects of deforestation depend on how the vegetation is removed and how the soil and water resources are subsequently developed and managed.

Clearing for agricultural development is the major initial step that influences sustained land productivity. The choice of appropriate methods of land clearing, therefore, is important both economically and ecologically. The choice of an appropriate method or a combination of methods depends on several factors (Table 13.6). Important among these are the vegetation, topography, soil type and intended land-use. No less important are the economics, infrastructure, and availability of required equipment and maintenace facilities.

Land-clearing operations are often initiated at the end of the rainy season when the soil water has been depleted enough to support heavy machinery without causing undue compaction and soil disturbance. The magnitude of the degradative effects of deforestation on soil also depend on soil wetness, time of clearing, method adopted, and choice of attachments used. The skill of the operator is also an important factor (Lal et al., 1985).

V. EFFECTS OF DEFORESTATION

'Taming' and 'domesticating' a tropical forest ecosystem is a misguided challenge. Whereas utilization of natural resources is permissable and desirable, exploitation for short-term and rapid economic gains is indeed unforgivable. In addition to the lack of appropriate resource data needed for proper planning, another difficulty in initiating a rational policy for forest resource management is the lack of quantitative data regarding the ecological effects of deforestation.

The ecological effects of deforestation are broadly classified into 'regional' and 'global' (Table 13.7). Among regional effects of deforestation are those on micro- and meso-climate, soil, hydrology, fauna and flora, and on crop production. Some important global effects are effects on CO_2 balance, and world water balance.

A. Regional and local effects

1. Micro- and meso-climate

The removal of vegetation cover increases the amount of solar radiation and rainfall directly reaching the soil surface. The wind velocity is also increased. The buffering effect of leaf litter and other biota against sudden fluctuations in temperatures is also removed. The quasi-stable equilibrium that exists in the soil-water-forest ecosystem is disturbed by forest removal. Consequently, the micro- and meso-climates and drastically influenced.

Table 13.7 Effects of deforestation and conversion to arable land-use

Regional					Global	
Climate	Soil	Hydrology	Fauna and Flora	Production		
(i) Increase in insolation (ii) Increase in temperature range (iii) Increase in relative humidity (iv) Decrease in vapour return to atmosphere (v) Increase in wind	(i) Increase in soil temperature range (ii) Compaction (iii) Tendency of the survey layer to dessicate (iv) Decrease in organic matter (v) Decrease in litter return (vi) Accelerated erosion (vii) Hardening of plinthite	(i) Decrease in infiltration (ii) Increase in run-off (iii) Decrease in avapotranspiration (iv) Increase in interflow (v) Decrease in soil-water storage (vi) Increase in fluctuation of ground-water table	(i) Decrease in soil fauna (ii) Decrease in floral diversity (iii) Shift in predominant species	(i) Decrease in gross production (ii) Increase in net primary production (iii) Increase in yield of grain crops (rice, wheat, soybean, sugarcane, sorghum) (iv) Decrease in drought stress (iii) & (iv) are results of increased CO_2 in the atmosphere	Increase in CO_2 return and general warming Rainfall patterns	Decrease in vapour return to atmosphere Changes in Evaporation World water balance

Table 13.8 Solar radiation received under forest and in the cleared area at Okumu during some days of dry season of 1984–85

Day	Solar radiation (g cal cm^{-2} day^{-1})		
	Forest	Cleared area	Forest/cleared
December 4	10	270	0.037
December 6	9	229	0.039
December 8	8	229	0.035
December 10	10	243	0.041
December 12	10	229	0.043
December 14	11	270	0.401
December 16	10	283	0.035
Average	9.7±0.9	250.4±21.6	0.039±0.003

(i) Temperature and relative humidity: The forest canopy effectively intercepts the solar radiation. The data in Table 13.8 show that only 10 g cal cm^{-2} day^{-1} of radiation reached the ground surface under forest as compard to 250 g cal cm^{-1} day^{-1} received in the cleared area. Consequently, the amplitude of diurnal and seasonal changes in temperature and relative humidity near the ground surface is increased by deforestation. Experiments conducted at IITA, Ibadan, Nigeria, showed that deforestation increased the maximum temperature by 5 to 8°C, and decreased the minimum temperature by 1 to 2 °C at 10, 50,

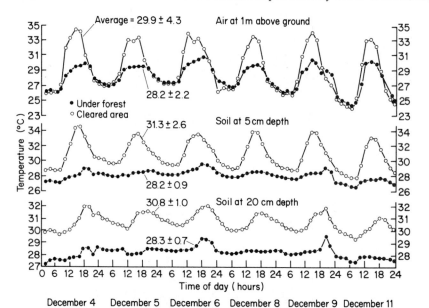

Figure 13.2 Soil and air temperatures under and outside (cleared area) the forest for six consecutive days during the dry season of 1984–85 at Okumu (Ghuman and Lal, 1985)

and 100 cm above the ground surface. The differences in air temperature between cleared and forested land were more pronounced near the ground than at 1 m above the ground surface. In another study at IITA, Lawson *et al.* (1981) observed smaller temperature variations under forest than in the cleared land. Forested land had lower maxima and higher minima than cleared land. In a more humid zone near Benin city, Nigeria, the maximum air temperatures were 37°C and 26°C on cleared and forested sites, respectively, within 20 metres of one another. There was also a notable phase difference in the temperature wave. The maximum temperature was attained earlier and was maintained for a longer time on cleared than on forested land (Fig. 13.2). Temperature distribution profiles from 20 cm below the soil surface to 10 m above are illustrated for a high forest near Benin, Nigeria, in Fig. 13.3. Temperature profiles at 0600 and 1400 hours are markedly different and coalesce into one at about 20 cm below the soil surface.

The high relative humidity commonly observed within a forest canopy also decreases in cleared land because of increased air temperature, higher wind velocity, and a reduction in biota-transpired water vapours. The data obtained from IITA, Ibadan, showed that differences in the minimum humidity in cleared and forested land were of the order of 15–20 per cent. The time of occurrence of minimum relative humidity, however, was 2–3 hours later on cleared than on forested land. Lawson *et al.* (1981) also observed more humid conditions in the forest in comparison with the cleared land. The differences in relative humidity between the cleared and forested sites were more pronounced for minimum rather than the maximum values. The lower minimum humidity on the cleared land reflected the greater proportion of incoming radiation reaching the soil surface. Near Benin City, Nigeria, the minimum relative humidities observed were 49 and 87 per cent for cleared and forested plots, respectively (IITA/UNU, 1984) (Fig. 13.4). Deforestation therefore, increases aridity of the micro-climate.

(ii) Evaporation: The evaporative demand is lower under forest than on cleared land. Lawson *et al.* (1981) reported total evaporation from March to September to be 794 mm and 1409 mm for forested and cleared areas, respectively. The data in Fig. 13.5 from a high rainforest near Benin City show a marked increase in evaporation from the cleared land. Decreases in minimum relative humidity, increases in insolation and maximum air temperatures, and relatively greater wind velocity on cleared land, increase the evaporative demand. The cleared areas, therefore, have a higher mean evaporation than the forested soil.

(iii) Rainfall: Also affected by deforestation is the proportion of the rainfall directly reaching the ground surface. Through-fall through the forest canopy ranges from 60 to 90 per cent depending on the tree density, canopy characteristics, and rainfall amount and intensity (Bernhard-Reversat *et al.*, 1972; UNESCO, 1978; Lawson *et al.*, 1981). The mean stem flow is usually less than

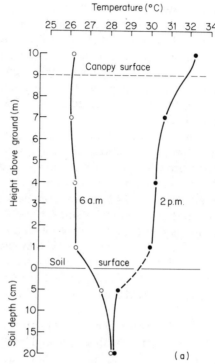

Figure 13.3a Temperature profiles in a forest canopy at Okomu. The data are an average of temperatures measured at six days during the dry season of 1984–85 at Okomu (Ghuman and Lal, 1985)

Table 13.9 Monthly rainfall (cm) from June to December 1984 at Okomu

Month	Cleared area	Rainfall (cm) Under forest	
		Throughfall	Stem flow + canopy intercepted
21–28 June	4.86	3.90	0.96
July	25.51	19.87	5.64
August	23.23	21.15	2.08
September	32.07	28.63	3.44
October	20.36	19.51	0.85
November	0.00	0.00	0.00
December	0.00	0.00	0.00
Total	106.03	93.06	12.97

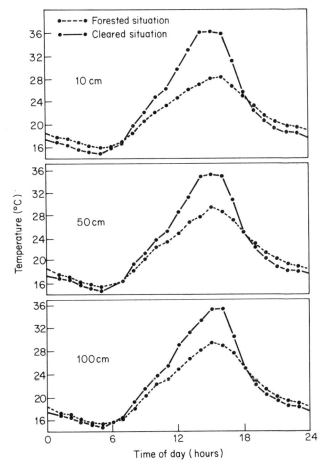

Figure 13.3b Effects of deforestation on ambient temperatures at IITA, Ibadan, Nigeria (Lal and Cummings, 1979)

10 per cent (Lawson *et al.*, 1981), although 18 to 28 per cent has also been reported (Kline *et al.*, 1968). The direct rainfall interception, depending on the canopy and rainfall characteristics, varies between 12 and 24 per cent (Kline *et al.*, 1968; Lawson *et al.*, 1981). The data on partitioning of rainfall into various components for a high rainforest in southern Nigeria are shown in Table 13.9. On an average, through-fall was about 12.2 per cent less than the rainfall in the open. Deforestation decreases the effective amount of rainfall by increasing evaporation and losses due to surface run-off.

Removal of forest cover, therefore, has a drastic effect on micro-climate especially near the ground surface. The implications of alterations on micro- and meso-climate on a regional scale have not been documented. The effect on a regional scale, however, would depend on the area and number of individual clearings.

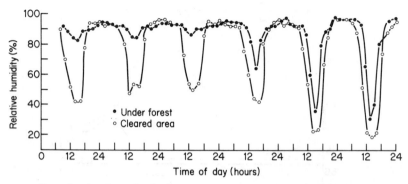

Figure 13.4 Relative humidity under and outside (cleared area) the forest for six consecutive days during the dry season of 1984–85 at Okomu (Ghuman and Lal, 1985)

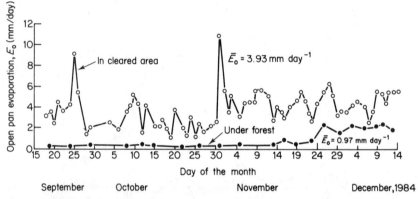

Figure 13.5 Open pan evaporation in the cleared area and under forest at Okomu during 1984, E_0 is E_0 (Ghuman and Lal, 1985)

2. Soil physical properties

Soil physical properties are also altered by deforestation, and the magnitude of the alterations depends on the methods of deforestation, prevalent climate, and antecedent soil properties. Mechanical clearings with heavy machinery often result in immediate and more drastic effects on soil physical properties than manual clearing or tree poisoning.

(i) Soil temperature: Increase in insolation reaching the soil surface, and decrease in rainfall effectiveness results in higher amplitude of the diurnal and seasonal fluctuations in soil temperature. The forested soil has less difference between the maximum and the minimum temperatures. In general, the maximum soil temperature is higher and the minimum temperature is lower in cleared than in the forested land. The fluctuations in soil temperature are also

enhanced by the lack of buffering effect of vegetation cover of litter fall, and due to the general decrease in soil moisture caused by higher evaporation.

In Africa d' Hoore (1954) reported significant effects of vegetation removal on soil temperature (Table 13.9a). Increases in the maximum soil temperature due to deforestation were reported from Ghana by Cunningham (1963). Whereas the minimum temperatures did not differ, the maximum temperature for the 7.5-cm depth was 11°C higher in the fully exposed than in the fully shaded treatment. The maximum soil temperatures were 27°, 32°, and 38°C under shade, half-exposure, and full-exposure treatments, respectively. The temperature under artificial shade was similar to that in the soil of the adjacent

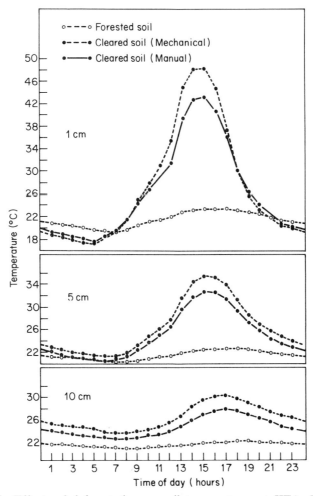

Figure 13.6 Effects of deforestation on soil temperatures at IITA, Ibadan. The measurements are averages of data obtained between 2 and 10 December, 1975 (Lal and Cummings, 1979)

Table 13.9a Effects of vegetation removal on fluctuation in soil temperature under different ecologies (d'Hoore, 1954)

Vegetation	Depth (cm)	Soil temperature (°C)		
		0800 h	1100 h	1400 h
No vegetation	1	26.8	44.0	53.5
	10	27.5	31.7	36.8
Savanna vegetation				
Under grass	1	24.7	41.5	50.4
	10	27.8	32.9	40.1
	20	27.5	33.4	41.1
Under shrubs	1	21.6	22.7	26.9
	10	23.0	22.9	24.8
	20	23.8	23.8	25.4
Air temperature		22.2	28.3	36.0

forest. Lal and Cummings (1979) reported that the maximum soil temperature was higher on mechanical than on manually cleared land. Their data in Fig. 13.6 show that the maximum soil temperature at 1 cm depth was 25° and 20°C higher for mechanically and manually cleared than forested land, respectively. Forest removal influenced soil temperature at 5 cm and 10 cm depths also. Lawson *et al.* (1981) reported 10° to 12°C higher maximum soil temperatures at 5 cm depth in cleared than in forested lands. The minimum temperature was, however, higher in forested land by 3° to 4°C. In western Nigeria, Swift *et al.* (1981) reported that temperature on the soil surface was 18°C on cleared with only 3°C in forested plots. That is why the surface soil of the cleared plot had a strong tendency to desiccate. The increase in the maximum soil temperature following clearing is indicative of the alterations in energy balance of the soil surface.

(ii) Soil texture and structure: Unless the surface soil is scraped off and carried to the windrows, deforestation *per se* does not have any immediate effect on textural properties of the soil. In the long run, however, soil textural properties on cleared land are altered due to accelerated soil erosion, decreased activity of soil fauna, and exposure-induced increases in surface soil temperatures. Accelerated soil erosion results in preferential removal of soil colloids leaving behind coarse sand and gravels on the surface. Declines in the biotic activity of earthworms and termites which bring clay-sized particles to the soil surface, can eventually decrease the clay content of the surface horizon. Decreases in the clay content of the cleared and cultivated land are also attributed to eluviation of clay sized particles from the surface to the subsoil horizons. In Ghana Cunningham (1963) observed significant differences in soil mechanical

Table 13.10 Effect of degree of soil exposure on soil textural properties (Cunningham, 1963). *Reproduced by permission of the* Journal of Soil Science

Horizon (cm)	Coarse sand (%) (0.20–2.00 mm)			Fine sand (%) (0.02–0.20 mm)			Silt + clay (%) (< 0.02 mm)		
	Shade	Half-exposure	Full exposure	Shade	Half-exposure	Full exposure	Shade	Half-exposure	Full exposure
0–5	33.1	35.0	43.8	35.1	34.6	30.4	26.3	27.6	23.6
5–15	44.8	42.8	34.0	30.9	33.5	36.6	22.8	22.5	28.6

Table 13.11 Effect of degree of exposure following clearing on soil porosity and water stable aggragates in the semi-deciduous forest zone of Ghana (Cunningham, 1963). *Reproduced by permission of the* Journal of Soil Science

Treatment	Porosity (volume %)			Water stable aggregates (% air dry soil, > 3 mm)	
	Capillary pore space	Non-capillary pore space	Total pore space	Total structure	Crumb structure
Shade	37.0	14.7	15.7	55.3	39.8
Half-exposure	35.0	16.4	51.4	50.2	28.9
Full exposure	32.6	10.1	42.7	48.7	29.0

analyses 3 years after imposing the clearing treatments. Whereas the textural composition had changed little in the 0 to 5 cm layer of the shaded and half-shaded plots, there was a measurable increase in coarse sand fraction and a decrease in the fine sand, silt, and clay contents of the exposed plots (Table 13.10). The clay and fine soil fractions were believed to have eluviated to the 5 to 15 cm layer.

Soil structure is more drastically affected by deforestation than soil texture, and the effect is generally greater from mechanical than manual clearing methods. Soil structure implies the degree of soil aggregation and aggregate stability to raindrop impact. The protective effects of vegetal cover, regular addition of leaf litter and other organic matter to the soil surface, and high activity of soil fauna encourage development of aggregated structure in forested soil. Depending on the textural and mineralogical composition, aggregates are relatively stable. Both inter- and intra-aggregate porosity is high. Drastic alterations in microclimate, decline in biotic activity, lack of the protective leaf litter and organic matter, result in decreased soil aggregation. Whatever aggregates exist are weakly developed and are unstable. Surface soil on even newly cleared land is prone to slaking, crusting, and surface-seal due to a collapse of structural units.

The decline in structural properties following deforestation is mostly due to reduced soil fauna activity, i.e., earthworms and termites. Slaking, crusting and surface seal are also enhanced by the ultra-desiccation that the exposed soil surface undergoes from high insolation. Cunningham (1963) observed that exposure of an Alfisol for 3 years in the semi-deciduous rain forest zone of Ghana decreased capillary and non-capillary porosity of the 0 to 7.5 cm layer. Furthermore, both full and half exposure of the soil for 3 years decreased both total and true crumb structure of the 0 to 7.5 cm layer (Table 13.11). Ollagnier *et al.* (1978) observed in Ivory Coast that the structural stability index of the bare soil was only 56 per cent of that of the forested land (Table 13.12). These authors attributed the decline in soil organic matter content from 2.85 to 1.24 per cent. Hulugalle *et al.* (1984) observed that mechanized land clearing resulted in a drastic decrease in macropores exceeding 14.3 µm in pore radius (Table 13.13). In contrast, however, pore volume of 2 to 14 µm radius was increased by the mechanized land clearing operations.

(iii) Soil compaction: Bulk density is generally low in forested soil, and the

Table 13.12 Effects of deforestation on soil structure and organic matter content at Adiopodoume, Ivory Coast (Ollagnier *et al.*, 1978). *Reproduced by permission of IRMO-Oleagineux*

Parameter	Forest	Bare soil
Henin index of structural stability	1350	750
Organic matter (%)	2.85	1.24

Table 13.13 Effect of clearing method on total porosity and apparent pore size distribution (0–100 mm depth interval) (Hulugalle et al., 1984). *Reproduced by permission of Williams & Wilkins Co.*

Clearing method	Total porosity	Apparent pore size distribution (%)		
		14.3 μm	14.3–2.0 μm	2.0 μm
Uncleared	0.62	79.1	2.9	17.4
Manual	0.61	76.1	3.9	20.0
Tree pusher	0.57	71.1	4.0	24.9
Tree pusher + root rake	0.56	52.7	28.6	18.7
Shearblade	0.56	59.1	23.8	17.1
LSD (0.05)	0.02	7.9	5.2	7.3

* Pore radius.

magnitude of its increase by deforestation depends on clearing methods, amount of leaf litter on the soil surface, antecedent textural and structural properties, soil-water content and mineralogical composition. Mechanical land clearing techniques often result in a greater increase in bulk density than manual land clearing methods (Van der Weert, 1974; Seubert et al., 1977). The degree of compaction also depends on the pressure exerted at the soil surface, vibrations in the soil, and the ability of the soil to resist compaction.

The pressure exerted by the machine depends on its weight and the forward veolicty. In Surinam Van der Weert (1974) observed that bulldozer clearing increased soil bulk density and decreased porosity up to about 80 cm depth (Figs. 13.7 and 13.8). Mechanical clearing decreased the volume of macro-

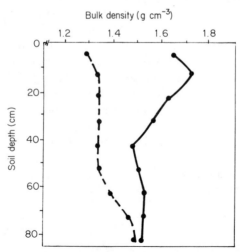

Figure 13.7 Effects of bulldozer clearing on bulk density of a soil in Surinam: (– –) after clearing (– – –) control (Van der Weert, 1974). *Reported by permission of Tropical Agriculture*

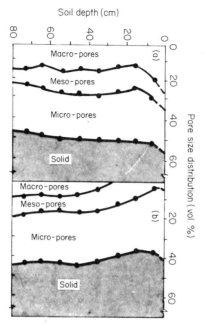

Figure 13.8 Effects of bulldozer clearing on total porosity and pore size distribution of a soil in Surinam: (a) control, (b) after clearing (Van der Weert, 1974). *Reproduced by permission of Tropical Agriculture*

(>180 μm) and meso-pores (30–180 μm) and increased that of the micro-pores (<30 μm). At Yurimaguas in the Upper Amazon region of Peru, Seubert *et al.* (1977) observed that the bulk density of the cleared Ultisol was 1.24 and 1.46 g/cm^3 for 0–2 cm depth, and 1.51 and 1.67 g/cm^3 for 8–10 depth for burned and bulldozed plots, respectively. In Bahia, Brazil, Silva (1981) reported that clearing with a D-7 track-tractor affected bulk density and porosity up to the 20 cm depth (Table 13.14). The maximum compaction occurred in the layer immediately beneath the soil surface i.e. at 5 cm depth.

Table 13.14 Effect of clearing with D-7 tractor on bulk density (g/cm^3) and porosity (%) of an Oxisol in Bahia, Brazil (Silva, 1981)

Depth (cm)	Forested control		Mechanically cleared	
	Bulk density	Porosity	Bulk density	Porosity
2.5	1.09	57	1.19	53
5.0	1.25	52	1.60	39
10.0	1.37	48	1.68	36
15.0	1.36	48	1.65	36
20.0	1.49	43	1.60	39
30.0	—		—	

Table 13.15 Effects of methods of deforestation on soil bulk density and penetrometer resistance of 0–5 cm layer (unpublished data of R. Lal)

Clearing method	Bulk density (g/cm^3)				Penetrometer resistance (kg/cm^2)			
	Pre-clearing 1978	1979	1980	1981	Pre-clearing 1978	1979	1980	1981
Traditional	0.64*	1.06	1.07	1.27	0.21	0.96	0.52	1.32
Manual	0.68	1.17	1.17	1.39	0.20	1.42	0.75	1.19
Shearblade	0.70	1.19	1.37	1.38	0.26	1.0	1.84	2.19
Tree pusher/ root rake	0.60	1.24	1.32	1.42	0.20	1.3	0.73	2.23

* Each value is a mean of 25 separate measurements.

In western Nigeria Lal and Cummings (1979) observed that a greater increase in bulk density occurred in the upper 0–3 cm layer. The percentage change in bulk density was 76, 60, and 28 for mechanical clearing, 49, 46, and 29 for slash and burn, and 43, 34 and 22 for slash clearing at 0–3, 3–5, and 5–10 cm depths, respectively. In a follow-up study on a similar soil, Lal (1984) observed a gradual increase in bulk density with time after land clearing (Table 13.15). Even with traditional clearing and farming, bulk density increased from a pre-clearing value of 0.64 g/cm^3 in 1978 to 1.27 g/cm^3 in 1981 when the land was returned to bush fallow. The relative increase in bulk density was even greater in intensively cultivated treatments. However, Van der Weert (1974), Seubert et al. (1977), and Silva (1981) observed more drastic compaction of some soils of tropical South America caused by mechanical land clearing treatments. In contrast, Hulugalle et al. (1984) observed that although clearing resulted in an overall increase in bulk density, differences in soil bulk density of plots cleared by different mechanical attachments were not significant. For example, their data showed that the bulk densities of the forested control, and the manual, tree pusher, tree pusher/root rake and shearblade cleared treatments were 1.00, 1.04, 1.15, 1.17, and 1.18 g/cm^3 (with LSD = 0.08), respectively.

Penetrometer resistance is also increased with mechanized methods of deforestation as a result of increased bulk density. Lal and Cummings (1979) observed that the relative increase to penetrometric resistance in the 0–4 cm layer was 10 to 20 times in mechanically cleared plots compared with 3- to 4-fold increase in the manually cleared treatments. Furthermore, the penetrometric resistance under the wheel tracks of the bulldozer was about twice as much as on the rest of the plot and the soil under the wheel tracks was compacted to about 15 cm depth because of the stress imposed by the heavy equipment. An assessment of the degree of soil compaction on the basis of penetrometer resistance without evaluating soil moisture and the textural properties is difficult. Because of the variable soil moisture content, the comparison of increase in bulk density with time following different methods of land clearing shown in Table 13.15 does not indicate a corresponding increase in penetrometer resistance.

(iv) Water transmission properties: Impaired soil structure and decrease in macro-porosity following clearing results in reduced water tansmission through the soil profile. Increased compaction of the surface layer, decreased stability of structural aggregates and development of surface seal and crust following exposure reduce infiltration rate and capacity. Regardless of the method of clearing, soil exposure *per se* decreases macroporosity, creates surface seal and crust, and drastically decreases rainfall acceptance of the surface layer. This degradative process is exaggerated by mechanized land clearing. For an Ultisol in South America Seubert *et al.* (1977) reported that the cumulative infiltration was 12 times more in manually cleared burned plots than in bulldozed plots. The equilibrium infiltration rates averaged 11.8 cm/hr in the burned plots and 0.9 cm/hr in those that were bulldozed. Silva (1981) studied the effects of different methods of land clearing on the water transmission properties of an Oxisol in Southern Bahia, Brazil. The equilibration infiltration rates were 3, 20 and 24 cm/hr for mechanical clearing, manual clearing and forested control, respectively.

In southwestern Nigeria Lal and Cummings (1979) reported the infiltration capacity to be 50, 140, 175 and 300 cm/160 minutes for clearing with bulldozer, slash and burn, slash, and uncleared control, respectively. With reference to the pre-clearing infiltration capacity, there was a decrease of 74, 66, and 45 per cent for mechanical clearing, slash and burn, and slash alone, respectively. The post-clearing equilibrium infiltration rates were 62, 44, and 17 cm/hr for slash, slash and burn, and mechanically cleared plots, respectively. The saturated hydraulic conductivity decreased from 16.1 to 1.3 cm min^{-1} for mechanical clearing, 15.2 to 5.0 cm min^{-1} for slash and burn, and 9.8 to 4.6 cm min^{-1} for slash clearing. Lal (1981) reported that the decrease in infiltration capacity and rate on mechanically cleared plots persisted for at least 2 years even with improved subsequent management following clearing (Table 13.16).

Table 13.16 Effects of land-clearing methods on infiltration rate of an Alfisol near Ibadan, Nigeria (Lal, 1981). *Reproduced by permission of Elsevier Science Publishers*

Infiltration	Methods of land clearance					
	Mechanical		Slash and burn		Slash	
	1976	1977	1976	1977	1976	1977
Capacity (cm/3h)	136	109	222	213	316	260
Rate (cm/h)	32	28	37	44	60	49
LSD (0.05)	Capacity		Rate			
(i) 1976	70		25			
(ii) 1977	85		16			

Table 13.17 Effect of clearing method on the infiltration characteristics (Philip's equation) and the saturated hydraulic conductivity (0–50 mm depth interval) of the soil (Hulugalle *et al.*, 1984). *Reproduced by permission of Williams & Wilkins Co.*

Clearing method	Infiltration characteristics			Saturated hydraulic conductivity (mm s^{-1})
	Sorptivity (mm s$^{-1/2}$)	Transmissivity	Cumulative infiltration in 3 hr (mm)	
Uncleared	17.8	3.48×10^{-2}	2220.2	4.1×10^{-1}
Manual	17.4	1.05×10^{-1}	2946.2	3.8×10^{-1}
Tree pusher	17.1	2.24×10^{-2}	2019.0	2.6×10^{-1}
Tree pusher/root rake	10.2	2.89×10^{-2}	1185.8	2.6×10^{-1}
Shearblade	6.9	3.84×10^{-2}	1122.6	5.8×10^{-1}
LSD (0.05)	2.9	2.39×10^{-2}	345.6	5.8×10^{-2}

In another study, for a similar Alfisol in southwest Nigeria Hulugalle *et al.* (1984) reported differences in soil-water sorptivity due to different clearing attachments used with mechanized clearing. The sorptivity of the tree pusher-root rake and shearblade clearing treatments were 57.3 and 38.8 per cent of the forested control, respectively. Also affected adversely was the saturated hydraulic conductivity (Table 13.17).

(v) Soil-water retention properties: Alterations in pore size distribution, total porosity, inter- and intra-aggregate pores, and in percentage of soil aggregated into stable crumb structure by land clearing and soil exposure also affect the soil-water retention properties. The moisture content of the exposed soil is influenced by increased insolation, high temperatures and increased wind movement. Cunningham (1963) reported differential soil moisture contents due to exposure—the mean soil moisture content of the upper 0–15 cm layer being 32.7, 28.2, and 21.0 per cent for the shaded, half-exposed and fully-exposed soils, respectively. For coarse-textured soils, the decrease in soil moisture retention is also associated with a very rapid decline in soil organic matter content after land clearing.

Lal and Cummings (1979) observed difference in soil moisture retention characteristics following different methods of land clearing. For example, moisture retention at zero suction measured *in situ* decreased from a preclearing value of 35 per cent to 22 per cent with mechanical clearing. There was only a slight decrease with manual clearing. Lal (1981) observed that methods of deforestation influenced moisture retention more at 0 and 0.1 bar suctions than at 0.3 or 0.5 bar suctions (Table 13.18). Once again, mechanical clearing decreased moisture retention at low suctions more than manual clearing. In a subsequent study Hulugalle *et al.* (1984) observed that at a soil-water potential of 0 kPa more water was retained in the surface soil of the manually cleared and uncleared plots than in the mechanically cleared treatments (Table 13.19). Compared with the uncleared control, the decrease in soil-water retention at

Table 13.18 Effect of land clearance methods on pF characteristics of 0–10 cm layer of an Alfisol near Ibadan, Nigeria (Lal, 1981). *Reproduced by permission of Elsevier Science Publishers*

Suction (bars)	Gravimetric moisture content (%) for different methods			
	Mechanical	Slash and Burn	Slash	LSD (0.05)
0	34.0	36.8	37.3	6.9
0.1	14.8	15.2	15.4	3.9
0.3	11.0	11.1	12.1	2.5
0.5	9.0	9.3	10.1	2.8

Table 13.19 Soil-water retention of the 0–100 mm depth as affected by land-clearing method (Hulugalle *et al.*, 1984). *Reproduced by permission of Williams & Wilkins Co.*

Clearing method	Volumetric water content ($m^3\ m^{-3}$) at specified soil water potentials (kPa)				
	0.0	−10.0	−33.0	−50.0	−70.0
Uncleared	0.398	0.126	0.115	0.111	0.108
Manual	0.436	0.146	0.131	0.126	0.122
Tree pusher	0.400	0.165	0.150	0.145	0.142
Tree pusher + root rake	0.369	0.265	0.183	0.143	0.105
Shearblade	0.315	0.232	0.168	0.132	0.100
LSD (0.05)	0.026	0.030	0.028	0.021	0.024

0 kPa was 7.3 per cent and 20.9 per cent for the three pusher-root rake and shearblade clearings, repectively. Effects on soil moisture retention properties due to heavy traffic are generally greater in the high than in the low suction range.

The review presented in this section indicates the detrimental effects of forest clearing on soil physical properties. Although machine-clearing methods have a more adverse effect than manual techniques, forest removal *per se* decreases clay and soil colloids, percentage volume occupied by transmission or macro-pores, structural aggregates and their stability, and water transmission and retention properties. Reduction in porosity (and water transmission and retention) is a direct effect of soil compaction. Although machine-induced soil compaction is observed even beyond 50 cm depth, the most drastic effect occurs in the layer immediately beneath the surface horizon.

3. *Hydrology and soil-water balance*

In the same way as for energy balance, the available literature indicates pronounced changes in water balance and in surface and subsurface flow following deforestation. The rainfall-interception effect of the forest vegetation is removed and the temperature and moisture regimes of the exposed soil

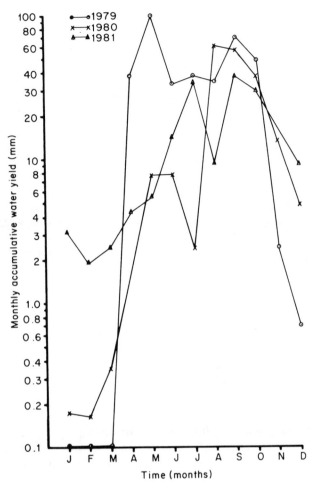

Figure 13.9 Change in total water yield from the cleared catchment with time after deforestation (Lal, 1983)

fluctuate widely. Rapid decline in structural stability causes slaking of the surface soil, reduction in its rainfall acceptance, and increase in surface run-off. Removal of deep-rooted trees decreases water utilization from subsoil horizons thereby increasing the interflow and percolation losses.

Deforestation causes an increase in stream flow, and the magnitude of increase is proportional to the percentage forest cover removed from a catchment (Hibbert, 1967). Increase in stream flow following deforestation is due to decrease in evapotranspiration, and to reduction in water extraction from the subsurface horizons (Pereira, 1973).

In southwestern Nigeria Lawson et al. (1981) observed that deforestation of a 44-hectare watershed drastically increased the total water yield compared with the pre-clearing level. Furthermore, the ground-water rose to 30 cm below the surface during the rainy season. Compared with a negligible water run-off from a forested control, 23 per cent of 1445 mm of rain received was lost as surface run-off from the cleared watershed. In addition, both the duration and the quantity of interflow water were increased by deforestation. Within 8 months after forest removal, about 1.5 per cent of the rain received was lost as interflow. The data in Fig. 13.9 indicate a steady increase in total water yield from the cleared catchment with time after deforestation. The interflow component increased from an unmeasurable trace in January during the dry season of 1978 before clearing to 0.1, 0.18 and 3.2 mm month^{-1} after clearing in January 1979, 1980, and 1981, respectively. This progressive increase in interflow was due to a gradual decrease in forest regrowth and in the non-utilization of subsoil water by the shallow-rooted annuals that succeeded the forest cover. A decrease in soil-water storage capacity of the profile due to decrease in the organic matter content and a decline in the relative proportion of retention pores is also responsible for the increase in interflow following deforestation.

Alterations in the micro- and meso-climate of a cleared watershed also influence diurnal patterns in interflow. The data in Fig. 13.10 show the effect of daily evapotranspiration over the cleared watershed on interflow measured with a 5:1 weir. The maximum rate of diurnal interflow measured after a prolonged dry spell occurs in the early hours of the day (0400 to 0600 hours), and the minimum is usually observed late in the afternoon (between 1600 and 1800 hours). The effects of evapotranspiration over the clearing watershed in reducing the interflow are usually observed 3–4 hours after the period of maximum evaporative demand. Due to the uniform micro- and meso-climate of forested watershed, there are only slight diurnal fluctuations in interflow.

Changes in stream flow following conversion depend on the land-use, i.e., conversion to forest/tree plantations, to seasonal food crops, or to grazed pastures. The main differences lie in the canopy cover, root system and crop-water use. There are also differences among tree crops depending upon their canopy architecture, spacing, etc. For example, the change in water balance caused by forest conversion to tea or coffee is different from that caused by rubber, coconuts or pine.

Pereira and Hosegood (1962) and Pereira (1973) compared the water use by forest, forest cleared, tea and shade trees planted, developing cover of tea and shade trees, and full canopy of tea and shade trees in a watershed study in Kenya. They observed that initially there was less water use by the newly established tea plantation than by the forest. With the canopy development of the tea plantation, however, the differences disappeared. the initial rise in the ground-water table also eventually disappeared. In New South Wales, Australia, Brown (1972) reported an abrupt increase in flow pattern following an accidental fire in the watershed. There were changes in the shape of the flood hydrographs with pronounced sharp secondary peaks on the rising side of many of the hydrographs. The water yield increased significantly for 4 years following the fire incident.

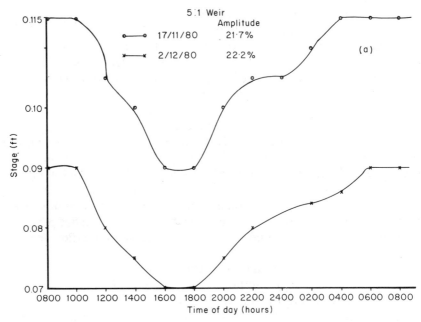

Figure 13.10 Diurnal fluctuations in interflow from a 44-hectare watershed at IITA, Ibadan, measured with a 5:1 weir (Lal, 1983). Fluctuations (a) changes in one day (b) changes in seven consecutive days

The increase in stream flow following conversion to tree or accidental fire also depends on the rotation cycle or the time required for regeneration. Stream flow will increase whenever the old plantation is replaced by a new one or is accidentally burned. Subsequent reforestation then decreases stream flow and surface run-off. One of the relevent examples of this is the impact of reforestation on stream flow of a 36 ha watershed in Tennessee (TVA, 1962).

4. Run-off and soil erosion

In addition to interflow, deforestation also increases the direct and surface run-off components of the hydrological balance. Soil detachement and splash by raindrop impact increase sediment transport in accelerated run-off. With a few exceptions in very humid regions, soil erosion from under a rain forest cover is usually negligible. The removal of protective vegetation increases the potential erosion hazard.

(i) Logging and erosion: Even partial clearing for logging and developing infrastructure can severely increase erosion hazard. In Indonesia, Rusland and Menan (1980) studied erosion differences between undisturbed forest and skidding roads. Soil erosion and water run-off were the most severe on newly constructed and used skid roads and declined on abandoned, unused roads

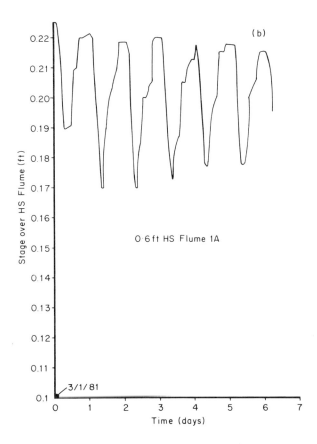

(Table 13.20). Severe erosion was also reported in the Philippines from the logged-over site and especially on the road fill-slopes (Lim Suan, 1980). In North Queensland, Australia, Gilmour (1971) reported that logging of virgin tropical rain forest increased the sediment load 6- to 12-fold. The principal sources of the sediment were the logging and skidding roads. The log landing at the stream also contributed to the sediment load (Table 13.21). In another study, Gilmour (1977) observed load increased from about 180 ppm before logging to about 320 ppm during the first year after logging and to about 520 ppm during the second year.

(ii) Perennial crops and erosion: Conversion of forest to plantation/perennial crops causes less erosion except during the initial establishment phase. Erosion is further lessened by the usual practice of seeding a cover crop between the trees, e.g. *Puereria, Mucuna Centrosema*, etc. Establishing ground cover in the plantation and along water ways is important. In the Cameron Highlands of Malaysia, Shallow (1956) compared sediment yields from three watersheds with different land-uses—forest, tea and vegetables. At high flow, 300 cusecs, the sediment load increased substantially. In Sumatra, Indonesia Gintings

Table 13.20 The effects of logging activities and skid roads on runoff and erosion in Indonesia (Ruslan and Manan, 1980)

Treatments	Erosion (t ha^{-1} month^{-1})	Run-off (m^3 ha^{-1} month^{-1})
Newly constructed and used skid roads	12.9	189.1
Newly constructed but unused skid road	10.8	148.6
Skid road abandoned for 2 years	6.2	42.7
Skid road abandoned for 3 years	3.2	19.2
Forest undisturbed	0.0	2.1

Slope 8–10%, rainfall 2430 mm/yr, soil latasol, plot size 4mm × 2m and observation period 5 months.

Table 13.21 Effect of a log landing on stream sedement load in Northern Queensland after 8 mm of rain between 1320 and 1340 (Gilmour, 1971)

Sampling location on stream	Time (hr)	Sediment concentration (ppm)
Upstream of landing	1345	48
Immediately downstream of landing	1345	2602
50 m downstream	1350	203
400 m downstream	1400	186
400 m downstream	1545	21

(1981) compared run-off and sediment load from undisturbed forest with coffee plantations and observed that run-off and erosion under coffee depended upon the age and maintenace of the plantation (Table 13.22a, b). The surface run-off increased with increase in the age of coffee plantation, and the run-off and erosion were higher under coffee than in undisturbed forest.

Establishment of improved forestry on cleared lands leads to reduced sediment yields. Once established, improved tree plantations are hydrologically almost similar to the undisturbed forest. The data of Hardjono (1980) from Indonesia showed that sediment yield from a reforested catchment was less than half of that from the catchment used for upland dry agriculture.

Accelerated erosion following deforestation has been reported from tropical Asia by Dugain (1953), Thijsse (1977), BIOTROP (1978), Kronfellner-Kraus (1980) and Junor (1981). Kellman (1969) studied surface run-off characteristics during 227 days in upland Mindanao in the Philippines. The rates of run-off in a 10-year-old abaca plantation were twice as high as in natural mixed dipterocarp forest, four times as high as in a young rice field, and about 50 times as high as in a 12-year-old rice field. Among the western coast of southern India, Chinnamani (1977) observed that erosion from a poorly managed tea plantation was 40 k to 50 t ha^{-1} yr^{-1} compared with only 0.06 t ha^{-1} yr^{-1} from the forested land. In Hongkong Lam (1978) reported that suspended sediment discharge increased with increase in the percentage area without forest cover. In Ivory

Table 13.22 Comparison of surface run-off and erosion between forest and coffee plantation in Sumatra (Gintings, 1981)

(a) January – April, 1980

Parameter	Undisturbed natural forest	16-year-old Coffee plantation
Slope (%)	52–65	46–49
Rainfall (mm)	926.5	926.5
Surface run-off (m^3/ha)	104.75	633.37
Percentage of rainfall	11.3	68.4
Erosion (t ha^{-1} 6 months^{-1})	0.28	1.18

(b) May – October, 1980

Parameter	Undisturbed natural forest	Coffee plantation		
		1 year	3 years	16 years
Slope (%)	52–65	59–63	62–66	46–49
Rainfall (mm)	1338.4	1338.4	1338.4	1338.4
Surface run-off (m^3/h)	101.42	237.4	453.98	837.57
Percentage of rainfall	7.6	17.8	33.0	62.6
Erosion (t ha^{-1} 6 months^{-1})	0.31	1.94	1.57	1.27

Coast Ollagnier *et al.* (1978) observed less run-off and erosion under forest and perennial crops than under maize or cassava (Table 13.23).

(iii) Grain crops: Soil erosion is the most severe under short-rotation grain crops. In West Africa the magnitude of accelerated erosion following deforestation was measured by Auberville (1959). In Ivory Coast Roose (1979) observed that run-off and erosion were 50 and 1000 times greater on cleared and cultivated plots than on forested control. Effects of method of deforestation on run-off an erosion have been reported from western Nigeria by Lal (1983). Among land development treatments involving complete clearing, manually cleared plots lost of total of 48 mm of run-off and 5.1 t ha^{-1} of soil over a period of 3 years, from 1979 to 1981. In comparison, plots mechanically cleared using shearblade lost 105 mm of run-off and 4.8 t ha^{-1} of soil (Table 13.24), whereas those cleared with tree pusher lost 250 mm of run-off and 20 t ha^{-1} of soil. There was more run-off and soil erosion from plots with ploughed and harrowed seedbeds than from those receiving no-till treatments.

Similar effects of deforestation on run-off and soil erosion have been reported from tropical America (Buch, 1978; Le Baron *et al.*, 1979; Roche, 1981; Hecht, 1981). Ramos and Merinho (1980) reported the effects of deforestation on run-off and soil erosion from northeastern Brazil. Data from plot measurements indicated 115 t ha^{-1} of soil loss from bare plot, 8.6 t ha^{-1} from shrub growth herbaceous vegetation, and only 1.2 t ha^{-1} from shrub growth

Table 13.23. Effect of land-use and vegetal cover on run-off and erosion in southern Ivory Coast (Ollagnier et al., 1978)

Vegetation	Slope (%)					
	4.5		7		23	
	E†	R	E	R	E	R
Undisturbed forest	—	—	0.1	1	0.1	1
Perennial crops (oil palm, coffee, cocoa with cover crops)	—	—	0.3	2	—	—
Annual crops (i) cassava	—	—	32	22	—	—
(ii) maize	—	—	92	30	—	—
Bare soil	57	37	125	33	520	25

†E = soil erosion in t ha^{-1} year^{-1}; R = run-off as % of rainfall.

Table 13.24 Effects of deforestation on run-off and soil erosion (modified from Lal, 1983)

Method of deforestation	Watershed area (ha)	Run-off (mm)		Soil erosion (t ha^{-1})	
		1979	1979–1981	1979	1979–1981
Forest control	15	T	T	T	T
Traditional	2.6	3.0	6.6	0.01	0.02
Manual	3.15	35.0	27.9	2.7	5.1
Shearblade	2.7	86.0	104.8	4.0	4.8
Tree pusher	3.6	201.5	250.3	17.5	20.0

with tree cover. The percentage run-off from these treatments was 52, 26 and 18, respectively. Smith (1981) reported soil erosion of up to 100 t ha^{-1} on land developed for a transmigration scheme in the Brazilian Amazon region.

(iv) Grazed pastures: Conversion to properly grazed and adequately stocked pastures creates less erosion than arable land use. Excessive grazing and high stocking rates, however, severely accelerate the erosion hazard. The effects of change in land-use and grazing following deforestation on run-off and erosion have not been widely studied. In Cameroon Hurault (1968) observed that even the low stocking rate of one head of cattle per 10 hectares accelerated soil erosion. In Kenya Dunne (1979) analysed sediment yields from 61 Kenyan watersheds and observed more sediment yield from grazed and agricultural watersheds than from those under partial or complete forest cover. The regression equations relating sediment yeild with run-off and slope indicate a lower run-off exponential for forested than grazed or agricultural catchments (Table 13.25). In the steeplands of Machakos, Kenya, Thomas et al. (1981)

Table 13.25 Sediment yield from some Kenyan catchments with different land uses (after Dunne, 1979). *Reproduced by permission of Elsevier Science Publishers*

Land-use		Regression equation	Correlation coefficient (R^2)
(a)	Forest catchment	$SY = 1.56\ Q^{0.46}\ S^{-0.03}$	0.98
		$SY = 2.67\ Q^{0.38}$	0.98
(b)	Forest > agriculture	$SY = 1.10\ Q^{1.28}\ S^{0.047}$	0.76
(c)	Agriculture > forest	$SY = 1.14\ Q^{1.48}\ S^{0.51}$	0.74
(d)	Range land	$SY = 4.26\ Q^{2.17}\ S^{1.12}$	0.87

SY = Sediment yield (t km^{-2} yr^{-1}).
Q = Mean annual run-off (mm).
S = Relief.

Table 13.26 Effect of grazing on soil erosion in Machakos, Kenya (Thomas *et al.* (1981)

Land-use	Soil loss (mm/yr)	Ratio
Degraded grazing land	4.5	50
Cultivated land	1.3	15
Good grazing land (bush and woodland)	0.7	1

observed that well maintained pastures had slight or tolerable levels of erosion. With uncontrolled grazing on degraded lands, however, erosion increased by 50 times (Table 13.26).

Although more quantatitive data are needed from well planned and adequately equipped experiments to determine the effects of deforestation on run-off and erosion from different soils and ecologies of the humid tropics, review of the available data indicates that run-off and erosion increase drastically with deforestation. Furthermore, the severity of erosion from cleared land depends on the method of deforestation and land-use. Erosion is more severe from mechanical clearing than from manually cleared land.

5. Soil chemical properties

If the claim by some researchers that most of the nutrient capital of the tropical forest is tied in the biomass is valid, then this ecosystem is more susceptible to man-made changes than those systems where in a greater proportion of the nutrient pool is stored in the soil. As was briefly discussed in the section dealing with the forest, however, the claim that the biomass contains most nutrient reserves is a debatable issue. Although only additional research data from well-designed experiments can test these popular and emotional myths, forest conversion is bound to effect the nutrient balance both directly and indirectly. Directly, the nutrients stored in the biomass are taken out of the ecosystem by logging, removing fuel wood, or simply by burning. Indirectly, deforestation disrupts the nutrient cycles of major elements (C, N, P, S), and some of the elements are taken out of the ecosystem through leaching, surface run-off, or volatilization.

In spite of these broad-based conclusions, it is difficult to generalize the effects of deforestation on the nutrient pool or the soil chemical properties. The effects differ for different soils, vegetation, rainfall pattern, and methods of deforestation. Similar to the response of soil physical properties, mechanized clearing affects soil chemical properties more adversely than manual and traditional clearing methods. The effects on soil chemical properties following deforestation are more drastic in rain forest ecologies than in the savanna. The effects are also more drastic for less fertile Oxisols and Ultisols than for relatively more fertile Alfisols and Inceptisols.

The effects of deforestation on soil chemical properties are immediate, medium-term and long-term. The immediate effects are due to the addition of ash, removal of nutrient-rich surface layer, and mechanical mixing of the surface and subsoil layers. The medium- and long-term effects are due to disruption in the biotic activity of soil fauna and in the supply of organic matter and leaf litter, as well as an enhanced rate of mineralization of organic matter content and losses due to leaching and surface run-off. The long-term effects of deforestation on soil chemical properties depend on the land-use and soil and crop management systems adopted. Obviously, the chemical properites are affected differently by different systems of soil and crop management, e.g., mulch farming, conventional versus no-tillage, agro-forestry versus seasonal crops, and mixed-farming versus plantation crops.

In contrast to the study of soil physical and hydrological properties, effects of deforestation and of change in land-use on soil chemical properties have been studied extensively. That is not to say, however, that no additional research is needed or that all answers are known and validated. Readers are referred for more detailed and specific information from tropical Africa to the following reports by Cunningham (1963), Nye and Greenland (1964), Diatta (1974), Godefroy and Jacquin (1975), Ayanaba *et al.* (1976), Blic (1976), Juo and Lal (1977), Ollagnier *et al.* (1978), Lal and Cummings (1979), Lal (1981), Swift *et al.* (1981), Kang *et al.* (1981), Nwoboshi (1981) and Martin (1983) among others. Similar information for tropical America is available from Seubert *et al.* (1977), Cassel *et al.* (1981), Sourabié (1980), Sanchez *et al.* (1982, 1983) and others. The effects of deforestation on chemical properties of soils in tropical Asia are reported by Hubert-Schoumann (1974), Webb (1975) and others.

The general conclusions from these studies are that: (i) chemical fertility and soil organic matter content decline rapidly following deforestation and cultivation; (ii) loss of bases decreases soil pH; (iii) the cation exchange capacity decreases because of the loss of soil colloids, etc. On the other hand, the rate of deterioration in soil chemical properties can be effectively curtailed by appropriate soil and crop management and balanced fertilizer application, e.g., mulch farming and cover crops, no-till and agroforestry, liming and the choice of appropriate nitrogenous fertilizers (the use of ammonium sulphate should be avoided), and by using suitable cropping sequences and crop combinations.

6. Soil Fauna

Deforestation influences the population and diversity of soil fauna directly and indirectly. The direct effects on faunal population include the physical injuries during clearing operations especially by fire and the use of motorized equipment. The indirect effects on soil fauna are due to alterations in soil moisture and temperature regime, decreases in food supply and diversity, and exposure to predators and parasites. The specific effects of deforestation on soil fauna are described in other chapters. Deforestation and change in land-use does not always decrease the total population, but it invariably alters the dominant species and decreases species diversity. The change in total population, however, is affected by the subsequent land-use and the use of agrochemicals. In Sarawak Collins (1983) observed that there were 25 species of termites in the forest but only six survived in the cleared land. Among the 25 species observed, *Macrotermitinae* were the least affected by deforestation. In general, deforestation decreases the population of the soil-feeding termites (Wood *et al.*, 1977). The effects of clearing on fauna are also related to litter input and its quality. Although the litter input is generally less on the cleared land, its nutritional level is generally superior.

Alterations in species diversity and dominations, however, influence soil physical properties and the rate of decomposition of organic residue (Wood *et al.*, 1977; Collins, 1983; Anderson and Wood, 1983). For example, the population of earthworms decreases following deforestation and subsequent cultivation, although the population of other soil fauna may have increased. Decline in the activity of earthworms can affect soil-water and air movement. Although there is voluminous literature dealing with earthworms *per se* and the effects of soil and crop management on earthworm population, little quantitative information exists establishing a cause-effect relationship between earthworm activity and soil physical properties following deforestation. Kirkham (1982) used an electric analogue to quantify the effects of earthworm channels on water and air conductance, and observed that as both the length and diameter of channels increased, conductance increased exponentially (Fig. 13.11). Large increases in conductance, however, are limited by the narrow channel diameter in the vertical and small channel length in the horizontal direction. Worm channels also support higher concentrations of oxygen and nutrients than the surrounding soil. These conclusions based on an analogue study were confirmed by Aina (1984) who observed that much of the high infiltrability associated with Alfisols in southwest Nigeria under forest cover is due to earthworm activity. Earthworm channels amounted to 0.37 per cent and

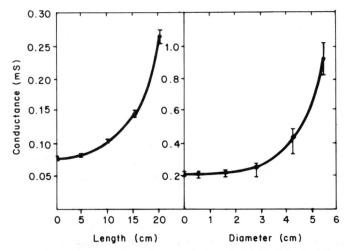

Figure 13.11 Conductance of earthworm holes. Left: Holes of different lengths oriented horizontally; Right: Holes of different diameter oriented vertically (Kirkham, 1982). *Reproduced by permission of Pedobiologia*

0.02 per cent of the soil volume under forest and cleared land, respectively (Fig. 13.12). The infiltration rate of soil under forest was 14 litre hr^{-1} without earthworm activity and 50 litre hr^{-1} with earthworm activity, respectively (Fig. 13.13). The minimum infiltrability of conducting channels was 0.91 litre hr^{-1} on forested land and 0.38 litre hr^{-1} on cleared land.

In a similar way to that of earthworms, the activity of other soil fauna, e.g., termites and ants, also influence soil physical properties and moisture and air conductance through the soil. Research data are needed, however, for differ-

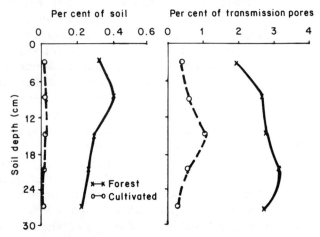

Figure 13.12 Total porosity of earthworm channels and percent of transmission pores in an Alfisol as affected by earthworm activity (Aina, 1984). *Reproduced by permission of Pedobiologia*

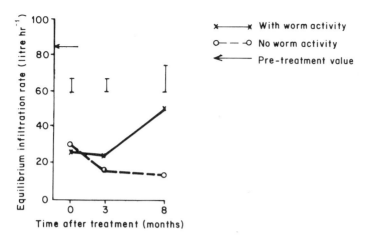

Figure 13.13 Effects of earthworm activity on equilibrium infiltration rate of an Alfisol (Aina, 1984). *Reproduced by permission of Pedobiologia*

ent soils and ecologies to establish quantitative relations between soil physical properties and soil fauna.

7. Crop yields for different land-uses following deforestation

The yield response to methods of deforestation differs among soils, crops, fertilizers applied, the prevailing weather conditions and the management intensity. Crop yield also depends on the vegetation and type of forest, i.e., virgin forest, high forest, secondary forest or short-duration fallow vegetation. There is also a strong interaction between the method of land clearing and inherent soil fertility. Differences in crop yield due to methods of land clearing are more pronounced in soils of low inherent fertility, e.g., Ultisols and Oxisols, than on soils of relatively high levels of antecedent nutrient reserves, i.e., Alfisols and Inceptisols. Yield is also affected more in crops with high nutrient requirements. Most of these interacting factors are listed in Table 13.27. Although the results are difficult to generalize, a few basic principles can be established:

(i) Short-duration fallow under shifting cultivation: In shifting cultivation and related bush fallow systems the existing vegetation cover is traditionally removed only by manual methods using native tools. When dried the felled vegetation is burnt *in situ*. The question of mechanical or chemical methods of land clearing are thus irrelevant. With traditional clearing where burning is an integral component, the crop yield depends on the duration of the fallow and vegetation management, if any, during the fallowing phase. Depending upon the antecedent level of soil fertility and the effective reestablishment of the native vegetation, each soil has an optimum length of fallow for productivity restoration. Crop yields increase with increase in the duration of fallow up to

Table 13.27 Factors affecting crop yield response to methods of land clearing

Soil	Native Vegetation	Actual yield			Soil and Crop management
		Crop	Micro-climate		
(i) Nutrient and organic matter profile	(i) Age of the forest	(i) Nutrient requirement	(i) Rainfall effectiveness and water balance		(i) Fertilizer level
(ii) Water-holding capacity	(ii) Predominant species	(ii) Growth duration	(ii) Air and soil temperature		(ii) Tillage methods
(iii) Effective rooting depth	(iii) Ecosystem (forest or savanna)	(iii) Root system	(iii) Evaporative demand		(iii) Cultivar
(iv) Structural stability		(iv) Water requirement and susceptibility to drought			(iv) Pest management
(v) Susceptibility to compaction		(v) Growth cycle			(v) Crop residue management

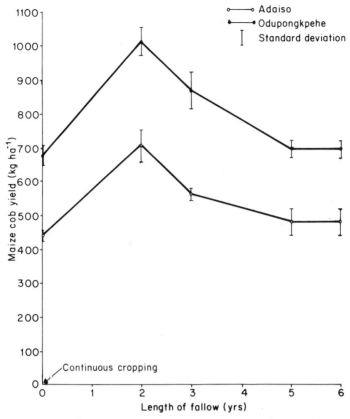

Figure 13.14   The effects of length of fallow on maize yield at two sites in the semi-deciduous forest zone of Ghana (Singh, 1961).

the optimum length. For an Alfisol in the semi-deciduous forest zone in Ghana Singh (1961) observed that the optimum duration of fallowing was 2 or 3 years (Fig. 13.14). The yield decline observed after 5 or 6 years of fallowing was attributed to the nutrient depletion through the removal of timber and firewood. That is why the vegetation management during the fallow period also affects yields during the cropping phase. The optimum fallow duration is generally longer for soils of low inherent fertility and for those whose physical properties are severely degraded.

(ii) Antecedent soil fertility: Mechanical clearing that results in scraping off the fertile surface layer and in mass-removal of the vegetation containing most nutrient capital has a greater effect on the following crop yield for soils with low inherent fertility than soils of higher nutrient reserves and favourable distribution of plant nutrients in the soil profile. That is why the crop yields are more drastically affected on Ultisols and Oxisols than, say, on Alfisols. These arguments are supported by the comparison of data in Table 13.28 for an

Ultisol and in Table 13.29 for an Alfisol. Seubert *et al.* (1977) observed for an Ultisol in Peru that yields in unfertilized plots averaged merely one-third of those obtained in burned plots (Table 13.28). In contrast Lal (1981) did not observed significant reductions in maize grain yield on mechanically cleared

Table 13.28 Effects of clearing methods on crop yield for an Ultisol in the Upper Amazon region of Peru (Seubert *et al.*, 1977). *Reproduced by permission of Tropical Agriculture*

Crop	Fertilizer level[†]	Grain yeild (t ha^{-1})	
		Manual/Burned	Bulldozed
Rice	0	1.33	1.70
	NPK	3.00	1.47
	NPKL	2.90	2.33
Maize	0	0.10	0.00
	NPK	0.44	0.04
	NPKL	3.11	2.36
Soybeans	0	0.70	0.15
	NPK	0.95	0.30
	NPKL	2.65	1.80
Cassava (fresh tubers)	0	15.4	6.5
	NPK	18.9	41.9
	NPKL	25.6	24.9
Guinea grass (dry straw)	0	12.3	8.3
	NPK	25.2	17.2
	NPKL	32.2	24.2

[†]$N = 50$ kg ha^{-1}; $P = 172$ kg ha^{-1}; $K = 40$ kg ha^{-1}; Lime (L) = 4 t ha^{-1}.

Table 13.29 Maize grain yield on an Alfisol for different methods of land clearing and with and without chemical fertilizers (Adapted from Lal, 1981). *Reproduced by permission of Elsevier Science Publishers*

Clearance method	Fertilizer level[†]	Maize grain yield (t ha^{-1})		
		First crop	Second crop	Third crop
Mechanical	F_0	4.42	1.21	2.41
	F_1	4.93	1.66	3.35
Slash and burn	F_0	4.78	1.62	4.04
	F_1	5.50	2.09	4.88
Slash	f_0	5.45	1.92	3.81
	F_1	5.45	1.92	3.81
LSD (0.05)				
(i) Clearance method		0.69	0.71	0.98
(ii) Fertilizer level		0.49	0.81	0.33

[†]F_0 = no fertilizer; F_1 = 120 kg N, 26 kg P and 30 kg K per ha.

plots, although maize grain yield was generally higher on manually cleared plots (Table 13.29). Whereas liming is an important input to obtain high yields on acidic Ultisols and Oxisols (Table 13.28), it is not required on Alfisols (Table 13.29). Similarly, Hulugalle *et al.* (1984) observed a reduction of 6 to 10 per cent with mechanical over manual clearing methods.

(iii) Crop species: The data from Peru by Seubert *et al.* (1977) indicate significant differences among crops between two methods of land clearing. With low fertilizer inputs, grain yields of maize and soybean were more drastically reduced than that of rice. Sanchez *et al.* (1983) analysed grain yields of 37 harvests of upland rice, 17 of corn, 24 of soybean, and 10 of peanuts grown on Ultisol with and without fertilizer. The yield decline without fertilizer was most rapid in acidity-sensitive crops, e.g., maize and soybeans. Whereas rice and maize were a complete failure on bulldozed unfertilized plots, cassava produced 6.4 t/ha of fresh tubers and guinea grass 8.3 t/ha of dry forage. Furthermore, the effects of clearing methods on yields of second or third crops after deforestation also depended on the preceding crop. A grain crop grown after a legume, tuber, or pasture would have less yield decline than that preceded by another grain crop.

(iv) Soil management: Cultural practices of soil and crop management after land clearing affect crop performance. Growing a legume/grass pasture for 1 or 2 years following land clearing as an ameliorative measure to restore physical, chemical and biological soil properties of mechanically cleared plots improves

Figure 13.15 Voluntary regrowth of mucuna can suffocate maize severely reduce maize grain yield (courtesy Hulugalle, 1984)

yields. Yield of a following grain crop may be adversely affected if the pasture, and especially the climbers, e.g., *Mucuna utilis*, is not adequately controlled. Hulugalle and Lal (unpublished data of 1984) observed that whereas *Mucuna utilis* improved soil physical properties of the mechanically cleared plots, maize grain yield was severely suppressed by voluntary regrowth (Fig. 13.15).

B. Global effects

The important global effects of deforestation are related to the role of forests in recycling carbon and water vapours. Possible alterations in the carbon cycle by deforestation have created considerable debate among environmentalists, climatologists, oceanographers, and the environmentalists. Climatologists are also concerned about the role of forests in recycling water vapour and the possible effects of deforestation on the rainfall regime and the hydrology of large river systems. The possibility of changes in the carbon cycle, with resulting global warming trends, has been debated by oceanographers, and the air pollution in relation to deforestation is a popular issue with the environmentalist. Most of these subjects, though interesting, are outside the scope of this book. Those aspects that are relevant to the physical edaphology are discussed here briefly.

1. The 'greenhouse gas' syndrome

The so called 'greenhouse gases' are those that are transparent to short-wave radiation but do not permit re-emission of longer wavelength infrared radiation back into space. An increase in the concentration of these gases in the atmosphere can lead to an increase in global temperatures. Important among the greenhouse gases are carbon dioxide and methane.

The forest ecosystem plays a significant role in regulating the concentration of CO_2 in the atmosphere; that is why the issue of deforestation and its effects on carbon dioxide has become a major concern in recent years. An increase in carbon dioxide concentration by about 15 per cent over the past century due both to fossil fuels and deforestation (Fig. 13.16, Revelle, 1982) has increased interest in the terrestrial biomass as a source of additional CO_2 in the atmosphere (Bolin, 1977; Woodwell *et al, 1978; Woodwell, 1983; Lovelock, 1983). It is estimated that doubling the atmospheric carbon dioxide would raise the mean temperature of the global surface by 2.8°C.* The global temperature is believed to have increased by 0.4°C between 1885 and 1940, corresponding with a mean increase in CO_2 concentration of 43 ppm (Revelle, 1982).

Tropical forests regulate the CO_2 in the atmosphere by three mechanisms: (i) through major river systems draining tropical rain forests by which a considerable amount of carbon is transported to the oceans; (ii) through resynthesis of carbon within the tropical biomass; and (iii) through storage in soil as soil organic matter. These recycling mechanisms are disrupted by deforestation. Deforestation influences the CO_2 level in the atmosphere in two ways. A

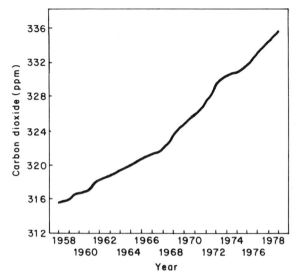

Figure 13.16 Trend since 1958 in the concentration of carbon dioxide in the atmosphere (parts per million by volume) (Revelle, 1982). *Reproduced by permission of W.H. Freeman & Co*

considerabe amount of carbon is stored within the tropical forest reserves (Table 13.30). Firstly, the carbon stored in the biota is released into the atmosphere following deforestation, burning, and decomposition. Various estimates of the amount of this release reported in the literature are inconsistent and are at the most rough guesses. Woodwell *et al.* (1978) estimated that clearing and harvest of tropical forest causes a total annual release of 4.4×10^{15}g of carbon, of which 0.9×10^{15}g is resynthesized into net productivity and the net release in the atmosphere is about 3.5×10^{15}g. Although it is a debatable issue, Woodwell and his colleagues argue that deforestation releases twice as much CO_2 as fossil fuel. Woodwell *et al.* (1982a, b) estimated the net global release of carbon to atmosphere from deforestation since 1860 at about

Table 13.30 Carbon stored in tropical forest reserves (Grainger, 1980). *Reproduced by permission of* The Ecologist

Vegetation	Carbon store (% of all plant C)
Tropical rainforest	41.5
Tropical seasonal forest	14.1
Temperate evergreen forest	9.5
Temperate deciduous forest	11.4
Boreal forest	13.0
Others	10.5

180×10^{15}g. The release in 1980 from deforestation was estimated to be between 1.8×10^{15}g and 4.7×10^{15}g C/yr. In comparison, the release from fossil fuel was about 5.2×10^{15}g. These authors estimate that net carbon flux from tropical forests has increased from about 100×10^{12}g C in 1920 to a maximum of 900×10^{12}g C/yr from Latin America and to 700×10^{12}g C/yr in tropical Africa (Fig. 13.17). In contrast, the net carbon flux is now negative in both Europe and North America. Deforestation also increases CO_2 release from the soil by increasing the rate of mineralization (Feller, 1977). These estimates are very tentative at best and certainly need more refinement through precise figures on annual rates of deforestation and a vegetation survey of the remaining forest reserves. Secondly, deforestation adds to the net release of CO_2 into the atmosphere because trees are estimated to resynthesize 10 to 20 times more carbon per unit area than land in crops or pastures (Revelle, 1982). Mass-scale deforestation over the last two to three centuries to develop agricultural land must, therefore, have contributed immensely to the CO_2 pool of the atmosphere.

In addition to deforestation, there are other indirect and compensatory effects of forest removal on production of greenhouse gases in the tropics. Agricultural activities such as clearing, burning, and cultivation influence the activity of termites. The magnitude of the effect depends on many factors including the termite species. Zimmerman *et al.* (1982) reported that termites have the potential to emit large quantities of methane, carbon dioxide and

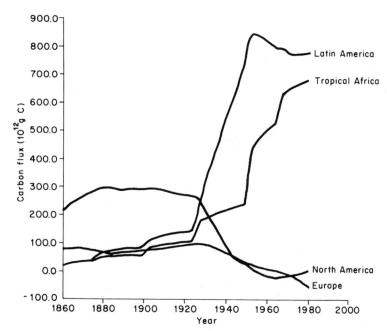

Figure 13.17 Carbon flux from different sources between 1860 and 1980 (Woodwell *et al.*, 1982a, b)

Table 13.31 Estimates of termite global biomass consumption and methane emission (after Zimmerman et al., 1982)

Ecology	Area (10^{12} m^2)	Net Primary Productivity (g m^{-2} yr^{-1})	Total produced (10^{15} g yr^{-2})	Termite density (number m^{-2})	Total termite consumption by termites	Percent of biomass consumed (10^{12})	Annual CH$_4$ production
Tropical wet forest	4.6	1200	5.5	1000	0.6	12	2.9
Tropical moist forest	6.1	1500	9.2	4450	3.8	41	17.3
Tropical dry forest	7.8	1200	9.4	3165	3.4	36	15.7
Temperate	12.0	1250	15.4	600	1.0	7	4.6
Wood/shrubland	8.5	700	6.0	431	0.5	9	2.3
Wet savannas	14.2	1200	17.0	4402	8.7	51	39.9
Dry savanna	4.3	900	3.9	861	0.5	13	2.3
Temperate grassland	9.0	600	5.4	2139	2.7	50	12.4
Cultivated land	11.9	650	7.7	2813	4.7	60	21.6
Desert scrub	18.0	90	1.6	229	0.6	38	2.8
Clearing burning	6.8		9.6	6850	6.5	68	29.8
Total	103.2		90.3	2.4×10^{17}*	33.0	37	151.6
Percentage of total	68		77		28		

* Total estimated population.

other gases. They estimated that 0.77 per cent of the carbon ingested by termites was re-emitted as methane (CH_4), 84.8 per cent as CO_2, 0.03 per cent as CO and 0.005 per cent as dimethyl disulphide (DMDS). Their estimates of the potential global emission of CH_4 shown in Table 13.31 indicate termite-induced methane production of 1.5×10^{14}g (or 1.1×10^{14}g of C). Other greenhouse gases emitted are estimated at 4.6×10^{16}g of CO_2 (1.3×10^{16}g of C), 10^{13}g of CO (0.4×10^{13}g of C) and 7×10^{11}g of DMDS. The estimated gross amount of CO_2 produced by termites is more than twice the net global input from fossil fuel combustion, believed to be 5.4×10^{15}g C/year. Some estimates indicate that termites process about 28 per cent of the earth's net primary productivity. Although this much biomass would eventually be oxidized, termites accelerate the normal decomposition process. If these estimates are correct, the regions that would have the largest emissions are tropical wet savannas. The use of fire for burning following deforestation is thought to be a major source of methyl chloride in the atmosphere (Lovelock, 1983).

In view of many uncertainties and unknowns it is difficult to provide any definite answer regarding the effects of depletion of tropical rainforests on concentration of 'greenhouse gases' in the atmosphere. Different sources and sinks of these gases are not adequately understood to remove uncertainties and replace speculation by facts. What is the role of transport of carbon by rivers for permanent storage in the oceans? What is the effectiveness of oceans in reabsorbing these gases from the atmosphere? The relevance and importance of deforestation in the tropics to the amount of CO_2 and other 'greenhouse gases' in the atmosphere necessitate a coordinated effort to determine precisely the rate of deforestation, and its effect on the carbon balance. The global effects of CO_2 are important enough to justify this topic as a high research priority.

2. Climate

Deforestation has both direct and indirect effects on global climate. The most direct effects are due to increased albedo (light and radiation reflection) on cleared land surface. Potter *et al.* (1975) estimated that albedo over forest is 0.07 compared with 0.25 on cleared land. Most of the voluminous literature available on the subject (Goodland and Irwin, 1975; Villa Nova *et al.*, 1976; Marques *et al.*, 1977; Lettau *et al.*, 1979; Salati *et al.*, 1983) is regrettably based more on speculation than on experimentally determined facts. Properly designed experiments are difficult to conduct except on a very small scale. However, computer simulation models have provided some insights of what could happen given certain assumptions and boundary conditions (were they to come true). One such model has recently been tested by Hansen *et al.* (1984).

Potter *et al.* (1975) used a computer simulation model to evaluate the effects of deforestation on global climate with two assumptions: (i) albedo change only, and (ii) complete deforestation effect including increased run-off rate and decrease in evaporation rate. There were no major differences in the results

obtained from the two assumptions and the authors summarized the global effect of deforestation as follows:
  (i) increased surface albedo
  (ii) reduced surface absorption of solar energy
  (iii) surface cooling
  (iv) reduced evaporation and sensible heat flux from the surface
  (v) reduced convective activity and rainfall
  (vi) reduced release of latent heat
  (vii) cooling in the middle and upper tropical troposphere
  (viii) increased tropical lapse rates
  (ix) increased precipitation in the latitude bands of 5 to 25°N and 5 to 25°S and a decrease in the equator–pole temperature gradient
  (x) reduced meridional transport of heat and moisture out of equatorial zones
  (xi) global cooling (0.2 – 0.3K) and decrease in precipitation between 45 and 85°N and at 40 and 60°S

The decrease in precipitation in high northern latitudes was estimated to be only 1.5 to 2.6 per cent, which amounts to 0.9 to 0.7 per cent when averaged over the globe. The data of Potter *et al.* (1975) on simulated water balance in relation to tropical deforestation shown in Table 13.32 indicate a slight effect on global cooling, reduced atmospheric water vapour content (reduced greenhouse effect), reduced global mean precipitation and decreased water vapour residence time in the atmosphere, which would accelerate the atmospheric water cycle.

An increase in the concentrations of 'greenhouse gases' in the atmosphere has indirect implications in terms of global climatic change. A measurable increase in global temperature has already been observed since 1885 due to increased atmospheric concentrations of CO_2. The global increase in temperature can alter rainfall distribution patterns. Similarly to the CO_2 question, deforestation-induced effects on global climatic changes are not well understood, and most projections are mere speculations. Many researchers, however, feel that major climatic shifts are imminent if mass-scale deforestation

Table 13.32 Effects of deforestation on global water balance (Potter *et al.*, 1975). *Reproduced by permission of Macmillan Journals Ltd.*

Parameter	Control	Completed deforestation	Albedo change only
Atmospheric water vapour content ($g\ cm^{-2}$)	2.4731	2.4032	2.394
Evaporation ($g\ cm^{-2}\ d^{-1}$)	0.33323	0.32983	0.33016
Rainfall ($g\ cm^{-2}\ d^{-1}$)	0.33305	0.32996	0.33169
Cloudiness	0.493	0.490	0.491
Atmospheric residence time (d)	7.4236	7.2847	7.2499

occurs in the Amazon and the Congo Basins (Mathew, 1970; Villa Nova et al., 1978; Baumgartner and Kirchner, 1980; Thompson, 1980; Dickinson, 1982). The envisaged increase in global temperature may influence the rates of precipitation and run-off in northern latitudes (Revelle, 1982). Flow rates in some rivers of northern latitudes may decrease substantially. Some climatologists speculate an appreciable decrease in precipitation and increases in temperature and evaporative demand in the bands of latitudes centered on 40°N and 10°S. In contrast, precipitation may increase between 10° and 20°N and in regions north of 50°N and south of 30°S. A direct consequence of the global warming trend may be the disintegration of the West Antarctic ice sheet, which could inundate many low-lying coastal regions.

Whatever scanty and preliminary information exists, it seems that deforestation-induced global changes have some compensating effects. The net results of these alterations are far from being well enough understood to make any justifiable conclusions.

3. World water balance

One of the immediate and debatable effects of deforestation on a regional scale is the increase in surface run-off during the rainy season and a possible decrease in the dry season. A relevant case for discussion is that of the Amazon. The area of the Amazon Basin is 6–7 million km^2. Its estimated water outflow is about 5.5×10^{12} m^3 year^{-1} which corresponds to approximately 80 cm of rainfall (Friedman, 1977; Salati et al., 1979). This flow constitutes about 15 per cent of the total fresh water flow on the earth. Mean annual rainfall in the Amazon Basin varies between 2000 to 2400 mm, and some estimates show that more than 50 per cent of the rainfall is returned to the atmosphere through evapotranspiration whereas about 45 per cent is drained into the ocean (Villa Nova et al., 1976). The major concern is that if water run-off increases by deforestation and the water vapours returned to the atmosphere decrease, what are the long term implications of the water balance in the region? Would the Amazon Basin in fact turn from 'Green Hell to Red Desert' (Goodland and Irwin, 1975)? The answers to these questions are not easy to come by in view of contradictory reports based on speculative data.

First, the change in run-off rate due to deforestation has stirred up considerable controversy. For example, in Venezuela Lamprecht (1977) observed higher fluctuations in river flow due to deforestation of the southern slopes of the coastal cordillera and northeastern Andean region. Also Gentry and Lopez-Parodi (1980) reported a pronounced and statistically significant increase in the Amazon flow between 1962 and 1978 measured at Iquitos (Fig. 13.18 a & b) indicating an increase in the run-off of water from Upper Amazon. Although this increase in flow was partly due to high rainfall received over the catchment, the effect of the large-scale deforestation that occurred during this period was thought to be responsible for the increase. These findings support the arguments by Sidi (1975) and others that deforestation may result in

Figure 13.18a Depths of annual high and low water marks of the Amazon at Iquitos, 1962 to 1978 (Gentry and Lopez-Parodi, 1980).

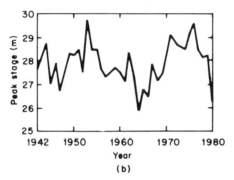

Figure 13.18b Peak stage of the Rio Negro at Manaus, 1942 to 1980 (Gentry and Lopez-Parodi, 1981)

irreversible changes in the Amazon water balance. The claims of increased run-off by deforestation made by Gentry and Lopez-Parodi have, however, been disputed by Nordin and Meade (1982), who present the flow records at Manaus from 1903 through 1980 and show no major deviation in flow during the last two decades in comparison with the long-term mean flood stage of 27.7 m.

Secondly, the contribution of recycled water vapour to rainfall over the Amazon basin is difficult to quantify because the supposedly homogenous hydrometeorological conditions over the 6–7 million km^2 area just do not exist. Salati *et al.* (1979) concluded on the basis of an isotope study that hydrometeorological conditions in the basin are inhomogeneous. Scientific answers to these questions, therefore, would require detailed studies conducted separately for small, relatively homogenous zones of the Amazon Basin during different seasons. The visual observations made from commercial flights on cloud formation over cleared and uncleared areas of the Amazon merely add to the utter confusion that already exists.

VI. CONCLUSIONS

The review presented leaves some important and relevant questions unanswered regarding deforestation in the tropics and its ecological consequences. The foremost open question is regarding the forest inventory and its conversion rates. The available estimates on forest resources and conversion rates are utterly confusing, obsolete, unreliable and often contradictory. Interpretations and judgements made on the basis of these unreliable estimates are liable to cause gross errors and serious, expensive and ecologically disastrous mistakes. The first research priority of the international community should be to obtain scientifically verifiable, routinely validated estimates of forest reserves in different ecologies and their conversion rates. These estimates should be based on standardized terminology.

The regional and local effects of deforestation are more easily quantifiable than global effects. Available data indicate that deforestation has immediate and drastically adverse effects on: (i) soil structure and pore size distribution, (ii) soil texture and organic matter content, and (iii) water retention and transmission properties. Furthermore, deforestation influences energy balance, increases fluctuations in soil and air temperature, relative humidity and soil moisture, and increases the amount of insolation reaching the soil surface. The soil-atmosphere interphase has a tendency to be more arid than it was under forest cover. Deforestation disrupts the normal pathways of major nutrients and organic carbon. Depending upon the antecedent level of nutrients and on the method of vegetation removal, deforestation can result in the escape of nutrient capital out of the ecosystem. The effects of deforestation on yields of succeeding crops are hard to generalize because response depends on a multitude of interacting factors, including the post-clearing soil and crop management.

Equally difficult to generalize are the global effects of deforestation. Most of the available literature is speculative, laced with emotions, and supported by few experimentally determined facts. The question of carbon dioxide and other 'greenhouse gases' remains an open issue. No doubt a large amount of carbon tied in the biomass is released upon deforestation, but its pathways and sinks are far from being understood. On the basis of the evidence available it is difficult to separate the effects from the noise. The problem of expected temperature rise due to deforestation-induced CO_2 increase in the atmosphere is further complicated by the compensatory cooling effect due to increase in albedo. The interaction of floral and faunal activity in carbon cycling is another intriguing issue awaiting further research.

The hydrometeorological effect of deforestation in terms of global water balance is another example of the difficult issues that can only be resolved by obtaining data on small homogeneous regions for many consecutive seasons. Long-term and reliable records of stream flow are needed to establish flow trends in relation to changing land-use. Computer models are no doubt important tools to indicate knowledge gaps and approximate long-term trends, but are no substitute for reliable experimental data.

And then there remains the question of 'dos' and 'don'ts'. Should the forests be removed considering their global consequences? On the basis of available evidence the scientific community would be hard put to provide a convincing argument against deforestation to those that need additional land to meet their pressing demands. To be on firm ground is to have reliable data on debatable issues. Emotional rhetorics produce short-term, if any, effects. Scientists should adopt a positive approach and should consider alternatives. The argument that effects of deforestation are really catastrophic is not supported by the results of the agricultural revolution over the past two centuries, which created productive farmland from the wilderness. Of course, an increase in farmland in the temperate latitudes may not have had the same ecological effect as a similar level of expansion would have in the tropics. Once again only with scientific data can we replace myths by facts. Will it be too late, by the time we have the data needed, to make a rational judgement? Maybe! That is why it is justifiable to adopt a conservative yet positive approach — gradual development of forest resources for effective utilization rather than rapid mass-scale transformation. The scientific community owes an answer on how best to manage an 'altered' ecology to those who have no alternative but to clear forest. Some answers already available are promising.

REFERENCES

Aina, P.O., 1984, Contribution of earthworm to porosity and water infiltration in a tropical soil under forest and long term cultivation, *Pedobiologia*, **26**, 131–137.

Anonymous, 1982, The vanishing jungle, *Economist*, Sept., 1982, 89–90.

Anderson, J.M., and Wood, T.B., 1983, Mound composition and soil modification by soil-feeding termite species (*Termitimae, Termitidae*) in a Nigerian riparian forest, *Pedobiologia*, (In Press).

Anderson, J.M., and Swift, M.J., 1983, Decomposition in tropical forests, in S.L. Sutton, T.C. Whitmore, and L.C. Chadwick, (eds.), *The Tropical Rainforest*, Blackwell Sci. Pub., New York, pp. 287–309.

Auberville, A., 1959, Erosion under forest cover and erosion on deforested land in the wet tropics, *Bois For. Trop.*, **68**, 3–14.

Ayanaba, A., Tuckwell, S.B., and Jenkinson, D.S., 1976, The effects of clearing and cropping on the organic reserves and biomass of tropical forest soils, *Soil Biol. Biochem.*, **8**, 519–525.

Bastos, T.X., Diniz, T.D., and De. A.S., 1974, Soil temperature in a humid equatorial forest, *Bol. Téc. IPEAN*, **64**, 73–83.

Baumgartner, A., and Kirchner, M., 1980, Impacts due to deforestation, in I–W. Bach (ed.), *Interactions of Energy and Climate*, Reidel, Dordrecht, pp. 305–316.

Bernhard-Reversat, F., Juttle, C., and Lemee, G., 1972, Quelques aspects de la périodicité écologique et de l'activité végétale saisonniere en forest ombrophile sempervivante de Cote-d'Ivoire, in F.B. Golley, and P.M. Golley, (eds), *Tropical Ecology with Emphasis on Organic Production*, Univ. Georgia Press, Athens, GA, pp. 214–234.

BIOTROP, Indonesia, 1978, *Long-Term Effects of Logging in South East Asia*, Proc. Symp. 24–27 June, 1975, Darmaga, BIOTROP Spec. Pub., No. 3, 177 pp.

Blancaneoux, P., and Araujo, J., 1982, The decisive action of man and of the present climate on the evolution of soils and savannas in south Venezuela, Amazon Federal Territory, *Cahiers, ORSTOM Pedol.*, **19**, 131–150.

Blic, P. De, 1976, Behaviour of ferrallitic soils in the Ivory Coast following mechanical clearing and pre-clearance environments, *Cahiers ORSTOM Pedol.*, **14**, 113–130.

Bolin, B., 1977, Changes of land biota and their importance for the carbon cycle, *Science*, **196**, 613–615.

Brown, J.A.N., 1972, Hydrological effects of a bushfire in a catchment in southeastern New South Wales, *J. Hydrol.*, **15**, 77–96.

Buch, M.W. Von, 1978, Degradation of ignimbrite soils and forest destruction in the pine-region of Honduras, *Mitteilungen der Bundesforschungsanstalt fur Forstund Holzwirtschaft*, **118**, 146–163.

Cassel, D.K., Bandy, D.E., and Alegre, J., 1980, in *Agronomic-economic Research on Soils of the Tropics*, 1978–1979 Report, Dept. of Soil Sci., North Carolina State Univ., Raleigh.

Chinnamani, S., 1977, Soil and water conservation in the hills of the Western Ghats, *Soil Cons. Digest*, **5**, 25–33.

Collins, N.M., 1983, The effect of logging on termite (Isoptera) diversity and decomposition processes in lowland dipterocarp forest, in J.J. Furtado (ed.) *Tropical Ecology and Development*, Int. Soc. Trop. Ecology and Ecology, Kuala Lumpur, Malaysia, pp. 113–121.

Cunningham, R.K., 1963, The effect of clearing a tropical forest soil, *J. Soil Sci.*, **14**, 334–345.

Daniel, J.G., and Kilasingam, A., 1974, Problems arising from large scale forest clearing for agricultural use, The Malaysian experience, *Malay. For.*, **37**, 152–160.

Daniel, J.G., and Kilasingham, A., 1975, Problems arising from large scale forest clearing for agricultural use; The Malaysian experience, *Planter*, **51**, 250–257.

Diatta, S., 1974, Effect of cultivation on the plateau soils of continental casamance. Results of two-year experiments, *Agron. Trop.*, **30**, 344–351.

Dickinson, R.E., 1982, Effects of tropical deforestation on climate, in *Studies in Third World Societies* No. 14, William and Mary College, Williamsburg, VA, pp. 411–441.

Dugain, F. 1953, First observations on erosion in New Caledonia, *Agron, Trop. Nugent*, **8**, 466–475.

Dunne, T., 1979, Sediment yield and landuse in tropical catchments, *J. Hydrol.*, **42**, 281–300.

Eckholm, E.P., 1976, *Losing ground*, Norton, New York.

FAO, 1979, *Report of a workshop on land development in outer islands of Indonesia*, FAO, Rome.

FAO, 1981, *Agriculture, Toward 2000*, FAO, Rome, 134 pp.

Feller, C., Soil changes in recent clearing in the Terres Neuves region (East Senegal) 2. Biological aspects and characteristics of the organic matter, *Cahiers, ORSTOM Pedol.*, **15**, 291–301.

Friedman, I., 1977, The Amazon Basin, another Sahel? *Science*, **197**, 4928.

Gentry, A.H., and Lopez-Parodi, J., 1980, Deforestation and increased flooding of the upper Amazon, *Science*, **210**, 1354–1356.

Gentry, A.H., and Lopez-Parodi, J., 1982, Deforestation and increased flooding of the upper Amazon, *Science*, **215**, 426.

Gilmour, D.A., 1971, The effects of logging on stream flow and sedimentation in a north Queensland rainforest catchment, *Commonw. For. Rev.*, **50**, 38–48.

Gilmour, D.A., 1977, Effects of rainforest logging and clearing on water yield and quality in a high rainfall zone of northeast Queensland, *Hydrol. Symp., Inst. of Engineers, Brisbane*, pp. 156–160.

Gintings, A.N., 1981, *Surface runoff and soil erosion on land covered by coffee plantation versus undisturbed natural forest in Sumberjaya, Lampung, Sumatra*, M.S. Disst., IPB, Bogor, Indonesia.

Godefroy, J., and Jacquin, F., 1975, Effect of vegetation on humification in a ferrallitic soil, *Cahiers, ORSTOM Pedol.*, **13**, 279–298.

Goodland, R.J.W., and Irwin, H.W., 1975, *Amazon jungle: Green Hell to Red Desert?* Elsevier, New York.

Grainger, A., 1980, The state of the World's tropical forests, *Ecologist*, **10**, 6–54.

Hamilton, L.S., and King, P.N., 1983, *Tropical Forested Watersheds: Hydrologic and Soils Response to Major Uses or Conversions*, Westview Press, Boulder, Colorado, 168 pp.

Hansen, J., Lacis, A., Rind, D., Russell, G., Stone, P., Fung, I., Ruedy, R., and Lerner, J., 1984, Climate sensitivity: analysis of feedback mechanisms, in *Climate Processes and Climate Sensitivity, Geo. Monogr.* 29, American Geophysical Union, pp. 13–163.

Hardjono, H.W., 1980, The effect of permanent vegetation and its distribution on stream flow of three sub-watersheds in Central Java, *Seminar Hydrol. Watershed Mgt.*, Surakarta, 5 June, 1980, Indonesia.

Hecht, S.B., Ph.D. Thesis, University of California, Berkley.

Hibbert, A.R., 1967, Forest treatment effects of water yield, *Proc. Int. Symp. Forests Hydrology, Penn. State Univ.*, Pergamon Press, New York, pp 537–543.

Hoore, J.D., 1954, Proposed classification of the accumulation zones of the free sesquioxides on a genetic basis. Afr. soils 3, 66–81.

Hubert-Schoumann, A., 1974, Bush clearing or cultivation on burnt fields in North Thailand, *J. Agric. Trop. Bot. Appl. (France)*, **22**, 251–255.

Hulugalle, N., Lal, R., and Ter Kuile, C.H.H., 1984, Soil physical changes and crop root growth following different methods of land clearing in western Nigeria, *Soil Sci.* (In Press).

Hurault, J., 1968, Effects of overgrazing on the high plateaux of Adamawa (Cameroons): Aerial Surveys and Integrated Studies, *Nat. Resour. Res.*, **6**, 463–468.

IITA/UNU, 1984, Effects of deforestation and land use on soil hydrology, microclimate and productivity in the humid tropics, *Prog. Rep., IITA*, Ibadan, 27 pp.

Jackson, P., 1983, The tragedy of our tropical rainforests, *Ambio*, **12**, (5), 252–254.

Junor, R.S., 1981, Impact of erosion on human activities, Bagmati Catchment Nepal, *J. Soil Cons. Serv. N.S.W.*, **37**, 41–50.

Juo, A.S.R., and Lal, R., 1977, The effect of fallow and continuous cultivation on the chemical and physical properties of an Alfisol in western Nigeria, *Plant Soil*, **47**, 567–584.

Kang, B.T., Wilson, G.F., and Sipkens, I., 1981, Alley cropping maize and *Leucaena leucocephala* in southern Nigeria, *Plant Soil*, **63**, 165–179.

Kellman, M.C., 1969, Some environmental components of shifting cultivation in upland Mindanao, *J. Trop. Geogr.*, **28**, 40–56.

Kirkham, M.B., 1982, Water and air conductance in soil with earthworms: an electric-analogue study, *Pedobiologia*, **23**, 367–371.

Kline, J.R., Jordan, C.F., and Drewry, G., 1968, Tritium movement in soil of a tropical rainforest (Puerto Rico), *Science*, **160**, 550–557.

Klinge, H., 1966, Verbreitung tropischer Tieflandspodsole, *Naturwissenschaften*, **17**, 442–443.

Klinge, H., 1978, Studies on the Ecology of Amazon Caatinga Forest in Southern Venezuela: Biomass Dominance of Selected Tree Species in the Amazon Ceatinga near Sau Carlos de Rio Negro, *Acta Cientifica Venezolana*, **29**, 258–269.

Kronfellner-Kraus, G., 1980, Erosion and torrent problems in Indonesia, *Centralblatt für das Gesamte Forestwesen*, **3**, 129–150.

Lal, R., 1981, Clearing a tropical forest II. Effects on crop performance, *Field Crops Research*, **4**, 345–354.

Lal, R., 1983, Soil erosion in the humid tropics with particular reference to agricultural land development and soil management, *IAHS Publ.*, **40**, 221–239.

Lal, R., 1984, Compaction, erosion and soil mechanical problems on tropical arable lands, *Proc. Int. Workshop on Soils*, ACIAR, Canberra, Australia, pp. 160–164.

Lal, R., and Cummings, D.J., 1979, Clearing a tropical forest I. Effects on soil and micro-climate, *Field Crops Res.*, **2**, 91–107.

Lal, R., Sanchez, P.A., and Cummings, R.W. Jr., 1985, *Land Clearing and Development in the Tropics*, A.A. Balkema, Rotterdam.

Lam, K.C., 1978, Soil erosion, suspended sediment and solute production in three Hong Kong catchments, *J. Trop. Geog.*, **47**, 51–62.

Lamprecht, H., 1977, quoted in Gentry and lopex-Parodi (1980).

Lanly, Jean Paul, 1982, *Tropical Forest Resources*, FAO Forestry Paper 30, 106 pp.

Lawson, T.L., Lal, R., and Oduro-Afriyie, K., 1981, Rainfall redistribution and microclimatic changes over a cleared watershed, in R. Lal and E.W. Russell, (eds), *Tropical Agricultural Hydrology*, Wiley & Sons, Chichester, pp. 141–151.

Lea, D.A., 1975, Human sustenance and tropical forest, *in Ecological effects of increasing human activities on tropical and sub-tropical forest ecosystems*, Australian/ENESCO Commision for Man and the Biosphere, Canberra, Australia, pp. 83–102.

Le Baron, A., Bond, L.K., Aitken, S.P., and Michallsen, L., 1979, An explanation of the Bolivian Highland grazing-erosion syndrome, *J. Range Mgt.*, **32**, 201–208.

Lettau, H., Lettau, K., and Molion, L.C.B., 1979, Amazonia's hydrological cycle and the role of atmospheric recycling in assessing deforestation effects, *Mon. Wea. Rev.*, **107**, 227–237.

Lim Suan, M.P., 1980, Technical feasibility and economic viability of selective logging in the Insular Lumber Company. *Annual Report*, For. Res. Inst. Laguna, Philippines, pp. 142–143.

Lovelock, J.E., 1983, *GAIA, a new look at life on Earth*, Oxford University Press, Oxford, 157 pp.

Marques, J., Santos, J.M., Villa Nova, N.A., and Salati, E., 1977, Precipitable water and water vapour flux between Belem and Manus, *Acta Amazonica*, **7**, 355–362.

Martin, G., 1983, Clearing and preparation of land for industrial oil palm plantation, *Oléagineux*, **38**, 219–232, 223–225.

Mathew, W.T., 1970, *Report of the Study of Critical Environmental Problems (SCEP), Man's Impact on the Global Environment*, MIT Press, Cambridge, MA.

Myers, N., 1980, *Conversion of tropical moist forests*, Nat. Acad. Sci., Washington, D.C., 205 pp.

Myers, N, 1981, The Hamburger Connection: How central Americas forests became North America's, Hamburger, Ambio X, 3–8.

Myers, N., 1983, Conversion rates in tropical moist forest: review of a recent survey, in F. Mergen, (ed), *Tropical Forests, Utilization and Conservation*, Yale For. Env. St., New Haven, CT, pp. 48–66.

Nordin, C.F., and Meade, R.H., 1982, Deforestation and increased flooding of the Upper Amazon, *Science*, **215**, 426–427.

Nwoboshi, L.C., 1981, Soil productivity aspects of agrisilviculture in the West African rainforest, *J. Agro. Ecosystems*, **7**, 263–270.

Nye, P.H., and Greenland, D.J., 1964, Changes in soil after clearing a tropical forest, *Plant Soil*, **21**, 101–112.

O'Keefe, P., and Kristofferson, L., 1984, The uncertain energy-path and third world development, *Ambio*, **13**, 168–170.

Ollagnier, M., Lauzeral, A., Olivin, J., and Ochs, R., 1978, Evolution des sols, sous palmeraie aprés défrichement de le foret, *Oleagineux*, **33**, 357–547.

Pereira, H.C., 1973, *Land Use and Water Resources*, Cambridge University Press, Cambridge 246 pp.
Pereira, H.C., and Hosegood, P.H., 1962, Comparative water-use of softwood plantations and bamboo forest, *J. Soil Sci.*, **13**, 299–313.
Postel, S., 1984, Protecting forests, in L.R. Brown et al. (eds) '*State of the World 1984*', Worldwatch Institute, W.W. Norton & Co., New York, 75–92.
Potter, G.L., Ellsaesser, H.W., MacCracken, M.C., and Luther, F.M., 1975, possible climatic effects of tropical deforestation, *Nature*, **258**, 697–698.
Persson, R., 1977, *Forest resources of Africa, Part II: Regional Analysis. Res. Note 22*, Dept. For. Survey, Royal College of Forestry, Stockholm, Sweden.
Persson, R., 1974, *World Forest Resources*. Royal College of Forestry, Stockholm.
Ramos, A.D., and Marinho, H.E., 1980, Erodibility of a lithosol without vegetation cover and under 2 vegetation types in the native scrub pasture of northern Brazil. EMBRAPA—Recife, Boletim de Pesquisa, No 2, 116 pp.
René, D., 1976, Symbiosis between earth and humankind, *Science*, **193**, 459–462.
Revelle, R., 1982, Carbon Dioxide and world climate, *Sci. Amer.*, **247**(2), 33–41.
Roche, M.A., 1981, Watershed investigation for development of forest resources of the Amazon region in French Guyana, in R. Lal, E.W. Russell, (eds) *Tropical Agricultural Hydrology*, Wiley & Sons, Chichester, pp 75–82.
Roose, E.J., 1979, Present dynamics of a highly desaturated ferrallitic soil derived from sandy clay deposits under cultivation and under dense subequatorial rainforest in southern Ivory Coast, Adiopodoumé 1964–1976, *Cahiers ORSTOM Pedol.*, **17**, 259–281.
Ross, M., 1984, *Forestry in land use policy for Indonesia*, Ph.D. Thesis, Green College, University of Oxford, 266 pp.
Ruslan, M., and Manan, S., 1980, The effect of skidding road on soil erosion and run-off in the forest concession of Pulan Laut Lakimantou, Indonesia, *Seminar Hydrol. Watershed Mgmt.*, Sura Karta, Indonesia.
Rubinoff, I., 1983, Strategy for preserving tropical rainforest, *Ambio*, **12**, 255–258.
Salati, E., Dall'Olio, A., Matsue, E., and Gat, J.R., 1979, Recycling of water in the Amazon Basin: an isotope study, *Water Resources Res.*, **15**, 1250–2158.
Salati, E., Lovejoy, T.E., and Vose, P.B., 1983, Precipitation and water recycling in tropical rain forests; Commision on Ecology, Occasional Paper No 2, *Environmentalist*, **3**(1), 67–74.
Salati, E. and Vose, P., 1983, Depletion of tropical rain forests, *Ambio*, **12**(5), 67–71.
Sanchez, P.A., Bandy, D.E., Villachaica, J.H., and Nicholaides J.J., III, 1982, Amazon Basin soils: Management and continuous crop production, *Science*, **216**, 821–827.
Sanchez, P.A., Villachica, J.H., and Bandy, D.E., 1983, Soil fertility dynamics after clearing a tropical rainforest in Peru, *Soil Sci. Soc. Amer. J.*, **47**, 1171–1178.
Seiler, W., and Crutzen, P.J., 1980, *Climatic change*, **2**, 207–247.
Seubert, C.E., Sanchez, P.A., and Valvarde, C., 1977, Effects of land clearing methods on soil properties and crop performance in an Ultisol of the Amazon Jungle of Peru, *Trop. Agric. (Trinidad)*, **54**, 307–321.
Shallow, P.G., 1956, River flow in the Cameron Highlands, *Hydroelectric Tech. Memo. No. 3*, Central Electricity Board, Kuala Lumpur.
Sidi, H., 1975, Tropical rivers as expressions of their terrestrial environments, in F.B. Golley, and E. Medina, (eds), *Tropical Ecological Systems*, Springer-Verlag, Berlin, pp. 275–288.
Silva, L.F. Da, 1981, Edaphic changes in 'tabuleiro' soils (Haplorthoxs) as affected by clearing, burning and management systems, *Revista Theobroma*, **11**, 5–19.
Singh, K. 1961, Value of bush, grass of legume fallow in Ghana, *J. Sci. Food Agric.*, **12**, 160–168.

Smith, N.J.H., 1981, Colonization lessons from a tropical forest, *Science*, **214**, 755–761.
Sommer, A., 1976, Attempts at an assesment of the world's tropical forests, *Unasylva*, **28**, 112–113 (FAO, Rome).
Sourabie, N., 1980, Soil evolution under irrigated sugercane in the Banfora region (Upper Volta), *Cahiers ORSTOM Biol.*, **42**, 25–41.
Swift, M.J., Russel-Smith, A., and Perfect, T.J., 1981, Decomposition and mineral nutrient dynamics of plant litter in regenerating bush-fallow in the sub-humid tropics, *J. Ecol.*, **69**, 981–995.
TVA (Tennessee Valley Authority), 1962, *Reforestation and erosion control influences upon the hydrology of the pine. Tree Branch Watershed, 1941 to 1960*, TVA, Knoxville, Tenn., USA.
Thijsse, J.P. 1977, Soil erosion in the humid tropics, *Landbouwkundig Tijdscrift*, **89**, 408–411.
Thomas, D.B., Barber, R.G., and Moore, T.R., 1981, Terracing of cropland in low rainfall areas of Machakos Dist., Kanya, *J. Agric. Eng. Res.*, **25**, 57–63.
Thompson, K., 1980, Forests and climate change in America: some early views, *Climate Change*, **3**, 47–64.
UNESCO/UNEP/FAO, 1978, *Natural Resources Research XIV, Tropical Forest Ecosystems*, UNESCO, Paris.
Van der Weert, 1974, The influence of mechanical forest clearing on soil conditions and resulting effects on root growth, *Trop. Agric. (Trinidad)*, **51**, 325–331.
Villa Nova, N.A., Salati, E., and Matsui, E., 1976, Estimativa de evapohauspiracao da Bacia Amazonica, *Acta Amazonica*, **6**, 215–228.
Villa Nova, N.A., Ribeiro, M.N.G., Nobre, C.A., and Salati, E., 1978, Radiacao Solar em Manus, *Acta Amazonia*, **8**, 417–421.
Webb, B.H., 1975, Objective planning of land clearing for mechanised agricultural development, *Planter*, **51**, 231–249.
Wood, T.G., Johnson, R.A., and Ohiagu, C.E., 1977, Populations of termites (Isoptera) in natural and agricultural ecosystems in Southern Guinea Savanna near Mokwa, Nigeria, *Geo-Eco-Trop*, **1**, 139–148.
Woodwell, G.M., 1978, The carbon dioxide question, *Sci. Amer.*, **238**(1), 34–43.
Woodwell, G.M., Whittaker, R.H., Reiners, W.A., Likens, E.G., Delwiche, C.C., and Botkin, D.B., 1978, The biota and the world carbon budget, *Science*, **199**, 141–146.
Woodwell, G.M., Hobbie, J.E., Houghton, R.A., Melillo, J.M., Moore, B., Palm, C.A., Peterson, B.J., and Shaver, G.R., 1982, Biotic contributions to the global carbon cycle, *Woods Hole Conf., 10—12 Feb., 1982*.
Woodwell, G.M., Houghton, R.A., Hobbie, J.E., Melillo, J.M., Moore, B., Peterson, B.J. and Shaver, G.R., 1983, Global deforestation: contribution to atmospheric carbon dioxide. *Science*, 222, 1081–1086.
Zimmerman, P.R., Grenbers, J.P., Wandiga, S.O., and Curzen, P.J., 1982, Termites: A potentially large source of atmospheric methane, carbon dioxide, and molecular hydrogen, *Science*, **218**, 563–565.

Chapter 14
Tillage

I	NEED FOR TILLAGE	566
II	TILLAGE AND SOIL PHYSICAL PROPERTIES	570
	A. Texture	571
	B. Structure	571
	C. Soil Compaction and Mechanical Properties	575
	1. Semi-arid regions	575
	2. Subhumid and humid regions	578
	D. Soil Moisture Retention	584
	E. Infiltration	585
	F. Run-off and Erosion	590
	G. Soil-water Conservation	594
III	TILLAGE AND SOIL MICROCLIMATE	598
	A. Soil Temperature	598
	B. Soil Air	599
IV	TILLAGE AND BIOLOGICAL ACTIVITY	604
V	TILLAGE AND SOIL CHEMICAL PROPERTIES	604
VI	CROP RESPONSE TO SOIL TILLAGE	606
VII	CONCLUSIONS	612

Mechanical soil tillage, used widely to facilitate seedbed preparation, has serious environmental and edaphological implications. Man has tilled the land ever since the beginning of permanent and established agriculture. In addition to its primary objective of establishing seed-soil contact, tillage is used to alleviate soil compaction and improve rainfall acceptance, dispose of pathogen-infested crop residue, incorporate fertilizer and minimize weed competition. Whether these objectives are achieved depends on the frequency and intensity of tillage, timing of operations and implements used, and soil and climate factors. In his quest to achieve these objectives, man has developed diverse and innovative implements to invert, loosen, mix, level and pulverize the soil. Heavy earth-moving and earth-ripping equipment are energy-intensive devices. Manual or animal-driven labour-intensive equipment is gradually being replaced by energy-intensive heavy machines.

Tillage operations affect soil, microclimate, flora and fauna, both directly and indirectly. The trends and magnitude of the alterations, however, depend on a multitude of interesting factors. The effects of manual tillage operations, for example, on fertile soil with well distributed nutrients in the soil profile are different from those of mechanized tillage operations on sloping land of shallow effective rooting depth. The magnitude of alterations also depends on the antecedent level of soil properties, vegetations cover prior to the tillage operation, and the climatic region.

I. NEED FOR TILLAGE

Is tillage necessary? If so, how much, how often, and with what equipment? These are some important questions of major concern to soil scientists, agronomists, environmentalists, and planners. There are two contrasting views—one believes that tillage is not indispensible and its role as a weed-control technique can be replaced by other equally effective measures, i.e. herbicides. In contrast, others argue that mechanical tillage is necessary to rapidly alleviate soil's mechanical and physical constraints. Both arguments are valid, depending on soil properties, crops to be grown, and socio-economic factors. It is difficult to generalize conclusions on tillage requirments because the need for tillage or lack of it is often soil and crop specific. Tillage requirements in relation to soil properties have been described for temperate

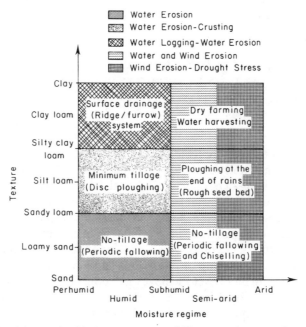

Figure 14.1 A general guide for assessment of tillage requirements for soils of different texture and moisture regimes (Lal, 1985). *Reproduced by permission of Elsevier Science Publishers*

zone soils by Soane (1975), Soane and Pidgeon (1975), Cannell *et al.* (1971, 1978), Galloway and Friffith (1978), and Pidgeon and Ragg (1979). Tillage requirements in relation to properties of soils in the tropics and the climatic regimes discussed by Lal (1985a,b) are based on variables including erosivity, erodibility, soil loss tolerance, compaction, soil temperature regime, available water holding capacity, cation exchange capacity, soil organic matter content and the quantity of crop residue mulch. A rating system was thus developed to assess the tillage requirements of a soil. Based on the available information for soil management and climatic problems, general guidelines for appropriate tillage systems are depicted in Figs. 14.1 to 14.3. It is apparent that there is no single tillage system universally applicable for all soils and crops. Economic crops can be grown without any tillage on some soils, whereas both primary and secondary tillage operations are required on others. A wide range of tillage operations of varying intensity, e.g. minimum tillage to intense tillage systems, are applicable for other soils and crops. In fact the tillage requirments depend on antecedent soil properties and vary as the soil contstraints to crop production vary. Tillage is a dynamic concept and should be conceptualized as a tool needed to alleviate soil constraints. This tool, as any other, should be used only if there exist specific constraints to crop production. The intensity and type of tillage operations, are, therefore, done with both short- and long-term objectives (Table 14.1). The short-term objectives are to optimize soil temperature and moisture regimes, minimize weed competition, stimulate root

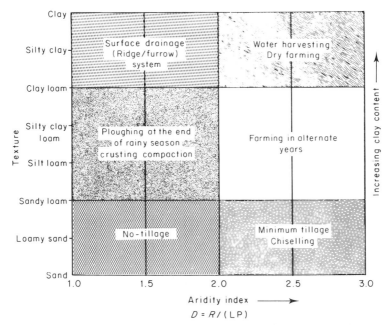

Figure 14.2 A general guide for assessment of tillage requirements for the semi-arid tropics (Lal, 1985). *Reproduced by permission of Elsevier Science Publishers*

Table 14.1 Objectives of seedbed preparation

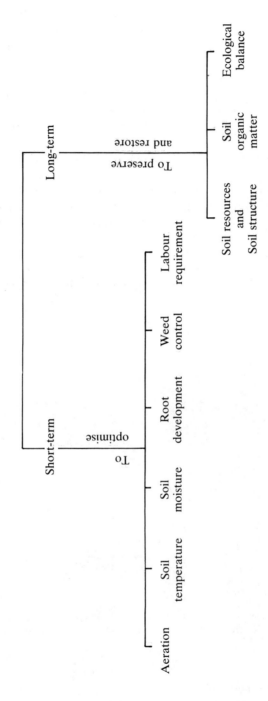

Table 14.2 Factors affecting choice of tillage methods

Climate	Soil	Crop	Socio-economic
1. Erosivity	1. Texture	1. Canopy characteristics	1. Farm size
2. Temperature	2. Structure	2. Duration	2. Infrastructure
3. Precipitation	3. Erodibility	3. Root system	3. Marketing facility
	4. Rooting depth	4. Water requirement	4. Labour
	5. Slope	5. Soil conserving or soil degrading crop	5. Technology
	6. Organic matter	6. Rotation	6. Resources
	7. Clay mineralogy	7. Susceptibility to pests and diseases	7. Tradition or culture
	8. Iron and aluminium oxides		
	9. Surface features including residue cover		

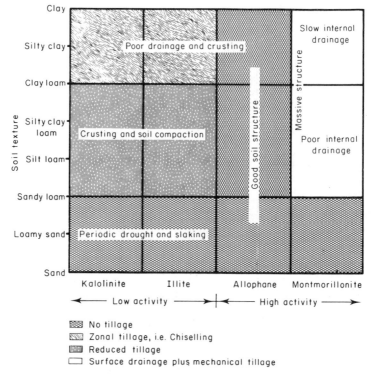

Figure 14.3 A general guide for assessment of tillage requirements for the humid tropics in relation to soil texture and mineralogical composition (Lal, 1985). *Reproduced by permission of Elsevier Science Publishers*

proliferation and development, and decrease labour constraints for seeding and harvesting. In the long run, however, tillage operations must preserve the resource base and ecological stability. It is important to make a distinction between exploitation of the limited and non-renewable land resources for short-term production gains and consideration of soil and climatic constraints to preserve this resource base. The choice of suitable tillage methods depends on a range of biophysical and socio-economic factors (Table 14.2). In addition to climatic factors, soil and crop characteristics are important determinants of tillage needs.

II. TILLAGE AND SOIL PHYSICAL PROPERTIES

Tillage operations are performed to alter the physical state of the soil. Although changes in physical properties are drastic immediately following the tillage operations, it is often the long-term effects of tillage operations that have been a major concern. The direct effects of tillage are due to shearing caused by tillage implements and the wheel traffic, i.e., loosening and compaction, breaking down of clods, increasing surface roughnesss, smearing and

Table 14.3 Effects of tillage methods on particle size distribution of 0–7 cm layer of an Alfisol at Ife, Nigeria, 36 months after initiating the experiment (Aina, 1982)

Tillage methods	Mulch	Sand (%)	Silt (%)	Clay (%)
Control (Initial level)		64.9	17.6	17.5
No-tillage	With	66.7	16.8	16.6
	Without	71.6	15.3	13.1
Ridges	With	69.6	15.8	14.6
	Without	75.6	12.4	12.0
Mounds	With	72.5	13.6	13.9
	Without	76.6	12.1	11.2

Table 14.3a Effects of cultivation on relative aggregation of an Oxisol and an Ultisol in Brazil (Grohmann, 1960)

Soil	Forest pasture	Cultivated annually
Oxisol	100	69
Ultisol	58	5

wheel rut. Indirect effects are alterations in microclimate due to exposure of the soil changes in surface sealing and crusting, variations in the susceptibility of the soil to erosion, and changes in population diversity of soil fauna.

A. Texture

The immediate effects of tillage on soil texture are due to mixing, homogenization, and soil inversion. In the long run, tillage influences the rate and intensity of clay migration by eluviation and movement in surface run off. Ahn (1968) reported significant downward movements of clay-sized particules in cultivated soil. In Southwestern Nigeria, Aina (1982) observed a measurable decrease in clay content of tilled compared with untilled soil. His data in Table 14.3 indicate 2 to 6 per cent decrease in clay, 2 to 5 per cent decrease in silt and 10 to 12 per cent increase in sand in the tilled compared with the untilled plots. The effect of tillage in decreasing clay and silt content was less drastic on mulched than on unmulched treatments.

Clay and silt content is also influenced indirectly by the tillage-induced alterations in the activity of soil fauna. The casting activity of earthworms commonly observed on untilled plots (Lal, 1983) brings clay and silt to the surface layer.

B. Structure

The effects of tillage are more drastic on soil structure than on soil texture. The magnitude and trends of alterations, however, depend on the initial structural properties. In semi-arid West Africa, soils are easily compacted, develop massive structure, crust, and form surface seals (Charreau and Nicou, 1971a,b; Nicou, 1974; 1977). Under these conditjions, tillage improves structural properties. However, the ameliorative effects of mechanical tillage are often short-lived, and the weakly developed structure deteriorates soon after. That is why repeated and more intensive tillage is needed to bring about temporary improvements in soil sturcture with every successive cultivation cycle. The ameliorative effects of tillage on soil structure are, therefore, a mixed blessing.

It is a common observation that virgin soils in equilibrium with climax vegetation often have the most favourable structural properties. In the long run, therefore, mechanical tillage operations result in degradation of soil structure. These long-term adverse effects are indirect and are due to tillage-induced decline in soil organic matter and in the activity of soil fauna. Decreases in clay and silt content also aggravate structural degradation. Some examples of structural deterioration due to tillage operations are described below.

Long-term effects of tillage operations on stuctural properties of some Pampas soils in Buenos Aires are reported by Tallarico *et al.* (1960). These reasearchers compared the state of soil aggregation after 12 years of tillage with that of 10 years of no-till seedbed preparation. Tillage decreased the proportion of large aggregates and of particles less than 100 μm. High stability of sturctural aggregates in untilled soils was associated with high organic matter content. In Brazil, Grohmann (1960) and Grohmann and Arruda (1961) reported that intensive ploughing caused structural degradation of a terra-roxa/legitima. The two-ploughing treatment resulted in mean aggregate diameter of 0.45 mm. In comparison, the mean aggregate diameter of one-ploughing and one-hoeing treatment was 0.70 and 0.95 mm, respectively (Table 14.3a). Sidiras *et al.* (1982) studied the effects of four tillage methods on size-distribution of water stable aggregates after 4 years of cultivation. For 0–10 cm depth, untilled soil had the highest percentage of 5.66 to 9.52 mm size range of aggregates. Smaller-sized aggregate dominated in tilled treatments. For 10–20 cm depth, the untilled soil had a higher proportion of aggregates, between 2.00 and 9.52 mm (Fig. 14.4). The organic matter content was positively correlated with the size distribution of water-stable aggregates.

That tillage reduces proportion and stability of macro-aggregates has also been demonstrated by research conducted in tropical Africa. In East Africa, Pereira and Jones (1954) observed that soil disturbance caused by clean weeding reduced the stability of aggregates to raindrop impact and total porosity (Table 14.4). The total porosity was 63.7, 65.1 and 65.8 per cent for clean-weeded, slashed weeds, and tall weeds treatment, respectively. In comparison, the volume of pores draining at 50 cm water tension was 19.9, 23.0 and 24.4 per cent, respectively. In West Africa, intensive tillage has also been proven to degrade soil structure. In Nigeria, Armon and Lal (1979) observed that the mean weight diameter of aggregates of the 0 to 5 cm layer was 3.6 and 2.6 mm for no-till and ploughed soil, respectively (Table 14.5). For an Alfisol at

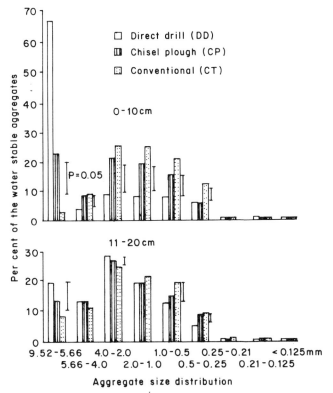

Figure 14.4 Distribution of the water stable aggregates for 0–10 and 11–20 cm soil depths after 4 years of conventional tillage, chisel plough, and direct drilling of an Oxisol in Londrina, Brazil (Sidiras et al., 1982)

Ife, Nigeria, Aina (1982) observed a greater progressive decrease in the mean weight diameter of aggregates with ridging and mound methods than with the no-till system (Table 14.6). After 36 months, the mean diameter of aggregates was 2.6 and 1.4 mm for no-till, 1.5 and 1.2 mm for ridges, and 1.4 and 1.1 mm for mounds with and without residue mulch, respectively. Both mulching and tillage had significant effects on the mean weight diameter of stable aggregates.

The effects of tillage on structural properties of some Australian soils were studied by Hamblin (1980). Untilled soils had significantly greater stability than ploughed soil. However, Hamblin reported that improvement in soil structue on untilled land takes place after several years and at a slow rate. The adverse effects of tillage on structural properties of temperate zone soils have been reported by Boone et al. (1976). That prolonged mechanical cultivation leads to morphological degradation of macro-aggregates and decreases their water stability, porosity and mechanical strength is also reported for fertile chernozems with relatively high organic matter content. Medvedev (1979) observed that water stability and structural coefficients were lower for ploughed than untilled chernozem (Table 14.7).

Table 14.4 Structure effects of weed-control cultivations on a Lateritic soil (Modified from Pereira and Jones, 1954). *Reproduced by permission of* Experimental Agriculture

Field treatment	Dry-sieving fractions (% of whole soil)		Water-stable aggregates greater than 0.5 mm (% of whole soil)		
	Clods over 1.25 cm diameter	Soil passing 0.5 mm diameter sieve	After vacuum wetting	After immersion wetting	After rain-impact wetting
W_0 Clean weeded	31.3	44.2	41.2	24.8	27.9
W_1 Slashed weeds	33.6	46.8	41.9	27.9	30.8
W_2 Tall weeds	19.8	45.3	40.3	28.9	31.2
LSD (0.05)	3.7	4.3	3.9	5.3	3.1

Table 14.5 Effect of tillage on mean weight diameter of aggregates of an Alfisol (Armon et al., 1982)

Year	Mean weight diameter (mm)		LSD (0.05)
	Untilled	Ploughed	
1977	3.46	2.62	0.40
1978	3.60	2.61	0.37

Table 14.6 Effects of tillage methods on mean weight diameter of water stable aggregates (> 2 mm) of the 0.7 cm layer of an Alfisol at Ife, Nigeria (Aina, 1982)

Tillage Methods	Mulch	Time (months) after initiating the experiment			
		0	12	24	36
No-till	With	3.1	2.8	2.4	2.6
	Without	2.8	2.2	1.7	1.4
Ridges	With	2.6	2.2	1.8	1.5
	Without	2.9	1.7	1.4	1.2
Mounds	With	2.9	1.8	1.5	1.4
	Without	2.6	1.5	1.3	1.1
LSD (0.05)		NS	0.4	0.5	0.2

C. Soil compaction and mechanical properties

As with the structural attributes, soil strength and mechanical properties are also in an improved condition immediately following mechanical tillage. The long-term tillage effects on mechanical properties are a debatable issue. For some soils, tillage is absolutely essential to alleviate soil compaction. For others, crops can be successfully grown without any soil disturbance. In semi-arid and arid regions with structurally inert soils where crop residue to protect the soil from desiccation is negligible, mechanical tillage has often been found to be useful.

1. Semi-arid regions

Tillage has been demonstrated to decrease soil bulk density and improve the ability of roots to penetrate deeper to exploit water and nutrients from the subsoil horizons (Feller and Milleville, 1977; Chopart, 1978; Mante, 1979; Nicou and Charreau, 1980; Klaij, 1983). For an Aridisol in Senegal Poulain and Tourte (1970) observed a significant decrease in the penetration resistance of a ploughed soil, compared with superficially ploughed soil (Table 14.8). The improvement in penetration resistance, however, did not last even until the end of a growing season. In semi-arid Tanzania, Macartney et al. (1971)

Table 14.7 Structural aggregate composition of chornozem as affected by ploughing (Medvedev, 1979)

Land-use	Depth (mm)	Amount of microaggregates upon dry sieving %, (mm)					Structure coefficient	Amount of microaggregates			Upon wet sieving %, mm	Water stability coefficient	Water intake (mm/hr)	
		>10	10–0.25	>1	>0.25	<0.25		>3	>1	>0.25	<0.025		Immediately after cultivation	At equilibrium density
Control	0–25	6.7	84.5	64.9	91.2	8.8	5.9	8.8	25.7	64.9	35.1	0.7	72	65
	30–40	6.3		61.5	84.4	15.6	3.6	1.7	10.9	56.4	43.6	0.7		
Ploughed	0–25	17.1	71.1	61.1	88.6	11.4	2.5	0.7	10.3	45.8	54.2	0.5	120	53
	30–40	9.2	77.4	60.2	86.6	13.4	3.4	0.1	9.6	51.3	48.7	0.6		

Although micro-aggregates do not change by planting, the water intake does.

Table 14.8 Effects of tillage methods on physical properties of some Aridisols in Senegal (Poulain and Tourte, 1970)

Depth (cm)	Superficial tillage (P$_1$)		Subsoiling to 40 cm depth (P$_2$) + Ploughing 25–30 cm + rotary hoe		Subsoiling to 70 cm depth (P$_3$) + Ploughing 35–40 cm + rotary hoe	
	Dior I	Dek II	I	II	I	II
(a) Penetration resistance (kg) (mean of 40 replications, CV 25–40%) recorded on 30 June 1966						
0–20	240	340	26	44	18	28
0–40	320	631	116	231	35	129
(b) Penetration resistance at the end of dry season (1969)						
0–20	323	442	298	266	320	328
0–40	348	445	420	394	363	357

Table 14.9 Bulk density, central Tanzania, as influenced by different methods of seedbed preparation (Macartney et al., 1971). *Reproduced by permission of* Tropical Agriculture

Treatment	Bulk density (g/cm^3)		Mean rooting depth (cm)
	0–3.8 cm	3.8–7.5 cm	
Uncultivated	1.49	1.44	—
Disc plough and harrow	1.19	1.22	84
Rip and disc harrow	1.39	1.47	53
Rip and ridge	1.25	0.98	79
Rip and no-till	1.38	1.11	43
Rip and tie ridge	1.23	1.03	76

observed a significant decrease in bulk density due to tillage (Table 14.9). Tillage-induced improvements in bulk density were observed even at the end of the first cropping cycle. Mean rooting depth in untilled plots was about 50 per cent of the intensively tilled treatment. Tillage investigations on Alfisol at Hyderabad, Central India, showed greater improvements in the penetrometer resistance of intensively cultivated treatments than in shallow cultivations.

The above findings are contradicted by the results of Dunham and his colleagues for a 4-year tillage investigation conducted at semi-arid Zaria, northern Nigeria (Dunham and Aremu, 1979; Aremu, 1980; Dunham, 1982, 1983). The data in Tables 14.10 and 14.11 show that soil bulk density measured 7 weeks after planting increased with increasing intensity of mechanical tillage. Accordingly, the percentage of water stable aggregates was more in untilled and less intensively tilled than in tilled treatments. Similar results were obtained from the arid region of India by Sharma et al. (1974) who observed that mechanized cultivation increased soil bulk density and decreased total porosity. Klaij (1983) did not observe any consistent trends in the effects of tillage on penetrometer resistance of an Alfisol in central India (Table 14.11).

These apparent contradictions are probably due to differences in soil and environmental factors. Alterations in soil bulk density due to mechanical disturbance depend on particle size distribution, organic matter and the antecedent soil moisture content. The amount, intensity and distribution of rainfall during the season and the soil and crop management systems adopted also play important roles in determining the tillage-induced effects on soil bulk density.

2. Subhumid and humid regions

Bulk density response to tillage for coarse-textured soils with low-activity clays in the humid and subhumid regions is different than in soils of the semi-arid regions. One or two high-intensity rainstorms on a ploughed soil often negate the loosening effects of ploughing. Consequently, the bulk density and

Table 14.10 Changes in bulk density of an Inceptisol at Zaria, Nigeria, for no-tillage and cultivated plots over a 4-year period (Dunham, 1982)

Year	Time of measurement	Depth (cm)	No-tillage	Cultivated	Significance of difference
1977	Initial	0–6		1.43	—
		10–16		1.47	—
1978	9 WAP	0–6	1.43	1.37	$P<0.1$
		10–16	1.44	1.52	$P<0.001$
1978	12 WAP	0–6	1.43	1.42	NS
		10–16	1.39	1.51	$P<0.05$
1979	8 WAP	0–6	1.51	1.45	$P<0.1$
		10–16	1.48	1.57	$P<0.05$
1980	9 WAP	0–6	1.35	1.34	NS
		10–16	1.43	1.48	NS

† WAP = Weeks after planting.

Table 14.11 Effect of tillage methods on soil penetrometer resistance (kg/cm^2) of two Alfisols during crop emergence at Hyderabad, India (Klaij, 1983)

Tillage treatment	1979		1980	
	RA–14	RW–2B	RA–14	RW–2B
1 Reshaping the old beds	2.8	2.6	4.2	2.2
2 Primary tillage	2.9	2.7	5.2	1.4
3 Chisel in the raw zone	2.8	2.5	5.0	1.2
4 Shallow cultivation	2.7	2.6	5.0	1.9
S. E.	±0.16	±0.28	±0.16	±0.24
CV (%)	26	33	14	43

RA = Red area No. 14 (Alfisol).
RW = Red watershed (Clayey Alfisol).

penetration resistance is often more on tilled than on untilled soil even a few days after the tillage operation.

At Ife, Nigeria, Aina (1979, 1982) observed lower bulk density and higher porosity in untilled than tilled soil (Table 14.12). Soil compaction and crusting were severe on plots receiving the most intense tillage treatment. The percentage of water stable aggregates was 5 times higher in untilled than in tilled soil. On a long-term tillage study on an Alfisol at Ibadan, Nigeria, Lal (1976, 1983) and Armon et al. (1982) also observed lower bulk density (Fig. 14.5) and penetrometer resistance (Fig. 14.6) on untilled than on tilled treatments. These data show that tillage resulted in soil compaction up to about 30 cm depth, corresponding with the depth of ploughing.

Contradictory results are also reported as to the effects of tillage in humid and subhumid ecologies. For a soil adjacent to that studied by Lal at Ibadan, Nigeria, Curfs (1976) reported that untilled plots had much higher soil bulk

Table 14.12 Physical properties of an Alfisol in Western Nigeria as influenced by tillage methods (Aina, 1979). *Reproduced from Soil Science Society of America Journal, Volume 43, 173–177, by permission of the Soil Science Society of America*

Tillage treatments	Bulk density (g/cm^3)	Porosity (%)	Air filled porosity at 60 cm H$_2$O suction (%)	Water stable aggregates > 2.36 mm (%)	Saturated hydraulic conductivity (cm/hr)
Uncultivated control	1.15	55.9	16.9	79	115
No-till	1.42	49.2	10.9	48	44
Ploughing	1.48	46.0	8.2	23	20
Ploughing and harrowing	1.51	44.1	6.8	15	15
LSD (0.05)	0.03	2.3	1.2	5	4

Table 14.13 Effect of a range of tillage methods on alteration in bulk density of an Alfisol at Ibadan, Nigeria (Curfs, 1976)

Tillage treatment	Years after land clearing								
	| |	2			3			4	
	Soil depth (cm)								
	0–10	10–20	0–10	10–20	0–10	10–20	0–10	10–20
Zero tillage*	1.41	1.48	1.53	1.56	1.58	1.62	1.65	1.68
Hand hoe	1.35	1.49	1.49	1.61	1.58	1.64	1.64	1.68
Rotary tiller (2-wheel tractor, 2WT)	1.40	1.51	1.44	1.60	1.49	1.60	1.57	1.66
Reversible plough (2WT)	1.37	1.49	1.44	1.60	1.45	1.63	1.52	1.66
Mould board plough shallow (4WT)	1.31	1.44	1.47	1.59	1.48	1.60	1.60	1.66
Mould board plough deep (4WT)	1.37	1.43	1.40	1.49	1.53	1.54	1.53	1.54
Disc plough shallow (4WT)	1.36	1.43	1.41	1.50	1.52	1.57	1.58	1.56
Rotovator (4WT)	1.29	1.36	1.39	1.41	1.41	1.53	1.56	1.58
Subsoiler of rotovator (4WT)	1.39	1.46	1.44	1.51	1.43	1.59	1.55	1.58
LSD (0.05)	N.S.	0.085	0.069	0.109	0.062	0.058	0.063	0.040
Mean pore volume	49.1	45.7	45.7	41.9	44.2	40.4	40.8	39.3

* Zero tillage without crop residue mulch.

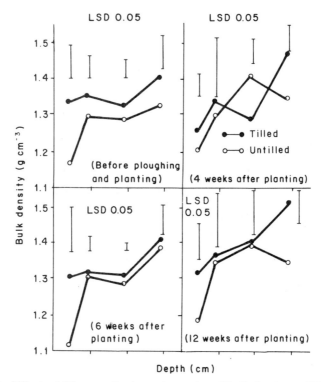

Figure 14.5 Effects of tillage methods on change in soil bulk density at different times after seeding maize on an Alfisol (Armon et al., 1982)

densities and lower total porosity and macroporosity than plots receiving tillage treatments (Table 14.13). In fact, the bulk density was often the lowest in the most intensively tilled plots. Furthermore, the rate of increase of soil strength and bulk density with years after land clearing and cultivation was greater on untilled than on various tillage treatments. This apparent discrepancy is once again due to different soil and crop management treatments. Mulching with crop residue is a very important determinant of soil properties. Soil physical, chemical and biological properties are drastically different with mulching and without it. Curfs' untilled (zero-tillage) treatments were without crop residue mulch, and that resulted in severe and rapid soil compaction. The beneficial effects of mulching on soil strength and mechanical properties have been widely demonstrated. As an example, the data of Aina (1982) shown in Table 14.14 indicate significantly greater bulk density in plots without than with mulch. The effect of mulching on decreasing bulk density was more pronounced for untilled than tilled treatments.

Mechanization of tillage operations also has profound effects on soil physical properties. Arable land-use and intensive cultivation increases bulk density and compaction in most soils. The rate of increase, regardless of tillage, is often more with mechanized than with manual tillage operations (Lal, 1985a).

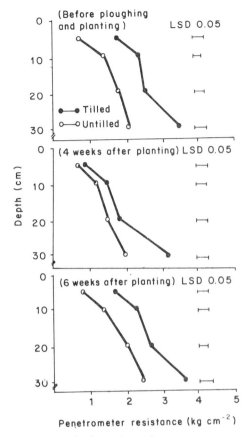

Figure 14.6 Effects of tillage methods on the resistance to penetrometer of an Alfisol measured at different times after plowing (Armon *et al.*, 1982)

Table 14.14 Effects of tillage methods and mulching on bulk density (g/cm³) of 0–7 cm layer of an Alfisol at Ife, Nigeria (Aina, 1982)

Tillage methods	Mulching	Time (months) after initiating the experiment			
		0	12	24	36
No-tillage	With	1.15	1.25	1.27	1.25
	Without	1.13	1.37	1.42	1.48
Ridges	With	1.09	1.36	1.39	1.41
	Without	1.05	0.45	1.48	1.50
Mounds	With	1.05	1.38	1.41	1.46
	Without	1.07	1.44	1.47	1.52
LSD (0.05)		0.06	0.08	0.05	0.06

D. Soil moisture retention

Drought stress is an important factor limiting crop production in the tropics. A wide range of tillage treatments is used to improve water-use efficiency and conserve rainwater within the root zone. Tillage is used to improve surface detention capacity, improve rainfall-acceptance by increasing infiltration, prolong time for water to infiltrate into the soil, and eliminate weeds to minimize competition. A wide range of tillage operations are used to achieve these objectives. The degree of success, however, varies among soils, implements, crops and climate environments.

The direct effects of tillage on moisture retention are due to alterations in total porosity, pore-size distribution and soil organic matter content. The duration of such an effect therefore depends on the subsequent status of these properties. In general, moisture retention at low suctions is more in untilled than in tilled soil. These conclusions are supported by the data of Lal (1983) in Nigeria and that of Sidiras *et al.* (1982) in Brazil. Opara-Nadi and Lal (1983) observed higher moisture retention in an untilled than tilled Alfisol between the pF ranges of 0.08 to 4.8 (Fig. 14.7). For an Oxisol in Brazil, Sidiras *et al.* observed significantly higher moisture retention at 0.06, 0.33 and 1 bar suction

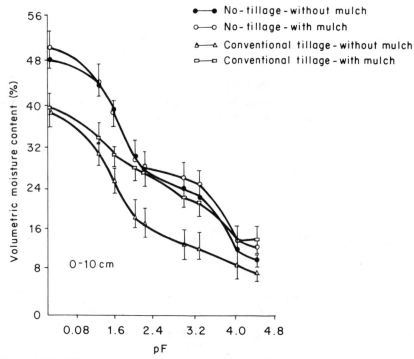

Figure 14.7 Effects of tillage methods on soil moisture retention properties of an Alfisol at IITA, Ibadan (Opara-Nadi and Lal, 1983)

in untilled than tilled soil (Table 14.15a Fig 14.8). Similar results have been reported for Alfisol in Nigeria (Table 14.15b).

Alterations in pF characteristics are just one factor responsible for soil-water storage in the root zone. The other factors responsible are infiltration, soil-water evaporation, and deep and profile root system development.

E. Infiltration

Infiltration or soil-water acceptance depends more on pore stability and continuity than on total porosity *per se*. The former is directly influenced by water stability of structural aggregates. A wide range of tillage operations are performed to alter water infiltration characteristics. Many studies have shown that the high infiltration rates commonly observed on uncultivated (virgin) soils in the tropics decline rapidly with arable land use and on intensively cultivated soils. Cultivation and intense grazing often result in compaction,

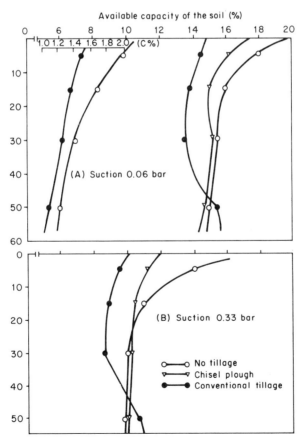

Figure 14.8 Effects of tillage methods on the release of plant available soil moisture in an Oxisol in Londrina, Brazil (Sidiras *et al.*, 1982)

Table 14.15a Effects of tillage methods on moisture retention characteristics of 3–10 cm layer of an Oxisol in Parana, Brazil (Sidiras et al., 1982)

Tillage method	Soil moisture (%) at different potential (bars)		
	0.06	0.33	1
Conventional plough	36.4	30.8	29.0
Chisel plough	39.2**	34.6**	32.6**
Untilled	44.7**	40.3**	38.2**

** Significantly different from conventional tillage at $0.01 > P$ level.

Table 14.15b Soil-water draining characteristics of surface 0–10 cm soil from ploughed and unploughed watersheds on Alfisols in Western Nigeria (Lal, 1984). *Reproduced by permission of Elsevier Science Publishers*

Tillage treatment	Soil water content (% w/w) at various matric potentials (bars)					
	0	0.03	0.1	1	3	15
Ploughed	28.1	21.6	18.5	8.9	6.5	5.6
Unploughed	31.3	23.7	21.2	9.9	7.3	6.3

Table 14.16 Relative efficiency of soil surface treatments in conserving rainfall (Lawes, 1961). *Reproduced by permission of* Experimental Agriculture

Treatment	Percentage of rainfall penetrating soil surface	Maximum infiltration rate (mm/hr)
Bare soil		
Undisturbed	31	12.7
Hoed at fortnightly intervals	49	22.9
Hoed after every storm	57	30.5
Mulches		
Dead grass	90	>127
Groundnut shells	89	>127
Sorghum stalks	98	>127

crusting and surface sealing (Fig. 14.9) thereby decreasing the soil's ability to receive torrential tropical rains. Tillage treatments are used to break the crust and surface seal and improve the infiltration rate.

Lawes (1961) observed in Zaria, northern Nigeria, that the effect of tillage on the infiltration rate was greatly altered by the presence or absence of mulch (Table 14.16). In an unmulched soil, mechanical disturbance increased the infiltration rate. However, the infiltration rate increased by the frequent tillage treatment was only one-quarter of that of the mulched plots. The presence of mulch obviously protects the soil from raindrop impact and prevents the

Figure 14.9 Crust development and surface sealing of an Alfisol

development of crust and surface seal (Lal, 1983). Subsequent studies by Dunham (1982) confirm the findings of Lawes that untilled soils with crop residue mulch are less prone to developing surface seal and maintain a higher infiltration rate than bare ploughed soils (Table 14.17). In semi-arid regions of Tanzania, Macartney et al. (1971) also observed that although initially infiltration was good in disc-plough and harrow treatments, subsequent surface capping reduced the rate of water penetration. The adverse effects of tillage

Table 14.17 Effects of tillage methods on equilibrium infiltration rate (mm/ha) of an Inceptisol at Samaru, Zaria (Aremu, 1980; Dunham, 1982)

Year	Time of measurement†	Position	Untilled	Tilled	Significance of difference
1978	8 WAP	Row	—	—	—
		Inter-row	22	10	$P < 0.01$
1978	14 WAP	Row	56	10	$P < 0.01$
		Inter-row	67	50	NS
1979	Before Planting	Row	58	17	$P < 0.05$
		Inter-row	59	25	$P < 0.05$
1980	1 WAP	Row	63	10	$P < 0.001$
		Inter-row	89	12	$P < 0.001$

† = Weeks after planting.

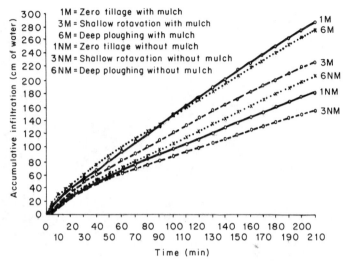

Figure 14.10 Infiltration rate in Egbeda soil following three tillage treatments under mulched and unmulched conditions (Curfs, 1976)

involving ploughing, discing, and harrowing as measured by reductions in the infiltration rates are also observed by Curfs (1976) and Lal (1984) in Nigeria (Figs. 14.10 and 14.11, respectively) and by Khatibu et al. (1984) in Zanzibar (Fig. 14.12). The data by Curfs indicate that any mechanical soil disturbance decreased the infiltration rate in comparison with the mulched-untilled control. The infiltration rate of untilled plots without mulch was significantly lower than some tillage treatments (Fig. 14.10). The significance of the data from Zanzibar (Fig. 14.12) lies in strong evidence of the importance of organic rather than inorganic mulches in increasing the infiltration rate. The lowest infiltration rate was recorded with the black and white polythene mulches. In

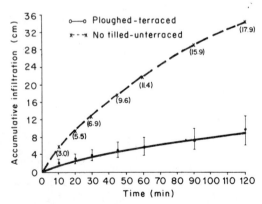

Figure 14.11 Cumulative infiltration into the ploughed and no-till treatments measured in January, 1980 after 5 years of continuous tillage treatments and maize production (Lal, 1984). *Reproduced by permission of Elsevier Science Publishers*

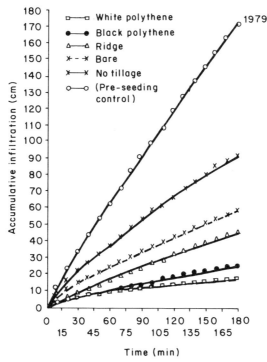

Figure 14.12 Effects of mulch and tillage treatments on accumulative water infiltration into an Ultisol in Zanzibar, Tanzania (Khatibu *et al.*, 1984). *Reproduced by permission of Elsevier Science Publishers*

Nigeria, Lal fitted the water-intake data into infiltration models and observed significant differences in soil-water sorptivity and transmissivity values of tilled and untilled soils.

Infiltration rate (untilled) = $82.4 \, t^{-\frac{1}{2}} + 60$ $r = 0.90$
Infiltrate rate (tilled) = $36.0 \, t^{-\frac{1}{2}} + 0.8$ $r = 0.92$

Where infiltration rate is in cm hr^{-1} and t is the time in hours.

The desirable physical properties for rice growth and yield are attained through a series of dry and wet tillage operations. The objective of tillage is to influence soil permeability and percolation, the water retention capacity of the surface layer, and total porosity and pore size distribution. Soil puddling, or tillage operations performed when soil is saturated, is traditionally done to deliberately destroy structural aggregates and to decrease macroporosity. In the Philippines, Sanchez (1973) observed a drastic decline in the drainage rates of puddled soils. The data in Fig. 14.13 by De Datta and Kerim (1974) show significant reduction in percolation losses by puddling. However, some soils are more easily puddled than others. Puddling has very little effect on the physical properties of coarse-textured soils. The effects of puddling are also less in soils that are easily dispersed, e.g., Vertisols.

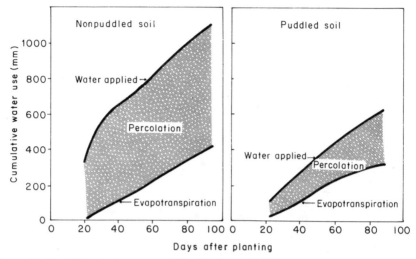

Figure 14.13 Water balance on puddled and unpuddled rice paddy soil in the Philippines (De Datta and Kerim, 1974). *Reproduced from Soil Science Society of America Proceedings, Vol. 38, 515–518, by permission of the Soil Science Society of America*

F. Run-off and erosion

The magnitude of run-off and soil erosion depends on interaction between soil and rainfall. Accelerated erosion is the detachment of particles and microaggregates caused by impacting raindrops and running water. The effectiveness of rainfall and run-off in detaching and transporting soil particles downstream is increased by several times if the soil is loose and unprotected by any vegetation cover. The cover not only reduces raindrop impact, it also increases the resistance to the flow of water run-off.

As with the effects on soil physical properties, the effects of tillage on run-off and soil erosion are hard to generalize. For some soils mechanical tillage is recommended to control erosion, and in others it is proven to be a decisively erosion-promoting practice. In addition to variations among soils, types of tillage and their interaction with residue management are important determinants of soil erosion. Antecedent soil properties also play a major role. If soil is compacted and has a well-developed surface seal, mechanical loosening and breaking the surface crust improve water infiltration and decrease run-off and erosion. On the other hand, if the soil surface is well structured and contains a large proportion of interconnected and stable water-conducting pores, tillage operations destroy these pores and render the soil more susceptible to erosion. The following are some examples of these contrasting but scientifically explainable phenomena.

Experiments conducted on structurally inert and easily compacted soils of West African Sahel show that mechanical tillage greatly decreases runoff and soil erosion in comparison with untilled or tilled treatments (Charreau, 1970).

Table 14.18 Effect of tillage on run-off erosion from a bare soil with 2% slope at Sefa in 1968 (Charreau and Seguy, 1969). *Reproduced by permission of IRAT*

Treatment	Erosion (t/ha)	Run-off (mm)	Percentage of rainfall
No-till	18.1	270.7	37.2
Ploughed	6.5	151.5	20.8

The data in Table 14.18 by Charreau and Seguy (1969) indicate that erosion in the Sahal zone of Senegal was reduced by 2 to 3 times by timely ploughing, i.e., 6.5 t ha^{-1} y^{-1} in comparison with 18.1 t ha^{-1} y^{-1}. These authors claim that ploughing reduced erosion directly through improving water infiltration and indirectly by promoting crop growth and ground cover. Regression equations developed relating run-off (mm), to rainfall (mm) for ploughed and unploughed treatments are:

$$Y_1 = 0.59 \ (\pm\ 0.09)\ X - 3.3\ (\pm\ 1.5) \qquad r = 0.91$$
$$Y_2 = 0.36 \ (\pm\ 0.11)\ X - 2.4\ (\pm\ 1.8) \qquad r = 0.74$$

Where Y_1 and Y_2 refer to run-off in no-till and ploughed treatments and X is the rainfall amount. Similar results have been reported from northwest India by Verma *et al.* (1979) who observed that deep tillage reduced water run-off. On an Alfisol at Hyderabad, India, Klaij (1983) observed less erosion and more run-off from plots receiving primary tillage operations (Table 14.19). Although drastically reduced, runoff and erosion were severe even on ploughed plots. Erosion on ploughed plots often exceeds the range of tolerable soil loss. That is why many researchers have recommended additional tillage techniques, e.g., contour and tied-ridge systems. The data in Tables 14.20 and 14.21 show the effectiveness of these systems in controlling erosion and reducing water run-off in semi-arid climates.

Tillage has a different effect on the soil erosion and run-off of biostructurally active soils in the humid and subhumid tropics. For these soils, structural

Table 14.19 Effects of tillage methods on run-off and soil erosion at Hyderabad, India (Klaij, 1983)

Date		Rainfall (mm)	Reshaping the old bed		Primary tillage	
			Run-off (mm)	Soil loss (kg/ha)	Run-off (mm)	Soil loss (kg/ha)
July	30	23.6	1.1	13	1.2	8
Aug.	14	16.4	2.2	5	3.1	6
	19	114.8	22.1	137	27.4	118
	20	72.6	17.2	172	21.8	87
Sept.	3	22.5	1.4	3	1.5	4
	6	31.1	4.4	6	5.2	9
	24	14.7	0.3	—	0.3	—
Total		295.7	48.7	336	60.5	232

Table 14.20 Effect of contour ridges on soil erosion at Nanokely, Madagastcar (Fournier, 1967)

Crop	Erosion (t/ha)	
	Contour ridges	Ploughed flat
Maize	3.9	19.1
Potatoes	4.2	5.1
Groundnuts	4.1	12.3

Table 14.21 Effect of tied ridges on run-off and soil erosion at Niangoloko, Upper Volta and on yield of groundnut and millet (Fournier, 1967)

Treatment	Run-off (%)	Erosion (t/km^2)	Yield (kg/ha)	
			Groundnut	Millet
Tied ridges	0.9	143.7	846	729
Sloping ridges	6.3	610.4	479	376
Ploughed flat	12.2	1318.6	658	352

Table 14.22 Soil and water losses from an Oxisol under soybean using simulated rainfall (Biscaia, 1982)

Tillage methods	Stage of crop growth	Run-off (%)	Erosion (kg/ha)
Conventional tillage	Seed bed	35	19 089
	100% canopy	8	953
Minimum tillage	Seedbed	25	12 018
	100% canopy	16	1656
No-tillage	Seedbed	16	1120
	100% canopy	5	151
Conventional tillage, bare soil	Dec. 1977	23	22 840
	April	21	13 934

Table 14.23 Effects of tillage methods on run-off and soil losses from soybean–wheat rotation on an Oxisol with 4% slope in Parana, Brazil (Sidiras et al., 1982)

Year	Rainfall (mm)	Conventional tillage		No-tillage	
		Run-off (%)	Erosion (kg/ha)	Run-off (%)	Erosion (kg/ha)
1978–79	671.9	2.9	552.6	1.0	183.3
1979–80	938.2	18.1	4067.0	12.1	474.9

Table 14.24 Effects of ploughing on run-off and soil erosion from watersheds growing maize in western Nigeria (Lal, 1984). *Reproduced by permission of Elsevier Science Publishers*

Year	Rainfall (mm)	Ploughed		Unploughed	
		Run-off (mm)	Erosion (t/ha)	Run-off (mm)	Erosion (t/ha)
1979	841.3	21.5	0.13	225.1	5.5
1980	900.0	34.4	0.33	153.0	1.9

stability and high rainfall acceptance are better achieved by leaving the soil alone and by protecting it from intense rains through stubble mulching and vegetation cover (Lal, 1983). Data in Tables 14.22 and 14.23 from Oxisols in Brazil indicate that ploughing accelerated soil erosion and increased losses due to water run-off. In western Nigeria Lal (1984) observed that during the year 1979 there was 10 times more run-off from the ploughed than unploughed watershed and 42.2 times more erosion (Table 14.24). Although there was some run-off from the unploughed watershed, the presence of crop residue mulch and the absence of soil disturbance caused by ploughing greatly reduced soil detachment and its transport in run-off. Analyses of the data of 1980 indicated that run-off and erosion from ploughed watershed were 4.44 and 5.25 times more respectively, than from unploughed watershed.

G. Soil-water conservation

Through its effect on infiltration, run-off, and water retention characteristics, tillage also modifies soil-water storage in the profile. Because tillage alters rooting depth and the ability of roots to penetrate into the subsoil layers, variable water utilization by crops is another consequence of the tillage operations. Once again responses to tillage for soil-water conservation and crop-water utilization differ among soils, ecologies and crops. Also important in defining the magnitude of these effects are soil and crop management, especially the crop residue management.

In semi-arid and arid regions intensive tillage has been demonstrated to improve soil-water conservation and crop-water use. In an arid region of Israel, Stibbe and Ariel (1970) observed that ploughing increased soil-water storage up to 90 cm depth (Table 14.25). In the 0–10 cm layer, however, unploughed soil had more water reserve than ploughed soil. During long dry seasons denuded soil surfaces are easily crusted and compacted. The loosening effect of tillage improves water infiltration and decreases losses due to run-off.

In semi-arid regions ploughing and tied-ridges are recommended to conserve soil water. In Tanzania Macartney *et al.* (1971) reported that tied-ridging following ripping conserved more water in the soil profile than disc ploughing, contour ridging or no cultivation (Table 14.26). In semi-arid regions with low,

Table 14.25 Alterations in soil moisture storage by different tillage methods in arid regions (Stibbe and Ariel, 1970). *Reproduced by permission of the* Netherlands Journal of Agricultural Science

Period	Tillage method	Soil moisture (%) at different depths (cm)			
		0–30		30–60	60–90
After rainfall, November	Ploughed	39.2		25.2	20.1
	Unploughed	34.4		18.6	18.5
After dry spell, December	Ploughed	24.2		23.5	22.0
	Unploughed	23.2		20.3	17.4
At seedling emergence, May	Ploughed + cultivated	0–10	16.8		
		10–20	29.6	31.4	31.9
		20–30	33.2		
	Unploughed	0–10	22.5		
		10–20	25.0	28.4	30.1
		20–30	27.1		

PWP at 15 bar = 15.0%; 1/3 bar = 26.0% by weight.

Table 14.26 Moisture storage in the root zone as influenced by tillage of a soil at Kongwa, Central Tanzania (Macartney *et al.*, 1971). *Reproduced by permission of* Tropical Agriculture

Tillage treatments	Available moisture (mm) in top 1.83 m			
	1966–67		1967–68	
	F	H	F	H
Uncultivated	—	—	—	—
Disc plough and harrow	29	61	131	118
Rip and disc harrow	39	82	166	129
Rip and ridge	—	—	152	130
Rip and no-till	35	110	156	123
Rip and tie ridge	42	98	194	157

F = Flowering; H = Harvest.

variable, and unpredictable rainfall, the tied-ridge system is often the only way to assure an economic yield from sorghum or millet. In Senegal, Nicou (1977) reported that available water reserves up to 2 m depth were 21, 64 and 79 mm for unploughed, fallow ploughed under, and millet straw ploughed under treatments, respectively (Table 14.27). Trends in yields of the following crops are an obvious consequence of the drastic tillage-induced differences in soil-water storage. In contrast Whiteman (1975) observed in Botswana that

Table 14.27 Effect of fallowing and tillage methods on water conservation and crop yield in Senegal (Nicou, 1977)

Surface conditions in 1971	Farming techniques in October 1971	Available water up to 2m depth on 27 July, 1971 (mm)	Yields in 1972 (kg/ha) Millets	Groundnuts
Fallow	Burning – Unploughed	21	1000	64
Fallow	Ploughed under	64	1800	1600
Millet	Straw ploughed under	79	2000	1600

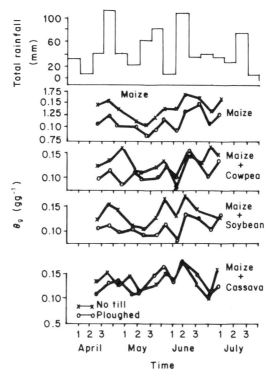

Figure 14.14 Soil moisture reserves for different crop mixtures on ploughed and no-till soil (Maurya and Lal, 1980). *Reproduced by permission of Experimental Agriculture*

weed-free summer fallow prior to ploughing had better water conservation than cropping or natural fallowing.

The conclusions drawn from the available data on tillage effects on soil-water storage in humid and subhumid regions are different from those in the semi-arid and arid climates. In Brazil, Kemper and Derpsch (1981) observed higher soil moisture storage in unploughed than ploughed soil. Similar observations have been made by Khatibu *et al.* (1984) in Zanzibar and Aina (1982) in Nigeria. In Western Nigeria Maurya and Lal (1980) observed higher soil moisture storage in unploughed than in ploughed treatments (Fig. 14.14). The reduced moisture conservation in ploughed plots is due to low infiltration, high run-off losses, and high evaporation losses.

The review of the literature on soil physical properties indicates that tillage has drastic effects on soil physical properties. These effects are positive for some soils and environments and negative in others. In arid and semi-arid tropics ploughing and intensive tillage increase porosity and decrease bulk density, increase infiltration and decrease run-off, increase soil-water storage and decrease evaporation losses, and have overall ameliorative effects on soils and crops. The opposite seems to be the case in the humid and subhumid regions. In humid ecologies, ploughing and intensive tillage operations increase compaction and decrease porosity, increase run-off and erosion by decreasing infiltration rate, increase evaporation and decrease soil-water storage, and have overall adverse effects on soil and crops. This differential response is partly due to different systems of soil and crop management. The ploughing in arid and semi-arid regions is a substitute for residue mulch because the latter is not available. Ploughing is obviously unnecessary in the humid regions with readily available mulch material. If cropping systems were developed to ensure the availability of mulch in drier climates, there might be no need for the intense and frequent use of mechanical tillage now recommended.

III. TILLAGE AND SOIL MICROCLIMATE

Tillage techniques are used to alter the microclimate in the vicinity of the soil surface. A range of tillage operations is performed to alter soil temperature and moisture regimes. Whereas suboptimal soil temperatures are not a limitation in lowland tropics, tillage methods can be used to decrease the maximum soil temperature if it exceeds the optimum range. Strongly interacting with soil temperature is the moisture regime, which has already been shown to be influenced by the mode, frequency, intensity and timing of tillage operations. Through its effects on porosity, crusting and surface seal, the tillage system also modifies the gaseous exchange between the soil and the atmosphere. The effect of tillage on soil temperature and aeration also varies among soils, moisture regimes and climate.

A. Soil temperature

Tillage affects soil temperature both directly and indirectly. Directly, tillage exposes the soil to more insolation and alters the albedo. Indirectly, tillage influences soil temperature through its effects on moisture regime, texture and structure. The latter properties modify thermal conductivity and diffusivity, thereby influencing the heat conductance through the soil profile. In temperate latitudes, where in early spring the soil is wet and has sub-optimal soil temperatures, soil is often ploughed to improve drainage and increase the maximum soil temperature. The orientation of ridges is deliberately selected to induce the warming trend. For the same reasons, ploughing and ridging increase the maximum soil temperatures in the tropics and, by increasing exposure, increase soil evaporation. Ploughed soil in the lowland tropics thus have higher maximum soil temperatures than unploughed soil. The soil temperatures on unploughed land are even less with residue mulch.

The effects of tillage systems on soil temperature have been studied for only a few soils and environments in the tropics. In Zanzibar, Khatibu *et al.* (1984) observed that ridged, ploughed flat, and white polythene mulch treatments had higher maximum soil temperatures than unploughed plots (Fig. 14.15). De-

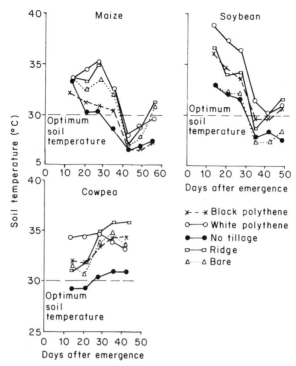

Figure 14.15 Effects of tillage methods and mulching on soil temperature regime at 5 cm depth measured in the Masika season in Zanzibar, Tanzania (Khatibu et al., 1984).
Reproduced by permission of Elsevier Science Publishers

pending on the season, soil temperature in the surface layer of the unploughed mulched treatments may have even been sub-optimal. In southwest Nigeria, Lal (1979) reported that the maximum soil temperatures on ridged and ploughed plots were higher and the minimum temperatures lower than on unploughed and straw-mulched treatments (Tables 14.28 and 14.29). During the second season, with frequent overcast skies and lower ambient temperatures, there were less differences in soil temperatures among treatments than in the first season.

B. Soil air

Tillage effects on soil aeration are due to alterations in total and air porosity, soil moisture retention and transmission, susceptibility to crusting and surface sealing, differences in biological activity and mineralization of crop residue and biomass. The latter two processes are related to soil respiration. The most direct effects of tillage on aeration and gaseous exchange are related to its effect on drainage and surface crust. Preliminary experiments conducted at IITA indicate that ploughing and intensive tillage operations of structurally

Table 14.28 Soil temperature (C°) at 5 cm depth under different crops one week after planting (first season, 1977) (Lal, 1979). *Reproduced by permission of Williams & Wilkins Co.*

Treatment	Time of day (hours)											
	0100	0300	0500	0700	0900	1100	1300	1500	1700	1900	2100	2300
Maize												
Black plastic	29.0	28.1	27.2	26.7	27.9	31.8	35.8	38.0	37.4	35.0	32.8	31.2
Clear plastic	31.0	30.0	29.0	28.4	29.0	32.8	38.4	42.5	42.0	38.0	34.9	33.0
Straw mulch	26.8	26.2	25.8	25.4	26.0	28.0	30.8	32.3	31.6	30.0	28.3	27.7
Ridges	26.2	25.0	23.8	23.0	25.3	32.5	40.8	45.0	42.5	36.7	32.0	29.3
Bare flat	27.7	26.6	25.8	25.0	26.3	30.8	36.2	39.8	38.7	35.0	32.0	30.1
Aluminium foil	26.7	26.6	25.3	25.0	26.6	29.9	33.4	34.9	33.6	31.0	28.9	28.0
Cowpea												
Black plastic	28.3	27.0	26.1	25.4	27.0	31.7	37.5	41.4	40.0	35.8	32.5	30.8
Clear plastic	30.1	29.0	28.0	27.4	28.9	33.7	39.7	44.0	42.0	37.5	34.2	32.4
Straw mulch	26.1	25.7	25.0	24.9	26.1	29.7	33.5	35.0	33.0	30.1	28.3	27.5
Ridges	26.0	24.8	23.7	22.9	25.2	32.8	40.3	44.0	4.6	36.1	31.9	29.2
Bare flat	27.5	26.4	25.5	24.9	26.1	30.7	36.0	39.6	38.5	34.9	31.9	30.0
Aluminium foil	27.0	26.6	26.2	26.0	26.6	28.9	30.5	31.5	31.0	29.8	28.6	28.0

Soybean												
Black plastic	29.1	28.2	27.3	26.8	27.5	31.5	36.8	39.5	38.1	35.0	32.5	31.0
Clear plastic	30.2	29.1	28.0	28.4	28.5	33.7	40.5	45.0	43.3	38.6	34.9	32.8
Straw mulch	26.2	25.8	25.0	24.9	25.8	28.7	33.0	35.1	33.6	30.8	28.8	27.8
Ridges	25.8	24.2	23.0	22.0	25.5	34.6	42.6	45.4	42.0	36.4	31.8	28.9
Bare flat	27.7	26.4	25.3	24.9	26.7	32.0	33.8	41.9	40.0	35.5	32.2	30.2
Aluminium foil	28.1	27.7	27.0	26.5	27.1	29.3	31.3	32.5	32.2	31.1	30.0	29.4
Cassava												
Black plastic	28.9	28.0	27.0	26.5	27.6	30.9	35.2	38.0	37.5	34.9	32.4	30.9
Clear plastic	29.2	28.1	27.2	26.7	29.5	37.0	45.5	47.5	43.4	37.1	33.3	31.5
Straw mulch	26.0	25.4	25.0	24.8	26.0	29.9	34.5	36.0	33.0	29.9	28.0	27.0
Ridges	26.0	24.2	23.0	22.0	25.5	34.6	42.6	45.4	42.0	36.4	31.8	28.9
Bare flat	27.2	26.0	25.0	24.0	26.9	33.8	41.1	44.4	41.5	36.0	32.2	30.1
Aluminium foil	27.2	26.4	26.0	25.4	27.0	29.5	32.2	33.0	32.0	30.5	29.2	28.5

Table 14.29 Soil temperature (C°) at 5 cm depth under different crops one week after planting (second season 1977) (Lal, 1979). *Reproduced by permission of Williams & Wilkins Co.*

Treatment	Time of day (hours)											
	0100	0300	0500	0700	0900	1100	1300	1500	1700	1900	2100	2300
Maize												
Black plastic	25.2	25.0	25.0	24.8	25.2	26.5	27.5	29.0	28.8	27.6	26.6	25.9
Clear plastic	26.1	26.0	25.6	25.4	26.5	28.2	29.0	30.4	30.1	28.6	27.7	26.8
Straw mulch	24.9	24.7	24.4	24.3	24.9	25.6	26.0	27.0	26.6	26.0	25.5	25.0
Ridges	23.7	23.3	23.0	23.1	24.0	25.7	26.3	27.6	27.4	25.6	24.8	23.6
Bare flat	24.8	24.4	24.1	24.0	25.0	26.3	26.8	27.9	27.4	26.1	25.5	24.9
Aluminium foil	24.7	24.6	24.4	24.3	24.4	24.9	25.2	25.5	26.6	25.6	25.3	25.1
Cowpea												
Black plastic	26.0	25.6	25.4	25.1	25.5	27.0	28.0	29.2	29.3	28.5	27.6	26.8
Clear plastic	24.8	24.5	24.0	23.9	24.5	26.0	26.6	28.0	27.9	26.6	25.8	25.1
Straw mulch	24.9	24.6	24.4	24.2	24.7	25.6	26.0	26.9	26.5	25.9	25.4	25.0
Ridges	23.7	23.4	23.1	23.1	24.1	25.9	26.4	27.7	27.0	25.6	25.0	23.9
Bare flat	24.7	24.3	24.0	24.0	24.9	26.2	26.6	27.7	27.2	26.0	25.3	24.8
Aluminium foil	24.4	24.1	24.0	24.4	25.3	25.6	26.4	26.4	26.4	25.0	25.2	25.0
Soybean												
Black plastic	25.2	25.0	26.0	24.9	25.3	26.7	27.5	28.4	28.4	27.4	26.6	25.8
Clear plastic	26.1	25.9	25.6	25.4	26.4	28.5	29.3	30.7	30.5	29.0	27.9	27.0
Straw mulch	25.2	25.1	25.0	25.0	25.0	25.2	25.5	26.0	26.3	26.1	26.0	25.5
Ridges	23.6	23.3	23.0	23.0	24.1	26.1	26.8	28.0	27.8	25.8	24.9	23.6
Bare flat	24.6	24.6	24.4	24.3	24.4	24.8	25.1	25.5	26.6	25.5	25.3	25.1
Aluminium foil	24.9	24.8	24.5	24.4	24.7	25.3	25.7	26.3	26.4	26.0	25.8	25.4
Cassava												
Black plastic	26.2	26.0	25.7	25.5	26.0	27.0	27.7	28.6	28.5	28.0	27.3	26.8
Clear plastic	26.3	26.0	25.9	25.6	26.4	28.0	28.5	30.0	29.6	28.4	27.7	27.0
Straw mulch	25.5	25.4	25.2	25.1	25.3	25.8	26.0	26.6	26.4	26.0	25.9	25.5
Ridges	24.3	24.0	23.7	23.6	24.9	26.4	27.0	28.2	27.7	26.1	25.1	24.4
Bare flat	23.8	23.6	23.5	23.8	24.9	27.5	28.0	28.2	27.4	25.6	24.7	23.6
Aluminium foil	24.9	24.7	24.5	24.4	24.7	25.3	25.7	26.3	26.4	26.0	25.8	25.4

Table 14.30 Effects of cultivation and drainage conditions on ethylene concentrations in soil air measured at IITA in 1975 (Unpublished data of Agbahungba and Lal, 1975)

Treatment	Ethylene concentration (ppm) v/v						
	10 Oct.	13 Oct.	15 Oct.	16 Oct.	17 Oct.	21 Oct.	
(a) Effect of ploughing and cultivation							
Cultivated	9.4	6.6	2.4	2.3	2.3	—	
Uncultivated control	0.0	0.0	0.0	0.0	0.0	—	
(b) Water table depth (cm)							
25	—	—	—	—	1.0	1.9	
15	—	—	—	—	1.9	4.2	
5	—	—	—	—	2.7	2.3	

unstable soils decrease oxygen concentration and increase that of CO_2 and ethylene in soil air. The ethylene concentrations in cultivated soils are higher than in uncultivated soil possessing better structure and higher in poorly-drained than well-drained soils (Table 14.30).

The effects of tillage-induced alterations in the oxygen diffusion rate (ODR) into an Alfisol at IITA, Ibadan, are shown in Table 14.31. These observations were made at different growth stages of maize during the first season of 1983. In general, the ODR was high for the unploughed treatments. Immediately after heavy rains on 47 days after seeding, the least ODR was measured in disc ploughed and rotovated and ridged treatments and the maximum in plots that had been chiselled to 50 cm depth. The low ODR in both ridged and rotovated treatments is due to the presence of a surface seal and crust. More data on the effect of tillage on soil aeration are needed for tropical environments.

IV. TILLAGE AND BIOLOGICAL ACTIVITY

Tillage affects biological activity both directly and indirectly: directly by destroying antecedant habitat and indirectly by influencing food diversity and availability, soil moisture and temperature regimes, and exposure to predators and parasites. The most notably affected by tillage are earthworms. Mechanical tillage invariably reduces earthworm activity. These effects have been discussed earlier in various chapters describing soil fauna.

V. TILLAGE AND SOIL CHEMICAL PROPERTIES

Tillage methods are also used to influence soil chemical properties. Soil inversion and mixing of crop residue into the upper soil layer significantly affects soil organic matter content. The mineralization rate of soil organic matter is increased by ploughing (Fig. 14.16a). Perhaps more important than the gross organic carbon is the biomass carbon content, which is often greater in unploughed than in ploughed soil. Ploughing and mixing render homogeneous concentrations of plant nutrients within the ploughed layer. In contrast, organic matter and plant nutrients are concentrated in the surface soil horizon in an unploughed culture (Fig. 14.16b) (Juo and Lal, 1979). Some of the relatively less soluble plant nutrients (i.e. lime and P) have a tendency to be concentrated in the top few centimetres of the unploughed soil. This is particularly so for soils with a high capacity to render immobile some plant nutrients (e.g., P in Oxisols).

Because of these differential concentrations of plant nutrients, fertilizer response can be different in ploughed and unploughed soils. In fact, the optimal rate, time and mode of application of fertilizers may be different for ploughed and unploughed cultures.

Table 14.31 Effects of tillage methods on oxygen diffusion rate into an Alfisol growing maize at IITA, Ibadan, Nigeria (Unpublished data of Lal, 1983)

Treatments	ODR ($\mu g\ cm^{-2}\ min^{-1}$) at different growth stages			
	I	II	III	IV
Unploughed and mulched	0.516a	0.652ab	0.722a	0.682ab
Unploughed and chiseled	0.599a	0.530b	0.674a	0.630ab
Moldboard ploughing and harrowing	0.458a	0.729a	0.637ab	0.653ab
Disc ploughing and rotovation	0.568a	0.588ab	0.696a	0.719a
Unploughed without mulch	0.441a	0.630ab	0.720a	0.665ab
Moldboard ploughing at rain ending	0.521a	0.577ab	0.565b	0.592ab
Moldboard ploughing harrowed and mulch	0.568a	0.604ab	0.703a	0.666ab
Moldboard ploughing and ridging	0.583a	0.637ab	0.704a	0.530a
LSD (0.05)	0.193	0.138	0.099	0.138

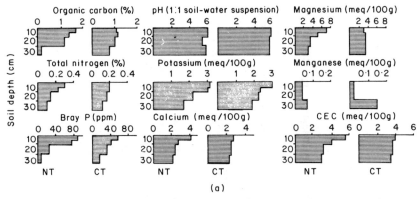

Figure 14.16a Nutrient profile of an Alfisol under different tillage systems (Armon et al., 1981) NT, no-tillage; CT, conventional

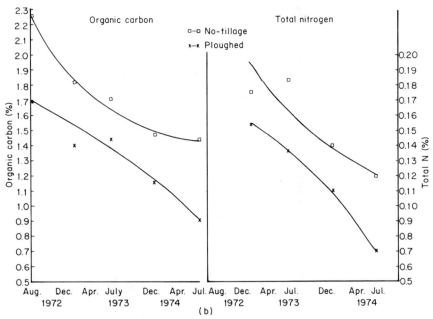

Figure 14.16b Effects of tillage methods on rate of decline of soil organic matter content (Lal, 1983)

VI. CROP RESPONSE TO SOIL TILLAGE

Just as the effects of tillage on soil physical properties differ among soils and environment, so also do crop response. If tillage alleviates some soil constraints that inhibit crop growth, yeilds are expected to be improved. If not, the effects of tillage on crop yields are often masked by other soil and crop management factors, and by environmental characteristics.

For easily compacted soils in arid and semi-arid regions, tillage and subsoiling usually have positive effects on crop yield. The favourable effects are due to many factors, foremost among them being water conservation. However, the results vary among soils. The most positive effects of tillage and subsoiling observed by Poulain and Tourte (1970) and Charreau and Nicou (1971) are attributed to better porosity, improved root penetration, and more soil-water storage in the root zone. The other extreme of this very favourable response observed in the Sahel region by Charreau and his colleagues is the research conducted in Angola by Almeida (1960) who compared yields of crops grown on plots ploughed in May or October to 25 or 10 cm depths or tilled by surface movement of soil into hills 20–25 cm high and 1 m apart. These variable tillage treatments had no effect on crop yields. In addition, there are other studies where yield response to tillage has either been inconclusive, or where similar yields were obtained with less intense tillage methods. For example, in Egypt Eldin and Maksoud (1977) observed that seedbed and subsequent crop yeilds were as satisfactory with a single pass of chisel plough to which a clod crusher was attached as those obtained by the customary two or more passes. Similarly, in Taiwan subsoilers used once, twice or three times with and without rotary tilling on various light and heavy soils produced inconclusive effects on growth and yield of sugarcane (Taiwan Sugar Research Institute, 1979). In northern Nigeria, Adeoye (1982) observed that yields of sorghum, maize and cotton were higher for ploughing depth of 15 than at 5 or 30 cm depths (Table 14.32a). The favourable effects on crop yields were due to increased macroporosity, water storage and rate of wetting of subsoil in treatments ploughed to 15 or 30 cm depth. The author observed, however, that the beneficial effects of increased porosity were short-lived and disappeared by the end of the first cropping cycle. In many semi-arid regions a tied-range system has produced better yields than deep ploughing (Marcartney et al., 1971; IITA, 1981; Fournier, 1967).

The negative effects of ploughing and of intensive tillage methods on crop yield are reported from semi-arid regions of Northern Territory and other tropical regions of Australia (Stonebridge et al., 1973; Tothill, 1974; Melville, 1978; McCown et al., 1980a, b; Bateman and Rowlings, 1980). the negative

Table 14.32a Crop grain yield (kg/ha) fro varying depths of soil tillage at Zaria, Nigeria (Adeoye, 1982). *Reproduced by permission of Elsevier Science Publishers*

Year	Crop	Tillage depth (cm)			LSD (0.05)
		5	15	30	
1974	Maize (cobs)	5120	5730	5760	565
1975	Maize (cobs)	4920	5700	5410	680
1975	Sorghum (heads)	2190	2290	2460	540
1976	Sorghum (heads)	2330	2710	2720	445
1976	Cotton (seed)	1680	1890	2170	480

Table 14.32b Effects of tillage depth on bulk density and porosity of a soil at Zaria, Nigeria (Adeoye, 1982). *Reproduced by permission of Elsevier Science Publishers*

Tillage depth (cm)	Depth of measurement (cm)	5 weeks after			15 weeks after		
		Bulk density (g/cm$_3$)	Total porosity (%)	Macro-porosity (%)	Bulk density (g/cm$_3$)	Total porosity (%)	Macro-porosity (%)
5	0–7	1.23	53.2	14.8	1.46	49.1	9.5
	10–20	1.49	47.3	9.1	1.50	46.9	8.5
15	0–7	1.20	53.6	16.1	1.41	49.4	9.4
	10–20	1.21	52.0	15.9	1.48	49.1	11.1
30	0–7	1.19	54.4	15.1	1.45	48.9	8.0
	10–20	1.18	53.9	15.6	1.47	49.7	10.3

Bulk density before ploughing = 1.45 and 1.54 g/cm^3 for 0–15 and 15–30 cm depth.

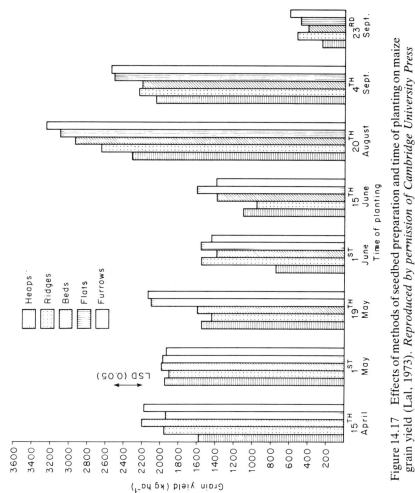

Figure 14.17 Effects of methods of seedbed preparation and time of planting on maize grain yield (Lal, 1973). *Reproduced by permission of Cambridge University Press*

Figure 14.18 Lodging of maize grown on ridges

response is due to crusting and higher soil temperatures on ploughed than on unploughed and mulched soil. In neighbouring Fiji, Chandra (1977) obtained favourable yields of sorghum without tillage.

In humid and subhumid regions of West Africa (Aina, 1982; Hayward *et al.*, 1980; Lal, 1983) favourable yields of grain crops have been obtained without primary or secondary tillage. The data in Fig. 14.17 are based on experiments conducted at IITA during 1970. Maize grown on ridged or mounded seedbeds invariably produced low yields due to supra-optimal soil temperatures, low soil moisture, and high incidence of lodging (Fig. 14.18). This does not imply, however, that all crops would respond the same way. The data in Table 14.33 show that in another study, in 1977, the yields of cowpea, soybean and cassava were also less on ridged than on flat or mulched seedbeds. The yield reduction by ridging was lower in cowpea than in maize, soybean or cassava because the cowpeas can tolerate higher temperatures and lower soil moisture. The differences among treatments were more marked in the first season than in the second season, which was milder.

Table 14.33 Crop grain yield (t/ha) effects of different mulches and methods of seedbed preparation on an Alfisol in western Nigeria (Lal, 1979). *Reproduced by permission of Williams & Wilkins Co.*

Treatment	Maize	Cowpea	Soybean	Cassava
(a) *First season, 1977*				
Black plastic mulch	5.4	0.64	1.9	—
Clear plastic mulch	4.7	0.67	1.2	—
Straw mulch	6.9	0.73	1.7	—
Ridges	4.9	0.46	0.2	—
Ploughed unmulched	5.5	0.62	1.6	—
Aluminium soil mulch	6.5	0.85	2.1	—
LSD (0.05)	1.3	0.26	0.7	
(b) *Second season, 1977*				
Black plastic mulch	2.2	0.60	1.3	7.9
Clear plastic mulch	2.3	0.76	1.3	9.7
Straw mulch	1.9	0.38	1.6	8.6
Ridges	1.7	0.45	1.4	2.3
Ploughed unmulched	1.6	0.60	1.5	3.2
Aluminium soil mulch	2.4	0.65	1.5	11.1
LSD (0.05)	1.0	0.38	0.4	2.9

The negative effects of tillage on easily compacted and eroded soils in the humid and subhumid regions are aggravated by the use of heavy machines. Lal (1984) reported that grain yield declined with continuous cropping of maize for 12 consecutive crops with mechanized farm operations. The rate of decline, however, was more on ploughed than unploughed soil. The grain yield of the sixth year was 3 and 1 t ha^{-1} for unploughed and ploughed watersheds, respectively. The reduction in maize yield on unploughed watersheds was due to soil compaction caused by combine harvesting and other vehicular traffic. On the other hand, the very significant decline in maize grain yield of the ploughed watershed was due to an overall degradation of soil productivity caused by erosion, compaction, decline in organic matter content, and increase in soil acidity. On a companion study with manual seeding and harvesting operations, however, the mean yield for 22 consecutive crops was 3.0 and 2.6 t ha^{-1} season^{-1} for unploughed and ploughed treatments, respectively (Fig. 14.19). The favourable crop response to elimination of tillage operations was due to better soil physical, chemical and biological environments (Lal, 1983).

Similar to the results obtained in the sahel, the data of Sar (1976) from Surinaam contradicts the conclusions by Lal regarding the favourable yields with elimination of tillage on soils of the humid tropics. Sar's data (Table 14.34) show that yields of various crops were greater on ploughed than on unploughed (minimum till) treatments. These results are similar to those of Curfs (1976) who also reported better crop yields on ploughed than on unploughed soils. One of the reasons for this discrepancy is the lack of standardized nomencla-

Table 14.34 Crop response to different tillage treatments in the humid region of Surinaam (Sar, 1976)

Crop	Grain yield (kg/ha) at 12% moisture content				
	Rotovation	Ploughing	Unploughed	Mean	S. E.
Cowpea	736	836	596	722	130
Maize	2728	2978	2414	2707	309
Groundnut	1901	1947	1675	1841	222
Sorghum	2734	3032	2513	2760	326
Soybean	1465	1517	1155	1379	210

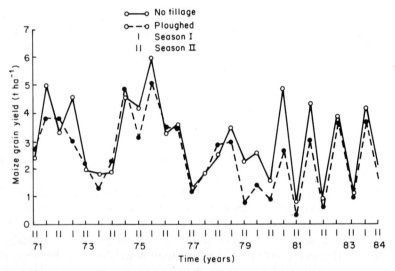

Figure 14.19 Effects of no-till and ploughing methods on maize grain yield for 27 consecutive crops of maize at IITA, Ibadan, Nigeria (unpublished data, Lal, 1984)

ture. Some researchers conduct experiments whereby the previous residue mulch is retained on the surface of the control unploughed plots. The soil and microclimate conditions are drastically different with crop residue mulch and without. It is important to indicate that most experiments where the crop response to ploughing has been overwhelmingly favourable, the unploughed control (no-till) had no crop residue mulch. This was the case in the studies by Charreau and Nicou in Senegal, Curfs in Nigeria and Sar in Surinaam. There thus is a need to standardize the terminology and define the package of agronomic practices used for different tillage systems.

VII. CONCLUSIONS

Tillage is an important tool by which man influences soil and microclimate with the objective of improving agricultural production. Ever since primitive man learned to manipulate soil by simple tools, tillage has been considered an essential component of farming and its usefulness had never been questioned. It is only within about the last 50 years that tillage and its consequences have been a concern to the scientific community. It is also only within the last half century that mechanized tillage operations with their drastic effects on soil and environments have been widely used. The effects of manual and animal-driven tillage implements are perhaps not serious enough to arouse scientific curiosity. In addition to its effects on soil, the high energy requirements of mechanized tillage operations have led to a closer examination of the need for this expensive, time-consuming technique.

The consequences of tillage on soils and environments are hard to generalize. The effects vary among soils, climates, antecedent properties, and the subsequent management. The short-term effects of tillage are often different from the long-term effects.

Mechanical loosening of soil by various tillage operations is often useful for arid and semi-arid regions where soils are easily compacted and there is little crop residue and stubble mulch to protect the soil. Although the beneficial effects are short-lived and disappear by the end of the first cropping cycle, deep ploughing and soil inversion improve porosity and infiltration, decrease run-off and erosion, and improve soil-water storage and crop yield. The beneficial effects of tillage are less in the presence of crop residue mulch than without it. The tied-ridge system is a useful and effective water conserving device on soils with stable structure and where the capacity of the ridge-furrow system is more than the rainfall received.

In the humid and subhumid regions, however, mechanical tillage often has more adverse than beneficial effects. Ploughing exposes structurally unstable soil to raindrop impact, decreases infiltration rate and increases run-off and soil erosion, increases the maximum soil temperature and evaporation, and often reduces crop yield. When there is an adequate quantity of crop residue mulch and chemical weed control, mechanical tillage for this region is economically wasteful and ecologically incompatible. For most soils and for grain crops, the intensity and frequency of tillage operations can be substantially reduced.

Tillage research in the tropics is still in its infancy. Most of the contradictory results are due to the lack of systematic investigations on this very important aspect of farming. The results of tillage research should always be linked to detailed analyses of soil properties, environmental factors, and crop characteristics. The results from soil and environments are not comparable without this supporting analytical data. For tillage inputs to be minimized, it is important to: (i) develop alternate systems for weed control, (ii) provide sources of mulch through appropriate cropping systems, (iii) alleviate soil compaction through biological means, (iv) develop suitable tools that can seed through mulch and on uneven ground, (v) and develop effective drainage and erosion control systems for soils prone to waterlogging and susceptible to erosion.

To describe the ecological effects of tillage is a complex question indeed. Tillage can be at least partially eliminated if alternative systems of weed control and compaction alleviation are developed. This is necessitated not only by high costs of fuel but also by the possibility of increased risks of soil erosion on ploughed lands in humid and subhumid environments. So far the development of sophisticated tillage tools has been a misdirected effort. Rather than curing the disease by understanding the causative factors, the scientific community has in the past been misled by the symptoms. With an understanding of the soil's constraints and potential and of crop requirements, it is possible to eliminate excessive, expensive, and often ecologically harmful tillage. This does not mean that ploughless agriculture would be possible on all soils and crops, but unnecessary soil manipulation and wasteful use of energy and resources should be minimized.

REFERENCES

Adeoye, K.B., 1982, Effect of tillage depth on physical properties of a tropical soil and on yield of maize, sorghum and cotton, *Soil Till. Res.*, **2**, 225–231.

Agbahungba, G.A., and Lal, R., 1975, Ethylene concentration in soil air after the rain, *Mimeo*, IITA, Ibadan, Nigeria, 30 pp.

Ahn, P.M., 1968, The effects of large-scale mechanized agriculture on the physical properties of West African soils, *Ghana J. Agric. Sci.*, **1**, 35–40.

Aina, P.O., 1979, Soil changes resulting from long-term management practices in western Nigeria, *Soil Sci. Soc. Amer. J.*, **43**, 173–177.

Aina, P.O., 1982, Soil and crop responses to tillage and seedbed configuration in southwestern Nigeria. *Proc. 9th ISTRO Conf.*, Osijek, Yugoslavia, pp.72–78.

Almeida, F.L.S. de, 1960, Effect of depth and time of ploughing on soil productivity, *Agron. Angol.*, No. 12, 31–60.

Aremu, J.A., 1980, Effect of different cultivation techniques on infiltration, *M.Sc. Thesis*, ABU, Zaria, Nigeria.

Armon, M., and Lal, R., 1979, Soil conditions and tillage systems in the tropics, *Proc. 8th ISTRO Conf.*, Hohenheim, Germany.

Armon, M., Lal, R., and Obi, M., 1982, Effect of tillage system on properties of an Alfisol in South-West Nigeria, *Ife J. Agric.*, **1982**, 1–15.

Bateman, R.J., and Rowlings, R.W., 1980, Conservation cropping . . . a new way of farming for the South Burnett?, *Queensland Agric. J.*, **106**(3): xiii-xvi.

Biscaia, R.C.M., 1982, Soil and water losses with the rotating-boom rainfall simulator for wheat-soybean rotation, *Proc. 9th ISTRO Conf.*, Osijek, Yugoslavia, pp.569–577.

Boone, F.R., Slager, S., Miedema, R., Eleveld R., 1976, Some influences of zero-tillage on the structure and stability of fine-textured river level soil, *Neth. J. Agric. Sci.*, **24**(2), 105–119.

Cannell, R.W., Davies, D.B., Mackney, D., and Pidgeon, J.D., 1971, The suitability of soils for sequential direct drilling of combine-harvested crops in Britain: a provisional classification, *Outlook Agric.*, **9**, 306–316.

Cannell, R.,Q., Davies, D.B., Mackney, D., and Pidgeon, J.D., 1978, The suitability of soils for sequential direct drilling of combine-harvested crops in Britain: a provisional classification. Outl. Agric., **9**, 306–316.

Chandra, S., 1977, Minimum tillage practices: possible applications in Fiji agriculture, *Fiji Agric. J.*, **39**, 39–46.

Charreau, C., 1970, Problems created by the growth and development of agriculture in the arid tropics, in *Traditional African Agricultural Systems and Their Improvement*, Ford Foundation/IRAT/IITA Seminar, 16–20 Nov., 1970, IITA, Ibadan, Nigeria.

Charreau, C., and Nicou, R., 1971a, Improvement of arable profile in sandy and sandy-clay soils of the West African dry tropical zone and its agronomic consequences. *Agron. Trop.*, **26**, 209–255.

Charreau, C. and Nicou, R., 1971b, Improvement of the arable profile in sandy and sandy-clay soils of the West African dry tropical zone and its agronomic consequences, *Agron. Trop.*, **26**, 903–978.

Charreau, C., and Seguy, L., 1969, Mesure de l' erosion et du ruissellement a sefe en 1969, *Agric. Trop.*, **XXIV 11**, 1055–1097.

Chopart, J.L., 1978, *Prolongation de la periode des labours de fin de cycle grace a des techniques d' economie de l'eau*, Institut Senegalais de Recherches Agricoles (I.S.A.R.), Bambey, Senegal, 65 pp.

Curfs, H.P.F., 1976, *System development in agricultural mechanization with special reference to soil tillage and weed control*, H. Veenman and Zonen, B.V., Wageningen, 179 pp.

De Datta, S.K., and Kerim, N.S.A.A.A., 1974, Water and nitrogen economy of rainfed rice as affected by soil puddling, *Soil Sci. Soc. Amer. Proc.*, **38**, 515–518.

Dunham, R.J., 1982, No-tillage crop production research at Samaru, *Proc. XI Ann. Conf. Weed Sci. Soc. Nigeria*, 1–6 March, 1982, Zaria, Nigeria.

Dunham, R.J., 1983, Soil management research in the Nigerian Savanna, *Mimeo*, IAR, Samaru, Zaria, Nigeria, 33 pp.

Dunham, R.J., and Aremu, A.J., 1979, Soil conditions under conventional and zero tillage at Samaru, *Proc. NSCC/SSSN Conference*, Kano, Nigeria.

Eldin, S. and Maksoud, A., 1977, A study on the performance of chisel plough/clod crusher combination. *Ann. Agric. Sci. Moshtohor*, **7**, 113–180.

Feller, C., and Milleville, P., 1977, Evolution of recently cleared soils in the region of Terres Neuves (eastern Senegal). Methods of study and evolution of the principal morphological and physico-chemical characteristics, *Cahiers ORSTOM, Biol.*, **12**(3), 199–211.

Fournier, F., 1967, Research on soil erosion and soil conservation in Africa, *African Soils*, **12**, 53–96.

Galloway, H.M., and Friffith, D.R., 1978, Tillage. Which is best for each soil type? Part 2, *Crops Soils Mag.*, **30**(9), 10–14.

Grohmann, F., 1960, Soil aggregate analysis, *Bragantia*, **19**, 201–213.

Grohmann, F., and Arruda, H.V. de., 1961, Effect of tillage practices on structure of terra-roxa legtima soil, *Bragantia*, **20**, 1203–1209.

Hamblin, A.P., 1980, Changes in aggregate stability and associated organic matter properties after direct drilling and ploughing on some Australian soils, *Aust. J. Soil Res.*, **18**(1), 27–36.

Hayward, D.M., Wiles, T.L., and Watson, G.A., 1980, Progress in the development of no-tillage systems for maize and soya beans in the tropics, *Outlook Agric.*, **10**(5), 255–261.

IITA, 1981, *IITA Research Highlights*, International Institute of Tropical Agriculture, Ibadan, Nigeria, pp.7–10.

Juo, A.S.R., and Lal, R., 1979, Nutrient profile in a tropical Alfisol under conventional and no-till systems, *Soil Sci.*, **127**, 168–179.

Kemper, B., and Derpsch, R., 1981, Results of studies made in 1978 and 1979 to control erosion by cover crops and no-tillage techniques in Parana, Brazil, *Soil Till. Res.*, **1**, 253–267.

Khatibu, A.I., Lal, R., and Jana, R.K., 1984, Effects of tillage methods and mulching on erosion and physical properties of a sandy clay loam in an equatorial warm humid region, *Field Crops Res.*, **8**, 239–254.

Klaij, M.C., 1983, Analaysis and evaluation of tillage on an Alfisol in a semi-arid tropical region of India, *Des Namiddags Te Vier UUr In De Aula Van De Landbouwhogechool*, Te Wageningen, The Netherlands.

Lal, R., 1976, No-tillage effects on soil properties under different crops in western Nigeria. Soil Sci. Soc. Amer. J., **40**, 762–768.
Lal, R., 1979, Soil and micro-climatic considerations for developing tillage systems in the tropics, in R. Lal, (ed), *Soil Tillage and Crop Production*, IITA Proc. Series 2, pp.48–62.
Lal, R., 1983, No-till Farming. *IITA Tech. Bull.* 2, Ibadan, Nigeria, 68 pp.
Lal, R., 1984, Mechanized tillage systems effects on soil erosion from an Alfisol in watersheds cropped to maize, *Soil Till. Res.*, **4**, 349–360.
Lal, R., 1985a, Mechanized tillage systems effects on properties of a tropical Alfisol in watersheds cropped to maize, *Soil Till. Res.* (In Press).
Lal, R., 1985b, A soil suitability guide for different tillage methods in the tropics, *Soil Till. Res.* (In Press).
Lawes, D.A., 1961, Rainfall conservation and the yield of cotton in Northern Nigeria, *Emp. J. Exp. Agric.*, **29**, 307–318.
Macartney, J.C., Northwood, P.J., Dagg, M. and Dawson, R., 1971, The effect of different cultivation techniques on soil moisture conservation and the establishment and yield of maize at Kongwa, Central Tanzania, *Trop. Agric.*, **48**, 9–23.
Mante, E.F.G., 1979, The effect of tillage depth and other farming practices on some physical properties of a Vertisol and on the yield of sorghum, *Proc. Appropriate Tillage Workshop*, 16–20 January 1979, IAR, Zaria, Nigeria. Commonwealth Secretariat, UK, pp. 95–106.
Maurya, P.R., and Lal, R., 1980, Effects of no-tillage and ploughing on roots of maize and leguminous crops, *Exp. Agric.*, **16**, 185–193.
McCown, R.L., Jones, R.K., and Peake, D.C.I., 1980a, Short-term benefits of zero-tillage in tropical grain production, in *Pathways to Productivity, Proc. Aust. Agron. Congr.* Queensland Agric. Coll., Lawes, April 1980, p.220
McCown, R.L., Jones, R.K. and Peake, D.C.I., 1980b, A ley farming system for the semi-arid tropics, in *Pathways to Productivity, Proc. Aust. Agron. Congr.*, Queensland Agric. Coll., Lawes, April 1980, p.188.
Medvedev, V.V., 1979, Some changes in the physical properties of chernozems upon cultivation, *Soviet Soil Sci.*, **11**(1), 70–78.
Melville, I.R., 1978, Conservation tillage for N.T.? A look at minimum tillage systems in North America and Queensland, *N.T. Rural News Mag.*, **3**(4), 14–15.
Nicou, R., 1974, The problem of caking with the drying out of sandy and sandy clay soils in the arid tropical zone, *Agr. Trop.*, **30**, 325–343.
Nicou, R., 1977, Le travail du sol dans les terres exondées du Sénégal: motivations; constraints, *Mimeo*, ISRA/CNRA, Bambey, Senegal, 50 pp.
Nicou, R., and Charreau, C., 1980, Mechanical impedance to land preparation as a constraint to food production in the tropics, *Proc. Soil Constraints Conference*, IRRI, Los Banos, Philippines, pp.372–388.
Opara-Nadi, O.A., and Lal, R., 1983, The effects of tillage methods on hydrological properties of a tropical Alfisol, *Proc. German Soc. Soil Sci. Conf.*, 4–10 Sept. 1983, Trier, Germany.
Pereira, H.J., and Jones, P.A., 1954, A tillage study in Kenya Coffee. Part I. The effects of tillage practices on coffee yields, *Emp. J. Exp. Agric.*, **22**, 231–240.
Pidgeon, J.D., and Ragg, J.M., 1979, Soil, climatic and management options for direct drilling cereals in Scotland, *Outlook Agric.*, **10**(1), 49–55.
Poulain, J.F., and Tourte, R., 1970, Effects of deep preparation of dry soil on yields from millet and sorghum to which nitrogen fertilizers have been added (sandy soil from a dry tropical area), *African Soils*, **15**, 553–586.
Sanchez, P.A., 1973, Puddling tropical rice soils. Parts I and II, *Soil Sci.*, **115**, 149–158.
Sar, T. van der, 1976, Tillage for dry annual crops in the humid tropics, *De Surinaamse Landbouw*, **24**, 93–98.
Sharma, D.L., Darra, B.L., Nathani, G.P., and Sharma, P.N., 1974, Effect of

continuous cultivation and irrigation under paddy-wheat rotation on the physical make-up of soils of chambal command area of Kota Rajsthan, *Indian J. Agric. Res.*, **8**(2), 77–82.

Sidiras, N., Henklain, J.C., and Derpsch, R., 1982, Comparison of three different tillage systems with respect to aggregate stability; The soil and water conservation and the yields of soybean and wheat on an Oxisol, *Proc. 9th ISTRO Conf.*, Osijek, Yugoslavia, pp.537–544.

Soane, B.D., 1975, Studies on some soil physical properties in relation to cultivations and traffic, in *Soil Physical Conditions And Crop Production*, Technical Bulletin, 29, Ministry of Agriculture, Fisheries and food, HMSO, London, pp.160–182.

Soane, B.D., and Pidgeon, J.D., 1975, Tillage requirements in relation to soil physical properties, *Soil Sci.*, **119**(5), 376–384.

Stibbe, E., and Ariel, D., 1970, No-tillage as compared to tillage practices in dryland farming of a semi-arid climate, *Neth. J. Agric. Sci.*, **18**, 293–307.

Stonebridge, W.C., Fletcher, I.C., and Lefroy, D.B., 1973, 'Spray-Seed': the Western Australian direct sowing systems, *Outlook Agric.*, **7**(4), 155–161.

Taiwan Sugar Research Institute, 1979, Studies on sub-soiling and earthing-in, *Taiwan Sugar*, **26**, 59–61.

Tallarico, L.A., Ferreiro, A.C., and Stillo, F.S., 1960, Effect of landuse on the state of aggregation of some pamps soils, *Rev. Invest. Agric. (Buenos Aires)*, **14**, 315–333.

Tothill, J.C., 1974, Experiences in sod-seeding Siratro into native speargrass pastures on granite soils near Mundubbera, *Trop. Grasslds*, **8**, 128–131.

Verma, H.N., Singh, R., Prihar, S.S., and Chaudhary, T.N., 1979, Runoff as affected by rainfall characteristics and management practices on gently sloping sandy loam, *J. Indian Soc. Soil Sci.*, **27**(1), 18–22.

Virgo, K.J., and Ysselmuiden, I.L.A., 1979, Cultivating soils of tropical steepland, *World Crops*, **31**(6), 210, 216–221.

Whiteman, P.T.S., 1975, Moisture conservation by fallowing in Botswana, *Exp. Agric.*, **11**, 305 314.

Chapter 15
Farming Systems

I	INTRODUCTION	618
II	SOILS UNDER SHIFTING CULTIVATION AND TRADITIONAL AGRICULTURAL SYSTEMS	619
III	SOILS UNDER INTENSIVE LAND-USE	629
IV	RECOMMENDED CULTURAL PRACTICES TO SUPPORT CONTINUOUS CROPPING	634
	A. Mulching	635
	1. Physical effects	635
	2. Chemical effects	654
	3. Biotic activity	655
	4. Crop response	657
	B. Crop Rotations and Fallowing	657
	1. Rotations	658
	2. Planted fallows	662
	C. Agroforestry	668
	D. Mechanization	669
	E. Pasture Management	671
	F. Fertilizers and Manures	672
	G. Management of Wetlands	672
V	CONCLUSIONS	680

I. INTRODUCTION

The terms tropical agricultural systems or tropical farming systems are used interchangeably and imply resource management strategies. Although the traditional tropical agricultural systems are invariably subsistence, based on low inputs, modernization strategies attempt to achieve high and economic agricultural production on a continuous basis. Improved agricultural systems should however, be sustainable, preserve the resource base, and maintain high environmental quality. New and improved systems must look beyond production and specifically address the issues of biological sustainability and ecological stability. The land resource is not only finite, it is also non-renewable. The objective of improved agricultural systems, therefore, lies in preserving and improving the land in productivity while maintaining biological diversity,

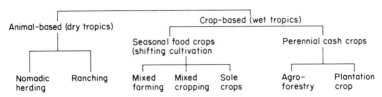

Figure 15.1 A generalized classification of tropical farming systems

environmental quality, and ecological stability. The emphasis on agricultural productivity alone is being short-sighted, and this narrow-minded approach has even more serious repercussions in tropical than in temperate zone ecologies.

The traditional tropical agricultural systems are complex, diverse, and, therefore, difficult to define and classify (Grigg, 1974; Webster and Wilson, 1980; Ruthenberg, 1980; Simmonds, 1984). Most traditional systems have built-in good security and ecological stability. A simplified classification of the traditional agricultural systems shown in Fig. 15.1 indicates varying degrees of intensity and diversity. Compound farming systems, as observed in densely populated regions of South-east Asia and Eastern Nigeria, are often developed to provide continuous and intensive land-use with a minimum of commercial inputs. Mixed cropping, growing more than one crop species simultaneously, is the common feature of the most traditional agricultural systems. It ensures a continuous flow of diverse food supply, and provides a continuous ground cover.

Agricultural systems involving the production of seasonal food crops are based on shifting cultivation and bush fallow rotations to restore soil fertility and preserve ecological stability. These systems have evolved over the millenia, and for subsistence landholders are the most desirable resource management strategies. The most important issue at hand is to develop sustainable food systems that have high production but do not upset the delicate soil - vegetation - climate balance.

II. SOILS UNDER SHIFTING CULTIVATION AND TRADITIONAL AGRICULTURAL SYSTEMS

Under shifting cultivation and the related bush fallow system soil is used for food production for a few years and then the land is fallowed for fertility restoration. This system has evolved throughout the tropics and has a built-in mechanism to restore soil physical, nutritional and biological properties. The use of land for seasonal crop production, either extensively by shifting cultivation or intensively by continuous land-use, results in decline in soil quality. The rate of decline, however, is less if the duration of cultivation is short in comparison with the fallow period. The rate of decline is also less with manual

Table 15.1 Effects of duration of cultivation on soil properties at 2 village sites in Senegal (Siband, 1972). *Reproduced by Permission of IRAT*

Soil property	Village Sare Bidji (1)			Village Diankancounda Oguneul (2)		
	5 Yrs	16 Yrs	80 Yrs	2 Yrs	15 Yrs	50 Yrs
(a) 0–10 cm layer						
Clay (%)	9.1	9.2	6.0	9.3	8.9	6.5
Silt (%)	5.5	4.3	2.9	4.7	4.2	3.3
Sand (%)	85.4	86.5	91.1	86.0	86.9	90.2
Available water (%)	5.7	4.5	4.4	5.6	5.8	5.0
Organic matter (%)	2.4	1.5	1.3	1.5	1.3	1.1
Total nitrogen (%)	0.73	0.54	0.43	0.67	0.49	0.41
pH (1:1 H_2O)	6.5	6.4	6.0	6.6	6.2	5.6
CEC (meq/100g)	4.7	3.5	2.6	3.5	3.2	2.3
Base saturation (%)	86	80	69	81	69	52
(b) 10–25 cm layer						
Clay (%)	12.3	12.0	15.8	12.5	15.6	12.5
Silt (%)	5.9	5.4	6.5	5.2	6.3	5.5
Sand (%)	81.8	82.6	77.7	82.3	78.1	82.0
Available water (%)	5.9	6.3	5.4	5.9	6.0	5.2
Organic matter (%)	1.1	0.8	0.9	0.9	0.9	0.7
Total nitrogen (%)	0.38	0.38	0.43	0.54	0.41	0.32
pH (1:1 H_2O)	5.9	6.3	5.4	5.9	6.0	5.2
CEC (meq/100g)	3.4	3.1	3.1	3.1	3.1	3.2
Base saturation (%)	58	70	40	56	67	37
(c) 25–40 cm layer						
Clay (%)	21.4	18.2	23.0	22.5	20.8	25.9
Silt (%)	8.6	7.3	8.8	8.7	8.3	10.0
Sand (%)	70.0	74.5	68.2	68.8	70.9	64.1
Available water (%)	7.3	4.2	5.0	6.5	5.5	5.2
Organic matter (%)	0.7	0.8	0.7	0.7	0.8	0.8
Total nitrogen (%)	0.37	0.41	0.44	0.44	0.41	0.44
pH (1:1 H_2O)	5.6	6.1	5.2	5.3	6.1	5.5
CEC (meq/100g)	3.6	3.0	3.7	4.1	3.6	4.5
Base saturation (%)	52	72	40	32	69	50

Soil of village 1 = Rouge; soil of village 2 = Beige.

than mechanized farming methods, and differs widely among soil and crop management practices, cropping systems and the land-use.

There are contradictory reports about the effects of shifting cultivation on soil properties and land degradation. Some reports indicate that shifting cultivators are destroying the forest resources and have accelerated soil erosion hazard. Others believe that shifting cultivation maintains ecological stability and is the only solution to the problem of rapid soil degradation. The controversy obviously exists because of the lack of quantitative data describing the effects of shifting cultivation on soils and environments. The effects vary widely among soils, rainfall, vegetation, durations of cultivation and fallow period, etc. The duration of cultivation is usually longer for soils of high inherent fertility. For soils of high fertility, the productivity decline to below the sustenance level occurs slowly, so that farmers can cultivate the same land for a number of years before returning it to the forest fallow.

The effects of shifting cultivation on soil physical properties have not been widely documented. Shifting cultivation in general has been described by many (Watters, 1971; FAO, 1974). The magnitude of change in soil properties obviously depends on the length of cultivation. In Casamance, Senegal, Siband (1972) characterized physical properties of farms on ferrallitic and tropical ferruginous soils under traditional cultivation systems for 2 to 5, 5 to 16 and 50 to 80 years of cultivation. The cultivation duration caused a marked deterioration of the surface soil with rapid fall in organic matter content, decrease in clay content, reduction in water-holding capacity and cation exchange capacity (Table 15.1). These studies were conducted in two villages (Sare Bidji and Diankancounda). Most changes in soil properties were observed in the 0–10 cm layer. In comparison with the 5 year cultivation cycle, the plots that had been cultivated for 80 years recorded a decrease of 34% in clay, 47% in silt, 45% in organic matter, 47% in nitrogen, 0.5 units in pH, 23% in available water capacity, 45% in CEC, and 17% in the base saturation. Similar changes were observed in the village Diankankounda (Table 15.1). Grain yield of upland rice was accordingly affected as a result of the degradation in soil quality (Table 15.2). For example, rice grain yield in plots cultivated for 16 years was only 60% of those cultivated for 5 years. The root density in the 30 cm layer in plots cultivated for 16 and 80 years was 83% and 47% of those cultivated for only 5 years. The rice straw yield in village Diankancounde with Beige soil were 2808, 2123, and 1135 kg/ha from plots cultivated for 2, 15 and 50 years, respectively. Similar results are reported by Charreau (1972) (Table 15.3). A watershed management experiment conducted at IITA compared soil properties under traditional farming with those under intensive mechanised agriculture. The rate of change in soil physical properties under traditional farming was much less than under intensive agriculture (Table 15.4). The most drastic change occurred with mechanized farming systems.

Table 15.2 Growth and yield of upland rice as affected by the duration of cultivation of a 'Rouge' soil in Senegal (Siband, 1972). *Reproduced by permission of IRAT*

Duration of cultivation (years)	Rice grain yield (kg ha^{-1})	Stover yield (kg ha^{-1})	Root growth 0–30 cm layer (mg/450 cm^3)
5	2078	2555	2286
16	1250	2698	1907
80	—	655	1067

Table 15.3 Evolution of soil properties (0–10 cm depth) with the duration of cultivation of a soil in southern Senegal (Charreau, 1972)

Duration of cultivation (years)	pH	Organic matter (%)	Total nitrogen (%)	C/N	Clay (%)	Silt (%)	Available water (%)	Exchangeable cations (meq/100g)			Base (%)
								Ca^{2+}	Mg^{2+}	K^+	
Forest control	6.3	16.5	0.90	18.3	11.1	5.0	4.1	5.0	1.7	0.07	87
3	6.0	13.8	0.79	17.5	10.2	4.7	4.7	2.7	1.2	0.07	81
12	5.9	11.6	0.68	17.0	10.5	4.7	3.7	2.2	1.0	0.04	86
46	6.0	6.8	0.43	15.8	9.0	4.3	2.8	1.4	0.5	0.04	53
90	5.9	5.0	0.35	14.3	7.4	4.2	3.3	1.0	0.5	0.04	64

Table 15.4 Effects of land clearing methods and post-clearing management on physical properties of an Alfisol (Unpublished data, Lal, 1981)

Treatment	Bulk density (g cm^{-3})				Penetrometer resistance (kg cm^{-2})				Equilibrium infiltration rate (cm hr^{-1})			
	1978	1979	1980	1981	1978	1979	1980	1981	1978	1979	1980	1981
Traditional farming	0.64	1.06	1.07	1.27	0.21	0.96	0.52	1.32	175	51	35	86
Manual clearing	0.68	1.17	1.17	1.39	0.20	1.40	0.75	1.19	383	20	10	7
Shearblade	0.70	1.19	1.37	1.38	0.26	1.0	1.84	2.19	102	11	14	5
Tree pusher/root rake	0.60	1.24	1.32	1.42	0.20	1.30	0.73	1.23	162	17	11	3

Table 15.5 Effects of shifting cultivation on soil bulk density (g/cm³) near Khon Kaen, Thailand (Takahashi et al., 1983)

Treatment	Depth (cm)	Jan. 1980	Aug. 1980	Jan. 1981	Oct. 1981	Jan. 1982
Shifting cultivation	0–5	1.148	1.065	1.146	1.045	1.145
	5–15	1.320	1.272	1.307	1.338	1.316
Continuous upland agriculture	0–5	1.148	1.042	1.077	1.082	1.131
	5–15	1.272	1.303	1.345	1.310	1.324
Bare fallow	0–5	1.058	1.031	1.158	1.169	1.247
	5–15	1.242	1.338	1.357	1.398	1.400
Forested control	0–5	0.989	0.910	0.874	0.948	0.988
	5–15	1.245	1.301	1.282	1.214	1.246

Shifting cultivation plot received no tillage operation.

In a 2 year study investigating the effects of shifting cultivation on physical properties of soils near Khon Kaen in Thailand, Takahashi et al. (1983) observed that in comparison with the forested control, soil bulk density increased in both shifting cultivation and intensive farming treatments (Table 15.5). The maximum increase was, however, observed in the bare plot, in which bulk density increased both in the surface and subsoil layers. Soil hardness was also the least in forested control. The surface soil compaction observed in the shifting cultivation treatment was attributed to the lack of tillage performed in these plots. It is not clear from the data in Table 15.5, however, as to the time after tillage operations in continuous cultivation and bare plot treatment that bulk density samples were taken. Soil compaction was caused by the clearing and burning operations. There were no differences in infiltration capacity and in equilibrium infiltration rate because of high soil

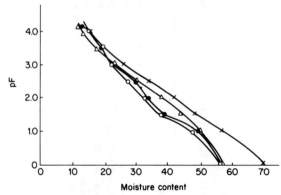

Figure 15.2 Soil moisture retention curves in shifting cultivation (○–○), intensive tillage, fertilizer and mulching (●–●), bare fallow (△–△), and natural forest cover (×–×) (Takahashi et al., 1983).

Table 15.6 Erosion and run-off from soil under shifting cultivation in the Philippines (Kellman, 1969). *Reproduced by permission of the* Singapore Journal of Tropical Geography

	Run-off (%)		Sediment loss (g day^{-1})	
	Cropping period	Post-harvest	Cropping period	Post-harvest
Logged-over forest	—	—	—	—
New maize swidden	1.52	0.86	3.03	0.65
New rice swidden	1.08	0.42	1.45	0.37
2-year-old maize swidden	1.78	0.69 (4.08)	12.05	9.81
12-year-old rice swidden	11.64	6.73 (14.15)	119.31	6.32

Numbers in parentheses refer to post cropping period.

variability. The soil moisture retention characteristics also indicated that forested soil had higher total porosity and macroporosity than the cultivated treatments (Fig. 15.2).

Soil erosion and water run-off are often less from land under shifting than intensive cultivation. This is especially true for the fallow phase. However, run-off and soil erosion increase with increase in duration of cultivation. In the Philippines Kellman (1969) studied water run-off and soil erosion during cultivation and fallow phases. The data in Table 15.6 show that water run-off increased from 1.08 per cent in the first year after clearing to 11.64 per cent in the 12th consecutive year of cultivation. The run-off losses were even more severe in the post-cropping period. Similarly, soil erosion losses increased from 1.45 g/day in the first year to 119.31 g/day in the 12th year. The estimated annual loss of N through erosion was 0.575 g/m^2 in the first year of cultivation of maize and rice. It increased to 2.205 g/m^2 in the second year under maize and to 23.55 g/m^2 in the 12th year under rice. Lal (1981) observed negligible run-off and soil erosion from a traditionally cultivated plot both during cropping and fallow phases. In this study conducted at Ibadan, Nigeria, the high tree cover and mulch from the slashed vegetation provided adequate protection against accelerated erosion (See Table 13.24 of Chapter 13). Takahashi *et al*. (1983) reported that in 1980 soil erosion under shifting cultivation was 81.9 and 56.8 per cent of that under continuous cultivation and bare fallow treatments, respectively (Table 15.7). However, soil erosion under shifting cultivation was 18.6 times greater in 1980 and 2.4 times greater in 1981 than under the forested control. Because of below average rainfall, soil erosion was less in 1981 than in 1980 (Table 15.7). In northeastern India, Mishra and Ramakrishnan (1983) observed that soil erosion increased with decrease in the length of the fallow cycle from 10 to 5 years. When the fallow cycle was shortened, even terraces had no effect in curtailing soil erosion hazard.

From the scanty literature available, it is difficult to quantify the reduction in crop yields by the deterioration in soil physical properties during the cropping

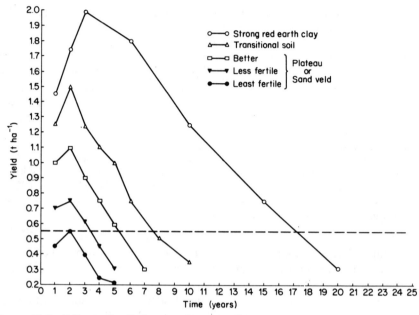

Figure 15.3 Effects of soil characteristics on yield decline under shifting cultivation (Allan, 1965).

Table 15.7 Soil erosion (m³ ha⁻¹) under shifting cultivation and continuous cropping at Khon Kaen, Thailand (Takahashi *et al.*, 1983)

Treatment	1980	1981
Shifting cultivation	87.6	6.9
Continuous cropping	107.0	6.4
Bare fallow	154.3	25.9
Forested control	4.7	2.9
Total rainfall (mm)	1542	1009

phase. The rate of decline in crop yield under shifting cultivation varies among soils, crops, and the climate. In East Africa Allan (1965) observed that whereas it took 17 years of continuous cultivation of a soil with good physical characteristics and gentle slopes for the maize yield to fall below the lower economic limit, it took only 1 to 2 years on soils of poor physical characteristics (Fig. 15.3). In Senegal Siband (1972) observed significant reductions in yield of upland rice with increasing number of years under cultivation (Fig. 15.4). Lal (1983) compared soil loss/grain yield ratio for shifting cultivation with other intensive methods of land management (Fig. 15.5). Furthermore, the soil loss/yield ratios were high in the first year of cultivation. Because the soil erosion was negligible in shifting cultivation treatment, the soil loss/grain yield ratio was lower by several orders of magnitude. Similar trends were observed in

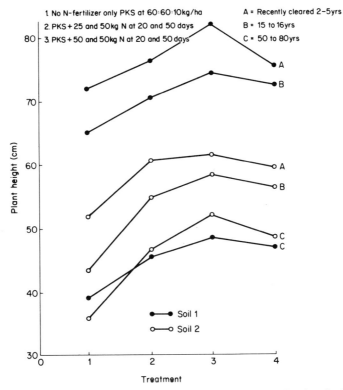

Figure 15.4 Yield decline in upland rice with duration of cultivation in Senegal (Siband, 1972).

Table 15.8 Maize grain yield at Khon Kaen, Thailand, under shifting cultivation and with different level of inputs (Kamanoi et al., 1983)

Treatment	Input			Grain yield (kg ha^{-1})	
	Tillage	Weeding	Fertilizer	1980	1981
Shifting cultivation	No	Yes	No	4001±37	3181±441
	No	No	No	3963±296	3263±193
Intensive cultivation	Yes	Yes	Yes	5134±298	6478±224
	Yes	No	Yes	4469±422	6448
	Yes	Yes	−N	4453±419	5634±5
	Yes	Yes	−P	4511±278	4809±238
	Yes	Yes	−K	5182±156	5969±156
	Yes	Yes	No	4123±203	4084±98

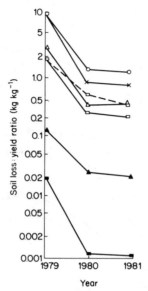

Figure 15.5 Soil loss: grain yield ratio for different management systems (Lal, 1983)

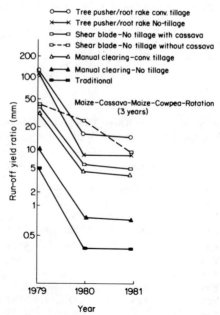

Figure 15.6 Run-off: grain yield ratio for different management systems (Lal, 1983)

the water run-off/yield ratio (Fig. 15.6). In Thailand Kamanoi *et al.* (1983) reported less yield from shifting cultivation than from intensively managed treatments (Table 15.8). Mechanical tillage increased maize yield by 120 kg/ha in 1980 and 900 kg/ha in 1981 because the soil compaction was greater in the second year. In the shifting cultivation treatment, grain yield declined by about 20 per cent in the second year.

III. SOILS UNDER INTENSIVE LAND-USE

Soil properties are more drastically influenced by continuous and intensive farming than by shifting cultivation. The collapse of many large-scale land development schemes is partly attributed to rapid deterioration of soil properties. Many researchers have quantified the effects of continuous cultivation on chemical and nutritional properties of the soil. Fauck *et al.* (1969) reported from Casamance, Senegal, that continuous cropping decreased soil organic matter content by 25 to 50% in the first 2 years. Soil pH decreased from 6.4 to 6.2, 5.9 and 5.0 after 2, 6 and 15 years of cultivation. Their data indicated that a new state of equilibrium is attained under cultivation (Table 15.9). Similar results on plateau soils of continental Casamance were observed by Diatta (1974). In Ivory Coast Buanec (1972) conducted a 10 year experiment on a 10 hectare watershed comprising ferrallitic soils and observed no apparent decrease in potential fertility, although crop yields varied markedly depending on the rainfall amount. Similar results were observed in Ghana by Ofori *et al.* (1969) and Ofori (1973). In Ivory Coast Feller and Milleville (1972) observed that continuous cropping resulted in a decline in moisture retention at pF 2.5, 3.0 and 4.2 (Fig. 15.7). The differences were more marked at pF 2.5 than at other pF values. These differences occurred in spite of relatively little change in the clay content, implying thereby a decrease in organic matter content and an overall collapse of soil structure. At Grimari experiment station in the Central African Republic, Cointepas and Makilo (1982) compared soil properties in plots under cultivation for 45 years to those of adjacent virgin plots. Reduced macroporosity (30%) and decreased structural stability were observed in the cultivated plots. In addition, exchangeable Ca and the Mg were found to be decreased by about 40% and soil pH by 0.7 units. These changes occurred mainly during the first few years of cultivation. In western Nigeria Aina (1979)

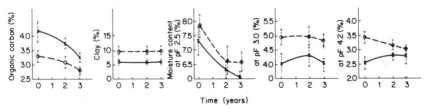

Figure 15.7 Effects of continuous and intensive landuse on ferruginous soils in Ivory Coast on organic carbon, clay, and moisture retention characteristics (Feller and Milleville, 1972)

Table 15.9 Effects of continuous cultivation on soil properties in Casamance, Senegal (Fauck et al., 1969). *Reproduced by permission of IRAT*

Property	Soil	Forest control	Cultivated
Structural Instability Index (i_s)	Beige	0.61–0.72	1.93–2.14
	Rouge	0.49–0.56	1.25–1.57
Soil permeability (cm h^{-1})	Beige	2.1–2.5	1.3–1.7
	Rouge	2.2–2.7	2.4–2.6
Exchangeable Ca^{2+} (meq/100g)	Beige	8.04–10.14	1.47–2.05
	Rouge	4.21–5.38	1.28–2.05

observed significant declines in the physical properties of cultivated soil in comparison with the forested control (Table 15.10). Measurable declines were recorded in water stable aggregates, aggregate stability, total and air-filled porosity, and hydraulic conductivity. On the other hand, cultivation increased soil bulk density.

The effects of continuous cultivation on soil physical properties for soils in western Nigeria have been reported by Lal (1985). The results show that soil physical properties and chemical constituents declined substantially in 6 years of continuous cultivation. The rate of decline, however, depend on soil and crop management. For example, the cumulative infiltration 2 hours after the beginning of the test for no-till and ploughed watersheds decreased from 77 and 65 cm in 1976 to 38 and 28 cm in 1978, 28 and 9 cm in 1979, and to 12 and 5 cm in 1980, respectively, (Fig. 15.8). This drastic decline in infiltration capacity

Table 15.10 Effects of cultivation of Iwo soil in western Nigeria on soil properties (Aina, 1979). *Reproduced from* Soil Science Society of America Journal, *Volume 43, 173–177, by permission of the Soil Science Society of America*

Soil properties	Iwo			LSD (0.05)
	Forest control	Cultivated		
		F_0	F_1	
Water stable aggreate (>2.36 mm, %)	85	12	9	12
Aggregate stability (N^{-1})	0.033	0.143	0.148	0.032
Sand (%)	69.2	75.7	74.0	4.5
Silt (%)	10.9	7.1	8.7	3.1
Clay (%)	19.9	17.3	17.3	6.2
Bulk density (g cm^{-3})	1.24	1.56	1.58	0.07
Total porosity (%)	53.3	43.2	41.3	3.1
Air filled porosity (60 cm suction)	20.5	8.1	7.6	2.9
Hudraulic conductivity (cm hr^{-1})	120.2	17.1	15.9	8.8

F_0 = no fertilizer; F_1 = with fertilizer.

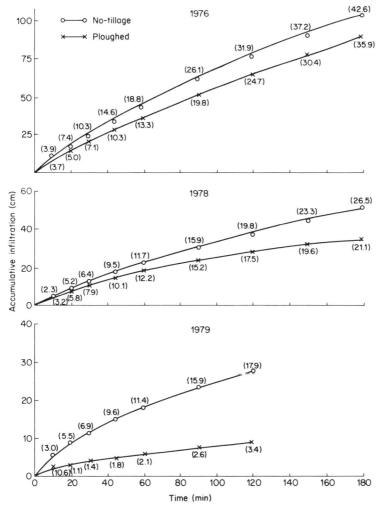

Figure 15.8 The effects of continuous cultivation on water infiltration rate of an Alfisol grown to 12 consecutive crops of maize. Figures in parentheses refer to the standard cover of the mean (Lal, 1984)

indicates structural collapse and elimination of transmission pores as a result of vehicle traffic and soil compaction. Similarly affected were the soil water retention characteristics. The data in Table 15.11 show that changes in water retention with the duration of cultivation were differently affected by the two tillage methods.

Erosion on some soils of the tropics is generally observed to be high during the first few years after land clearing. That is why soil erodibility peaks 2 to 3 years after clearing and then declines with subsequent reduction in the amount of easily erodible material. During a 6-year comparison Lal (1981) observed the maximum soil erodibility to be between the second and the third years after

Table 15.11 Changes in soil water retention in the 0 to 10-cm layer at different times after tillage treatments were imposed

Matric potential (MPa)	Initial data, 1975		2 years		5 years		6 years	
	No-tillage	Ploughed	No-tillage	Ploughed	No-tillage	Ploughed	No-tillage	Ploughed
0	34.7±6.3	39.5±3.5	—	—	36.2±3.0	35.1±2.9	41.3±4.8	30.9±7.4
0.03	32.1±5.9	35.2±2.8	—	—	27.5±2.9	27.0±2.3	26.7±6.8	23.1±2.6
0.1	14.7±3.7	17.7±2.1	21.3±4.2	17.6±4.3	24.6±3.2	23.2±1.8	17.5±2.7	13.8±1.8
1	8.4±3.8	11.2±1.1	11.0±3.0	9.2±2.1	11.5±1.7	11.1±1.6	10.6±1.3	9.2±1.3
3	7.0±3.0	9.7±1.2	9.4±2.4	7.8±3.1	8.5±1.5	8.1±1.1	—	—
15	5.4±2.8	8.1±1.0	7.2±3.0	6.8±2.0	7.3±1.2	7.0±0.8	—	—

There were 20 measurements per stated mean.

Table 15.12 Effect of cultivation on soil erosion in the Sahel (Charreau, 1972)

Country	Location	Period of study	Slope (%)	Rainfall (mm)	Soil erosion (t ha^{-1})		
					Forest	Arable land	Bare soil
Upper Volta (Burkina-Fasso)	Ouagadouogou	1967–1970	0.5	850	0.1	0.6–0.8	10–20
Senegal	Séfa	1954–1968	1–2	1300	0.2	7.3	21.3
Ivory Coast	Bouaké	1960–1970	4.0	1200	0.1–0.2	0.1–26.0	81–30
Ivory Coast	Abidjan	1954–1970	7.0	2100	0.03	0.1–90.0	108–170

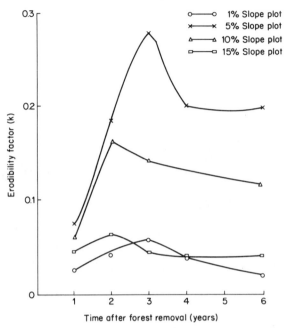

Figure 15.9 Changes in erodibility of an Alfisol with time after land clearing and development (Lal, 1981). *Reproduced by permission of Elsevier Science Publishers*

land clearing, followed by a measurable decline (Fig. 15.9). The decline in erodibility observed on bare plots may take a long time in cultivated plots with protective vegetation cover. This implies an initial increase in erosion with continuous cultivation for a variable period of time depending on soil properties, rainfall patterns and soil and crop management systems. The data in Table 15.12 from Charreau (1972) support the argument of lower erosion from cropped than bare land.

IV. RECOMMENDED CULTURAL PRACTICES TO SUPPORT CONTINUOUS CROPPING

Improved farming systems to enable continuous cropping are based on those cultural practices, component systems or subsystems that alleviate resource/production constraints due to biophysical and socio-economic factors. Appropriate subsystems for different ecologies are recommended to maintain a favourable level of soil physical properties and microclimatic conditions. Some relevant cultural practices/subsystems are mulching, crop rotations and planted fallows, fertilizer and manuring, pasture management and controlled grazing etc. Methods of deforestation and tillage, important subsystems, have been discussed in separate chapters. In addition to the cultural practices *per se*, the mode of their implementation also affects soil physical properties and microclimate. Mechanized farm operations have a greater effect on soil physical properties than manual methods.

A. Mulching

'Mulch' means a layer of dissimilar material separating the soil surface from the atmosphere. The word dissimilar implies that another layer of soil cannot be considered as mulch as is mistakenly used in the term 'dust mulch'. Mulch may be organic (crop residue, straw, etc.) or inorganic (plastic sheets), and it may be grown *in situ* or brought-in. Some common examples of the *in situ* produced mulch are the previous crop residue mulch left on the soil surface or that produced by a specially grown cover crop. A large variety of materials are used as brought-in mulch, e.g., straw, sawdust, petroleum and plastic products, and even gravels. The sub-surface application of organic or inorganic materials, even as a uniform layer, does not fall under the category of mulch.

The practice of 'mulching' has been widely used as a management tool for centuries in many ancient civilizations, and its effects on soils, crops and environments have been extensively studied (Lal, 1979). Mulch has a buffering effect and dampens the influence of environmental factors on soil. The magnitude of this buffering effect depends on the quantity, quality, and durability of the material. Also important are the ambient climatic conditions and the soil properties. The effects of mulch on soil, therefore, depend on the climate, soil and the mulch material interaction and differ from region to region. Crop residue mulches are of special importance in the tropics and affect physical, chemical and biological soil properties. These are interactive effects, often difficult to quantify separately. For example, physical effects on soil structure and hydrological properties are influenced by alterations in soil biological properties. Mulching influences soil and microclimate directly by intercepting rainfall and solar radiation, and by adding plant nutrients. Indirectly, mulching influences soil physical and chemical properties by altering soil moisture and temperature regimes, and the activity of soil flora and fauna.

1. Physical effects

(i) Texture and structure: The recurrent use of mulch for many years can lead to some alterations in soil texture through its indirect effects on run-off, soil erosion and clay eluviation, and through alterations in soil biological properties. Lal *et al.* (1980) observed higher clay and silt content in the 0 to 5 cm layer of plots receiving repeated applications of 2 to 12 t ha^{-1} of mulch applied twice in a year (Table 15.13). These differences in soil texture were observed within 18 months after initiating the experiment. The geometric mean of the particle size distribution within 18 months was 0.330 mm for the soil from the control plots and 0.191 mm for the plots receiving 12 t ha^{-1} season^{-1} mulch rate.

Soil structural properties are influenced more readily than textural properties by mulch. Mulch influences structural properties directly by preventing

Table 15.13 Effects of mulch rate on particle and pore-size distribution of 0 to 5 cm layer of Alfisol 18 months after initiating the experiment (Lal et al., 1980). *Reproduced from Soil Science Society of America Journal, Volume 44, 827–833, by permission of the Soil Science Society of America*

Mulch rate (t ha^{-1})	Texture (%)			Porosity (%)				Mean weight diameter (mm)
	Clay	Silt	Sand	Total	Macro	Meso	Micro	
0	7.4	8.9	83.7	48.1	17.9	7.1	23.1	1.2
2	8.2	12.5	79.3	50.2	20.6	7.2	22.4	1.4
4	9.1	12.7	78.2	54.8	27.7	8.4	18.7	2.1
6	11.1	11.3	76.9	55.1	28.9	8.3	17.9	2.0
12	11.2	12.7	76.1	59.0	37.0	8.2	13.2	2.4
LSD (0.05)	0.8	2.1	2.8	3.8	5.3	n.s.	4.2	—

raindrop impact and indirectly by promoting biotic activity. Measurable effects of mulch are observed on total porosity and macroporosity, and on percentage of water stable aggregates and on aggregate size distribution. The data in Table 15.13 also show that macroporosity (>30 µm pores) was higher in the surface layer of the soil receiving 4, 6 and 12 t ha^{-1} mulch. The reserve was the trend with the microporosity. The mean weight diameter of water stable aggregates also increased with increase in mulch rate (Table 15.13). Lal et al. (1980) observed that whereas 0.25 to 0.5 mm was the dominant aggregate size for unmulched control and 2 t ha^{-1} treatment, 1 to 4 mm was the predominant aggregate size for the 4 to 6 t ha^{-1} and 4 to 8 mm for the 12 t ha^{-1} mulch rates. The following empirical relationships were developed between mulch rate and soil structural properties:

Per cent water stable aggregates (>0.05 mm) = $42 + 7.36x - 0.41x^2$ $r = 0.98^{**}$
   Dispersion ration = $26.9 \exp(-0.09x)$ $r = 0.97^{**}$
   Erosion ratio = $71.9 \exp(-0.09x)$ $r = 0.96^{**}$

where x is the mulch ratio in t ha^{-1} season^{-1}.

Petroleum products are also used in the form of emulsions on the soil surface to influence structural properties (Soong, 1980). Use of such products, being expensive, is limited to plantation and high-value cash crops. De Vleeschauwer et al. (1978) reported measurable effects of different petroleum emulsions on soil structural and textural properties while chemical properties were only slightly changed. The comparison of synthetic soil conditioners with organic mulches (Table 15.14) shows that the latter are either more or equally effective in stabilizing soil aggregates and increasing macroporosity as the petroleum by-products. The largest mean weight diameter of aggregates was observed for the crop residue mulch treatment.

(ii) Water transmission: By influencing the macroporosity and stability of structural aggregates, mulching improves water movement within the soil profile. However, the quantity of mulch required for maintenance of favourable infiltration capacity and structural stability depends on soil and environmental factors. Field experiments conducted at IITA and elsewhere in the tropics have shown that both the soil water 'sorptivity' and the 'transmissivity' parameters of Philip's infiltration equation depend on mulch rate and duration (Table 15.15). One year after applying the residue mulch, the equilibrium infiltration rate was 32, 52, 60, 73 and 97 cm hr^{-1}, respectively, for 0, 2, 4, 6 and 12 t ha^{-1} season^{-1} mulch rates (Lal et al., 1980). Similarly affected was the saturated hydraulic conductivity. Soil conditioners and other petroleum by-products are also used to alter water transmission properties. De Vleeschauwer et al. (1978) observed that although the equilibrium infiltration rate was increased by some synthetic soil conditioners (PAM), the maximum rate was observed in the straw mulch treated plot.

(iii) Moisture regime: Moisture regime is influenced by the effects of mulch

Table 15.14 Effects of synthetic soil conditions on textural and structural properties of an Alfisol (De Vleeschauwer et al., 1978). *Reproduced by permission of Catena Verlag*

	Texture			Geometric mean particle size (mm)	Porosity (%)					Mean weight diameter of aggregates (mm)
	Sand	Silt	Clay		Total	Macro	Meso	Micro		
Control	79.7	10.4	9.9	0.265	53.5	23.5	19.0	11.0		2.5
Straw mulch	71.6	16.3	12.1	0.165	62.3	33.5	16.5	12.3		3.7
No-till	78.9	11.4	9.7	0.236	47.1	19.6	17.5	10.0		3.3
Soil penetration	71.4	16.9	11.7	0.172	54.6	15.6	14.2	14.8		2.0
Bitumen	74.1	14.4	11.5	0.184	61.2	29.2	16.0	16.0		3.5
Polyacrylamide	71.7	16.5	11.8	0.166	62.0	33.5	14.0	14.5		3.2
LSD (0.05)	5.4	3.9	2.1	0.051	6.7	9.2	3.8	3.4		0.31

Soil penetrant is a blend of 50% polyoxyethylene esters of mixed alcohol. Bitumen is applied as 50% emulsion of bitumen in water.

Table 15.15 Effects of mulch rate on water transmission properties of an Alfisol (Lal et al., 1980) *Reproduced from* Soil Science Society of America Journal, *Volume 44, 827–833, by permission of the Soil Science Society of America*

Mulch rate (t ha^{-1})	Soil water sorptivity	Soil water tranmissivity	Saturated hydraulic conductivity (cm h^{-1})
0	0.32	5.56	30a*
2	0.57	7.81	45a
4	0.67	7.50	70b
6	0.84	10.21	132c
12	1.05	15.36	129c

* Duncan's Multiple Range Test.

on soil moisture retention (pF) properties, infiltration rate, soil evaportion and water condensation at night due to temperature reversals. The effects of mulch on soil evaporation depend on the mulch material and the soil moisture status. Plastic sheets (Fig. 15.10) used as mulch reduce soil evaporation more than straw mulch, especially during the first stage of soil-water evaporation when the soil is very moist.

Mulch also influences soil moisture retention properties through its effect on pore size distribution and soil structure. The data in Table 15.16 show that a high mulch rate increases soil-water retention at low suctions through increases in macro-pores and inter-aggregate pores. A comparison between plastic and straw mulches on soil moisture reserves under maize is shown in Fig. 15.11 (Maurya and Lal, 1981). Less difference in soil moisture reserves during the first season was due to frequent rainless periods. In the second season, however, plastic mulched plots had significantly more soil moisture reserves than straw mulch or unmulched treatments (Maurya and Lal, 1980). If, however, plastic mulches are not fixed properly and the rainwater does not enter into the soil, those plots may have less moisture than straw mulch or

Table 15.16 Effects of mulch rate on moisture retention properties of 0 to 5 cm layer of an Alfisol (Unpublished data of Lal and Akinremi, 1983)

Mulch rate (t ha^{-1})	Moisture retention (% w/w) at different suctions (bars)			
	0.03	0.06	1	15
0	27.3	26.1	5.5	3.2
2	29.1	26.1	6.5	5.1
4	29.6	28.8	5.2	3.7
6	29.6	28.5	5.4	4.2
8	31.2	30.5	7.3	5.8

Figure 15.10 The use of plastic sheet as mulch is an effective water conservation technique

Table 15.17 Slope–soil erosion relationship for different mulch rates for slopes ranging between 1 and 15 per cent (Lal, 1976)

Mulch rate (t ha^{-1})	Equation regression	r	Mean soil loss (t ha^{-1})
0	$Y = 11.8\ S^{1.13}$	0.81	76.6
2	$Y = 0.5\ S^{0.87}$	0.35	2.4
4	$Y = 0.07\ S^{1.05}$	0.57	0.37
6	$Y = 0.01\ S^{1.0}$	0.46	0.09

Y = estimated soil erosion (t ha^{-1}).
S = slope (%).

Table 15.18 Effects of crop residue mulch on oxygen diffusion rate (μg cm^{-1} mm^{-1}) under maize

Date	No-till		Ploughed	
	Mulch	Unmulched	Mulch	Unmulched
5.5.1983	0.690	0.456	0.409	0.378
14.5.1983	0.749	0.597	0.716	0.564
24.5.1983	0.739	0.512	0.639	0.514

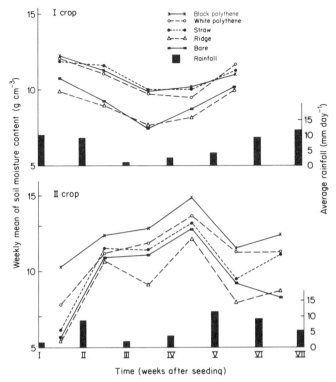

Figure 15.11 Soil moisture reserves under maize for straw and plastic surface mulches at IITA, Ibadan (Maurya and Lal, 1978)

unmulched treatments (Lal, 1979). The data in Tables 15.17 and 15.18 on neutron count ratio under different mulch treatments indicate higher soil moisture reserves under straw than under plastic mulch or aluminium treatment. In Zanzibar, on the other hand, Khatibu *et al.* (1984) observed higher soil moisture reserves under plastic mulch treatments (Figs. 15.12 and 15.13) for both long and short rainy seasons.

A comparison of the effect of straw mulch or synthetic soil conditioners on soil moisture reserves for an Alfisol was made by De Vleeschauwer *et al.* (1978). Their data (Fig. 15.14) show that soil moisture content in the surface layer of the straw mulch treatments was consistently more than in other treatments. The mean percentage gravimetric moisture content was 14.4 and 14.7 for the mulch and no-till treatments as compared to 10.4, 10.0, 9.2 and 9.0 per cent for soil penetrant, polyacrylamide, control and bitumen treated plots, respectively.

The literature survey of the effects of mulching on soil moisture reserves indicates seemingly contradictory results. It seems that the effects of mulch on soil moisture reserves vary among soils, mulch material, rainfall amount and frequency, and the crops grown. Vigorous crop growth, often caused by mulch

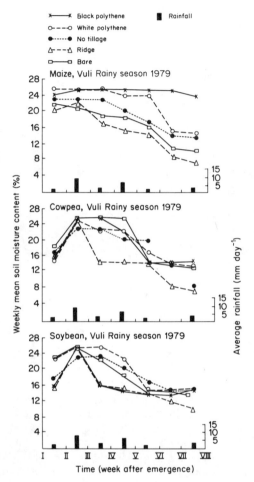

Figure 15.12 Effects of mulch materials and tillage methods on soil moisture reserves in the Vuli season in Zanzibar (Khatibu *et al.*, 1984). *Reproduced by permission of Elsevier Science Publishers*

treatments, extracts more soil water than a crop at poor stand. There are thus often no systematic trends in soil moisture reserves due to mulch treatments.

(iv) Soil temperature: Similarly to soil moisture, the soil temperature regime is variably influenced by mulch treatments. Mulch increases soil temperature during cooler weather and decreases it during hot spells. The effects on soil temperature also vary among soils, the mulch material used and rate of its application. In general, mulching has a damping effect on the amplitude of the diurnal fluctuations in soil temperature. The data in Figs. 15.15 and 15.16 by Lal (1974) show that straw mulch significantly decreased maximum soil temperature. Temperature differences of 8 to 10°C at 5 cm depth are not uncommon. The minimum soil temperature, on the other hand, was lower for

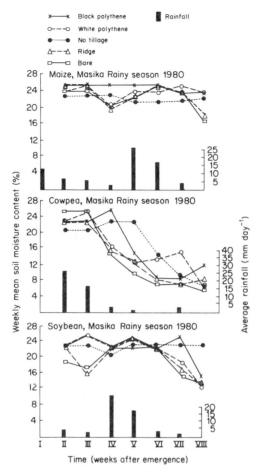

Figure 15.13 Effects of mulch materials and tillage methods on soil moisture reserves in the Masika season in Zanzibar (Khatibu et al., 1984). *Reproduced by permission of Elsevier Science Publishers*

the unmulched than mulched treatments. The data on Alfisols by Lal et al. (1980) shows that both the amplitude and the periodicity of soil temperature were affected by the rate of mulch application. The differences in the maximum temperature were 3.3., 4.1., 4.5 and 5.4°C for 2, 4, 6 and 12 t ha^{-1} mulch rates treatments, respectively, compared with the unmulched control. In semi-arid Australia, experiments conducted by Jones and McCown (1983) on Alfisols at Katherine in Northern Territory show that crop residue mulch decreased soil surface temperature drastically, and significantly increased maize and sorghum emergence (Figs. 15.17a, b, c).

The effect of plastic-sheet mulch on soil temperature is different from that of straw or chopped-up plastic. The data in Fig. 15.18 by Maurya and Lal (1981) show the mean weekly records of soil temperature at 5 cm depths measured on

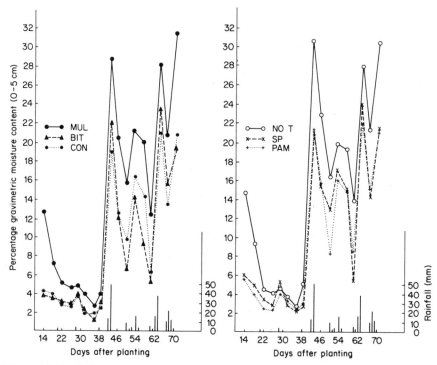

Figure 15.14 Soil moisture regime under maize during the second season of 1976 for different mulch materials e.g. straw mulch (MUL), bitumen (BIT), no-tillage (NOT), soil penetrant (SP) and Polyanylamide (PAM) and bare ploughed (CON) (De Vleeschauwer et al., 1978). *Reproduced by permission of Catena Verlag*

a sunny day at Ibadan, Nigeria. There were obvious differences in soil temperature with respect to mulching, time of the day, and methods of seedbed preparation. Compared with the flat bare ground surface, the straw mulch decreased the maximum soil temperature and the clear plastic mulch increased it. The effect of black polythene mulch was similar to that of the ridge treatment. The amplitude of diurnal fluctuations in soil temperature was 7, 10, 12, 12 and 15°C for straw mulch, bare soil surface, ridged soil, black polythene and white polythene mulch, respectively. Because of the air gap between the soil and the plastic sheet, white polythene mulch creates a greenhouse effect.

Synthetic soil conditioners influence soil temperature through their effect on the soil moisture regime and the albedo. Soil conditioners that darken the soil surface increase the maximum soil temperature. The data by De Vleeschauwer et al. (1978) show that the effects of soil conditioners on soil temperature regime also depend on the season, time of the day and the per cent crop cover. The shading effect of the growing crop decreases the differences in soil temperature among treatments. Generally, the bitumen-treated plots had significantly higher maximum temperatures during the first 7 weeks (Figs. 15.19 and 15.20) than other treatments. The minimum temperature at 0700

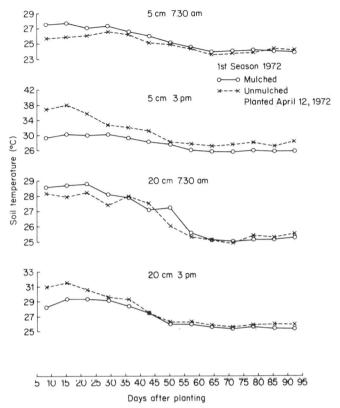

Figure 15.15 Effects of straw mulch on soil temperature regime during the first season maize at IITA (Lal, 1974). *Reproduced by permission of Plant and Soil*

hours, on the other hand, was lower than in mulched plots. The differences in soil temperature between straw mulch and bitumen increased with increasing solar radiation or air temperature according to the following equation:

$$Y = 274.9 \, X^{0.311} \qquad r = 0.78^{***}$$

where Y is the temperature difference between the mulch and the bitumen-treated plots in °C and X is the solar radiation in g cal cm^{-2} day^{-1}. Comparisons of the diurnal fluctuations in soil temperature between straw mulch and the soil conditioner treatments for a sunny and cloudy day are shown in Figs. 15.21 and 15.22 (De Vleeschauwer *et al.*, 1978). As expected, the amplitude of the fluctuations is much less in the mulch treatment than in bitumen and other soil conditioner treated plots, and the differences are more pronounced on a sunny day.

(v) Aeration: Mulching improves soil aeration by promoting free exchange of gases between the soil and the atmosphere. This is facilitated by improvements in total and macro-porosity, by decrease in splash and crusting, and by

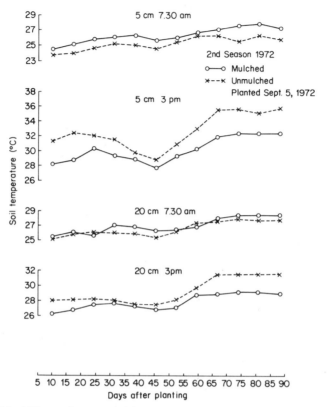

Figure 15.16 Effects of straw mulch on soil temperature regime during the second season maize at IITA (Lal, 1974). *Reproduced by permission of Plant and Soil*

improving the overall soil drainage. The high oxygen diffusion rates (Table 15.18) in mulched compared with unmulched treatments, even after heavy rains, are an indication of the better soil drainage conditions and of the presence of freely conducting macropores in mulched soils. Prevention of soil crust by mulch is the important factor influencing soil aeration.

The gaseous composition of soil air also depends on the nature of the mulch material (C:N ratio, etc.), its rate of decomposition, the soil moisture regime and the climatic environment. The data in Fig. 15.23 show that the ethylene concentration in soil air was generally higher in decomposing mulch than in unmulched plots.

(vi) Run-off and soil erosion: In contrast to the variable effects on soil temperature and moisture regimes, mulch invariably decreases soil erosion and often reduces water run-off rate and its amount. Mulch cover protects the soil against raindrop impact, increases the infiltration rate by improving soil structure and macroporosity, and decreases run-off velocity through physical resistance to water flow. The effects of mulch on soil and water conservation in

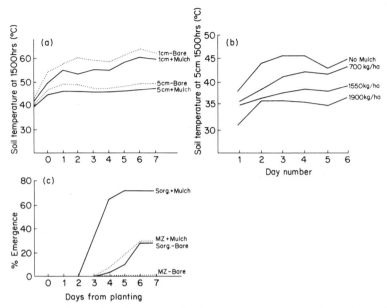

Figure 15.17 Effects of mulching on (a,b) soil temperature regime of a soil at Katherine, N.T., Australia and on (c) maize (MZ) and sorghum (Sorg.) emergence (Jones and McCown, 1983)

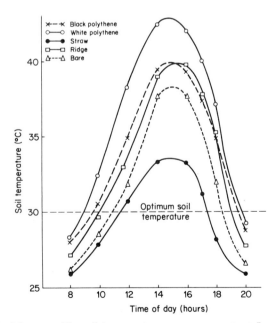

Figure 15.18 Mean weekly soil temperature measurements at 5 cm depth on a sunny day in the first growing season of 1976 for different mulch materials (Maurya and Lal, 1978)

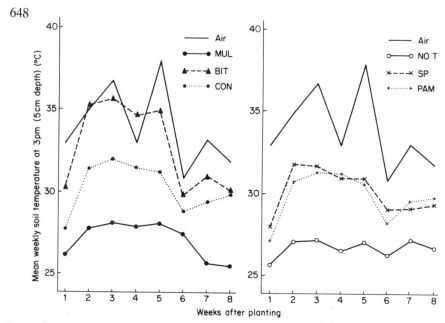

Figure 15.19 Weekly soil temperature measurements at 1500 hours during the second season 1976 for straw mulch (MUL), bitumen (BIT), bare ploughed (CON), no-tillage (NOT), soil penetrant (SP) and polyacrylamide (PAM) (De Vleeschauwer et al., 1978). *Reproduced by permission of Catena Verlag*

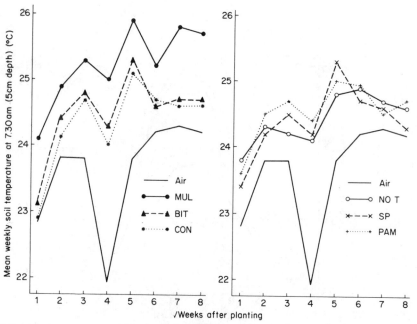

Figure 15.20 Weekly soil temperature measurements at 0730 hours during the second season 1976; see Fig. 19 for legend (De Vleeschauwer et al., 1978). *Reproduced by permission of Catena Verlag*

649

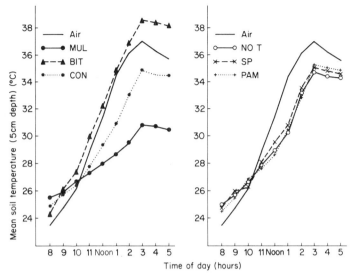

Figure 15.21 Diurnal fluctuations of the 5cm soil temperature for a clear day as affected by the various mulch treatments; see Fig. 19 for legend (De Vleeschauwer et al., 1978). *Reproduced by permission of Catena Verlag*

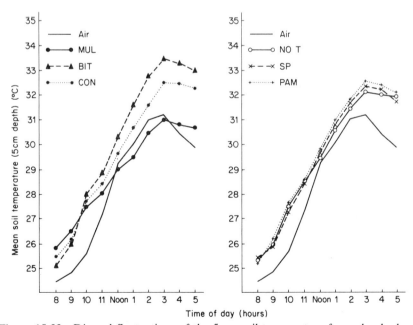

Figure 15.22 Diurnal fluctuations of the 5 cm soil temperature for a cloudy day as affected by the various mulch treatments (see Fig. 19 for legend) (De Vleeschauwer et al., 1978). *Reproduced by permission of Catena Verlag*

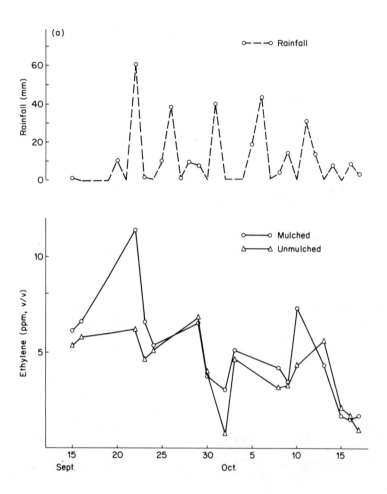

Figure 15.23 Mean concentration of ethylene in soil air (a) in mulched and unmulched plots (b) different mulch materials and (c) maize stover mulch and unmulched plot

the tropics have been extensively reviewed (Lal, 1975, 1977, 1983, 1984). For some soils mulch does not substantially decrease run-off amount and yet drastically reduces soil erosion. The reduction in soil erosion in these cases is due to prevention of soil splash and to decrease in run-off velocity. The water run-off from mulched plots is often clear with little sediment. The effect of mulch in reducing run-off and erosion is due to:
- reduction of raindrop impact,
- improving soil structure and porosity,
- increasing soil infiltration rate by decreasing surface sealing,

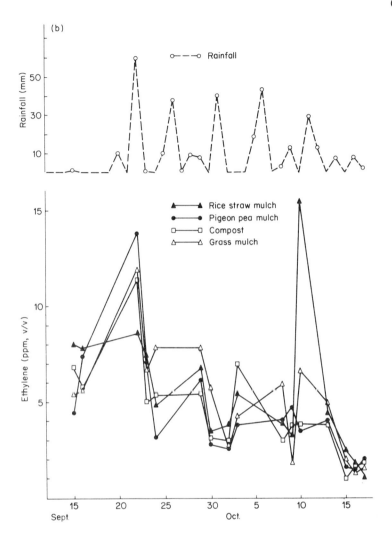

- increasing surface storage and detention capacity,
- decreasing run-off velocity,
- improving the biological activity related to soil cover and its influence on porosity.

Soil conditioners are often not as effective as residue mulches. This difference is partly due to the lack of protective cover that decreases raindrop impact, increases surface detention component, and decreases run-off velocity. In general, losses in water run-off decrease exponentially with increase in mulch rate (Fig. 15.24). In comparison with unmulched bare soil a marked

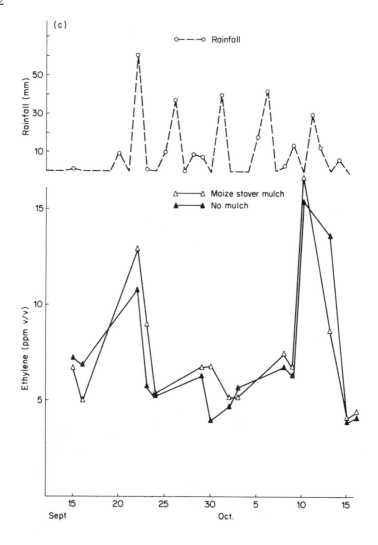

decrease in run-off for mulch rates of 4 to 6 t ha^{-1} occurs (Table 15.19). Lal (1976) reported that the mean annual run-off loss (average of 4 slopes of an Alfisol in southwest Nigeria) was 393.1, 80.7, 30.1 and 12.9 mm for mulch rates of 0, 2, 4, and 6 t ha^{-1}, respectively. The crop residue mulch is also effective in preventing soil erosion. Lal (1976) reported that a mulch of 4 to 6 t ha^{-1} drastically reduced soil erosion of 10 to 15 per cent. He measured soil erosion of 110, 3.5, 0.5 and 0.3 t ha^{-1} season^{-1} from plots mulched with 0, 2 and 4 t ha^{-1} of rice straw, respectively. For the slopes ranging between 1 and 15 per cent, soil erosion decreased exponentially with increase in mulch rate (Table 15.20).

Table 15.19 Mulching effects on water run-off for mulch rates of 0 to 6 t ha^{-1} (Lal, 1976)

Slope (%)	Regression equation	r
1	$Y = 0.39\ M^{-9.73}$	0.78
5	$Y = 1.16\ M^{-0.36}$	0.80
10	$Y = 5.53\ M^{-0.27}$	0.86
15	$Y = 5.26\ M^{-0.55}$	0.75

M = mulch rate (t ha^{-1})

Table 15.20 Mulching effects on soil erosion for mulch rates of 0 to 6 t ha^{-1} (Lal, 1976)

Slope (%)	Regression equation	r
1	$Y = 0.19\ M^{-0.53}$	0.85
5	$Y = 1.25\ M^{-0.71}$	0.85
10	$Y = 1.09\ M^{-0.67}$	0.96
15	$Y = 0.98\ M^{-0.24}$	0.72

M = mulch rate (t ha^{-1})

Furthermore, with an adequate quantity of crop residue mulch, the degree of slope steepness had less effect on soil erosion than without the residue mulch (Table 15.20). Lal *et al.* (1980) observed that run-off losses were 18, 10, 4, 1 and 0 per cent of the rainfall, for 0, 2, 4, 6 and 12 t ha^{-1} mulch rates respectively. The corresponding soil losses were 10, 2, 0.5, 0.1 and 0 t ha^{-1} for the respective mulch rates.

The effectiveness of mulch in reducing run-off and soil erosion has also been demonstrated in Zanzibar and Tanga in Tanzania. Khatibu *et al.* (1984) evaluated the effects of 6 t ha^{-1} of straw mulch on run-off and erosion. Their data in Table 15.21 show that mulch at 6 t ha^{-1} provided effective erosion control even on steep slopes of up to 22 per cent. The annual soil erosion from bare plots was 315, 488, and 490 times more and run-off 18, 23, and 29 times

Table 15.21 Effects of mulching on run-off and soil erosion observed in Zanzibar (Khatibu *et al.*, 1984). *Reproduced by permission of Elsevier Science Publishers*

Observations	Unmulched	Mulched
Run-off (mm)	108.6	6.9
Run-off (% of rainfall)	10.2	0.01
Soil erosion (t ha^{-1})	5.5	0.2
Nutrient loss in run-off (kg ha^{-1})	28.0	2.0

Table 15.22 Mulching effects on run-off and soil loss at Tanga, Tanzania (Ngatunga et al., 1984). *Reproduced by permission of* Goederma

Season	Slope (%)	Run-off (mm)		Soil erosion (t ha^{-1})	
		Bare fallow	Mulched	Bare fallow	Mulched
(a) Vuli	10	9.6	0.7	10.1	0.04
	19	10.8	1.2	26.1	0.06
	22	15.1	0.8	21.2	0.08
(b) Masika	10	41.0	2.1	27.8	0.08
	19	48.9	1.7	66.7	0.11
	22	40.2	1.1	62.9	0.10
(c) Annual Total	10	50.6	2.8	37.8	0.12
	19	65.7	2.9	92.8	0.19
	22	55.2	1.9	88.2	0.18

greater than that from the mulched plots for 10%, 19% and 22% slopes respectively. Obviously mulch is more effective in controlling soil erosion than in reducing water run-off. Similar effects of mulch on run-off and soil erosion control for slopes of up to 22 per cent were reported for Tanga, Tanzania, by Ngatunga *et al.* (1984) (Table 15.22).

2. Chemical effects

Soil chemical properties are more affected by differences among mulch materials than are soil physical properties. Obviously, the addition of plant nutrients in organic mulches is one major difference in comparison with unmulched or those mulched with chemically inert materials, e.g. plastic mulches. The composition of mulch materials, thus, has an important effect on soil chemical properties. Soil chemical properties are also influenced through the effects of mulches on biotic activity i.e., both micro- and macro-organisms. Soil biotic activity responds to the alterations in soil temperature and moiture regimes and on the availability of food supply and energy substrates supplied through the mulch.

One of the most important effects of mulching is on soil organic matter content. Keeping the soil organic matter content at high levels is an important aspect of productivity maintenance for soils in the tropics. Rather than the uphill task of increasing soil organic matter content, it may suffice if the rate of decline can be controlled to a manageable level. Experiments conducted in Africa and elsewhere have shown that regular and substantial additions of crop residue mulch decrease the rate of decline of soil organic matter content (Lal and Kang, 1982). For example, Lal *et al.* (1980) observed that rate of decline of organic carbon during the 18 months after land clearing was 0.103, 0.100, 0.092, 0.083 and 0.078% per month for mulch rates of 0, 2, 4, 6 and 12 t ha^{-1}

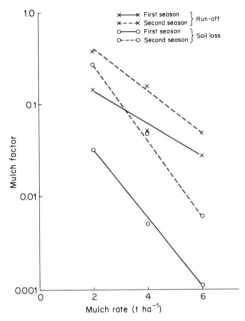

Figure 15.24 Effect of mulch rate on water run-off loss (Lal, 1976)

season^{-1}, respectively (Fig. 15.25). Mulching at high rates can be impractical and perhaps uneconomic if the corresponding yield increments are small.

The cation exchange capacity, an important factor in nutrient retention and availability, is substantially influenced by organic matter content in soils containing predominantly low activity clays. With relatively high soil organic matter content mulched soils have often higher ECEC than those without mulch (Lal, 1973; Lal et al., 1980; Lal and Kang, 1982). The magnitude of the addition of plant nutrients by mulch depends on the quality, quantity and frequency of the material added. The most noticeable effect is on the K status.

3. Biotic activity

Mulches affect soil biological properties both directly and indirectly. Direct effects include addition of food substrate and indirect effects are due to changes in soil temperature and moisture regimes. The use of inert materials as mulch causes only indirect effects on soil biotic activity. Microorganisms (bacteria, fungi, actinomycetes) and soil fauna (earthworms, termites, millipedes etc.) play an important role in decomposition of soil organic matter. The activity and diversity of these organisms is substantially affected by mulching, the quality of mulch material, and rate of its application. The rate of decomposition of crop residue depends, other factors being the same, on the quantity of residue supplied as mulch. Lal et al. (1980) reported that the rate of decomposition of mulch increased with increasing rate of mulch application (Table 15.23)

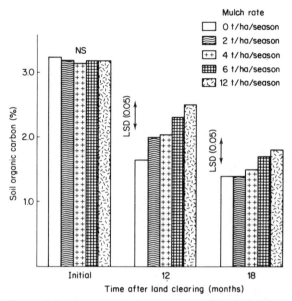

Figure 15.25 Rate of decline in organic carbon for different mulch rates during 18 months following land clearing (Lal and Kang, 1982)

Table 15.23 Decomposition rate of rice straw and guinea grass for different mulch rates (Lal *et al.*, 1980). *Reproduced from* Soil Science Society of America Journal, Volume 44, 827–833, *by permission of the Soil Science Society of America*

Mulch rate (t ha^{-1})	Regression equation	Correlation coefficient (R^2)
	Guinea grass	
2	$Y = -4.74 + 0.93\,t - 0.0022\,t^2$	0.97
4	$Y = -4.00 + 0.81\,t - 0.0020\,t^2$	0.97
6	$Y = -4.00 + 0.75\,t - 0.0015\,t^2$	0.98
12	$Y = -3.09 + 0.66\,t - 0.0011\,t^2$	0.99
	Rice straw	
2	$Y = -5.66 + 1.05\,t - 0.0024\,t^2$	0.95
4	$Y = -4.93 + 0.89\,t - 0.0017\,t^2$	0.95
6	$Y = -4.54 + 0.79\,t - 0.0012\,t^2$	0.95
12	$Y = -4.37 + 0.69\,t - 0.0009\,t^2$	0.95

Y = per cent decomposition.
t = time in days after mulch application.

because the latter increased species diversity and population density. In fact the activity of earthworm *Hyperiodrilus africanus* was linearly related to the mulch rate (Table 15.24). Intense biotic activity created by mulch has important implications in improving soil physical as well as nutritional properties.

Table 15.24 Regression relating mulch rate with activity of earthworm *Hyperiodrilus* spp. at IITA, Ibadan (Lal *et al.*, 1980). *Reproduced from* Soil Science Society of America Journal, *Volume 44, 827–833, by permission of the Soil Science Society of America*

Month	Regression equation	Correlation coefficient (r)
April	$Y = 1.15X + 3.30$	0.96**
May	$Y = 0.94X + 2.44$	0.96**
June	$Y = 0.90X + 4.33$	0.93**
July	$Y = 1.81X + 0.45$	0.99**
August	$Y = 1.45X + 2.84$	0.97**
September	$Y = 1.62X + 1.93$	0.97**
October	$Y = 1.02X + 3.29$	0.97**
Annual	$Y = 1.41X + 2.66$	0.98**

** Significant at the 0.01 leval.
Y = Worm casts m^{-2} month^{-1}.
M = Mulch rate, t ha^{-1}.

4. Crop response

Crop response to mulching is difficult to generalize because it depends on climate, soil and crop interaction. Yield responses to mulch differ among regions, soils and crops. A detailed literature survey of crop response to mulching has been reported earlier (Lal, 1979). A positive response has been reported in Africa (Jurion and Henry, 1969; Lal, 1975; Maurya and Lal, 1981), in India (Prihar *et al.*, 1979) and in South America (Wade and Sanchez, 1983). From their studies on an Oxisol in the eastern Amazon region of Brazil, Schöningh and Alkämper (1984) observed that grain yield of maize and cowpea was significantly increased by mulching (Table 15.24a)., Mulches with high decomposition rate (low C:N ratio) increased yield more than those with low decomposition rate (high C:N ratio). This was particularly true for plots that did not receive NPK fertilizer. The negative response commonly reported with the application of a mulch material with high C:N ratio is often due to competition for N caused by microbial immobilization. The favourable crop response in the tropics is due in part to improvements of moisture and temperature regimes, improvements in soil structure, soil and water conservation and partly due to addition of some plant nutrients.

B. Crop rotations and fallowing

Crop rotations and planted fallows are important tools to influence soil properties. Changes in soil properties depend on crops grown and the sequences and combinations adopted. In general legumes and deep-rooted crops have a more favourable effect on soil properties than shallow-rooted annuals. That is why open-row cereals are commonly termed soil-degrading crops in contrast to the soil-conserving effects of closed-canopy legumes and woody

Table 15.24a Yield response (kg/ha) of maize and cowpea to different mulch materials and mineral fertilizer in an eastern Amazon Oxisol, Capitao Poco (CPATU), Para, Brazil, 1983 (Schöningh and Alkämper, 1984)

Mulches used (10t/ha of DM)	First crop (Maize)*		Second crop (Cowpea)**	
	NPK 120–80–60	NPK 0–0–0	NPK 30–80–60	NPK 0–0–0
Elephant grass	4646	2144	1227	80
Pueraria	5697	3342	1187	114
Weeds	4911	2215	1394	105
Sec. forest (2–3 years)	4462	1560	1191	35
Sec. forest (4–5 years)	4479	1807	1397	95
Rice husks	4398	1146	1487	163
Maize cops + maize husks	4863	2101	1302	41
Bare soil	3539	78	1169	7
L.S.D. 5%	987		281	
L.S.D. 1%	1321		376	
L.S.D. 0.1%	1729		493	

* = grain moisture content 14.5%.
** = grain moisture content 13%.

perennials. The rate of development and the duration of the vegetal cover, and the nature of the root system are important factors that influence soil properties. Some considerations on suitable crop rotations are discussed by Henin (1960).

1. Rotations

Considerable research on the effect of crop rotations on soil properties has been done in semi-arid West Africa by IRAT. Nicou (1975) observed significant differences in penetrometer resistance and the coefficient of cohesion due to crops. The data in Fig. 15.26 show that the coefficient of cohesion in soil varied with crop rotations and locations. In western Nigeria Juo and Lal (1977) observed significant differences in soil properties due to crop rotations. The soil under crops that produce more residue (stover) contained more organic matter than those producing less. Similar differences existed in the concentrations of exchangeable cations.

Crops affect soil physical properties through the action of their root systems. Tuberous roots of cassava have loosening effects on the soil (Fig. 15.27), decrease the bulk density and improve water infiltration (Table 15.25). Long duration crops that provide a prolonged ground cover e.g., cassava, often have ameliorative effects on soil physical properties.

Soil erosion and water run-off are also differentially affected by different crops. Crops that establish an early and close canopy cover protect the soil against the impact of rain drops. Soil erosion is often severe in crops that have

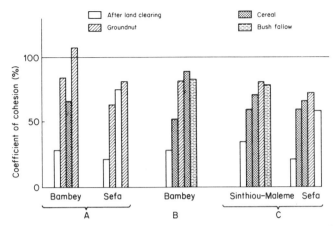

Figure 15.26 Variation in coefficient of cohesion of soil due to crop rotation. A = Regeneration, groundnut, millet, B = Regeneration, millet, groundnut, millet, bush fallow, C = Regeneration, maize, millet (or sorghum or rice), groundnut, bush fallow (Nicou, 1975). *Reproduced by permission of IRAT*

Table 15.25 Effects of cassava versus maize cultivation on water infiltration rate on plots (unpublished data, Lal, 1981)

Time (minutes)	Cumulative infliltration (m)	
	Cassava	Maize
	Jan. 1981	Jan. 1981
10	6.5	3.5
20	10.2	5.7
30	14.4	7.3
45	19.6	10.5
60	23.6	12.2
90	29.0	17.6
120	36.8	20.4

slow initial vigour and require a long time to establish a good cover. In Ivory Coast Berger (1964) observed that erosion from plots under maize was different depending on whether the previous crop was maize or cowpeas. Similar observations were made by Lal (1976) in Nigeria. (Table 15.26). Aina *et al.* (1979) observed that soil erosion from a maize + cassava mixed cropping sytem was less than from monoculture cassava. The data in Table 15.27 show the low values of crop management factor 'C' for maize + cassava mixed cropping and for soybean. Soil erosion decreases exponentially with increases in the canopy cover. The regression equations relating soil erosion with vegetal cover for different crop rotations are shown in Table 15.28. The exponents of the regression equations were larger for soybeans and pigeon peas based rotations than for root crops. Run-off and soil erosion are influenced by the

Figure 15.27   Loosening effects of the development of root tubers of cassava on soil

Table 15.26   Effects on erosion (t ha^{-1}) of maize following corn versus corn following cowpea (Lal, 1976)

Slope (%)	Corn–Corn		Cowpea–Corn	
	1973	1972	1973	1972
1	0.4	0.1	0.3	0.3
5	2.8	0.4	4.0	1.1
10	2.8	0.6	3.0	0.9
15	17.1	1.3	35.4	4.4

canopy cover and height, and crop duration. Crops with dense canopy close to the ground surface provide better protection against erosion than open-row crops. In support of these conclusions Lal (1983) reported different relationships between soil erosion and leaf area index for different crops (Table

Table 15.27 Influence of cropping systems on crop management factor C (Aina et al., 1979). Reproduced from Soil Science Society of America Journal, Volume 43, 173–177, by permission of the Soil Science Society of America

Cropping system	Soil erosion	
	First season	Second season
Cassava (monoculture)	0.72	0.39
Maize–cassava (mixed cropping)	0.43	0.05
Soybean	0.19	0.02

Table 15.28 Simple regression of soil loss Y (in metric ton ha^{-1} cm of rain^{-1}) on percent vegetal cover (X) under different cropping systems (Aina et al., 1979). Reproduced from Soil Science Society of America Journal, Volume 43, 173–177, by permission of the Soil Science Society of America

Cropping system	r	Regression equation
Soybean–soybean	0.63**	$Y = 5.38\ e^{-0.04X}$
Pigeonpea–pigeonpea	0.94**	$Y = 3.27\ e^{-0.01X}$
Maize–cassava (mix-cropping)	0.84**	$Y = 2.20\ e^{-0.01X}$
Cassava (monoculture)	0.90**	$Y = 2.71\ e^{-0.01X}$

Table 15.29 Regression equations relating soil splash with leaf area index, rainfall amount and 30-min maximum intensity (unpublished data of S. Huke and R. Lal)

Cropping system	Regression equation	r
Cassava	$E = 0.56 + 0.70\ I_{30} + 0.43\ A - 0.46$ LAI	0.93
Yam	$E = 0.08 + 0.88\ I_{30} + 0.51\ A - 0.18$ LAI	0.76
Maize	$E = 0.15 + 0.06\ I_{30} + 0.53\ A - 0.05$ LAI	0.77
Sweet potato	$E = 0.91 + 0.30\ I_{30} + 0.38\ A - 0.91$ LAI	0.53
Cassava + sweet potato	$E = 0.45 + 0.18\ I_{30} + 0.08\ A - 0.28$ LAI	0.63
Maize + sweet potato	$E = 0.31 + 0.03\ I_{30} + 0.54\ A - 0.07$ LAI	0.58
Yam + maize	$E = 0.14 + 0.32\ I_{30} + 0.65\ A - 0.07$ LAI	0.80
Cassave + maize	$E = 0.02 + 0.11\ I_{30} + 0.23\ A - 0.01$ LAI	0.87

E = splash (kg m^{-2}); I_{30} = maximum 30-min intensity (h^{-1}); LAI = leaf area index; A = rainfall amount per storm (mm).

15.29). In India Prahbakara and Dakshinamurti (1976) observed that soil structure was improved by the practice of relay cropping owing to the minimum soil disturbance and continuous addition of root residues.

Nutrient and water use efficiency of compatible crop mixtures is often more than component crops grown as monoculture. Higher water use efficiency has been reported in India for maize/soybean and maize/mungbean intercrops and other crop mixtures in relation to their respective monocrops (De and Singh,

1981; Natarajan and Willey, 1980; Steiner, 1982). In Nigeria Hulugalle and Lal (1985) studied the soil-water balance of intercropped maize and cowpea and reported high water use efficiency of mixed versus monocultures. During seasons of drought stress the water use efficiency was more for intercropped maize and cowpea planted in the same than in alternate rows.

2. Planted fallows

The use of planted fallows and cover crops has long been advocated for restoration of soil physical and nutritional properties, and for build-up of soil organic matter. Planted fallows are more effective in increasing soil organic matter and in improving soil structure than natural fallowing used in shifting cultivation and related bush fallowing systems. In Kenya Pereira and Beckley (1952) investigated the effects of various lengths of fallowing by grass and legume mixtures on soil physical properties and the yield of the fallowing arable crops. They reported more rapid improvement in soil physical properties by fallowing with grass than by legume covers, and observed that the structural improvements thus caused are rapidly lost during arable cultivation. The data in Table 15.30 show significant improvements in pore space and soil moisture retention in cropped or grass plots in comparison with those without cover. Clean cultivation, including clean weeding, is detrimental to the soil in tropical climates. Pereira and Jones (1954) reported that clean weeded plots had lower porosity and infiltration rates than those under tall weeds or where weeds had been merely slashed to provide an extra ground cover (Table 15.31). The most effective method of establishing grass leys for fertility restoration was by undergrowing it through a standing crop by the relay system. The use of fallowing for fertility restoration is recommended by Jameson and Kerkham (1960) in Uganda and by Barnes (1981) in Zimbabwe.

Earlier conducted research in Nigeria (Vine, 1953; Dennison, 1959) has conclusively demonstrated the ameliorative effects of fallowing on soil properties. In northern Nigeria Wilkinson (1975) observed a linear increase in soil infiltration rates with square root of the time the land had been kept under fallow: Equilibrium infiltration rate (cm/hr)$=0.76+1.982$ (Fallow period, years)$^{1/2}$ Similar to the observations of Pereira and colleagues in East

Table 15.30 Effects of cropping systems on free-draining pore space of surface soil (Pereira *et al.*, 1958). *Reproduced by permission of Experimental Agriculture*

Cropping system	Pore space (% by volume) at different tensions (cm of water)			Soil temperature at 2.5 cm depth (F°)
	10	50	100	
Bare fallow	7.4	22.0	25.7	121.6
Crops and grasses	11.5	26.1	28.1	86.8–95.4
LSD (0.05)	4.1	5.2	4.9	

Table 15.31 Effects of methods of weed control on physical properties of lateritic soil (Modified from Pereira and Jones, 1954). *Reproduced by permission of* Experimental Agriculture

Treatment	Pore space (%)		Percolation (cm hr^{-1})		Dry sieving (%)		Water stable aggregates 0.5 mm (%)		
	Total	Pores drained at 50 cm suction	Under 1 cm static head	Artificial rain at 15 cm hr^{-1}	>1.25 cm	<0.5 mm	After vacuum wetting	After immersion wetting	After rain impact wetting
Clean weeded	63.7	19.9	24.9	11.2	31.3	44.2	41.2	24.8	27.9
Slash weeded	65.1	23.0	26.2	12.7	33.6	46.8	41.9	27.9	30.8
Tall weeds	65.8	24.4	39.6	13.0	19.3	45.3	40.3	28.9	31.2
LSD (0.05)	0.7	1.6	14.2	1.3	3.7	4.3	3.9	5.3	3.1

Table 15.32 Effects of cover crops on soil moisture characteristics of an Alfisol in western Nigeria (Lal et al., 1979). *Reproduced by permission of Williams & Wilkins Co.*

Cover crops	Soil moisture retention (%) at different suctions (bars)															
	0		0.1		0.3		0.5		1		2		3		15	
	B	A	B	A	B	A	B	A	B	A	B	A	B	A	B	A
Brachiaria	32.0	37.5	17.5	18.7	13.0	12.6	11.8	11.8	11.1	11.3	10.5	9.4	9.0	9.1	7.7	8.3
Paspalum	35.7	40.9	17.5	19.5	14.2	14.3	12.4	12.7	11.8	11.1	10.7	11.2	9.9	10.0	8.7	9.3
Cynodon	36.1	4.3	18.3	19.7	14.3	13.5	13.6	12.5	12.4	10.5	10.3	9.3	9.9	8.5	8.8	9.9
Pueraria	31.7	39.7	16.4	18.4	14.4	14.0	12.7	12.9	12.0	12.1	10.9	10.8	9.8	9.2	8.9	9.2
Stylosanthes	31.5	40.3	18.3	19.1	14.6	14.6	13.6	13.0	12.9	11.1	10.9	10.2	9.9	9.2	9.4	9.5
Stizolobium	32.9	40.2	17.4	19.0	15.1	14.2	13.8	13.7	12.8	11.5	11.7	10.6	10.7	9.6	8.8	9.7
Psophocarpus	34.2	38.3	16.2	18.8	14.8	14.2	13.1	13.7	12.6	11.2	11.6	11.4	10.3	10.0	9.2	9.7
Centrosema	34.1	38.1	17.3	18.4	15.1	15.9	13.9	14.2	13.0	11.3	11.9	11.2	10.4	9.9	9.0	9.9
Control	30.6	37.6	15.3	15.7	15.2	14.5	13.9	12.9	13.1	11.4	11.7	10.9	10.3	8.4	8.6	8.9
LSD (0.05)	4.7	3.9	9.7	2.7	3.0	2.8	2.5	2.6	2.7	2.0	1.9	1.8	2.2	2.1	2.5	1.9

B = 1974 before seeding cover crops; A = 1976, 2 years after.

Africa, Wilkinson also reported that most of the benefits in infiltration rate by fallowing were eliminated by the end of the first cropping season. For example, the initial infiltration rate of 5.84 cm hr^{-1} decreased to 0.97 cm hr^{-1} at the end of the first cropping cycle. Wilkinson and Aina (1976, 1977) confirmed these observations for soils of the humid regions of southwestern Nigeria.

In western Nigeria Lal *et al.* (1978, 1979) used a range of grass and leguminous cover crops to restore physical properties of an eroded, compacted and degraded Alfisol. In comparison with the natural fallow and ambient soil conditions, fallowing for 2 years with grass and legume covers improved soil moisture retention capacity particularly at the low suction ranges (Table 15.32). Also improved were the organic matter content, cation exchange capacity, infiltration rate and the field moisture capacity (Table 15.33). The lowest soil bulk density and the highest infiltration rates were observed in plots sown to *Psophocarpus palustris*. In another study, Hulugalle *et al.* (1985) observed that the deleterious effects of land clearing by heavy machines were alleviated by fallowing with *Mucuna utilis* grown for about 1 year immediately after the land clearing. These authors observed significant improvements in infiltration rate and in cumulative infiltration of the plots growing mucuna in comparison with those sown to maize (Table 15.34). The increase in cumulative infiltration by mucuna cover was 134% in annual clearing, 55% in shearblade clearing, 15% in treepusher clearing, and 189% in clearing by treepusher/root rake combinations. The improvements in total porosity ranged from 2% to 4%. However, there was no increase in the porosity of the severely compacted plots of treepusher/root clearing by 1 year fallowing. In such cases a longer duration of fallowing is necessary to render substantial improvements in soil structure. The degree of improvement in structural properties of severely compacted soils increases with increased duration of the fallow period (Aweto, 1981).

Improvements in structural properties of easily crusted and compacted soils of the semi-arid regions of Australia by fallowing with appropriate grass/legume mixtures have also been reported (Mehanni, 1974; Bridge *et al.*, 1983). It is now well established that seasonal crops can be grown through the sod produced by suppressed covers. The killed-sod mulch thus produced creates years)^{1}. In agreement with the observations of Pereira and colleagues in East favourable soil temperature and moisture regimes (Table 15.35) for seedling establishment and prevents the development of surface seals. Many studies have shown that deterioration in soil structure and infiltration rate during the cropping phase can be drastically reduced by managing the fallow cover through a no-till system. Sod-systems have been successfully used for growing seasonal crops in Ghana (Kannegieter, 1967; 1969); Nigeria (Lal *et al.*, 1979), Australia (McCown *et al.*, 1979) and South America (Wade and Sanchez, 1983). Although attempts have also been made to grow seasonal crops through the unkilled sod (Akobundu, 1982), competition by unsuppressed covers can drastically reduce crop yield. Some climbers (e.g. *Mucuna utilis*) climb up the standing maize crop and cause serious yield reductions (Fig. 15.28).

Table 15.33 Effects of legume and grass covers on physical properties of a tropical Alfisol (Lal et al., 1979). *Reproduced by permission of Williams & Wilkins Co.*

Cover crop	Bulk density g (cm⁻³)	Infiltration* capacity (cm (3 hr)⁻¹)	Infiltration rate (cm hr⁻¹)	Organic carbon (%) B	Organic carbon (%) A	CEC (meq/100g) B	CEC (meq/100g) A	Equilibrium soil moisture content (%) 0–5 cm	6–10 cm	11–15 cm
Brachiaria	1.34	65	19	1.21	1.57	6.1	8.5	10.1	10.6	10.3
Paspalum	1.35	65	14	1.23	1.45	7.1	8.2	9.7	11.4	10.3
Cynodon	1.30	66	18	1.30	1.70	6.9	8.9	14.8	11.5	8.9
Pueraria	1.32	76	16	1.27	1.50	6.6	7.7	20.1	12.7	12.1
Stylosanthes	1.33	74	16	1.30	1.63	6.3	8.8	18.5	15.9	15.6
Stizolobium	1.33	77	21	1.30	1.57	6.7	10.5	14.7	14.2	11.7
Psophocarpus	1.14	124	42	1.20	1.57	5.9	10.9	21.2	15.8	13.6
Centrosema	1.33	72	18	1.30	1.53	6.5	10.0	15.9	15.0	15.5
Control	1.42	71	13	1.33	1.37	8.2	8.4	11.0	12.7	12.0
LSD (0.05)	0.04	66	9	0.50	0.23	2.3	3.5	6.2	2.3	3.1

* Preceding infiltration capacity was 44 cm (3hr)⁻¹ and infiltration rate was 17 cm hr⁻¹
B = in 1974 seeding cover crops;   A = 1976 2 years after.

Table 15.34 Effects of *mucuna* fallowing on infiltration characteristics (Hulugalle *et al.*, 1985)

Clearing treatment	Cropping system	Porosity (%)	Moisture retention (% v/v) a= different suctions (kPa)			Infiltration rate (mm s^{-1})		Cumulative infiltration in 3 hr (mm)
			0	10		5 min	180 min	
Manual	Maize	55.0b	35.3b	15.9a		0.22b	0.02cd	608bc
Manual	*Mucuna*	59.0a	42.1a	15.0a		0.28a	0.08a	1422a
Shearblade	Miaze	54.0c	29.5c	14.9a		0.15c	0.02cd	492cd
Shearblade	*Mucuna*	57.0b	34.9b	16.4a		0.20bc	0.04b	764b
Treepusher	Maize	57.0b	31.1c	17.4a		0.08d	0.02cd	287de
Treepusher	*Mucuna*	59.0a	39.3a	12.7a		0.16bc	0.004d	329de
Treepusher/ root rake	Maize	53.0c	29.2c	14.4a		0.09d	0.002d	193e
Treepusher/ root rake	*Mucuna*	52.0c	33.0b	15.7a		0.13cd	0.03bc	557bcd

Figure 15.28 The climbing legume cover crops suffocate crops and drastically reduce economic yields

Table 15.35 Effect of killed sod-mulch on soil temperature and soil-water potential on a Blain sand at 1500 hours on 13th January, 1979 at Katherine, N.T., Australia (McCown et al., 1980)

Depth (cm)	Without mulch				With mulch			
	Inter-row		Row		Inter-row		Row	
	°C	Bars	°C	Bars	°C	Bars	°C	Bars
1	51	<−0.04	50	<−0.40	35	<−0.20	46	<−0.30
5	44	−0.20	44	−0.30	34	−0.18	39	−0.22
10	41	−0.18	40	−0.18	33	−0.10	36	−0.20

C. Agroforestry

The term agroforestry refers to the practice of growing deep-rooted woody shrubs and perennials in association with seasonal food crops. If the mixtures are compatible, the practice may combine the benefits of bush fallowing while enabling the production of seasonal crops. The deep-rooted perennials, e.g., pigeon pea (*Cajanus cajan*) or *Leucaena leucocephala* improve the physical properties of the subsoil layers and minimize environmental degradation by decreasing the soil exposure. This special case of mixed-cropping is termed alley cropping (Kang *et al.*, 1981), hedgeway cropping (Mongi and Huxley, 1979) or avenue cropping (Wijewardene and Waidyanatha, 1984). In addition to Africa and Asia, the system has also been found to be useful in the Pacific

Table 15.36 Effects of alley-cropping of maize on run-off and soil erosion from 70 × 10 m plots established on slopes of about 5%. Rainfall of 30th May, 1984 amount = 44.5 mm (Unpublished data of R. Lal)

Treatment	Run-off		Soil erosion (kg ha^{-1})
	(mm)	%	
Ploughed	35.1	78.9	3715
No-till	1.6	3.6	30
Leucaena 4 m	2.6	5.8	113
Leucaena 2 m	0.5	1.1	12
Glyricidia 4 m	13.3	29.9	1576
Glyricidia 2 m	8.5	19.1	851

and tropical America (Vergara, 1982). Some examples of commonly used tree species include *Leucaena leucocephala*, *Gliricidia maculata*, and *Tephrosia candida*. Other useful species suitable for agroforestry are *Sesbania*, *Parkia*, *Calliandra*, *Albizia*, *Acacia*, *Cassia*, *Pithecellobium*, *Mimosa*, *Prosopis* and *Samanea*. The choice of an appropriate specie depends on soil, rainfall regime and the climate (Vergara, 1982). The root system of the woody perennials should not be competitive with the seasonal crops.

There is little research information describing the effects of agroforestry practices on soil physical properties and microclimatic environment. The additional ground cover and mulch materials produced by woody perennials may improve soil physical properties and reduce run-off and soil erosion. For the system to be effective, however, the spacing should be close and there should be frequent additions of pruning to the soil in the form of crop residue mulch. Studies are now in progress at IITA to quantify the effects of various tree species on soil properties.

In addition to providing mulch material, closely-spaced tree rows planted on contour may retard run-off velocity and allow more time for water to infiltrate the soil. The time required for hedge establishment that effectively retards run-off velocity varies among species, soils and rainfall regimes. In the humid and subhumid region it takes about 3 years for *Leucaena* planted at 2 m intervals to reduce run-off and soil erosion. The data in Table 15.36 show the effects of 2-year-old hedges of *Leucaena* and *Glyricidia* on run-off and soil erosion. It is apparent that under the soil and climatic conditions in Ibadan, *Leucaena* is more effective than *Glyricidia* and that 2 m spacing is more effective in run-off and erosion control than 4 m spacing. Suitable tree species are, however, different for different ecologies.

D. Mechanization

Motorized farm operations and wheel traffic are a major factor responsible for severe soil compaction observed in tropical arable lands. Soil compaction-

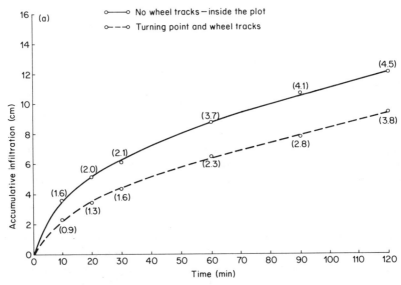

Figure 15.29a Water infiltration rate in mechanized no-till plots for headlands (turning points) and inside the plot (Lal, 1985)

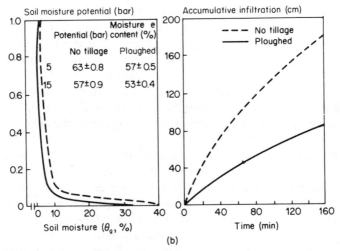

Figure 15.29b Water infiltration rate and moisture retention curves of no-till plots after growing 32 consecutive crops (Lal, 1983)

induced alterations lead to accelerated soil erosion and the overall degradation of the natural resource. For example, in Nigeria Lal (1985) reported significant compaction of the headland soil where the farm equipment is turned. The crop stand growth and yield at the boundaries near the turning points of farm equipment were noticeably poor in comparison with the rest of the fields inlands. The data in Fig.15.29 compare the infiltration rate of headlands with that of 50 metres inside. The accumulative water infiltration 2 hours after the test began was 28 per cent more inside than near the turning point. Maize grain yield for the compacted region was 1.2 ± 0.4 t ha^{-1} in comparison with $1.9 \pm 0,5$ t ha^{-1} for the uncompacted soil, a reduction of 27%. In contrast soil compaction has not been observed in plots cultivated manually in spite of growing 32 consecutive crops of maize (Lal, 1985) (Fig. 15.29a and b). The infiltration rate on no-till plots is high in spite of intensive cultivation for 15 consecutive years.

Mechanized harvesting and cart movements on the soil increase soil compaction, especially if the soil is wet. In the quest to shorten the turn-around time to grow a second crop, farmers often try to speed up the harvesting procedure. Mechanized harvesting is particularly damaging if the soil is already wet, as is often the case in regions of bimodal rainfall distribution. Significant compaction from load carts was detected in surgercane fields in Guyana (Howson, 1977) and in Puerto Rico (Shukla and Ravalo, 1976). In a Puerto Rican field with a soil moisture content of 27.4%, the zone of significant compaction was 15 cm. When the soil moisture at harvesting in an adjacent field was 48.3%, the zone of significant compaction was increased to 45 cm depth.

E. Pasture management

Another factor that leads to adverse changes in soil physical properties is uncontrolled and excessive grazing. Excessive grazing depletes the vegetation cover, exposes soil to high intensity rains, compacts the surface soil layer, and decreases infiltration rate of the surface soil layer as a result of trampling (Pereira and Hosegood, 1961; Lundgren and Lundgren, 1972). Pereira et al. (1967) concluded from investigations of grass leys on a lateritic red soil in Kenya that the trampling caused by 20 yearling beasts on 1 acre for 2 days produced severe run-off, even from a paddock that was completely covered by a dense mat of stolans and foliage. Paddock grazed 2 days prior to a storm lost half as much, and no flow was observed from a paddock grazed 5 days earlier. Dunne (1979) observed that the sediment yield from some Kenyan catchments

Table 15.37 The regression equations relating sediment yield to run-off and relief for different land-uses in Kenya (Dunne, 1979)

Land-use	Regression equation	Correlation coefficient (R^2)
1 Forested catchment	SY = 1.56 $Q^{0.46}$ $S^{-0.03}$	0.98
	SY = 2.67 $Q^{0.38}$	0.98
2 Forest > agriculture	SY = 1.10 $Q^{1.28}$ $S^{0.047}$	0.76
3 Agriculture > forest	SY = 1.14 $Q^{1.48}$ $S^{0.51}$	0.74
4 Rangeland	SY = 4.26 $Q^{2.17}$ $S^{1.12}$	0.87

SY = mean annual sediment yield (t km^{-2} y^{-1}).
Q = mean annual run-off (mm).
S = relief (dimensionless).

Table 15.38 Effect of grazing on soil erosion in Machakes, Kenya (Thomas et al., 1980)

Land-use	Soil loss (mm yr^{-1})	Ratio
Degraded grazing land	4.5	50
Cultivated land	1.3	15
Good grazing land (bush and woodland)	0.07	1

was related exponentially to the annual run-off. The exponent of the mean annual run-off was highest for the grazed land (Table 15.37). In the steeplands of Machakos, Kenya, Thomas et al. (1980) observed that well maintained pasture had slight or tolerable levels of erosion. With uncontrolled grazing on degraded lands, however, erosion increased by 50 times (Table 15.38). In Cameroon Hurault (1968) observed that even the low stocking rate of one head of cattle per 10 hectares accelerated soil erosion on the clayey soils of the high plateau of Adamawa.

F. Fertilizers and manures

Alterations in soil physical properties during intensive cultivation are partly due to a rapid decline in soil organic matter and a reduction in the biotic activity of soil fauna. Good crop growth produces more biomass, which adds to the soil's organic matter capital. Balanced nutrient application to stimulate good crop growth is thus important to soil physical conditions. All soil processes being interlinked, good soil and crop husbandry is as necessary to the soil's physical and nutritional properties as it is to good crop growth. Farmyard manure and compost are better ameliorative measures than chemical fertilizers.

G. Management of wetlands

The total area of marshes and swamps in the world equals about $2 \times 10^6 km^2$, of which 75% is located in the tropics (Crutzen, 1985). In the tropics of South and Southeast Asia these swamps have been utilized intensively. In fact the lowland rice culture is a sustainable system that has supported the high population density in Southeast Asia without causing the severe ecological imbalance and soil degradation problems that have limited the intensive utilization of tropical uplands. To meet the increasing demand, however, additional wetlands, marshes and swamps will have to be brought under rice production (Fig. 15.30). In fact, the area sown to rice is expected to increase from 98 million hectares in 1975 to 126 million hectares by the year 2000 (Dudal and Hrabovrzky, 1981).

Wetlands in tropical Africa have hitherto been un-utilized and constitute a

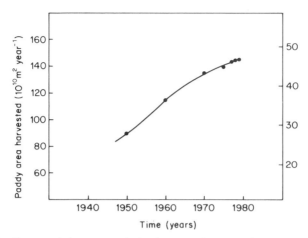

Figure 15.30 Temporal increase of the harvested area of rice paddies and the corresponding global CH_4 emission. The calculated Ch_4 emission rates do not take into account the possible influence of mineral nitrogen fertilizer on the CH_4 production rates in paddy soils (Seiler et al., 1984)

potentially useful land resource for intensive cropping. Wetlands and hydromorphic soils can be intensively utilized for double or even triple cropping provided that the problems of their development, water management, and health-hazard-based social taboos are overcome. A recent survey by Hekstra et al. (1983) showed that in the sub humid and humid regions of West Africa (regions with growing seasons > 165 days) covering a total area of 2.2 million km², the total wetland area ranges from 12.6 to 28.5%. Out of this, at least one-third of the wetland is occupied by streamflow valleys. These wetlands are the likely target for development for meeting the pressing demand for increasing food production in tropical Africa.

What are the ecological implications of converting the un-utilized wetlands to rice paddies? Some researchers believe that the methane concentration in the atmosphere is increasing at a fast rate of 1.5% per year (Crutzen, 1985; Seiler and Conrad, 1985). Various sources of methane production are believed to be enteric fermentation in domestic ruminants, biomass burning in the tropics, termites, and rice paddies and fresh water swamps (Tables 15.39 and 15.40). Crutzen (1985) estimated that about 2.7×10^{14}g of methane is released annually from all sources listed above. However, an average emission of methane from rice paddies during the rice growing season is estimated to range from 12 g/m² (Seiler, 1984) in Spain, 25 g/m² (Cicerone et al., 1983) in California, and 54 g/m² (Holzapfel-Pschorn and Seiler, 1985) in Italy. The tentative estimates of methane emission for different rice producing regions of the world were made by Seiler et al. (1984) (Table 15.41). Most of these figures quoted are tentative because few observations have been made from rice paddies in the tropics.

Table 15.39 Global production rates (10^{12} g/yr) of individual sources for atmospheric CH_4 (After Seiler and Conrad, 1985)

	1950	1960	1970	1975
Sources				
Ruminants	49 – 69	56 – 81	65 – 92	72 – 99
Paddy fields	43 – 106	56 – 135	65 – 158	69 – 167
Swamps	11 – 57	11 – 57	11 – 57	11 – 57
Other biogenic sources	8 – 20	8 – 21	9 – 22	9 – 22
Biomass burning	41 – 74	47 – 84	51 – 91	53 – 97
Leakage of natural gas	3 – 4	7 – 10	14 – 20	19 – 29
Coal mining	20	24	28	30
Other nonbiogenic sources	1	1	1	1 – 2
Total production	176 – 351	210 – 413	244 – 469	264 – 503
Sinks				
Reaction with OH	210	230	270	290
Flux into stratosphere	44	48	56	60
Soils	24	26	30	32
Total decomposition	278	304	356	382

Table 15.40 Contribution of tropical areas to the global flux (10^{12} g/yr) of CH_4, N_2O, H_2, and CO between different ecosystems and the atmosphere (After Seiler and Conrad, 1985)

	Total	Tropical areas
CH_4 emission by rice paddies	70–170	67–162
CH_4 emission by wetlands	11–57	9–46
CH_4 emission by ruminants	72–99	20–50
CH_4 emission by termites	2–5	2–5
CH_4 emission by biomass burning	53–97	42–78
CH_4 uptake by soils	32	25
N_2O emission from soils	4.5–17	3–15
N_2O emission from fertilized rice paddies	<0.1	<0.1
H_2 uptake by soils	70–110	30–40
CO uptake by soils	190–580	70–40
CO emission by plants	50–100	40–75
CO production by photochemical oxidation of CH_4 and NMHC	700–2200	560–1800

Crutzen (1985) estimated that the organic matter production in natural tropical wetlands is about 4 kg m^{-2} y^{-1} or 6×10^{15}g for the whole tropics. This much organic matter content decomposed anaerobically can cause an annual methane emission of $1.2 - 1.8 \times 10^{14}$g (Blake, et al. 1982). Firstly, no one knows for sure whether these figures and extrapolation are correct. Even if they were, we do not know the global implications of increased concentrations of these greenhouse gases in the upper atmosphere. It is, therefore, desirable to increase mean on-farm rice yield rather than to increase area under production.

It is important to identify rice production constraints in relation to soil and water management and to develop technological systems to alleviate these constraints. Production from wetlands already under cultivation can be increased by developing technology that facilitates multiple cropping. The soil and water resources of the wetlands should be so managed as to meet edaphic requirements for both semi-aquatic rice and the following upland crops grown on the residual soil moisture. In this connection the use of appropriate tillage methods of seedbed preparation is vital in terms of managing soil structure.

H. Forestry

That the intensive arable land-use based on motorized farm operations and mechanical tillage leads to soil degradation is well recognized. It is a general belief, however, that the establishment of perennial crops, plantations, and man-made forests following the removal of natural forests is an ecologically better alternative because the latter improves soil structure and restores soil fertility. The belief in restorative effects of man-made forests is an over-generalization because, with improper management, severe soil compaction and accelerated soil erosion and land degradation are also observed under plantation crops. Undoubtedly with good management, however, the chances of attaining economical success and ecologically compatible systems are better with establishment of plantation crops such as cocoa, coffee, tea, oil palm, banana, rubber than with annuals. In addition to being native to the humid tropics, perennial crops protect the soil against raindrop impact and high insolation due to continuous vegetation cover.

Soil erosion and erosion-related degradation can be severe in plantations of semi-deciduous trees, whenever cover crops are not adequately established, and uncontrolled burning is frequent. In Kenya Maher (1950) advocated intensive soil conservation measures to control erosion under wattle. In Trinidad, for example, Bell (1973) observed severe erosion of 152 t/ha/year in frequently burned teak plantation. In India, Jose and Koshy (1972) observed decrease in organic matter content under teak in comparison with natural forest. Similar observations were made under teak plantation in Senegal (Maheut and Dommergues, 1960).

Lundgren (1978) reviewed the literature that brought to attention the yield decline of man-made forests and other plantation crops due to progressive soil deterioration. It is well established, for example, that the growth of second rotation *Pinus radiata* in Australia is poorer than the first rotation (Lewis and Harding, 1963; Keeves, 1966). Lower production of the second rotation was due to soil compaction, loss in soil organic matter and nitrogen, and decrease in bases (Hamilton, 1965; Florence, 1967; Muir, 1970; Boardman, 1973). Similar observations of decline in soil quality under tree crops have been reported from Kenya (Robinson et al., 1966; Robinson, 1967), Swaziland (Evans, 1975) and elsewhere in the tropics (Chaffey, 1973; Lawton, 1973; Evans, 1976; Reynolds

Table 15.41 Methane production rates using the CH_4 flux rates measured by Cicerone and Shetter (1982) and corrected for different temperature (Seiler et al., 1984)

Region	Area harvested (m^2)	Vegetation period (days)	Average temperature (°C)	CH_4 release rate (g/m^2/h)	CH_4 emission (10^{12} g/yr)
USA, Japan, Europe	5×10^{10}	150	25–28	6.6–11.2	1.2–2.0
Asia, Africa, South America (wet season)	98×10^{10}	120	27–30	9.7–14.3	27.4–40.4
Asia, Africa, South America (dry season)	42×10^{10}	150	22–28	4.0–11.2	6.1–16.9
Total	145×10^{10}	—	—	—	34.7–59.3

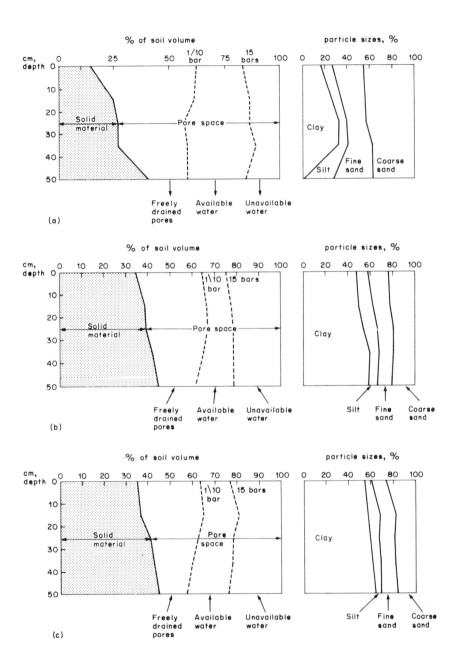

Figure 15.31 Partitioning of total soil volume into solid and voids. (a) Soil under natural forest, (b) plantations of *Pirus patula* and (c) under *Cypressus Insitanica* (Lundgren, 1978). *Reproduced by permission of the author*

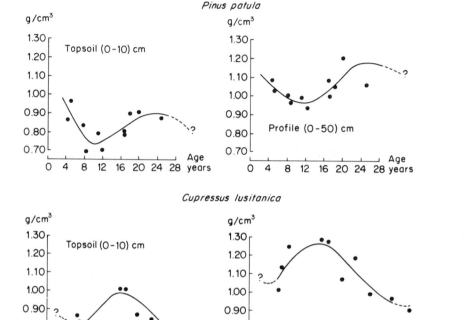

Figure 15.32 Changes in bulk density of the surface soil with time in plantations of *Pinns patula* and *Cypressus Insitanica* in Tanzania (Lundgren, 1978). *Reproduced by permission of the author*

and Wood, 1977; and Lundgren, 1978). Growth of tree crops such as *Gmelina arborea* are very sensitive to soil compaction (Johnson, 1976).

Lundgren (1978) conducted a detailed study regarding the effects of plantation crops on physical and nutritional properties of soils in Tanzania. The soil

Table 15.42 Effects of *Pinus patula* and *Cupressus lusitanica* in Shume on soil bulk density in comparison with the natural forest (Lundgren, 1978). *Reproduced by permission of the author*

Treatment	Soil bulk density (g/cm^3) at different depths (cm)				
	0–10	10–20	20–30	30–40	40–50
Natural forest	0.58	0.64	0.73	0.79	1.26
P. patula	0.85±0.08	1.01±0.08	1.07±0.08	1.10±0.10	1.13±0.11
C. lusitanica	0.85±0.14	1.08±0.13	1.21±0.14	1.26±0.14	1.28±0.16

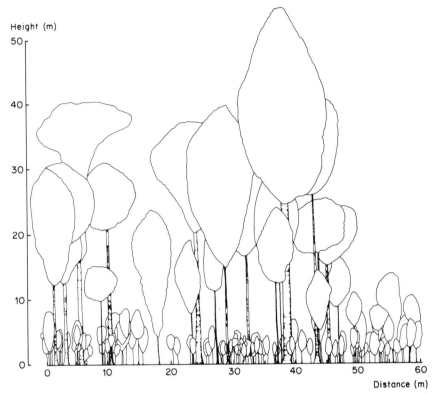

Figure 15.33 Profile diagram, 60 × 7.5m, along the western side of the sample plot under natural forest (Lundgren, 1978). *Reproduced by permission of the author*

under natural forest had low bulk density in comparison with that under plantation crops (Table 15.42). Consequently, the total pore volume was much higher with high percentage of free draining pores under natural forest than under *Pinus* or *Cupressus* (Fig. 15.31a, b and c). Lundgren's data showed that soils under *Cupressus* became progressively more compact than the pine soils (Fig. 15.32). As a result of alterations in soil structure and the pore size distribution, the water retention at field capacity and at wilting point was also adversely affected (Fig. 15.31a, b and c). As a result of this decrease in the available water capacity, the drought is a major growth limiting factor even in plantation crops. In addition to changes in soil physical properties, Lundgren (1978) also observed decrease in soil chemical fertility under plantation in comparison with the natural forest.

Based on these studies Lundgren (1978) concluded that without special soil conservation measures, the change in landuse from natural forest to plantation crops will inevitably result in soil deterioration in the form of decreased soil organic matter content and nutrient levels and loss of topsoil structure and porosity. The magnitude and rate of deterioration, however, will depend on soil conditions, climatic environments, and management. The replacement of multi-storey canopy (Fig. 15.33) of natural forest by monoculture plantation affects soil fauna, the amount of leaf litter, root biomass, and water and energy balance. These ecological alterations lead to similar effects on soil as do annual grain crops. The difference is more in degree than in kind.

It is important to realise, therefore, that replacement of forest by perennials or by annuals also leads to decline in soil quality as do arable landuse. The degree of deterioration, in both cases, depends on soil and crop management systems. More important than the growth habit of a given crop, therefore, is the soil and crop management practices adopted. High soil quality can be preserved even by growing seasonal crops as long as effective soil conservation measures are adopted. On the other hand, rapid soil deterioration is inevitable consequence if soil management is inadequate even when natural vegetation is replaced to estabish man-made forests and perennial crops. The most important aspect of soil management is not *what* is grown but *how* it is being grown.

V. CONCLUSIONS

Soil physical properties are more favourably regulated by the traditional farming systems than by intensive agricultural systems based on high stocking rate or monocropping. Mechanization of farm operations has pronounced adverse effects on soil physical properties. All systems that provide a continuous ground cover, maintain regular addition of organic matter to the soil, and promote biotic activity have favourable effects on soil physical properties e.g. mulching, mixed cropping, cover crops, agro-forestry, etc.

Sustainable agricultural systems that preserve the resource base for continuous use must, by necessity, be based on evolutionary change from the traditional system by gradual improvements rather than by rapid transformations. These evolutionary changes must consider both bio-physical and socio-economic circumstances of farmers. The term 'stability' in this context, therefore, implies a gradual change rather than a static system. Mulching, agro-forestry, mixed cropping, planted fallows, etc., are important subsystems of these evolving sustainable agricultural systems of the tropics.

REFERENCES

Aina, P.O., 1979, Soil changes resulting from long-term management practices in western Nigeria, *Soil Sci. Soc. Amer. J.*, **43**, 173–177.

Aina, P.O., Lal, R., and Taylor, G.S., 1979, Effects of vegetal cover on soil erosion on an Alfisol. In R. Lal, and D.J. Greenland, (eds), *Soil Physical Properties and Crop Production in the Tropics*, J. Wiley & Sons, Chichester, UK, pp. 501–507.

Akobundu, I.O., 1982, Live mulch crop production in the tropics, *World Crops*, **1982**, 125–126, 114–145.

Allan, W., 1965, *The African Husbandman*, Oliver and Boyd, Edinburgh.

Aweto, A.O., 1981, Organic matter build-up in fallow soil in a part of south-western Nigeria and its effects on soil properties, *J. Biogeography*, **8**, 67–74.

Barnes, D.L., 1981, Residual effects of grass leys on the productivity of sandy granite-derived soils. 1. Harvested leys, *Zimbabwe J. Agric. Res.*, **19**, 51–67.

Bell, T.I.W., 1973, Erosion in the Trinidad teak plantation. *Commonw. For. Rev.*, **52**, 223–233.

Berger, J.M., 1964, Cultural soil profiles in Central Ivory Coast, *Cahiers ORSTOM Pédol.*, **2**, 41–69.

Blake, D.R., Mayer, E.W., Tyler, St. C., Makide, D.C., Montagne, D.C., and Rowland, F.S., 1982, Global increase in atmospheric methane concentration between 1978 and 1980. *Geophys. Res. Lett.* **9**, 447–480.

Boardman, R., 1973, Longterm productivity. Position paper. Forest Research Working Group No. 3. Third Meeting, Caloundra, Qld., 13 p.

Bridge, B.J., Mott, J.J., Winter, W.H., and Hartigan, R.J., 1983, Improvement in soil structure resulting from sown pastures on degraded areas in the dry savannah woodlands of Northern Australia, *Aust. J. Soil Res.*, **21**, 83–90.

Buanec, B. Le, 1972, Ten years of mechanised cultivation on a Central Ivory Coast watershed. Evolution of fertility and productivity, *Agron. Trop.*, **27**, 1191–1211.

Chaffey, D.R., 1973, Decline in productivity under successive rotations of forest monoculture. Land Resources Div., ODA, U.K., Mimeo.

Charreau, C., 1972, Problèmes posés par l' utilisation agricole des sols tropicaux par des cultures annuelles, *L' Agron. Trop.*, **27**, 905–929.

Cicerone, R.J., Shetter, J.D., and Delwiche, C.C., 1983, Seasonal variation of methane flux from a California rice paddy, J. Geophys. Res. 88, 11022–11024.

Cointepas, J.P., and Makilo, R., 1982, Evaluation of soil evolution under intensive cultivation in an experimental station under humid tropical conditions, *Cahiers ORSTOM Pédol.*, **19**, 271–282.

Crutzen, P.J., 1985, The role of the tropics in atmospheric chemistry. Symp. Proc. 'Climatic, Biotic, and Human Interactions in the Humid Tropics: Vegetation and Climate Interactions in Amazonia', 25 Feb.–1 March, 1985, INPE, Sao Jose Dos Campos, SP, Brazil.

De, R., and Singh, S., 1981, Management practices for intercropping systems. *Proc. Int.' Workshops on Intercropping, 10–13 January, 1979*, ICRISAT, Hyderabad, India, pp. 17–21.

Dennison, E.B., 1959, The maintenance of soil fertility in the southern Guinea Savanna Zaria of northern Nigeria, *Trop. Agric.*, **36**, 171–178.

De Vleeschauwer, D., Lal, R., and De Boodt, M., 1978, The comparative effects of surface applications of organic mulch versus chemical soil conditions on physical and chemical properties of the soil and on plant growth, *Catena*, **5**, 337–349.

Diatta, S., 1974, Effect of cultivation on the plateau soils of continental Casamance. Results of 2-year experiments, *Agron. Tropicale*, **30**, 344–351.

Djokoto, R.K., and Stephens, D., 1961, Thirty long-term fertilizer experiments under continuous cropping in Ghana, I., *Exp. J. Exp. Agric.*, **29**, 181–195.

Dudal, R., and Hrabovszky, J., 1981, Rice soils for food production. p. 31–41. In *Proc. Symp. on Paddy Soils, Inst. Soil Sci. Academia Sinica*, Nanjing, China.

Dunne, T., 1979, Sediment yield and land-use in tropical catchments, *J. Hydrology*, **42**, 281–300.

Evans, J., 1975, Two rotations of *Pinus patula* in the Usutu Forest, Swaziland. *Commonw. For. Rev.*, **54**, 69–81.

Evans, J., 1976, Plantations: productivity and prospects. *Aust. For.*, **39**, 150–163.

FAO, 1974, Shifting cultivation and soil conservation in Africa, *Soils Bull.*, **24**, 248 pp.

Fauck, R., Moureau, C., and Thomanu, C., 1969, Changes in soils of Séfa (Casamance, Senegal) after fifteen years of continuous cropping, *Agron. Trop. (Paris)*, **24**, 263–301.

Feller, C., and Milleville, P., 1977, Evolution of recently cleared soils in the region of

Terres Neuves (eastern Senegal). Methods of study and evolution of the principal morphological and physico-chemical characteristics, *Cahiers ORSTOM Biol.* **12**(3), 199–211.

Florence, R.G., 1967, Factors that may have a bearing upon decline of productivity under forest monoculture. *Aust. For.*, **31**, 50–71.

Grigg, D.B., 1974, *The Agricultural System of the World: an Evolutionary Approach*, Cambridge University Press, London.

Hamilton, C.D., 1965, Changes in soils under *Pinus radiata*. *Aust. For.*, 275–289.

Hekstra, P., Andriesse, W., de Vries, C.A., and Bus, G., 1983, Wetland utilization research project in West Africa, ILRI, Wageningen, The Netherlands, p. 25.

Henin, S., 1960, Some considerations on rotations, *Oléagineux*, **15**, 9–12.

Holzapfel-Pschorn, A., and Seiler, W., 1985, Methane emission during a vegetation period from an Italian rice paddy. *J. Geophys. Res.*, (In Press).

Howson, D.F., 1977, A recording cone penetrometer for measuring soil resistance, *J. Agr. Eng. Res.*, **22**, 209–212.

Hulugalle, N., Lal, R. and Ter Kuile, C.H.H., 1985, Amelioration of soil physical properties by mucuna following mechanized land clearing of a tropical rainforest, *Soil Sci.* (In Press).

Hulugalle, N.R. and Lal, R., 1985, Soil water balance of intercropped maize and cowpea grown in a tropical hydromorphic soil in western Nigeria, *Agron. J.* (In Press).

Hurault, J., 1968, Effects of overgrazing on the high plateaux of Ademawa (Cameroons), *Nat. Resources Res.*, **6**, 463–468.

Jameson, J.D., and Kerkham, R.K., 1960, The maintenance of soil fertility in Uganda, I. Soil fertility experiments at Serere, *Emp. J. Exp. Agric.*, **28**, 179–192.

Johnson, N.E., 1976, Biological opportunities and fast growing plantations. *J. of Forestry*, **74**, 206–211.

Jones, R.K., and McCown, R.L., 1983, Research on a no-till, tropical legume–ley farming strategy, in J.G. Ryan (ed.), *Eastern Africa ACIAR Consultation on Agricultural Research*, ACIAR, Canberra, pp.108–121.

Jose, A.I., and Koshy, M.M., 1972, A study of the morphological, physical and chemical characteristics of soils as influenced by teak vegetation. *Indian For.*, **96**, 338–348.

Juo, A.S.R., and Lal, R., 1977, The effect of fallow and continuous cultivation on the chemical and physical properties of an Alfisol in western Nigeria, *Plant Soil*, **47**, 567–584.

Jurion, F., and Henry, J., 1969, *Can primitive farming be modernized?* INEAC Hons. Series 1964, 457 pp.

Kamanoi, M., Hayashi, S., and Bunpromma, K., 1983, Maize cultivation, in K. Kyuma, and C. Pairintra, (eds), *Shifting Cultivation*, Ministry of Science and Technology, Bangkok, Thailand, pp.144–153.

Kang, B.T., Wilson, G.F., and Sipkens, L., 1981, Alley cropping maize (*Zea mays*) and Leucaena in southern Nigeria, *Plant Soil*, **63**, 165–179

Kannegieter, A., 1967, Zero cultivation and other methods of reclaiming Pueraria fallowed land for food crop production in the forest zone of Ghana, *Trop. Agric.* (Ceylone), 123, 1–23.

Kannegieter, A., 1969, The combination of a short term *pueraria* fallow, zero cultivation and fertilizer application: its effect on a following maize crop, *Trop. Agric.* (Ceylone), **125**, 1–18.

Keeves, A., 1966, Some evidence of loss of productivity with successive rotations of *Pinus radiata* in the southeast of South Australia, *Aust. For.*, **30**, 51–63.

Kellman, M.C., 1969, Some environmental components of shifting cultivation in upland Mindanao, *J. Trop. Geog.* **28**, 40–56.

Khatibu, A.I., Lal, R., and Jana, R.K., 1984, Effects of tillage methods and mulching on erosion and physical properties of a sandy clay loam in an Equatorial warm humid region, *Field Crops Res.* **8**, 239–254.

Lal, R., 1973, Soil temperature, soil moisture, and maize yield from mulched and unmulched tropical soils, *Plant Soil*, **40**, 129–143.

Lal, R., 1975, Role of mulching techniques in tropical soil and water management, *IITA Tech. Bull.* 1, Ibadan, 38 pp.

Lal, R., 1976, Soil erosion problems on an Alfisol in western Nigeria and their control, *IITA Monograph 1*, Ibadan, 208 pp.

Lal, R., 1977, Soil management systems and erosion control, In D.J. Greenland and R. Lal (eds), *Soil Conservation and Management in the Tropics*, J. Wiley & Sons, Chichester, UK, pp. 81–86.

Lal, R., 1979, Mulch, in R.W. Fairbridge, and C.W. Finkl, Jr., *The Encyclopaedia of Soil Science*, Encyclopaedia of Earth Sciences, Vol. XII, Dowden, Hutchinson & Ross, Inc., Stroudsburg, Pennsylvania, pp.314–319.

Lal, R., 1981, Soil erosion problems on an Alfisol in western Nigeria VI. Effects of erosion on experimental plots, *Geoderma*, **25**, 215–230.

Lal, R., 1983, Soil erosion in the humid tropics with particular refernece to agricultural land development and soil management, *IAHS Publ.*, **140**, 221–239.

Lal, R., 1984, Mechanized tillage systems effects on soil erosion from an Alfisol in watersheds cropped to maize, *Soil Till. Res.*, **4**, 349–360.

Lal, R., 1985, Mechanized tillage systems effects on properties of a tropical Alfisol in watersheds cropped to maize, *Soil Till. Res.* (In Press).

Lal, R., and Kang, B.T., 1982, Management of organic matter in soils of the tropics and subtropics, *Proc. XII Int. Cong. Soil Sci.*, **IV**, 152–178.

Lal, R., Wilson, G.F., and Okigbo, B.N., 1978, No-till farming after various grasses and leguminous cover crops in tropical Alfisol I. Crop performance, *Field Crops Res.*, **1**, 71–84.

Lal, R., Wilson, G.F., and Okigbo, B.N., 1979, Changes in properties of an Alfisol produced by various crop covers, *Soil Sci.*, **127**, 377–382.

Lal, R., De Vleeschauwer, D., and Malafa Nganje, R., 1980, Changes in properties of a newly cleared tropical Alfisol as affected by mulching, *Soil Sci. Soc. Amer. J.*, **344**, 827–833.

Lawton, R.M., 1973, A review of the possible causes of yield decline in second rotation conifer plantations and some suggested lines of investigation. Land Resources Division, ODA, Mimeo, 5 pp.

Lewis, N.B., and Harding, J.H., 1963. Soil factors in relation to pine growth in South Australia. *Aust. For.*, **27**, 27–34.

Lundgren, B., 1978, Soil conditions and nutrient cycling under natural and plantation forests in Tanzanian Highlands, Dept. of Soils Swedish University of Agricultural Sciences, Uppsala, Sweden, 429 pp.

Lundgren, B., and Lundgren, L., 1972, Comparison of some soil properties in one forest and two grasslands eco-systems on Mount Mer, Tanzania, *Geografiska Annaler*, **54**(9), 227–240.

Maheut, J. and Dommergues, Y., 1960, Les teckeraise de Casamance. Capacité de production des peuplements; charactéristiques biologiques et maintien de potentiel productif des sols. *Bois For. Trop.*, **70**, 25–42.

Maurya, P.R., and Lal, R., 1980, Effects of no-tillage and ploughing on roots of maize and leguminous crops, *Exp. Agric.*, **16**, 185–193.

Maurya, P.R., and Lal, R., 1981, Effects of different mulch materials on soil properties and on the root growth and yield of maize (*Zea mays*) and cowpea (*Vigna unguiculata*), *Field Crops Res.*, **4**, 33–45.

McCown, R.L., Haaland, G., and de Haan,C., 1979, The interaction between cultivation and livestock production in a semi-arid Africa, pp. 297–332, in A.E. Hall et al. (eds), *Agriculture in Semi-Arid Environments*, Berlin, Springer-Verlag.

Mehanni, A.H., 1974, Short term effect of some methods of improving soil structure in red-brown earth soils of the northern irrigation areas, Victoria, *Aust. J. Exp. Agric. Anim. Husb.*, **14**, 689–693.

Mishra, B.K., and Ramakrishnan, P.S., 1983, Slash and burn agriculture at higher elevations in north-eastern India. I. Sediment, water and nutrient losses, *Agric. Ecosyst. Env.*, **9**, 69–82.

Mongi, H.O., and Huxley, P.A., 1979, *Soils Research in Agroforestry*, ICRAF, Nairobi, 584 pp.

Muir, W.D., 1970, The problems of maintaining site fertility with successive croppings, *Aust. J. Sci.*, **32**, 316–324.

Natarajan, M., and R.W. Willey, 1980, Sorghum–pigeonpea intercropping and the effects of plant population density 2. Resource use, *J. Agric. Sci., Camb.*, **95**, 5–65.

Ngatunga, E.L.N., Lal, R., and Uriyo, A.P., 1984, Effects of surface management on runoff and soil erosion from plots at Mlingano, Tanzania, *Geoderma*, **33**, 1–12.

Nicou, R., 1975, Contribution to the study and improvement of the porosity of sandy and sandy-clay soils in the dry tropical zone, *Agron. Trop.*, **29**, 110–1127.

Ofori, C.S., and Nandy, S., 1969, The effect of method of seed cultivation on yield and fertilizer response of maize grown on a forest ochrosol. *Ghana J. Agric. Sci.* **2**, 19–24.

Ofori, C.S., 1973, Decline in fertility status of a tropical ochrosol under continuous cropping, *Exp. Agric.*, **9**, 15–22.

Peers, A.W., 1962, Establishment of grass leys by undergrowing to cereals, *E. Afr. Agric. For. J.*, **27**, 145–149.

Pereira, H.C., and Beckley, V.R.S., 1952, Grass establishment on eroded soil in a semi-arid African reserve, *Emp. J. Exp. Agric.*, **21**, 1–15.

Pereira, H.C., and Jones, P.A., 1954, A tillage study in Kenya Coffee II. The effects of tillage practices on the structure of the soil. *Emp. J. Expl. Agric.*, **22**, 323–331.

Pereira, H.C., Chenery, E.M., and Mills, W.R., 1954, The transient effects of grasses on the structure of tropical soils, *Emp. J. Exp. Agric.*, **22**, 148–160.

Pereira, H.C., Wood, R.A., Brzostowski, H.W., and Hosegood, P.A., 1958, Water conservation by fallowing in semi-arid tropical East Africa, *Emp. J. Exp. Agric.*, **26**, 213–228.

Pereira, H.C., and Hosegood, P.H., 1961, The productivity of semi-arid thorn shrub country under intensive management, *Emp. J. Exp. Agric.*, **29**, 269–286.

Pereira, H.C., Hosegood, P.H., and Dagg, M., 1967, Effects of tied ridges, terraces and grass leys on a lateritic red soil in Kenya, *Exp. Agric.*, **3**, 89–98.

Prahbakara, J., and Dakshinamurti, C., 1976, Dynamics of soils structure in relay cropping: intensive cultivation technique. *Mededelingen van de Faculteit Landbouwwetenschappen*, Rijksuniversiteit Gent **41**(1), 459–462.

Prihar, S.S., Singh, R., Singh, N., and Sandhu, K.S., 1979, Effects of mulching previous crops or fallow on dryland maize and wheat, *Exp. Agric.*, **15**(2), 129–134.

Robinson, J.B.D., 1967a, The effects of exotic softwood crops on the chemical fertility of a tropical soils. *E. Afr. Agric. For. J.*, **33**, 175–191.

Robinson, J.B.D., 1967b, The preservation unaltered of mineral nitrogen in tropical soils and soil extracts. *Plant and Soil* **27**, 53–80.

Robinson, J.B.D., Hosegood, P.H., and Dyson, W.G., 1966, Note on a preliminary study of the effects of an East African softwood crop on the physical and chemical conditions of a tropical soil. *Commonw. For. Rev.*, **45**, 359–365.

Ruthenberg, H., 1980, *Farming systems in the tropics*, Oxford University Press, Oxford, 3rd Edn.

Schöningh, E., and Alkämper, J., 1984, Effects of different mulch materials on soil properties and yield of maize and cowpea in an eastern Amazon Oxisol, *Proc. 1st Symp. on the Humid Tropics, November 1984*, Belem, Brazil.

Seiler, W., 1984, Contribution of biological processes to the global budget of methane in the atmosphere. In *Current Perspective in Microbial Ecology* (M.J. Klug and C.A. Reddy, eds.) pp. 468–477, ASM, Washington, D.C., 1984.

Seiler, W., and Conrad, R., 1985, Exchange of atmospheric trace gases with anoxic and oxic tropical ecosystems. Symp. Proc. 'Climatic, Biotic, and Human Interactions in the Humid Tropics : Vegetation and Climate Interactions in Amazonia'. 25 Feb.–1 March, 1985, INPE, Sao Jose Dos Campos, SP, Brazil.

Shukla, L.N., and Ravalo, E.J., 1976, Soil compaction in sugarcane fields due to transit casts, *Syar Y Azucar*, 7, 26–36.

Siband, P., 1972, Soil evolution under traditional cultivation in Haute-Casamance (Senegal). Main results, *Agron. Trop.*, 27, 574–591.

Simmonds, N.W., 1984, The state of the art of farming systems research, Edinburgh School of Agric. U.K. *Mimeo*, 135 pp.

Soong, N.K., 1980, Influence of soil organic matter on aggregation of soils in Peninsular Malaysia, *J. Rubber Res. Inst. Malaysia*, 28(1), 32–46.

Steiner, K.G., 1982, *Intercropping in tropical smallholder agriculture with special reference to West Africa*, GTZ, Eschborn, FDR.

Takahashi, T., Nagahori, K., Mongkolsawat, C., and Losirikul, M., 1983, Run-off and Soil Loss, in K. Kyuma, and C. Pairintra, (eds), *Shifting Cultivation*, Ministry of Science and Technology, Bangkok, pp. 84–109.

Thomas, D.B., Barber, R.G., and Moore, T.R., 1980, Terracing of cropland in low rainfall areas of Machakos Dist., Kenya, *J. Agric. Eng. Res.*, 25, 57–63.

Vergara, N.T., 1982, New directions in agro-forestry: the potentials of tropical legume trees, *I.P.I.*, East-West Centre, Honolulu, 36 pp.

Vine, H., 1953, Experiments on the maintenance of soil fertility at Ibadan, Nigeria, 1922–1951, *Emp. J. Exp. Agric.*, 21, 65–85.

Wade, M.K., and Sanchez, P.A., 1983, Mulching and green manure applications for continuous crop production in the Amazon Basin, *Agron. J.*, 75, 39–44.

Watters, R.F., 1971, Shifting cultivation in Latin America, *FAO Forestry Dev. Paper*, 17, 305 pp.

Webster, C.C., and Wilson, P.N., 1980, *Agriculture in the Tropics*, Longman, London, 2nd Edn.

Wijewardene, R., and Waidyanatha, P., 1984, *Systems, Techniques and Toos: Conservation Farming*, Dept. of Agric. Sri Lanka, Colombo, 38 pp.

Wilkinson, G.E., 1975, Effect of grass fallow rotations in Northern Nigeria, *Trop. Agric.*, 52, 97–103.

Wilkinson, G.E., and Aina, P.O., 1976, Infiltration of water into 2 Nigerian soils under forest and subsequent arable cropping, *Geoderma*, 15, 51–59.

Wilkinson, G.E., and Aina, P.O., 1977, Shifting tropical forest soils in Nigeria from bush to arable crops: the effect on the infiltration of water, *Geoderma*, 15, 51–59.

PART IV

Towards Improvements In Tropical Agriculture

Chapter 16
An Ecological Approach To Tropical Agriculture

I	INTRODUCTION	689
II	EDAPHIC AND ECOLOGICAL CONSTRAINTS OF HIGH-INPUT MECHANIZED AGRICULTURE	691
	A. Soil Compaction	692
	B. Soil Erosion	693
	C. Drought Stress	693
III	SOIL RESOURCES: CONSTRAINTS AND POTENTIALS	694
IV	REQUIREMENTS OF A SUSTAINABLE, IMPROVED AGRICULTURAL SYSTEM	695
V	ECOLOGICAL AGRICULTURE: A SUSTAINABLE SYSTEM	697
VI	PROMISING TECHNOLOGICAL PACKAGES	698
	A. Mulch farming	698
	(i) No-tillage or reduced tillage	699
	(ii) Planted fallows	699
	B. Mixed cropping and crop rotations	702
	C. Agro-forestry	703
	D. Mixed farming	703
VII	A CASE STUDY FOR THE HUMID TROPICS	703
VIII	RESEARCH AND DEVELOPMENT PRIORITIES	705

I. INTRODUCTION

The high-input temperate zone agriculture of Western Europe and North America has broken the yield barriers and dramatically increased food production over the past 50 years. A combination of suitable inputs and improved varieties has also been the basis of the so called 'Green Revolution' in South and Southeast Asia. Elsewhere, however, attempts at revolutionizing the traditional agriculture by high-input mechanized farming have often been ecological and economic disasters (Pagel, 1969; Gretzmacher, 1977; Ormerod, 1978; Fisher and Matter, 1980; Bauer, 1977; Henzell, 1975). Agriculture in tropical Africa is at the crossroads and urgently needs direction in adopting a system that guarantees a substantial increase in food production to avert mass starvation and famine. The important issue to be resolved by agricultural scientists and planners is the approach to be adopted in transforming traditional agriculture into a sustainable enterprise that can provide the basic necessities of life and meet the aspirations of a modern society.

High-input agriculture has left many questions unanswered pertaining to ecological, economic, social and political issues even in relatively stable temperate zone environments. For example, the inputs of fertilizers and agricultural chemicals amount to 2.5 million kcal., e.g., 38.9 per cent of the total inputs needed to produce one hectare of corn in the United States of America (Pimental and Pimental, 1979). The output/input ratio for corn production in the U.S.A. is 2.93 in comparison with 16.67 for Mexico (Table 16.1). As it now stands Third World agriculture is much more energy efficient than Western agriculture (Le Pape and Mercier, 1983). But efficient as it is, the net output of Third World agriculture and particularly that of tropical Africa leaves much to be desired. For example, Bayliss-Smith (1982) studied the energy flow of a traditional farm in New Guinea and observed that the net farm output was 10.4 MJ per person per day. In comparison the output in North American agriculture for 1979 was estimated at 1500 MJ per person per day, about 150 times more. These are the two extreme examples, on the one hand of a self-sufficient agricultural society dependent on its own human energy resources, and on the other hand of a mass-consumption society heavily dependent on fossil fuels. The constraint lies in the low net output of the traditional tropical agricultural systems in which large areas are kept under forest/bush fallow for fertility restoration. With extensive, low-input but self-sustained agriculture, one hectare can barely provide subsistence living for two persons (Bayliss-Smith, 1982). Does it mean that the energy consumption patterns of Third World agriculture will have to be increased to raise its net production? It is hoped that there are alternative strategies.

An abrupt break-away from low-input, self-sustaining, subsistence tropical

Table 16.1 The energy output/input ratio for some grain crops (Le Pape and Mercier, 1983)

Crop	Third World	Western Countries
Wheat	6.25 (India)	2 (USA)
Rice	7.7 (Philippines)	1.35 (USA)
Corn	16.67 (Mexico)	2.7 (USA)
Sorghum	20	3.1 (France)
Traditional tropical farming*	14.2–13.0 (New Guinea, South India)	
Collective mechanized farm	1.3 (USSR)	

* Bayliss-Smith (1982).

agricultural systems would create environmental problems of pollution, erosion, etc., of grossly aggravated dimensions because of the constraints of the harsh/fragile environments, and non-availability of the needed inputs. The undesirable outcome of the isolated examples of high-input mechanized agricultural schemes adopted in Africa and elsewhere in the humid and subhumid tropics warrants a critical evaluation of the approach needed to bring about lasting improvements in African agriculture.

II. EDAPHIC AND ECOLOGICAL CONSTRAINTS OF HIGH-INPUT MECHANIZED AGRICULTURE

Failure of large-scale mechanized agricultural schemes in Africa and elsewhere in the tropics has been attributed to both exogeneous and soil factors. Some important exogeneous factors include socio-economic conditions and lack of infrastructure, insufficient baseline data to enable adequate planning for resource development and management, and the much publicized failures of monsoons. The soil factors are, however, basic to all planning and their importance has often been underestimated.

The reasons for rapid declines in the fertility of soils in the tropics with intensive and continuous cultivation vary among soils, ecological conditions, crops grown, and management strategies. Among soil chemical and nutritional properties are the decline in soil organic matter content and pH. Deterioration of soil physical properties influences the mechanical, biological and nutritional properties of soil. There exists in some tropical soils a close relationship between crop yields and soil physical properties. Dabin (1962) computed three indices:

Structural Index $(A) = S_t \times P_u \times E_u$

Moisture Index $(B) = P_{up10} \times E_u / S_t$

Index of drying $(C) = A_c \times \log K$

Where S_t is structural stability, P_u is useful porosity, E_u is the available moisture, A_c is air capacity and K is water permeability. Dabin observed that the Structural Index (A) was highly correlated with rice yield in Nigeria, cacao yield in Ivory Coast, cotton production in Togo, and banana growth on well-drained soils. In addition the index of drying was correlated with banana growth on poorly drained soils and with cotton growth on heavy-textured soils containing montmorillonitic clay, i.e., Vertisols. Lal (1981) developed multiple regression equations and observed that maize grain yield was significantly correlated with accumulative soil erosion, organic carbon content, moisture retention at zero suction, and with the equilibrium infiltration rate:

$$Y = 1.79 - 0.007 \, (E) + 0.70 \, (OC) + 0.07 \, (M_o) + 0.002 \, I_c$$
$$r = 0.90$$

when Y is maize yield (t ha^{-1}), E is the cumulative soil erosion (t ha^{-1}), OC is organic carbon content (%), M_o is moisture content at zero suction (%), and I_c is infiltration capacity (cm). Similarly, cowpea grain yield was found to be correlated with erosion (E), organic carbon (OC), and the available water-holding capacity (AWC) as follows:

$$Y = 0.10 - 0.0005\ E + 0.044\ (OC) + 0.02\ (AWC) \qquad r = 0.72$$

Some major soil physical processes of edaphic importance are compaction, erosion and drought:

A. Soil compaction

Soil compaction is a major concern in large-scale motorized agriculture because of the increasing use of large and heavy equipment. Motorized tillage systems and other farm operations performed by heavy machines increase soil bulk density, decrease total porosity and macroporosity, reduce water retention and transmission characteristics, and increase the overall soil compaction. Soil under wheels supporting heavy loads is compacted within a few seasons of cultivation for grain crop production. Wheel-induced soil compaction is a severe constraint for most soils. Martin (1963) reported that the clearing of savanna vegetation in Ivory Coast for arable land-use changed the soil structure in the surface layer from niciform to polythedric and turned it into a 'powdery' structure within 2 years. This deterioration of soil structure was severe, irrespective of the type of agricultural equipment, and neither green manuring nor liming could restore the original structural attributes. For Alfisols in a subhumid region Lal (1985b) observed rapid declines in infiltration rate brought about by motorized farm operations. The decline in infiltration capacity was due to structural collapse and elimination of transmission pores caused by vehicle traffic. The soil was particularly compacted near the turning points.

Coarse-textured soils with low activity clays are easily compacted. Khan (1968) observed that abundance of sand fraction, especially 0.1–0.25 mm fraction, and low contents of clay and silt, render some West African Alfisols highly prone to soil compaction. Loess-derived soils containing high silt and fine sand fractions and low organic matter content are also easily compacted (Nicou and Chopart, 1979). In contrast, fertility of soils with high sand content is affected more by the lack of available nutrients than by soil compaction *per se* (Willatt, 1967a, b).

The combined effect of high bulk density and low total porosity is impaired seedling establishment, inhibited root development, and low fertilizer and water-use efficiency (Lugo-Lopez, 1960; Ghildyal and Satyanarayana, 1969; Lal, 1985b). The root growth of banana trees is also inhibited in compacted soils, as was reported for some soils of Cameroon, Ivory Coast and Madagascar (Godefroy, 1967). In Taiwan, Yang *et al.* (1978) observed that mechanical compaction affected root growth of surgarcane. In Andosols of low bulk density, predominance of retention over transmission pores decreases oxygen diffusion rate and adversely affects growth of perennial crops, e.g., coffee, cacao, etc. (Gavande, 1969a; b).

B. Soil erosion

Deterioration in soil physical properties causes accelerated soil erosion, and the latter aggravates the degradation of soil physical, chemical and biological properties. Obi and Asiegbu (1980) reported that the physical properties of eroded soils of eastern Nigeria were more degraded than those of uneroded soils. The topsoil of the world is being depleted at the rate of 0.7% per year (Brown, 1984). FAO (1984) estimated that if soil erosion continues unchecked between 1975 and the year 2000, about 18 per cent of the rainfed cropland in developing tropical regions will be lost and the rainfed crop productivity will fall by about 29 per cent. Soil erosion-induced desertification affects vast areas of semi-arid and arid lands. Dregne (1978) estimated that the world's arid lands affected by desertification range from 18.0, 53.6, 28.3 and 0.1 per cent, respectively, for slight, moderate, severe and very severe degrees of desertification.

Although the possible ecological implications of severe soil erosion in the tropics have caused much concern among scientists and planners around the world, there is a conspicuous lack of quantitative data regarding its magnitude and impact. The voluminous literature on the erosion hazard in the tropics is, regrettably, emotional rhetoric containing few facts. Most data on sediment transport in major tropical rivers are out-dated by 15 to 20 years, and the reliability of these data is questionable. Nevertheless, the isolated field plot measurements indicate erosion rates exceeding 300 to 500 t ha^{-1} yr^{-1} in Bangladesh, Java, Trinidad, Guatemala and several locations in tropical Africa. In order to assess the impact of different levels of soil erosion, however, one has to define (i) the rate of soil erosion, (ii) the rate of new soil formation, (iii) the economic and environmental impact of on-site and off-site damage, and (iv) the extent to which the degraded soils can be easily and economically restored.

The effects of accelerated soil erosion on crop yields have not been quantified for a wide range of soils, crops and ecological environments. A few reports (Lal, 1981, 1983b, c; El-Swaify et al., 1982; Mbagwu et al., 1984; Stocking, 1984) indicate that erosion can drastically reduce yields on soils with shallow effective rooting depth, on those where most nutrient capital is confined to the surface horizon, and where the subsoil is not favourable to root growth. In contrast, there are a few deep and fertile soils where even the high rates of erosion now observed have either no or only slight adverse effects.

C. Drought stress

The term 'drought stress' implies crop response to the integrated effects of low available water-holding capacity, high evaporative demand, and high soil and ambient temperatures. Compaction and high water run-off cause severe and frequent drought stress even in regions of high annual rainfall. The adverse effects of drought stress are aggravated by high soil temperatures, especially for soils of low available water reserves. Severe reductions in crop yields attributable to drought stress have been expressed for semi-arid (Kovda et al., 1976; Fisher, 1980) and for humid regions (Wolf, 1975; Hsiao et al., 1980). Drought stress has been shown to severely reduce yields of upland rice (Ahmadi, 1983a; b), cowpea (Kamara and Godfrey-Sam-Aggrey, 1979; Turk et al., 1980, Turk and Hall, 1980a,-c), cacao (Machado and Alvim, 1981), maize (Lal, 1985), soybeans (Sivakumar et al., 1981), sorghum (Rathore et al., 1981) and groundnut (Gatreau, 1977). Drought stress affects growth of leguminous crops by inhibiting nodulation (Willatt, 1967). Plants with deep root systems (some perennial grasses) are less affected than those with shallow root systems (Dagg, 1969). Bigort (1977) observed in Ivory Coast that yam, cotton, maize and rice yielded less in years of insufficient or infrequent rainfall. However, maize and rice were more susceptible to drought than cotton or yam. Crops susceptible to drought do not respond to fertilizers and other chemical amendments.

III. SOIL RESOURCES: CONSTRAINTS AND POTENTIALS

Both soil physical and nutritional constraints inhibit yields in the humid and subhumid tropics with soils containing predominantly low activity clays. Physically, these soils are easily compacted, have low available water-holding capacity, and are susceptible to accelerated soil erosion. Once the forest cover is removed, grain crop establishment and growth is inhibited by crusting, drought stress, high soil temperatures, and accelerated erosion. Chemically, soils of the humid and subhumid tropics have low nutrient reserves, especially of nitrogen, phosphorus and some trace elements. Organic matter content is generally low, and losses due to leaching, volatilization, run-off and erosion are high. Intensive use of these soils for grain crop production leads to very rapid depletion of the nutrient reserves.

Alfisols of the semi-arid tropics have additional problems of low and erratic rainfall distribution, high surface soil temperatures, and development of surface seal and crust (Arndt, 1965; Mott et al., 1979). Seedling establishment is hampered by mechanical impedance, supra-optimal soil temperatures, and drought. Heavy textured Vertisols, with high swell-shrink capacities, have severe water management and tillage problems. These soils are structurally unstable, and are highly susceptible to water erosion even on gentle slopes. Timely seedbed preparation is hindered by poor trafficability of these soils both when dry and when wet. Some soils of the semi-arid regions of West Africa are underlain by hardened plinthite (McNeil, 1964; Kellogg and Orvedal, 1969). The shallow layer above the hardened 'laterite' can be easily lost to accelerated soil erosion.

In addition to the serious lack of water in rainfed agriculture, soils of the

African Sahel and arid tropics are prone to wind erosion, are easily compacted, and are notoriously poor in organic matter content and essential plant nutrients. Historically, these soils have been used for nomadic herding and extensive farming for food crop production.

Ash-derived soils, alluvial flood-plains, and soils derived from basic rocks have generally favourable nutritional, physical and hydrological conditions for intensive food crop production. Only a small proportion of land surface in the tropics is covered by these soils.

In spite of soils of low inherent fertility with severe physical constraints, research station experimentation has shown that high and sustainable crop yields are achievable with 'appropriate systems of soils and crop management' (Lal, 1983a; Sanchez et al., 1982). If conditions are right, these research findings can be adapted under farmer's conditions to boost agricultural production. In spite of higher gross productivity, the net economic productivity of tropical ecosystems is often lower than that of temperate regions (Lieth, 1976).

IV. REQUIREMENTS OF A SUSTAINABLE, IMPROVED AGRICULTURAL SYSTEM

Creative agricultural technology should promote symbiosis between man and the tropical environments. The technology must conform to and address the specific ecological constraints and long-range consequences of the simplification of the complex, diverse and stable natural ecosystems. A strategy for increasing food supply must look beyond production economics and consider the vital issues of ecological stability and sustainability. For an improved agricultural system to succeed, it must have the following characteristics.

(i) Low-input and self-sustaining. The overall energy input must be regulated so that output:input ratio is high. The system must endeavour to reduce nutrient losses by effectively containing leaching and erosion, and improve nutrient capital through natural processes. Nitrogen deficiency being a major constraint in most soils of the tropics, several alternate sources will have to be tapped, e.g., symbiotic nitrogen fixation by legumes, and use of organic manures and compost, and enhancing use of Azolla. The effective use of phosphatic fertilizers can be increased through mycorrhizal infection. Nonetheless, it does not mean that synthetic fertilizers are to be avoided, but that soil and crop management systems should be so developed as to minimize dependence on costly synthetic inputs. The energy output:input ratio of even the West European and North American agriculture is low, and definitely a bad example to be followed. Hoping to increase and sustain agricultural output by dumping chemicals is bound to cause frustrations and disappointments. The energetics of a watershed management experiment at IITA comparing the output:input data for five systems of intensive land-use with that of traditional farming in the 1979, 1980, and 1981 cropping seasons are shown in Table 16.2. The traditional farming treatments were based on incomplete clearing by native tools (e.g. axe,

Table 16.2 Energetics* (GJ/ha) of different land management systems for maize–cassava–maize–cowpea rotation on an Alfisol in Nigeria

Treatments	1979			1980			1981		
	I	O	O:I	I	O	O:I	I	O	O:I
Traditional farming	3.6	8.4	2.4	33.7	125.6	3.7	3.7	16.9	4.7
Manual/No-till	7.8	25.9	3.3	36.9	235.5	6.4	11.3	52.7	4.5
Manual/Conventional tillage	7.8	26.4	3.4	36.4	188.4	5.2	12.6	56.2	4.5
Shearblade/No-till	3.7	32.9	8.9	36.9	219.8	6.0	11.3	66.5	5.9
Tree pusher, root rake/No-till	3.0	23.0	7.7	36.9	312.9	8.5	11.3	55.3	4.9
Tree pusher, root rake/Conventional tillage	3.0	29.1	9.7	36.4	282.6	7.8	12.6	36.8	2.9

* I, input; O, output.

machette), followed by *in situ* burning, and received no inputs of herbicides, fertilizers or any other amendments. The intensive land-use treatments also received no fertilizers for the appropriate chemicals. The uniform cropping system adopted was maize for 1979, cassava for 1980, and maize followed by cowpea for 1981.

When the labour and seed/cuttings inputs for traditional land clearing are fully accounted for, the low output:input ratio for the traditional farming systems is due to the low yields. The low crop yields were due to the effects of shading and competition by weeds. These results of low efficiency of the traditional farming methods are contradictory to those reported by Rappaport (1973) for New Guinea. For the study at IITA, the manual clearing methods proved to be the least efficient because of the slow and time-consuming clearing operations in comparison with rapid but ecologically less desirable machine-clearing. During the second and third years, however, the yields and efficiency of the machine-clearing treatments decreased due to soil compaction and accelerated run-off and soil erosion. The soil degradation problems were less severe with manual clearing and no-till system based on mulch farming.

(ii) Land restorative. The objective is not to maximize the yield but to increase and sustain the net output by preserving the natural resource. The desired high net output must be achieved with a minimum of soil degradation. The soil quality and its productive capacity must be preserved and improved by preventing soil erosion, promoting high biological activity of soil fauna, improving soil organic matter content, and by continuously replacing the nutrients removed in the harvested produce through effective recycling mechanisms. The production efficiency of a system must be evaluated in terms of its effect on the natural resource, i.e., yield per unit soil erosion, per unit loss in soil organic matter content, and per unit decline in effective cation exchange capacity or pH. The fertility-mining systems must be stopped.

(iii) Ecologically compatible. Simplified ecosystems are less stable than diversified systems. 'Ecosystem fragility' refers to the environmental instability created by simplification of a 'community'. The problems of accelerated soil erosion, rapid nutrient depletion, shift in floral and faunal population are the results of over-simplification of an ecosystem.

V. ECOLOGICAL AGRICULTURE: A SUSTAINABLE SYSTEM

The traditional tropical agricultural system based on fertility-restorative measures of shifting cultivation and bush fallow rotation are examples of ecological agriculture on the subsistence farming level. The basic concept is to retain their ecological stability while improving net output. The sustainable ecologically stable agricultural system should seek to increase net production without causing serious environmental changes and preserve soil productivity. Kiley-Worthington (1981) defined ecological agriculture as 'the establishment and maintenance of an ecologically self-sustaining low input, economically viable, small farming system managed to maximize production without causing large or long-term changes to the environment, or being ethically or aesthetically unacceptable'. Although misconceptions regarding tropical ecosystems still persist (Janzen, 1973) due to the lack of basic data, the literature surveyed in previous chapters indicates that many proven sub-systems or concepts are available which can be used as building-blocks for developing sustained-yield tropical agroecosystems. The basic components of these sustainable systems include:

(i) Vegetative cover: Continuous ground cover is necessary to provide a buffer against sudden fluctuations in micro- and mesoclimate, and prevent degradative effects of raindrop impact. Vegetative cover is an effective soil conserving measure.

(ii) Regular supply of organic matter: A regular and sizeable addition of organic material is essential to maintain a favourable level of soil organic matter content and to stimulate biotic activity. Structural collapse can be avoided by maintaining high organic matter content and by enhancing activity of soil fauna.

(iii) Nutrient recycling mechanisms: The nutrients leached out of the root zone must be captured and recycled for crop utilization. There are also obvious advantages to substituting biological N for synthetic fertilizers.

VI. PROMISING TECHNOLOGICAL PACKAGES

The basic concepts of a self-sustaining, low-input, diversified and efficient agricultural system must be sythesized into practical technological packages to suit the specific needs of farming communities in different agroecological regions of the tropics (Fig. 16.1). That is a tall order indeed, and the task of national and international research organizations should be to achieve just that. The following sections describe some potentially useful packages.

A. Mulch farming

The requirements of a continuous vegetative cover and a regular supply of organic matter can be met by adapting a range of mulch farming techniques (Jurion and Henry, 1969; Lal, 1975; Wade and Sanchez, 1983; Uexkull, 1982). The rate and frequency of mulch required is often different for different soils. Some of the cultural practices compatible with mulch farming principles are:

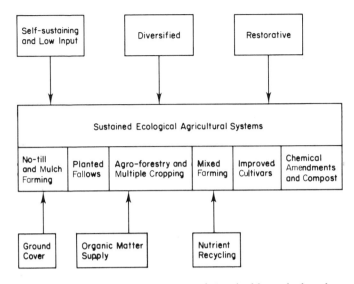

Figure 16.1 Concepts and components of sustainable agricultural systems

(i) No-tillage or reduced tillage: By eliminating or minimizing mechanical seedbed preparation, the previous crop residue is left on the soil surface as mulch. Research and development priorities must address the issues of how to adapt the no-till systems for food crop production in the tropics (Krause *et al.*, 1978; Lal, 1983a).

(ii) Planted fallows: Frequent use of appropriate grass or legume covers is necessary to restore soil structure, improve soil organic matter content, and for pest control (Talineau *et al.*, 1979). Suitable species of cover crops and planted fallows can restore soil structure and organic matter content more efficiently than natural bush fallow systems (Morel and Quantin, 1964; Robinson *et al.*, 1966; Stephens, 1967; Radwanski, 1969; Aweto, 1981). The suitable covers are different for different ecologies. If the concept of incorporating planted fallows in the rotation system is to be accepted, the cover crops must also have some immediate economic benefits for the farming community. As for example, some pasture and browse legumes, identified for tropical Australia, have the potential to accumulate a sizeable quantity of N (Vallis and Gardener, 1985). Seeding food crops through the dead or suppressed sod mulch must be done by the principles of no-till farming, in which the crop establishment is achieved by mechanical mowing (Wilson *et al.*, 1982) by chemical suppression (Akobundu, 1982) or after light grazing (McCown *et al.*, 1984). Some covers also have allelopathic effects on perennial weeds otherwise difficult to control.

While planted fallows have definite ameliorative effects on soil physical,

Table 16.3 Effects of 1 year of seeding cover crops and soil surface management on infiltration and soil bulk density (Unpublished data, R. Lal)

Treatment	Cumulative infiltration (cm)	Bulk density (g cm^3)	
		Overall	Fine earth
Bare soil	44	1.38	1.31
Straw mulch	119	1.39	1.27
Indigofera spp.	93	1.44	1.32
Desmodium triflorum	66	1.49	1.40
Paspalum notatum	33	1.43	1.35
Centrosema pubscens	73	1.49	1.41
Axonopus campestris	93	1.44	1.30
Arachis prostrate	86	1.47	1.36
Psophocarpus palustris	62	1.37	1.25

Table 16.4 Gravimetric soil moisture measurements (0–5 cm depth) under different cover crops and surface mulches during the dry season

Cover crop	11/11/76	18/11/76	24/11/76	30/11/76	6/12/76	14/12/76
0–5 cm						
Straw mulch	17.0	12.4	7.6	7.5	8.1	7.5
Bare	16.0	10.9	7.2	6.5	6.7	6.9
Desmodium	9.8	7.5	3.4	3.4	2.8	2.0
Arachis	16.6	7.9	5.4	5.3	4.3	3.5
Indigofera	16.5	10.8	5.3	3.9	5.5	5.3
Psophocarpus	13.2	9.4	6.9	5.8	5.3**	5.5**
Paspalum	12.8	8.7	6.9	2.9	2.7	2.4
Centrosema	12.9	8.8	6.3	6.5	6.2	5.5
Axonopus	15.1	7.8	3.0	3.5	3.9	3.9
5–10 cm						
Straw mulch	14.0	12.2	8.2	7.3	8.7	6.1
Bare	12.4	10.5	7.9	7.5	7.5	5.4
Desmodium	13.0	8.4	4.0	4.7	2.4	1.9
Arachis	14.8	8.4	5.2	5.4	3.5	3.2
Indigofera	10.2	6.1	3.9	5.7	6.1	4.9
Psophocarpus	11.3	10.3	4.7	8.0	4.2	5.4
Paspalum	12.0	8.3	5.0	4.0	4.6	2.7
Centrosema	10.2	8.1	7.2	6.5	5.8	5.4
Axonopus	11.1	6.2	3.6	3.2	3.8	3.9

**Mowed

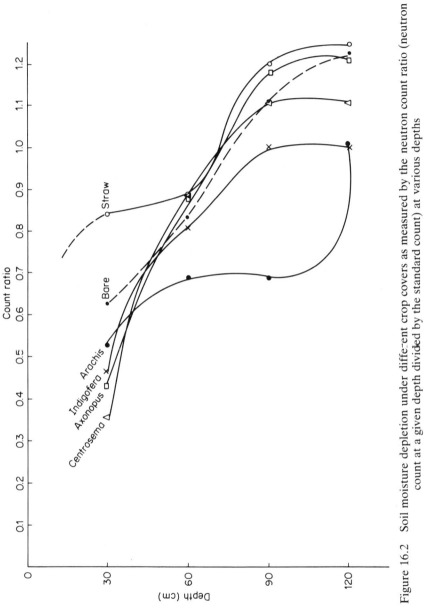

Figure 16.2 Soil moisture depletion under different crop covers as measured by the neutron count ratio (neutron count at a given depth divided by the standard count) at various depths

chemical and biological properties, fallow management for grain crop production remains an open question. Crops can be grown through the chemically killed sod provided that regrowth, either vegetative or through seeds, is not a severe problem. Light controlled grazing is an economically desirable practice for regions where mixed farming is possible provided that planted cover and weeds are effectively controlled. The practices of live mulch, green bed, etc, are definitely conservation-effective measures and have been extensively used for plantation crops, e.g., oil palm, rubber, etc. This technique is also potentially useful for growing seasonal grain crops provided that the planted fallow does not excessively compete for nutrients, water and light. Ideally, the planted fallow should be a low-growing, non-climbing legume with deep and tap root system. There is a scope for improvement.

An experiment was conducted at IITA, Ibadan, Nigeria to determine the alterations in soil physical properties under different systems of soil surface management. The soil had been cultivated for about 7 years and the surface layer was compacted and practically devoid of soil macrofauna. Some planted fallows were slow to establish, and therefore did not influence cumulative infiltration and bulk density during the first year as much as the residue mulch (Table 16.3). Once the cover had been fully established, the biotic activity and soil physical properties were more favourable under planted fallows than in ploughed bare or straw mulch treatments.

The root system of live covers influences soil moisture regimes. If not effectively suppressed, live covers can compete aggressively with grain crops. The data in Table 16.4 compare the soil moisture content of the 0–5 and 5–10 cm depths during the dry season of November – December 1976 of different planted covers and surface management treatments. In general, the soil moisture content was higher in unplanted mulched or unmulched plots than under live covers. This is expected because actively transpiring green foliage deplete soil moisture reserves more rapidly than the vegetation-free soil. The deep root system of some covers can deplete soil moisture reserves to considerable depths. The data of neutron count ratio with depth shown in Fig. 16.2 indicate that *Arachis* and *Indigofera* can deplete soil moisture reserves up to 120 cm depths. In contrast *Centrosema* and *Axonopus* utilize moisture reserves in the top 30 cm layer. All other factors remaining the same, shallow-rooted annuals are more compatible with *Indigofera* and *Arachis* than with *Centrosema* and *Axonopus*.

B. Mixed cropping and crop rotations

Growing mixtures of crops of different agronomic and morphologic characteristics is a useful concept in diversifying the plant community, providing continuous ground cover, and optimizing the use of a limited resource (Wiley, 1979; Okigbo, 1978). Traditional mixed cropping systems growing as many as 15 crops at random in a small field can be improved upon by selecting the 2 or 3 most compatible mixtures and growing them in an orderly manner. This method facilitates mechanization of farm operations, e.g., seeding, weed control, fertilizer application, harvesting, etc. The spatial arrangements of crop mixtures can be so arranged to rotate legumes and cereals.

C. Agro-forestry

The concept of mixtures to create diversity is appropriately extended by growing woody and herbaceous perennials in association with seasonal annuals. The root system of perennials is deeper and the canopy higher than the annuals, and the appropriate mixtures can be so managed as to optimize the utilization of both above- and below-ground resources in both space and time (Nwoboshi, 1974; King, 1979; Budowski, 1982; Kang et al., 1981). Suitable combinations of species of perennials and annuals and cultural practices of their management are different for different soils and environments. While leguminous shrubs and trees fix some atmospheric nitrogen, their deep root system utilizes water and nutrient reserves in the subsoil. The tap root system creates macro-channels and voids to facilitate water transmission. The foliage prunings applied as mulch to the soil surface provide nutrients for the shallow-rooted annuals grown in association with the shrubs.

D. Mixed farming

Integrating livestock with seasonal and perennial crops, where feasible, is a useful concept to enhance production while preserving soil resources and maintaining ecological stability. Mixed farming is also an appropriate concept to utilize grass and/or legume covers as planted fallow for immediate economic gains (McCown et al., 1979), although the effects of grazing on soil properties may be detrimental (Bayer and Ochere, 1984). For a smallholder with a few cattle or small ruminants, the cut-and-feed system may be a possibility to avoid the adverse effects of animal trampling on soil structure and water transmission properties. A low stocking rate is an obviously conservation-effective approach.

VII. A CASE STUDY FOR THE HUMID TROPICS

The dominant farming system in the lowland humid tropics is the extensive land-use of shifting cultivation and related bush-fallow rotation. Upland soils of the ecology are low in inherent fertility, are easily compacted, and are susceptible to erosion by water. Productivity declines rapidly with mechanized intensive arable land-use.

The conceptual basis of improved technology for management of uplands in the humid tropics comprises: (i) preservation of soil resource and ecological stability; (ii) maximization of output per unit input and sustaining it, and (iii) adaptation of crops and cropping systems to soil constraints. Keeping these principles in mind the main components of improved technology for the humid

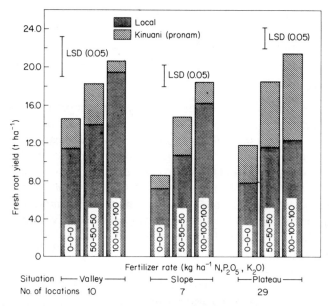

Figure 16.3 Yield of variety Kinuani and Local in three farming niches at three rates of fertilizers in Bas Zaire 1983–84 (Unpublished data of S.J. Pandey)

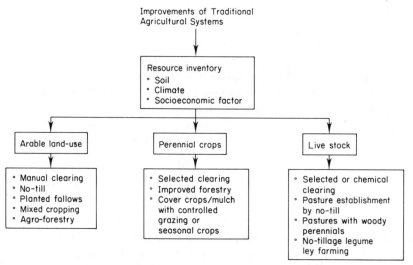

Figure 16.4 Alternate sustainable agricultural systems

tropics would consist of (a) ecologically compatible land clearing methods, (b) mulch farming with no-till or minimum tillage, (c) legume-based cropping systems including agro-forestry and perennials, (d) integrating crops with livestock, and (e) alleviation of severe labour constraints through appropriate mechanization including the use of draft animals.

These improvements should be incorporated into the traditional system step-by-step. This gradual improvement should be preferred over the rapid transformation and sudden replacement by a new system. An example of this method of improvement is illustrated below:

A. Traditional
B. (A) + improved cultivars, cropping systems and agroforestry
C. (B) + land clearing methods and minimum tillage
D. (C) + appropriate labour-saving tools
E. (D) + integration with livestock
F. (E) + use of appropriate chemicals

The first and the easiest step to improve the traditional system is to introduce high yielding cultivars and incorporate them into compatible and organized cropping systems including integration with woody perennials. For example, the data from Bas-Zaire in Fig. 16.3 show a drastic increase in cassava tuber yields by introducing an improved variety especially for the low-fertility Plateau soil. The next step is to improve land clearing and tillage methods to facilitate mechanization, to alleviate labour constraints, and to improve soil and water conservation. Soil fertility restoration is primarily achieved through biological means supplemented with fertilizer and other amendments.

For small- or medium-size land holders, three examples of sustainable systems are outlined in Fig. 16.4: (i) arable land-use based on mulch farming, (ii) perennial crops or improved forestry with cover crops and mulches, (iii) livestock raising with improved pastures and browse herbs. Land clearing and development is a major factor responsible for soil degradation (Sioli, 1980; Lal, 1981; Lal et al., 1983). The system of growing cash crops with mulches is exemplified in eastern Africa, e.g., Kenya (Othieno and Laycock, 1977) and in Rwanda. The livestock-based systems are more applicable for Latin America. For the semi-arid tropics, some promising no-tillage legume ley farming systems are described by McCown et al. (1979). Most of these systems have been proven effective for some representative ecologies in improving production while minimizing soil degradation. A flow chart depicting a sequence of steps for land development and soil management is outlined in Fig. 16.5.

VIII. RESEARCH AND DEVELOPMENT PRIORITIES

The available information must be validated and adapted by practically oriented research on representative soils and environments. There is also a need to evaluate economic feasibility and social acceptability of the basic concepts, subsystems, and components involved. It is relevant to encourage

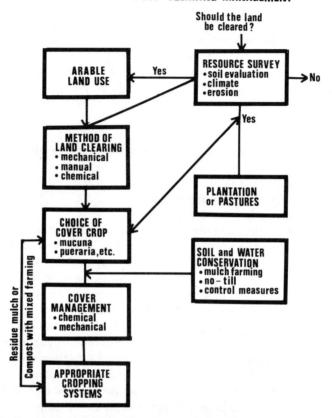

Figure 16.5 Sequence of steps for land development and soil management for high and sustained production

conceptual understanding of local agricultural problems, and develop socially acceptable solutions.

Mulch farming is the basic ingredient of the most sustainable components and subsystems; and yet, it is a scarce ingredient. Agronomic research is needed to develop cropping systems for different ecologies to ensure adequate production of mulch materials. Alternative systems of providing fodder, fuel and fencing material have to be developed so that crop residues can be spared for mulching the soil. This is not an unachievable agronomic task.

Although socio-economic issues are important in adaptation of improved technologies, it is also the right combination of various subsystems on which success or failure of the improved system depends. Farming systems are site-specific and the combination of various subsystems varies among soils and ecologies. The final and workable recipe has to be fine-tuned by local adaptive research.

Basic soil management research is needed on representative soils and ecologies for (a) developing energy-efficient soil management and tillage systems, (b) adapting conservation-effective farming systems, and (c) studying environmental impact of agro-chemicals and cultural practices.

REFERENCES

Ahmadi, N., 1983a, Genetic variability and inheritance of drought tolerance mechanisms in rice *Oryza sativa* L. 1. Development of the root system, *Agron. Trop. (Paris)*, **38**, 110–117.

Ahmadi, N., 1983b, Genetic variability and inheritance of drought tolerance mechanisms in rice *Oryza sativa*, L. 2. Susceptibility of the stomata to water deficit, *Agron. Trop. (Paris)* **38**, 118–122.

Akobundu, I.O., 1982, Live mulch crop production in the tropics, *World Crops*, **1982**, 125–126, 144–145.

Arndt, W., 1965, The impedance of soil seals and forces of emerging seedlings, *Aust. J. Soil Res.*, **3**, 55–68.

Aweto, A.O., 1981, Secondary succession and soil fertility restoration in south-western Nigeria. II. Soil fertility restoration, *J. Ecol.*, **69**, 609–614.

Bauer, F.M., 1977, *Cropping in North Australia: Anatomy of Success and Failure*, ANU Press, Canberra.

Bayer, W., and Ochere, E.O., 1984, Effect of livestock-crop integration on grazing time of cattle in a subhumid African savanna. *Proc. Int. Savanna Symp.*, CSIRO, 28–31 May, 1984, Brisbane, Australia.

Bayliss-Smith, T.P., 1982, *The Ecology of Agricultural Systems*, Cambridge University Press, Cambridge, U.K.

Bigort, Y., 1977, Fertilizing, ploughing and cultivated species under conditions of irregular rainfall in central Ivory Coast. Account of the main results of a trial on cropping systems in 1967–74, *Agron. Trop.*, **32**, 242–247.

Brown, L.R., 1984, Conserving soils, in *State of the World, 1984*, A Woldwatch Institute Report on Progress Towards Sustainable Society, W.W. Norton & Co., New York, pp. 53–73.

Budowski, G., 1982, Applicabililty of agro-forestry systems. In L.H. MacDonald (ed) *Agroforestry in the African Humid Tropics*, The United Nations University, Tokyo, Japan, 6–12.

Dabin, B., 1962, Relationships between the physical properties and fertility of tropical soils, *Ann. Agron. (Paris)*, **13**, 111–140.

Dagg, M., 1969, Hydrological implications of grass roots studies at a site in East Africa, *J. Hydrol.*, **9**, 438–444.

Dregne, H.E., 1978, Desertification: Man's abuse of land, *J. Soil Water Cons.*, **33**, 11–14.

El-Swaify, S.A., Dangler, E.W., and Armstrong, C.L., 1982, *Soil erosion by water in the tropics*, Res. Ext. Series 624, Univ. of Hawaii, Honolulu, 173 pp.

FAO, 1984, *Land, Food and People*, FAO, Rome, 96 pp.

Fisher, H., and Matter, H.E., 1980, Trends in development in tropical agriculture today. *Giessener Beitrage Zür Entwicklungsforschung*, Reihe I, **6**, 131 pp.

Fisher, R.A., 1980, Influence of water stress on crop yield in semi-arid regions, in N.C. Turner and P.S. Kramer (eds) *Adaptation of plants to Water and High Temperature Stress*, J Wiley & Sons, New York, pp. 323–339.

Gautreau, J., 1977, Levels of intercultivar leaf potentials and adaptation of the groundnut to drought in Senegal, *Oléagineux*, **32**, 323–332.

Gavande, S.A., 1969a, Effect of soil moisture tension on air-filled porosity and oxygen diffusion in six Costa Rican soils under coffee, *Turrialba*, **19**, 39–48.

Gavande, S.A., 1969b, Influence of soil moisture on oxygen diffusion and water use by cacao, *Proc. 2nd Int. Congr. Cacao Res., Brazil 1967*, pp. 431–435.

Ghildyal, B.P., and Satyanarayana, T., 1969, Influence of soil compaction on shoot and root growth of rice (*Oryza sativa*), *Indian J. Agron.*, **14**, 187–192.

Godefroy, J., 1967, Subsoiling in banana plantations, *Fruits*, **22**, 341–350.

Gretzmacher, R.,1977, Studies on the feasibility of a large-scale mechanised farm in the East Central State, Nigeria, *Bodenkultur*, **28**(3), 303–324.

Henzell, E.F., 1975, Agricultural development in tropical Australia, *Etruscan*, **24**, 18–23.

Hsiao, T.C., O'Toole, J.C., and Toma, V.S., 1980, Water stress as a constraint to crop production in the tropics, in *Priorities for Alleviating Soil-Related Constraints to Food Production in the Tropics*, IRRI, Los Banos, pp. 339–369.

Janzen, D.H., 1973, Tropical agroecosystems, *Science*, **182**, 1219–1219.

Jurion, F., and Henry, I., 1969, *Can primitive farming be modernized?* INEAC, Brussels, 457 pp.

Kamara, C.S., and Godfrey-Sam-Aggrey, W., 1979, Time of planting, rainfall and soil moisture effects on cowpea in Sierra Leone, *Exp. Agric.*, **15**, 315–320.

Kang, B.T., Wilson, G.F., and Sikpens, I., 1981, Alley cropping maize and *Leucaena leucocephala* in southern Nigeria, *Plant Soil*, **63**, 164–179.

Kellogg, C.E., and Orvedal, A.C., 1969, Potentially arable soils of the world and critical measures for their use, *Adv. Agron.*, **21**, 109–170.

Khan, S.H., 1968, Mechanized rice production in the Accra Plains. *The Ghana Farmer* XII 167–171.

Kiley-Worthington, M., 1981, Ecological Agriculture: What it is and how it works, *Agric. Env.*, **6**, 349–168.

King, K.F.S., 1979, Agroforestry and the utilization of fragile ecosystems, *Forest Eco. Mgt.*, **2**, 161–168.

Kovda, V.A., Onishchenko, S.K., and Rosanov, B.G., 1976, On the probability of droughts and secondary salinization of soils of the world, *Nauchnye Doklady Vysshei Shkoly, Biologicheskie* Nauki, **2**, 7–20.

Krause, R., Lorenz, F., and Wieneke, F., 1978, Tillage in tropical and subtropical climates, *Berichte über Landwirtschaft*, **59**, 308–828.

Lal, R., 1975, Role of mulching techniques in tropical soil and water management, *IITA Tech. Bull.* 1, Ibadan, 38 pp.

Lal, R., 1981a, Soil erosion problems on Alfisols in western Nigeria. VI. Effects of erosion on experimental plots, *Geoderma*, **25**, 215–230.

Lal, R., 1981b, Land clearing and hydrological problems, in R. Lal, and E.W. Russell, (eds.), *Tropical Agricultural Hydrology*, J. Wiley & Sons, Chichester, U.K., pp. 131–140.

Lal, R., 1983a, No-till farming, *IITA Monograph* 2, Ibadan, 68 pp.

Lal, R., 1983b, Erosion-caused productivity decline in soils of the humid tropics, *Soil Tax. News*, **5**, 4–5, 18.

Lal, R., 1983c, Soil erosion and its relation to productivity in tropical soils, *Malama Aina Conference*, January 1983, Honolulu, Hawaii.

Lal, R., 1984, Mechanized tillage systems effects on soil erosion from Alfisol in watersheds cropped to maize, *Soil Till. Res.*, **4**, 349–360.

Lal, R., 1985a, Soil erosion from tropical arable lands and its control, *Adv. Agron.*, **37** (In Press).

Lal, R., 1985b, Mechanized tillage systems effects on properties of a tropical Alfisol in watersheds cropped to maize, *Soil Till. Res.* (In Press).

Lal, R., Wilson, G.F., and Okigbo, B.N., 1979, Changes in properties of an Alfisol produced by various crop covers, *Soil Sci.*, **127**, 377–382.

Lal, R., Sanchez, P.A., and Cummings, R.W., Jr., 1983, *Land Clearing and Development in the Tropics*, A.A. Balkema, Rotterdam (In Press).

Le Pape, Y., and Mercier, J.R., 1983, Energy and agricultural production in Europe and the third World, in D. Knorr, (ed.), *Sustainable Food Systems* Ellis Horwood Ltd., Chichester, U.K., pp. 104–123.
Lieth, H., 1976, Biological productivity of tropical lands, *FAO Unasylva*, **28**, 24–31.
Lugo-Lopez, M.A., 1960, Pore size and bulk density as mechanical soil factors impeding root development, *J. Agric. Univ. (Puerto Rico)*, **44**, 40–44.
Machado, R.C.R., and Alvim, P. De T., 1981, Effects of soil water deficit on the flushing, flowering, and water status of cocoa trees, *Revista Theobroma*, **11**, 183–191.
Martin, G., 1963, The degradation of soil structure under mechanized cultivation in the Niari valley, *Cahiers ORSTOM, Pedol.*, **2**, 8–14.
Mbagwu, J.S.C., Lal, R., and Scott, T.W., 1984, Effects of artificial desurfaing on Alfisols and Ultisols in southern Nigeria. Parts I & II, *Soil Sci. Soc. Amer. J.*, **48**, 828–838.
McCown, R.L., Haaland, G., and de Haan, C., 1979, The interaction between cultivation and livestock production in semi-arid Africa, in A.E. Hall, G.H. Cannell and H.W. Lawton (eds.) *Agriculture in Semi-arid Environments*, Springer-Verlag, Berlin, pp. 297–332.
McCown, R.L., Hones, R.K., and Peake, D.C.I., 1984, Evaluation of a no-till, tropical legume ley farming strategy, in R.C. Muchow (ed), *Agro-Research for Australia's semi-arid tropics*, University of Queensland Press, Brisbane, Australia.
McNeil, M., 1964, Lateritic soils, *Sci. Am.*, **211**(5), 96–102.
Mott, J., Bridge, B.T., and Arndt, W., 1979, Soil seals in tropical tall grass pastures of Northern Australia, *Aust. J. Soil Res.*, **30** 483–494.
Morel, R., and Quantin, P., 1964, Fallowing and soil regeneration under the Sudan-Guinea climate in Central Africa, *Agron. Trop.*, **19**, 105–136.
Nicou, R., and Chopart, J.L., 1979, Water management in sandy soils of Senegal, in R. Lal (ed.) *Soil Tillage and Crop Production*, IITA, Proc. Series 2, IITA, Ibadan, Nigeria, p. 248–257.
Nwoboshi, L.C., 1974, Soil productivity aspects of agri-silviculture in the West African tropical moist forest zone, *African Soils*, **19**, 1–13.
Obi, M.E., and Asiegbu, B.O., 1980, The physical properties of some eroded soils of southeastern Nigeria, *Soil Sci.*, **130**(1), 39–48.
Okigbo, B.N., 1978, Cropping systems and related research in Africa, *AAASA Occasional Publication Series* – OT-1, 81 pp.
Ormerod, W.E., 1978, The relationship between economic development and ecological degradation: how degradation has occurred in West Africa and its progress might be halted, *J. Arid Env.*, **1**, 357–379.
Othieno, C.O., and Laycock, D.H., 1977, Factors affecting soil erosion under tea field, *Trop. Agric.*, **54**, 323–331.
Pagel, H., 1969, Problem of the fertility of ferrallitic soils, *Albrecht-Thaer-Arch.*, **13**, 339–350.
Pimental, D., and Pimental, M., 1979, *Food, Energy and Society*, Edward Arnold Ltd., London.
Radwanski, S.A., 1969, Improvement of red acid sands by the neem tree (*Azadirachta indica*) in Sokoto, North-Western State of Nigeria, *J. Appl. Ecol.*, **6**, 507–511.
Rappaport, R., 1971, The flow of energy in an agricultural society. *Sci. Amer.*, Sept. **1971**, 117–132.
Rathore, T.R., Chhonkar, P.K., Sachan, R.S., and Ghildyal, B.P., 1981, Effect of soil moisture stress on legume-Rhizobium symbiosis in soybeans, *Plant Soil*, **6**, 445–450.
Robinson, J.B.D., Hosegood, P.H., and Dyson, W.G., 1960, Note on a preliminary study of the effects of an East African softwood crop on the physical and chemical conditions of a tropical soil, *Commonw. For. Rev.*, **45**, 359–365.
Sioli, H., 1980, Foreseable consequences of actual development and alternative ideas, in F. Barbira-Scazzocchio, (ed) *Land, People and Planning in Contemporary*

Amazonia, Cambridge University Press, Cambridge, U.K., pp. 257–268.
Sivakumar, M.V.K., Seetharama, N., Gill, K.S., and Sachan, R.C., 1981, Response of sorghum to moisture stress using line source sprinkler irrigation. I. Plant-water relations, *Agric. Water Mgt.*, **3**, 279–289.
Sanchez, P.A., Bandy, D.E., Villachica, J.H., and Nicholaides, J.J., 1982, Amazon Basin soils: management for continuous crop production, *Science*, **216**, 821–827.
Stephens, D., 1967, Effects of grass fallow treatments in restoring fertility of Bugaanda clay loam in South Uganda, *J. Agric. Sci. Camb.*, **68**, 391–463.
Stocking, M.A., 1984, Erosion and soil productivity: a review of its effects and recent research with suggestions for further investigation, *Consultant Report*, FAO, Rome, 74 pp.
Talineau, J.C., Bonzon, B., Fillonneau, C., and Hainnaux, G., 1979, Contribution to the study of a grassland agrosystem inthe humid tropical environment of the Ivory Coast. I. Analysis of some soil physical parameters, *Cahiers ORSTOM Pédol.*, **17**(2), 141–157.
Turk, K.J., Hall, A.E., and Asbell, C.W., 1980, Drought adaptation of cowpea I. Influence of drought on seed yield, *Agron. J. (Calif.)*, **72**, 413–420.
Turk, K.J., and Hall, A.E., 1980a, Drought adaptations of cowpea. II. Influence of drought on plant water status and relations with seed yield. *Agron. J.*, **72**, 421–427.
Turk, K.J., and Hall, A.E., 1980b, Drought adaptation of cowpea. III. Influence of drought on plant growth and relations with seed yield, *Agron. J.*, **72**, 428–433.
Turk, K.J., and Hall, A.E., 1980c, Drought adaptation of cowpea. IV. Influence of drought on water use, and relations with growth and seed yields, *Agron. J.*, **72**, 434–439.
Uexkull, H.R. Von, 1982, Suggestion for the management of 'problem soils' for food crops in the humid tropics, *Trop. Agric. Res. Ser.*, **15**, 139–152.
Vallis, I., and Gardener, C.J., 1985, Nitrogen inputs into agricultural systems by Stylosanthes, in H.M. Stace and L.A. Edye (eds) *The Biology and Agronomy of Stylosanthes*, Academic Press, Sydney, Australia (In Press).
Wade, M.K. and Sanchez, P.A., 1983, Mulching and green manure applications for continuous crop production in the Amazon Basin, *Agron. J.* **75**, 39–45.
Waddell, E., 1977, The return to traditional agriculture, *Ecologist*, **7**, 444–447.
Willatt, S.T., 1967a, Soil moisture studies under soyabeans, *Rhod. Zamb. Mal. J. Agric. Res.*, **5**, 229–232.
Willatt, S.T., 1967b, The fertility of sandveld soils under continuous cultivation. III. Changes in physical properties. *Rhod. Zamb. Mal. J. Agric. Res.*, **5**, 129–131.
Wiley, R.W., 1979, Inter-cropping, its importance and research needs, *Field Crop Abst.*, **32**, 2–10.
Wilson, G.F., Lal, R., and Okigbo, B.N., 1982, Effects of cover crops on soil structure and on yield of subsequent arable crops grown under strip tillage on an eroded Alfisol. *Soil & Tillage Research*, **2**, 233–250.
Wolf, J.M., 1975, Soil-water relations in Oxisols of Puerto Rico and Brazil, in E. Bornemisza and A. Alvarado (eds), *Soil Management in Tropical America, Proc. Seminar CIAT*, Cali, Colombia, pp. 145–154.
Yang, P.C., Yang, S.J., and Ho, F.W., 1978, Effects of mechanical cultivation on soil compaction, root development and sugarcane yield, *Taiwan Sugar*, **25**, 117–121.

Author Index

Abaturov, B.D., 276
Abbott, I., 313, 316
Abe, T., 339–340, 351, 405
Abrahamsen, G., 290
Adams, J.A., 182
Adamson, A.M., 371, 409
Adeoye, K.B., 607–608
Afolayan, T.A., 487
Agafonov, O., 118
Agarwal, G.W., 287, 313
Ahlgren, C.E., 271, 465
Ahlgren, I.F., 271, 465
Ahmad, N., 84
Ahmadi, N., 694
Ahn, P.H., 117, 132
Ahn, P.M., 183, 571, 573, 575
Aina, P.O., 77, 151, 543–545, 579, 582–583, 597, 610, 629–630, 659, 661, 665
Ajayi, S.S., 487
Ajuwon, S., 492
Akinremi, 639
Akobundu, I.O., 321, 665, 699
Alban, D.H., 205, 217
Alfred, J.R.B., 267–268, 270–271, 657–658
Alkampo, J., 280
Allan, W., 626
Almeida, F.L.S., 607
Alvarado, A., 427, 432–433
Alvim, P.T., 104, 694
Amezquita, E.C., 135, 137, 138, 148, 152
Anderson, J.M., 232, 543
Anderson, K.E., 204
Anderson, M.G., 204
Andrews, D.J., 63
Araragi, M., 279
Aremu, J.A., 578, 587
Ariel, D., 594–595
Aristeguieta, L., 469
Aritajat, U., 323
Arle, R., 287

Armon, M., 572, 575, 579, 582–583, 606
Arnason, T., 453
Arndt, W., 144, 694
Arnold, R.W., 179
Arruda, H.V. de, 572
Arshad, M.A., 338, 395, 399–411
Arya, L., 427
Asamoa, G.K., 120
Asiegbu, B.O., 693
Asseline, J., 484
Aston, A.R., 204
Athias, F., 266, 287, 323, 459–460, 474
Atlavinyte, O., 315, 324
Aubert, G., 168, 239, 309, 485
Auberville, A., 485, 539
Aurangebadkar, R.K., 470, 472, 474, 493
Aweto, A.O., 252, 665, 699
Ay, P., 11
Ayanaba, A., 277–278, 542
Ayoade, J.P. 62
Ayres, I., 290

Babalola, O., 135, 185, 188, 193, 198
Babin, B., 168
Bachelier, G., 404, 423
Baena, A.R.C., 118, 120
Bagnouls, F., 53
Bahl, K.N., 287
Bailey, H.P., 32, 34–35
Baker, S., 37
Bakr, A.A., 204, 219
Balasubrumanian, A., 277
Ball, D.F., 214
Balogun, C., 56
Banage, W.B., 157, 288
Banerjee, S.P., 144
Banfield, C.F., 182
Barley, K.P., 315
Barnes, B.T., 319
Barnes, D.L., 662
Barney, G.O., 7

Bascomb, C.L., 182
Batchelder, R.B., 453
Bateman, R.J., 607
Bates, J.A.R., 305–306
Batz, 187
Bauer, A., 188, 218
Bauer, F.M., 689
Baumgartner, A., 556
Baweja, K.D., 287
Baxter, F.P., 429
Bayer, W., 703
Baylis, H.A., 292
Bayliss-Smith, T.P., 690
Beard, J.S., 100
Beaudou, A.G., 182
Beckett, P.H.T., 179
Beckley, V.R.S., 662
Belfield, W., 267
Bell, K.R., 188
Bell, T.I.W., 676
Bellairs, V., 266
Belobrov, V.P., 118
Ben-Asher, J., 188
Benmema, J., 89
Benoit, P., 63–65, 67
Berger, J.M., 659
Bernhard-Reversat, F., 519
Bertrand, R., 129, 140–142
Betsch, J.M., 287
Bharat, R., 314
Bhatt, J.V., 315
Bhattachrya, T., 267
Bhatti, H.K., 295
Bidwell, O.W., 446
Biggar, J.W., 214
Bigort, Y., 694
BIOTROP, 538
Biscaia, R.C.M., 592
Biswas, A.K., 48
Bjorkman, O., 242
Black, A.S., 214
Blic, P. de, 542
Block, W., 266, 288
Blyth, J.F., 218
Boardman, R., 676
Bodot, P., 346, 407
Bogdan, A.V., 468
Bolin, B., 550
Bonell, M., 251
Bonnet, J.A., 249
Bonsu, M., 126, 135, 139, 152–154
Boone, F.R., 573
Bornemisza, E., 146
Bosser, J., 280

Bouche, M.B., 287, 315
Bouchev, 52
Boughey, A.S., 453, 469
Bouillon, A., 340, 346
Boulad, A.O., 309
Bourliere, F., 101–102, 265
Bowden, D.J., 60
Bowden, J.M., 116
Boyer, P., 352, 359, 371, 384, 389, 394–395
Boyle, J.R., 187, 487
Brechner, I., 84
Bredler, E., 198
Bregt, A.K., 219
Bresler, E., 217
Breznak, J.A., 395
Brian, M.V., 340
Bridge, B.J., 665
Bridge, B.T., 150
Briggs, D.J., 182
Brinkman, W.L.F., 487
Brockington, N.R., 485
Bronchart, R., 249
Brown, C.H., 40
Brown, J.A.N., 535
Brown, L.R., 693
Brown, S.M., 325
Brown, W.L., 427
Bruce-Okine, E., 162, 164, 475
Bryan, R.B., 204
Buanec, B. le, 120, 629
Buch, M.W. von, 539
Buck, L.V., 25
Budowski, G., 464, 485, 703
Budyko, M.I., 31–33
Buffington, J.D., 271
Bunning, E., 242
Buol, S.W., 4, 89, 96, 168, 215, 427
Buringh, P., 3–4, 115
Burrough, P.A., 217–219
Byram, G.M., 456

Cabanis, Y., 484
Cachan, P., 240, 242
Cadima, A.A., 147, 149
Calaby, J.H., 347
Calvin, L.D., 218
Camarche, A., 217–218
Cannell, R.W., 567
Canpolle, A.P., 26
Carbonnel, J.P., 252
Carroll, C.R., 423
Carson, R., 260
Carter, A., 313

Carter, D.B., 30
Carter, G.S., 242
Carvallo, H.O., 198
Caseley, J.C., 325
Cassel, D.K., 187–188, 218, 542
Cavalcanti, L.H., 461
Caveness, F.E., 270, 316
Chaffey, D.R., 676
Chand, S., 144
Chandra, S., 610
Chang, J.H., 42
Chapman, A.L., 54–55
Charley, J.L., 205
Charreau, C., 142, 207, 572, 575, 590–591, 607, 612, 621–622, 632, 634
Charter, C.F., 385
Chauvel, A., 132
Cheema, P.S., 380–381
Cheney, N.P., 456
Cherrett, J.M., 427
Chevalier, A., 384, 464
Chinnamani, S., 538
Chopart, J.L., 575, 692
Chou, M., 7
Cicerone, R.J., 673
Clark, J.D., 453
Cocheme, J., 39–40, 54, 56, 59
Coe, R., 68
Cointepas, J.P., 629
Collinet, J., 182–183
Collins, J.F., 182
Collins, N.M., 340, 345, 352, 404, 543
Collis-George, N., 162, 479
Commonwealth Bureau of Soils, 182
Conrad, R., 673–675
Conseil Scientifique pour l'Afrique, 100–101
Cook, L., 467
Corbet, A.S., 487
Cornforth, I.S., 314
Correa, J.C., 135
Coster, C.H., 37
Courtney, F.M., 219
Couteaux, M., 266, 273, 287, 323
Coutinho, L.M., 453, 459, 461, 469, 487
Couto, W., 215
Covolo, G., 315
Critchley, B.R., 270, 275, 289, 317, 407, 423, 424, 437
Crosson, P.R., 6, 12
Crowther, L.T., 155
Crutzen, P.J., 489, 506, 662, 673, 675
CSIRO, 125, 469
CTFT, 84

Cummings, D.J., 317, 466, 480, 485, 487–488, 524, 530–532, 542
Cummings, R.W., 17–18
Cunningham, R.K., 523–527, 532, 542
Curfs, H.P.F., 579, 581–582, 588, 611–612

D'Hoore, J., 523–524
Da Silveira, P.M., 118, 132, 156
Dabek-Szreniawska, M., 278
Dabin, B., 691
Dabral, B.G., 248
Dagg, M., 248, 251, 694
Dancette, C., 142, 207–215, 240
Dantas, M., 267
Darlong, V.T., 267–268, 270–271
Darwin, C.F., 340
Dash, M.C., 288, 316, 322
Daubenmire, R., 467, 495
Davies, D.M., 290, 324
Davol, F.D., 485
Day, G.M., 315
De, R., 661
De Ban, L.F., 463
De Datta, S.K., 165–166, 589–590
De Heinzelin, J., 384
De Izara, D.C., 273
de la Rosa, D., 219
De Nascimento, J.C., 487
De Ploey, J., 384, 386
De Rham, P., 486
De Vleeschauwer, D., 299–315, 637–638, 641, 644–645, 648–649
De Vries, P.F.W.T., 17
De Wit, C.T., 17
Deiomande, T., 424
Delye, G., 424
Dennison, E.B., 662
Derpsch, R., 597
Deshpande, T.L., 116
Dexter, A.R., 324
Dhawan, C.L., 305
Diatta, S., 542, 629
Dickinson, R.E., 506, 556
Dimov, D.I., 120
Dinchev, D., 214
Dindal, D.L., 270
Dommergues, Y., 280, 676
Donahue, R.L., 463
Dow, H., 59
Drees, L.R., 179–180, 214
Dregne, H.E., 160, 693
Drosdoff, M., 118, 145, 149, 151
Drummond, H., 340, 387
Du Pleiss, M.C.F., 484

Dudal, R., 92, 95, 181, 205, 672
Dugain, F., 538
Dunham, R.J., 578–579, 587
Dunin, F.X., 204
Dunne, T., 540–541, 671
Duval, J., 240
Duweini, A.K. el, 289, 291, 307

Easton, E.G., 287
Ebert, H., 326
Edelman, C.H., 276
Edmisten, J., 249
Edwards, C.A., 287, 299, 301, 319, 325
Edwards, D.C., 474–475
Ehlers, W., 313, 319
Ehrlich, P.R., 445
El-Swaify, S.A., 116, 132, 161, 693
Eldin, S., 607
Ellis, F.B., 319
Elvim, P., 147, 149
Elwell, H.A., 84
Emberger, L., 26
Eno, C.F., 325
Erhart, H., 387, 389, 411
Escolar, R.P., 146
Ettershank, G., 361, 432
Evans, G.C., 240, 242
Evans, J., 676
Ezeta, F.N., 212

Falesi, I.C., 168
FAO, 3–11, 13–17, 19, 88, 234, 505–506, 510, 621, 693
Farming in South Africa, 404
Farnworth, E., 231
Fassbender, H.W., 487
Fauck, R., 629–630
Feller, C., 552, 575, 629
Ferrar, P., 340, 347
Ferrari, T.J., 181, 218
Finlayson, D.G., 325
Fisher, H., 689
Fisher, M.J., 479
Fisher, R.A., 694
Fitzpatrick, E.A, 50, 66–67
Flohn, H., 25
Florence, R.G., 676
Fluhler, H., 217
Fogliata, F.A., 472
Folster, H., 385
Foth, H.D., 92, 95–98, 100
Fournier, F., 592, 607
Fox, R.L., 218
Frank, N.R., 427

Frankart, R., 168
Franquin, P., 39–40, 54, 56, 59
Frederick, K.D., 6, 12
Freeze, R.A., 179–180
Freise, F.W., 485
Frere, M., 14–15
Fridland, V.M., 168
Friedman, I., 556
Friffith, D.R., 567
Frometa, B.L., 118
Furch, K., 314
Furley, P.A., 204
Furusaka, C., 214

Gajem, Y.M., 182, 187
Galloway, H.M., 567
Gardener, C.J., 699
Garnier, B.J., 31
Garren, K.H., 453, 468
Gates, G.E., 287, 292
Gatreau, J., 694
Gaussen, H., 26
Gaussens, H., 53
Gavande, S.A., 147, 167–169, 693
Gay, F.J., 347, 382, 387
Geidel, H., 217
Geissert, D., 219
Gelger, R., 26–27
Gentry, A.H., 556–557
Gerakis, P.A., 465
Gerard, B.M., 319, 322
Ghabbour, S.I., 288–289, 291, 307
Ghilarov, M.S., 367, 371, 375
Ghildyal, B.P., 692
Ghosh, R.C., 240
Ghuman, B.S., 157–159, 232, 347
Gifford, G.F., 198
Gillison, A.N., 99
Gillon, D., 266, 270, 453, 458, 460, 464, 489
Gillon, Y., 266, 270
Gilmour, D.A., 251, 537–538
Gintings, A.N., 537, 539
Giovannini, G., 472–473
Glavis, C., 427
Glover, J., 59, 63, 252
Glover, P.E., 338, 346, 404
Godefroy, J., 132, 244, 542, 692
Godfrey-Sam-Aggrey, 694
Goffinet, G., 404
Gokhale, N.G., 395
Golledge, A., 487
Golley, B., 99
Golley, F.B., 231–232

Gomez-Pompa, A., 99
Gomma, A.A., 216
Gonzalez, M.A., 118, 147, 167–169
Gonzalez Abreau, A., 214
Goodland, R.J.A., 375, 395, 399, 411
Goodland, R.J.W., 554, 556
Goosen, D., 96–102, 427
Gopalswamy, A., 215
Graff, O., 265, 315, 319
Grainger, A., 510–511, 551
Granier, P., 484
Grasse, P.P., 351, 378, 381, 389
Greaves, T., 367
Greenland, D.J., 4, 117, 120, 157, 232, 252, 466, 542
Greenslade, P.J.M., 423–424
Gregory, S., 56, 59–60
Gretzmacher, R., 689
Grieve, I.C., 205
Griffith, G., 339, 389
Grigg, D.B., 619
Grisi, B.M., 489
Grohmann, F., 252, 572
Grubb, P.J., 232, 242, 461
Guedez, J.E., 132
Guerra, R.A.T., 290
Guilloteau, J., 456
Gupta, M.L., 305
Gupta, S.R., 339, 349, 405
Gurovich, L.A., 198
Gwynne, M.D., 252

Haantjens, H.A., 297, 311
Hack, H.R., 215, 217
Hadley, M., 101, 265
Hall, A.E., 694
Hamblin, A.P., 573
Hamilton, C.D., 676
Hamilton, L.S., 504
Hansen, J., 554
Harding, J.H., 676
Hardjono, H.W., 538
Hargreaves, G.H., 26, 32, 36
Harker, W., 287
Harris, W.V., 338, 340, 354, 370, 387, 389, 404, 409
Hartge, K.H., 187
Hatfield, J.L., 203
Haug, R.M., 472
Hay, R.K.M., 319
Hayward, D.M., 610
Hazra, A.K., 267
Heath, G.W., 309, 384
Hebbert, R.H.B., 219

Hecht, S.B., 539
Hekstra, P., 673
Henin, S., 658
Henry, J., 657, 698
Henzell, E.F., 689
Hermann, H.R., 340
Herrera, R., 231, 253–254, 287–288
Hesse, P.H., 338, 346, 374, 381, 390–391, 394–395
Hesse, P.R., 409
Hettner, A., 25
Heuveldop, J., 248, 251
Hewitt, P.H., 378
Hibbert, A.R., 534
Higgins, G., 17
Hingston, F.J., 489
Hodgkins, E.J., 453, 468
Hoeksema, K.J., 276
Holdaway, F.G., 382
Holdren, J.P., 445
Holdridge, L.R., 37, 240, 246, 249
Hole, F.D., 211, 265, 275–276, 429, 446
Holldobler, B., 427
Holt, J.A., 266–269, 339, 347, 345, 365, 384
Holzapfel-Schorn, A., 673
Hopkins, B., 456–457, 468
Hosegood, P.H., 535, 671
Howell, F.C., 453
Howse, P.E., 341
Howson, D.F., 671
Hrarovrzky, J., 672
Hsiao, T.C., 694
Huang, F.Z., 301
Hubert-Schoumann, A., 542
Hudson, N.W., 76, 84
Huijbregts, 219
Huke, R.E., 38
Huke, S., 661
Hulugalle, N., 527, 530, 532, 549–550, 662, 665, 667
Hundal, S.D., 165–166
Hurault, J., 540, 662
Huxley, P.A., 668

IAR, 151
ICRISAT, 36, 75, 78, 123, 146, 150
Ivanvi, K.P., 40
IITA, 121, 127, 189, 519, 607
Irmler, 290, 314
Irwin, H.W., 554, 556
Isbell, A.F., 96
Izara, D.C. DE, 489

Jackson, I.J., 69–71
Jackson, P., 505–507
Jacomine, P.K.T., 168
Jacquin, F., 542
Jameson, J.D., 662
Jansen, I.J., 179
Janzen, D.H., 698
Jarvis, M.G., 182
Jatzold, R., 31, 37, 39
Jenik, J., 242
Jenkinson, D.S., 278
Jenny, H., 314
Joachim, A.W.R., 132, 339, 369, 394, 485
Johannes, W., 463
Johnson, N.E., 677
Johnson, O.H., 71
Jones, M.J., 63, 88, 120
Jones, P.A., 572, 574, 662–663
Jones, R.K., 643, 647
Jordan, C.F., 103, 246–248, 251, 253
Jose, A.I., 676
Josens, G., 427
Joshi, N.V., 297, 299
Julka, J.M., 287
Junor, R.K., 538
Juo, A.S.R., 211, 542, 604, 658
Jurion, F., 667, 698

Kaiser, P., 277, 280
Kakshinamurti, C., 661
Kale, R.D., 287, 315
Kalla, S.E. el, 216
Kamanoi, M., 627, 629
Kamara, C.S., 694
Kandiah, S., 339, 394, 485
Kang, B.T., 181, 183, 187, 205–207, 214–215, 345, 347, 354, 390, 399, 411, 486, 542, 654–656, 668, 703
Kannegieter, A., 665
Kanwar, J.S., 297, 299, 305, 315
Karpacehevskii, L.O. 187
Kassam, A.H., 48, 59–60, 63–64, 66, 84, 105, 135
Katz, B., 277
Kautilva, A., 48
Kayani, S.A., 404
Kayll, A.J., 467–468
Keeves, A., 676
Keisling, T.C. 217
Kelkar, B.V., 297, 299
Kellman, M.C., 538, 625
Kellogg, C.E., 207, 485, 694
Kelly, R.D., 151
Kelly, W.P., 473

Kemp, P.B., 374
Kemper, B., 597
Kenworthy, J.M., 59, 63
Keogh, R.G., 325
Kerim, N.S.A., 589–590
Kerkham, R.K., 662
Kevan, D.K.McE., 265
Kevan, K., 427
Keyfitz, N., 10
Khambata, S.R., 315
Khan, A.W., 287, 295, 316
Khan, S.H., 692
Khanna, S.S., 144
Khatibu, A.I., 588–589, 597–599, 641–643, 653
Kidieri, S., 346
Kiefer, R.W., 217
Kilbertus, G., 278
Kilewe, A.M., 122
Kiley-Worthington, M., 698
Killick, D.J.B., 469
King, K.F.S., 703
King, P.N., 504
Kininmonth, W.R., 54–55
Kirchner, M., 556
Kirkham, M.B., 543–544
Kisu, M., 166
Klaer, W., 484
Klages, K.H.W., 37
Klaij, M.C., 575, 578–579, 591
Kline, J.R., 249, 521
Klinge, H., 231–232, 251
Knabe, D.T., 63
Koenigs, F.F.R., 116
Komarek, E.V., 453
Konandreas, P., 19–20
Kool, J.B., 217
Koon, C.K., 84
Koppen, W., 26–28, 37
Koshy, M.M., 676
Kovda, V.A., 694
Kowal, J.M., 17, 48, 59–60, 63–64, 84, 102–103, 105, 135, 138, 232, 252, 313
Kozlova, A.V., 347, 394, 402
Krantz, B.A., 162
Kraus, E.B., 48
Krause, R., 699
Krishna, K., 340, 404
Krishnamoorthy, R.V., 287
Krishnan, A., 40
Kristofferson, L., 507
Kpoger, M., 291
Kronfellner-Kraus, G., 538
Kruzinga, S., 75

Kurtywa, J.C., 48
Kusnezov, N., 423
Kyuma, K., 144

Lacey, C.J., 453, 461, 470
Laird, J.M., 291
Laker, M.C., 359, 395
Lal, R., 63, 76–78, 117, 126, 132, 135, 139, 151, 152–154, 155, 157–159, 162–164, 186, 192–194, 214, 232, 251, 278, 299, 301–302, 309, 313, 315, 317, 318, 321, 324, 347, 395, 466, 473, 479–483, 485–488, 492, 513, 516, 524, 530–534, 539, 542, 548, 550, 556–557, 570–572, 579–582, 584, 586–589, 594, 597, 599–606, 609–611, 623, 625–626, 628, 630–631, 634–637, 639–643, 645–647, 650–662, 664–666, 669–671, 691–695, 698–700, 705
Lam, K.C., 538
Lambert, J.D.H., 453
Lamprecht, H., 556
Langohr, R., 132
Lappartient, J.R., 309
Lascano, R.J., 198
Lasebikan, B.A., 266, 317
Laudelout, N., 487
Lauer, W., 30
Laurie, M.V., 248
Lavelle, P., 265, 268, 271, 273, 287, 289, 291, 295, 297–298, 317, 424
Lawes, D.A., 586–587
Lawson, T.L., 40, 239, 519, 521, 524, 534
Lawton, R.M., 676
Laycock, D.H., 705
Le Baron, A., 539
LeHouerou, H.N., 53, 55–56, 94
Le Pape, Y., 690
Lea, D.A., 504
Lebrun, P., 325
Lee, K.E., 287, 339–341, 346–347, 354, 357, 369–370, 383–384, 389, 391–392, 394–395, 399, 404, 423
Lee, R., 214
Leger, R.G., 203
Lehri, L.K., 369, 374, 391, 394, 399
Leibundgut, H., 325
Leita, J. del, 212
Lemon, P.C., 464, 468
Leneuf, B., 309
Lepage, M.G., 384
Leprun, J.C., 349, 399
Leprun, L.C., 339, 375, 380, 385, 391
Lettau, H., 31, 554

Leveque, A., 120, 183, 384, 386
Levieux, J., 423–424, 427
Levings, S.C., 427
Lewis, N.B., 676
Lewis, T., 427
Li, C.K., 248, 251
Libera, C.L.F., 151
Lierop, W. van, 214
Lieth, H., 695
Lim Suan, M.P., 537
Ling, A.H., 467
Ljungstron, P.O., 287, 290
Loffler, E., 484
Lofty, J.R., 287, 299, 301, 319
Longhurst, C., 423
Longman, K.A., 242
Loos, R., 378
Loots, G.C., 273
Lopez-Parodi, J., 556–557
Loring, S.J., 319
Loveday, J., 219
Lovelock, J.E., 550, 554
Lucchesi, S., 472–473
Ludlow, M.M., 242
Lugo-Lopez, M.A., 146, 151, 692
Luk, S.H., 204
Lundgren, B., 671, 676–679
Lundgren, L., 671
Luscher, M., 378, 380–381
Lutz, J.F., 118, 183
Lynch, J.M., 277
Lyonga, S.N., 492
Lyons, C.H., 325

Macartney, J.C., 575, 578, 587, 594–595, 607
Machado, R.C.R., 694
Mackenzie, A.F., 214
Macleod, D.A., 218
Madge, D.S., 265, 287–289, 292, 295, 314, 429
Madhawan, S.K., 473
Maeda, T., 167
Maher, 676
Maheut, J., 676
Mahlhot, G., 346
Maignien, R., 168
Mainstone, B.J., 467
Majer, J.D., 291
Makilo, R., 629
Maksoud, A., 607
Malaisse, F., 248
Malajczuk, M., 489

Maldague, M., 266, 352, 392, 394–395, 404
Maldague, M.E., 340, 346, 350, 367
Malkomes, H.P., 314
Manning, H.L., 59
Mante, E.F.G., 575
Marques, J., 554
Martin, G., 542, 692
Martin, N.A., 287, 296, 325
Martinez Cruz, A., 277
Martonne, E., 30, 53
Masefield, G.B., 467
Masson, H., 456, 458
Matheron, G., 219
Mathew, W.T., 556
Mathys, G., 275
Matsumoto, T., 339–340, 351
Matter, H.E., 689
Maureaux, C., 280
Maurya, P.R., 597
Maurya, P.R., 639, 641, 643, 647, 657
Mba, C.C., 313, 315
Mbagwu, J.S.C., 121, 129, 144–145, 151–152, 693
Mcarthur, A.G., 456
McColl, H.P., 267, 316
McCown, R.L., 145, 150, 211, 607, 643, 647, 665, 668, 699, 703, 705
McCulloch, J.S.G., 59, 248, 251
McGeorge, W., 473
McNeil, M., 694
Meade, R.H., 557
Medina, E., 237, 465, 467–469, 472, 474, 477
Medina, H.P., 252
Medts, A.D., 325
Medvedev, V.V., 573, 376
Mehanni, A.H., 665
Mehr-Homji, V.M., 37
Meigs, P., 30
Meiklejohn, J., 394, 464, 487
Melville, I.R., 607
Menan, S., 536, 538
Menaut, J.C., 99
Menk, J.R.F., 218
Mercier, J.R., 690
Merinho, H.E., 539
Merino, J.F., 266
Messenger, A.S., 315
Meyer, B., 118
Meyer, J.A., 268, 287, 341, 345, 351, 387, 412
Miedema, R., 211, 339, 390
Miehlich, G., 179

Mielke, H.W., 402, 408–409
Milfred, C.J., 217
Miller, D., 31
Millette, G.J.F., 203
Milleville, P., 575, 629
Millington, A.C., 157
Minnich, J., 286–287, 295
Mire, P.B.de, 383
Mishra, B.K., 625
Mistra, R., 99
Mmkonga, A.A., 168
Moberg, J.P., 168
Modena, A.C., 315
Mohr, E.C.J., 37, 48, 54, 88–90, 248, 485
Moira, 17
Mollitor, A.V., 205
Moltham, H.D., 181, 218
Monasterio, M., 464
Mongi, H.O., 668
Montgomery, R.J., 183
Moore, A.W., 467, 487
Moormann, F.R., 136, 151, 162, 181, 183, 187, 205–207, 214–215, 249
Morais, F.I., 214
Moran, E., 231, 253–254
Morel, R., 699
Morgan, C., 204
Morgolis, E., 492
Morth, H.T., 71
Mostert, J.S., 54
Mostert, J.W.C., 484
Mott, J., 125–126, 134–135, 145, 694
Mott, J.J., 424
Moureaux, C., 459–460
Muir, W.D., 676
Murillo, B., 301
Murphy, P.G., 103–104
Murphy, P.W., 265
Murray, J.M., 347
Myero, N., 231, 233
Myers, J.G., 453
Myers, N., 505, 508–511, 513

Nagai, V., 218
Nahon, D., 309
Nakamura, Y., 309
Natarajan, M., 662
Natarajan, T., 275
Nel, J.J., 378
Nemeth, A., 287–288
Neumeyer, K., 118
Ngatunga, E.L.N., 654
Nicou, R., 572, 575, 595–596, 607, 612, 658–659, 692

Nielsen, D.R., 187–188, 218
Nieuwolt, S., 48–49, 248
Nijhawan, S.D., 297, 299, 305, 315
Nishita, H., 472
Nix, H.A., 102
Noirot, C.H., 341, 369, 377–378, 380–382, 389
Nordin, C.F., 557
Norman, M.J.T., 467, 470, 486, 491
Norris, J.M., 219
Norse, D., 17
Norse, O., 3, 5, 106
Nortcliff, S., 219
Northcote, K.H., 125
Northrup, M.L., 187
Nwa, E.U., 74–75
Nwoboshi, L.C., 542, 703
Nye, P.H., 276, 287, 295, 297, 305, 307, 309–311, 314, 338, 354, 383–385, 390–391, 399, 429, 432, 466, 542

O'Keefe, P., 507
Obeng, H.B., 121
Obi, A.O., 135
Obi, M.E., 693
Ochere, E.O., 703
Ochse, J.J., 37, 50, 54
Odum, H.T., 99, 232, 251
Odum, P.P., 445
Ofori, C.S., 629
Ogawa, H., 232
Oguntoyinbo, J.S., 56
Ohiagu, C.E., 341, 347, 350
Ojo, O., 53, 84
Okigbo, B.N., 4, 321, 702
Okoro, G.E., 486, 492
Okwakol, M.J.N., 404
Oliveira, J.B. de, 182
Oliveira, L.B., 118
Ollagnier, M., 527, 539–540, 542
Olson, J.S., 314
Oluwole, A., 301–302, 318
Omayuli, A.P.O., 277
Omo Malaka, S.M., 338, 364, 371, 399
Opara-Nadi, O.A., 584
Oreshkina, N.S., 187
Ormerod, W.E., 689
Orvedal, A.C., 694
Oswall, N.C., 144
Othieno, C.O., 705
Otremba, E., 40
Owens, L.B., 309
Oyebande, O., 56

Pagel, H., 132, 689
Pairintra, C., 144
Panabokke, C.R., 59, 123, 183
Pandey, S.J., 704
Pandilteskera, D.G., 132, 369
Papadakis, J., 39–40
Parker, C.A., 313
Pathak, A.N., 369, 374, 391, 394, 399, 435
Patra, U.C., 288
Patterson, G.T., 217
Peck, A.J., 198
Pecrot, A., 446
Pedro, G., 132
Pendleton, R.L., 374, 402, 408–409
Pereira, H.C., 84, 251, 534–535, 662–663, 665, 671
Pereira, H.J., 572, 574
Perfect, T.J., 275
Pernes, J., 439
Persson, R., 512–513
Peterson, R.G., 218
Philip, J.R., 179, 204
Philippson, A., 25
Philipson, W.R., 118
Phillips, J., 236
Phillips, J.F.V., 453, 469, 495
Pidgeon, J.D., 567
Pierce, T.G., 301
Pigeon, R.F., 99
Pilgrim, D.H., 204
Pillai, K.S., 266, 269, 275
Pimental, D., 690
Pimental, M., 690
Pinho, A.F. des, 214
Pitot, A., 456, 458
Pla Sentis, I., 118, 120, 145, 148, 155
Plowman, K.P., 269, 315
Polyakov, I.S., 187
Pomeroy, D.E., 266–267, 269, 273, 345, 354, 384, 387
Popenoe, H.L., 252
Popov, G.F., 53, 55–56, 94
Postel, S., 15, 17, 233–234, 505
Potter, G.L., 554–555
Poulain, J.F., 207, 215, 240, 575, 577, 607
Prahbakara, J., 661
Prihar, S.S., 657
Primavesi, A.M., 315
Pullan, R.A., 341, 409
Puttarudriah, M., 314

Quantin, P., 699

Radwanski, S.A., 699

Ragg, J.M., 567
Rai, B., 280
Raignier, A., 429
Rains, A.B., 456–457
Raison, R.J., 453, 456–457, 470, 486, 491
Raj, D., 215
Rajagopal, D., 369
Ramakrishnan, P.S., 625
Ramamohan, R.V., 157
Rambelli, A., 464
Ramirez, A.C., 214
Ramos, A.D., 539
Rao, B.K.S., 248
Rao, K.S., 473
Rao, N.A.N., 157
Rao, P.S.C., 198
Rao, P.V., 218
Rappaport, R., 697
Rathore, T.R., 694
Raunet, M., 389
Ravalo, E.J., 671
Reddy, M.V., 287
Reichart, K., 147
Reid, I., 202
Reinecke, A.J., 287, 289
Rengeswamy, P., 150
Resck, D.V.S., 118
Revelle, R., 4, 550–552, 556
Reynolds, S.G., 180, 188, 218, 676
Rhee, J.A. van., 326
Ricaud, R., 214
Richards, P.W., 99, 240
Riehl, H., 48, 51, 57
Riquier, J., 183
Risch, S.J., 423
Ritter, M., 319
Robertson, L.S., 214
Robinson, J.B.D., 338, 390–391, 395, 399, 676, 699
Robinson, P., 59
Roche, M.A., 539
Roeder, M., 146
Rogowski, A.S., 204
Roonwal, M.L., 341, 399
Roose, E.J., 84, 244, 251, 384, 387, 539
Roose, J., 484
Ross, M., 512
Rosswall, T., 486
Rotini, O.T., 473
Rougerie, G., 251
Rounce, N.V., 409, 411
Rowlings, R.W., 607
Roynoel, J., 349, 399
Rubinoff, I., 507

Ruelle, J.E., 340, 378, 382
Rundel, P.W., 465
Ruslan, M., 536, 538
Russo, D., 198
Ruthenberg, H., 619
Rwakaikara, D., 266–267, 269, 273
Ryke, P.A.J., 270, 273, 287

Sague Diaz, H., 151
Saha, A.K., 150
Sakal, R., 305
Salati, E., 507, 554, 556–557
Salem, M.Z., 429
Salt, G., 266
San Jose, J.J., 467, 469, 472, 474, 477
Sanchez, 4, 17, 88–89, 96–97, 231, 253, 463, 472, 542, 549, 589, 657, 665, 695, 698
Sands, W.A., 346–347, 402
Sands, 407
Sans, W.W., 326
Santamaria, F., 183
Santos, O.M., 489
Sar, T. van der, 611–612
Sarlin, P., 252
Sarmiento, G., 99, 464
Sartz, R.S., 188, 198
Sastry, K.S.S., 314
Satchell, J.E., 265, 291
Satyanarayana, T., 692
Saxena, G.S., 157
Schaefer, R., 280
Schafer, J.W., 92, 95–98, 100
Schafer, P., 217
Scharringa, M., 201
Schimper, A.F.W., 50
Schoch, P.G., 248
Scholl, D.G., 463, 473
Schoningh, E., 280, 657–658
Schubart, H.O.R., 267
Schulz, J.P., 242
Schumann, J.E., 54
Schwerdtle, F., 319
Scott, D., 211, 214
Scott, J.D., 467
Segalen, P., 183
Segun, A.O., 287
Seguy, L., 591
Seiler, W., 407, 506, 673–676
Selvakumari, G., 144
Senapati, B.K., 314, 322
Serafino, A., 266
Sertsu, S.M., 463, 472
Seubert, C.E., 466, 528–531, 542, 548–549

Shaler, N.S., 429
Shallow, P.G., 557
Shantz, H.L., 453, 464, 468
Sharma, D.L., 578
Sharma, G.D., 265, 287, 429
Sharon, D., 71
Sharpley, A.N., 298, 307, 324
Sheikh, K.H., 404
Shrikhande, J.G., 435
Shukla, J., 60–61
Shukla, L.N., 671
Siband, 620–621, 626–627
Sibi, H., 556
Sidiras, N., 572–573, 584–586, 592
Sien, C.L., 84
Silva, L.F. DA, 119, 529–531
Simeon, F.R., 151
Simmonds, N.W., 619
Singh, J., 266, 269, 275
Singh, J.S., 339–341, 345, 347, 380, 382
Singh, K., 547
Singh, M., 40
Singh, S., 661
Singh, U.R., 339–341, 345, 347, 380, 382
Sioli, H., 705
Sisson, J.G., 198
Sivakumar, M.V.K., 59, 694
Skaife, S.H., 378
Slavov, D., 214
Smith, E.L., 470
Smith, L.P., 215
Smith, N.J.H., 540
Smith, R.E., 219
Smyth, A.J., 183
Snider, R.J., 275, 325
Soane, B.D., 567
Sommer, A., 508–510
Soong, N.K., 132, 637
Soota, T.D., 287
Sourabie, N., 542
Southwell, L.T., 291
Soyer, J., 346, 484
Spears, B.M., 369–371
Spratt, M., 253
Springer, E.P., 198
Springett, J.A., 273, 287, 290, 489
Sreenivasan, A., 470, 472, 474, 493
Sreenivasen, P.S., 74
Srivastava, A.K., 314
Stafford, C.J., 325
Stark, N., 252–253
Steiner, K.G., 662
Stenersen, J., 325
Stent, H.B., 487

Stephens, D., 699
Stern, J., 198
Stern, R.D., 68
Stibbe, E., 594–595
Stockdill, S.M.J., 316
Stocker, G.C., 467
Stocking, M.A., 84, 693
Stockinger, K.R., 63
Stone, J., 156
Stone, L.F., 118, 132
Stonebridge, W.C., 607
Stoops, G., 352, 384–386, 392, 395
Story, R., 309
Stout, J.D., 277
Strang, R.M., 252
Stringer, A., 325
Sturtz, J.D., 467
Suarez, F.D., 218
Subagja, J., 275, 325
Sudan, A., 25
Sudres, R., 485
Swaminathan, M.S., 20
Swift, M.J., 232, 251, 524, 542
Syers, J.K., 298, 307
Sys, C., 168, 346, 351, 356, 383

Taiwan Sugar Research Institute, 607
Takahashi, T., 124, 128, 133–134, 624–626
Talineau, J.C., 699
Tallarico, L.A., 572
Taltasse, P. 385
Tanaka, A., 96, 245, 250
Tavernier, R., 446
Ten Berge, H.F.M., 219
Teotia, S.P., 299
Tergas, L.E., 252
Tessema, T., 463
Tessier, F., 389
Thijsse, J.P., 538
Thomas, D.B., 540–541, 672
Thomas, M.F., 183
Thompson, A.R., 326
Thompson, K., 556
Thomson, A.J., 290, 324
Thornthwaite, C.W., 29
Thorp, J., 265, 429
Thum, J., 217
Tinley, K.L., 469
Tiwan, V.K., 280
Tomlin, A.D., 326
Torres, V., 216
Tosi, J.A., 37
Tothill, J.C., 467, 607
Tourte, R., 575, 577, 607

Tracey, J.G., 240
Tran-Vinh-An, 297, 306–307, 309
Trapnell, C.G., 469, 485, 487
Tricart, J., 389
Troll, C., 25–26, 39
Trollope, W.S.W., 453, 461, 469
Trudgill, S.T., 182
Turk, K.J., 694
Turner, B.L., 453, 464
TVA, 536

US Forest Service, 465
US President's Science Advisory Committee, 3–4
Ueckert, D.N., 339, 349, 351
Uexkull, H.R. von, 698
Uhl, C., 253
UNESCO, 88, 99–100, 231, 233–235, 237–238, 248–249
UNU, 519
Uribe, A.H.F., 487
USDA, 14, 88, 295

Valdes, A., 19–20
Valiachmedov, B.V., 345, 347, 384–385
Vallis, I., 214, 699
Van Baren, F.A., 48, 248, 485
Van Bavel, C.H.M., 198
Van de Graff, R.H.M., 144
Van den Berg, R.A., 270
Van der Weert, 528–530
Van Rensburg, H.J., 467
Van Vuura, W., 211, 339, 390
Varazashvili, L.I., 187
Vareschi, V., 459
Vergara, N.T., 669
Verheye, W., 168
Verma, H.N., 591
Vermeulen, F.H.B., 181, 218
Viani, R., 457
Vickery, M.L., 100
Vidal, R., 207
Villa Nova, N.A., 554, 556
Villecourt, P., 486
Vincent, C., 467
Vincent, J.M., 277
Vine, H., 120, 135, 662
Vintola, I., 217–218
Virmani, S.M., 59, 63
Viro, P.J., 480
Visser, S.A., 157, 287, 309–310
Voertman, R.F., 37
Vogl, R.J., 453, 461, 464, 480, 493
Voisin, A., 286

Vos, A. de, 101
Vose, P., 507
Vreeken, W.J., 204

Wade, M.K., 657, 665, 698
Waidyanatha, P., 668
Wakao, N., 214
Walker, B.H., 151
Walker, H.O., 54–55, 84
Walker, P.H., 182, 214
Wall, G.J., 217
Wallwork, J.A., 265, 423
Walter, H., 253
Waltson, A.B., 464
Wambeke, A. van, 89, 181, 205
Wang, C.M., 216
Wanner, H., 251
Waring, S.A., 214
Warkentin, B.P., 167
Wasawo, D.P.S., 287, 309–310
Watnabe, H., 270–271, 275
Watson, J.A.L., 387
Watson, J.P., 309, 338, 345, 354, 372, 384–385, 391, 395, 402, 432
Watters, R.F., 621
Webb, B.H., 542
Weber, N.A., 427
Webster, C.C., 619
Webster, C.D., 54
Webster, R., 179, 219
Weesner, F.M., 340, 404
Wehrmann, J., 463
Wells, C.G., 486
West, N.E., 205
West, O., 467, 484
Wetselaar, R., 486
Wheeler, W.M., 423
Whitehead, P.H., 325
Whiteman, P.T.S., 595
Whitmore, T.C., 99, 239, 242
Wiecek, C.S., 315
Wielemaker, W.G., 277, 338, 346, 352, 385, 387, 391, 399, 414–415, 427
Wierenga, P.J., 198
Wijewardene, R., 668
Wild, A., 88, 120
Wild, H., 402, 409
Wilde, R.H., 182
Wilding, L.P., 179–180, 214
Wiley, R.W., 702
Wilkinson, G.E., 151, 311, 662, 665
Wilkinson, W., 338, 409
Willatt, S.T., 692, 694
Willey, R.W., 662

Williams, E.C., 424
Williams, J.B., 129–130, 142, 151, 412
Williams, M.A.J., 384, 386
Williams, W.M., 214
Wilson, G.F., 699
Wilson, P.N., 54, 619
Wilson, S.R., 179
Wolf, J.M., 145, 149, 151, 694
Wood, T.G., 267, 291, 330–341, 346–347, 354, 357, 369–370, 383–384, 389, 391–392, 394–395, 399, 404–405, 407, 409, 423, 543, 677
Woodwell, G.M., 550–552
World Bank, 14
World Forestry Congress, 510
Worthington, E.B., 54
Wright, R.L., 179

Yaalon, D.H., 446

Yadav, M.R., 157
Yakushev, V.M., 389
Yang, P.C., 692
Yang, T.W., 278
Yao, A.Y.M., 68
Yaron, B., 446
Yates, G.W., 316
Yirgou, D., 463
Yoshino, 52
Yost, R.S., 218–219
Young, A., 88
Young, R.S., 487
Yperlaan, G.J., 75

Zablotskii, V.R., 187
Zachariae, G., 326
Zimmerman, P.R., 405–407, 552–553
Zonn, S.V., 248, 251
Zusevics, J.A., 145, 151

Subject Index

Agricultural development, planning and projects, 21, 24
Agricultural research, 21, 215–220
Agrochemicals, 22, 260, 273–275, 314, 324–326, 690
Agro-ecology, 47, 697–698
Agroforestry, 668–671, 703
Air pollution, 489, 550–554
 termites, 405–406
Atmospheric chemistry, 489

Biomass, 461, 486, 489
Biomes, tropical, 99–100, 102
Biosphere, 47

Climate, 554–556
 agroclimatological surveys, 54
 evaporation, 519
 evapotranspiration, 248
 humidity, 238–242, 518–519
 insolation, 248
 microclimate, 267
 rainforest microclimate, 516–521
 temperature, 48, 67, 238–242, 245, 248, 268, 275, 518–519
 termites, 345–350
 wind intensity, 248
 wind velocity, 239
Climatology
 energy balance, 242
 inter-tropical convergence zone, 51, 237
 microclimate, dew formation, 242
 microclimate instrumentation, 238
 moisture regime, 17
 photosynthesis, 105
 radiation, 103–105, 242
 radiation budget, 17, 42
 rainforest microclimate, 235–242
  light intensity, 242
  sunfleck light, 242

Countries
 Afghanistan, 6
 Angola, 14, 607
 Argentina, 6, 273
 Australia, 57, 66, 90, 96, 100, 102, 123, 125, 204, 211, 214, 240, 251, 266–269, 340, 354, 361, 365, 369, 384, 386–387, 392, 399, 412, 423–424, 432, 453, 461, 467, 470, 479, 489, 507, 535, 537, 573, 607, 643, 665
 Bangladesh, 6
 Barbados, 505
 Belize, 453
 Benin, 14
 Botswana, 469, 595
 Brazil, 6–7, 90, 97, 100, 102, 118–119, 135, 146–147, 151–152, 212, 233, 244, 252, 385, 453, 459, 461, 469, 485, 489, 492, 505, 507, 529, 531, 539–540, 572, 584, 594, 597, 657
 Bulgaria, 214
 Burma, 94, 292
 Cambodia, 94, 244, 252
 Cameroon, 76, 92, 234–235, 340, 384, 492, 540, 672
 Canada, 217
 Chad, 6
 Chile, 6–7, 198, 218
 China, 12, 94, 248, 301, 446
 Colombia, 6, 97, 102, 118
 Congo, 234–235, 266
 Costa Rica, 167, 245, 249, 266, 432, 505, 510
 Cuba, 118, 151, 214, 216, 505
 Dominican Republic, 151
 Ecuador, 507
 Egypt, 6, 216, 289, 307, 383, 607
 England, 102–103
 Ethiopia, 6, 14, 92, 340, 463, 472
 Fiji, 610
 French Guyana, 266, 273, 278

725

Gabon, 14
Gambia, 129, 142, 151, 157, 412
Germany, 217, 315
Ghana, 6, 14, 92, 120, 239, 322, 523–524, 527, 547, 629, 665
Guinea, 120
Guyana, 242, 375, 399
Honduras, 507
Hong Kong, 538
India, 6, 57, 68, 74–76, 84, 90, 94, 100, 123, 126, 144, 150, 157, 160, 162, 215, 240, 266–275, 280, 297, 301, 305, 314–315, 322, 340, 349, 369, 374, 395, 399, 435, 470, 474, 493, 505, 538, 578, 591, 625, 657, 660–661
Indonesia, 56, 94, 100, 116, 242, 505, 537–538
Iran, 6
Israel, 188, 198, 594
Ivory Coast, 92, 239–240, 242, 244, 266, 268, 278, 280, 295, 297, 323, 382, 424, 457, 459, 474, 484, 486, 507, 527, 539, 629, 659, 691, 694
Japan, 17, 279
Kenya, 6, 122, 352, 387, 399, 404, 427, 474, 535, 540, 671–672
Liberia, 56, 92
Malagasy, 280, 459, 507
Malawi, 409, 468
Malaysia, 71, 248, 340, 351, 405, 505, 507, 537
Mexico, 214, 273, 424, 690
Mozambique, 60
New Guinea, 311
New Zealand, 214, 273, 307, 316
Nicaragua, 507
Niger, 6, 18
Nigeria, 6, 56, 60, 63, 67, 74–88, 90, 92, 102–103, 121, 135, 144, 151, 157–159, 198–199, 208, 211, 215, 238, 249, 266, 270, 278, 295–298, 305, 309–313, 317, 326, 350–356, 364, 371, 375, 389, 399, 402–407, 412, 424, 429, 432, 435, 457, 466–468, 485–487, 492, 505, 519–521, 524, 530–532, 534, 539, 543, 571–573, 578–579, 585–588, 597, 607, 612, 629–630, 643–644, 658, 662–665, 691
Pakistan, 6, 295, 316, 404
Panama, 424
Papua New Guinea, 484, 504, 690
Peru, 6–7, 118, 247, 529, 549

Philippines, 6, 315, 537–538, 589, 625
Rwanda, 507
Senegal, 14, 129, 142, 207–208, 215, 240, 270, 384, 399, 458, 575, 591, 595, 612, 621, 626, 629
Sierra Leone, 14, 60, 92, 211
South Africa, 270, 273, 309, 340, 378, 384, 404, 461, 469
Spain, 301
Sri Lanka, 94, 123, 369, 485
Sudan, 6, 56, 92, 215, 295, 385, 387, 456
Surinam, 90, 242, 529, 611–612
Syria, 6
Taiwan, 216
Tanzania, 6, 18, 68, 374, 409, 411, 575, 587–588, 594, 597–598, 641, 653
Thailand, 6, 90, 94, 124, 126, 144, 150, 207, 270–271, 275, 279, 374, 408, 507, 624
Togo, 120, 691
Trinidad, 279
Uganda, 14, 157, 266–267, 273, 310, 345, 354, 384, 387, 404, 507
UK, 205, 278
United States, 17, 68, 349, 369, 371, 429
Upper Volta, 14, 375, 384–385, 387
Uruguay, 102
USA, 188, 199, 204–205, 217, 690
USSR, 187
Venezuela, 6, 118, 120, 145, 151, 237, 248, 254, 288, 459, 468–469, 474, 477, 556
Zaire, 6, 17, 234, 249, 297, 307, 340, 350–352, 384–386, 392, 404
Zambia, 412, 469, 485, 487
Zimbabwe, 151, 354, 372, 385, 432, 469
Crop calendar, 41
Crop growth
  earthworms, 315–316
  termites, 409–413
Crop mulches, 152, 278–279, 319–321, 635–657, 698
  crop residue, 492
Crop residue management, 468, 472–473
Crop suitability, 37–39, 41
Crop yield and productivity, 3–4, 13–20, 42, 67, 217, 489–493, 545–550, 606–612, 657, 691–694
  beans, 18
  cassava, 18
  effect of earthworms, 314–316
  rice, 17
Cropping systems, 273–275, 280, 618–680, 697–703

fallow, 321–322, 505, 662–669, 699–702
mixed cropping, 702–703
no-till, 699
rotation, 658–662, 702–703
shifting cultivation, 21, 212–214, 504–505, 507, 513, 545–547, 619–629
Crops
  banana, 275
  cassava, 207, 275, 539, 549
  cereals, 12
  citrus, 275
  cocoa, 200, 207
  coconut, 535
  coffee, 535
  cola, 200
  cotton, 105
  cover crops, 662–669
    Arepens, 321
    Brachiaria, 321
    Centrosema, 537
    D triflorum, 321
    I. spicata, 321
    Melinis, 321
    Mucuna, 537, 665
    Paspalum notatum, 321
    Psophocarpus palustris, 665
    Pueraria, 537
    Stylosanthes, 321
  cowpea, 105, 321
  fuel crops, 15, 17
  gmelina, 151
  grain, 105, 411, 539
  groundnut, 68, 105, 151, 549
  maize, 105, 142, 207, 275, 321, 492, 539
  millet, 207
  oil palm, 205–207, 505
  okra, 135
  perennial, 212, 321, 537–539
  pigeon pea, 211
  pineapple, 484
  plantation, 200
  rice, 38, 92, 105, 166, 535, 549
  root crops, 200
  rubber, 535
  sorghum, 37
  soybean, 105, 549
  sugarcane, 505
  sunflower, 135
  tea, 535
  tomato, 105
  wheat, 315
Cultural practices, 269–275, 317–326
  ants, 437
  termites, 407–409

Developing countries, 5, 8, 10–12

Ecological change, 445–451, 454–465
Ecological instability, 445, 450, 464
Ecological stability, 22, 445, 697–698
Ecology, savanna, 454–455
Ecology, soil, 260
  research, 260–264
Ecosystems, 46
  fire-dependent, 453–455, 461, 464, 480, 485, 493
  fire-free, 453–455, 464, 480, 485, 493
  forest, 253–254
  fragile, 255
  nutrient recycling, 22, 253–254, 445, 485, 698
  savanna, 484
  tropical, 102–106, 246–247
  tropical rainforest, 231, 445, 470–489, 516–521
Energy consumption, agriculture, 690, 695–697
Environmental damage, 445–450, 454–465, 485–489, 493–495
Erosion, 90, 97, 151–152, 160, 204–205, 248, 251–252, 277, 280, 324, 359–361, 386–389, 463–465, 472, 480–484, 536–541, 590–594, 631–634, 646–654, 692
  control (mulching), 646, 655
  raindrop impact, 152, 247, 251, 463
  slope wash, 252
  soil creep, 252
  splash, 252, 480–484
Evapotranspiration, 29–30, 37–42, 48, 63, 66–67

Farm planning, 41
Farming systems, 618–680
  continuous cropping, 634–672
  high-input, 21, 689–694
  intensive, 97, 492, 629–634
  low-input, 24, 97, 690–691, 695
  mechanized, 22, 98, 248, 539, 689–694
  mixed farming, 703
  multiple cropping, 4
  requirements, 695–697
  traditional, 489–491, 504–505, 619–629
Food supply, problems, 12–21
Forage crops, 275
Forage grasses
  Macroptilium atropurpureum, 479
  Petrie, 479
Forest soils, tropical, 505

Forestry and woodlands management, 467
Forests and woodlands
 classification, 511–512
 evergreen, 251
 forest conversion, 504–513
 leaf area index, 242, 252
 logging, 505, 507, 536–537
 primary forest, 238–239, 246, 251–252
 rooting patterns, 252
 savanna, 26, 37–39, 99–101, 251, 453–493
 semi-deciduous forest, 249, 251, 454
 tree density, 252
 tree size, 252
 tropical, 99–100
 tropical rainforest, 26, 37–39, 231, 453
Fuel-wood, 505

Geography
 tropical river systems, 237, 245, 251
 tropics, definition, 40

Hydrology, 533–536
 river systems, sediment load, 251
 run-off, 67, 204–205, 248, 251–252, 254, 463, 472, 480–484, 521, 536–541, 590–594, 646–654, 694
  mulching, 646–655
 water balance, 40, 235–236
 water supply, 66
 world water balance, 556–559

Inputs
 fertilizers, 324, 672
 herbicides, 325
 pesticides, 325–326
Institutions and organizations
 BIOTROP, 538
 Conseil Scientifique pour l'Afrique, 100–101
 CSIRO, 125, 479
 CTFT, 84
 FAO, 3–11, 13–17, 19, 88, 505, 510
 IAR, 151
 IRAT, 142, 658
 Sudan Meteorological Survey, 56
 Unesco, 88, 99–100, 233–235, 252
 USDA, 14, 88, 446
 World Bank, 14
 World Forestry Congress, 510
International Agricultural Research Centres
 ICRISAT, 123, 146, 150

IITA, 63, 71, 121, 127, 313, 518–519, 604–610, 621, 637, 669
Land capability, 38, 102–106, 160
Land clearing, 254, 270–273, 706
 deforestation, 248, 270, 317–318, 407–409, 437, 504–511
Land clearing methods
 burning, 271–273, 280, 322–323, 453–493, 513
 manual, 466, 513, 527
 mechanized, 513, 527, 532, 539
 tree poisoning, 513
Land clearing results, global effects, 550–559
Land degradation, 21, 461–465, 470–485, 507, 691
Land development and reclamation
 projects, 21, 270
 planning, 41
Land management, pasture renovation, 271, 467–470, 479, 489–491
Land restoration, 697
Land types
 Cerrados, 96, 102, 106, 118, 126, 152, 252, 453, 459, 461, 469, 489
 Llanos, 96, 102, 106, 469, 472, 474, 477
 Pampas, 102, 106, 489, 572
 Terra Rossa, 254
 Terra Roxa Estruturada, 90, 97
Land use, 3–12, 41
 grazing, 322, 469, 540, 671–672
 pasture management, 6–9, 273, 467–470, 489–491, 671–672
 ranching, 505
 statistics, 3–12

Marketing, 12–17

Plant adaptability, 37–38
Plant diseases, 18
Plant growth, 42, 67, 197, 215
 drought stress, 41, 67–68, 97–98, 106, 135, 237, 248–249, 252, 492, 693–694
 root development, 207
 rooting depth, 98, 135, 248
Plant pests, 18
Population problems, 7–13, 505

Rainfall, 48–88, 248–249, 251, 519–521
 evaporation, 248
 growing season, 17, 42, 62–67
 intensity, 76–88, 98, 248
 monsoon, 66, 89, 96

spatial variability, 71
stemflow, 248, 519–521
termites, 345–350, 352
throughfall, 248, 521
transpiration, 248
Regions
  Africa, 4, 6–7, 9–11, 13–15, 18, 21, 39–40, 54–56, 59, 84, 91, 100–102, 120, 151, 233–235, 249, 292, 309, 340, 352, 354, 384–385, 387, 389, 424, 453, 464, 467, 484–485, 505, 507, 510–511, 523, 657, 690
  Africa, Central, 41, 90, 92, 116, 125, 170, 356, 484
  Africa, East, 68–71, 91–92, 116, 123, 126, 248, 374, 394, 411, 468, 572, 625–626, 662–665
  Africa, Southern, 92, 251
  Africa, West, 37, 41, 56, 63–65, 76, 89–92, 96, 116, 120–121, 123, 125, 135, 142, 146, 159–160, 244, 247, 248, 251, 267, 295, 311, 351, 505, 510, 572, 590, 610, 658
  Amazon, 4, 41, 56, 89, 96–97, 100, 116, 135, 146, 233, 244, 248, 250–252, 254, 267, 280, 423, 489, 529, 540, 556–557
  America, Central, 91, 116, 151, 167, 245, 249, 254, 453, 505, 510
  America, North, 4, 10, 21, 91, 214
  America, South, 4, 6, 39, 84, 91, 100–101, 151, 472, 489, 505, 510, 530–531, 657, 665
  America, Tropical, 9, 96–97, 102, 106, 117–118, 145, 454, 464, 507, 539
  Arid and Semi-Arid Tropics, 37, 40, 54, 66, 386, 389, 454, 575–578
  Asia, 4, 7, 12, 17, 91, 93–94, 123, 233, 340, 453, 464, 507, 538
  Asia, Central, 345, 375, 384, 394
  Asia, South, 116, 123, 144, 151, 159, 689
  Asia, Southeast, 38, 84, 91, 116, 151, 167, 245, 446, 510, 689
  Australia, Tropical, 4, 7, 41, 50, 91, 96, 116, 125, 135, 144–145, 211, 467
  Borneo, 252
  Caribbean, 84, 151, 427, 505
  Congo, 56, 170
  Europe, 4, 12, 21
  Far East, 6, 8, 10–11, 13–16, 239–240
  Hawaii, 57, 218
  Humid and Semi-Humid Tropics, 37, 248, 251–252, 578–583, 703
  Java, 116, 254
  Lake Chad, 92
  Latin America, 7, 10–11, 13–17, 116, 510–511
  Malay Peninsula, 94
  Near East, 6, 10–11, 13, 15–16
  Nile, 295, 383
  Oceania, 7, 507
  Puerto Rico, 118, 145, 151, 248–249, 251
  Sahel, 92, 129
  Sarawak, 252, 543
  Sumatra, 41, 100, 505, 537
  temperate, 21
Research and development priorities, 705–707
Resources planning, 505
Resources
  forest, 4–7, 99–106, 233–235, 504–559
  land, 3–7, 17
  soil, 694–695

Savanna, 453–493
  definition, 100–101
  derived, 464, 474, 485, 493
  vegetation, 468–470, 484
Soil biology, 21
  ant nests, 427
  earthworm casts, 291–309
  mulching, 655–656
  termitaria, 340–377, 404
  tillage, 604
Soil characteristics
  acidity, termites, 395–402
  CN ratio, termites, 394–395
  exchangeable bases, termites, 395–402
  high activity clays, 115–116, 160–166
  low activity clays, 114–115, 117, 250, 351, 474
  organic matter content, 207, 249, 268–269, 275, 474, 698
  termites, 392–395
  self-mulching, 164
  slope, 212
Soil chemistry, 192–193, 214–215, 252–254, 275–277, 392–402, 435, 466, 486–487, 541–543, 604, 691
Soil classification, 168–170
Soil conservation, mulching, 152
soil degradation, 392, 691
  leaching, 464
  termites, 399, 402
  organic matter loss, 98
Soil erodibility, termites, 386–389

soil fauna, 22, 151, 199–200, 211, 247, 260–281, 463–464, 487–489, 527, 543–545, 655–656
 ants, 423–439
  nests, 427
 arthropods, 489
 classification, 265–266
 earthworms, 286–327, 524, 527, 655–656
  adverse effects, 313–314, 316
  casting activity, 291–296, 571
  soil properties, 288–315
 ecology, 316–317
 humus formation, 275, 314–315, 383
 mesofauna, 265
 microfauna, 265
 nematodes, 316–317
 soil aggregation, 313, 369
 soil formation, 309–311, 383–384
 soil pore creation, 276, 311–313, 367, 370, 383, 433–435
 soil profile inverstion, 276, 352–364, 369, 383–385, 392, 427–429
 soil turnover, 264, 275–276, 299, 383, 392
 termites, 151, 340–415, 487, 524, 527
  fungus-feeding, 351, 404
  ground-dwellers, 351, 370, 374
  humus-feeding, 404
  litter consumption, 351–352
  litter decomposition, 404–405
  mound-builders, 351, 370, 374
  nests, 340–377, 404
Soil fertility and mineral nutrition, 97–99, 114, 160, 214–215, 252–254, 315–316, 463, 466, 485–487
 ants, 435
 fertility decline, 691
 nutrient deficiencies, 106, 135
 nutrient loss, 485–487
 termites, 392–402, 409–413
Soil flora, 260–281, 487–489
 ectomycorrhizae, 489
 megaflora, 265
 mesoflora, 265
 microflora, 265, 277–280, 487
Soil horizon, 98, 446
Soil humidity, termitaria, 379–381
Soil management, 18, 98, 549–550
Soil mechanics, 182–187
 compaction, 97, 160, 200, 248, 323–324, 463–465, 485, 527–531, 573–583, 692
Soil microbiology, 277–280

Soil microclimate, 201–204, 598–604
 termitaria, 377–382
Soil physics, 22, 97–99, 114–170, 179–220, 242–252, 275–280, 461–463, 466, 470–485, 522–533, 570–598, 635–637, 691
 aeration, 155, 167, 645–646
 mulching, 645–646
Soil processes
 aggregation, 276, 278, 313
 gaseous exchange, 276, 377–378
 humification, 276
 laterization, 389–392
 water movement, 276
 weathering, 276
Soil productivity, 22, 102–106
Soil properties, 22, 114, 178–220, 242–252, 275–280, 461–463, 466, 470–485, 522–533, 541–543, 570–598, 604, 691
 ants, 427–435
 bulk density, 125–129, 159–160, 185–186, 194–197, 207, 246–248, 527–530
  ants, 432–433
  termites, 364–367
 changes, 446, 450
 gravel content, 182–185, 193–194
 particle size distribution, 117–125, 182–187, 470–472
 pore size, 268
 porosity, 247, 250, 463, 472, 474, 485, 532–533
 soil air, 599–604, 645–646
 swell–shrink, 135, 165, 351
 termites, 351–377
 texture, 192–193, 244, 252, 463, 472, 524–527, 571
  ants, 429–432
  mulching, 635–637
  termites, 351–364, 385–386
 variability, 178–220
 wettability, 461–463, 473–474, 480–484
Soil resources, 694–695
Soil restoration, 326, 697
Soil structure, 129–134, 162, 168–170, 246–248, 250, 278, 461–463, 472–473, 484, 524–527, 571–573
 mulching, 635–637
 termites, 367–370
Soil temperature, 41, 42, 63, 68, 98, 157, 162, 168, 201, 238–239, 267–268, 457–461, 474, 522–524, 598–599, 642–645
 earthworms, 288–289
 mulching, 642–645
 termitaria, 381–382

Soil water, 62–66, 165–167, 187–200, 207, 267
  conservation, 594–598
  deficit, 248
  drainage, 249
  earthworms, 289–290
  evaporation, 248
  hydraulic conductivity, 152
  infiltration, 150–152, 479–484, 531, 585–589
  infiltration rate, 249
  moisture balance, 474–479
  moisture content, 240, 249
  moisture deficiency, 252
  moisture regime, 252, 637–642
    mulching, 637–642
  moisture retention, 250, 474, 532–533, 584–585
    ants, 433–435
    termites, 374–377
  moisture transmission, 150–155, 479–480, 637
    mulching, 637
    termites, 370–374
  movement, 249
  water balance, 248–249, 533–536
  water repellency, 463, 473–474, 480–484
  water-holding capacity, 135
Soils
  Alfisol, 90–98, 116, 118, 120, 123–126, 135, 144–145, 151, 159–160, 162, 164, 170, 182, 211, 247, 309, 313, 356, 384, 454, 473–474, 479–480, 492, 527, 532, 542–549
  Andosol, 91, 116, 167–168, 170, 242, 309
  Aridisol, 91–96, 115, 142, 454
  Entisol, 92, 115–116, 245, 249
  Inceptisol, 92, 96, 97, 115–116, 118, 155, 242, 245, 249, 542, 545
  Kraznozem, 479
  Laterite, 89–90
  Latosol, 135, 151
  Lithosol, 93, 492
  Mollisol, 115–116, 160, 165, 170
  Nitosol, 116
  Oxisol, 88–93, 97–98, 116, 119–120, 151, 170, 182, 244, 247–250, 254, 280, 384, 454, 479, 531, 542, 545, 547, 549
  Plinthite, 89–90, 98
  Regosol, 93
  Ultisol, 90, 92–98, 116, 118, 126, 144, 151–152, 159–160, 170, 182, 199, 244, 247–250, 254, 454, 479, 531, 542, 545, 547–549
  Vertisol, 91–96, 115–116, 160–165, 170, 182, 351, 470–472, 474, 479
  Yermosol, 93
Soil definitions, 264
Soils research, 215–220
  analysis, 265
  design of experiments, 215–216
  geostatistical techniques, 219
  kriging, 219–220
  measurements and instrumentation, 186, 188, 193–197, 199, 203–204
  sampling, 217–220
  statistical methods, 216–220
  tropical soil biology, 277
  pF measurements, 135–150, 250
Soils (tropical savanna), 106, 135

Tillage, 212–214, 566–614
  crop response, 606–612
  seedbed preparation, 164, 275
Tillage systems, 318–319
  minimum tillage, 606–612
  mounds, 212–214
  no till, 275, 606–612
  ploughing, 166, 264
  puddling, 166, 589
Tropical/temperate comparisons
  leaf litter decomposition, 314
  soil tillage, 264
  soils, 182

Vegetation, 88–89, 99–102, 231–255, 460–461, 468–470
  A. canescens, 469
  abaca, 538
  acacia, 275, 669
  Acacia karroo, 469
  Brachystegia specififormis, 469
  Bridelia ferruginea, 469
  Burkea africana, 469
  Butyrospermum paradoxum, 469
  changes, 460–461, 464, 468–470, 493
  Chrusopogon fallacx, 470
  climbers, 242
  epiphytes, 242
  eragrostis, 275
  grasses, 211, 469
    Guinea grass, 278
    Hyparrhenia, 252
    Imperata, 464, 513
    Panicum maximum, 211

Hyparrhenia, 469
Isoberlinia angolensis, 469
Julbernaridia paniculata, 469
Kikuyu, 275
leaf litter, 247, 251, 266–267
Lophira lanceolata, 469
Melinis minutiflora, 470
Parinari curatellifolia, 469
Parkis clappertoniana, 469
Peterocarpus erinaceous, 469
savanna, 248
Sorghum plumosum, 470
Syzygium guineense, 469
termites, 402–405
Themedea triandra, 469
Trachypogon montufari, 469
Trachypogon plumosus, 468–469
trees, 248
    Acacia albida, 207, 669
    Acio barteri, 207
    Albizia, 669
    bamboo, 249
    Calliandra, 669
    Cassia, 669
    Chorophora excelsa, 207
    Geronniera plycnemia, 249
    Gliricidia maculata, 669
    Leucaena leucocephala, 669
    Mimosa, 669
    Parkia, 669
    Phyllostachys, 207
    pine, 270
    Pithecellobium, 669
    Prosopis, 669
    rainforest species, 232, 242, 252
    Samanea, 669
    Sesbania, 669
    Tephrosia candida, 669
Vegetation zones, 233–235
Vegetative canopy, 238–242, 248
Vegetative cover, 205–211, 232, 239, 248, 252, 280, 484–485, 698

Water conservation and reclamation, 41
Water management, 92–93